DESIGN
AND
CONSTRUCTION
OF
CONCRETE SHELL ROOFS

DESIGN
AND
CONSTRUCTION
OF
CONCRETE SHELL ROOFS

G. S. Ramaswamy

United Nations Expert,
Saudi Arabia
Formerly Visiting Professor,
University of Arizona

CBS

CBS Publishers & Distributors Pvt. Ltd.

New Delhi • Bengaluru • Chennai • Kochi • Kolkata • Mumbai
Hyderabad • Nagpur • Patna • Pune • Vijayawada

ISBN: 81-239-0990-X

First Indian Edition: 1986
Reprint: 1999, 2002, 2003, 2005

This edition has been reprinted in India by arrangement with
Robert E. Kraiger Publishing Company, Inc., USA

Published by **Satish Kumar Jain** and produced by **Varun Jain** for
CBS Publishers & Distributors Pvt. Ltd.,
4819/XI Prahlad Street, 24 Ansari Road, Daryaganj, New Delhi - 110002
delhi@cbspd.com, cbspubs@airtelmail.in • www.cbspd.com
Ph.: 23289259, 23266861, 23266867 • Fax: 011-23243014

Corporate Office: 204 FIE, Industrial Area, Patparganj, Delhi - 110 092
Ph: 49344934 • Fax: 011-49344935
E-mail: publishing@cbspd.com • publicity@cbspd.com

Branches:
• *Bengaluru:* 2975, 17th Cross, K.R. Road, Bansankari 2nd Stage,
  Bengaluru - 70 • Ph: +91-80-26771678/79 • Fax: +91-80-26771680
  E-mail: cbsbng@gmail.com, bangalore@cbspd.com
• *Chennai:* No. 7, Subbaraya Street, Shenoy Nagar, Chennai - 600030
  Ph: +91-44-26681266, 26680620 • Fax: +91-44-42032115
  E-mail: chennai@cbspd.com
• *Kochi:* Ashana House, 39/1904, A.M. Thomas Road, Valanjambalam,
  Ernakulum, Kochi • Ph: +91-484-4059061-65
  Fax: +91-484-4059065 • E-mail: cochin@cbspd.com
• *Kolkata:* 6-B, Ground Floor, Rameshwar Shaw Road, Kolkata - 700014
  Ph: +91-33-22891126/7/8 • E-mail: kolkata@cbspd.com
• *Mumbai:* 83-C, Dr. E. Moses Road, Worli, Mumbai - 400018
  Ph: +91-9833017933, 022-24902340/41 • E-mail: mumbai@cbspd.com

Representatives:

• Hyderabad: 0-9885175004	• Nagpur: 0-9021734563
• Patna: 0-9334159340	• Pune: 0-9623451994
• Jharkhand: 0-9811541605	• Uttarakhand: 0-9716462459

Printed at:
Neekunj Print Process, Delhi (India)

To my wife Seetha.

PREFACE TO REVISED EDITION

When the first edition of this book was conceived and written in the early sixties, the large-scale use of electronic digital computers in structural analysis was just round the corner. In the first version of the manuscript, written in 1961, the 4 × 4 matrices figuring in Chapter 8 were inverted by tedious and time-consuming methods, using a mechanical calculator. Fortunately, by the middle of 1964, when the manuscript was being finalized, the author gained access to an IBM 1620 Computer. This led to a thorough revision of the manuscript to orient the presentation to suit digital computers.

In the thirteen years that have elapsed since the book first appeared, structural analysis techniques have undergone a near revolution, thanks to the burgeoning use of digital computers. During the same period, equally impressive progress has been made to develop matrix methods of structural analysis tailor-made to match the power and versatility of these amazingly fast calculating tools. Of these, the finite-element technique has more or less dominated the scene. At the latest tally, it has been estimated that no less than 12,000 papers and reports on the finite element technique have appeared in print. With such powerful aids available to him, the structural analyst can now state with confidence that any structure at all that can be conceived can also be analyzed—at a cost. But this unprecedented spurt in publication activity has not led any spectacular breakthroughs in creativity.

Authors for the most part appear to have directed their energies to the pedestrian goals of refining structural analysis to the point of diminishing returns. They have been looking for problems to suit their particular techniques rather than forging techniques to solve problems of practical interest. Gem polishers have been many; but the gem-finders have been few and far between. These statements will be illustrated with special reference to thin shells used in Civil Engineering practice. A close examination of the papers published during the past decade on finite element techniques applied to thin shells reveals that a large majority of them involved the papers published during the past decade on finite element techniques applied to thin shells reveals that a large majority of them involved the development of refined elements and the demonstration of their merits by applying them to carefully selected standard test problems. There have hardly been any efforts to explore optimum shapes or deal with realistic boundary conditions. Interest in thin shells among architects and practising engineers is bound to gradually decline, unless efforts are made to open new frontiers, involving the search for new shapes, innovative applications and the use of new construction materials. The four new chapters added to the second enlarged edition have been chosen with the object of stimulating interest among those concerned for initiating further development of thin shells on these lines.

In Chapter 23, a state of the art survey of finite element analysis applied to thin shells is presented. The possibilities that thin shells offer for building intermediate floors are examined in Chapter 24 with the aid of a fully worked example. Shell foundations have been used to a limited extent in the past for individual footings. The use of shell rafts and the assessment relative merits of different shell forms for this purpose are still to be fully explored. Those interested in this subject will find Chapter 25 to be of special interest. One of the obvious infirmities of concrete is its notoriously low tensile strength. For this reason, the effort in the past has been to design concrete shells as compression forms. Ferrocement, hailed sometimes as superreinforced concrete, liberates us from this restriction because of its higher tensile strength and ductility. It opens the door to the imaginative use of tension forms. Some of these possibilities are briefly explored in the concluding chapter by means of examples.

The author acknowledges his gratitude to Dr. T.V.S.R. Appa Rao and Mr. Nagesh Iyer of the Structural Engineering Research Centre, Madras and Dr. David Gunaratnam, his colleague at the University of the West Indies, for their advice and assistance in writing Chapter 23. He thanks his colleagues Dean R. H. Gallagher and Professor C. S. Desai for making several useful comments on Chapter 23. He is indebted to many publishers, authors and others who permitted him to quote from their publications or use photographs supplied by them. In particular he would like to record his grateful appreciation to Mr. Gautam and Miss Gira Sarabhai and Dr. Ramaiah, the Director of the Structural Engineering Research Centre, Madras and his colleagues and Mr. Ashoke Chatterjee, the Executive Director of the National Institute of Design, Ahmedabad, India. Miss Angela Carrington and Mr. Austin Rodriguez deserve a special word of praise for a meticulously typed manuscript and neatly executed drawings.

G. S. Ramaswamy

CONTENTS

PART IV   OTHER ASPECTS

PART V   NEW DEVELOPMENTS

DESIGN
AND
CONSTRUCTION
OF
CONCRETE SHELL ROOFS

INTRODUCTION

Each age produces its own architecture out of its own materials and technology.

Eero Saarinen

Ours is the age of concrete. According to the Italian architect Ponti, it has liberated us from the rectangle. It has also freed us from the restrictions imposed by the ubiquitous post-and-lintel type of construction devised to suit older construction materials such as stone, timber, and steel. It is only in recent years that the special properties of concrete have been fully appreciated and exploited. In the words of architect-engineer Nervi, "Here in effect is a material which by its monolithicity and plastic mouldability has widened beyond imagination the scope of architecture." No structural form perhaps does greater justice to the special attributes of concrete than thin-shell construction. During the past four decades, the development of thin shells as a structural form has added an exciting chapter to contemporary architecture.

Shell as a Structural Form

Thin shells are an example of strength through *form* as opposed to strength through *mass*. The effort in design is to make the shell as thin as practical requirements will permit so that the dead weight is reduced and the structure functions as a membrane free from large bending stresses. By this means, a minimum of materials is used to the maximum structural advantage. Shells of double curvature are among the most efficient of known structural forms. Most shells occurring in nature are doubly curved. Shells of eggs, nuts, and the human skull are commonplace examples. These naturally occurring shells are hard to crack or break.

Historical Notes

The thin reinforced-concrete shell, as we know it today, had its beginnings in Germany in the 1920s. Most of the early shells built were cylindrical barrels. Two German engineers—Finsterwalder and

1

Dischinger—were the first to develop a theoretical analysis applicable to reinforced-concrete cylindrical shells around the year 1930. An early American contribution was a paper by Schorer (1936), which attempted a very much simplified analysis. His method is remarkable for the physical intuition underlying it. Until about 1940, the cylindrical shell more or less dominated the scene. Other forms of shells were rarely tried out. An exception perhaps is the thin-shell dome.

The modern thin-shell dome may be regarded as an evolution of a structural form known and used by man from very ancient times. The Pantheon, believed to have been built in A.D. 125, and several medieval domes such as the one over St. Peter's at Rome show that this structural form was favored by architects of all ages. Being the most economical

PHOTO I-1. Model of the Sports Stadium at Havana. (*Courtesy of the Preload Company, Inc., New York. Architect: Arroyo & Menendez.*)

means of covering large column-free areas, many variations of the dome continue to be employed by contemporary architects. But the modern thin-shell dome, such as the one roofing the Sports Stadium at Havana, is a far cry from the massive masonry dome of the Middle Ages. A comparison of relative weights would bring home this difference. The 137-ft-diameter dome over St. Peter's built in 1590 weighs approximately 12,500 lb/sq yd. Its modern counterpart, the dome over the Sports Stadium at Havana, weighs only 648 lb/sq yd although it is 290 ft in diameter. The shell roof over the auditorium at Urbana, Ill., may be cited as an interesting variation of the traditional dome.

It was only in the mid-thirties and early forties that designers began looking for other forms of shells for roofs. Around 1935, two French engineers—Aimond and Laffaile—published studies on the properties and potentialities of the hyperbolic paraboloid. It was, however, left to Felix Candela, the Mexican architect-engineer, to promote and popularize the hyperbolic paraboloid for roofing factories, churches, clubs, grandstands, and whatnot. The hyperbolic paraboloid has two interesting properties. Though it is a warped surface, it can be cast on straight forms, being a ruled surface. Another property is that its exposed edges can be completely freed of normal and tangential stresses. This means that the shell edge needs no thickening or the provision of stiffening members. This property has been exploited to dramatic advantage by Candela in a number of imaginative designs. The roofs over the restaurant at Xochimilco and the open chapel at Cuernavaca are the most striking examples. Other ruled surfaces which may be

PHOTO I-2. Dome at the University of Illinois. (*Courtesy of Dr. A. L. Parme. Architects: Harrison & Abramovitz. Structural Engineers: Amman & Whitney. General Contractors: Felmley-Dickerson Co.*)

cast on straight forms are the conoid and the one-sheet hyperboloid. Of these, the conoid had its beginnings in France. The one-sheet hyperboloid appears to have been first employed for roofing in Germany in the form of Silberkuhl shells. The last ten years have witnessed an unprecedented spurt of activity in the field of thin shells. Many new forms are continuously being experimented upon and added to the growing vocabulary of shells. Suspensed shells or hanging roofs may be cited as examples.

Economics of Shells

Our enthusiasm for shell roofs needs to be tempered by the hard facts of economics. Shell structures tend to be expensive when only a few units of a kind are to be built. Thin-shell roofs work out to be competitive or economical only if several identical units are involved so that many reuses of the forms are ensured. The cost of forms tends to make shell roofs sometimes expensive in spite of their low consumption of materials—cement and steel. Form costs can be brought down by either resorting to mobile forms or precasting.

Analysis and Design

In the analysis of shell roofs, the structure is regarded as homogeneous and isotropic. However, in the design of reinforcement, the concrete is assumed to be cracked and steel is provided to take care of the full direct and diagonal tension. This dichotomy is inescapable as long as we continue to use the elastic theory for stress analysis of thin shells.

PHOTO I-3. Restaurant in Xochimilco, Mexico D.F. (*Courtesy of Felix Candela. Architect: Joaquin Alvarez Ordonez. Structural design and construction: Felix Candela.*)

Shell roofs of complex shapes do not always lend themselves to calculation by analytical means. There is therefore a growing trend toward the use of scale models as aids to design.

Books on thin shells tend to place undue emphasis on stress analysis and pay little or no attention to designing and detailing. The student is thus left with the impression that analysis is all. Translating an analysis into a workable design and the design into an actual structure are equally important steps.

There are many secondary effects such as those due to shrinkage and temperature for which exact calculations are not possible in our present state of knowledge. In tropical countries, the stresses due to temperature changes may be even more severe than those caused by loading.

The secret of avoiding cracks in such shells lies in the provision of closely spaced small-diameter reinforcement. Such reinforcement also imparts greater ductility to the structure. This is the underlying principle of ferrocement successfully employed by Nervi. He uses many layers of small-diameter mesh reinforcement embedded in rich mortar to build up the required thickness of the shell.

Selection of Shell Type

Shell shapes are so numerous and their possible applications are so varied that it is almost impossible to lay down any general rules for the

PHOTO I-4. Prestressed hanging shell roof of the "Schwarzwaldhalle" at Karlsruhe, West Germany. (*Courtesy of Dycherhoff & Widmann, Kommanditgesellschaft, Germany. Design: Prof. Schelling, Karlsruhe. Photographed by Schlesijer, Karlsruhe.*)

selection of the proper type of shell for any given situation. If good daylighting is not a requirement, long multiple cylindrical shells with featheredge beams may serve the purpose. Where the width of the building to be roofed over is very large, a short shell may be preferred with arch traverses spaced at one-sixth to one-third of the chord width. Folded plates are generally competitive with shells in the span range of 40 to 80 ft. The limiting span of long reinforced-concrete cylindrical shells is about 100 ft. For larger spans, poststressing the shell would generally be worthwhile. If good daylighting is desired, several alternatives are available. A northlight cylindrical shell, a northlight folded plate, a conoid, and a tilted hyperbolic paraboloid are some of the possible solutions.

Acoustics

Acoustics of shell roofs does not usually receive the attention it deserves at the planning stage. The result is that expensive correction measures are usually found necessary after the building is commissioned. Acoustic treatment of shell roofs is a job for a specialist. At the same time, the designer of shell roofs must have some appreciation of the factors involved.

For any enclosure used for purposes of hearing, the reverberation time (RT) is an important consideration. It is the time during which a sound of a given intensity decays to 10^{-6} fraction of its initial value

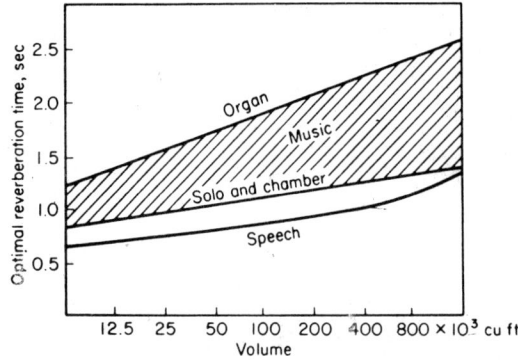

Fig. I-1. (*From Vern O. Knudsen, "Architectural Acoustics," 3d ed., copyright © 1947, John Wiley & Sons, Inc., Courtesy of John Wiley & Sons, Inc.*)

after the source is shut off. The requirements of the so-called optimal RT depend on the volume of the enclosure. The variation of this optimum RT for speech and music with the volume of the enclosure is shown in Fig. I-1. When the enclosure has to serve more than one purpose, a compromise is effected and usually the lower RT is chosen.

The other problems that arise in doubly curved shell roofs are those of flutter echoes and sound foci. These are of considerable importance when acoustical demands are severe. Normally in large enclosures used for storage or sports, the covering up of floor space by an audience provides sufficient relief from these two effects. For enclosures used specifically for speech or music, reference will have to be made to a specialist. Flutter echo is a rapid succession of reflected pulses resulting from a single initial pulse. Wherever such problems are anticipated, it is best to consult an expert at the planning stage itself.

Planning and Coordination

If a shell-roof project is to be successfully carried out, it is essential that the architect and engineer concerned should work together right from the very beginning. All fixtures to be installed in the shell should be decided upon in advance. Subsequent puncturing of the shell after it is built is to be avoided. A successful shell project is an enterprise in which the closest coordination is necessary among the three partners involved—the architect, the engineer, and the builder.

PART I

BACKGROUND MATHEMATICS

CHAPTER 1

SOLUTION OF EIGHTH-DEGREE POLYNOMIAL EQUATIONS

1-1 Characteristic Equations of Cylindrical Shells

The various bending theories of cylindrical shells developed in Chapter 7 lead to homogeneous, linear partial differential equations of the eighth order known as characteristic equations. The auxiliary equations of the characteristic equations are algebraic equations of the eighth degree. The aim of this chapter is to develop techniques for the solution of these equations. Elementary operations with complex numbers are reviewed as a preliminary to this development.

COMPLEX NUMBERS

1-2 Definitions

A number $z = x + iy$, where $i = \sqrt{-1}$, is known as a complex number whose real and imaginary parts are x and y. It may be observed that x and y are real numbers themselves. A complex number may be represented graphically as shown in Fig. 1-1. The point P represents the complex number $z = x + iy$. Such a graphical representation is known as an Argand diagram. The number $r = \sqrt{x^2 + y^2}$, which is always taken as positive, is known as the *modulus* of the complex number and is usually written as $|z|$. The angle θ is known as the *argument* of the complex number.

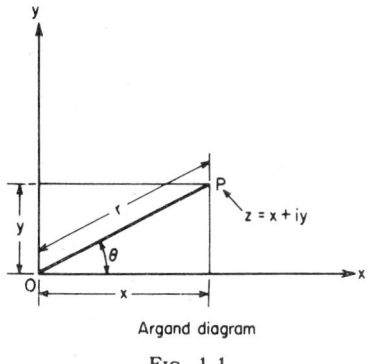

Argand diagram

Fig. 1-1

From Fig. 1-1,

$$x = r \cos \theta \quad \text{and} \quad y = r \sin \theta \tag{1-1}$$

The argument θ is not unique because $\tan^{-1}(y/x) = \theta + 2\pi k$, where k is any integer. The angle θ corresponding to $k = 0$ is known as the *principal argument* of z. Making use of the relations in (1-1), we may write

$$z = r(\cos \theta + i \sin \theta) = re^{i\theta} \tag{1-2}$$

This is another way of representing complex numbers. If two complex numbers z_1 and z_2 are such that $z_1 = x + iy$ and $z_2 = x - iy$, z_1 and z_2 are known as *conjugates* of each other.

1-3 Properties of Complex Numbers

The following are some of the properties of complex numbers:

(i) A complex number $z = x + iy$ is zero if and only if both the real part x and the imaginary part y are equal to zero.

(ii) Complex numbers obey the rules of algebra.

(iii) The sum, difference, and quotient of two complex numbers are themselves complex numbers.

(iv) Two complex numbers are equal when and only when the real part of one is equal to the real part of the other and the imaginary part of one is equal to the imaginary part of the other.

1-4 Operations with Complex Numbers

Let $z_1 = x_1 + iy_1$ and $z_2 = x_2 + iy_2$ be two complex numbers. Their sum

$$z_1 + z_2 = (x_1 + x_2) + i(y_1 + y_2) \tag{1-3}$$

If z_2 is subtracted from z_1, the result is

$$z_1 - z_2 = (x_1 - x_2) + i(y_1 - y_2) \tag{1-4}$$

Multiplication and division of complex numbers are best carried out by representing them as $z_1 = r_1 e^{i\theta_1}$ and $z_2 = r_2 e^{i\theta_2}$. The product

$$\begin{aligned} z_1 \cdot z_2 &= r_1 r_2 e^{i(\theta_1 + \theta_2)} \\ &= r_1 r_2 [\cos(\theta_1 + \theta_2) + i \sin(\theta_1 + \theta_2)] \end{aligned} \tag{1-5}$$

The quotient

$$\begin{aligned} \frac{z_1}{z_2} &= \frac{r_1}{r_2} e^{i(\theta_1 - \theta_2)} \\ &= \frac{r_1}{r_2} [\cos(\theta_1 - \theta_2) + i \sin(\theta_1 - \theta_2)] \end{aligned} \tag{1-6}$$

1-5 *n*th Root of a Complex Number

To find the nth root of the complex number $z = x + iy$, it is convenient to use the form $z = re^{i\theta}$. It is easily verified that

$$(z)^{1/n} = (r)^{1/n}\, e^{i(\theta/n)} = (r)^{1/n} \left[\cos\left(\frac{\theta + 2\pi k}{n}\right) + i\sin\left(\frac{\theta + 2\pi k}{n}\right)\right] \tag{1-7}$$

k taking on values of 0, 1, 2,..., $(n-1)$ to give the n roots.

Example 1-1

Find the square root of the complex number $z = a + ib$.

Solution

Let $z = a + ib = re^{i(\theta + 2\pi k)}$. The two roots are found by putting $k = 0$ and 1. They are

$$z_1 = (r)^{1/2}\left(\cos\frac{\theta}{2} + i\sin\frac{\theta}{2}\right)$$

and

$$z_2 = -(r)^{1/2}\left(\cos\frac{\theta}{2} + i\sin\frac{\theta}{2}\right)$$

Hence

$$\sqrt{a + ib} = \pm(r)^{1/2}\left(\cos\frac{\theta}{2} + i\sin\frac{\theta}{2}\right)$$

Noting that $r\cos\theta = a$ and $r\sin\theta = b$, we may write

$$\cos\theta = \frac{a}{\sqrt{a^2 + b^2}}$$

Making use of the relations $\cos\theta = 2\cos^2(\theta/2) - 1 = 1 - 2\sin^2(\theta/2)$

$$\cos\frac{\theta}{2} = \frac{1}{\sqrt{2}}\frac{1}{(r)^{1/2}}(\sqrt{a^2 + b^2} + a)^{1/2}$$

and

$$\sin\frac{\theta}{2} = \frac{1}{\sqrt{2}}\frac{1}{(r)^{1/2}}(\sqrt{a^2 + b^2} - a)^{1/2}$$

Hence

$$\sqrt{a + ib} = \pm\frac{1}{\sqrt{2}}[(\sqrt{a^2 + b^2} + a)^{1/2} + i(\sqrt{a^2 + b^2} - a)^{1/2}] \tag{1-8}$$

It is possible to show in a similar manner that

$$\sqrt{a - ib} = \pm\frac{1}{\sqrt{2}}[(\sqrt{a^2 + b^2} + a)^{1/2} - i(\sqrt{a^2 + b^2} - a)^{1/2}] \tag{1-9}$$

These results are important and find frequent application.

Example 1-2

Solve the equation

$$p^4 + a^4 = 0$$

Solution

The equation may be rewritten as

$$p = a \sqrt[4]{-1}$$

Thus the problem is that of finding the four roots of -1. It is easily verified that we may write

$$-1 = e^{i(\pi + 2\pi k)}$$

the modulus of the number being unity. Hence

$$p = ae^{i(\pi/4 + 2\pi k/4)}$$

The four roots may now be written down by assigning values of 0, 1, 2, and 3 to k. The roots are

$$p_1 = ae^{i(\pi/4)} = a\left(\cos\frac{\pi}{4} + i\sin\frac{\pi}{4}\right) = \frac{a}{\sqrt{2}}(1+i)$$

$$p_2 = ae^{i(3\pi/4)} = a\left(-\cos\frac{\pi}{4} + i\sin\frac{\pi}{4}\right) = \frac{a}{\sqrt{2}}(-1+i)$$

$$p_3 = ae^{i(5\pi/4)} = -a\left(\cos\frac{\pi}{4} + i\sin\frac{\pi}{4}\right) = \frac{a}{\sqrt{2}}(-1-i)$$

$$p_4 = ae^{i(7\pi/4)} = a\left(\cos\frac{\pi}{4} - i\sin\frac{\pi}{4}\right) = \frac{a}{\sqrt{2}}(1-i)$$

We may collect the four roots together and write

$$p = \pm\frac{a}{\sqrt{2}}(1 \pm i)$$

The four roots may be represented on an Argand diagram by the points 1, 2, 3, and 4, respectively (Fig. 1-2). This result finds application in the theory of beams on elastic foundations.

Example 1-3

Solve the equation

$$(m^2 - \lambda_n^2)^4 + \frac{\lambda_n^4}{k} = 0$$

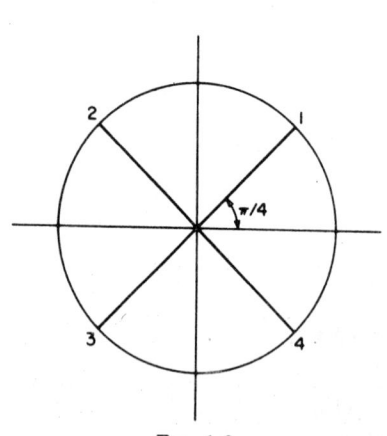

FIG. 1-2

Solution

This equation may be recast in the same form as the equation in Example 1-2 by letting $p = m^2 - \lambda_n^2$ and $\lambda_n/k^{1/4} = a$. The eight roots of m may then be written as $m = \pm[\lambda_n^2 \pm (\lambda_n/k^{1/4})(1/\sqrt{2})(1 \pm i)]^{1/2}$

This result is used later in Chapter 7 to find the roots of the D-K-J characteristic equation.

Example 1-4

Solve the equation

$$m^8 + \rho^8 = 0$$

Solution

Proceeding as in the previous example,

$$m = \rho \sqrt[8]{-1}$$

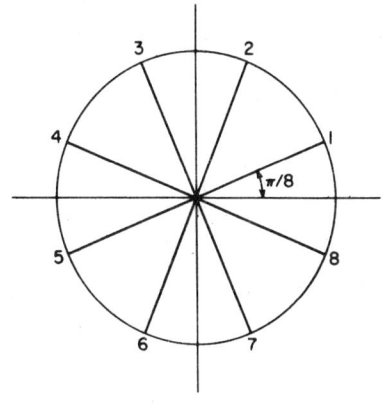

FIG. 1-3

Noting that $-1 = e^{i(\pi + 2\pi k)}$, the eight roots may be written down as follows by assigning values to k ranging from 0 to 7:

$$m_1 = \rho e^{i(\pi/8)} = \frac{\rho}{8^{1/4}}[+(2^{1/2} + 1)^{1/2} + i(2^{1/2} - 1)^{1/2}]$$

$$m_2 = \rho e^{i(3\pi/8)} = \frac{\rho}{8^{1/4}}[+(2^{1/2} - 1)^{1/2} + i(2^{1/2} + 1)^{1/2}]$$

$$m_3 = \rho e^{i(5\pi/8)} = \frac{\rho}{8^{1/4}}[-(2^{1/2} - 1)^{1/2} + i(2^{1/2} + 1)^{1/2}]$$

$$m_4 = \rho e^{i(7\pi/8)} = \frac{\rho}{8^{1/4}}[-(2^{1/2} + 1)^{1/2} + i(2^{1/2} - 1)^{1/2}]$$

$$m_5 = \rho e^{i(9\pi/8)} = \frac{\rho}{8^{1/4}}[-(2^{1/2} + 1)^{1/2} - i(2^{1/2} - 1)^{1/2}]$$

$$m_6 = \rho e^{i(11\pi/8)} = \frac{\rho}{8^{1/4}}[-(2^{1/2} - 1)^{1/2} - i(2^{1/2} + 1)^{1/2}]$$

$$m_7 = \rho e^{i(13\pi/8)} = \frac{\rho}{8^{1/4}}[+(2^{1/2} - 1)^{1/2} - i(2^{1/2} + 1)^{1/2}]$$

$$m_8 = \rho e^{i(15\pi/8)} = \frac{\rho}{8^{1/4}}[+(2^{1/2} + 1)^{1/2} - i(2^{1/2} - 1)^{1/2}]$$

Points 1 to 8 on the Argand diagram (Fig. 1-3) represent the roots $m_1, m_2, ..., m_8$. The roots derived in this example are in fact those of the auxiliary equation associated with the Schorer characteristic equation developed in Chapter 7.

ITERATIVE SOLUTION OF EIGHTH-DEGREE POLYNOMIAL EQUATIONS

1-6 Method in Outline

The auxiliary equations relating to bending theories of cylindrical shells—other than the Schorer and D-K-J theories—do not have explicit roots. In this section, a simple iterative procedure is developed for extracting the roots of such equations to any desired degree of accuracy. This method, due to Ramaswamy and Ramaiah [1], is an extension of a rule given by Steinman[1] for finding the real roots of higher-degree equations. The auxiliary algebraic equation associated with the bending theories of cylindrical shells is of the eighth degree. It contains only the even powers of the variable. The eight roots of this equation are symmetrical about the x and y axes on the Argand diagram and occur in four complex conjugate pairs. This eighth-degree equation is first recast as a biquadratic equation by means of a suitable substitution. The relevant resolvent cubic equation is then formed. It is known from the experience gained from the analysis of a large number of cylindrical shells that the two roots of this cubic are nearly $+2$ and -2, the third root being nearly zero. The third root may be ignored without loss of accuracy. The initial trial values of $+2$ and -2 for the two roots are next refined by iteration using the Steinman rule. Knowing the roots of the cubic, the roots of the biquadratic are found. From these, the roots of the eighth-degree equation are determined.

1-7 Steinman Rule

If the given polynomial equation is

$$mx^n = ax^{n-1} + bx^{n-2} + cx^{n-3} + dx^{n-4} + \cdots$$

and if x_0 is the approximate value of one of its real roots, then its exact value is given by

$$x_1 = \frac{a + \dfrac{2b}{x_0} + \dfrac{3c}{x_0{}^2} + \dfrac{4d}{x_0{}^3} + \cdots}{m + \dfrac{b}{x_0{}^2} + \dfrac{2c}{x_0{}^3} + \dfrac{3d}{x_0{}^4} + \cdots} \qquad (1\text{-}10)$$

[1] Steinman, D. B., "Simple Formula Solves All Higher Degree Equations," *Civil Engineering*, vol. 21, no. 2, February, 1951.

This rule can be applied successively to improve an assumed trial value for the root.

1-8   Application to an Example

An example would best illustrate this procedure.

Example 1-5

The following data relate to a cylindrical shell:
Span $l = 83.25$ ft, radius $a = 25.0$ ft, thickness $d = 0.25$ ft.

$$\lambda_n = \frac{\pi a}{l} = 0.9434212174$$

$$\rho^8 = \frac{12\lambda_n^4 a^2}{d^2} = 95,061.311784$$

The auxiliary equation of the Finsterwalder theory, developed in Chapter 7, takes the form

$$m^8 + 2m^6(1 - \lambda_n^2) + m^4(1 - 4\lambda_n^2 + \lambda_n^4) + m^2(\lambda_n^4 - 2\lambda_n^2) + 12\frac{\lambda_n^4 a^2}{d^2} = 0$$

Making the substitution $\bar{m} = m/\rho$, the following equation results:

$$\bar{m}^8 + \frac{2(1 - \lambda_n^2)}{\rho^2}\bar{m}^6 + \frac{1 - 4\lambda_n^2 + \lambda_n^4}{\rho^4}\bar{m}^4 + \frac{\lambda_n^4 - 2\lambda_n^2}{\rho^6}\bar{m}^2 + 1 = 0$$

Substituting the data given for the shell, the equation becomes

$$\bar{m}^8 + 0.0125241882\bar{m}^6 - 0.0057342889\bar{m}^4 - 0.0001824796\bar{m}^2 + 1 = 0 \quad (1\text{-}11)$$

The substitution $y = \bar{m}^2$ reduces it to the biquadratic

$$y^4 + 0.0125241882y^3 - 0.0057342889y^2 - 0.0001829796y + 1 = 0 \quad (1\text{-}12)$$

The equation is of the form $y^4 + py^3 + qy^2 + ry + s = 0$.
To eliminate the term y^3, a substitution $y = (x - p/4)$ is next made. The equation now assumes the form

$$x^4 - 0.0057931096x^2 - 0.0001463254x + 1.0000005149 = 0 \quad (1\text{-}13)$$

Equation (1-13) is now of the form $x^4 + ax^2 + bx + c = 0$.
The resolvent cubic equation is $z^3 + 2az^2 + (a^2 - 4c)z - b^2 = 0$. Noting that $a = -0.0057931096$, $b = -0.0001463254$, and $c = 1.0000005149$, the resolvent cubic equation becomes

$$z^3 - 0.0115862192z^2 - 3.9999684995z - 0.0000000214 = 0 \quad (1\text{-}14)$$

To apply the Steinman rule, Equation (1-14) may be rewritten as

$$z^3 = 0.0115862192z^2 + 3.9999684995z + 0.0000000214 \qquad (1\text{-}15)$$

The First Root

Assume $z_0 = +2$. Inserting this in the Steinman formula, $z_1 = 2.0057852̇60$. The second cycle gives $z_1 = 2.005793627$. The third cycle gives the same value, showing that there is no scope for further improvement.

The Second Root

Let us start with a trial value $z_0 = -2$. Inserting this value in the Steinman value, we get $z_2 = -1.994198990$. A second trial gives $z_2 = -1.994207403$. A third trial results in $z_2 = -1.994207402$. Hence the three roots of the resolvent cubic equation are $z_1 = +2.005793627$, $z_2 = -1.994207402$, and $z_3 = 0$.

The roots of Equation (1-13) may now be written down as follows, using rules given by Descartes:

$$x_1 = \tfrac{1}{2}[+\sqrt{z_1} + i(\sqrt{z_2} + \sqrt{z_3})]$$

$$x_2 = \tfrac{1}{2}[+\sqrt{z_1} - i(\sqrt{z_2} + \sqrt{z_3})]$$

$$x_3 = \tfrac{1}{2}[-\sqrt{z_1} + i(\sqrt{z_2} - \sqrt{z_3})]$$

$$x_4 = \tfrac{1}{2}[-\sqrt{z_1} - i(\sqrt{z_2} - \sqrt{z_3})]$$

Hence

$$x_1 = \tfrac{1}{2}(+1.416260437 + i\,1.412164085)$$

$$x_2 = \tfrac{1}{2}(+1.416260437 - i\,1.412164085)$$

$$x_3 = \tfrac{1}{2}(-1.416260437 + i\,1.412164085)$$

$$x_4 = \tfrac{1}{2}(-1.416260437 - i\,1.412164085)$$

We know that $y = (x - p/4) = (x - 0.00313104705)$. Hence, adding (-0.00313104705) to the real parts of the roots,

$$y_1 = \tfrac{1}{2}(+1.413129390 + i\,1.412164085)$$

$$y_2 = \tfrac{1}{2}(+1.413129390 - i\,1.412164085)$$

$$y_3 = \tfrac{1}{2}(-1.419391484 + i\,1.412164085)$$

$$y_4 = \tfrac{1}{2}(-1.419391484 - i\,1.412164085)$$

Now

$$\bar{m}^2 = y$$

or

$$\bar{m} = \pm\sqrt{y}$$

The square roots are extracted by using the rules derived in Example 1-1 to give

$$\bar{m}_{1,2,3,4} = \pm(0.923432954 \pm i\,0.38231364805)$$

and

$$\bar{m}_{5,6,7,8} = \pm(0.3817155951 \pm i\,0.924879742)$$

Since $m = \rho\bar{m}$, the values of m are

$$m_{1,2,3,4} = \pm(3.869508850 \pm i\,1.602028646)$$
$$m_{5,6,7,8} = \pm(1.599522594 \pm i\,3.875571401)$$

Check

The product of the roots must be equal to ρ^8, the constant term in the eighth-degree equation in m.

The product of the roots is

$$[(3.869508850)^2 + (1.602028646)^2]^2 \times [(1.599522594)^2 + (3.875571401)^2]^2$$
$$= 95,061.35927$$

The value of $\rho^8 = 95,061.31178$. Hence the agreement is close.

The iteration procedure was explained with reference to the Finsterwalder theory only by way of illustration. This procedure may be employed to determine the roots relating to all other theories as well.

1-9 Other Methods

A brief reference will now be made to some of the other methods in use. In a method described by Lundgren [2] the eighth-degree equation is successively transformed into a biquadratic and later into a coupled quadratic equation. In the next step, the coupled quadratic is reduced to an ordinary quadratic equation and solved. Working backward, the eight roots of the original equation are found. Since it is an iterative method, at least two cycles are required to get reasonably accurate results. The procedure is involved and several operations with complex numbers are necessary. The method given in ASCE Manual 31 [3] essentially consists in reducing the eighth-degree equation in the first instance to a quartic equation whose resolvent cubic equation is solved by trigonometric methods. The method is tedious as it involves many operations with complex numbers. A rapid but approximate method for writing down the roots is that due to Carbone [4]. Let

$$x^8 + a_2x^6 + a_4x^4 + a_6x^2 + a_8 = 0$$

be the eighth-degree equation whose eight roots $x_1, x_2, ..., x_8$ are required. Let p be the real positive value of $(a_8)^{1/8}$. The eight roots are

$$x_{1,2,3,4} = \pm\alpha_1 \pm i\beta_1$$

and

$$x_{5,6,7,8} = \pm\alpha_2 \pm i\beta_2$$

where

$$\alpha_1 = 0.92388 \left(p + \frac{a_2}{8p}\right) + \frac{0.0478 \left(\dfrac{5a_2^2}{16} - a_4\right)}{p^3}$$

$$\beta_1 = 0.38268 \left(p - \frac{a_2}{8p}\right) - \frac{0.1155 \left(\dfrac{5a_2^2}{16} - a_4\right)}{p^3}$$

$$\alpha_2 = 0.38268 \left(p + \frac{a_2}{8p}\right) - \frac{0.1155 \left(\dfrac{5a_2^2}{16} - a_4\right)}{p^3}$$

and

$$\beta_2 = 0.92388 \left(p - \frac{a_2}{8p}\right) + \frac{0.0478 \left(\dfrac{5a_2^2}{16} - a_4\right)}{p^3}$$

CHAPTER 2

FOURIER SERIES

2-1 Representation of Periodic Functions

A function $f(x)$ is said to be periodic, if $f(x) = f(x + k)$, k being known as the period. In this chapter, methods are described for developing a periodic function in infinite trigonometric series known as Fourier series. Subject to a few restrictions, known as Dirichlet conditions,[1] almost any arbitrary function, not necessarily periodic, can be developed in this manner, provided the function is defined as periodic outside the interval in which we are interested.

2-2 Expansion in a Fourier Series

Let us suppose that we are interested in representing the function $f(x)$ in a Fourier series in the interval $-\pi$ to $+\pi$. The period in this case is 2π. We may write

$$f(x) = \frac{a_0}{2} + \sum_{n=1}^{n=\infty} (a_n \cos nx + b_n \sin nx) \tag{2-1}$$

It only now remains to determine the coefficients a_0, a_n, and b_n. Multiplying both sides by dx and integrating term by term, it will be seen that no terms except a_0 will contribute anything on the right-hand side as

$$\int_{-\pi}^{+\pi} \cos nx \, dx = \int_{-\pi}^{+\pi} \sin nx \, dx = 0$$

for $n = 1, 2,..., \infty$. Hence

$$a_0 = \frac{1}{\pi} \int_{-\pi}^{+\pi} f(x) \, dx \tag{2-2}$$

[1] Sokolnikoff, Ivan S. and E. S. Sokolnikoff, "Higher Mathematics for Engineers and Physicists," chap. 2, art. 8, McGraw-Hill Book Co., New York, 1941.

21

Next, multiply both sides by cos nx and integrate term by term, observing that

$$\int_{-\pi}^{+\pi} \sin mx \cos nx \, dx = 0$$

for all integral values of m and n and that

$$\int_{-\pi}^{+\pi} \cos mx \cos nx \, dx = 0$$

for all integral values of m and n except $m = n$. We get

$$a_n \int_{-\pi}^{+\pi} \cos^2 nx \, dx = \int_{-\pi}^{+\pi} f(x) \cos nx \, dx$$

Simplifying,

$$a_n = \frac{1}{\pi} \int_{-\pi}^{+\pi} f(x) \cos nx \, dx \qquad (2\text{-}3)$$

Equation (2-2) is only a special case of (2-3) corresponding to $n = 0$. In a similar manner,

$$b_n = \frac{1}{\pi} \int_{-\pi}^{+\pi} f(x) \sin nx \, dx \qquad (2\text{-}4)$$

We are now ready to apply the technique to an example.

Example 2-1

Expand $f(x) = x$ in Fourier series in the interval $-\pi$ to $+\pi$.

Solution

The function $y = f(x)$ to be expanded in Fourier series is shown plotted in Fig. 2-1. The function is shown outside the interval in

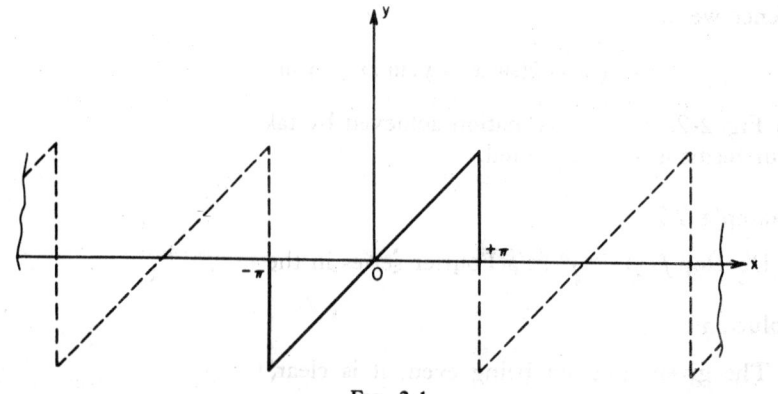

FIG. 2-1

dotted lines to indicate its periodicity.   Using (2-2),

$$a_0 = \frac{1}{\pi} \int_{-\pi}^{+\pi} x \, dx = 0$$

This is to be expected as a_0, physically interpreted, stands for the average value of the function over the interval.   This value is clearly zero, as the area of the diagram between $-\pi$ and $+\pi$ is zero. a_0, *thus, will be zero, whenever the given function is odd.*   Using (2-3),

$$a_n = \frac{1}{\pi} \int_{-\pi}^{+\pi} x \cos nx \, dx = 0$$

This could also have been foreseen, if it is recalled that

$$\cos x = 1 - \frac{x^2}{2!} + \frac{x^4}{4!} - \frac{x^6}{6!} + \cdots$$

Hence, the cosine terms cannot occur in the expansion for an odd function such as $f(x) = x$.   Only the sine terms will appear.   From (2-4),

$$b_n = \frac{1}{\pi} \int_{-\pi}^{+\pi} f(x) \sin nx \, dx$$

$$= \frac{1}{\pi} \left[\left(-\frac{x}{n} \cos nx \right)_{-\pi}^{+\pi} + \frac{1}{n} \int_{-\pi}^{+\pi} \cos nx \, dx \right]$$

$$= \frac{1}{\pi} \left[\left(-\frac{x}{n} \cos nx \right)_{-\pi}^{+\pi} + \frac{1}{n^2} (\sin nx)_{-\pi}^{+\pi} \right]$$

$$= -\frac{2}{n} \cos n\pi = -\frac{2}{n} (-1)^n \tag{2-5}$$

Hence we may write

$$f(x) = 2(\sin x - \tfrac{1}{2} \sin 2x + \tfrac{1}{3} \sin 3x - \cdots) \tag{2-6}$$

In Fig. 2-2, the approximation achieved by taking one, two, three, and four terms is shown plotted.

Example 2-2

Develop $f(x) = x^2$ in a Fourier series in the interval $-\pi \leqslant x \leqslant \pi$.

Solution

The given function being even, it is clear that sine terms will not appear in the expansion.

We need evaluate only a_0 and a_n.

$$a_0 = \frac{1}{\pi} \int_{-\pi}^{+\pi} x^2 \, dx = \frac{2}{3} \pi^2$$

$$\frac{a_0}{2} = \frac{1}{3} \pi^2$$

$$a_n = \frac{1}{\pi} \int_{-\pi}^{+\pi} x^2 \cos nx \, dx$$

$$= \frac{1}{\pi} \left[\left(\frac{x^2}{n} \sin nx \right)_{-\pi}^{+\pi} - \int_{-\pi}^{+\pi} \frac{2x}{n} \sin nx \, dx \right]$$

$$= \frac{1}{\pi} \left(\frac{1}{n} x^2 \sin nx + \frac{2}{n^2} x \cos nx - \frac{2}{n^3} \sin nx \right)_{-\pi}^{+\pi}$$

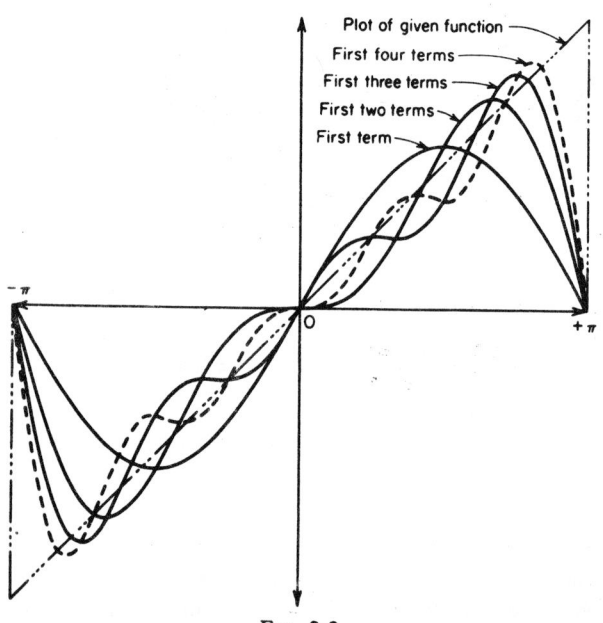

Plot of given function

First four terms

First three terms

First two terms

First term

FIG. 2-2

Simplifying,

$$a_n = \frac{4}{n^2} \cos n\pi = \frac{4}{n^2} (-1)^n \qquad (2\text{-}7)$$

Hence

$$x^2 = \frac{\pi^2}{3} + 4 \sum_{n=1}^{n=\infty} (-1)^n \frac{1}{n^2} \cos nx \qquad (2\text{-}8)$$

In Fig. 2-3, the approximation achieved by taking one, two, three, and four terms is shown plotted.

2-3 Expansion in the Interval 0 to π

In many applications, it is convenient to expand a given function in terms of sines only or cosines only, irrespective of whether the given function is odd or even.   This is possible if the interval is from 0 to π. The expansion is obtained by employing the following artifice.   An

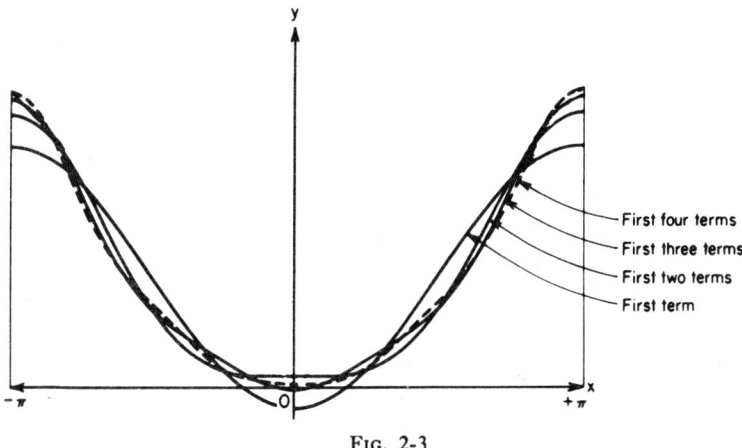

First four terms
First three terms
First two terms
First term

Fig. 2-3

auxiliary function $F(x)$ is introduced.   If expansion in terms of cosines only is desired, $F(x)$ is defined such that

$$F(x) \equiv f(x) \qquad 0 \leqslant x \leqslant +\pi$$

and

$$F(x) \equiv f(-x) \qquad -\pi \leqslant x \leqslant 0$$

By this means we ensure that the auxiliary function is an even function which, within the range in which we are interested, is identical with the function whose expansion is desired. $F(x)$ being an even function, we need look only for cosine terms.   From (2-3),

$$a_n = \frac{1}{\pi} \int_{-\pi}^{+\pi} F(x) \cos nx \, dx$$

$$= \frac{1}{\pi} \left[\int_{-\pi}^{0} F(x) \cos nx \, dx + \int_{0}^{+\pi} F(x) \cos nx \, dx \right]$$

$$= \frac{1}{\pi} \left[\int_{-\pi}^{0} f(-x) \cos n(-x)(dx) + \int_{0}^{+\pi} f(x) \cos nx \, dx \right]$$

$$= \frac{2}{\pi} \int_{0}^{+\pi} f(x) \cos nx \, dx \tag{2-9}$$

The constant term a_0 may be obtained by putting $n = 0$ in (2-9) to give

$$a_0 = \frac{2}{\pi} \int_0^{+\pi} f(x)\, dx \tag{2-10}$$

If, now, we wish to expand the given function in terms of sines only, $F(x)$ is defined such that

$$F(x) \equiv f(x) \qquad 0 \leqslant x \leqslant \pi$$

and

$$F(x) \equiv -f(-x) \qquad -\pi \leqslant x \leqslant 0$$

This implies that $F(x)$ is odd and only sine terms will appear in the expansion. Again from (2-4),

$$\begin{aligned}
b_n &= \frac{1}{\pi} \int_{-\pi}^{+\pi} F(x) \sin nx\, dx \\
&= \frac{1}{\pi} \left[\int_{-\pi}^{0} -f(-x) \sin n(-x)(-dx) + \int_0^{+\pi} f(x) \sin nx\, dx \right] \\
&= \frac{2}{\pi} \int_0^{+\pi} f(x) \sin nx\, dx \tag{2-11}
\end{aligned}$$

We may now turn to an example to illustrate this procedure.

Example 2-3

Expand $f(x) = x$ in the interval 0 to π in terms of cosines only.

Solution

Using (2-9),

$$\begin{aligned}
a_n &= \frac{2}{\pi} \int_0^{+\pi} x \cos nx\, dx \\
&= \frac{2}{\pi} \left[\left(\frac{x}{n} \sin nx \right)_0^{\pi} - \frac{1}{n} \int_0^{+\pi} \sin nx\, dx \right] \\
&= \frac{2}{\pi n^2} (\cos n\pi - 1) = \frac{2}{\pi n^2} [(-1)^n - 1] \tag{2-12}
\end{aligned}$$

Making use of (2-10),

$$a_0 = \frac{2}{\pi} \int_0^{+\pi} x\, dx = \pi \tag{2-13}$$

Hence the expansion desired is

$$x = \frac{\pi}{2} - \frac{4}{\pi} \left(\frac{1}{1^2} \cos x + \frac{1}{3^2} \cos 3x + \frac{1}{5^2} \cos 5x + \cdots \right) \tag{2-14}$$

2-4 Expansion in the Interval −*l* to +*l*

The problem is to expand a function $f(x)$ in Fourier series in the interval $-l$ to $+l$. Introduce a new variable z which takes on values from $-\pi$ to $+\pi$ as x varies from $-l$ to $+l$ to cover the given interval. Clearly,

$$x = \frac{l}{\pi} z \tag{2-15}$$

and

$$dx = \frac{l}{\pi} dz \tag{2-16}$$

It is now possible to write $f(x) = f[(l/\pi)z]$. The function of z thus defined may be expanded in a Fourier series as follows in the interval $-\pi$ to $+\pi$ by using the results of Art. 2-2. Thus

$$f\left(\frac{l}{\pi} z\right) = \frac{a_0}{2} + \sum_{n=1}^{n=\infty} (a_n \cos nz + b_n \sin nz) \tag{2-17}$$

Using (2-2),

$$a_0 = \frac{1}{\pi} \int_{-\pi}^{+\pi} f\left(\frac{lz}{\pi}\right) dz \tag{2-18}$$

Substituting the relations (2-15) and (2-16) in (2-18),

$$a_0 = \frac{1}{l} \int_{-l}^{+l} f(x)\, dx \tag{2-19}$$

Similarly,

$$a_n = \frac{1}{\pi} \int_{-\pi}^{+\pi} f\left(\frac{lz}{\pi}\right) \cos nz\, dz$$

$$= \frac{1}{l} \int_{-l}^{+l} f(x) \cos \frac{n\pi x}{l}\, dx \tag{2-20}$$

and

$$b_n = \frac{1}{l} \int_{-l}^{+l} f(x) \sin \frac{n\pi x}{l}\, dx \tag{2-21}$$

An example would best illustrate the use of these formulas.

Example 2-4

Develop $f(x)$ in a Fourier series in the interval $-l$ to $+l$ given that

$$f(x) = 0 \qquad -l < x < 0$$
$$f(x) = p \qquad 0 < x < +l$$

Solution

Using (2-19),

$$a_0 = \frac{1}{l} \int_{-l}^{+l} f(x) \, dx = \frac{1}{l} \int_0^{+l} p \, dx = p$$

$$\frac{a_0}{2} = \frac{p}{2}$$

From (2-20),

$$a_n = \frac{1}{l} \int_{-l}^{+l} f(x) \cos \frac{n\pi x}{l} \, dx = \frac{1}{l} \int_0^{+l} p \cos \frac{n\pi x}{l} \, dx$$

$$= \frac{p}{n\pi} \left(\sin \frac{n\pi x}{l} \right)_0^{+l} = 0$$

Making use of (2-21),

$$b_n = \frac{1}{l} \int_{-l}^{+l} f(x) \sin \frac{n\pi x}{l} \, dx$$

$$= \frac{1}{l} \int_0^{+l} p \sin \frac{n\pi x}{l} \, dx = \frac{p}{n\pi} (1 - \cos n\pi)$$

Hence

$$f(x) = \frac{p}{2} + \frac{2p}{\pi} \left(\sin \frac{\pi x}{l} + \frac{1}{3} \sin \frac{3\pi x}{l} + \frac{1}{5} \sin \frac{5\pi x}{l} + \cdots \right) \qquad (2\text{-}22)$$

2-5   Half-range Series in the Interval 0 to *l*

By introducing an auxiliary function as in Art. 2-3, it is possible to develop a function $f(x)$ in a Fourier series in the interval 0 to *l*. It is easily verified that the formulas for a_0, a_n, and b_n that are applicable to such series are the following:

$$a_0 = \frac{2}{l} \int_0^{+l} f(x) \, dx \qquad (2\text{-}23)$$

$$a_n = \frac{2}{l} \int_0^{+l} f(x) \cos \frac{n\pi x}{l} \, dx \qquad (2\text{-}24)$$

and

$$b_n = \frac{2}{l} \int_0^{+l} f(x) \sin \frac{n\pi x}{l} \, dx \qquad (2\text{-}25)$$

In this interval, it is possible to expand a function in terms of cosines only or sines only. In the three examples that follow, this procedure is illustrated.

Example 2-5

Expand the function $f(x) = p$, where p is a constant, in a sine series in the interval 0 to l. Using formula (2-25),

$$b_n = \frac{2}{l} \int_0^{+l} p \sin \frac{n\pi x}{l} \, dx = \frac{2p}{n\pi}(1 - \cos n\pi) = \frac{2p}{n\pi}[1 - (-1)^n]$$

Even sine terms will not appear in the expansion. We may write the series as

$$p = \frac{4p}{\pi}\left(\sin \frac{\pi x}{l} + \frac{1}{3}\sin \frac{3\pi x}{l} + \cdots\right)$$

or

$$p = \frac{4p}{\pi} \sum_{n=1,3,5}^{n=\infty} \frac{1}{n}\sin \frac{n\pi x}{l} \tag{2-26}$$

The approximation achieved by taking one, two, three, and four terms of the series is indicated in Fig. 2-4.

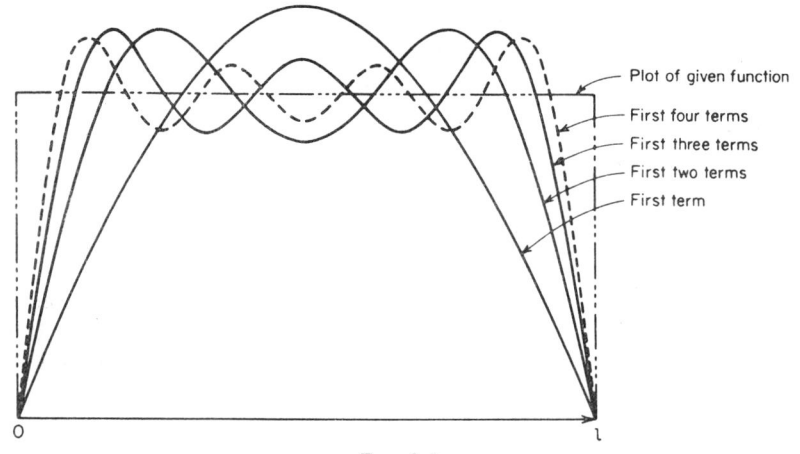

Plot of given function
First four terms
First three terms
First two terms
First term

FIG. 2-4

Example 2-6

$$f(x) = p \qquad 0 < x < \frac{l}{2}$$

$$f(x) = -p \qquad \frac{l}{2} < x < l$$

Develop $f(x)$ in a Fourier series in terms of cosines only in the interval 0 to l.

Solution

$$a_0 = \frac{2}{l} \int_0^{+l} f(x)\, dx$$

$$= \frac{2}{l} \left(\int_0^{+l/2} p\, dx + \int_{+l/2}^{+l} -p\, dx \right) = 0$$

$$a_n = \frac{2}{l} \int_0^{+l} f(x) \cos \frac{n\pi x}{l}\, dx$$

$$= \frac{2}{l} \left(\int_0^{+l/2} p \cos \frac{n\pi x}{l}\, dx - \int_{+l/2}^{+l} p \cos \frac{n\pi x}{l}\, dx \right) = \frac{4p}{n\pi} \sin \frac{n\pi}{2}$$

$$= \frac{4p}{n\pi} (-1)^{(n-1)/2} \qquad \text{with} \quad n = 1, 3, 5, \ldots$$

Hence

$$f(x) = \frac{4p}{\pi} \left(\cos \frac{\pi x}{l} - \frac{1}{3} \cos \frac{3\pi x}{l} + \frac{1}{5} \cos \frac{5\pi x}{l} - \frac{1}{7} \cos \frac{7\pi x}{l} + \cdots \right) \qquad (2\text{-}27)$$

The approximation of the function by one, two, three, and four terms of the series is indicated in Fig. 2-5. It may be observed in passing that the series within the brackets becomes $(1 - \frac{1}{3} + \frac{1}{5} - \frac{1}{7} + \cdots)$ if we put $x = 0$. The sum of this series is known to be $\pi/4$. It is thus confirmed that the series converges to p at $x = 0$.

Example 2-7

$$f(x) = 0 \qquad 0 < x < d - \frac{\epsilon}{2}$$

$$f(x) = p \qquad d - \frac{\epsilon}{2} < x < d + \frac{\epsilon}{2}$$

where p is a constant, and

$$f(x) = 0 \qquad d + \frac{\epsilon}{2} < x < l$$

Find the Fourier expansion of the function in sine series.

Solution

$$b_n = \frac{2}{l} \int_0^{+l} f(x) \sin \frac{n\pi x}{l}\, dx$$

$$= \frac{2}{l} \int_{d-\epsilon/2}^{d+\epsilon/2} p \sin \frac{n\pi x}{l}\, dx$$

$$= -\frac{2p}{n\pi} \left(\cos \frac{n\pi x}{l} \right)_{d-\epsilon/2}^{d+\epsilon/2}$$

$$= \frac{4p}{n\pi} \sin \frac{n\pi d}{l} \sin \frac{n\pi \epsilon}{2l}$$

Let us examine the value of b_n when $\epsilon \to 0$. In limit $\sin(n\pi\epsilon/l) = n\pi\epsilon/l$.

$$b_n = \frac{2p\epsilon}{l} \sin \frac{n\pi d}{l}$$

Let

$$p\epsilon = P$$

Hence

$$b_n = \frac{2P}{l} \sin \frac{n\pi d}{l}$$

The series expansion of $f(x)$ is

$$f(x) = \frac{2P}{l} \sum_{n=1}^{n=\infty} \sin \frac{n\pi d}{l} \sin \frac{n\pi x}{l} \qquad (2\text{-}28)$$

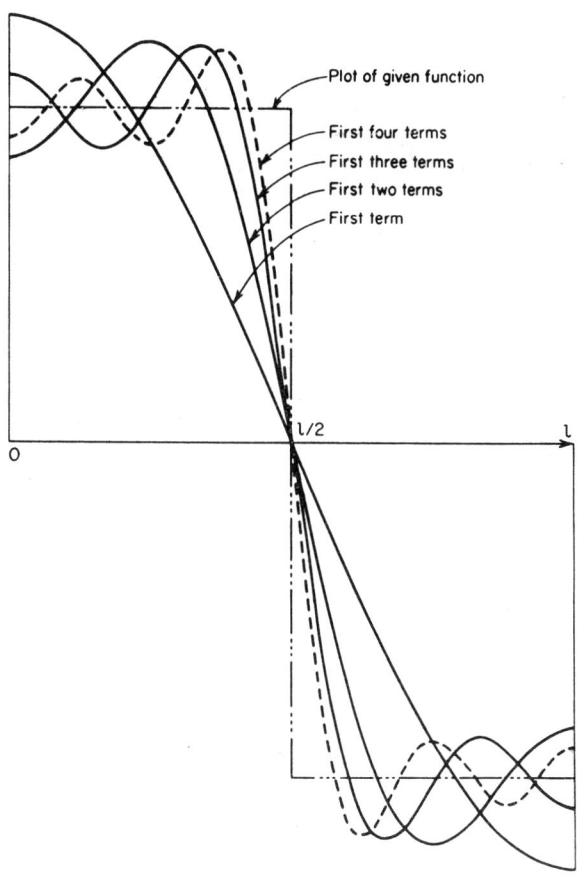

Plot of given function

First four terms

First three terms

First two terms

First term

FIG. 2-5

The above series, though divergent, is often used in structural analysis. Such use is justified by Hobson.[1]

The results developed in Examples 2-5, 2-6, and 2-7 find many applications in structural analysis. The expansions derived in Examples 2-5 and 2-6 are employed to represent a uniformly distributed load in the theory of cylindrical shells. The function expressed in series form in Equation (2-28) may be used to represent a concentrated load on a beam.

2-6 Double Fourier Series

Problems on plates and shells sometimes involve the expansion of a given function in double Fourier series. Let it be required to develop $f(x,y)$ in Fourier series in the interval $0 < x < a$ and $0 < y < b$ (Fig. 2-6). We may write

$$f(x, y) = \sum_{m=1}^{m=\infty} \sum_{n=1}^{n=\infty} a_{mn} \sin \frac{m\pi x}{a} \sin \frac{n\pi y}{b}$$

To determine any particular coefficient a_{pq} we multiply both sides by $\sin(p\pi x/a)$ and integrate both sides with respect to x, noting that

$$\int_0^{+a} \sin \frac{m\pi x}{a} \sin \frac{p\pi x}{a} = 0$$

when $m \neq p$

and

$$\int_0^{+a} \sin \frac{m\pi x}{a} \sin \frac{p\pi x}{a} = \frac{a}{2}$$

when $m = p$

Hence we have

$$\int_0^{+a} f(x, y) \sin \frac{p\pi x}{a} = \frac{a}{2} \sum_{n=1}^{n=\infty} a_{pn} \sin \frac{n\pi y}{b}$$

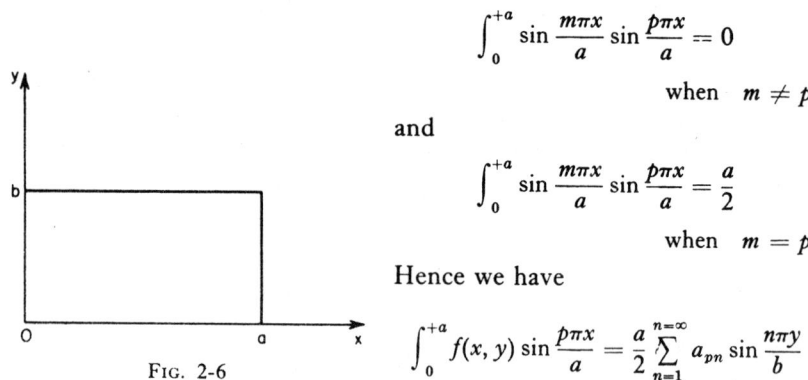

FIG. 2-6

Similarly, we may now multiply both sides by $\sin(q\pi y/b)$ and integrate with respect to y to get

$$a_{pq} = \frac{4}{ab} \int_0^{+b} \int_0^{+a} f(x, y) \sin \frac{p\pi x}{a} \sin \frac{q\pi y}{b} \, dx \, dy \qquad (2\text{-}29)$$

[1] Hobson, E. W., "The Theory of Functions of a Real Variable and the Theory of Fourier Series," vol. II, second revised enlarged edition, p. 493, Dover Publications, Inc., New York, 1957.

In the special case where $f(x,y) = q_0$, a constant, it is easy to verify that

$$a_{pq} = \frac{4q_0}{ab} \int_0^{+b} \int_0^{+a} \sin\frac{p\pi x}{a} \sin\frac{q\pi y}{b}\, dx\, dy = \frac{16q_0}{\pi^2 pq} \tag{2-30}$$

where p and q are odd integers. If p, q, or both are even integers, $a_{pq} = 0$. This result will be used for expanding a uniformly distributed load on shells of double curvature.

2-7 Applications in Structural Analysis

The application of Fourier series to structural analysis is best explained by means of an example.

Example 2-8

Derive an expression for the deflection of a simply supported beam of span l carrying a uniformly distributed load of intensity p.

Solution

Choosing the origin at the end A (Fig. 2-7), the Fourier expansion for p may be written as $p = (4p/\pi)\,[\sin{(\pi x/l)} + \frac{1}{3}\sin{(3\pi x/l)} + \cdots]$ using

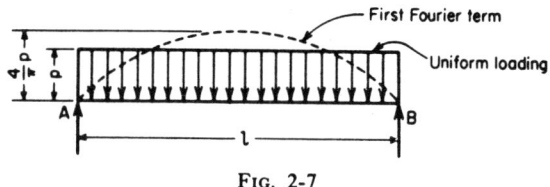

Fig. 2-7

the result derived in Example 2-5. Hence we may write

$$EI\frac{d^4y}{dx^4} = \frac{4p}{\pi}\left(\sin\frac{\pi x}{l} + \frac{1}{3}\sin\frac{3\pi x}{l} + \frac{1}{5}\sin\frac{5\pi x}{l} + \cdots\right)$$

Integrate twice and introduce two arbitrary constants to get

$$-EI\frac{d^2y}{dx^2} = \frac{4pl^2}{\pi^3}\left(\sin\frac{\pi x}{l} + \frac{1}{3^3}\sin\frac{3\pi x}{l} + \frac{1}{5^3}\sin\frac{5\pi x}{l} + \cdots + Ax + B\right)$$

Because the bending moment is zero at $x = 0$ and $x = l$, $A = B = 0$. Integrating twice again and noting that $y = 0$ at $x = 0$ and $x = l$,

$$EIy = \frac{4p}{\pi}\left(\frac{l}{\pi}\right)^4\left(\sin\frac{\pi x}{l} + \frac{1}{3^5}\sin\frac{3\pi x}{l} + \frac{1}{5^5}\sin\frac{5\pi x}{l} + \cdots\right) \tag{2-31}$$

Deflection at midspan may be found by putting $x = l/2$. This would result in the series

$$EIy_{(x=l/2)} = \frac{4p}{\pi}\left(\frac{l}{\pi}\right)^4\left(1 - \frac{1}{3^5} + \frac{1}{5^5} - \cdots\right) \qquad (2\text{-}32)$$

Even a single term of the series gives a very accurate value for the deflection. Taking only the first term, the deflection at the center works out to $0.01307\ pl^4/EI$ as against the exact value of $0.01302\ pl^4/EI$. The error is only 0.38 %. Thus it is seen that, although the load component given by the first term of the Fourier series departs considerably from a uniform distribution, it gives a very accurate result for the deflection. This explains why it is often adequate to consider one term of the series only. The elastic lines corresponding to the uniformly distributed load p and its first Fourier component will hardly be distinguishable from each other.

CHAPTER 3

ELEMENTS OF MATRIX ALGEBRA

3-1 Matrix Algebra in Structural Analysis

Matrix methods offer an elegant and systematic means of formulating complex problems of structural analysis. Structural engineers need only master a few simple rules set out in this chapter to be able to make use of this powerful tool.

3-2 Definitions

A matrix is a square or rectangular array of numbers which are usually enclosed in square brackets thus:

$$[A] = \begin{bmatrix} a_{11} & a_{12} & a_{13} \\ a_{21} & a_{22} & a_{23} \\ a_{31} & a_{32} & a_{33} \end{bmatrix}$$

Each number of the array, such as a_{23}, is known as an *element* of the matrix. Thus, there are nine elements in matrix $[A]$. The first subscript carried by the element stands for the row and the second subscript for the column in which it occurs. For example, it is clear that a_{23} occurs in the second row and the third column. If a matrix has m rows and n columns, it is said to be of the *order* $(m \times n)$. A *square matrix* has as many rows as it has columns. The elements lying on the diagonal connecting the top left-hand corner and the bottom right-hand corner of a *square matrix* define its leading *diagonal* or *principal diagonal*.

3-3 Addition and Subtraction

If a matrix $[A]$ is added to a matrix $[B]$ to give a matrix $[C]$, the elements of the matrix $[C]$ are found by adding the corresponding elements of the matrices $[A]$ and $[B]$. For example, if

$$[A] = \begin{bmatrix} a_{11} & a_{12} & a_{13} \\ a_{21} & a_{22} & a_{23} \\ a_{31} & a_{32} & a_{33} \end{bmatrix}$$

and

$$[B] = \begin{bmatrix} b_{11} & b_{12} & b_{13} \\ b_{21} & b_{22} & b_{23} \\ b_{31} & b_{32} & b_{33} \end{bmatrix}$$

$$[C] = \begin{bmatrix} (a_{11} + b_{11}) & (a_{12} + b_{12}) & (a_{13} + b_{13}) \\ (a_{21} + b_{21}) & (a_{22} + b_{22}) & (a_{23} + b_{23}) \\ (a_{31} + b_{31}) & (a_{32} + b_{32}) & (a_{23} + b_{33}) \end{bmatrix} \quad (3\text{-}1)$$

If matrix $[B]$ is subtracted from matrix $[A]$ to give matrix $[D]$, the elements of $[D]$ may be found by subtracting the corresponding elements of $[B]$ from those of $[A]$. Using the matrices $[A]$ and $[B]$ for purposes of illustration,

$$[D] = \begin{bmatrix} (a_{11} - b_{11}) & (a_{12} - b_{12}) & (a_{13} - b_{13}) \\ (a_{21} - b_{21}) & (a_{22} - b_{22}) & (a_{23} - b_{23}) \\ (a_{31} - b_{31}) & (a_{32} - b_{32}) & (a_{33} - b_{33}) \end{bmatrix} \quad (3\text{-}2)$$

3-4 Multiplication

Suppose it is desired to multiply matrix $[A]$ by matrix $[B]$. *It has to be clearly specified whether $[A]$ is to be postmultiplied or premultiplied by $[B]$.* This is because these two operations do not, in general, give the same result. Thus, $[A][B]$ is not always equal to $[B][A]$. *For postmultiplication of $[A]$ by $[B]$ to be possible, there must be as many columns in $[A]$ as there are rows in $[B]$.* Such matrices are said to be *conformable*. Let us suppose that the order of the matrix $[A]$ is $(m \times p)$ and that of $[B]$ is $(p \times n)$; then the order of the matrix $[A][B]$ is $(m \times n)$. Multiplication of two matrices is carried out by the "row by column" rule. An example would best illustrate the operation.

Let it be required to postmultiply $[A]$ by $[B]$ given that

$$[A] = \begin{bmatrix} a_{11} & a_{12} & a_{13} \\ a_{21} & a_{22} & a_{23} \\ a_{31} & a_{32} & a_{33} \end{bmatrix}$$

and

$$[B] = \begin{bmatrix} b_{11} & b_{12} & b_{13} \\ b_{21} & b_{22} & b_{23} \\ b_{31} & b_{32} & b_{33} \end{bmatrix}$$

Let

$$[A][B] = [C]$$

where

$$[C] = \begin{bmatrix} c_{11} & c_{12} & c_{13} \\ c_{21} & c_{22} & c_{23} \\ c_{31} & c_{32} & c_{33} \end{bmatrix}$$

The elements of $[C]$ are found as follows: To determine c_{11}, for instance, the first-row elements of $[A]$ are multiplied by the first-column elements of $[B]$ and the results are added. Thus

$$c_{11} = (a_{11}b_{11} + a_{12}b_{21} + a_{13}b_{31})$$

In a similar manner,

$$c_{12} = (a_{11}b_{12} + a_{12}b_{22} + a_{13}b_{32})$$

Generalizing this procedure,

$$c_{ij} = \sum_{p=1}^{p=3} a_{ip}b_{pj} \tag{3-3}$$

We have already seen that, for matrix multiplication to be possible at all, p must be the same for a and b. Let us apply this rule to determine c_{32}. For this element, $i = 3, j = 2, p = 1, 2$, and 3. Hence

$$c_{32} = (a_{31}b_{12} + a_{32}b_{22} + a_{33}b_{32})$$

An example will clarify the operation involved.

Example 3-1

Given that

$$[A] = \begin{bmatrix} 2 & 1 & 1 \\ 3 & 2 & 3 \\ 4 & 1 & 2 \end{bmatrix}$$

and

$$[B] = \begin{bmatrix} 2 & 2 \\ 1 & 2 \\ 1 & 1 \end{bmatrix}$$

Evaluate the product $[A]\,[B]$.

Solution

Because the matrix $[A]$ has three columns and the matrix $[B]$ has three rows, multiplication is possible. The order of the matrix $[A]$ $[B] = (3 \times 3) \times (3 \times 2) = (3 \times 2)$.

$$[A]\,[B] = \begin{bmatrix} 6 & 7 \\ 11 & 13 \\ 11 & 12 \end{bmatrix}$$

It is of interest to observe that premultiplication indicated by $[B]\,[A]$ is impossible as the matrix $[B]$ has two columns and the matrix $[A]$ has three rows.   As in ordinary algebra, the associative law of multiplication holds good for continued products.   Thus

$$[A] [B] [C] = [A]([B] [C]) = ([A] [B]) [C]$$

For the continued products to have meaning, the adjacent matrices of the chain must be conformable.   The elements of the product matrix $[D] = [A] [B] [C]$ are found by double summation.   Thus

$$d_{ij} = \sum_q \sum_p a_{ip} b_{pq} c_{qj}$$

3-5   Special Matrices

A matrix with one column only is known as a *column matrix* or *column vector*.   The elements of a column vector are usually enclosed within brackets as indicated below:

$$\{A\} = \begin{bmatrix} a_{11} \\ a_{21} \\ a_{31} \\ a_{41} \end{bmatrix}$$

A matrix with elements equal to unity along the principal diagonal and zero everywhere else is known as a *unit matrix*.   It is usually designated by $[I]$, and its elements are

$$[I] = \begin{bmatrix} 1 & 0 & 0 & 0 \\ 0 & 1 & 0 & 0 \\ 0 & 0 & 1 & 0 \\ 0 & 0 & 0 & 1 \end{bmatrix}$$

The unit matrix has the property that any matrix $[A]$ remains unaltered when premultiplied or postmultiplied by the unit matrix.   Or

$$[A] [I] = [I] [A] = [A] \tag{3-4}$$

The unit matrix in matrix algebra plays a role similar to unity in algebra.

3-6   Inverse or Reciprocal Matrix

The *inverse* or *reciprocal matrix* of a given matrix is such that, when it is multiplied by the given matrix, the result is a unit matrix.   The inverse of the matrix $[A]$ is usually written as $[A]^{-1}$.   From definition, it is clear $[A]^{-1} [A] = [A][A]^{-1} = [I]$.

3-7   Methods of Inverting Matrices

The process of finding the inverse of a given matrix is known as *inversion*.   The inverse matrix can always be found, unless the given matrix is singular. A matrix is said to be *singular* if the determinant formed out of its elements vanishes. Several methods are available for inverting matrices [5]. The standard rule for inverting matrices is that given by Pipes.[1] But this procedure is seldom convenient in practice for inverting matrices of high order.

3-8 The Modified Gauss-Doolittle Method

In this short introduction to matrix algebra, it is not possible to deal with all the available methods for inverting matrices. Only one method —the Gauss-Doolittle scheme—has been selected for detailed presentation. It follows directly from the definition of an inverse matrix and does not involve the memorizing of any rules.

In this method, the given matrix [A] and a unit matrix [I] are written side by side. A selected sequence of operations is performed on both the matrices so that the matrix [A] is reduced, in gradual stages, to a unit matrix. Since the same operations are carried out on the matrix [I], it would have been reduced to [A]⁻¹. The steps involved are enumerated below.

Operation 0

Write matrices [A] and [I] side by side. Each matrix may be provided with a check column in which the sums of the elements occurring in each row of [A] and [I] are indicated. All operations carried out on the matrices are carried out on the check columns as well and any arithmetical error made is easily detected.

Operation 1A

Select one of the leading diagonal elements of [A] as a *pivot*. The row and column in which this element occurs are known as the pivotal row and column. Divide all elements of each row by the element that occurs in the pivotal column of that row. It is clear that the effect of this step would be to reduce all elements in the pivotal column of [A] to unity.

Operation 1B

Keeping the pivotal row unaltered, the elements in the other rows are replaced by new elements which are obtained by subtracting the elements in the pivotal row from the corresponding elements of the other rows. The object of this step is to reduce all elements in the pivotal column of [A] except the pivot itself to zero.

---

[1] Pipes, Louis A., "Applied Mathematics for Engineers and Physicists," 1st ed., p. 81, McGraw-Hill Book Company, New York, 1946.

Operations 2A and 2B

Selecting a new pivot for the matrix, operations identical with those described under 1A and 1B are carried out with the object of reducing all elements except the pivot in the new pivotal column to zero.

The process is repeated as many times as is necessary, with a new pivot each time.   At the end of these operations, all elements except those on the leading diagonal of [A] would have been reduced to zero.   Along the leading diagonal the elements will have some numerical values.   The last step consists in reducing these elements to unity by dividing each row by the diagonal element of that row.   The same sequence of operations enumerated for [A] is simultaneously applied on [I], reducing it, gradually, to [A]$^{-1}$.   An example will serve to illustrate the scheme.

Example 3-2

Invert the matrix [A] given below.

$$[A] = \begin{bmatrix} 4 & 2 & 3 & 1 \\ 1 & 1 & 1 & 2 \\ 2 & 3 & 2 & 3 \\ 3 & 4 & 4 & 2 \end{bmatrix}$$

Inversion of Matrix [A]

Opera-tion No.	[A]				Check column	[I]				Check column
0	4	2	3	1	10	1	0	0	0	1
	1	1	1	2	5	0	1	0	0	1
	2	3	2	3	10	0	0	1	0	1
	3	4	4	☐2	13	0	0	0	1	1
1A	4	2	3	1	10	1	0	0	0	1
	1/2	1/2	1/2	1	5/2	0	1/2	0	0	1/2
	2/3	1	2/3	1	10/3	0	0	1/3	0	1/3
	3/2	2	2	1	13/2	0	0	0	1/2	1/2
1B	5/2	0	1	0	7/2	1	0	0	−1/2	1/2
	−1	−3/2	−3/2	0	−4	0	1/2	0	−1/2	0
	−5/6	−1	☐−4/3	0	−19/6	0	0	1/3	−1/2	−1/6
	3/2	2	2	1	13/2	0	0	0	1/2	1/2
2A	5/2	0	1	0	7/2	1	0	0	−1/2	1/2
	2/3	1	1	0	8/3	0	−1/3	0	1/3	0
	5/8	3/4	1	0	19/8	0	0	−1/4	3/8	1/8
	3/4	1	1	1/2	13/4	0	0	0	1/4	1/4
2B	15/8	−3/4	0	0	9/8	1	0	1/4	−7/8	3/8
	1/24	☐1/4	0	0	7/24	0	−1/3	1/4	−1/24	−1/8
	5/8	3/4	1	0	19/8	0	0	−1/4	3/8	1/8
	1/8	1/4	0	1/2	7/8	0	0	1/4	−1/8	1/8

Opera- tion No.	[A]				Check column	[I]				Check column
3A	-5/2	1	0	0	-3/2	-4/3	0	-1/3	7/6	-1/2
	1/6	1	0	0	7/6	0	-4/3	1	-1/6	-1/2
	5/6	1	4/3	0	19/6	0	0	-1/3	1/2	1/6
	1/2	1	0	2	7/2	0	0	1	-1/2	1/2
3B	|-8/3|	0	0	0	-8/3	-4/3	4/3	-4/3	4/3	0
	1/6	1	0	0	7/6	0	-4/3	1	-1/6	-1/2
	2/3	0	4/3	0	2	0	4/3	-4/3	2/3	2/3
	1/3	0	0	2	7/3	0	4/3	0	-1/3	1
4A	1	0	0	0	1	1/2	-1/2	1/2	-1/2	0
	1	6	0	0	7	0	-8	6	-1	-3
	1	0	2	0	3	0	2	-2	1	1
	1	0	0	6	7	0	4	0	-1	3
4B	1	0	0	0	1	1/2	-1/2	1/2	-1/2	0
	0	6	0	0	6	-1/2	-15/2	11/2	-1/2	-3
	0	0	2	0	2	-1/2	5/2	-5/2	3/2	1
	0	0	0	6	6	-1/2	9/2	-1/2	-1/2	3
5	1	0	0	0	1	1/2	-1/2	1/2	-1/2	0
	0	1	0	0	1	-1/12	-5/4	11/12	-1/12	-1/2
	0	0	1	0	1	-1/4	5/4	-5/4	3/4	1/2
	0	0	0	1	1	-1/12	3/4	-1/12	-1/12	1/2

Hence the inverse matrix $[A]^{-1}$ is

$$[A]^{-1} = \begin{bmatrix} 1/2 & -1/2 & 1/2 & -1/2 \\ -1/12 & -5/4 & 11/12 & -1/12 \\ -1/4 & 5/4 & -5/4 & 3/4 \\ -1/12 & 3/4 & -1/12 & -1/12 \end{bmatrix}$$

3-9 Solution of Linear Simultaneous Equations

Let it be required to solve the following set of linear simultaneous equations:

$$a_{11}x_1 + a_{12}x_2 + a_{13}x_3 + \cdots + a_{1n}x_n = b_1$$

$$a_{21}x_1 + a_{22}x_2 + a_{23}x_3 + \cdots + a_{2n}x_n = b_2$$

$$\begin{matrix} . & . & . & \cdots & . \\ . & . & . & \cdots & . \end{matrix}$$

$$a_{n1}x_1 + a_{n2}x_2 + a_{n3}x_3 + \cdots + a_{nn}x_n = b_n$$

In matrix form, this set of equations may be written as

$$[A]\{X\} = \{B\} \tag{3-5}$$

The matrix $[A]$ is usually known as the coefficient matrix.   To solve for $x_1, x_2, \ldots, x_n$, premultiply both sides by $[A]^{-1}$ to get

$$[A]^{-1}[A]\{X\} = [A]^{-1}\{B\}$$

Noting that

$$[A]^{-1}[A] = [I]$$

we have

$$[I]\{X\} = [A]^{-1}\{B\}$$

But, we know that

$$[I]\{X\} = \{X\}$$

Hence

$$\{X\} = [A]^{-1}\{B\} \qquad (3\text{-}6)$$

It is clear from (3-6) that the solution of a set of linear simultaneous equations involves the inversion of the coefficient matrix $[A]$.

Example 3-3

Solve the equations

$$\begin{aligned}
4x_1 + 2x_2 + 3x_3 + x_4 &= 16 \\
x_1 + x_2 + x_3 + 2x_4 &= 4 \\
2x_1 + 3x_2 + 2x_3 + 3x_4 &= 11 \\
3x_1 + 4x_2 + 4x_3 + 2x_4 &= 21
\end{aligned}$$

The coefficient matrix $[A]$ of this problem is the same as the matrix of Example 3-2.   Its inverse has already been found as

$$[A]^{-1} = \begin{bmatrix}
1/2 & -1/2 & 1/2 & -1/2 \\
-1/12 & -5/4 & 11/12 & -1/12 \\
-1/4 & 5/4 & -5/4 & 3/4 \\
-1/12 & 3/4 & -1/12 & -1/12
\end{bmatrix}$$

$$\{B\} = \begin{bmatrix} 16 \\ 4 \\ 11 \\ 21 \end{bmatrix}$$

$$\{X\} = [A]^{-1}\{B\}$$

$$= \begin{bmatrix} 1 \\ 2 \\ 3 \\ -1 \end{bmatrix}$$

Hence

$$x_1 = 1 \qquad x_2 = 2 \qquad x_3 = 3 \qquad \text{and} \qquad x_4 = -1$$

Example 3-4

Solve

$$x_1 + 2x_2 + x_3 = 1$$
$$2x_1 + x_2 + 2x_3 = -1$$
$$3x_1 - 2x_2 = -2$$

The above equations can be written in the matrix form as below:

$$\begin{bmatrix} 1 & 2 & 1 \\ 2 & 1 & 2 \\ 3 & -2 & 0 \end{bmatrix} \begin{bmatrix} x_1 \\ x_2 \\ x_3 \end{bmatrix} = \begin{bmatrix} 1 \\ -1 \\ -2 \end{bmatrix}$$

Inversion of the Matrix $[A] = \begin{bmatrix} 1 & 2 & 1 \\ 2 & 1 & 2 \\ 3 & -2 & 0 \end{bmatrix}$

Operation No.	Given matrix			Check column	Unit matrix			Check column
0	1	2	1	4	1	0	0	1
	2	1	2	5	0	1	0	1
	3	-2	|0|	1	0	0	1	1

In the given matrix, zero occurs in the pivot itself. To get a nonzero element in the pivot, change the rows as below.
The rows of the unit matrix are also changed correspondingly. Now the matrix is written as below:

	Given matrix			Check column	Unit matrix			Check column
	3	-2	0	1	0	0	1	1
	2	1	2	5	0	1	0	1
	1	2		1		4	1	0	0	1
	3	-2	0	1	0	0	1	1
1A	1	1/2	1	5/2	0	1/2	0	1/2
	1	2	1	4	1	0	0	1
	3	-2	0	1	0	0	1	1
1B	0	|-3/2|	0	-3/2	-1	1/2	0	-1/2
	1	2	1	4	1	0	0	1

2A	−3/2	1	0	−1/2	0	0	−1/2	−1/2
	0	1	0	1	2/3	−1/3	0	1/3
	1/2	1	1/2	2	1/2	0	0	1/2
2B	−3/2	0	0	−3/2	−2/3	1/3	−1/2	−5/6
	0	1	0	1	2/3	−1/3	0	1/3
	1/2	0	1/2	1	−1/6	1/3	0	1/6
3A	1	0	0	1	4/9	−2/9	1/3	5/9
	0	1	0	1	2/3	−1/3	0	1/3
	1	0	1	2	−1/3	2/3	0	1/3
3B	1	0	0	1	4/9	−2/9	1/3	5/9
	0	1	0	1	2/3	−1/3	0	1/3
	0	0	1	1	−7/9	8/9	−1/3	−2/9

Hence

$$[A]^{-1} = \begin{bmatrix} 4/9 & -2/9 & 1/3 \\ 2/3 & -1/3 & 0 \\ -7/9 & 8/9 & -1/3 \end{bmatrix}$$

Thus

$$\begin{bmatrix} x_1 \\ x_2 \\ x_3 \end{bmatrix} = \begin{bmatrix} 4/9 & -2/9 & 1/3 \\ 2/3 & -1/3 & 0 \\ -7/9 & 8/9 & -1/3 \end{bmatrix} \begin{bmatrix} 1 \\ -1 \\ -2 \end{bmatrix} = \begin{bmatrix} 0 \\ 1 \\ -1 \end{bmatrix}$$

Therefore

$$x_1 = 0$$
$$x_2 = 1$$
$$x_3 = -1$$

3-10 Matrix Methods in Structural Analysis

In this introductory treatment it is not possible to deal with the applications of matrix methods to structural analysis. The interested reader may consult references 6, 7, and 8. The aim of this chapter is merely to introduce matrix operations employed in the later chapters.

CHAPTER 4

PROPERTIES OF CURVES

4-1 Plane Curves Used as Directrices

A knowledge of the properties of plane curves commonly employed as directrices of cylindrical shells is essential for developing the membrane theory of Chapter 6.   The curves considered are the arc of a circle, the parabola, the cycloid, the catenary, and the semiellipse.   Of these, the first four belong to one family and their curvature properties can be expressed by a single formula.   The semiellipse alone needs to be considered separately.

4-2 Radius of Curvature

It may be recalled that the radius of curvature at any point of a curve $y = f(x)$ described in cartesian coordinates may be expressed as

$$R = \frac{\left[1 + \left(\frac{dy}{dx}\right)^2\right]^{3/2}}{\frac{d^2y}{dx^2}} \qquad (4\text{-}1)$$

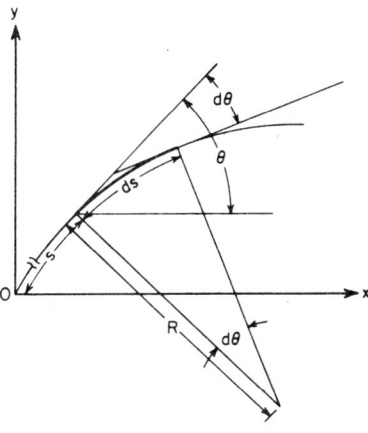

Referring to Fig. 4-1, the radius of curvature may also be expressed in the form

$$R = \frac{ds}{d\theta} \qquad (4\text{-}2)$$

where θ is the angle made by the tangent to the horizontal.   It is proposed to demonstrate that $R = R_0 \cos^n \theta$ is the equation to the family

FIG. 4-1

of curves comprising the arc of a circle, the parabola, the cycloid, and the catenary.   In this equation, R is the radius of curvature at any point and R_0 the radius of curvature at the crown where $\theta = 0$ and the tangent is horizontal.   It will also be shown that if

(i) $n =$ 0,   the curve is a circle
(ii) $n =$ 1,   the curve is a cycloid
(iii) $n = -2$,   the curve is a catenary

and

(iv) $n = -3$,   the curve is a parabola

4-3   The Arc of a Circle

If the curve is an arc of a circle, the radius of curvature R is constant. Hence $R = R_0 = R_0 \cos^0 \theta$.   It is thus evident that when $n = 0$, the equation $R = R_0 \cos^n \theta$ describes an arc of a circle.

4-4   The Parabola

The cartesian coordinates x and y of a point on the parabola shown in Fig. 4-2 may be conveniently expressed in parametric form as $x = 2at$ and $y = at^2$ so that they satisfy the equation to the parabola $x^2 = 4\,ay$. Differentiating,

$$dy = 2at\,dt$$

and

$$dx = 2a\,dt$$

Hence

$$\frac{dy}{dx} = t = \tan \theta \qquad (4\text{-}3)$$

Differentiating once again,

$$\frac{d^2y}{dx^2} = \frac{d}{dx}\left(\frac{dy}{dx}\right) = \frac{dt}{dx} = \frac{1}{2a} \qquad (4\text{-}4)$$

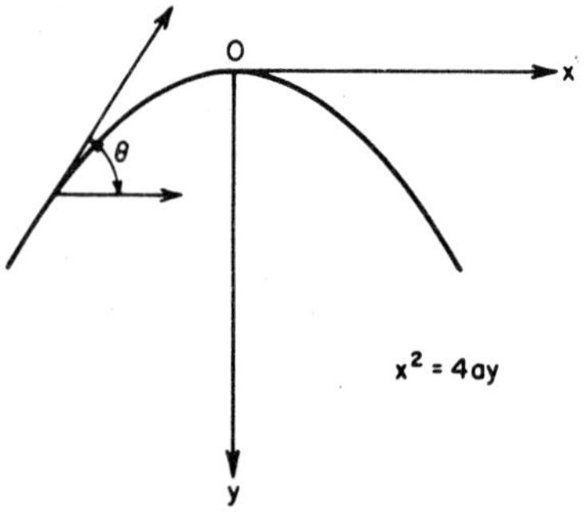

$x^2 = 4ay$

Parabola

FIG. 4-2

Substituting (4-3) and (4-4) in Equation (4-1), we get

$$R = 2a(1 + t^2)^{3/2} = 2a(1 + \tan^2 \theta)^{3/2} = 2a \sec^3 \theta \qquad (4\text{-}5)$$

R_0 corresponding to the crown which is also the origin is found by putting $t = 0$ in (4-5) to give

$$R_0 = 2a \qquad (4\text{-}6)$$

From (4-5) and (4-6) it is clear that for a parabola

$$R = R_0 \cos^{-3} \theta \qquad (4\text{-}7)$$

4-5 The Cycloid

It may be recalled that a cycloid is generated by a point P (Fig. 4-3) on the circumference of a circle as it rolls along a straight line without sliding. The equation of the cycloid in parametric form is given by

$$x = a(t - \sin t)$$
$$y = a(1 - \cos t)$$
(4-8)

Differentiating the relations in (4-8),

$$dy = a \sin t \, dt$$

and

$$dx = a(1 - \cos t) \, dt$$

Hence

$$\frac{dy}{dx} = \frac{\sin t}{1 - \cos t} = \cot \frac{t}{2} = \tan \theta$$
(4-9)

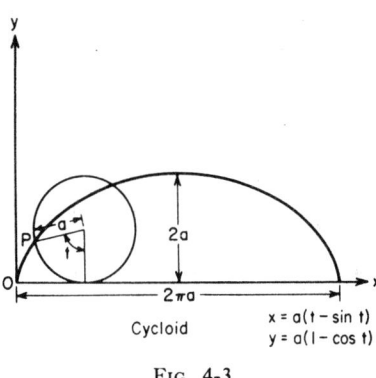

Cycloid

$x = a(t - \sin t)$
$y = a(1 - \cos t)$

Fig. 4-3

Differentiating (4-9) once again with respect to x, we get

$$\frac{d^2y}{dx^2} = \frac{d}{dx}\left(\frac{dy}{dx}\right) = -\frac{1}{2} \operatorname{cosec}^2 \frac{t}{2} \frac{dt}{dx} = -\frac{\operatorname{cosec}^2 \dfrac{t}{2}}{2a(1 - \cos t)}$$

$$= -\frac{1}{4a} \operatorname{cosec}^4 \frac{t}{2}$$
(4-10)

Substituting (4-9) and (4-10) in (4-1),

$$R = -4a \sin \frac{t}{2}$$
(4-11)

But from (4-9) it is evident that

$$\theta = \left(\frac{\pi}{2} - \frac{t}{2}\right)$$

Hence

$$\sin \frac{t}{2} = \cos \theta$$

Substituting this result in (4-11), we have

$$R = -4a \cos \theta$$
(4-12)

At the crown, $t = \pi$ and hence from (4-11),

$$R_0 = -4a \qquad (4\text{-}13)$$

From (4-12) and (4-13),

$$R = R_0 \cos\theta$$

Hence $R = R_0 \cos^n\theta$ represents a cycloid if $n = 1$.

4-6 The Catenary

The equation to a catenary (Fig. 4-4) in cartesian coordinates is

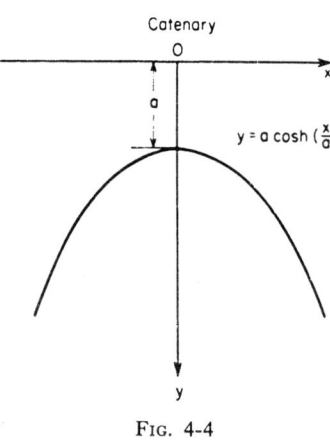

Catenary

$y = a \cosh\left(\frac{x}{a}\right)$

Fig. 4-4

$$y = a \cosh\left(\frac{x}{a}\right) \qquad (4\text{-}14)$$

$$\frac{dy}{dx} = \tan\theta = \sinh\left(\frac{x}{a}\right) \qquad (4\text{-}15)$$

Differentiating (4-15) once again with respect to x,

$$\frac{d^2y}{dx^2} = \frac{1}{a}\cosh\left(\frac{x}{a}\right) \qquad (4\text{-}16)$$

Hence from (4-11),

$$R = \frac{\left[1 + \sinh^2\left(\frac{x}{a}\right)\right]^{3/2}}{\frac{1}{a}\cosh\left(\frac{x}{a}\right)} = a\cosh^2\left(\frac{x}{a}\right) \qquad (4\text{-}17)$$

But from relation (4-15),

$$\sinh^2\left(\frac{x}{a}\right) = \tan^2\theta$$

Hence

$$\cosh^2\left(\frac{x}{a}\right) = 1 + \tan^2\theta = \sec^2\theta$$

Substituting this value in (4-17),

$$R = a\sec^2\theta = a\cos^{-2}\theta \qquad (4\text{-}18)$$

At the origin, $x = 0$ and $\cosh(x/a) = 1$. Hence from (4-17),

$$R_0 = a \qquad (4\text{-}19)$$

From (4-18) and (4-19), we deduce the relation

$$R = R_0 \cos^{-2} \theta \qquad (4\text{-}20)$$

4-7 The Semiellipse

The parametric coordinates of a point on the semiellipse shown in Fig. 4-5 may be written as

$$x = a \cos t \qquad y = b \sin t \qquad (4\text{-}21)$$

Hence

$$dy = b \cos t \, dt \qquad dx = -a \sin t \, dt$$

$$\frac{dy}{dx} = -\frac{b}{a} \cot t = \tan \theta \qquad (4\text{-}22)$$

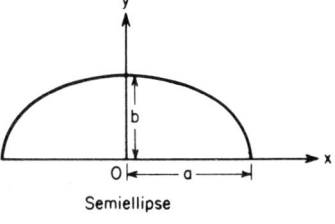

Semiellipse

FIG. 4-5

Differentiating once more with respect to x,

$$\frac{d^2y}{dx^2} = \frac{b}{a} \operatorname{cosec}^2 t \frac{dt}{dx} = -\frac{b}{a^2} \operatorname{cosec}^3 t \qquad (4\text{-}23)$$

Substituting for $\operatorname{cosec}^3 t$ from (4-22), we may rewrite (4-23) as

$$\frac{d^2y}{dx^2} = -\frac{b}{a^2} \left(1 + \frac{a^2}{b^2} \frac{\sin^2 \theta}{\cos^2 \theta}\right)^{3/2} \qquad (4\text{-}24)$$

Substituting (4-22) and (4-24) in (4-1), we arrive at the relation

$$R = \frac{(1 + \tan^2 \theta)^{3/2}}{-\dfrac{b}{a^2} \left(1 + \dfrac{a^2}{b^2} \dfrac{\sin^2 \theta}{\cos^2 \theta}\right)^{3/2}}$$

Simplifying,

$$R = \frac{-a^2 b^2}{(a^2 \sin^2 \theta + b^2 \cos^2 \theta)^{3/2}} \qquad (4\text{-}25)$$

CHAPTER 5

ELEMENTS OF DIFFERENTIAL GEOMETRY
AND CLASSIFICATION OF SHELLS

In this chapter, it is proposed to review some concepts of differential geometry as a preliminary to the classification of shells. Because of its elegance and brevity, vector analysis is employed in the treatment. We begin with the elementary operations of vector algebra.

5-1 Vector Addition, Subtraction, and Multiplication

A *vector* may be defined as a quantity which has both a magnitude and a direction. Displacement, velocity, and force are familiar examples. A vector **a** can therefore be represented by a directed line segment AB (Fig. 5-1), the length of the line representing the magnitude and the

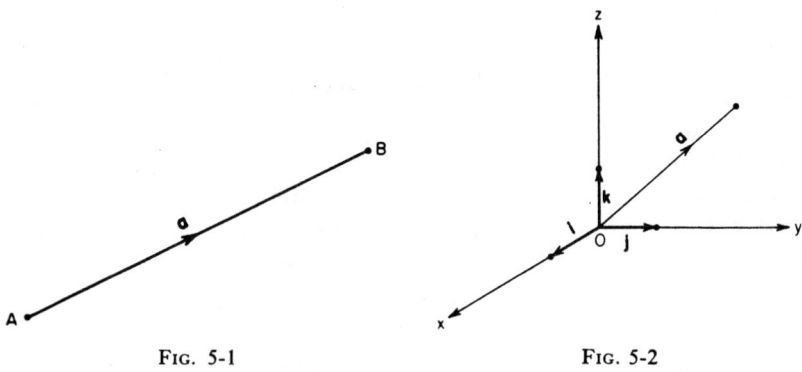

FIG. 5-1   FIG. 5-2

direction of the line indicating the direction of the vector. A *scalar* quantity has only magnitude but no direction. The mass of a body is an example. The magnitude of a vector is usually indicated by $|\mathbf{a}|$. Consider a right-handed system of cartesian coordinates x, y, and z (Fig. 5-2). Let **i**, **j**, and **k** be unit vectors directed along the x, y, and z

directions. Any other vector can now be represented in terms of the base vectors **i**, **j**, and **k** as follows:

$$\mathbf{a} = a_x\mathbf{i} + a_y\mathbf{j} + a_z\mathbf{k} \tag{5-1}$$

where a_x, a_y, and a_z are the projections of the vector **a** in the x, y, and z directions. It is easily verified that $|\mathbf{a}| = \sqrt{a_x{}^2 + a_y{}^2 + a_z{}^2}$. Let us suppose that it is required to find the sum of two vectors **a** and **b**. Let them be expressed in terms of the unit base vectors as

$$\mathbf{a} = a_x\mathbf{i} + a_y\mathbf{j} + a_z\mathbf{k}$$

and

$$\mathbf{b} = b_x\mathbf{i} + b_y\mathbf{j} + b_z\mathbf{k}$$

Clearly,

$$\mathbf{a} + \mathbf{b} = (a_x + b_x)\,\mathbf{i} + (a_y + b_y)\,\mathbf{j} + (a_z + b_z)\,\mathbf{k} \tag{5-2}$$

Vector addition is diagrammatically illustrated in Fig. 5-3.

$$\mathbf{a} + \mathbf{b} = \mathbf{PQ} + \mathbf{QR} = \mathbf{PR}$$

This is the familiar parallelogram law of addition used in physics for adding forces. A simple physical explanation of vector addition may be given as follows: If a person proceeds from P to Q and thence to R, the net result is that he goes from the initial point P to the terminal point R. Vector subtraction $\mathbf{a} - \mathbf{b}$ may be thought of as equivalent to the vector addition $\mathbf{a} + (-\mathbf{b}) = (-\mathbf{b} + \mathbf{a})$. The directed line segment **PS** stands

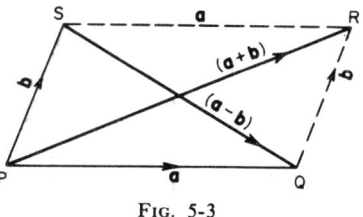

Fig. 5-3

for **b**. Hence the directed line segment **SP** stands for $(-\mathbf{b})$. The vector **a** is represented by the directed line segment **PQ**. Hence $(-\mathbf{b} + \mathbf{a}) = \mathbf{SP} + \mathbf{PQ} = \mathbf{SQ}$. It is also easily seen that

$$\mathbf{a} - \mathbf{b} = (a_x - b_x)\,\mathbf{i} + (a_y - b_y)\,\mathbf{j} + (a_z - b_z)\,\mathbf{k} \tag{5-3}$$

There are two kinds of vector multiplication—*scalar* multiplication resulting in what is called a *dot product* and *vector* multiplication resulting in a *cross product*. The dot product of two vectors **a** and **b**, written as **a** · **b**, is, by definition, the product of the length of one of the vectors and the scalar projection of the other on it.

The result is a *scalar*, and its magnitude is given by $|\mathbf{a}|\,|\mathbf{b}|\cos\theta$ where θ is the angle between the two vectors. It is easily verified that

$\mathbf{a} \cdot \mathbf{b} = \mathbf{b} \cdot \mathbf{a}$.   If the vectors $\mathbf{a}$ and $\mathbf{b}$ are perpendicular to each other $\cos \theta = 0$ and hence $\mathbf{a} \cdot \mathbf{b} = 0$.   It is easily seen that $(\mathbf{i} \cdot \mathbf{i}) = (\mathbf{j} \cdot \mathbf{j}) = (\mathbf{k} \cdot \mathbf{k}) = 1$ and $(\mathbf{i} \cdot \mathbf{j}) = (\mathbf{j} \cdot \mathbf{i}) = (\mathbf{i} \cdot \mathbf{k}) = (\mathbf{k} \cdot \mathbf{i}) = (\mathbf{j} \cdot \mathbf{k}) = (\mathbf{k} \cdot \mathbf{j}) = 0$. These results enable the dot products of two vectors, whose cartesian components are known, to be written down.   Observing, moreover, that the distributive law holds for scalar multiplication, we may write

$$\mathbf{a} \cdot \mathbf{b} = (a_x\mathbf{i} + a_y\mathbf{j} + a_z\mathbf{k}) \cdot (b_x\mathbf{i} + b_y\mathbf{j} + b_z\mathbf{k})$$

$$= (a_xb_x + a_yb_y + a_zb_z) \tag{5-4}$$

The cross product of two vectors $\mathbf{a}$ and $\mathbf{b}$ is a vector $\mathbf{c}$ which is normal to the plane on which $\mathbf{a}$ and $\mathbf{b}$ lie and is so directed that the vectors $\mathbf{a}, \mathbf{b}$, and $\mathbf{c}$ form a right-handed system.   The magnitude of the vector $\mathbf{c} = |\mathbf{a}|\,|\mathbf{b}|\sin\theta$, where $\sin\theta$ is the absolute magnitude of the sine of the angle θ between the two vectors.   The area of the shaded parallelogram in Fig. 5-4 represents the magnitude of the vector $\mathbf{c}$.   By applying the distributive law, it can be shown that $\mathbf{a} \times \mathbf{b}$ may be written in determinant form as

$$\mathbf{a} \times \mathbf{b} = \begin{vmatrix} \mathbf{i} & \mathbf{j} & \mathbf{k} \\ a_x & a_y & a_z \\ b_x & b_y & b_z \end{vmatrix}$$

$$= (a_yb_z - a_zb_y)\,\mathbf{i} + (a_zb_x - a_xb_z)\,\mathbf{j} + (a_xb_y - a_yb_x)\,\mathbf{k} \tag{5-5}$$

For a more detailed treatment of vector analysis, the reader may consult reference 9.

5-2   Parametric Representation of a Surface

A surface may be defined as the locus of a point whose position vector $\mathbf{r}$ may be expressed as a function of two variables.

A surface may thus be defined by the following three parametric equations in a cartesian system:

$$x = x \tag{5-6a}$$

$$y = y \tag{5-6b}$$

$$z = f(x, y) \tag{5-6c}$$

x and y are known as the curvilinear coordinates of a point on the surface.   The position vector $\mathbf{r}$ may be written as

$$\mathbf{r} = x\mathbf{i} + y\mathbf{j} + z\mathbf{k} \tag{5-7}$$

5-3   The First Quadratic Form

Consider two adjacent points P and Q on the surface with position vectors $\mathbf{r}$ and $\mathbf{r} + d\mathbf{r}$, respectively (Fig. 5-5).

$$d\mathbf{r} = \frac{\partial \mathbf{r}}{\partial x}\, dx + \frac{\partial \mathbf{r}}{\partial y}\, dy \tag{5-8}$$

Let the arc length PQ be equal to ds.

$$ds^2 = d\mathbf{r} \cdot d\mathbf{r}$$

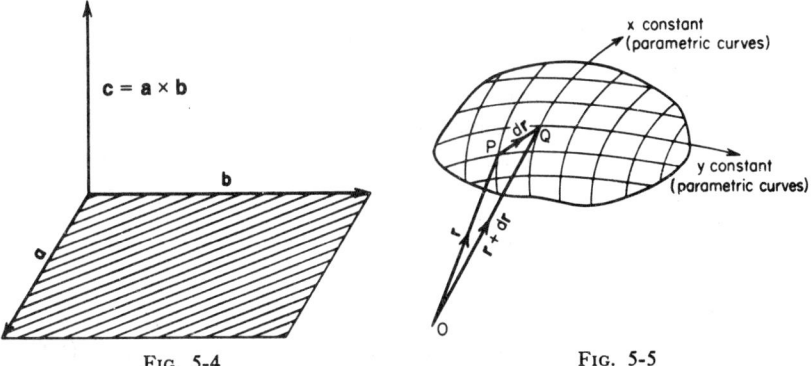

| FIG. 5-4 | FIG. 5-5 |

Using relation (5-8) and noting that the distributive law holds good for scalar products,

$$ds^2 = \frac{\partial \mathbf{r}}{\partial x} \cdot \frac{\partial \mathbf{r}}{\partial x}\, dx^2 + 2\,\frac{\partial \mathbf{r}}{\partial x} \cdot \frac{\partial \mathbf{r}}{\partial y}\, dx\, dy + \frac{\partial \mathbf{r}}{\partial y} \cdot \frac{\partial \mathbf{r}}{\partial y}\, dy^2$$

But

$$\frac{\partial \mathbf{r}}{\partial x} = \mathbf{i} + \frac{\partial z}{\partial x}\mathbf{k}$$

and

$$\frac{\partial \mathbf{r}}{\partial y} = \mathbf{j} + \frac{\partial z}{\partial y}\mathbf{k}$$

Substituting these values in the expression for ds, we arrive at the relation

$$ds^2 = \left[1 + \left(\frac{\partial z}{\partial x}\right)^2\right] dx^2 + 2\,\frac{\partial z}{\partial x}\frac{\partial z}{\partial y}\, dx\, dy + \left[1 + \left(\frac{\partial z}{\partial y}\right)^2\right] dy^2 \tag{5-9}$$

We may rewrite this expression more compactly by using Monge's notation for the derivatives of z with respect to x and y. According to this notation,

$$\frac{\partial z}{\partial x} = p \qquad \frac{\partial z}{\partial y} = q \qquad \frac{\partial^2 z}{\partial x^2} = r \qquad \frac{\partial^2 z}{\partial x\,\partial y} = s \qquad \frac{\partial^2 z}{\partial y^2} = t$$

Using this notation, (5-9) may be recast as

$$ds^2 = (1 + p^2)\, dx^2 + 2pq\, dx\, dy + (1 + q^2)\, dy^2 \tag{5-10}$$

Relation (5-10), which enables the elemental arc length to be calculated, is known as the *first quadratic form* of the surface. In differential geometry, this relation is more usually written in the form

$$ds^2 = E\, dx^2 + 2F\, dx\, dy + G\, dy^2 \tag{5-11}$$

In the cartesian system

$$E = (1 + p^2) \qquad F = pq \qquad \text{and} \qquad G = (1 + q^2)$$

E, F, and G are known as the *fundamental magnitudes* of the surface. If the parametric curves are orthogonal, $F = 0$.

5-4   Equation of the Normal to a Surface

The equation of the unit normal to a surface may be written as

$$\mathbf{n} = \frac{\dfrac{\partial \mathbf{r}}{\partial x} \times \dfrac{\partial \mathbf{r}}{\partial y}}{\left| \dfrac{\partial \mathbf{r}}{\partial x} \times \dfrac{\partial \mathbf{r}}{\partial y} \right|} \tag{5-12}$$

$$\mathbf{n} = -\frac{p}{\sqrt{1 + p^2 + q^2}}\mathbf{i} - \frac{q}{\sqrt{1 + p^2 + q^2}}\mathbf{j} + \frac{1}{\sqrt{1 + p^2 + q^2}}\mathbf{k} \tag{5-13}$$

5-5   The Second Quadratic Form

Being the vector product of $\partial \mathbf{r}/\partial x$ and $\partial \mathbf{r}/\partial y$, $\mathbf{n}$ is at right angles to both of them.   Hence

$$\mathbf{n} \cdot \frac{\partial \mathbf{r}}{\partial x} = 0 \tag{5-14}$$

$$\mathbf{n} \cdot \frac{\partial \mathbf{r}}{\partial y} = 0 \tag{5-15}$$

Differentiating (5-14) with respect to x,

$$\frac{\partial \mathbf{n}}{\partial x} \cdot \frac{\partial \mathbf{r}}{\partial x} = -\mathbf{n} \cdot \frac{\partial^2 \mathbf{r}}{\partial x^2} = -L \tag{5-16}$$

Similarly,

$$\frac{\partial \mathbf{n}}{\partial y} \cdot \frac{\partial \mathbf{r}}{\partial x} = -\mathbf{n} \cdot \frac{\partial^2 \mathbf{r}}{\partial x \, \partial y} = -M \qquad (5\text{-}17)$$

Differentiating (5-15) with respect to y,

$$\frac{\partial \mathbf{n}}{\partial y} \cdot \frac{\partial \mathbf{r}}{\partial y} = -\mathbf{n} \cdot \frac{\partial^2 \mathbf{r}}{\partial y^2} = -N \qquad (5\text{-}18)$$

We also know that the unit tangent vector is given by

$$\frac{\partial \mathbf{r}}{\partial s} = \frac{\partial \mathbf{r}}{\partial x} \cdot \frac{dx}{ds} + \frac{\partial \mathbf{r}}{\partial y} \cdot \frac{dy}{ds} \qquad (5\text{-}19)$$

A new definition will now be introduced. *A normal section at a point on a surface is defined as the plane curve obtained by cutting the surface by a plane, containing the normal to the surface at that point.* The principal normals to such curves will be parallel to the normal $\mathbf{n}$ to the surface. Let us set up an expression for the curvature of one such curve. The curvature of a normal section at a point P is known as the normal curvature. Let it be denoted by $\varkappa_n$. An expression for $\varkappa_n$ may be derived by differentiating relation (5-19). We may also note that

$$\frac{\partial^2 \mathbf{r}}{\partial s^2} = \varkappa_n \cdot \mathbf{n} \qquad \text{(Frenet-Seret formula)} \qquad (5\text{-}20)$$

This is because $\partial \mathbf{r}/\partial s$, the unit tangent vector, and its derivative $\partial^2 \mathbf{r}/\partial s^2$ have to be at right angles. Hence $\partial^2 \mathbf{r}/\partial s^2$ will be directed along the unit normal. Differentiating (5-19) and making use of (5-20),

$$\frac{\partial^2 \mathbf{r}}{\partial s^2} = \varkappa_n \cdot \mathbf{n} = \left(\frac{\partial^2 \mathbf{r}}{\partial x^2}\right) \cdot \left(\frac{dx}{ds}\right)^2 + 2 \frac{\partial^2 \mathbf{r}}{\partial x \, \partial y} \cdot \left(\frac{dx}{ds}\right)\left(\frac{dy}{ds}\right)$$

$$+ \left(\frac{\partial^2 \mathbf{r}}{\partial y^2}\right)\left(\frac{dy}{ds}\right)^2 + \frac{\partial \mathbf{r}}{\partial x} \cdot \frac{d^2 x}{ds^2} + \frac{\partial \mathbf{r}}{\partial y} \cdot \frac{d^2 y}{ds^2} \qquad (5\text{-}21)$$

To find $\varkappa_n$ we form the dot product of the two vectors $\partial^2 \mathbf{r}/\partial s^2$ and $\mathbf{n}$ to give

$$\frac{\partial^2 \mathbf{r}}{\partial s^2} \cdot \mathbf{n} = \varkappa_n = \mathbf{n} \cdot \frac{\partial^2 \mathbf{r}}{\partial x^2} \left(\frac{dx}{ds}\right)^2$$

$$+ 2\mathbf{n} \cdot \frac{\partial^2 \mathbf{r}}{\partial x \, \partial y} \left(\frac{dx}{ds}\right)\left(\frac{dy}{ds}\right) + \mathbf{n} \cdot \frac{\partial^2 \mathbf{r}}{\partial y^2} \left(\frac{dy}{ds}\right)^2 \qquad (5\text{-}22)$$

The last two terms in (5-21) do not yield anything, as

$$\mathbf{n} \cdot \frac{\partial \mathbf{r}}{\partial x} = \mathbf{n} \cdot \frac{\partial \mathbf{r}}{\partial y} = 0$$

Equation (5-22) may be more compactly recast as

$$x_n = L\left(\frac{dx}{ds}\right)^2 + 2M\left(\frac{dx}{ds}\right)\left(\frac{dy}{ds}\right) + N\left(\frac{dy}{ds}\right)^2$$

or

$$x_n = \frac{L\, dx^2 + 2M\, dx\, dy + N\, dy^2}{ds^2}$$

$$x_n = \frac{L\, dx^2 + 2M\, dx\, dy + N\, dy^2}{E\, dx^2 + 2F\, dx\, dy + G\, dy^2} \tag{5-23}$$

The numerator of this expression is known as the *second fundamental form*. One may easily verify that

$$L = \frac{r}{\sqrt{1 + p^2 + q^2}} \qquad M = \frac{s}{\sqrt{1 + p^2 + q^2}}$$

and

$$N = \frac{t}{\sqrt{1 + p^2 + q^2}}$$

Hence

$$x_n = \frac{r\, dx^2 + 2s\, dx\, dy + t\, dy^2}{ds^2 \sqrt{1 + p^2 + q^2}} \tag{5-24}$$

5-6 Principal Curvatures, Gauss Curvature, and Lines of Curvature

We may rewrite (5-23) as

$$(L - Ex_n)\, dx^2 + 2(M - Fx_n)\, dx\, dy + (N - Gx_n)\, dy^2 = 0 \tag{5-25}$$

or

$$(N - Gx_n)\left(\frac{dy}{dx}\right)^2 + 2(M - Fx_n)\frac{dy}{dx} + (L - Ex_n) = 0 \tag{5-26}$$

Equation (5-26) being a quadratic, there are in general two directions corresponding to a given x_n. If there is to be only one direction corresponding to a given x_n, the quadratic must have repeated roots. The condition for this to happen is

$$4(M - Fx_n)^2 - 4(L - Ex_n)(N - Gx_n) = 0 \tag{5-27}$$

or

$$(F^2 - EG)\, x_n^2 + (GL + EN - 2MF)\, x_n + (M^2 - LN) = 0 \tag{5-28}$$

This gives two values of x_n known as the *principal curvatures* at that point. The product of the two principal curvatures is defined as the *Gauss curvature* at the point. Its value is easily seen to be $(LN - M^2)/(EG - F^2)$. It is clear that Gauss curvature will be positive, zero, or negative according as

$$LN - M^2 \gtreqqless 0 \qquad (5\text{-}29)$$

This condition is also equivalent to

$$rt - s^2 \gtreqqless 0 \qquad (5\text{-}30)$$

The surface is called *synclastic, developable,* or *anticlastic* at a point according as the Gauss curvature is positive, zero, or negative at that point. If the quadratic equation (5-25) has repeated roots,

$$\frac{dy}{dx} = -\frac{(M - F x_n)}{(N - G x_n)} \qquad (5\text{-}31)$$

and

$$\frac{dx}{dy} = -\frac{(M - F x_n)}{(L - E x_n)} \qquad (5\text{-}32)$$

Hence

$$(N - G x_n)\, dy + (M - F x_n)\, dx = 0 \qquad (5\text{-}33)$$

$$(L - E x_n)\, dx + (M - F x_n)\, dy = 0 \qquad (5\text{-}34)$$

Eliminating x_n from (5-33) and (5-34), the two directions satisfy the relation

$$(EM - FL)\, dx^2 + (EN - GL)\, dx\, dy + (FN - GM)\, dy^2 = 0 \quad (5\text{-}35)$$

These two directions are known as the *principal directions*. The normal curvatures in these directions are known as the *principal curvatures* at that point. They are also the directions in which the normal curvatures at the point are a maximum and minimum.

Surface curves the tangents to which at any point coincide with one of the principal directions are known as *lines of curvature*. The curvature of a line of curvature is *not* a principal curvature, since the line of curvature need not be a normal curve. Two conditions are to be satisfied for the parametric curves to be lines of curvature.

(i) They have to be orthogonal. We have already seen that this is so if $F = 0$.

(ii) Let us suppose that we move along the parametric curve for which $x =$ constant. Hence $dx = 0$. From Equation (5-35), it is seen that this will be so if $FN = GM$. Similarly along the parametric curve $y =$ constant, $dy = 0$. Hence $EM = FL$. It is to be noted that $(EN - GL) \neq 0$. Multiplying the first relation by L and the second by N, we get

$$FLN = GML \tag{5-36}$$

$$ENM = FLN \tag{5-37}$$

From (5-36) and (5-37),

$$M(EN - GL) = 0$$

Noting that $(EN - GL) \neq 0$, $M = 0$. Hence $F = 0$ and $M = 0$ are the two conditions to be satisfied if the parametric curves are to be lines of curvature.

5-7 Choice of Curvilinear Coordinates

It is not necessary always to choose x and y as the curvilinear coordinates. We might as well have chosen u and v so that

$$x = f_1(u, v)$$

$$y = f_2(u, v)$$

$$z = f_3(u, v)$$

Relations derived in the previous articles can easily be modified for this choice of curvilinear coordinates. This will be illustrated with the aid of examples later. The choice of x and y as curvilinear coordinates does not result in any loss of generality.

5-8 Surfaces of Revolution

A *surface of revolution* is obtained by rotating a plane curve called the *meridian* about an axis lying in the plane of the curve. This plane is known as the *meridian plane*. The concepts developed in the previous articles will now be applied to find the principal curvatures of a surface of revolution. Referring to Fig. 5-6, we may write the coordinates of the surface, noting that $r_0 = f(z)$,

$$x = f(z) \cos \theta \tag{5-38a}$$

$$y = f(z) \sin \theta \tag{5-38b}$$

$$z = z \tag{5-38c}$$

z and θ are obviously the curvilinear coordinates employed to describe the surface.   The position vector **r** of a point P on the surface may be written as

$$\mathbf{r} = f(z) \cos \theta \, \mathbf{i} + f(z) \sin \theta \, \mathbf{j} + z\mathbf{k} \tag{5-39}$$

$$\frac{\partial \mathbf{r}}{\partial \theta} = -f(z) \sin \theta \, \mathbf{i} + f(z) \cos \theta \, \mathbf{j}$$

$$\frac{\partial \mathbf{r}}{\partial z} = f'(z) \cos \theta \, \mathbf{i} + f'(z) \sin \theta \, \mathbf{j} + \mathbf{k}$$

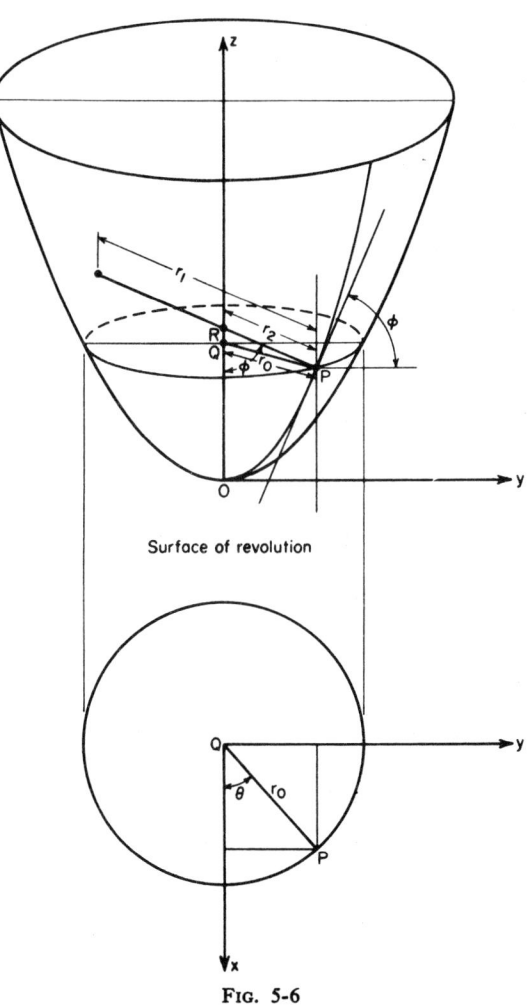

Surface of revolution

FIG. 5-6

$$\frac{\partial \mathbf{r}}{\partial \theta} \cdot \frac{\partial \mathbf{r}}{\partial \theta} = (f(z))^2 \sin^2 \theta + (f(z))^2 \cos^2 \theta = (f(z))^2$$

$$\frac{\partial \mathbf{r}}{\partial z} \cdot \frac{\partial \mathbf{r}}{\partial z} = (f'(z))^2 \cos^2 \theta + (f'(z))^2 \sin^2 \theta + 1 = [1 + (f'(z))^2]$$

$$\frac{\partial \mathbf{r}}{\partial \theta} \cdot \frac{\partial \mathbf{r}}{\partial z} = 0$$

Hence

$$E = (f(z))^2 \qquad F = 0 \qquad \text{and} \qquad G = [1 + (f'(z))^2]$$

$$\mathbf{n} = \frac{\dfrac{\partial \mathbf{r}}{\partial \theta} \times \dfrac{\partial \mathbf{r}}{\partial z}}{\left| \dfrac{\partial \mathbf{r}}{\partial \theta} \times \dfrac{\partial \mathbf{r}}{\partial z} \right|}$$

$$\frac{\partial \mathbf{r}}{\partial \theta} \times \frac{\partial \mathbf{r}}{\partial z} = \begin{vmatrix} \mathbf{i} & \mathbf{j} & \mathbf{k} \\ -f(z)\sin\theta & f(z)\cos\theta & 0 \\ f'(z)\cos\theta & f'(z)\sin\theta & 1 \end{vmatrix}$$

$$= f(z)\cos\theta\, \mathbf{i} + f(z)\sin\theta\, \mathbf{j} - f(z) f'(z)\, \mathbf{k}$$

$$\left| \frac{\partial \mathbf{r}}{\partial \theta} \times \frac{\partial \mathbf{r}}{\partial z} \right| = \sqrt{(f(z))^2 + (f(z))^2 (f'(z))^2}$$

$$= f(z)\sqrt{1 + (f'(z))^2}$$

Hence

$$\mathbf{n} = \frac{\cos\theta}{\sqrt{1 + (f'(z))^2}} \mathbf{i} + \frac{\sin\theta}{\sqrt{1 + (f'(z))^2}} \mathbf{j} - \frac{f'z}{\sqrt{1 + (f'(z))^2}} \mathbf{k} \qquad (5\text{-}40)$$

$$\frac{\partial^2 \mathbf{r}}{\partial \theta^2} = -f(z)\cos\theta\, \mathbf{i} - f(z)\sin\theta\, \mathbf{j}$$

$$L = \mathbf{n} \cdot \frac{\partial^2 \mathbf{r}}{\partial \theta^2} = -\frac{f(z)}{\sqrt{1 + (f'(z))^2}}$$

$$\frac{\partial^2 \mathbf{r}}{\partial z^2} = f''(z)\cos\theta\, \mathbf{i} + f''(z)\sin\theta\, \mathbf{j}$$

$$\mathbf{n} \cdot \frac{\partial^2 \mathbf{r}}{\partial z^2} = \frac{f''(z)}{\sqrt{1 + (f'(z))^2}} = N$$

$\mathbf{n} \cdot \partial^2 \mathbf{r}/\partial\theta\, \partial z$ is zero; hence $M = 0$. We may now substitute the values of E, G, L, and N in (5-28) to get

$$f(z)[1 + (f'(z))^2]^{1/2} \kappa_n{}^2 - \left[\frac{f(z) f''(z)}{1 + (f'(z))^2} - 1 \right] \kappa_n - \frac{f''(z)}{[1 + (f'(z))^2]^{3/2}} = 0$$

$$(5\text{-}41)$$

The two principal curvatures are the two roots of this quadratic equation. We may observe that $F = M = 0$. This means that the parametric curves are the lines of curvature for a surface of revolution. $z =$ constant gives circles which are known as *parallels of latitude* and $\theta =$ constant gives meridians. Solving (5-41), the two roots are

$$\kappa_1 = \frac{f''(z)}{[1 + (f'(z))^2]^{3/2}} \tag{5-42}$$

and

$$\kappa_2 = -\frac{1}{f(z)[1 + (f'(z))^2]^{1/2}} \tag{5-43}$$

κ_1 is easily identified as the curvature $1/r_1$ of the generating plane curve. Referring to Fig. 5-6, we know that

$$\tan \phi = \frac{1}{f'(z)}$$

$$r_2 = \frac{r_0}{\sin \phi} = r_0[1 + (f'(z))^2]^{1/2} = f(z)[1 + (f'(z))^2]^{1/2}$$

$$\frac{1}{r_2} = \frac{1}{f(z)[1 + (f'(z))^2]^{1/2}}$$

This is equal to κ_2 found in (5-43). Hence

$$\kappa_2 = \frac{1}{r_2} \tag{5-44}$$

5-9 Some Definitions

Before we can classify shell surfaces, we need a few more definitions. A *ruled* surface may be defined as a surface formed by the motion of a straight line which is known as the *generator* or *ruling*. A surface is said to be *singly ruled* if at every point only a single straight line can be ruled and *doubly ruled* if at every point two straight lines can be ruled. Conical shells, conoids, and cylinders are examples of singly ruled surfaces; the hyperbolic paraboloid and the hyperboloid of revolution of one sheet are examples of *doubly ruled* surfaces (Fig. 5-7). Ruled surfaces have the practical advantage that they may be cast on straight forms. A *surface of translation* is generated by the motion of a plane curve parallel to itself over another curve, the planes containing the two curves being at right angles to each other. One of the curves may be a straight line as in a cylinder. The elliptic paraboloid, generated by a convex parabola moving over another convex parabola or by a concave parabola moving over another concave parabola, is a surface of translation (Fig. 5-8).   A special case of this is the paraboloid of revolution for which both the parabolas involved are identical.   If a convex parabola moves over a concave parabola or vice versa, a hyperbolic paraboloid is generated.

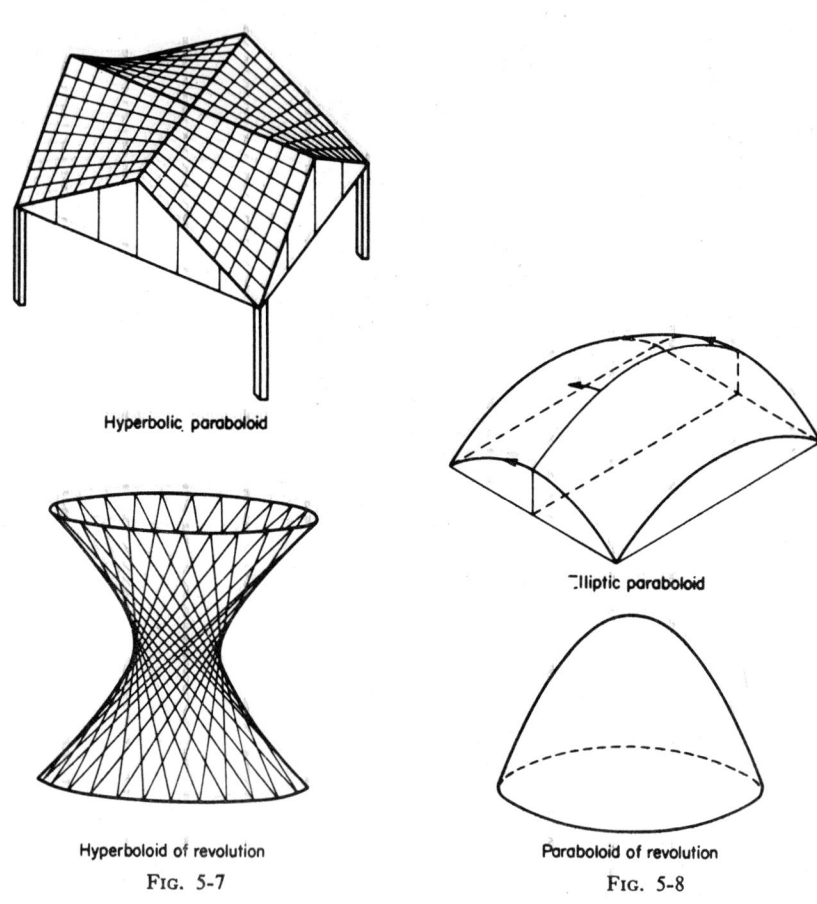

Hyperbolic paraboloid

Elliptic paraboloid

Hyperboloid of revolution

Fig. 5-7

Paraboloid of revolution

Fig. 5-8

The equation of a translational surface can always be written in the form

$$z = f_1(x) + f_2(y) \qquad (5\text{-}45)$$

5-10   Classification of Shell Surfaces

Shell surfaces may be broadly classified as *singly curved* and *doubly curved*.   Cylinders and cones are examples of singly curved surfaces. Singly curved surfaces are *developable*.   Thus a cylinder can be developed

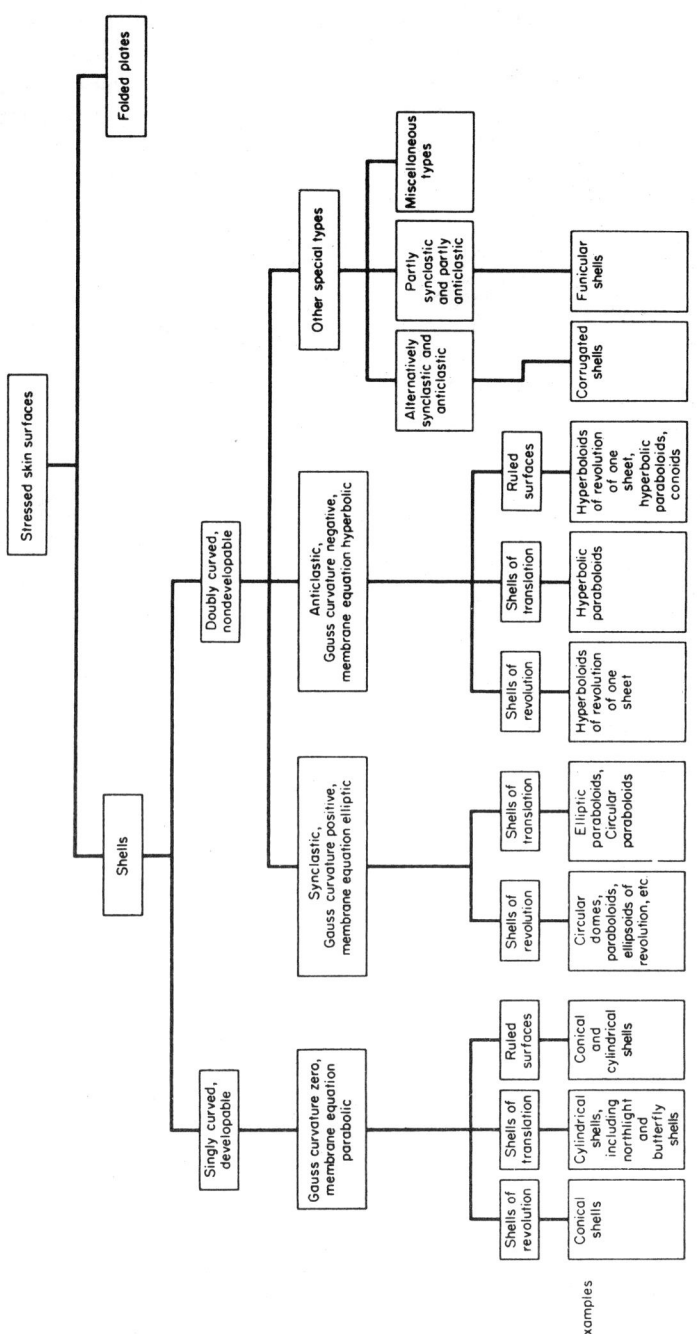

CHART 5-1. [*From Indian Standard Criteria for the Design of Reinforced Concrete Shells and Folded Plates (I.S. 2210-1962), Indian Standards Institution, New Delhi, April, 1963.   Used by permission of I.S.I., New Delhi.*]

into a plane rectangle *without stretching, shrinking, or tearing.* Similarly, a cone may be developed into a sector of a circle. Doubly curved surfaces are nondevelopable. Hence they will not tend to flatten out under loads. This explains their superior performance. Further classification of shell surfaces can be attempted on the basis of Gauss curvature. Subclassification is possible depending upon whether a shell is a translational surface, a ruled surface, or a surface of revolution. It may so happen that a shell is both a translational surface and a ruled surface, e.g., cylinders and hyperbolic paraboloids. Again, the surface may be a surface of revolution and also a ruled surface, e.g., a hyperboloid of revolution of one sheet. One possible classification of shell surfaces is that given in the Indian Standard Criteria for the Design of Reinforced Concrete Shell Structures and Folded Plates.[1] Classification Chart 5-1 on page 63 is reproduced from that publication.

The reader interested in studying differential geometry in greater detail may consult references 10 and 11.

---

[1] Indian Standard Criteria for the Design of Reinforced Concrete Shell Structures and Folded Plates (I.S. 2210–1962), Indian Standards Institution, New Delhi, April, 1963.

PART II

CYLINDRICAL SHELLS AND FOLDED PLATES

CHAPTER 6

MEMBRANE THEORY OF CYLINDRICAL SHELLS

6-1 Thin Shells

Shells and folded plates belong to the class of stressed-skin structures which by virtue of their geometry and small flexural rigidity tend to

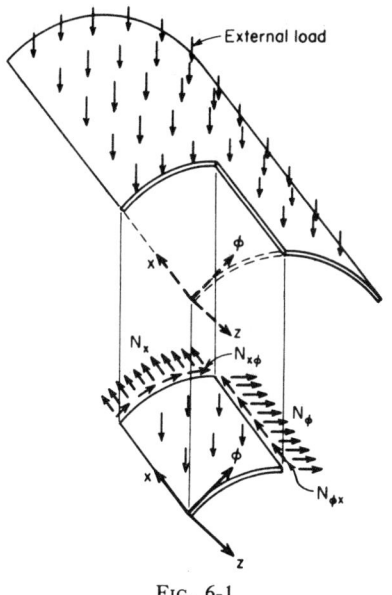

Fig. 6-1

carry applied loads primarily by direct stresses lying in their plane accompanied by little or no bending (Fig. 6-1). Their structural action is thus in marked contrast with that of a slab which carries loads by flexure (Fig. 6-2). This explains why it is possible to span openings as large as a 100 ft with a concrete shell which is hardly 3 in. thick. A shell which carries loads entirely by direct stresses lying in its plane is known as a *membrane*. For membrane action to be possible, the shell

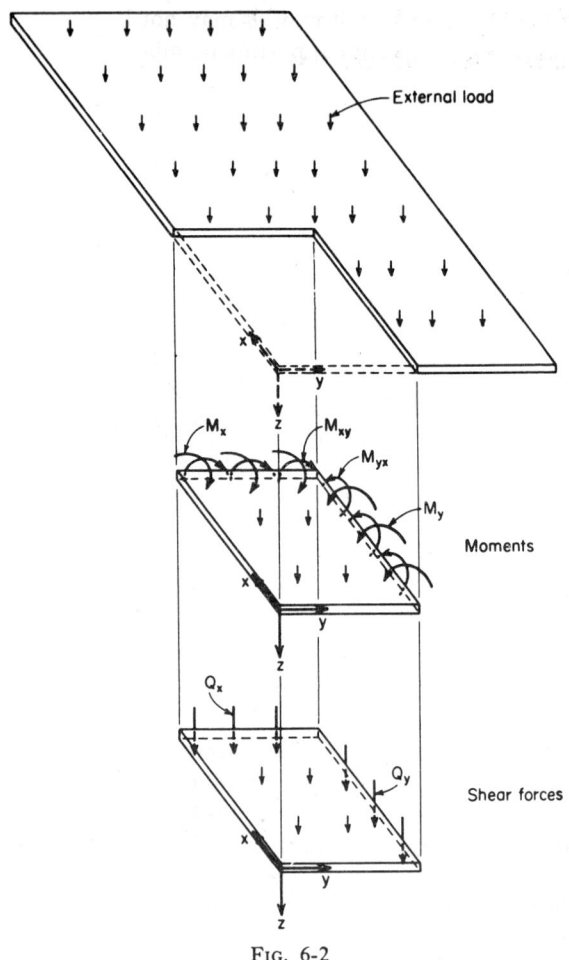

Fig. 6-2

has to be thin. According to Novozhilov [12], a shell may be regarded as thin if the ratio of its thickness d to its radius of curvature R at any point, i.e., $d/R \leqslant \frac{1}{20}$.

We shall be concerned only with thin shells in this book.

6-2 Parts of a Cylindrical Shell

A cylindrical shell may be thought of as a surface generated by a straight line moving over a plane curve. The straight line generating the surface is known as the *generator* and the plane curve that guides it is known as the *directrix* (Fig. 6-3). The directrices usually employed are the arc of a circle, the semiellipse, the parabola, the cycloid, and the catenary. A cylindrical shell may or may not be provided with an edge beam or edge member. The supporting members at the two ends of a shell

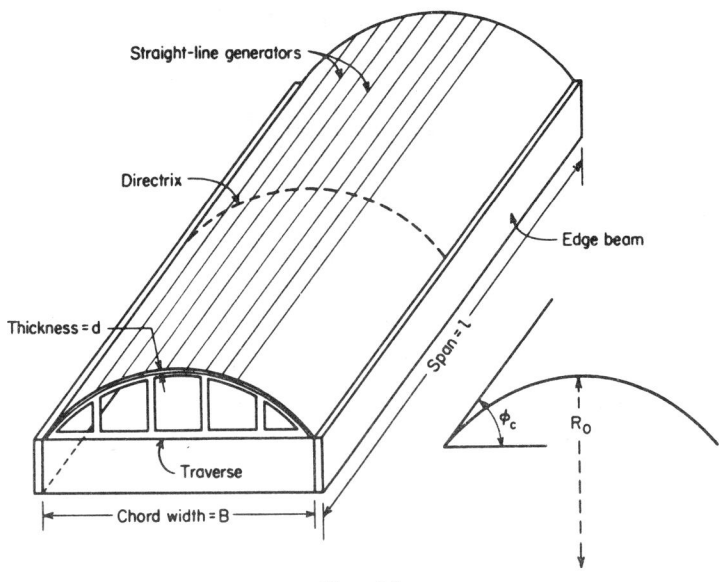

Fig. 6-3

are known as the *traverses*. The traverse may be a solid diaphragm, a tied arch, a trussed arch, or a rigid frame (Fig. 6-4). It is usually assumed that the shell is simply supported on the traverses. A further assumption is that the traverses are rigid in their own planes but flexible out of their planes so that they cannot receive any loads applied normally to them. The distance between adjacent traverses is known as the *span* of the shell. The projection of the arc of the shell is generally known as its *chord width*.

6-3 Loads

The loads usually considered in the design of cylindrical shells include the dead weight of the shell g and a snow load p_0 in regions subject to snowfall. The snow load is assumed to be uniform over the horizontal projection of the shell. In tropical countries, a live load of 10 to 15 psf of the shell surface is usually assumed instead of the snow load. The effects of wind cause only a suction on a cylindrical shell as long as the semicentral angle does not exceed 40°. This is usually the case.

6-4   Notes on the Membrane Theory

In the membrane theory, the shell is idealized as a membrane incapable of resisting bending stresses. For such a state of stress to exist, it is essential that the shell be a closed surface if it is circular, elliptical, or cycloidal. If the directrix is a parabola or catenary, the surface must

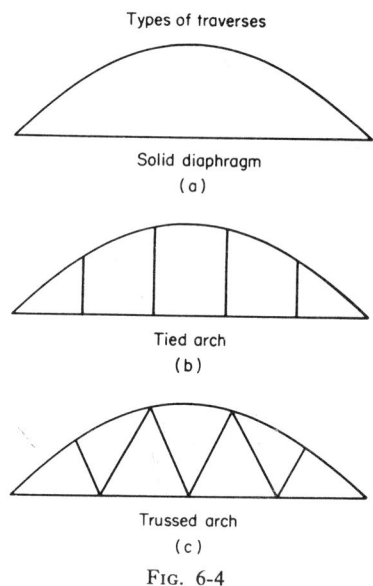

Types of traverses

Solid diaphragm
(a)

Tied arch
(b)

Trussed arch
(c)

Fig. 6-4

extend to infinity. An example of a pure membrane state of stress is furnished by a circular pipe subjected to fluid pressure. Shells in practice are of finite size terminated at their straight edges with or without stiffening edge members. In such shells, it is not possible to maintain a membrane state of equilibrium. Bending stresses are set up. These can be accounted for only by means of a bending theory to be developed in the next chapter. Bending stresses also develop in the vicinity of concentrated loads, cutouts, and stiffening ribs. The membrane theory is nevertheless useful in many practical cases in gaining some insight into the structural behavior of a shell. As we shall see later, it can also be used as a particular integral in the bending theory.

6-5 Equations of Equilibrium

Let $z = z(y)$ be a plane curve forming the directrix of a cylindrical shell (Fig. 6-5a). Let the origin be at the apex of the shell directrix at midspan. The coordinate x is measured along the crown generator, the coordinate y along the directrix, and the coordinate z along the

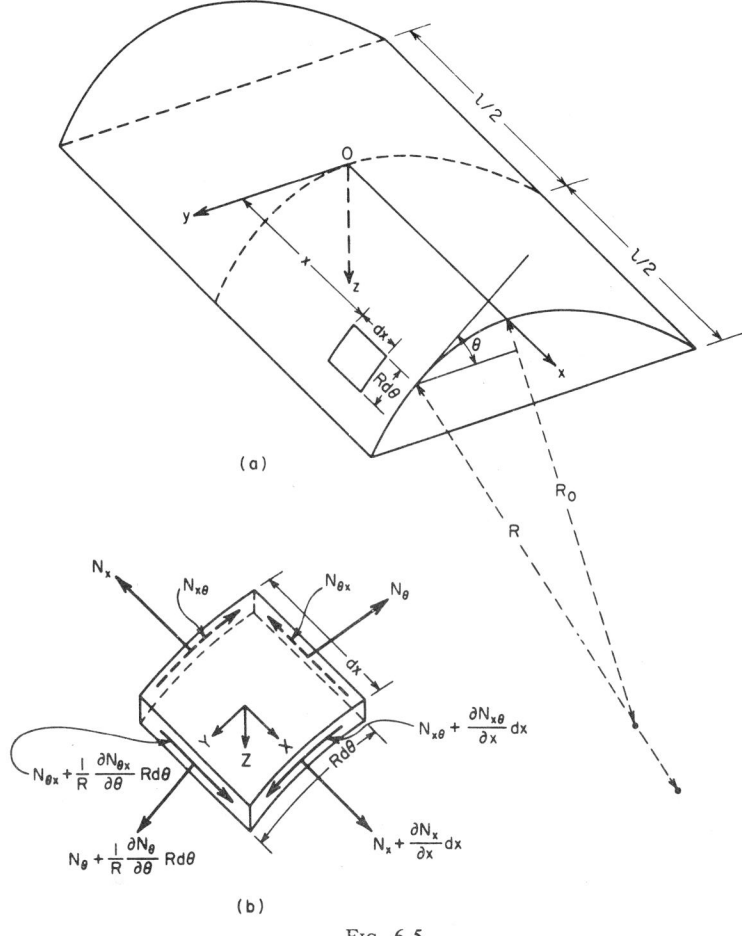

FIG. 6-5

inward normal. Figure 6-5b shows an element of the shell with the forces acting on it. N_x, N_θ, and $N_{x\theta}$ are forces developed per unit length. X, Y, and Z denote the components of the external load applied per unit area in the x, y, and z directions. It is clear that $dy = R\,d\theta$.

Summing up forces in the x direction and equating them to zero we get

$$\frac{\partial N_x}{\partial x} dx\, R\, d\theta + \frac{1}{R}\frac{N_{x\theta}}{\partial \theta} R\, d\theta\, dx + X\, dx\, R\, d\theta = 0$$

or

$$\frac{\partial N_x}{\partial x} + \frac{1}{R}\frac{\partial N_{x\theta}}{\partial \theta} + X = 0 \tag{6-1}$$

In like manner, we may write the equation of equilibrium in the y direction as

$$\frac{1}{R}\frac{\partial N_\theta}{\partial \theta} + \frac{\partial N_{x\theta}}{\partial x} + Y = 0 \tag{6-2}$$

Resolving forces along the inward normal to the element (Fig. 6-6),

$$2N_\theta\, dx\, (d\theta/2) + ZR\, d\theta\, dx = 0.$$

Simplifying,

$$N_\theta + ZR = 0 \tag{6-3}$$

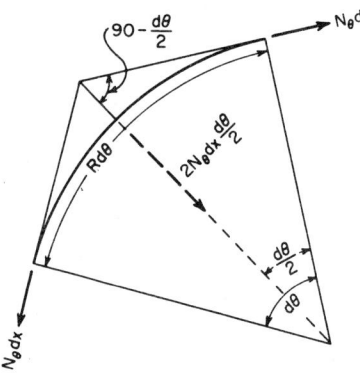

FIG. 6-6

Equations (6-1), (6-2), and (6-3) represent the three equations of equilibrium. It is to be noted that R, the radius of curvature, occurring in these equations, is a function of θ. N_θ is found from Equation (6-3), knowing Z and R. We may find $N_{x\theta}$ and N_x by integration as follows:

$$N_{x\theta} = -\int \frac{1}{R}\frac{\partial N_\theta}{\partial \theta} dx - \int Y\, dx + F_1(\theta) \tag{6-4}$$

where $F_1(\theta)$ is an arbitrary function of θ only. Similarly,

$$N_x = -\int \frac{1}{R}\frac{\partial N_{x\theta}}{\partial \theta} dx - \int X\, dx + F_2(\theta) \tag{6-5}$$

where $F_2(\theta)$ is again a function of θ only. The arbitrary functions $F_1(\theta)$ and $F_2(\theta)$ are to be found from the boundary conditions. In many cases of practical interest, X, Y, and Z are functions of θ only and do not vary along the x direction. If this is the case, it is evident from (6-3) that N_θ is a function of θ only. We may therefore write

$$N_{x\theta} = -\left(\frac{1}{R}\frac{dN_\theta}{d\theta} + Y\right) x + F_1(\theta) = -Kx + F_1(\theta) \tag{6-6}$$

where

$$K = \left(\frac{1}{R} \frac{dN_\theta}{d\theta} + Y \right)$$

Substituting for $N_{x\theta}'$ from (6-6) in (6-5) and integrating,

$$N_x = \left[\frac{x^2}{2R} \frac{dK}{d\theta} - \frac{1}{R} \frac{dF_1(\theta)}{d\theta} x - Xx + F_2(\theta) \right] \tag{6-7}$$

6-6 Stresses in a Simply Supported Shell

For a shell simply supported on the traverses, we have the following boundary conditions:

(i) $N_x = 0$ at $x = \pm l/2$. This boundary condition follows from the assumption that the traverses will not receive any loads applied normal to their planes.

(ii) $N_{x\theta} = 0$ at $x = 0$. This follows from the symmetry of the problem. For most shells occurring in practice, $X = 0$. From the second boundary condition and Equation (6-6) it is evident that $F_1(\theta) = 0$. Inserting the first boundary condition in Equation (6-7),

$$F_2(\theta) = - \frac{l^2}{8} \frac{1}{R} \frac{dK}{d\theta}$$

Hence we may now write the expressions for the stresses as

$$N_\theta = -ZR \tag{6-8a}$$

$$N_{x\theta} = -Kx \tag{6-8b}$$

$$N_x = -\frac{1}{2} \left(\frac{l^2}{4} - x^2 \right) \frac{1}{R} \frac{dK}{d\theta} \tag{6-8c}$$

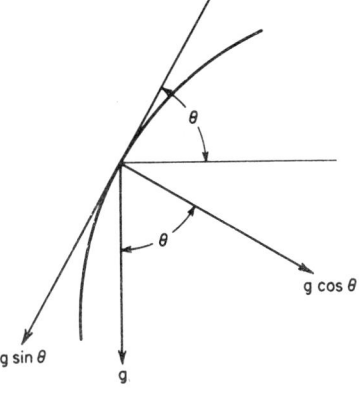

Fig. 6-7

6-7 Value of K for Dead and Snow Loads

Let the dead load be g per unit area of the surface. Referring to Fig. 6-7,

$$Y = g \sin \theta$$

and

$$Z = g \cos \theta$$

From relation (6-8a),

$$N_\theta = -gR \cos \theta \tag{6-9}$$

$$K = \left(\frac{1}{R} \frac{dN_\theta}{d\theta} + Y \right) = 2g \sin \theta - g \frac{1}{R} \frac{dR}{d\theta} \cos \theta \tag{6-10}$$

This value of K may be substituted in Equations (6-8b) and (6-8c) to find $N_{x\theta}$ and N_x.

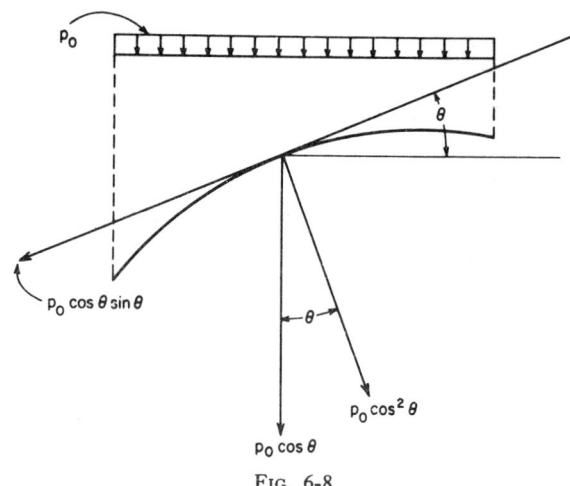

p_0

θ

$p_0 \cos \theta \sin \theta$

θ

$p_0 \cos^2 \theta$

$p_0 \cos \theta$

FIG. 6-8

Next consider a snow load of p_0 uniform over the horizontal projection of the shell surface (Fig. 6-8). The intensity of the vertical load $= p_0 \cos \theta$

Hence

$$Y = p_0 \sin \theta \cos \theta$$

and

$$Z = p_0 \cos^2 \theta$$

From (6-8a),

$$N = -p_0 R \cos^2 \theta \tag{6-11}$$

$$K = \left(\frac{1}{R} \frac{dN_\theta}{d\theta} + Y \right) = \left(3p_0 \sin \theta \cos \theta - p_0 \cos^2 \theta \frac{1}{R} \frac{dR}{d\theta} \right) \tag{6-12}$$

Knowing K, expressions for $N_{x\theta}$ and N_x may be written down using (6-8b) and (6-8c).

6-8 Expressions for Stresses under Dead and Snow Loads for Circular, Parabolic, Catenary, and Cycloidal Directrices

It has already been shown in Chapter 4 that the equations to these directrices may be written in the form $R = R_0 \cos^n \theta$.

(i) *Stresses under Dead Weight*

From (6-8a),

$$N_\theta = -gR \cos \theta = -gR_0 \cos^{n+1} \theta \tag{6-13}$$

Using (6-10), the value of K for this loading may be written as

$$K = 2g \sin \theta + gn \sin \theta = (n + 2) g \sin \theta$$

Hence

$$N_{x\theta} = -(n + 2) gx \sin \theta \tag{6-14}$$

Substituting the value of K in Equation (6-8c),

$$N_x = -\frac{n + 2}{2} g \left(\frac{l^2}{4} - x^2 \right) \frac{1}{R_0 \cos^{n-1} \theta} \tag{6-15}$$

(ii) *Stresses under Snow Load*

From relation (6-11),

$$N_\theta = -p_0 R_0 \cos^{n+2} \theta \tag{6-16}$$

The value of K appropriate to this load is found from Equation (6-12).

$$K = 3p_0 \sin \theta \cos \theta + \frac{p_0 \cos^2 \theta}{R_0 \cos^n \theta} nR_0 \cos^{n-1} \theta \sin \theta$$

$$= (n + 3) p_0 \sin \theta \cos \theta$$

Substituting this value of K in (6-8b) and (6-8c),

$$N_{x\theta} = -(n + 3) p_0 x \sin \theta \cos \theta \tag{6-17}$$

$$N_x = -\frac{(n + 3) p_0 \left(\frac{l^2}{4} - x^2 \right)}{2R_0} \frac{\cos^2 \theta - \sin^2 \theta}{\cos^n \theta} \tag{6-18}$$

By substituting the appropriate value of n in formulas (6-13) to (6-18), the stresses under dead and snow loads corresponding to any of the four directrices can be found.

6-9 Stresses in Shells with a Semielliptic Directrix

Following the procedure outlined in Arts. 6-7 and 6-8, the stresses in a cylindrical shell with a semielliptic directrix under dead and snow loads may be worked out. The expressions for the stresses are given below.

(i) *Stresses under Dead Weight*

$$N_\theta = -g \left[\frac{a^2 b^2 \cos\theta}{(a^2 \sin^2\theta + b^2 \cos^2\theta)^{3/2}} \right] \tag{6-19}$$

$$N_{x\theta} = -gx \left[2 + \frac{3(a^2 - b^2)\cos^2\theta}{a^2 \sin^2\theta + b^2 \cos^2\theta} \right] \sin\theta \tag{6-20}$$

$$N_x = -\frac{g}{2}\left(\frac{l^2}{4} - x^2\right)\left[\frac{2ab}{\alpha^3} + \frac{3(a^2-b^2)}{ab\alpha}\left(\cos^2\theta - \frac{2a^2}{b^2}\sin^2\theta\right)\right]\cos\theta \tag{6-21}$$

where

$$\alpha = \frac{ab}{(a^2 \sin^2\theta + b^2 \cos^2\theta)^{1/2}}$$

(ii) *Stresses under Snow Load*

$$N_\theta = -p_0 \frac{a^2 b^2 \cos^2\theta}{(a^2 \sin^2\theta + b^2 \cos^2\theta)^{3/2}} \tag{6-22}$$

$$N_{x\theta} = -3p_0 x \left(\frac{a^2 \sin\theta \cos\theta}{a^2 \sin^2\theta + b^2 \cos^2\theta} \right) \tag{6-23}$$

$$N_x = -\frac{3}{2} p_0 \left(\frac{l^2}{4} - x^2\right)\left[\frac{-a^2\sin^2\theta + b^2\cos^2\theta}{b^2(a^2\sin^2\theta + b^2\cos^2\theta)^{1/2}}\right] \tag{6-24}$$

It may be noted that, under dead weight, the entire cross section is under compression, i.e., N_x is negative, for certain values of b/a. For certain other values of b/a, N_x is compressive at the top and tensile at the bottom and tapers off to zero at $\theta = 90°$, corresponding to the bottom edge. Under dead load, N_x and N_θ are both zero for all values of b/a at $\theta = 90°$.

Under snow load, N_x is compressive at the top and tensile at the bottom for all values of b/a. It is not zero at $\theta = 90°$.

6-10 Tables for Stresses in Elliptic Shells

The formulas developed in the last article are not convenient for practical use as they involve tedious calculations. Tables such as those due to Krall [13] are available for use in the design office. An extract from Krall's tables is given in the Appendix. For making use of this table, formulas (6-19) to (6-24) need to be recast as follows:

(i) *Stresses under Dead Weight*

$$N_\theta = gat_2 \tag{6-25}$$

$$N_{x\theta} = gxs \tag{6-26}$$

$$N_x = \frac{g}{2a}\left(\frac{l^2}{4} - x^2\right)t_1 \tag{6-27}$$

(ii) *Stresses under Snow Load*

$$N_\theta = p_0 at_2' \tag{6-28}$$

$$N_{x\theta} = p_0 xs' \tag{6-29}$$

$$N_x = \frac{p_0}{2a}\left(\frac{l^2}{4} - x^2\right)t_1' \tag{6-30}$$

In these expressions the symbols used have the following meanings:

$$t_2 = -\frac{k^2}{\beta^3}\cos\theta \qquad s = -\left[2 + \frac{3(1 - k^2)\cos^2\theta}{\beta^2}\right]\sin\theta$$

$$t_1 = -\frac{1}{\beta}\left[3 + \sin^2\theta(1 + 2\sin^2\theta) - k^2\cos^4\theta - \frac{\sin^2\theta(4 - \cos^2\theta)}{k^2}\right]\cos\theta$$

$$t_2' = -\frac{k^2}{\beta^3}\cos^2\theta$$

$$s' = -\frac{3}{\beta^2}\sin\theta\cos\theta$$

$$t_1' = -\frac{3}{\beta}\left(\cos^2\theta - \frac{\sin^2\theta}{k^2}\right)$$

where

$$k = \frac{b}{a} \qquad \text{and} \qquad \beta = (\sin^2\theta + k^2\cos^2\theta)^{1/2}$$

6-11 Cylindrical Shell with Circular Directrix

The formula derived in Art. 6-8 may be specialized for the cylindrical shell with a circular directrix by putting $n = 0$ and noting that $R = R_0 = a$.

(i) *Stresses under Dead Weight*

$$N_\theta = -ga\cos\theta \tag{6-31}$$

$$N_{x\theta} = -2gx\sin\theta \tag{6-32}$$

$$N_x = -\frac{g}{a}\left(\frac{l^2}{4} - x^2\right)\cos\theta \tag{6-33}$$

The stresses N_θ and N_x are compressive everywhere. Figure 6-9 shows a free-body diagram of the edge beam with the shear forces transferred to it by the shell. The axial force P that develops in the

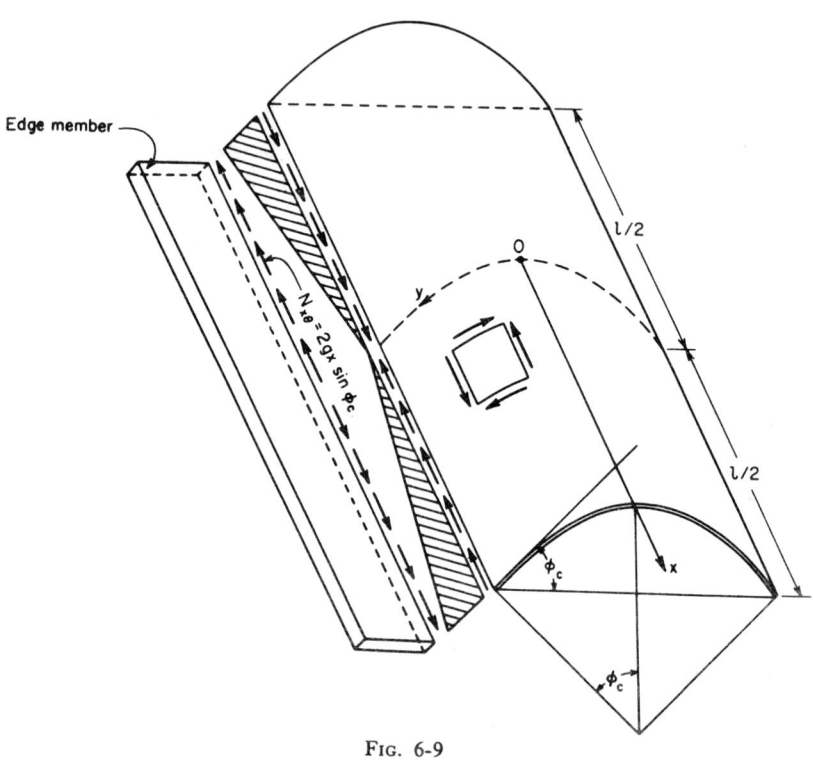

Fig. 6-9

edge beam at a distance x from the center of the span is given by

$$P = -\int_x^{l/2} (N_{x\theta})_{\theta=\phi_c}\, dx$$

$$= -\int_x^{l/2} -2gx \sin \phi_c \, dx$$

$$= +g \left(\frac{l^2}{4} - x^2\right) \sin \phi_c \tag{6-34}$$

The plus sign indicates that the force developed is a tension. The maximum tensile force in the edge beams occurs at $x = 0$, i.e., at midspan, and its value is $(gl^2/4) \sin \phi_c$.

(ii) *Stresses under Snow Load*

$$N_\theta = -p_0 a \cos^2 \theta \tag{6-35}$$

$$N_{x\theta} = -1.50 p_0 x \sin 2\theta \tag{6-36}$$

and

$$N_x = -1.50 \frac{p_0}{a} \left(\frac{l^2}{4} - x^2 \right) \cos 2\theta \tag{6-37}$$

The stress N_θ remains compressive for all values of θ. The stress N_x is compressive for $\theta < 45°$ and tensile for $\theta > 45°$. In the special case where the shell arc is semicircular, $N_{x\theta}$ vanishes at the shell edge. Consequently the axial force in the edge beam is zero.

6-12 Cylindrical Shell with Cycloidal Directrix

The stresses in a cylindrical shell with a cycloid as the directrix are found by putting $n = 1$ in the formulas derived in Art. 6-8.

(i) *Stresses under Dead Weight*

$$N_\theta = -g R_0 \cos^2 \theta \tag{6-38}$$

$$N_{x\theta} = -3gx \sin \theta \tag{6-39}$$

$$N_x = -\frac{3}{2} \frac{g}{R_0} \left(\frac{l^2}{4} - x^2 \right) \tag{6-40}$$

The stress N_x is found to be constant at all points on any cross section. At the edges, corresponding to $\theta = \pi/2$, N_θ vanishes and $N_{x\theta} = -3g x$.

The axial force P in the edge beam at a section distant x from the midspan is found as before. Thus

$$P = -\int_x^{l/2} -3gx \, dx = \frac{3}{2} g \left(\frac{l^2}{4} - x^2 \right)$$

It is again found to be a tension with a maximum value of $\frac{3}{8} gl^2$ at midspan.

(ii) *Stresses under Snow Load*

$$N_\theta = -p_0 R_0 \cos^2 \theta \tag{6-41}$$

$$N_{x\theta} = -4 p_0 x \sin \theta \cos \theta \tag{6-42}$$

$$N_x = -2 \frac{p_0}{R_0} \left(\frac{l^2}{4} - x^2 \right) \frac{\cos^2 \theta - \sin^2 \theta}{\cos \theta} \tag{6-43}$$

At the edge, where $\theta = \pi/2$,

$$N_x = \infty \qquad N_\theta = 0 \qquad \text{and} \qquad N_{x\theta} = 0$$

Because $N_{x\theta} = 0$, the axial force in the edge beam is zero. This means that the N_x forces form a couple, with their algebraic sum over the cross section adding up to zero. The sum of the N_x forces is

$$-\frac{4p_0}{R_0}\left(\frac{l^2}{4} - x^2\right)\int_0^\pi \frac{\cos^2\theta - \sin^2\theta}{\cos\theta} R_0 \cos\theta \, d\theta$$

This integral can be shown to be equal to zero.

6-13 Cylindrical Shell with Catenary Directrix

The stresses in a cylindrical shell with a catenary directrix are arrived at by setting $n = -2$ in the formulas derived in Art. 6-8.

(i) *Stresses under Dead Weight*

$$N_\theta = -\frac{gR_0}{\cos\theta} \tag{6-44}$$

$$N_{x\theta} = 0 \tag{6-45}$$

$$N_x = 0 \tag{6-46}$$

This leads us to the important conclusion that a catenary shell will degenerate into a series of independent arches (because $N_x = 0$) and transfer the entire load to their edge beams (because $N_{x\theta} = 0$) and transmit no load to the traverses. Such behavior is to be expected whenever the shell directrix is the funicular curve of the applied loading. This action of the catenary is taken advantage of in the so-called Ctesiphon or corrugated shells.

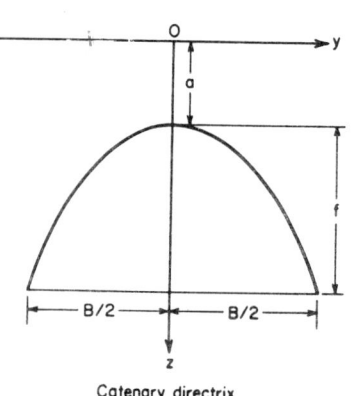

Catenary directrix

Fig. 6-10

(ii) *Stresses under Snow Load*

$$N_\theta = -p_0 R_0 \tag{6-47}$$

$$N_{x\theta} = -p_0 x \sin\theta \cos\theta \tag{6-48}$$

$$N_x = -0.50\frac{p_0}{R_0}\left(\frac{l^2}{4} - x^2\right)\cos 2\theta \cos^2\theta \tag{6-49}$$

The following relations are useful in dealing with the catenary. Referring to Fig. 6-10, the equation to the catenary may be written as

$$z = a \cosh\left(\frac{y}{a}\right) \tag{6-50}$$

At $y = B/2$, $z = (f + a)$. Hence $(f + a) = a \cosh (B/2a)$ or

$$f = a \left(\cosh \frac{B}{2a} - 1 \right) \tag{6-51}$$

Let us denote $B/2a$ by α and f/B by β. Inserting these in (6-51),

$$\beta = \frac{1}{2\alpha} (\cosh \alpha - 1) \tag{6-52}$$

This transcendental equation may be solved by trial and error. The values of α corresponding to various values of β are given in Table 6-1.

Table 6-1

β	0.50	0.40	0.30	0.20	0.10
α	1.616	1.371	1.088	0.762	0.395

It is also possible to tabulate $R_0 = a = B/2\alpha$ corresponding to various values of β (Table 6-2).

Table 6-2

β	0.50	0.40	0.30	0.20	0.10
R_0	0.309B	0.365B	0.460B	0.656B	1.266B

6-14 Cylindrical Shell with Parabolic Directrix

By inserting $n = -3$ in the formulas derived in Art. 6-8, the stresses in a cylindrical shell with a parabolic directrix are obtained as follows:

(i) *Stresses under Dead Weight*

$$N_\theta = - \frac{gR_0}{\cos^2 \theta} \tag{6-53}$$

$$N_{x\theta} = +gx \sin \theta \tag{6-54}$$

$$N_x = 0.50 \frac{g}{R_0} \left(\frac{l^2}{4} - x^2 \right) \cos^4 \theta \tag{6-55}$$

From (6-55), it is evident that the entire cross section is in tension. Consequently the edge member will be in compression. The axial compression P in the edge member is found as before.

$$P = -\int_{x}^{l/2} + gx \sin \phi_c \, dx$$

$$= -\frac{g}{2}\left(\frac{l^2}{4} - x^2\right) \sin \phi_c \qquad (6\text{-}56)$$

where ϕ_c is the slope of the tangent to the horizontal at the shell edge.

(ii) *Stresses under Snow Load*

$$N_\theta = -\frac{p_0 R_0}{\cos \theta} \qquad (6\text{-}57)$$

$$N_{x\theta} = 0 \qquad (6\text{-}58)$$

$$N_x = 0 \qquad (6\text{-}59)$$

The parabolic directrix being the funicular curve of the snow load, it is only to be expected that $N_x = N_{x\theta} = 0$. The shell degenerates into a series of independent arches and no load is transmitted to the traverses.

It can be shown that, if the directrix is a parabola, unit rings of the shell transmit to the edge beam stresses N_θ whose vertical component exceeds the dead weight of the shell. This may be proved as follows: Referring to Fig. 6-11, the equation to the parabola may be written as

$$z = \frac{4fy}{B^2}(B - y)$$

The slope of the tangent at the origin is evidently $4f/B$. It follows that $OC = OD = f$. It is easily shown that $R_0 = B^2/8f$ for a parabola. The vertical component of N_θ transferred to the two edge beams is

$$2\frac{gR_0}{\cos^2 \phi_c} \sin \phi_c = \frac{2g}{\cos \phi_c}\frac{B^2}{8f}\frac{4f}{B} = \frac{gB}{\cos \phi_c} = g(\overline{AC} + \overline{CB})$$

Hence the vertical reaction transferred to the two edge beams is seen to be proportional to $(\overline{AC} + \overline{CB})$. The dead weight of the shell is proportional to $\overparen{AOB}$. Because $(\overline{AC} + \overline{CB}) > \overparen{AOB}$, the result is proved. A simpler physical explanation is that, unlike other cylindrical shells, a shell with a parabolic directrix develops a positive shear stress $N_{x\theta}$ (Fig. 6-12). Observing the sense of these shear stresses it is clear that they add to the weight of the shell.

6-15 Some Comments on the Membrane Theory

Stresses in cylindrical shells with various directrices are summarized in Table 6-3 for ready reference. A critical study of this table leads to the following conclusions:

(i) According to the membrane theory, a thin shell acts partially as an arch and partially as a beam. The arch action is responsible for the transfer of reactions to the edge beams, and the beam action for the transfer of reactions to the traverses through the medium of shear stresses that develop between adjacent rings of the shell. When the tangent is vertical at the ends, pure beam action results. If the

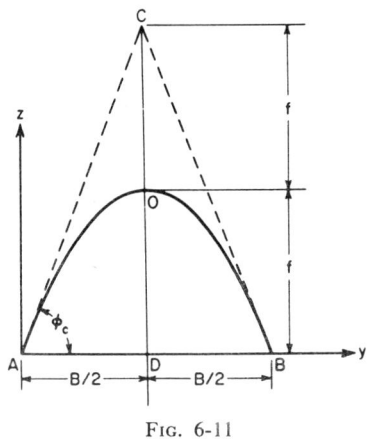

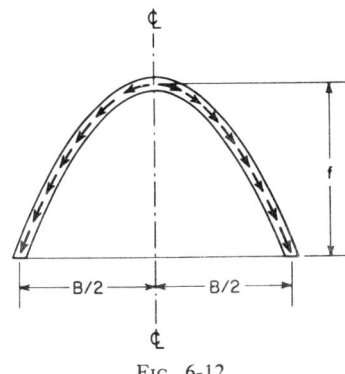

Fig. 6-11   Fig. 6-12

directrix chosen is the funicular curve of the applied loading, the shell degenerates into a series of independent arches and beam action completely disappears.

(ii) The external bending moment computed at any section treating the shell as a simply supported beam is resisted by

(a) The resultant of the N_x forces.

(b) The axial force P in the edge beams.

(c) The vertical component of the bending moment to which the edge beams are subjected. In the particular case where the end tangent of the shell is vertical, the external bending moment is resisted by a combination of (a) and (b). If the directrix is the funicular curve of the loading, the external bending moment is entirely resisted by (c). It may also happen that the external moment is resisted by (a) only. A case in point is the cycloidal directrix under snow load.

(iii) The value and variation N_θ is the same at all cross sections. N_θ is independent of the boundary conditions at the traverses.

(iv) In the membrane state, equilibrium of the shell is maintained by the in-plane stresses N_θ, $N_{x\theta}$, and N_x. Bending moments and transverse shears are absent. In an arch, bending moments will be absent if and only if the shape of the arch chosen corresponds to the funicular curve of the applied loading. Herein lies the basic difference between shell and arch action.

Table 6-3. Membrane Stresses in

Directrix	R	Due to dead weight g		
		N_θ	N_x	$N_{x\theta}$
Arc of a circle	$R = R_0 = a$	$-ga\cos\theta$	$-\dfrac{g}{a}\left(\dfrac{l^2}{4}-x^2\right)\cos\theta$	$-2gx\sin\theta$
Cycloid	$R = R_0\cos\theta$	$-gR_0\cos^2\theta$	$-\dfrac{1.5g}{R_0}\left(\dfrac{l^2}{4}-x^2\right)$	$-3gx\sin\theta$
Catenary	$R = R_0\cos^{-2}\theta$	$-\dfrac{gR_0}{\cos\theta}$	0	0
Parabola	$R = R_0\cos^{-3}\theta$	$-\dfrac{gR_0}{\cos^2\theta}$	$+\dfrac{0.50g}{R_0}\left(\dfrac{l^2}{4}-x^2\right)\cos^4\theta$	$+gx\sin\theta$
Ellipse $R_0=\dfrac{a^2}{b}$	$R=\dfrac{R_0\alpha^3}{a^3}$ $=\dfrac{\alpha^3}{ab}$	$-\dfrac{g\alpha^3}{ab}\cos\theta$	$-\dfrac{g}{2}\left(\dfrac{l^2}{4}-x^2\right)\left[\dfrac{2ab}{\alpha^3}\right.$ $+\dfrac{3(a^2-b^2)}{ab\alpha}\left(\cos^2\theta\right.$ $\left.-\dfrac{2\alpha^2}{b^2}\sin^2\theta\right)\Big]\cos\theta$	$-gx\left[2\right.$ $\left.+\dfrac{3(a^2-b^2)\cos^2\theta}{a^2\sin^2\theta+b^2\cos^2\theta}\right]\sin\theta$

R_0 = radius of curvature at crown
R = radius of curvature at any point
$\alpha = \dfrac{ab}{(a^2\sin^2\theta + b^2\cos^2\theta)^{1/2}}$ for elliptic shells

6-16 Membrane Theory of a Circular Cylindrical Shell for Fourier Loading

In Art. 6-4, mention has already been made of the possibility of using the membrane stresses as a particular integral in the bending theory to be developed in the next chapter. To be able to do this, the load on the shell has to be expressed as a Fourier series. This is necessary because, in the bending theory, the equations of equilibrium and stress-strain relations are combined to yield a single linear partial differential equation of the eighth order. The object of expressing the loading in Fourier series is to ensure that the differential equation developed for the midspan

Shells with Various Directrices

	Due to snow load p_0	
N_θ	N_x	$N_{x\theta}$
$-p_0 a \cos^2 \theta$	$-\dfrac{1.5p_0}{a}\left(\dfrac{l^2}{4} - x^2\right)\cos 2\theta$	$-1.5p_0 x \sin 2\theta$
$-p_0 R_0 \cos^3 \theta$	$-\dfrac{2p_0}{R_0}\left(\dfrac{l^2}{4} - x^2\right)\dfrac{\cos^2 \theta - \sin^2 \theta}{\cos \theta}$	$-2p_0 x \sin 2\theta$
$-p_0 R_0$	$-\dfrac{0.50p_0}{R_0}\left(\dfrac{l^2}{4} - x^2\right)\cos 2\theta \cos^2 \theta$	$-0.50p_0 x \sin 2\theta$
$-\dfrac{p_0 R_0}{\cos \theta}$	0	0
$-\dfrac{p_0 \alpha^3}{ab}\cos^2 \theta$	$-3p_0\left[\dfrac{(l^2/4 - x^2)}{\cdot \; 2}\right]\left[\dfrac{-a^2 \sin^2 \theta + b^2 \cos^2 \theta}{b^2(a^2 \sin^2 \theta + b^2 \cos^2 \theta)^{1/2}}\right]$	$-3p_0 x\left(\dfrac{a^2 \sin \theta \cos \theta}{a^2 \sin^2 \theta + b^2 \cos^2 \theta}\right)$

section of the shell may apply to other sections as well. In the membrane theory, the angle θ is measured from the crown to the edge of the shell. In developing the bending theory the angle ϕ is measured from the left edge. It is therefore necessary to recast the results of the membrane theory by replacing θ by ϕ. It may be observed that $\theta = (\phi_c - \phi)$. Consequent on replacing θ by ϕ, it is better to redesignate the stresses N_θ, $N_{x\theta}$, and N_x as N_ϕ, $N_{x\phi}$, and N_x, respectively, to be consistent with the bending theory. In what follows, these changes are introduced in deriving stresses from the membrane theory. It is best to rewrite the equations of equilibrium afresh. The following equations of equilibrium are easily derived with the aid of Fig. 6-13:

$$\frac{\partial N_x}{\partial x} + \frac{1}{a}\frac{\partial N_{x\phi}}{\partial \phi} + X = 0 \tag{6-60}$$

$$\frac{1}{a}\frac{\partial N_\phi}{\partial \phi} + \frac{\partial N_{x\phi}}{\partial x} + Y = 0 \tag{6-61}$$

$$N_\phi + aZ = 0 \tag{6-62}$$

These are identical with Equations (6-1), (6-2), and (6-3) except for the changes in notation. It is also to be noted that the directions of the shear stresses acting on the element are opposite to those of the shear stresses shown in Fig. 6-5b.

(i) *Stresses under Dead Weight*

We have already shown in Chapter 2 that a uniform load g may be developed in a Fourier series as

$$g = \frac{4}{\pi}g\left(\cos\frac{\pi x}{l} - \frac{1}{3}\cos\frac{3\pi x}{l} + \frac{1}{5}\cos\frac{5\pi x}{l} - \cdots\right)$$

Whenever the load is uniform in the x direction, it is usually adequate to consider the first term of this series.

Taking only the first term into account,

$$Z = \frac{4}{\pi}g\cos\frac{\pi x}{l}\cos(\phi_c - \phi)$$

and

$$Y = -\frac{4}{\pi}g\cos\frac{\pi x}{l}\sin(\phi_c - \phi)$$

Now from (6-62),

$$N_\phi = -\frac{4}{\pi}ag\cos\frac{\pi x}{l}\cos(\phi_c - \phi) \tag{6-63}$$

From (6-61),

$$N_{x\phi} = - \int \left(\frac{1}{a} \frac{\partial N_\phi}{\partial \phi} + Y \right) dx = + \frac{8l}{\pi^2} g \sin \frac{\pi x}{l} \sin (\phi_c - \phi) \qquad (6\text{-}64)$$

Again, from (6-62),

$$N_x = - \frac{1}{a} \int \frac{\partial N_{x\phi}}{\partial \phi} \, dx = - \frac{8gl^2}{a\pi^3} \cos \frac{\pi x}{l} \cos (\phi_c - \phi) \qquad (6\text{-}65)$$

(ii) *Stresses under Snow Load*

As before, we may represent p_0 as

$$p_0 = \frac{4}{\pi} p_0 \left(\cos \frac{\pi x}{l} - \frac{1}{3} \cos \frac{3\pi x}{l} + \frac{1}{5} \cos \frac{5\pi x}{l} - \cdots \right)$$

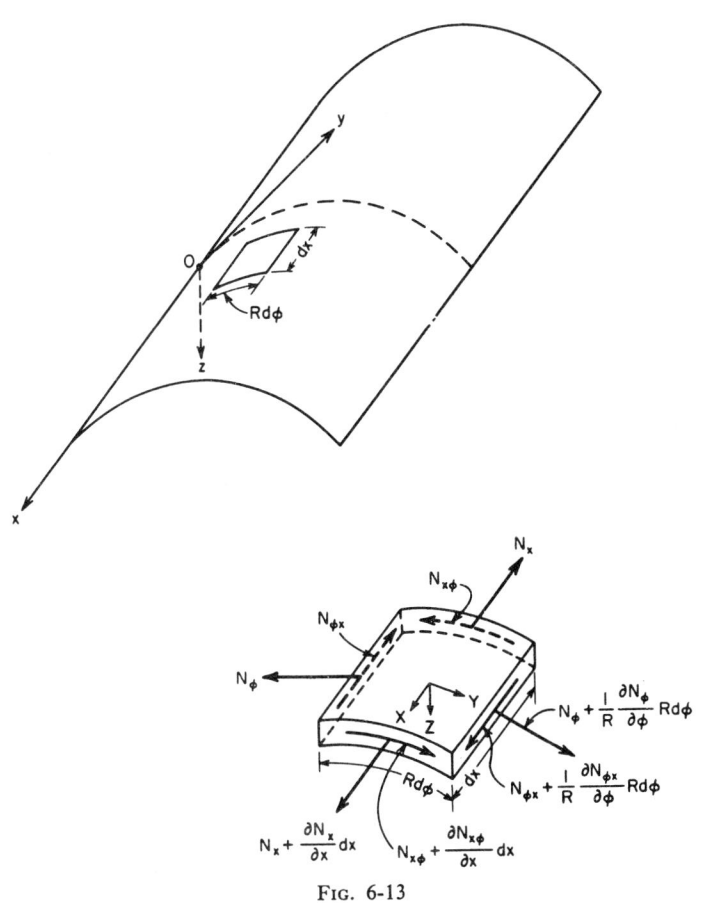

FIG. 6-13

Proceeding in the same manner as for the dead weight we may easily arrive at the following expressions:

$$N_\phi = -\frac{4p_0 a}{\pi} \cos \frac{\pi x}{l} \cos^2 (\phi_c - \phi) \tag{6-66}$$

$$N_{x\phi} = +\frac{6p_0 l}{\pi^2} \sin \frac{\pi x}{l} \sin 2(\phi_c - \phi) \tag{6-67}$$

$$N_x = -\frac{12p_0 l^2}{a\pi^3} \cos \frac{\pi x}{l} \cos 2(\phi_c - \phi) \tag{6-68}$$

CHAPTER 7

BENDING THEORY OF CYLINDRICAL SHELLS

7-1 The Need for a Bending Theory

Most reinforced-concrete cylindrical shells used in practice do not behave as membranes. Along the edges of the shell, stresses and displacements, different from those given by the membrane theory, usually exist. These depend on the manner in which the shell is supported or, in other words, on the type of *physical boundary conditions* that exist along the supporting edges. For example, consider a shell with free edges AB and CD. Along these edges, the membrane theory would indicate the presence of stresses N_ϕ and $N_{x\phi}$ (Fig. 7-1). But it is evident, from the boundary conditions, that these stresses cannot exist, the edges being free. The actual boundary conditions can be realized by applying *corrective line* loads. But the application of such *line loads* would cause the shell to bend and depart from its membrane state. The shell now seeks a new equilibrium and in that process brings into play bending moments, twisting moments, and radial shears. A bending theory is essential to account for these effects.

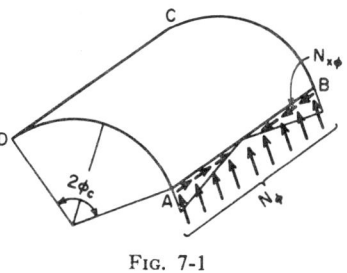

FIG. 7-1

7-2 Stress Analysis of Cylindrical Shells

The stress analysis of cylindrical shells may thus be carried out in three stages:

(*a*) A membrane analysis with the surface loads acting on the shell

(*b*) A bending analysis of the unloaded shell

(c) Superposition of the results of (a) and (b) to realize the actual boundary conditions that exist along the straight edges of the shell

The step described under (b) is often referred to as edge-disturbance analysis. Usually, the edge disturbances emanating from the straight edges alone are considered. The disturbances emanating from the curved edges are disregarded, in most cases, being less significant.

The process of stress analysis described above essentially consists in correcting the membrane stresses by superimposing the edge effects on them. This is not strictly rigorous, as we shall see later. But it is adequately accurate for most practical purposes.

EXPRESSIONS FOR STRAIN AND CHANGE IN CURVATURE

7-3 Uniaxial State of Stress

As a preliminary to setting up expressions for strains and change in curvature of a cylindrical shell, let us consider a few elementary states of stress. We begin with a uniaxial state of stress. Consider two adjacent points A and B on a bar subjected to tension (Fig. 7-2). Let the dis-

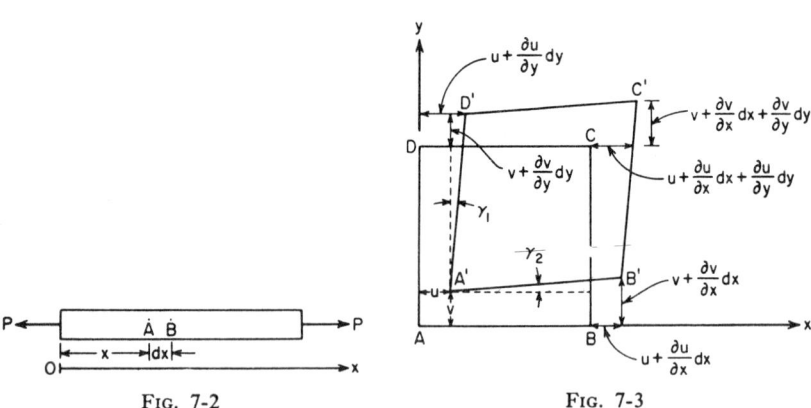

FIG. 7-2   FIG. 7-3

placements of A and B in the positive direction of the x axis be u and $u + (\partial u/\partial x)\, dx$, respectively. Hence the change in length of $AB = (\partial u/\partial x)\, dx$. The strain in the bar at A as B moves infinitely close to $A = \partial u/\partial x$.

7-4 State of Plane Stress

Next, consider an element $ABCD$ taken out of a body undergoing a plane state of stress (Fig. 7-3). $A'B'C'D'$ is the deformed element.

Let u and v be the components of the displacement of the point A in the x and y directions. The deformation of $ABCD$ into $A'B'C'D'$ may be visualized as taking place in three stages:

(a) A dilation in the x direction (Fig. 7-4)
(b) A dilation in the y direction (Fig. 7-5)
(c) An angular distortion (Fig. 7-3)

From Fig. 7-4, the strain in the x direction is found to be $\partial u/\partial x$. Similarly, from Fig. 7-5 it is clear that the strain in the y direction is

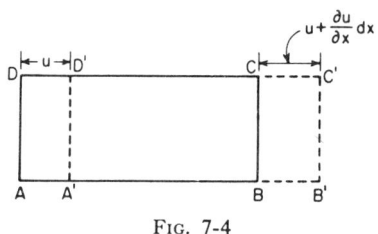

Fig. 7-4

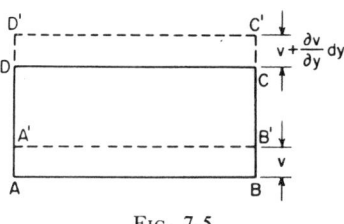

Fig. 7-5

$\partial v/\partial y$. The angular distortion $\gamma = (\gamma_1 + \gamma_2) = (\partial u/\partial y + \partial v/\partial x)$. We may now write the expressions for strain as follows:

$$\text{Strain in the } x \text{ direction} = \epsilon_x = \frac{\partial u}{\partial x} \tag{7-1}$$

$$\text{Strain in the } y \text{ direction} = \epsilon_y = \frac{\partial v}{\partial y} \tag{7-2}$$

$$\text{Shear strain} = \gamma_{xy} = \left(\frac{\partial u}{\partial y} + \frac{\partial v}{\partial x}\right) \tag{7-3}$$

7-5 Strains in a Circular Cylindrical Shell

It is convenient to use coordinates x and $a\phi$ as shown. u, v, and w are, respectively, the displacements of a point on the shell in the x direction, in the direction of the tangent at that point and in the direction of the inward-directed normal (Fig. 7-6). Let the origin be at midpoint of the left edge. The expressions for strain in the x direction and the shear strain would remain the same as for plane stress with $a\phi$ taking the place of y. Hence

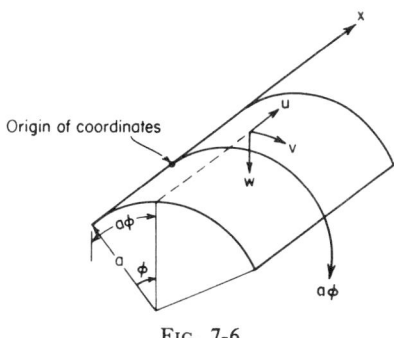

Fig. 7-6

$$\epsilon_x = \frac{\partial u}{\partial x} \quad \text{and} \quad \gamma_{x\phi} = \left(\frac{1}{a}\frac{\partial u}{\partial \phi} + \frac{\partial v}{\partial x}\right)$$

The strain in the ϕ direction will consist of two parts:

(a) Strain corresponding to a plane state of stress $= (1/a)(\partial v/\partial \phi)$

(b) Circumferential strain caused by w

The second effect may be calculated as follows: The circumference shrinks by $2\pi w$ when the radius changes from a to $(a - w)$. Hence

$$\text{Circumferential strain} = \frac{(a - w)\,d\phi - a\,d\phi}{a\,d\phi} = -\left(\frac{w}{a}\right)$$

Hence the total strain in the ϕ direction $\epsilon_\phi = [(1/a)(\partial v/\partial \phi) - w/a]$. We may now collect together the expressions for strain in a cylindrical shell:

$$\epsilon_x = \frac{\partial u}{\partial x} \tag{7-4}$$

$$\epsilon_\phi = \left(\frac{1}{a}\frac{\partial v}{\partial \phi} - \frac{w}{a}\right) \tag{7-5}$$

$$\gamma_{xy} = \left(\frac{1}{a}\frac{\partial u}{\partial \phi} + \frac{\partial v}{\partial x}\right) \tag{7-6}$$

These expressions ignore the effects of changes of curvature on the strains and are hence only approximately correct. But they are good enough for the analysis of reinforced-concrete shells.

7-6 Rotation of the Tangent

First, consider the differential radial displacements of two adjacent points A and B on the shell (Fig. 7-7). The tangent at A would rotate by

$$\frac{\dfrac{\partial w}{\partial \phi}\,d\phi}{a\,d\phi}$$

on this account. Because of the circumferential displacement v of the point A which moves from A to A', the tangent would rotate by an additional angle (v/a). Hence the tangent at A will rotate through a total angle of

$$\vartheta = \frac{1}{a}\left(v + \frac{\partial w}{\partial \phi}\right) \tag{7-7}$$

7-7 Change in Circumferential Curvature

The curvature before deformation is, by definition, $d\phi/(a\,d\phi) = 1/a$. The change in curvature is brought out by the change in the arc length of the element as well as the central angle that it subtends. The change in the latter is caused by the rotation of the tangents at A and B.

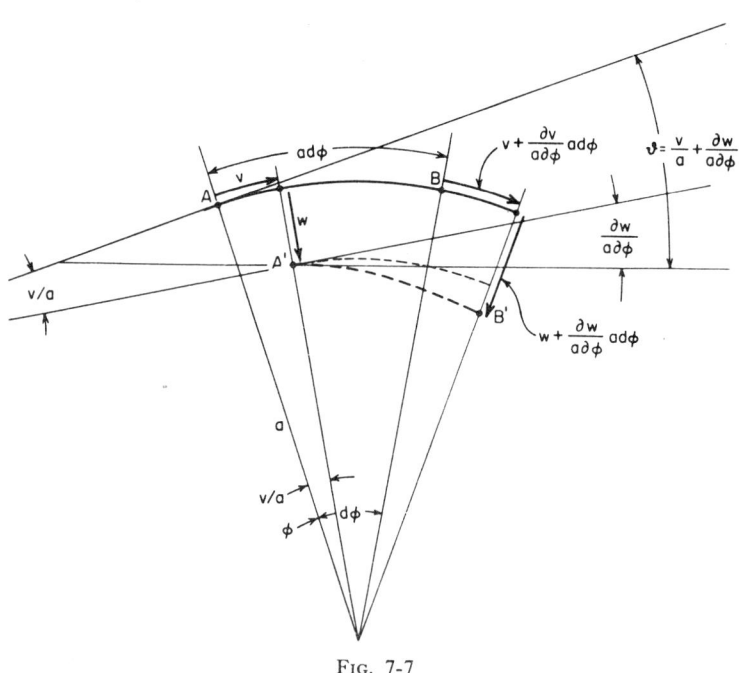

Fig. 7-7

Change in curvature = (Curvature after deformation − original Curvature)

Hence,

$$\chi_\phi = \frac{d\phi + \frac{d\vartheta}{d\phi}\, d\phi}{ds\,(1 + \epsilon_\phi)} - \frac{1}{a}$$

$$= \frac{1}{a}\left[\frac{d\vartheta}{d\phi}\,\frac{1}{(1 + \epsilon_\phi)} - 1\right]$$

Substituting the value of ϵ_ϕ from (7-5),

$$\chi_\phi = \frac{1}{a^2}\left(w + \frac{\partial^2 w}{\partial \phi^2}\right)$$

$$(7\text{-}8)$$

7-8   Stress Resultants

The forces and moments acting on a shell per unit length are usually known as the *stress resultants*.   In a cylindrical shell under bending, there are, in general, 10 stress resultants to be determined (Figs. 7-8 and 7-9).   The stress resultants may be related to the corresponding stresses as follows:

$$N_x = \int_{-d/2}^{+d/2} \sigma_x \frac{a+z}{a} \, dz \qquad (a)$$

$$M_x = \int_{-d/2}^{+d/2} \sigma_x z \frac{a+z}{a} \, dz \qquad (b)$$

$$M_{x\phi} = \int_{-d/2}^{+d/2} \tau_{x\phi} z \frac{a+z}{a} \, dz \qquad (c)$$

$$Q_x = \int_{-d/2}^{+d/2} \tau_{xz} \frac{a+z}{a} \, dz \qquad (d)$$

$$N_{x\phi} = \int_{-d/2}^{+d/2} \tau_{x\phi} \frac{a+z}{a} \, dz \qquad (e)$$

$$N_\phi = \int_{-d/2}^{+d/2} \sigma_\phi \, dz \qquad (f)$$

$$M_\phi = \int_{-d/2}^{+d/2} \sigma_\phi z \, dz \qquad (h)$$

$$M_{\phi x} = \int_{-d/2}^{+d/2} \tau_{\phi x} z \, dz \qquad (k)$$

$$Q_\phi = \int_{-d/2}^{+d/2} \tau_{\phi z} \, dz \qquad (l)$$

$$N_{\phi x} = \int_{-d/2}^{+d/2} \tau_{\phi x} \, dz \qquad (m)$$

(7-9)

The factor $(a + z)/a = (1 + z/a)$ appearing in some of these expressions takes care of the trapezoidal shape of the element of the shell in cross section.   If the shell is thin, $z/a \ll 1$ and can be ignored.   If this is done, it is easily verified that $M_{x\phi} = M_{\phi x}$ and $N_{x\phi} = N_{\phi x}$, because $\tau_{x\phi} = \tau_{\phi x}$.

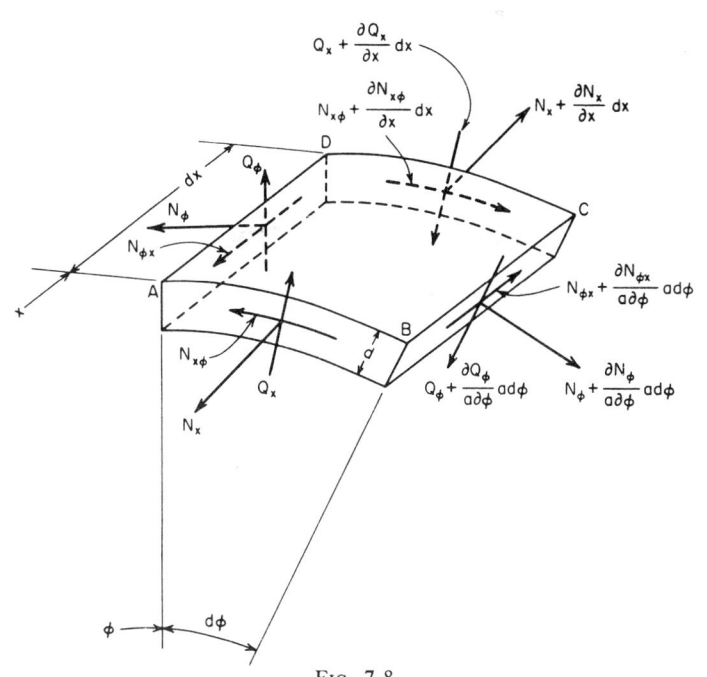

FIG. 7-8

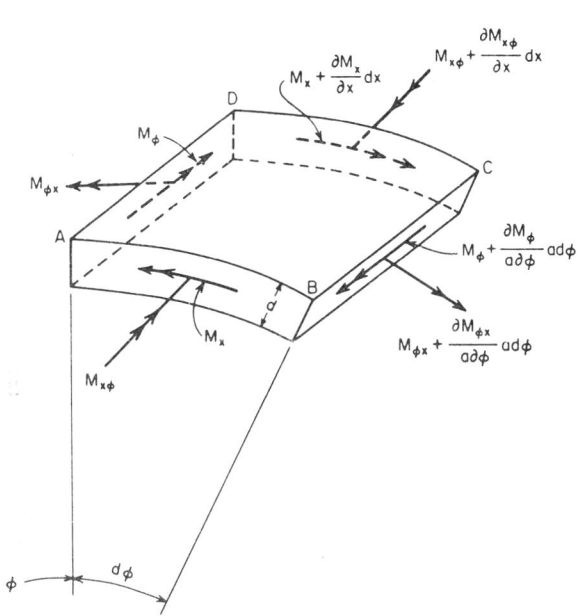

FIG. 7-9

7-9   Stress-Strain Relations

Making use of the expressions (7-4), (7-5), and (7-6) for the strains, the following stress-strain relations may be written down:

$$\frac{N_x}{Ed} - \frac{\nu N_\phi}{Ed} = \epsilon_x = \frac{\partial u}{\partial x} \qquad (a)$$

$$\frac{N_\phi}{Ed} - \frac{\nu N_x}{Ed} = \epsilon_\phi = \frac{1}{a}\left(\frac{\partial v}{\partial \phi} - w\right) \qquad (b) \qquad \text{(7-10)}$$

$$\frac{N_{x\phi}}{Gd} = \frac{2(1+\nu)\,N_{x\phi}}{Ed} = \gamma_{x\phi} = \left(\frac{1}{a}\frac{\partial u}{\partial \phi} + \frac{\partial v}{\partial x}\right) \qquad (c)$$

7-10   Moment-Curvature Relation

The transverse moment M_ϕ and the change in circumferential curvature χ_ϕ are related in the same way as the curvature and bending moment in a beam undergoing simple bending.   Hence we may write

$$M_\phi = -D\chi_\phi \qquad \text{(7-11)}$$

where $D = Ed^3/[12(1-\nu^2)]$ is the flexural rigidity of the shell.   The minus sign is affixed to indicate that a positive M_ϕ tends to flatten out the shell and reduce its curvature.

7-11   Membrane Displacements

The stress-strain relations derived in Art. 7-9 enable us to find the displacements u, v, and w corresponding to the membrane theory.   To simplify the expressions, it is assumed that $\nu = 0$.

Differentiating (7-10c) once with respect to x, we get

$$\frac{1}{Ed}\frac{\partial N_{x\phi}}{\partial x} = \frac{1}{2}\left(\frac{1}{a}\frac{\partial^2 u}{\partial x\,\partial \phi} + \frac{\partial^2 v}{\partial x^2}\right) \qquad \text{(7-12)}$$

Next, differentiate (7-10a) once with respect to ϕ to get

$$\frac{1}{Ed}\frac{\partial N_x}{\partial \phi} = \frac{\partial^2 u}{\partial x\,\partial \phi} \qquad \text{(7-13)}$$

Substituting for $\partial^2 u/\partial x\,\partial \phi$ from (7-13) in (7-12), we arrive at the relation

$$\frac{\partial^2 v}{\partial x^2} = \frac{1}{Ed}\left(2\frac{\partial N_{x\phi}}{\partial x} - \frac{1}{a}\frac{\partial N_x}{\partial \phi}\right) \qquad \text{(7-14)}$$

(i) *Dead Weight*

$N_{x\phi}$ and N_x corresponding to the dead weight are already available from the membrane theory in expressions (6-64) and (6-65).   These may be substituted in Equation (7-14) and integrated twice with respect to x to give

$$v = -\frac{8g}{\pi Ed} \sin(\phi_c - \phi) \cos kx \left(\frac{2}{k^2} + \frac{1}{a^2 k^4}\right) \qquad (7\text{-}15)$$

where $k = \pi/l$, corresponding to the first Fourier term. No arbitrary constant is involved as the expression satisfies the boundary condition $v = 0$ at $x = \pm \, l/2$.

$$w \approx \frac{\partial v}{\partial \phi} = \frac{8g}{\pi Ed} \cos(\phi_c - \phi) \cos kx \left(\frac{2}{k^2} + \frac{1}{a^2 k^4}\right) \qquad (7\text{-}16)$$

From (7-10a), we have $\partial u/\partial x = N_x/Ed$. Substituting for N_x from (6-65),

$$\frac{\partial u}{\partial x} = -\frac{1}{Ed} \frac{8g}{\pi a k^2} \cos(\phi_c - \phi) \cos kx$$

Integrating with respect to x and noting that the arbitrary constant involved is zero,

$$u = -\frac{1}{Ed} \frac{8g}{\pi a k^3} \cos(\phi_c - \phi) \sin kx \qquad (7\text{-}17)$$

(ii) *Snow Load*

Proceeding as before and making use of relations (6-67) and (6-68) for the stresses due to snow load we arrive at the following expressions:

$$v = -\frac{12p_0 a^2}{Ed\pi^5} \left(\frac{l}{a}\right)^4 \left[\left(\frac{a\pi}{l}\right)^2 + 2\right] \sin 2(\phi_c - \phi) \cos \frac{\pi x}{l} \qquad (7\text{-}18)$$

$$w = \frac{24p_0}{Ed} \frac{a^2}{\pi^5} \left(\frac{l}{a}\right)^4 \left[\left(\frac{a\pi}{l}\right)^2 + 2\right] \cos 2(\phi_c - \phi) \cos \frac{\pi x}{l} \qquad (7\text{-}19)$$

$$u = \frac{12p_0 l^3}{Ed\pi^4 a} \cos 2(\phi_c - \phi) \sin \frac{\pi x}{l} \qquad (7\text{-}20)$$

BENDING THEORIES OF CYLINDRICAL SHELLS

Starting from 1932, several rigorous and approximate bending theories have been put forward for the analysis of reinforced-concrete cylindrical shells. The earliest of these was due to Finsterwalder [14, 15].

THE FINSTERWALDER THEORY

7-12   Notes on the Finsterwalder Theory

By making a few simplifying assumptions, Finsterwalder was able to develop, for the first time, a theory that the engineer could use for the analysis of shell roofs. The assumptions underlying all shell theories and the additional simplifying assumption made by the Finsterwalder theory are listed below:

(*a*) The material is homogeneous and isotropic and obeys Hooke's law.

(*b*) Stresses normal to the shell surface are neglected.

(*c*) A rectilinear element normal to the middle surface of the shell remains rectilinear and normal even after deformation.

(*d*) All displacements of the shell surface are small.

(*e*) M_x, Q_x, and $M_{x\phi}$ are neglected in the analysis.

Of these, assumptions (*a*) to (*d*) are common ground to all bending theories of cylindrical shells. Assumption (*e*) was introduced by Finsterwalder to simplify the problem.

7-13   Equations of Equilibrium

It is possible to derive four equations of equilibrium for an element of the unloaded shell acted upon by the stress resultants shown in Figs. 7-8 and 7-9.

It is to be noted that M_x, Q_x, and $M_{x\phi}$ are not to be considered.

Equating all forces acting in the x direction to zero,

$$\frac{\partial N_x}{\partial x} + \frac{1}{a}\frac{\partial N_{x\phi}}{\partial \phi} = 0 \qquad (7\text{-}21a)$$

Summing up all forces in the ϕ direction, i.e., the direction of the tangent to the shell element at its midpoint pointing in the direction of increasing ϕ and equating them to zero, we get

$$\frac{1}{a}\frac{\partial N_\phi}{\partial \phi} a\,d\phi\,dx + \frac{\partial N_{x\phi}}{\partial x}dx\,a\,d\phi - 2Q_\phi\,dx\frac{d\phi}{2} = 0$$

The first two terms of this equation are the same as those appearing in the membrane theory. The additional term is the resolved component of the shear forces Q_ϕ in the ϕ direction (Fig. 7-8).

Simplifying, we get

$$\frac{\partial N_\phi}{\partial \phi} + a\frac{\partial N_{x\phi}}{\partial x} - Q_\phi = 0 \qquad (7\text{-}21b)$$

Equating to zero the sum of all forces in the direction of the inward normal drawn at the midpoint of the shell element (Fig. 7-8)

$$2N_\phi \, dx \, \frac{d\phi}{2} + \frac{\partial Q_\phi}{\partial \phi} \, d\phi \, dx = 0$$

On simplification, we get

$$N_\phi + \frac{\partial Q_\phi}{\partial \phi} = 0 \tag{7-21c}$$

Another equation of equilibrium results from equating the sum of moments of all forces about the generatrix AD. This equation is

$$\frac{\partial M_\phi}{\partial \phi} \, d\phi \, dx - Q_\phi \, a \, d\phi \, dx = 0$$

or

$$\frac{1}{a} \frac{\partial M_\phi}{\partial \phi} - Q_\phi = 0 \tag{7-21d}$$

7-14   The Finsterwalder Differential Equation

Let us introduce a function $f(\phi)$ which is such that

$$M_\phi = -f(\phi) \cos \frac{n\pi x}{l}$$

If $\lambda_n = (n\pi a/l)$, we may write

$$M_\phi = -f(\phi) \cos \frac{\lambda_n x}{a} \tag{7-22a}$$

From the equation of equilibrium (7-21d), Q_ϕ may be found

$$Q_\phi = \frac{1}{a} \frac{\partial M_\phi}{\partial \phi} = -\frac{1}{a} f^{\cdot} \cos \frac{\lambda_n x}{a} \tag{7-22b}$$

where $f^{\cdot}$ stands for $(\partial/\partial\phi) f(\phi)$. From (7-21c),

$$N_\phi = -\frac{\partial Q_\phi}{\partial \phi} = \frac{1}{a} f^{\cdot\cdot} \cos \frac{\lambda_n x}{a} \tag{7-22c}$$

We may now find $N_{x\phi}$ from (7-21b),

$$\frac{\partial N_{x\phi}}{\partial x} = -\frac{1}{a^2} f^{\cdot} \cos \frac{\lambda_n x}{a} - \frac{1}{a^2} f^{\cdot\cdot\cdot} \cos \frac{\lambda_n x}{a}$$

$$N_{x\phi} = -\frac{1}{\lambda_n a} (f^{\cdot} + f^{\cdot\cdot\cdot}) \sin \frac{\lambda_n x}{a} \tag{7-22d}$$

From (7-21*a*),

$$\frac{\partial N_x}{\partial x} = -\frac{1}{a}\frac{\partial N_{x\phi}}{\partial \phi} = \frac{1}{\lambda_n a^2}(f^{\cdot\cdot} + f^{::})\sin\frac{\lambda_n x}{a}$$

or

$$N_x = -\frac{1}{\lambda_n^2 a}(f^{\cdot\cdot} + f^{::})\cos\frac{\lambda_n x}{a} \qquad (7\text{-}22e)$$

Expressions for *u*, *v*, and *w* may now be derived by making use of the stress-strain relations (7-10). Before we do so, we may set *v*, the Poisson's ratio, to zero. Such an assumption is quite justified for reinforced-concrete structures. We thus have

$$\frac{N_x}{Ed} = \frac{\partial u}{\partial x}$$

Hence from (7-22*e*)

$$E\,d\,\frac{\partial u}{\partial x} = -\frac{1}{\lambda_n^2 a}(f^{\cdot\cdot} + f^{::})\cos\frac{\lambda_n x}{a}$$

or

$$E\,d\,u = -\frac{1}{\lambda_n^3}(f^{\cdot\cdot} + f^{::})\sin\frac{\lambda_n x}{a} \qquad (7\text{-}23a)$$

From relation (7-10*c*),

$$\frac{\partial v}{\partial x} = \frac{2N_{x\phi}}{Ed} - \frac{1}{a}\frac{\partial u}{\partial \phi}$$

Hence

$$E\,d\,\frac{\partial v}{\partial x} = 2N_{x\phi} - \frac{Ed}{a}\frac{\partial u}{\partial \phi}$$

$$= \left[-\frac{2}{\lambda_n a}(f^{\cdot} + f^{\cdot\cdot\cdot}) + \frac{1}{a\lambda_n^3}(f^{\cdot\cdot\cdot} + f^{::\cdot})\right]\sin\frac{\lambda_n x}{a}$$

Integrating,

$$E\,d\,v = \left[\frac{2}{\lambda_n^2}(f^{\cdot} + f^{\cdot\cdot\cdot}) - \frac{1}{\lambda_n^4}(f^{\cdot\cdot\cdot} + f^{::\cdot})\right]\cos\frac{\lambda_n x}{a} \qquad (7\text{-}23b)$$

Making use of (7-10*b*),

$$E\,d\,w = -\left[aN_\phi - E\,d\,\frac{\partial v}{\partial \phi}\right]$$

$$= -\left[f^{\cdot\cdot} - \frac{2}{\lambda_n^2}(f^{\cdot\cdot} + f^{::}) + \frac{1}{\lambda_n^4}(f^{::} + f^{::\cdot})\right]\cos\frac{\lambda_n x}{a} \qquad (7\text{-}23c)$$

Knowing v and w, we may write

$$\vartheta = \frac{1}{a}\left(v + \frac{\partial w}{\partial \phi}\right)$$

$$= \frac{1}{E\,d}\left[\frac{2}{a\lambda_n^2}(f^{\cdot} + f^{\cdots}) - \frac{1}{a\lambda_n^4}(f^{\cdots} + f^{:::})\right.$$

$$\left. - \frac{1}{a}f^{\cdots} + \frac{2}{a\lambda_n^2}(f^{\cdots} + f^{:::}) - \frac{1}{a\lambda_n^4}(f^{:::} + f^{::::})\right]\cos\frac{\lambda_n x}{a}$$

Simplifying,

$$E\,d\,\vartheta = \left[\frac{2}{a\lambda_n^2}(f^{\cdot} + 2f^{\cdots} + f^{:::})\right.$$

$$\left. - \frac{1}{a\lambda_n^4}(f^{\cdots} + 2f^{:::} + f^{::::}) - \frac{1}{a}f^{\cdots}\right]\cos\frac{\lambda_n x}{a} \qquad (7\text{-}24)$$

We may next find χ_ϕ.

$$\chi_\phi = \frac{1}{a^2}\left(w + \frac{\partial^2 w}{\partial\phi^2}\right)$$

$$= -\frac{1}{E\,da^2}\left[f^{\cdot\cdot} + f^{::} - \frac{2}{\lambda_n^2}(f^{\cdot\cdot} + 2f^{::} + f^{:::})\right.$$

$$\left. + \frac{1}{\lambda_n^4}(f^{::} + 2f^{::::} + f^{::::})\right]\cos\frac{\lambda_n x}{a} \qquad (7\text{-}25)$$

By substituting (7-22a) and (7-25) in the moment-curvature relation (7-11),

$$\left[f^{\cdot\cdot} + f^{::} - \frac{2}{\lambda_n^2}(f^{\cdot\cdot} + 2f^{::} + f^{:::}) + \frac{1}{\lambda_n^4}(f^{::} + 2f^{::::} + f^{::::})\right]$$

$$+ 12\frac{a^2}{d^2}f = 0$$

or

$$(f^{::::} + 2f^{:::} + f^{::}) - 2\lambda_n^2(f^{\cdot\cdot} + 2f^{::} + f^{:::}) + \lambda_n^4(f^{\cdot\cdot} + f^{::})$$

$$+ 12\lambda_n^4\frac{a^2}{d^2}f = 0 \qquad (7\text{-}26)$$

This is the eighth-order Finsterwalder differential equation. By inspection it is found that $-f(\phi) = He^{m\phi}$ will satisfy the equation. Substituting this solution, the following eighth-degree algebraic equation results:

$$(m^8 + 2m^6 + m^4) - 2\lambda_n^2(m^2 + 2m^4 + m^6) + \lambda_n^4(m^2 + m^4) + 12\lambda_n^4\frac{a^2}{d^2} = 0$$

Rearranging,

$$m^8 + 2m^6(1 - \lambda_n{}^2) + m^4(1 - 4\lambda_n{}^2 + \lambda_n{}^4) + m^2(\lambda_n{}^4 - 2\lambda_n{}^2) + 12\frac{\lambda_n{}^4 a^2}{d^2} = 0$$

(7-27)

This eighth-degree algebraic equation associated with the eighth-order partial differential equation governing the bending theory of cylindrical shells is known as the characteristic equation. Methods of finding the roots of the characteristic equation have been discussed in Chapter 1. The equation has eight complex roots m_1, m_2, m_3, m_4, . . . , m_8 which occur in conjugate pairs. They may be represented as follows:

$$
\begin{array}{ll}
m_1 = \alpha_1 + i\beta_1 & m_5 = -m_1 \\
m_2 = \alpha_1 - i\beta_1 & m_6 = -m_2 \\
m_3 = \alpha_2 + i\beta_2 & m_7 = -m_3 \\
m_4 = \alpha_2 - i\beta_2 & m_8 = -m_4
\end{array}
$$

(7-28)

7-15 Expression for M_ϕ

From (7-22a), we may write

$$
M_\phi = -f(\phi)\cos\frac{\lambda_n x}{a}
$$

$$
= [H_1 e^{(\alpha_1 + i\beta_1)\phi} + H_2 e^{(\alpha_1 - i\beta_1)\phi} + H_3 e^{(-\alpha_1 + i\beta_1)\phi} + H_4 e^{(-\alpha_1 - i\beta_1)\phi}
$$

$$
+ H_5 e^{(\alpha_2 + i\beta_2)\phi} + H_6 e^{(\alpha_2 - i\beta_2)\phi} + H_7 e^{(-\alpha_2 + i\beta_2)\phi} + H_8 e^{(-\alpha_2 - i\beta_2)\phi}]\cos\frac{\lambda_n x}{a}
$$

$$
= [H_1 e^{\alpha_1\phi}(\cos\beta_1\phi + i\sin\beta_1\phi) + H_2 e^{\alpha_1\phi}(\cos\beta_1\phi - i\sin\beta_1\phi)
$$

$$
+ H_3 e^{-\alpha_1\phi}(\cos\beta_1\phi + i\sin\beta_1\phi) + H_4 e^{-\alpha_1\phi}(\cos\beta_1\phi - i\sin\beta_1\phi)
$$

$$
+ H_5 e^{\alpha_2\phi}(\cos\beta_2\phi + i\sin\beta_2\phi) + H_6 e^{\alpha_2\phi}(\cos\beta_2\phi - i\sin\beta_2\phi)
$$

$$
+ H_7 e^{-\alpha_2\phi}(\cos\beta_2\phi + i\sin\beta_2\phi) + H_8 e^{-\alpha_2\phi}(\cos\beta_2\phi - i\sin\beta_2\phi)]\cos\frac{\lambda_n x}{a}
$$

Four out of the eight terms in the above expression are multiplied by an exponential with a negative index and the other four by an exponential with a positive index. The moment M_ϕ being in the nature of a disturbance emanating from the edge $\phi = 0$, one would expect its value to decay exponentially as we move away from the edge or, in other words, with increasing values of ϕ. This condition is satisfied by the four terms multiplied by an exponential with a negative index. The other four terms increase as ϕ increases. Hence they are to be omitted in considering the edge disturbance emanating from the left edge. They may be physically interpreted as emanating from the right edge. Their influence on the stress resultants would be accounted for later on by considering a similar edge disturbance emanating from the right edge. Hence we may write

$$M_\phi = [H_3 e^{-\alpha_1\phi}(\cos\beta_1\phi + i\sin\beta_1\phi) + H_4 e^{-\alpha_1\phi}(\cos\beta_1\phi - i\sin\beta_1\phi)$$
$$+ H_7 e^{-\alpha_2\phi}(\cos\beta_2\phi + i\sin\beta_2\phi) + H_8 e^{-\alpha_2\phi}(\cos\beta_2\phi - i\sin\beta_2\phi)]\cos\frac{\lambda_n x}{a}$$

or

$$M_\phi = [e^{-\alpha_1\phi}(H_3 + H_4)\cos\beta_1\phi + e^{-\alpha_1\phi}i(H_3 - H_4)\sin\beta_1\phi$$
$$+ e^{-\alpha_2\phi}(H_7 + H_8)\cos\beta_2\phi + e^{-\alpha_2\phi}i(H_7 - H_8)\sin\beta_2\phi]\cos\frac{\lambda_n x}{a}$$

It is to be noted that the constants H_3, H_4, H_7, and H_8 are complex numbers. Since M_ϕ is real, it follows that $(H_3 + H_4)$, $i(H_3 - H_4)$, $(H_7 + H_8)$, and $i(H_7 - H_8)$ have to be real. This means that the numbers H_3, H_4 and H_7, H_8 have to be conjugate pairs. To simplify the further work involved we may introduce constants A_n, B_n, C_n, and D_n which are such that

$$A_n = H_3 + H_4 \qquad B_n = i(H_3 - H_4)$$
$$C_n = H_7 + H_8 \qquad D_n = i(H_7 - H_8)$$

Hence

$$M_\phi = [e^{-\alpha_1\phi}(A_n\cos\beta_1\phi + B_n\sin\beta_1\phi)$$
$$+ e^{-\alpha_2\phi}(C_n\cos\beta_2\phi + D_n\sin\beta_2\phi)]\cos\frac{\lambda_n x}{a} \qquad (7\text{-}29)$$

It is to be noted that the solution for M_ϕ is really an infinite series with n taking on values $1, 3, 5, \ldots, \infty$. The function $\cos(\lambda_n x/a)$ would assume values of $\cos(\pi x/l)$, $\cos(3\pi x/l)$, $\cos(5\pi x/l)$, and so on. The suffix n attached to the constants A, B, C, and D denotes that they are associated with the nth term of the series.

7-16 Recurrence Formulas for Differentiation with Respect to ϕ

Differentiating M_ϕ once with respect to ϕ, we get

$$\frac{\partial M_\phi}{\partial \phi} = \{e^{-\alpha_1\phi}[(-\alpha_1 A_n + \beta_1 B_n)\cos\beta_1\phi + (-\beta_1 A_n - \alpha_1 B_n)\sin\beta_1\phi]$$
$$+ e^{-\alpha_2\phi}[(-\alpha_2 C_n + \beta_2 D_n)\cos\beta_2\phi + (-\beta_2 C_n - \alpha_2 D_n)\sin\beta_2\phi]\}\cos\frac{\lambda_n x}{a}.$$

It may be noted that differentiating M_ϕ once with respect to ϕ does not alter the form of the expression but only changes the coefficents

A_n, B_n, C_n, and D_n. We may denote the values of these constants after one differentiation by new constants $A_n^{①}$, $B_n^{①}$, $C_n^{①}$, and $D_n^{①}$ which are such that

$$A_n^{①} = -\alpha_1 A_n + \beta_1 B_n$$

$$B_n^{①} = -\beta_1 A_n - \alpha_1 B_n$$

$$C_n^{①} = -\alpha_2 C_n + \beta_2 D_n$$

and

$$D_n^{①} = -\beta_2 C_n - \alpha_2 D_n$$

The same rule can be successively applied for repeated differentiation with respect to ϕ. Let $A_n^{(k)}, \ldots, D_n^{(k)}$ be the values of the constants after the kth differentiation with respect to ϕ. They may be written as follows:

$$A_n^{(k)} = -\alpha_1 A_n^{(k-1)} + \beta_1 B_n^{(k-1)}$$

$$B_n^{(k)} = -\beta_1 A_n^{(k-1)} - \alpha_1 B_n^{(k-1)}$$

$$C_n^{(k)} = -\alpha_2 C_n^{(k-1)} + \beta_2 D_n^{(k-1)}$$

and

$$D_n^{(k)} = -\beta_2 C_n^{(k-1)} - \alpha_2 D_n^{(k-1)}$$

$$(7\text{-}30)$$

A more convenient set of recurrence formulas which directly connect $A_n, \ldots, D_n$ with $A_n^{(k)}, \ldots, D_n^{(k)}$ are given below.

$$A_n^{(k)} = [\text{real part of } (-\alpha_1 + i\beta_1)^k] A_n + [\text{imaginary part of } (-\alpha_1 + i\beta_1)^k] B_n$$

$$B_n^{(k)} = -[\text{imaginary part of } (-\alpha_1 + i\beta_1)^k] A_n + [\text{real part of } (-\alpha_1 + i\beta_1)^k] B_n$$

Similar formulas can be written down for $C_n^{(k)}$ and $D_n^{(k)}$. Making use of this rule, the following formulas may be derived:

$$A_n^{②} = (\alpha_1^2 - \beta_1^2) A_n - 2\alpha_1\beta_1 B_n$$

$$B_n^{②} = 2\alpha_1\beta_1 A_n + (\alpha_1^2 - \beta_1^2) B_n$$

$$A_n^{③} = (3\alpha_1\beta_1^2 - \alpha_1^3) A_n + (3\alpha_1^2\beta_1 - \beta_1^3) B_n$$

$$B_n^{③} = (\beta_1^3 - 3\alpha_1^2\beta_1) A_n + (3\alpha_1\beta_1^2 - \alpha_1^3) B_n$$

$$A_n^{④} = (\alpha_1^4 - 6\alpha_1^2\beta_1^2 + \beta_1^4) A_n + 4(\alpha_1\beta_1^3 - \alpha_1^3\beta_1) B_n$$

$$B_n^{④} = 4(\alpha_1^3\beta_1 - \alpha_1\beta_1^3) A_n + (\alpha_1^4 - 6\alpha_1^2\beta_1^2 + \beta_1^4) B_n$$

and so on.

7-17   Expressions for Stress Resultants, Displacements, and Rotation

Starting with the expression (7-29) for M_ϕ and making use of relations (7-22) to (7-24), the following expressions may be derived for the other stress resultants, displacements, and the rotation of the tangent.

$$Q_\phi = \frac{1}{a}\left[e^{-\alpha_1\phi}(A_n^① \cos\beta_1\phi + B_n^① \sin\beta_1\phi)\right.$$

$$\left. + e^{-\alpha_2\phi}(C_n^① \cos\beta_2\phi + D_n^① \sin\beta_2\phi)\right] \cos\frac{\lambda_n x}{a} \qquad (7\text{-}31a)$$

$$N_\phi = -\frac{1}{a}\left[e^{-\alpha_1\phi}(A_n^② \cos\beta_1\phi + B_n^② \sin\beta_1\phi)\right.$$

$$\left. + e^{-\alpha_2\phi}(C_n^② \cos\beta_2\phi + D_n^② \sin\beta_2\phi)\right] \cos\frac{\lambda_n x}{a} \qquad (7\text{-}31b)$$

$$N_{x\phi} = \frac{1}{\lambda_n a}\left\{e^{-\alpha_1\phi}\left[(A_n^① + A_n^③)\cos\beta_1\phi + (B_n^① + B_n^③)\sin\beta_1\phi\right]\right.$$

$$\left. + e^{-\alpha_2\phi}\left[(C_n^① + C_n^③)\cos\beta_2\phi + (D_n^① + D_n^③)\sin\beta_2\phi\right]\right\} \sin\frac{\lambda_n x}{a} \qquad (7\text{-}31c)$$

$$N_x = \frac{1}{\lambda_n^2 a}\left\{e^{-\alpha_1\phi}\left[(A_n^② + A_n^④)\cos\beta_1\phi + (B_n^② + B_n^④)\sin\beta_1\phi\right]\right.$$

$$\left. + e^{-\alpha_2\phi}\left[(C_n^② + C_n^④)\cos\beta_2\phi + (D_n^② + D_n^④)\sin\beta_2\phi\right]\right\} \cos\frac{\lambda_n x}{a} \qquad (7\text{-}31d)$$

$$E\,d\,u = \frac{1}{\lambda_n^3}\left\{e^{-\alpha_1\phi}\left[(A_n^② + A_n^④)\cos\beta_1\phi + (B_n^② + B_n^④)\sin\beta_1\phi\right]\right.$$

$$\left. + e^{-\alpha_2\phi}\left[(C_n^② + C_n^④)\cos\beta_2\phi + (D_n^② + D_n^④)\sin\beta_2\phi\right]\right\} \sin\frac{\lambda_n x}{a} \qquad (7\text{-}32a)$$

$$E\,d\,v = \left\{ e^{-\alpha_1\phi}\left[\left(\frac{A_n^⑤ + A_n^③}{\lambda_n^4} - \frac{2A_n^① + 2A_n^③}{\lambda_n^2}\right)\cos\beta_1\phi\right.\right.$$

$$\left. + \left(\frac{B_n^⑤ + B_n^③}{\lambda_n^4} - \frac{2B_n^① + 2B_n^③}{\lambda_n^2}\right)\sin\beta_1\phi\right]$$

$$+ e^{-\alpha_2\phi}\left[\left(\frac{C_n^⑤ + C_n^③}{\lambda_n^4} - \frac{2C_n^① + 2C_n^③}{\lambda_n^2}\right)\cos\beta_2\phi\right.$$

$$\left.\left. + \left(\frac{D_n^⑤ + D_n^③}{\lambda_n^4} - \frac{2D_n^① + 2D_n^③}{\lambda_n^2}\right)\sin\beta_2\phi\right]\right\} \cos\frac{\lambda_n x}{a}$$

or

$$E\,dv = \left\{ e^{-\alpha_1\phi} \left[\frac{1}{\lambda_n{}^4} (A_n{}^{\textcircled{5}} + \overline{1 - 2\lambda_n{}^2}A_n{}^{\textcircled{3}} - 2\lambda_n{}^2 A_n{}^{\textcircled{1}}) \cos \beta_1\phi \right. \right.$$

$$\left. + \frac{1}{\lambda_n{}^4} (B_n{}^{\textcircled{5}} + \overline{1 - 2\lambda_n{}^2}B_n{}^{\textcircled{3}} - 2\lambda_n{}^2 B_n{}^{\textcircled{1}}) \sin \beta_1\phi \right]$$

$$+ e^{-\alpha_2\phi} \left[\frac{1}{\lambda_n{}^4} (C_n{}^{\textcircled{5}} + \overline{1 - 2\lambda_n{}^2}C_n{}^{\textcircled{3}} - 2\lambda_n{}^2 C_n{}^{\textcircled{1}}) \cos \beta_2\phi \right.$$

$$\left. \left. + \frac{1}{\lambda_n{}^4} (D_n{}^{\textcircled{5}} + \overline{1 - 2\lambda_n{}^2}D_n{}^{\textcircled{3}} - 2\lambda_n{}^2 D_n{}^{\textcircled{1}}) \sin \beta_2\phi \right] \right\} \cos \frac{\lambda_n x}{a} \qquad (7\text{-}32b)$$

$$E\,dw = \left\{ e^{-\alpha_1\phi} \left[\frac{1}{\lambda_n{}^4} (A_n{}^{\textcircled{6}} + \overline{1 - 2\lambda_n{}^2}A_n{}^{\textcircled{4}} + \overline{\lambda_n{}^4 - 2\lambda_n{}^2}A_n{}^{\textcircled{2}}) \cos \beta_1\phi \right. \right.$$

$$\left. + \frac{1}{\lambda_n{}^4} (B_n{}^{\textcircled{6}} + \overline{1 - 2\lambda_n{}^2}B_n{}^{\textcircled{4}} + \overline{\lambda_n{}^4 - 2\lambda_n{}^2}B_n{}^{\textcircled{2}}) \sin \beta_1\phi \right]$$

$$+ e^{-\alpha_2\phi} \left[\frac{1}{\lambda_n{}^4} (C_n{}^{\textcircled{6}} + \overline{1 - 2\lambda_n{}^2}C_n{}^{\textcircled{4}} + \overline{\lambda_n{}^4 - 2\lambda_n{}^2}C_n{}^{\textcircled{2}}) \cos \beta_2\phi \right.$$

$$\left. \left. + \frac{1}{\lambda_n{}^4} (D_n{}^{\textcircled{6}} + \overline{1 - 2\lambda_n{}^2}D_n{}^{\textcircled{4}} + \overline{\lambda_n{}^4 - 2\lambda_n{}^2}D_n{}^{\textcircled{2}}) \sin \beta_2\phi \right] \right\} \cos \frac{\lambda_n x}{a}$$

$$(7\text{-}32c)$$

$$E\,d\vartheta = \frac{1}{\lambda_n{}^4 a} \left\{ e^{-\alpha_1\phi}[(A_n{}^{\textcircled{7}} + \overline{2 - 2\lambda_n{}^2}A_n{}^{\textcircled{5}} + \overline{1 - 4\lambda_n{}^2 + \lambda_n{}^4}A_n{}^{\textcircled{3}} \right.$$

$$- 2\lambda_n{}^2 A_n{}^{\textcircled{1}}) \cos \beta_1\phi + (B_n{}^{\textcircled{7}} + \overline{2 - 2\lambda_n{}^2}B_n{}^{\textcircled{5}}$$

$$+ \overline{1 - 4\lambda_n{}^2 + \lambda_n{}^4}B_n{}^{\textcircled{3}} - 2\lambda_n{}^2 B_n{}^{\textcircled{1}}) \sin \beta_1\phi]$$

$$+ e^{-\alpha_2\phi}[(C_n{}^{\textcircled{7}} + \overline{2 - 2\lambda_n{}^2}C_n{}^{\textcircled{5}} + \overline{1 - 4\lambda_n{}^2 + \lambda_n{}^4}C_n{}^{\textcircled{3}}$$

$$- 2\lambda_n{}^2 C_n{}^{\textcircled{1}}) \cos \beta_2\phi + (D_n{}^{\textcircled{7}} + \overline{2 - 2\lambda_n{}^2}D_n{}^{\textcircled{5}}$$

$$\left. + \overline{1 - 4\lambda_n{}^2 + \lambda_n{}^4}D_n{}^{\textcircled{3}} - 2\lambda_n{}^2 D_n{}^{\textcircled{1}}) \sin \beta_2\phi] \right\} \cos \frac{\lambda_n x}{a} \qquad (7\text{-}33)$$

7-18 Terms Originating at the Right Edge

In calculating functions such as M_ϕ in a symmetric shell at an angle ϕ from the left edge, the influence of terms originating at the right edge can be calculated by replacing ϕ by $(2\phi_c - \phi)$ in the relevant expression. These effects are to be superimposed over the corresponding terms originating from the left edge. For symmetric functions such as u, w, N_x, N_ϕ, and M_ϕ, involving the even derivatives of ϕ, these two effects are cumulative and are to be added. But for antisymmetric functions such as $N_{x\phi}$, v, ϑ, and Q_ϕ which involve the odd derivatives of ϕ, these two effects are of opposite sign and hence the effect of the right edge is to be added with an opposite sign to that of the left edge.

THE D-K-J THEORY

7-19 Notes on the D-K-J Theory

The simplest among the so-called "exact" theories which take into account M_x, $M_{x\phi}$, and Q_x ignored in the Finsterwalder theory is the D-K-J theory in which the three displacements u, v, and w appear in uncoupled form. The theory appears to be due to Donnell, who first used it in connection with his studies on the stability of thin-walled circular cylinders in 1933–1934 [16,17]. Kármán and Tsien also employed the same theory in 1941 in their investigations on the buckling of cylindrical shells [18]. Its presentation in a form suitable for the analysis of cylindrical shell roofs appeared in a book by Jenkins published in 1947 [19]. The theory is appropriately known as the Donnell-Kármán-Jenkins theory.

7-20 Shell as a Combination of a Disk, Plate, and Membrane

The structural action of a cylindrical shell, under bending, can be approximated by combining the structural actions of a disk, corresponding to the developed shell loaded in its own plane; of a plate, formed by the developed shell loaded at right angles to its plane; and of the shell regarded as a flexible membrane. These three actions may be described as "disk action," "plate action," and "membrane action." Omitting all terms not figuring in the equations for the disk, the plate, or the membrane offers an elegant approach for the derivation of the D-K-J equation. This artifice appears to have been first suggested by Csonka.[1]

7-21 Equations of Equilibrium

The equilibrium equations already derived in Art. 7-13 will now be modified to take into account M_x, $M_{x\phi}$, and Q_x. It will be assumed that $M_{x\phi} = M_{\phi x}$ and that $N_{x\phi} = N_{\phi x}$. Referring to Figs. 7-8 and 7-9, the following equations of equilibrium may be set up. Equating all forces acting on the elements in the x direction to zero,

$$\frac{\partial N_x}{\partial x} + \frac{1}{a} \frac{\partial N_{x\phi}}{\partial \phi} = 0 \tag{7-34a}$$

[1] This statement appears in P. Csonka, "On the Stress-function of the Circular Cylindrical Shell," *Acta Technica Academiae Scientiarum Hungaricae*, Budapest, 1960 (in English).

Equating the sum of all forces acting on the element in the ϕ direction to zero,

$$\frac{\partial N_\phi}{\partial \phi} - Q_\phi + a \frac{\partial N_{\phi x}}{\partial x} = 0$$

The term Q_ϕ appearing in this equation has to be dropped, as it does not occur in the corresponding equations of equilibrium of the disk, the plate, or the membrane shell. Hence this equation is rewritten as

$$\frac{\partial N_\phi}{\partial \phi} + a \frac{\partial N_{\phi x}}{\partial x} = 0 \qquad (7\text{-}34b)$$

The third equation of equilibrium is derived by equating all forces acting on the element in the normal direction to zero. This equation takes the form

$$a \frac{\partial Q_x}{\partial x} + \frac{\partial Q_\phi}{\partial \phi} + N_\phi = 0 \qquad (7\text{-}34c)$$

Equating the sum of all moments acting on the element about the generator AD to zero, we get

$$a \frac{\partial M_{x\phi}}{\partial x} + \frac{\partial M_\phi}{\partial \phi} - aQ_\phi = 0 \qquad (7\text{-}34d)$$

Similarly, the sum of all moments, acting on the element, taken about AB is equated to zero to get

$$\frac{\partial M_{\phi x}}{\partial \phi} + a \frac{\partial M_x}{\partial x} - aQ_x = 0 \qquad (7\text{-}34e)$$

The sum of all the moments acting on the element about its normal when equated to zero leads to

$$(N_{x\phi} - N_{\phi x}) a + M_{\phi x} = 0 \qquad (7\text{-}34f)$$

It may be explained that the last term in the above equation represents the component of the vector $M_{x\phi}$ in the direction of the normal to the element. As we have assumed $N_{x\phi} = N_{\phi x}$ to start with, this equation of equilibrium is violated unless $M_{x\phi} = 0$. This is an inherent drawback of this theory.

7-22 Stress-Strain Relations

The stress-strain relations are the same as those given by relations (7-4), (7-5), and (7-6) developed in Art. 7-5.

7-23 Moment-Curvature Relations

Because the Donnell theory takes into account the bending moment M_x and the twisting moment $M_{x\phi}$, ignored in developing the Finsterwalder theory, two additional moment-curvature relations are now necessary. The three moment-curvature relations used are the same as those appearing in plate theory.[1] They are

$$M_x = -D \frac{\partial^2 w}{\partial x^2} \qquad (7\text{-}35a)$$

$$M_\phi = -\frac{D}{a^2} \frac{\partial^2 w}{\partial \phi^2} \qquad (7\text{-}35b)$$

and

$$M_{x\phi} = -\frac{D}{a} \frac{\partial^2 w}{\partial x\, \partial \phi} \qquad (7\text{-}35c)$$

It may be noted that the moment-curvature relation (7-35b) for M_ϕ is different from the expression (7-11) of Art. 7-10 developed for the Finsterwalder theory. The difference consists in dropping the term w in the expression for χ_ϕ in the relation (7-8). This is consistent with the assumption made in Art. 7-20 that only terms appearing in the corresponding expressions for the disk, the plate, and the membrane shall be retained.

7-24 Flügge's Simultaneous Equations

The next step consists in substituting the stress-strain relations and the moment-curvature relations in the equations of equilibrium to eliminate the stress resultants from these equations and to get three simultaneous equations involving the three displacements u, v, and w. Substituting stress-strain relations (7-10) in equilibrium equations (7-34a) and (7-34b), we get

$$a^2 \frac{\partial^2 u}{\partial x^2} + \frac{1}{2}\left(\frac{\partial^2 u}{\partial \phi^2} + a \frac{\partial^2 v}{\partial x\, \partial \phi} \right) = 0 \qquad (7\text{-}36a)$$

and

$$\left(\frac{\partial^2 v}{\partial \phi^2} - \frac{\partial w}{\partial \phi} \right) + \frac{1}{2}\left(a \frac{\partial^2 u}{\partial x\, \partial \phi} + a^2 \frac{\partial^2 v}{\partial x^2} \right) = 0 \qquad (7\text{-}36b)$$

From Equations (7-34d) and (7-34e) we may write

$$Q_x = \frac{1}{a}\left[\frac{\partial M_{x\phi}}{\partial \phi} + a \frac{\partial M_x}{\partial x} \right] = -\frac{D}{a}\left[\frac{1}{a} \frac{\partial^3 w}{\partial x\, \partial \phi^2} + a \frac{\partial^3 w}{\partial x^3} \right]$$

---

[1] Timoshenko, S., and S. Woinowsky-Krieger, "Theory of Plates and Shells," 2d ed., McGraw-Hill Book Company, New York, 1959.

and

$$Q_\phi = -\frac{D}{a}\left[\frac{\partial^3 w}{\partial x^2\,\partial\phi} + \frac{1}{a^2}\frac{\partial^3 w}{\partial\phi^3}\right]$$

Also

$$a\frac{\partial Q_x}{\partial x} = -D\left[\frac{1}{a}\frac{\partial^4 w}{\partial x^2\,\partial\phi^2} + a\frac{\partial^4 w}{\partial x^4}\right]$$

and

$$\frac{\partial Q_\phi}{\partial\phi} = -\frac{D}{a}\left[\frac{\partial^4 w}{\partial x^2\,\partial\phi^2} + \frac{1}{a^2}\frac{\partial^4 w}{\partial\phi^4}\right]$$

Substituting these values in Equation (7-34c),

$$N_\phi - \frac{D}{a^3}\left(a^4\frac{\partial^4 w}{\partial x^4} + 2a^2\frac{\partial^4 w}{\partial x^2\,\partial\phi^2} + \frac{\partial^4 w}{\partial\phi^4}\right) = 0 \qquad (7\text{-}36c)$$

But

$$N_\phi = \frac{Ed}{a}\left(\frac{\partial v}{\partial\phi} - w\right)$$

Substituting this value of N_ϕ, we have

$$\left(\frac{\partial v}{\partial\phi} - w\right) - \frac{d^2}{12a^2}\left(a^4\frac{\partial^4 w}{\partial x^4} + 2a^2\frac{\partial^4 w}{\partial x^2\,\partial\phi^2} + \frac{\partial^4 w}{\partial\phi^4}\right) = 0 \qquad (7\text{-}36d)$$

Equations (7-36a), (7-36b), and (7-36d) are the simplified versions of the well-known simultaneous differential equations due to Flügge [20] in the three displacements u, v, and w. It is to be noted that Poisson's ratio has been assumed to be zero from the very beginning.

7-25 The D-K-J Equation

We shall next proceed to derive a single differential equation in the displacement w by eliminating u and v from the above set of equations. Differentiating Equation (7-36b) twice with respect to x,

$$(v'''' - w''') + \tfrac{1}{2}(au'''' + a^2 v'''') = 0 \qquad (7\text{-}37)$$

where $'$ represents $\partial/\partial x$ and $\cdot$ stands for $\partial/\partial\phi$. Now apply the operator $a\,(\partial^2/\partial x\,\partial\phi)$ on (7-36a) to get

$$a^3\frac{\partial^4 u}{\partial x^3\,\partial\phi} + \frac{1}{2}\left(a\frac{\partial^4 u}{\partial\phi^3\,\partial x} + a^2\frac{\partial^4 v}{\partial x^2\,\partial\phi^2}\right) = 0 \qquad (7\text{-}38)$$

Next apply the operator $\partial^2/\partial\phi^2$ on (7-36b) to get

$$(v^{\cdot\cdot} - w^{\cdot\cdot\cdot}) + \tfrac{1}{2}(au^{\cdot\cdot\cdot\prime} + a^2 v^{\prime\cdot\cdot}) = 0 \qquad (7\text{-}39)$$

From (7-38) and (7-39) we may write

$$a^3 u''''^{\cdot} = v^{\cdot \cdot} - w^{\cdot \cdot \cdot} \qquad (7\text{-}40)$$

Substituting this value of u''' in (7-37),

$$(v''''^{\cdot} - w''^{\cdot}) + \frac{1}{2}\left[\frac{1}{a^2}(v^{\cdot \cdot} - w^{\cdot \cdot \cdot}) + a^2 v''''\right] = 0$$

Simplifying,

$$\left[a^2 \frac{\partial^2}{\partial x^2} + \frac{\partial^2}{\partial \phi^2}\right]^2 v = w^{\cdot \cdot \cdot} + 2a^2 w''^{\cdot} \qquad (7\text{-}41)$$

Hence

$$\left[a^2 \frac{\partial^2}{\partial x^2} + \frac{\partial^2}{\partial \phi^2}\right]^2 v^{\cdot} = \left[a^2 \frac{\partial^2}{\partial x^2} + \frac{\partial^2}{\partial \phi^2}\right]^2 w - a^4 w''''$$

or

$$\left[a^2 \frac{\partial^2}{\partial x^2} + \frac{\partial^2}{\partial \phi^2}\right]^2 (v^{\cdot} - w) = -a^4 w'''' \qquad (7\text{-}42)$$

Apply operator $[a^2(\partial^2/\partial x^2) + \partial^2/\partial \phi^2]^2$ on Equation (7-36d) to obtain

$$\left[a^2 \frac{\partial^2}{\partial x^2} + \frac{\partial^2}{\partial \phi^2}\right]^2 (v^{\cdot} - w) - \frac{d^2}{12a^2}\left[a^2 \frac{\partial^2}{\partial x^2} + \frac{\partial^2}{\partial \phi^2}\right]^4 w = 0 \quad (7\text{-}42a)$$

Substituting the value of $[a^2(\partial^2/\partial x^2) + \partial^2/\partial \phi^2]^2[v^{\cdot} - w]$ from (7-42) in (7-42a), we finally get

$$\left[a^2 \frac{\partial^2}{\partial x^2} + \frac{\partial^2}{\partial \phi^2}\right]^4 w + 12\frac{a^6}{d^2} w'''' = 0$$

or, letting $k = (d^2/12a^2)$,

$$\left[a^2 \frac{\partial^2}{\partial x^2} + \frac{\partial^2}{\partial \phi^2}\right]^4 w + a^4 \frac{w''''}{k} = 0 \qquad (7\text{-}43)$$

This is Donnell's equation in w.

7-26 The D-K-J Characteristic Equation

Let us seek a solution for w in the form

$$w = He^{m\phi} \cos \frac{\lambda_n x}{a} \qquad (7\text{-}44)$$

Expanding the operator in (7-43), we may write

$$\left[a^8 \frac{\partial^8}{\partial x^8} + 4a^6 \frac{\partial^8}{\partial x^6 \partial \phi^2} + 6a^4 \frac{\partial^8}{\partial x^4 \partial \phi^4} + 4a^2 \frac{\partial^8}{\partial x^2 \partial \phi^6} + \frac{\partial^8}{\partial \phi^8}\right] w + a^4 \frac{w''''}{k} = 0$$

Substituting for w from (7-44),

$$(\lambda_n{}^8 - 4\lambda_n{}^6m^2 + 6\lambda_n{}^4m^4 - 4\lambda_n{}^2m^6 + m^8) + \frac{\lambda_n{}^4}{k} = 0$$

or

$$(m^2 - \lambda_n{}^2)^4 + \frac{\lambda_n{}^4}{k} = 0 \qquad (7\text{-}45)$$

This is the D-K-J characteristic equation whose roots can be found explicitly.

7-27 Roots of the Characteristic Equation

Solution of (7-45) gives

$$m = \pm \left[\lambda_n{}^2 \pm \lambda_n \left(\frac{1}{k}\right)^{1/4} \frac{1}{\sqrt{2}} (1 \pm i) \right]^{1/2}$$

This result has already been derived in Chapter 1.
To simplify the further work, we introduce the Aas-Jakobsen parameters ρ and κ defined as

$$\rho = \frac{\lambda_n^{1/2}}{(k)^{1/8}} \qquad \text{and} \qquad \kappa = \left(\frac{\lambda_n}{\rho}\right)^2$$

The eight roots m_1, m_2, m_3, ..., m_8 may now be written down as

$$m_1 = \alpha_1 + i\beta_1 \qquad m_5 = -m_1$$

$$m_2 = \alpha_1 - i\beta_1 \qquad m_6 = -m_2$$

$$m_3 = \alpha_2 + i\beta_2 \qquad m_7 = -m_3$$

$$m_4 = \alpha_2 - i\beta_2 \qquad m_8 = -m_4$$

where

$$\alpha_1 = \frac{\rho}{8^{1/4}} \left[\sqrt{(1 + \kappa\sqrt{2})^2 + 1} + (1 + \kappa\sqrt{2}) \right]^{1/2}$$

$$\alpha_2 = \frac{\rho}{8^{1/4}} \left[\sqrt{(1 - \kappa\sqrt{2})^2 + 1} - (1 - \kappa\sqrt{2}) \right]^{1/2}$$

$$\beta_1 = \frac{\rho}{8^{1/4}} \left[\sqrt{(1 + \kappa\sqrt{2})^2 + 1} - (1 + \kappa\sqrt{2}) \right]^{1/2} \qquad (7\text{-}46)$$

and

$$\beta_2 = \frac{\rho}{8^{1/4}} \left[\sqrt{(1 - \kappa\sqrt{2})^2 + 1} + (1 - \kappa\sqrt{2}) \right]^{1/2}$$

7-28 Expressions for the Stress Resultants and Displacements

All the stress resultants and displacements can be expressed in terms of w by making use of the equations of equilibrium, the stress-strain relations, and the moment-curvature relations. The expressions for the three moment resultants M_x, M_ϕ, and $M_{x\phi}$ in terms of w are already available in Equations (7-35). From Equation (7-36c),

$$N_\phi = \frac{D}{a^3}\left(a^4 \frac{\partial^4 w}{\partial x^4} + 2a^2 \frac{\partial^4 w}{\partial x^2 \, \partial\phi^2} + \frac{\partial^4 w}{\partial\phi^4}\right) = \frac{D}{a^3}\left(a^2 \frac{\partial^2}{\partial x^2} + \frac{\partial^2}{\partial\phi^2}\right)^2 w \quad (7\text{-}47)$$

Expressions have been derived already in Art. 7-24 for Q_x and Q_ϕ in terms of w, and these are listed below for convenient reference.

$$Q_x = -\frac{D}{a}\left[\frac{w^{'\cdot\cdot}}{a} + aw'''\right] \quad (7\text{-}48)$$

$$Q_\phi = -\frac{D}{a}\left[w''^{\cdot} + \frac{1}{a^2}w^{\cdot\cdot\cdot}\right] \quad (7\text{-}49)$$

Using Equation (7-34b), we may write

$$N_{\phi x}{}' = -\frac{1}{a}N_\phi{}^{\cdot} = -\frac{D}{a^4}(a^4 w''''^{\cdot} + 2a^2 w''^{\cdot\cdot\cdot} + w^{\vdots}) \quad (7\text{-}50)$$

From (7-34a), we note that

$$\frac{\partial N_x}{\partial x} = -\frac{1}{a}\frac{\partial N_{x\phi}}{\partial\phi} \quad \text{or} \quad \frac{\partial^2 N_x}{\partial x^2} = -\frac{1}{a}\frac{\partial^2 N_{x\phi}}{\partial x \, \partial\phi}$$

Hence substitution from (7-50) leads to

$$\frac{\partial^2 N_x}{\partial x^2} = \frac{D}{a^5}(a^4 w''''^{\cdot\cdot} + 2a^2 w''^{\vdots} + w^{\vdots}) \quad (7\text{-}51)$$

Since $N_x/Ed = u'$,

$$E\,du''' = \frac{\partial^2 N_x}{\partial x^2}$$

Hence

$$u''' = \frac{D}{E\,da^5}(a^4 w\,''''^{\cdot\cdot} + 2a^2 w''^{\vdots} + w^{\vdots}) \quad (7\text{-}52)$$

From the stress-strain relation (7-10c) it follows that

$$\frac{2N_{x\phi}{}'''}{Ed} = \frac{1}{a}u^{\cdot\cdot''} + v''''$$

Hence from (7-50) and (7-52),

$$v'''' = -\frac{D}{E\,da^6}[w^{:::} + 4a^2 w^{::''} + 5a^4 w^{\cdots''''} + 2a^6 w''''''\,] \quad (7\text{-}53)$$

7-29 Expression for w

Using the same reasoning as in Art. 7-15, an expression analogous to that of M_ϕ may be written taking into consideration the disturbances emanating from the left edge only. The expression for w will have the same form as that for M_ϕ given by Equation (7-29). We may write

$$w = [e^{-\alpha_1 \phi}(A_n \cos \beta_1 \phi + B_n \sin \beta_1 \phi) + e^{-\alpha_2 \phi}(C_n \cos \beta_2 \phi + D_n \sin \beta_2 \phi)] \cos \frac{\lambda_n x}{a}$$

$$(7\text{-}54)$$

Making use of the expressions developed in Art. 7-28, it is now possible to write down expressions for all stress resultants and displacements. The roots of the characteristic equation being explicitly known, these expressions assume relatively simple forms.

7-30 Explicit Expressions for the Stress Resultants and Displacements

The expressions for the three moment resultants given below are derived by making use of the moment-curvature relations given in Equations (7-35).

$$M_x = \frac{D\lambda_n^2}{a^2}[e^{-\alpha_1 \phi}(A_n \cos \beta_1 \phi + B_n \sin \beta_1 \phi)$$

$$+ e^{-\alpha_2 \phi}(C_n \cos \beta_2 \phi + D_n \sin \beta_2 \phi)]\cos \frac{\lambda_n x}{a} \quad (7\text{-}55)$$

$$M_\phi = -\frac{D}{a^2}[e^{-\alpha_1 \phi}(A_n^{②} \cos \beta_1 \phi + B_n^{②} \sin \beta_1 \phi)$$

$$+ e^{-\alpha_2 \phi}(C_n^{②} \cos \beta_2 \phi + D_n^{②} \sin \beta_2 \phi)]\cos \frac{\lambda_n x}{a} \quad (7\text{-}56)$$

$$M_{x\phi} = \frac{D\lambda_n}{a^2}[e^{-\alpha_1 \phi}(A_n^{③} \cos \beta_1 \phi + B_n^{③} \sin \beta_1 \phi)$$

$$+ e^{-\alpha_2 \phi}(C_n^{③} \cos \beta_2 \phi + D_n^{③} \sin \beta_2 \phi)]\sin \frac{\lambda_n x}{a} \quad (7\text{-}57)$$

Making use of Equation (7-48) we may write Q_x as

$$Q_x = -\frac{D\lambda_n}{a^3}\{e^{-\alpha_1 \phi}[(A_n \lambda_n^2 - A_n^{②})\cos \beta_1 \phi + (B_n \lambda_n^2 - B_n^{②})\sin \beta_1 \phi]$$

$$+ e^{-\alpha_2 \phi}[(C_n \lambda_n^2 - C_n^{②})\cos \beta_2 \phi + (D_n \lambda_n^2 - D_n^{②})\sin \beta_2 \phi]\}\sin \frac{\lambda_n x}{a}$$

The expression may be further simplified. We know that

$$A_n^{②} = (\alpha_1^2 - \beta_1^2) A_n - 2\alpha_1\beta_1 B_n$$

$$B_n^{②} = 2\alpha_1\beta_1 A_n + (\alpha_1^2 - \beta_1^2) B_n$$

From the values of the roots given in (7-46), it is easily verified that

$$(\alpha_1^2 - \beta_1^2) = \frac{\rho^2}{\sqrt{2}}(1 + \kappa\sqrt{2}) = \frac{\rho^2}{\sqrt{2}} + \lambda_n^2$$

Also

$$2\alpha_1\beta_1 = \frac{\rho^2}{\sqrt{2}}$$

Hence

$$A_n\lambda_n^2 - A_n^{②} = \left(-\frac{\rho^2}{\sqrt{2}}A_n + \frac{\rho^2}{\sqrt{2}}B_n\right) = -\frac{\rho^2}{\sqrt{2}}(A_n - B_n)$$

Similarly,

$$B_n\lambda_n^2 - B_n^{②} = -\frac{\rho^2}{\sqrt{2}}(A_n + B_n)$$

Proceeding in this manner, we may now write

$$Q_x = \frac{D\lambda_n\rho^2}{a^3\sqrt{2}}\{e^{-\alpha_1\phi}[(A_n - B_n)\cos\beta_1\phi + (A_n + B_n)\sin\beta_1\phi]$$
$$+ e^{-\alpha_2\phi}[-(C_n + D_n)\cos\beta_2\phi + (C_n - D_n)\sin\beta_2\phi]\}\sin\frac{\lambda_n x}{a} \quad (7\text{-}58)$$

From Equation (7-49), we may write

$$Q_\phi = -\frac{D}{a^3}\{e^{-\alpha_1\phi}[(A_n^{③} - \lambda_n^2 A_n^{①})\cos\beta_1\phi + (B_n^{③} - \lambda_n^2 B_n^{①})\sin\beta_1\phi]$$
$$+ e^{-\alpha_2\phi}[(C_n^{③} - \lambda_n^2 C_n^{①})\cos\beta_2\phi + (D_n^{③} - \lambda_n^2 D_n^{①})\sin\beta_2\phi]\}\cos\frac{\lambda_n x}{a}$$

This expression may be recast into a more compact form as follows:

$$A_n^{③} = (\alpha_1^2 - \beta_1^2) A_n^{①} - 2\alpha_1\beta_1 B_n^{①}$$

Hence

$$A_n^{③} - \lambda_n^2 A_n^{①} = \frac{\rho^2}{\sqrt{2}}(A_n^{①} - B_n^{①})$$

Similarly

$$B_n^{③} - \lambda_n^2 B_n^{①} = \frac{\rho^2}{\sqrt{2}}(A_n^{①} + B_n^{①})$$

Dealing with the other coefficients in a similar manner, the expression may be cast into the form

$$Q_\phi = -\frac{D}{a^3}\frac{\rho^2}{\sqrt{2}}\{e^{-\alpha_1\phi}[(A_n^{①} - B_n^{④})\cos\beta_1\phi + (A_n^{④} + B_n^{②})\sin\beta_1\phi]$$

$$+ e^{-\alpha_2\phi}[-(C_n^{①} + D_n^{②})\cos\beta_2\phi + (C_n^{④} - D_n^{②})\sin\beta_2\phi]\}\cos\frac{\lambda_n x}{a} \quad (7\text{-}59)$$

From Equation (7-47) it is found that N_ϕ can be obtained by applying the operator $[a^2(\partial^2/\partial x^2) + (\partial^2/\partial\phi^2)]^2$ on $(D/a^3)w$. Expanding, the operator takes the form $[a^4(\partial^4/\partial x^4) + 2a^2(\partial^4/\partial x^2\,\partial\phi^2) + \partial^4/\partial\phi^4]$. On the application of this operator on w, the coefficient of $e^{-\alpha_1\phi}\cos\beta_1\phi$ becomes

$$(A_n\lambda_n^4 - 2\lambda_n^2 A_n^{②} + A_n^{④})$$

This quantity, on simplification, becomes $-\rho^4 B_n$. It can be shown in a similar manner that the coefficient of the term $e^{-\alpha_1\phi}\sin\beta_1\phi$ becomes $\rho^4 A_n$. Coefficients of the other terms can be found in the same way. Hence we may build up the expression for N_ϕ as

$$N_\phi = \frac{D\rho^4}{a^3}[e^{-\alpha_1\phi}(A_n\sin\beta_1\phi - B_n\cos\beta_1\phi)$$

$$- e^{-\alpha_2\phi}(C_n\sin\beta_2\phi - D_n\cos\beta_2\phi)]\cos\frac{\lambda_n x}{a} \quad (7\text{-}60)$$

The expression for $N_{x\phi}$ may be derived by integrating (7-50) once with respect to x. This is easily found to give

$$N_{x\phi} = -\frac{D\rho^4}{a^3\lambda_n}[e^{-\alpha_1\phi}(A_n^{①}\sin\beta_1\phi - B_n^{①}\cos\beta_1\phi)$$

$$- e^{-\alpha_2\phi}(C_n^{①}\sin\beta_2\phi - D_n^{①}\cos\beta_2\phi)]\sin\frac{\lambda_n x}{a} \quad (7\text{-}61)$$

Integrating (7-51) twice with respect to x, the following expression is obtained for N_x:

$$N_x = -\frac{D\rho^4}{a^3\lambda_n^2}[e^{-\alpha_1\phi}(A_n^{②}\sin\beta_1\phi - B_n^{②}\cos\beta_1\phi)$$

$$- e^{-\alpha_2\phi}(C_n^{②}\sin\beta_2\phi - D_n^{②}\cos\beta_2\phi)]\cos\frac{\lambda_n x}{a} \quad (7\text{-}62)$$

Integrating (7-52) thrice with respect to x gives the following expression for u:

$$u = \frac{\lambda_n}{\rho^4}[e^{-\alpha_1\phi}(A_n^{②}\sin\beta_1\phi - B_n^{②}\cos\beta_1\phi)$$

$$- e^{-\alpha_2\phi}(C_n^{②}\sin\beta_2\phi - D_n^{②}\cos\beta_2\phi)]\sin\frac{\lambda_n x}{a} \quad (7\text{-}63)$$

To determine v, we start with the expression (7-53). If the operations indicated on w are carried out, the coefficient of the term $e^{-\alpha_1\phi}\cos\beta_1\phi$ becomes

$$[A_n^{\textcircled{7}} - 4\lambda_n^2 A_n^{\textcircled{5}} + 5\lambda_n^4 A_n^{\textcircled{3}} - 2\lambda_n^6 A_n^{\textcircled{1}}]$$

But

$$A_n^{\textcircled{7}} = (\alpha_1^2 - \beta_1^2)\,A_n^{\textcircled{5}} - 2\alpha_1\beta_1 B_n^{\textcircled{5}} = \left(\frac{\rho^2}{\sqrt{2}} + \lambda_n^2\right) A_n^{\textcircled{5}} - \frac{\rho^2}{\sqrt{2}} B_n^{\textcircled{5}}$$

Hence

$$A_n^{\textcircled{7}} - 4\lambda_n^2 A_n^{\textcircled{5}} = \left(\frac{\rho^2}{\sqrt{2}} - 3\lambda_n^2\right) A_n^{\textcircled{5}} - \frac{\rho^2}{\sqrt{2}} B_n^{\textcircled{5}}$$

Again

$$A_n^{\textcircled{5}} = \left(\frac{\rho^2}{\sqrt{2}} + \lambda_n^2\right) A_n^{\textcircled{3}} - \frac{\rho^2}{\sqrt{2}} B_n^{\textcircled{3}}$$

and

$$B_n^{\textcircled{5}} = \frac{\rho^2}{\sqrt{2}} A_n^{\textcircled{3}} + \left(\frac{\rho^2}{\sqrt{2}} + \lambda_n^2\right) B_n^{\textcircled{3}}$$

Making these substitutions,

$$A_n^{\textcircled{7}} - 4\lambda_n^2 A_n^{\textcircled{5}} + 5\lambda_n^4 A_n^{\textcircled{3}} = 2\lambda_n^2\left(\lambda_n^2 - \frac{\rho^2}{\sqrt{2}}\right) A_n^{\textcircled{3}} + \rho^2(\sqrt{2}\lambda_n^2 - \rho^2) B_n^{\textcircled{3}}$$

We next substitute for $A_n^{\textcircled{3}}$ and $B_n^{\textcircled{3}}$ in terms of $A_n^{\textcircled{1}}$ and $B_n^{\textcircled{1}}$ to get

$$[A_n^{\textcircled{7}} - 4\lambda_n^2 A_n^{\textcircled{5}} + 5\lambda_n^4 A_n^{\textcircled{3}} - 2\lambda_n^6 A_n^{\textcircled{1}}] = -\frac{\rho^6}{\sqrt{2}}[A_n^{\textcircled{1}} + (1 - \kappa\sqrt{2})\,B_n^{\textcircled{1}}]$$

Similarly, the coefficients of $(e^{-\alpha_1\phi}\sin\beta_1\phi)$, $(e^{-\alpha_2\phi}\cos\beta_2\phi)$, and $(e^{-\alpha_2\phi}\sin\beta_2\phi)$ are found. The resulting expression for v'''' is integrated four times with respect to x to give

$$v = -\frac{1}{\rho^2\sqrt{2}}\{-e^{-\alpha_1\phi}[A_n^{\textcircled{1}} + (1 - \kappa\sqrt{2})\,B_n^{\textcircled{1}}]\cos\beta_1\phi$$

$$+ e^{-\alpha_1\phi}[(1 - \kappa\sqrt{2})\,A_n^{\textcircled{1}} - B_n^{\textcircled{1}}]\sin\beta_1\phi$$

$$+ e^{-\alpha_2\phi}[C_n^{\textcircled{1}} - (1 + \kappa\sqrt{2})\,D_n^{\textcircled{1}}]\cos\beta_2\phi$$

$$+ e^{-\alpha_2\phi}[(1 + \kappa\sqrt{2})\,C_n^{\textcircled{1}} + D_n^{\textcircled{1}}]\sin\beta_2\phi\}\cos\frac{\lambda_n x}{a} \qquad (7\text{-}64)$$

Making use of the expressions for v and w, the expression for $\vartheta = (1/a)(v + \partial w/\partial\phi)$ may be written down. It is easily verified that

$$
\begin{aligned}
\vartheta = -\frac{1}{a\rho^2\sqrt{2}} \{ & -e^{-\alpha_1\phi}[A_n^{\oplus}(1+\rho^2\sqrt{2}) + (1-\kappa\sqrt{2})B_n^{\oplus}]\cos\beta_1\phi \\
& + e^{-\alpha_1\phi}[(1-\kappa\sqrt{2})A_n^{\oplus} - B_n^{\oplus}(1+\rho^2\sqrt{2})]\sin\beta_1\phi \\
& + e^{-\alpha_2\phi}[(1-\rho^2\sqrt{2})C_n^{\oplus} - (1+\kappa\sqrt{2})D_n^{\oplus}]\cos\beta_2\phi \\
& + e^{-\alpha_2\phi}[(1+\kappa\sqrt{2})C_n^{\oplus} + (1-\rho^2\sqrt{2})D_n^{\oplus}]\sin\beta_2\phi \} \cos\frac{\lambda_n x}{a}
\end{aligned}
\tag{7-65}
$$

The use of the explicit expressions derived in this article for the stress analysis of shells will be explained in the next chapter with the aid of an example.

THE SCHORER THEORY

7-31 Notes on the Schorer Theory

The Schorer theory published in 1936 [21] has the merit of extreme simplicity. Like Finsterwalder before him, Schorer assumed $M_x = Q_x = M_{x\phi} = 0$. Another assumption implicit in his theory is that the tangential strain

$$
\epsilon_\phi = \frac{1}{a}\left(\frac{\partial v}{\partial\phi} - w\right)
$$

and the shear strain

$$
\gamma_{xy} = \left(\frac{1}{a}\frac{\partial u}{\partial\phi} + \frac{\partial v}{\partial x}\right)
$$

are both small in comparison with the longitudinal strain $\epsilon_x = \partial u/\partial x$. This assumption leads to the following relations:

$$
\frac{\partial v}{\partial x} = -\frac{1}{a}\frac{\partial u}{\partial\phi}
\tag{7-66}
$$

and

$$
w = \frac{\partial v}{\partial\phi}
\tag{7-67}
$$

The Schorer theory is applicable only to "long shells." What this means will be examined later.

The transverse moment-curvature relation will be assumed to have the same form as in the Donnell theory, i.e.,

$$
M_\phi = -\frac{D}{a^2}\frac{\partial^2 w}{\partial\phi^2}
$$

7-32 The Schorer Differential Equation

From Equation (7-21*d*),

$$Q_\phi = \frac{1}{a} M_\phi^{\cdot} = -\frac{D}{a^3} w^{\cdots}$$

Again, from relation (7-21*c*),

$$N_\phi = -Q_\phi^{\cdot} = +\frac{D}{a^3} w^{::}$$

Making use of Equation (7-34*b*)

$$\frac{\partial N_{\phi x}}{\partial x} = -\frac{1}{a} N_\phi^{\cdot} = -\frac{D}{a^4} w^{:::}$$

From (7-34*a*),

$$\frac{\partial^2 N_x}{\partial x^2} = E\, du''' = +\frac{D}{a^5} w^{::::}$$

or

$$u''' = \frac{D}{E\, da^5} w^{::::}$$

From relation (7-66),

$$\frac{\partial v}{\partial x} = -\frac{1}{a}\frac{\partial u}{\partial \phi}$$

Hence

$$v'''' = -\frac{1}{a} u''''^{\cdot} = -\frac{D}{E\, da^6} w^{::::\cdot}$$

From relation (7-67),

$$w = \frac{\partial v}{\partial \phi}$$

Hence

$$w'''' = -\frac{D}{E\, da^6} w^{:::::}$$

Rearranging terms,

$$w^{::::} + \frac{12a^6}{d^2} w'''' = 0$$

or

$$w^{::::} + \frac{a^4}{k} w'''' = 0 \qquad (7\text{-}68)$$

where $k = d^2/12a^2$. This is the Schorer form of the differential equation for cylindrical shells.

7-33 The Schorer Characteristic Equation

As before, the solution for w may be sought in the form

$$w = He^{m\phi} \cos \frac{\lambda_n x}{a}$$

Substituting this solution in (7-68), we get the characteristic equation

$$m^8 + \frac{\lambda_n^4}{k} = 0$$

7-34 Roots of the Schorer Equation

The eight roots of the equation have already been derived in Chapter 1. They are

$$
\begin{aligned}
m_1 &= \alpha_1 + i\beta_1 & m_5 &= -m_1 \\
m_2 &= \alpha_1 - i\beta_1 & m_6 &= -m_2 \\
m_3 &= \alpha_2 + i\beta_2 & m_7 &= -m_3 \\
m_4 &= \alpha_2 - i\beta_2 & m_8 &= -m_4
\end{aligned}
$$

where

$$\alpha_1 = \frac{\rho}{8^{1/4}} (2^{1/2} + 1)^{1/2}$$

$$\beta_1 = \frac{\rho}{8^{1/4}} (2^{1/2} - 1)^{1/2}$$

$$\alpha_2 = \frac{\rho}{8^{1/4}} (2^{1/2} - 1)^{1/2}$$

$$\beta_2 = \frac{\rho}{8^{1/4}} (2^{1/2} + 1)^{1/2}$$

$$(7\text{-}69)$$

7-35 Expression for w

The first term of the infinite series for w may now be written as

$$
\begin{aligned}
w = [&e^{-\alpha_1 \phi}(A_n \cos \beta_1 \phi + B_n \sin \beta_1 \phi) \\
&+ e^{-\alpha_2 \phi}(C_n \cos \beta_2 \phi + D_n \sin \beta_2 \phi)] \cos \frac{\lambda_n x}{a}
\end{aligned}
$$

7-36 Expressions for Other Stress Resultants and Displacements

Starting with w, the expressions for the other stress resultants and displacements may be derived in the same manner as in the D-K-J theory. The constants A_n, A_n^Φ, ..., A_n^Φ; B_n, B_n^Φ, ..., B_n^Φ;

C_n, $C_n^\circled{1}$, ..., $C_n^\circled{5}$; D_n, $D_n^\circled{1}$, ..., $D_n^\circled{5}$ required for this purpose are given in Table 7-1. The expressions for the stress resultants and displacements are given below.

$$w = [e^{-\alpha_1\phi}(A_n \cos \beta_1\phi + B_n \sin \beta_1\phi)$$
$$+ e^{-\alpha_2\phi}(C_n \cos \beta_2\phi + D_n \sin \beta_2\phi)] \cos \frac{\lambda_n x}{a}$$

$$M_\phi = -\frac{D}{a^2}\frac{\rho^2}{2^{1/2}}\{e^{-\alpha_1\phi}[(A_n - B_n)\cos\beta_1\phi + (A_n + B_n)\sin\beta_1\phi]$$
$$- e^{-\alpha_2\phi}[(C_n + D_n)\cos\beta_2\phi - (C_n - D_n)\sin\beta_2\phi]\} \cos\frac{\lambda_n x}{a}$$

$$Q_\phi = -\frac{D}{a^3}\frac{\rho^2}{2^{1/2}}\{e^{-\alpha_1\phi}[(A_n^\circled{1} - B_n^\circled{1})\cos\beta_1\phi + (A_n^\circled{1} + B_n^\circled{1})\sin\beta_1\phi]$$
$$- e^{-\alpha_2\phi}[(C_n^\circled{1} + D_n^\circled{1})\cos\beta_2\phi - (C_n^\circled{1} - D_n^\circled{1})\sin\beta_2\phi]\} \cos\frac{\lambda_n x}{a}$$

$$N_\phi = -\frac{D\rho^4}{a^3}\{e^{-\alpha_1\phi}[B_n \cos\beta_1\phi - A_n \sin\beta_1\phi]$$
$$- e^{-\alpha_2\phi}[D_n \cos\beta_2\phi - C_n \sin\beta_2\phi]\} \cos\frac{\lambda_n x}{a}$$

$$N_{x\phi} = -\frac{D\rho^4}{\lambda_n a^3}\{e^{-\alpha_1\phi}[(A_n^\circled{1} \sin\beta_1\phi - B_n^\circled{1} \cos\beta_1\phi)]$$
$$- e^{-\alpha_2\phi}[(C_n^\circled{1} \sin\beta_2\phi - D_n^\circled{1} \cos\beta_2\phi)]\} \sin\frac{\lambda_n x}{a}$$

$$N_x = -\frac{D\rho^6}{2^{1/2}\lambda_n^2 a^3}\{e^{-\alpha_1\phi}[(A_n - B_n)\sin\beta_1\phi - (A_n + B_n)\cos\beta_1\phi]$$
$$+ e^{-\alpha_2\phi}[(C_n + D_n)\sin\beta_2\phi - (D_n - C_n)\cos\beta_2\phi]\} \cos\frac{\lambda_n x}{a} \qquad (7\text{-}70)$$

$$u = -\frac{D\rho^6}{2^{1/2}E\,da^2\lambda_n^3}\{e^{-\alpha_1\phi}[(A_n - B_n)\sin\beta_1\phi - (A_n + B_n)\cos\beta_1\phi]$$
$$+ e^{-\alpha_2\phi}[(C_n + D_n)\sin\beta_2\phi - (D_n - C_n)\cos\beta_2\phi]\} \sin\frac{\lambda_n x}{a}$$

$$v = \frac{D\rho^6}{2^{1/2}\lambda_n^4 E\,da^2}\{e^{-\alpha_1\phi}[(A_n^\circled{1} + B_n^\circled{1})\cos\beta_1\phi - (A_n^\circled{1} - B_n^\circled{1})\sin\beta_1\phi]$$
$$- e^{-\alpha_2\phi}[(C_n^\circled{1} - D_n^\circled{1})\cos\beta_2\phi + (C_n^\circled{1} + D_n^\circled{1})\sin\beta_2\phi]\} \cos\frac{\lambda_n x}{a}$$

$$\vartheta^\dagger = \frac{1}{a}\frac{\partial w}{\partial \phi} = \frac{1}{a}\{e^{-\alpha_1\phi}[A_n^\circled{1} \cos\beta_1\phi + B_n^\circled{1} \sin\beta_1\phi]$$
$$+ e^{-\alpha_2\phi}[C_n^\circled{1} \cos\beta_2\phi + D_n^\circled{1} \sin\beta_2\phi]\} \cos\frac{\lambda_n x}{a}$$

---

† The term v is ignored in comparison with $\partial w/\partial\phi$ in the expression $\vartheta = (1/a)(v + \partial w/\partial\phi)$.

Table 7-1. Table of Constants in Schorer's Theory

n	$A_n^{(n)}$	$B_n^{(n)}$	$C_n^{(n)}$	$D_n^{(n)}$
1	$-\alpha_1 A_n + \beta_1 B_n$	$-\beta_1 A_n - \alpha_1 B_n$	$-\alpha_2 C_n + \beta_2 D_n$	$-\beta_2 C_n - \alpha_2 D_n$
2	$\dfrac{\rho^2}{\sqrt{2}}(A_n - B_n)$	$\dfrac{\rho^2}{\sqrt{2}}(A_n + B_n)$	$-\dfrac{\rho^2}{\sqrt{2}}(C_n + D_n)$	$\dfrac{\rho^2}{\sqrt{2}}(C_n - D_n)$
3	$\dfrac{\rho^2}{\sqrt{2}}(A_n^{①} - B_n^{①})$	$\dfrac{\rho^2}{\sqrt{2}}(A_n^{①} + B_n^{①})$	$-\dfrac{\rho^2}{\sqrt{2}}(C_n^{①} + D_n^{①})$	$\dfrac{\rho^2}{\sqrt{2}}(C_n^{①} - D_n^{①})$
4	$-\rho^4 B_n$	$\rho^4 A_n$	$\rho^4 D_n$	$-\rho^4 C_n$
5	$-\rho^4 B_n^{①}$	$\rho^4 A_n^{①}$	$\rho^4 D_n^{①}$	$-\rho^4 C_n^{①}$
6	$-\dfrac{\rho^6}{\sqrt{2}}(A_n + B_n)$	$\dfrac{\rho^6}{\sqrt{2}}(A_n - B_n)$	$\dfrac{\rho^6}{\sqrt{2}}(C_n - D_n)$	$\dfrac{\rho^6}{\sqrt{2}}(C_n + D_n)$
7	$-\dfrac{\rho^6}{\sqrt{2}}(A_n^{①} + B_n^{①})$	$\dfrac{\rho^6}{\sqrt{2}}(A_n^{①} - B_n^{①})$	$\dfrac{\rho^6}{\sqrt{2}}(C_n^{①} - D_n^{①})$	$\dfrac{\rho^6}{\sqrt{2}}(C_n^{①} + D_n^{①})$

Note: $\alpha_2 = \beta_1$ and $\beta_2 = \alpha_1$.

7-37 Comments on the Schorer Theory

The Schorer characteristic equation (Art. 7-33) may be derived from the Finsterwalder characteristic equation (7-27) if all the lower derivatives with respect to ϕ are ignored in comparison with the eighth-order derivative. If this is done, we are left only with the first and last terms in the Finsterwalder characteristic equation, which then assumes the same form as the Schorer equation. This is, in fact, how Schorer arrived at his equation in his paper. To be able to discard the lower derivatives compared with the highest derivative in Equation (7-27), it is evident that $\lambda_n^2 = (n\pi a/l)^2$ must be equal to or less than 1. Confining ourselves to the first term of the Fourier series, this condition may be stated as

$$\frac{l}{a} \geqslant \pi \tag{7-71}$$

Or, in other words, according to Schorer, his theory is applicable only to shells whose $l/a \geqslant \pi$.

The Schorer theory may also be deduced as a special case of the D-K-J theory when $\kappa \to 0$. It is easily verified that, if this is the case, the D-K-J roots reduce themselves to the Schorer roots. Moreover, the expressions for the stress resultants and displacements derived from both the theories become identical.

7-38 Membrane Solution as the Particular Integral

It was stated in an earlier section that for purposes of design it is sufficiently accurate to superimpose the stress resultants and displacements derived from a bending theory of the unloaded shell, i.e., the homogeneous solution of the differential equation, on the stress resultants and displacements calculated from the membrane theory. This is generally satisfactory for loads which are uniformly distributed. This amounts to treating the membrane solution as the particular integral For other types of loadings, it may be necessary to use the particular integrals corresponding to the given loading instead of the solutions derived from the membrane theory. This procedure will be illustrated with reference to the Schorer theory.

7-39 Schorer's Differential Equation for a Loaded Shell

Let X, Y, and Z be the forces per unit area acting on the shell in the longitudinal, tangential, and radial directions. The derivation proceeds in the same manner as in Art. 7-32 except that the equations of equilibrium (7-21a), (7-21b), and (7-21c) made use of in the derivation have to be modified to include X, Y, and Z. Otherwise, what follows is self-explanatory. We begin, as before, with $M_\phi = -(D/a^2)\,w^{\cdot\cdot}$.

$$Q_\phi = \frac{1}{a} M_\phi^{\cdot} = -\frac{D}{a^3} w^{\cdot\cdot\cdot}$$

$$N_\phi = -Q_\phi^{\cdot} - aZ = -\frac{D}{a^3} w^{\vdots\vdots} - aZ$$

$$\frac{\partial N_{x\phi}}{\partial x} = -\frac{1}{a} N_\phi^{\cdot} - Y = -\frac{D}{a^4} w^{\vdots\vdots\cdot} + Z^{\cdot} - Y$$

$$\frac{\partial^2 N_x}{\partial x^2} = -\frac{1}{a} \frac{\partial^2 N_{x\phi}}{\partial x\,\partial\phi} - X'$$

$$E\,du''' = +\frac{D}{a^5} w^{\vdots\vdots\vdots} - \frac{Z^{\cdot\cdot}}{a} + \frac{Y^{\cdot}}{a} - X'$$

$$v'''' = -\frac{1}{a} u''''^{\cdot} = \frac{1}{Ed}\left[-\frac{D}{a^6} w^{\vdots\vdots\vdots\vdots} + \frac{1}{a^2} Z^{\cdot\cdot\cdot} - \frac{1}{a^2} Y^{\cdot\cdot} + \frac{1}{a} X'^{\cdot\cdot} \right]$$

$$w'''' = \frac{1}{Ed}\left[-\frac{D}{a^6} w^{\vdots\vdots\vdots\vdots} + \frac{1}{a^2} Z^{\vdots\vdots} - \frac{1}{a^2} Y^{\cdot\cdot\cdot} + \frac{1}{a} X'^{\cdot\cdot} \right]$$

or

$$\frac{D}{a^6} w^{\vdots\vdots\vdots\vdots} + E\,dw'''' - \frac{1}{a^2}(Z^{\vdots\vdots} - Y^{\cdot\cdot\cdot}) - \frac{1}{a} X'^{\cdot\cdot} = 0 \qquad (7\text{-}72)$$

7-40 Particular Integral for Uniform Vertical Load *g* due to Dead Weight

Referring to Fig. 6-7 and considering only the load component corresponding to the first Fourier term,

$$Z = \frac{4g}{\pi} \cos (\phi_c - \phi) \cos \frac{\pi x}{l}$$

$$Y = -\frac{4g}{\pi} \sin (\phi_c - \phi) \cos \frac{\pi x}{l}$$

and

$$X = 0$$

Substituting these values in the differential equation (7-72), we get

$$\frac{D}{a^5} w^{:::::} + E\,d\,a\,w'''' - \frac{1}{a}\left[\frac{8g}{\pi} \cos (\phi_c - \phi) \cos \frac{\pi x}{l}\right] = 0$$

Let us seek a particular integral of the form

$$w = C \cos (\phi_c - \phi) \cos \frac{\pi x}{l}$$

Substituting in the differential equation above and solving,

$$C = \frac{8g}{\pi a \left[E\,d\,a \left(\frac{\pi}{l}\right)^4 + \frac{Ed^3}{12a^5}\right]}$$

We may now write down the values of the stress resultants and displacements corresponding to the particular integral. Q_ϕ may be ignored in calculating N_ϕ because its influence is not significant.

$$N_\phi = -Za = -\frac{4ga}{\pi} \cos (\phi_c - \phi) \cos \frac{\pi x}{l} \tag{7-73a}$$

Hence

$$\frac{\partial N_{x\phi}}{\partial x} = -\frac{1}{a} N_\phi{}^{\cdot} - Y$$

$$N_{x\phi} = \frac{8gl}{\pi^2} \sin (\phi_c - \phi) \sin \frac{\pi x}{l} \tag{7-73b}$$

Hence

$$\frac{\partial N_x}{\partial x} = -\frac{1}{a} N_{x\phi}{}^{\cdot}$$

$$N_x = -\frac{8gl^2}{a\pi^3} \cos (\phi_c - \phi) \cos \frac{\pi x}{l} \tag{7-73c}$$

The expression for u follows from the relation

$$\frac{N_x}{Ed} = \frac{\partial u}{\partial x}$$

$$u = -\frac{8gl^3}{\pi^4 aEd} \cos(\phi_c - \phi) \sin\frac{\pi x}{l} \qquad (7\text{-}73d)$$

$$w = \frac{8g}{\pi a \left[E\,d\,a\left(\dfrac{\pi}{l}\right)^4 + \dfrac{Ed^3}{12a^5} \right]} \cos(\phi_c - \phi) \cos\frac{\pi x}{l} \qquad (7\text{-}73e)$$

The expression for v is next derived from the relation

$$w = \frac{\partial v}{\partial \phi}$$

$$v = -\frac{8g}{\pi a \left[E\,d\,a\left(\dfrac{\pi}{l}\right)^4 + \dfrac{Ed^3}{12a^5} \right]} \sin(\phi_c - \phi) \cos\frac{\pi x}{l} \qquad (7\text{-}73f)$$

It is easily verified that

$$\vartheta = \frac{1}{a}\left(v + \frac{\partial w}{\partial \phi} \right) = 0$$

$$M_\phi = -\frac{D}{a^2} \frac{\partial^2 w}{\partial \phi^2}$$

$$= +\frac{D}{a^2} \frac{8g}{\pi a \left[E\,d\,a\left(\dfrac{\pi}{l}\right)^4 + \dfrac{Ed^3}{12a^5} \right]} \cos(\phi_c - \phi) \cos\frac{\pi x}{l}$$

$$= \frac{8gD}{\pi a^3} \left[\frac{1}{E\,d\,a\left(\dfrac{\pi}{l}\right)^4 + \dfrac{Ed^3}{12a^5}} \right] \cos(\phi_c - \phi) \cos\frac{\pi x}{l} \qquad (7\text{-}73g)$$

$$Q_\phi = \frac{1}{a} M_\phi^{\cdot} = \frac{8gD}{\pi a^4} \left[\frac{1}{E\,d\,a\left(\dfrac{\pi}{l}\right)^4 + \dfrac{Ed^3}{12a^5}} \right] \sin(\phi_c - \phi) \cos\frac{\pi x}{l} \qquad (7\text{-}73h)$$

The expressions derived from the particular integral for N_x, N_ϕ, $N_{x\phi}$ and u, v, w are seen to be identical with those given by the membrane theory. It may be easily verified that M_ϕ and Q_ϕ tend to be negligible in most practical cases. We may therefore conclude that, where the loading on a shell is uniform, no serious error is involved in using the membrane solution in place of the particular integral.

7-41   "Long" and "Short" Shells

These terms, which often figure in shell literature, are best avoided. They have no precise meaning and it is difficult to demarcate the boundary between "long" and "short" shells. Several criteria have been proposed, from time to time, by different authors [3, 22, 23]. These will be discussed here briefly. Broadly speaking, a classification into long and short may be based on one of the following considerations:

(i) It is an advantage to be able to treat a shell as a beam of curved cross section so that the familiar theory of simple bending can be used to find the longitudinal and shear stresses in the shell. This is the basis of the Lundgren beam theory to be developed in a later chapter. Hence writers on shell theory have tried to define the limiting proportions of a shell in which the distribution of the longitudinal stress would approach very nearly the linear distribution in beams. This has been found to be the case if $l/a \geqslant 5$ for shells without edge beams. The provision of edge beams may bring down this limit to about three. Shells whose l/a ratios exceed these limits are sometimes described as long. The distribution of longitudinal stress in a shell depends on the deformation of the shell which, in turn, is influenced by the manner in which it is loaded and supported along its edges. Hence the size and type of edge member provided and the loading on the shell are also factors to be taken into account in defining this limit.

(ii) Several writers have tried to simplify the theory of cylindrical shells by ignoring certain stress resultants, by dropping secondary terms in the expressions for some of the stress resultants and displacements, or by omitting relatively less significant terms in the characteristic equation. For example, Finsterwalder assumed that $M_x = Q_x = 0$ and $M_{x\phi} = 0$. Schorer went a step further and made a more thoroughgoing approximation by dropping all except the term contributed by the highest derivative in the Finsterwalder characteristic equation. It now becomes important to define the range of validity of such approximations and the class of shells to which they are applicable. The terms long and short are useful in this context to provide some guidance to designers. The Finsterwalder and Schorer approximations are generally applicable only to long shells. As already noted, in Art. 7-37, Schorer has himself fixed a limit of $l/a \geqslant \pi$ for his theory to be valid. This limit is also sometimes employed to classify shells into long and short, as studies have shown that the range of validity of the Schorer and Finsterwalder theories is roughly the same. It may be noted in passing that when $l/a \geqslant \pi$, the stress distribution becomes nearly linear for most shells with edge beams. Hence this criterion is closely related to the one developed in (i) above.

(iii) Classification of shells into long and short has also been attempted on the basis of the extent to which the disturbances emanating from the straight edges penetrate into the body of the shell. Based on this consideration, the ASCE Manual [3] classifies shells as short if $l/a < 1.60$. In such shells, the edge disturbances will seldom penetrate beyond the crown and, even if they do, their values will be insignificant. Hence the disturbances emanating from the two edges do not overlap. Consequently, in formulating boundary conditions at a given straight edge, the disturbances emanating from the other edge may be ignored. Moreover, in the computation of stress resultants and displacements at any point on the shell, only the disturbances emanating from the nearer edge need be considered. Hence the stress analysis gets considerably simplified. A criterion which gives more or less the same results as that given in the ASCE Manual has also been proposed by Rabich [22]. Whether a shell is short or long is decided by the parameter $B/(l^2ad)^{1/4}$. If $B/(l^2ad)^{1/4} < 3$, the shell is regarded as long; and if $B/(l^2ad)^{1/4} > 5$, the shell is considered short. Shells with a parameter between 3 and 5 are classed as "quasi-short."

Aas-Jakobsen [23] was perhaps the first to classify shells into "long," "short," and "intermediate." He introduced two parameters ρ and κ, which have already been defined in Art. 7-27. Shells with ρ values between 4 and 7, with corresponding κ values of 0.03 and 0.12, are regarded as long. Those with ρ values between 10 and 20, with corresponding κ values of 0.15 and 0.30, are considered short. Shells with ρ values between 7 and 10, known as intermediate shells, are very rarely used in practice. The limits proposed by Aas-Jakobsen closely correspond to those set by the ASCE Manual and Rabich.

COMMENTS ON THE BENDING THEORIES OF CYLINDRICAL SHELLS

7-42 Characteristic Equations

In the foregoing articles, we have developed, in some detail, the Finsterwalder, D-K-J, and Schorer theories. These are adequate for the design of most reinforced-concrete thin shells met with in practice. To make the presentation complete, brief reference will now be made to some of the other bending theories, which are regarded as more exact. The characteristic equation of each theory is given, followed by brief comments. To make comparisons more meaningful, all the characteristic equations have been recast in uniform notation making use of the Aas-Jakobsen parameters ρ and κ. Poisson's ratio has been assumed to be zero. The notation $\bar{m} = m/\rho$ is introduced. As the roots of $\bar{m}$

are of the order of unity, this substitution permits the comparison of coefficients for their relative values. The characteristic equations of the Finsterwalder, D-K-J, and Schorer theories are also given for purposes of comparison (Table 7-2). The table is based on a similar comparison made by Arya [24].

7-43 "Exact" and "Approximate" Theories

The first seven theories presented in Table 7-2 may be termed "exact" in the sense that they may be applied to cylindrical thin shells of all proportions. Differences among them are only in certain minor details. The last three theories in the list are usually regarded as "approximate." They cannot be applied to shells of all proportions.

7-44 The Flügge Theory

Flügge [20] was the first to derive three simultaneous differential equations in u, v, and w. For most reinforced-concrete shells, $k < 10^{-5}$. Hence it may be neglected in comparison with unity. If this is done, the Flügge differential equations lead to the characteristic equation given in Table 7-2. The Flügge characteristic equation may be treated as exact and used as a yardstick to assess the accuracy of other theories.

7-45 The Dischinger Theory

Dischinger [25] arrived at the same simultaneous differential equations as Flügge. But the characteristic equations of the two theories are not identical. The differences between them are only in certain minor terms which are of no consequence. The Dischinger theory is thus nearly as accurate as Flügge's.

7-46 Aas-Jakobsen's Theory

The equations of equilibrium and force-displacement relations used are the same as in the Flügge theory. The characteristic equation obtained differs from that of Flügge only in certain minor terms.

7-47 Lundgren's Theory

Lundgren [2] arrives at his characteristic equation using the same equations of equilibrium as Flügge. Simplified but less accurate expressions are employed for the force-displacement relations. Nevertheless, the resulting characteristic equation differs from Flügge's only in insignificant respects.

Table 7-2. Characteristic Equations of Cylindrical Shells

Theory	Coefficient of				Constant term
	$\bar{m}^8$	$\bar{m}^6$	$\bar{m}^4$	$\bar{m}^2$	
1. Flügge	1.00	$\dfrac{2}{\rho^2} - 4\kappa$	$\dfrac{1}{\rho^4} - \dfrac{8\kappa}{\rho^2} + 6\kappa^2$	$-\dfrac{4\kappa}{\rho^4} + \dfrac{6\kappa^2}{\rho^2} - 4\kappa^3$	$\kappa^4 + 1$
2. Dischinger	1.00	$\dfrac{2}{\rho^2} - 4\kappa$	$\dfrac{1}{\rho^4} - \dfrac{8\kappa}{\rho^2} + 6\kappa^2$	$-\dfrac{4\kappa}{\rho^4} + \dfrac{6\kappa^2}{\rho^2} - 4\kappa^3$	$\dfrac{4\kappa^2}{\rho^4} + \kappa^4 + 1$
3. Aas-Jakobsen	1.00	$\dfrac{2}{\rho^2} - 4\kappa$	$\dfrac{1}{\rho^4} - \dfrac{8\kappa}{\rho^2} + 6\kappa^2$	$-\dfrac{8\kappa}{\rho^4} + \dfrac{13\kappa^2}{\rho^2} - 4\kappa^3$	$\kappa^4 + 1$
4. Lundgren	1.00	$\dfrac{2}{\rho^2} - 4\kappa$	$\dfrac{1}{\rho^4} - \dfrac{6\kappa}{\rho^2} + 6\kappa^2$	$-\dfrac{2\kappa}{\rho^4} + \dfrac{3\kappa^2}{\rho^2} - 4\kappa^3$	$\kappa^4 + 1$
5. ASCE Manual, 1	1.00	$\dfrac{2}{\rho^2} - 4\kappa$	$\dfrac{1.5}{\rho^4} - \dfrac{6\kappa}{\rho^2} + 6\kappa^2$	$\dfrac{1}{2\rho^6} - \dfrac{3\kappa}{\rho^4} + \dfrac{6\kappa^2}{\rho^2} - 4\kappa^3$	$\dfrac{1}{16\rho^8} - \dfrac{4\kappa}{8\rho^6} + \dfrac{1.5\kappa^2}{\rho^4} - \dfrac{2\kappa^3}{\rho^2} + \kappa^4 + 1$
6. Ivar Holand	1.00	$\dfrac{2}{\rho^2} - 4\kappa$	$\dfrac{1.5}{\rho^4} - \dfrac{6\kappa}{\rho^2} + 6\kappa^2$	$\dfrac{1}{2\rho^6} - \dfrac{3\kappa}{\rho^4} + \dfrac{6\kappa^2}{\rho^2} - 4\kappa^3$	$\dfrac{1}{16\rho^8} - \dfrac{4\kappa}{8\rho^6} + \dfrac{1.5\kappa^2}{\rho^4} - \dfrac{2\kappa^3}{\rho^2} + \kappa^4 + 1$
7. ASCE Manual, 2 D-K-J, Vlasov	1.00	-4κ	$+6\kappa^2$	$-4\kappa^3$	$+1$
8. Finsterwalder	1.00	$\dfrac{2}{\rho^2} - 2\kappa$	$\dfrac{1}{\rho^4} - \dfrac{4\kappa}{\rho^2} + \kappa^2$	$-\dfrac{2\kappa}{\rho^4} + \dfrac{\kappa^2}{\rho^2}$	$+1$
9. Schorer	1.00	0	0	0	$+1$

129

7-48 Methods Given in the ASCE Manual

The Manual develops a rigorous method for very long shells with $a/l < 0.20$. For shells with $a/l > 0.20$ two approximations are given.

Approximation 1

Although the method is described as approximate, a very high degree of accuracy is still maintained. The characteristic equation is simplified as

$$\left[\bar{m}^2 - \left(\kappa - \frac{1}{2\rho^2}\right)\right]^4 + 1 = 0$$

and the roots are found.

Approximation 2

If Poisson's ratio is taken as zero, the characteristic equation that results is not different from that of the D-K-J theory.

7-49 Ivar Holand's Method

Ivar Holand [26] arrives at the same characteristic equation as that given in the ASCE Manual, Approximation 1. But his force-displacement relations differ from those assumed in the ASCE Manual. Unlike in the ASCE method, he does not use the membrane solution as the particular integral.

7-50 ASCE Approximation 2 and the Vlasov and D-K-J Theories

If Poisson's ratio is set to zero, all these theories lead to the D-K-J characteristic equation of Art. 7-26. As explained in the ASCE Manual, these theories are accurate enough for short shells whose $l/a \leqslant 1.60$ or $a/l \geqslant 0.60$. Donnell derives his differential equation in w, while Jenkins chooses N_ϕ as the independent variable. That these theories would give accurate results for short shells can be inferred from Table 7-2. It is seen that the coefficients of the powers of $\bar{m}$ in the D-K-J equation are more or less the same as in the other exact characteristic equations. The only significant difference is that terms with powers of ρ in the denominator have been dropped. But this will not matter, as ρ is large for short shells.

7-51 The Finsterwalder Theory

The Finsterwalder theory is applicable only to long shells. Moe [27] reports that it is unsatisfactory for short shells with $\rho > 5$. McNamee [28] has drawn attention to two inconsistencies in the Finsterwalder formulation. He questions the retention of the term w/a^2 in the expression (7-8) and the term Q_ϕ in the equilibrium equation (7-21*b*). If these inconsistencies are removed, the Finsterwalder theory becomes almost identical with that of Schorer.

7-52 The Schorer Theory

The range of validity of the Schorer approximation is practically the same as that of the Finsterwalder theory. But, unlike the Finsterwalder characteristic equation, the Schorer characteristic equation has explicit roots which do not depend on the shell dimensions. Hence the calculations involved are less laborious and less time-consuming.

7-53 Application of Theories to Design

Summing up the discussion on the bending theories of cylindrical shells, the conclusion may be drawn that it is sufficiently accurate for purposes of design to employ the Schorer theory, if the shell is long, i.e., if $l/a \geqslant \pi$, and the D-K-J theory if the shell is short, i.e., $l/a \leqslant 1.60$. For intermediate shells, use may be made of one of the exact theories.

CHAPTER 8

DESIGN OF CYLINDRICAL SHELLS

8-1 Organization of Computations

It is essential to organize the lengthy calculations involved in the design of a cylindrical shell systematically so that they may be easily checked. The computations are best done in matrix form. To this

Table 8-1. Multipliers M in the Schorer Theory

Quantity	Whether odd or even	Multiplier
$\dfrac{\partial N_{x\phi}}{\partial x}$	Odd	$-\dfrac{D\rho^4}{a^4}\cos\dfrac{\lambda_n x}{a}$
Q_ϕ	Odd	$-\dfrac{D}{a^3}\dfrac{\rho^2}{\sqrt{2}}\cos\dfrac{\lambda_n x}{a}$
N_ϕ	Even	$-\dfrac{D\rho^4}{a^3}\cos\dfrac{\lambda_n x}{a}$
M_ϕ	Even	$-\dfrac{D}{a^2}\dfrac{\rho^2}{\sqrt{2}}\cos\dfrac{\lambda_n x}{a}$
$\dfrac{\partial u}{\partial x}$	Even	$-\dfrac{\lambda_n{}^2}{\sqrt{2}\,\rho^2 a}\cos\dfrac{\lambda_n x}{a}$
w	Even	$\cos\dfrac{\lambda_n x}{a}$
v	Odd	$+\dfrac{1}{\sqrt{2}\,\rho^2}\cos\dfrac{\lambda_n x}{a}$
ϑ	Odd	$+\dfrac{1}{a}\cos\dfrac{\lambda_n x}{a}$

end, it is necessary to rewrite the expressions for the stress resultants and displacements in a different form. Let H represent any shell action, be it a stress resultant or a displacement. We may, then, write

$$
\begin{aligned}
H = M\{ & e^{-\alpha_1\phi}[B_1(A_n \cos \beta_1\phi + B_n \sin \beta_1\phi) \\
& + B_2(B_n \cos \beta_1\phi - A_n \sin \beta_1\phi)] \\
& + e^{-\alpha_2\phi}[B_3(C_n \cos \beta_2\phi + D_n \sin \beta_2\phi) \\
& + B_4(D_n \cos \beta_2\phi - C_n \sin \beta_2\phi)]\}
\end{aligned}
\tag{8-1}
$$

where M is a multiplier.

8-2 Schorer's Theory in Matrix Notation

Rewriting the expressions for the stress resultants and displacements in the form indicated in Art. 8-1, we may arrive at Tables 8-1 and 8-2 containing the multipliers M and the constants B_1, B_2, B_3, and B_4 for all shell actions according to the Schorer theory. Making use of these two tables, we may proceed to write the stress resultants and displacements of the Schorer theory in matrix form as follows:

Table 8-2. Coefficients B_1, B_2, B_3, and B_4 in the Schorer Theory

Quantity	B_1	B_2	B_3	B_4
$\dfrac{\partial N_{x\phi}}{\partial x}$	$+\beta_1$	$+\alpha_1$	$-\alpha_1$	$-\beta_1$
Q_ϕ	$(-\alpha_1 + \beta_1)$	$(\alpha_1 + \beta_1)$	$(\alpha_1 + \beta_1)$	$(\beta_1 - \alpha_1)$
N_ϕ	0	$+1$	0	-1
M_ϕ	$+1$	-1	-1	-1
$\dfrac{\partial u}{\partial x}$	-1	-1	$+1$	-1
w	$+1$	0	$+1$	0
v	$-(\alpha_1 + \beta_1)$	$(\beta_1 - \alpha_1)$	$(\beta_1 - \alpha_1)$	$-(\alpha_1 + \beta_1)$
ϑ	$-\alpha_1$	$+\beta_1$	$-\beta_1$	$+\alpha_1$

$$
\begin{bmatrix} \dfrac{\partial N_{x\phi}}{\partial x} \\[2mm] Q_\phi \\[2mm] N_\phi \\[2mm] M_\phi \end{bmatrix}
= - \dfrac{D\rho^2}{a^2}\cos\dfrac{\lambda_n x}{a}
\begin{bmatrix} \dfrac{\rho^2}{a^2} & \cdot & \cdot & \cdot \\[2mm] \cdot & \dfrac{1}{a\sqrt{2}} & \cdot & \cdot \\[2mm] \cdot & \cdot & \dfrac{\rho^2}{a} & \cdot \\[2mm] \cdot & \cdot & \cdot & \dfrac{1}{\sqrt{2}} \end{bmatrix}
$$

$$
\times
\begin{bmatrix} \beta_1 & \alpha_1 & -\alpha_1 & -\beta_1 \\ (\beta_1 - \alpha_1) & (\alpha_1 + \beta_1) & (\alpha_1 + \beta_1) & (\beta_1 - \alpha_1) \\ 0 & +1 & 0 & -1 \\ +1 & -1 & -1 & -1 \end{bmatrix}
\begin{bmatrix} f_1 & f_2 & \cdot & \cdot \\ -f_2 & f_1 & \cdot & \cdot \\ \cdot & \cdot & f_3 & f_4 \\ \cdot & \cdot & -f_4 & f_3 \end{bmatrix}
\begin{bmatrix} A_n \\ B_n \\ C_n \\ D_n \end{bmatrix}
$$

$$(8\text{-}2)$$

$$
\begin{bmatrix} \dfrac{\partial u}{\partial x} \\[2mm] w \\[2mm] v \\[2mm] \vartheta \end{bmatrix}
= \cos\dfrac{\lambda_n x}{a}
\begin{bmatrix} -\dfrac{\lambda_n{}^2}{\sqrt{2}\,\rho^2 a} & \cdot & \cdot & \cdot \\[2mm] \cdot & 1 & \cdot & \cdot \\[2mm] \cdot & \cdot & \dfrac{1}{\sqrt{2}\,\rho^2} & \cdot \\[2mm] \cdot & \cdot & \cdot & \dfrac{1}{a} \end{bmatrix}
$$

$$
\times
\begin{bmatrix} -1 & -1 & +1 & -1 \\ +1 & 0 & +1 & 0 \\ -(\alpha_1 + \beta_1) & (\beta_1 - \alpha_1) & (\beta_1 - \alpha_1) & -(\alpha_1 + \beta_1) \\ -\alpha_1 & +\beta_1 & -\beta_1 & +\alpha_1 \end{bmatrix}
\begin{bmatrix} f_1 & f_2 & \cdot & \cdot \\ -f_2 & f_1 & \cdot & \cdot \\ \cdot & \cdot & f_3 & f_4 \\ \cdot & \cdot & -f_4 & f_3 \end{bmatrix}
\begin{bmatrix} A_n \\ B_n \\ C_n \\ D_n \end{bmatrix}
$$

$$(8\text{-}3)$$

The functions f_1, f_2, f_3, and f_4 are defined as indicated below

$$f_1 = e^{-\alpha_1 \phi}\cos \beta_1 \phi$$

$$f_2 = e^{-\alpha_1 \phi}\sin \beta_1 \phi$$

$$f_3 = e^{-\alpha_2 \phi}\cos \beta_2 \phi$$

$$f_4 = e^{-\alpha_2 \phi}\sin \beta_2 \phi$$

The matrices on the right-hand side of Equations (8-2) and (8-3) are each composed of the product of four matrices. The product may be written in the following form:

$$\{H\} = [M][B][F]\{A\} \qquad (8\text{-}4)$$

The matrix $[M]$ comprises the multipliers given in Table 8-1; the matrix $[B]$ consists of the array of coefficients B_1, B_2, B_3, and B_4 given in Table 8-2. The matrix $[F]$ is formed of the functions f_1, f_2, f_3, and f_4. The column matrix $\{A\}$ is made up of the arbitrary constants A_n, B_n, C_n, and D_n. The stress resultants and displacements given by (8-2) and (8-3) are those at a point at an angle ϕ from the left edge due to the disturbances emanating from that edge. The shell actions at the point due to the disturbances emanating from the right edge can be written down in matrix form as

$$\{H^*\} = [M]\,[B]\,[F^*]\,\{A\} \tag{8-5}$$

All the matrices on the right-hand side, except $[F^*]$, are the same as in (8-4)

$$[F^*] = \begin{bmatrix} f_1^* & f_2^* & \cdot & \cdot \\ -f_2^* & f_1^* & \cdot & \cdot \\ \cdot & \cdot & f_3^* & f_4^* \\ \cdot & \cdot & -f_4^* & f_3^* \end{bmatrix}$$

The functions f_1^*, f_2^*, f_3^*, and f_4^* are defined as

$$f_1^* = e^{-\alpha_1(2\phi_c-\phi)} \cos \beta_1(2\phi_c - \phi)$$

$$f_2^* = e^{-\alpha_1(2\phi_c-\phi)} \sin \beta_1(2\phi_c - \phi)$$

$$f_3^* = e^{-\alpha_2(2\phi_c-\phi)} \cos \beta_2(2\phi_c - \phi)$$

and

$$f_4^* = e^{-\alpha_2(2\phi_c-\phi)} \sin \beta_2(2\phi_c - \phi)$$

where ϕ_c is the semicentral angle.

The resultant shell action at any point is found by adding the contributions due to the left and right edges if the shell action is even; and by subtracting the contribution of the right edge from that of the left edge if the shell action is odd. Those shell actions, the expressions for which contain even-order derivatives of ϕ, are even and those shell actions the expressions for which involve the odd-order derivatives of ϕ are odd, as already explained in Art. 7-18. Whether a shell action is odd or even is indicated in Table 8-1.

8-3 The D-K-J Theory in Matrix Form

The shell actions of the D-K-J theory may also be presented in matrix form. The steps involved, being the same as those outlined in Art. 8-2, are not repeated. The multipliers given in Table 8-3 and the coefficients given in Table 8-4 are used to build up the matrix as before. Matrix equations similar to Equations (8-2) and (8-3) for the stress resultants and displacements at an angle ϕ from the left edge may be built up as before. Only those shell actions which would figure in the formulation of boundary conditions along the straight edges need be written down. $\bar{Q}_\phi$ included in Tables 8-3 and 8-4 is the effective shear at the edge. It is obtained by combining Q_ϕ and the shear contributed by the twisting moment. Hence it is easily verified that $\bar{Q}_\phi = Q_\phi + \partial M_{x\phi}/\partial_x$.

Table 8-3. Multipliers M in the D-K-J Theory

Quantity	Whether odd or even	Multiplier
$\dfrac{\partial N_{x\phi}}{\partial x}$	Odd	$-\dfrac{D\rho^4}{a^4}\cos\dfrac{\lambda_n x}{a}$
$\bar{Q}_\phi$	Odd	$\dfrac{D\rho^2}{a^3\sqrt{2}}\cos\dfrac{\lambda_n x}{a}$
N_ϕ	Even	$\dfrac{D\rho^4}{a^3}\cos\dfrac{\lambda_n x}{a}$
M_ϕ	Even	$-\dfrac{D\rho^2}{a^2\sqrt{2}}\cos\dfrac{\lambda_n x}{a}$
$\dfrac{\partial u}{\partial x}$	Even	$-\dfrac{\lambda_n{}^2}{a\sqrt{2}\,\rho^2}\cos\dfrac{\lambda_n x}{a}$
w	Even	$\cos\dfrac{\lambda_n x}{a}$
v	Odd	$\dfrac{1}{\rho^2\sqrt{2}}\cos\dfrac{\lambda_n x}{a}$
ϑ	Odd	$\dfrac{1}{a}\cos\dfrac{\lambda_n x}{a}$

Table 8-4. Coefficients B_1, B_2, B_3, and B_4 in the D-K-J Theory

Quantity	B_1	B_2	B_3	B_4
$\dfrac{\partial N_{x\phi}}{\partial x}$	β_1	α_1	$-\beta_2$	$-\alpha_2$
Q_ϕ	$-[\beta_1 + \alpha_1(\sqrt{2}\,\kappa - 1)]$	$[\beta_1(\sqrt{2}\,\kappa - 1) - \alpha_1]$	$-[(\sqrt{2}\,\kappa + 1)\alpha_2 + \beta_2]$	$[\beta_2(\sqrt{2}\,\kappa + 1) - \alpha_2]$
N_ϕ	0	-1	0	$+1$
M_ϕ	$(1 + \kappa\sqrt{2})$	-1	$-(1 - \kappa\sqrt{2})$	-1
$\dfrac{\partial u}{\partial x}$	-1	$-(1 + \kappa\sqrt{2})$	$+1$	$-(1 - \kappa\sqrt{2})$
w	$+1$	0	$+1$	0
v	$-[\alpha_1 + \beta_1(1 - \kappa\sqrt{2})]$	$[\beta_1 - \alpha_1(1 - \kappa\sqrt{2})]$	$[\alpha_2 - \beta_2(1 + \kappa\sqrt{2})]$	$-[\alpha_2(1 + \kappa\sqrt{2}) + \beta_2]$
$\vartheta^\dagger$	$-\alpha_1$	$+\beta_1$	$-\alpha_2$	$+\beta_2$

† Here, the approximate relation $\vartheta = (1/a)(\partial w/\partial\phi)$ has been used.

BOUNDARY CONDITIONS

8-4 Symmetric and Asymmetric Problems

Some of the common boundary conditions met with in practice and the approximations used in formulating them will now be discussed. We may, for convenience, distinguish between symmetric and asymmetric problems. A shell with a directrix symmetrical about its vertical axis and loaded symmetrically will lead to a symmetric problem if it has identical support conditions at both the edges.

8-5 Single Shell without Edge Beams

It is sometimes possible to dispense with edge beams, especially if the shell is short. Let us consider a single shell, symmetrically loaded and without edge beams. The boundary conditions for such a shell are obviously the following:

At $\phi = 0$,

$$
\begin{array}{lll}
\text{(i)} & M_\phi = 0 & \\[6pt]
\text{(ii)} & Q_\phi = 0 & \\[6pt]
\text{(iii)} & N_\phi = 0 & \\[6pt]
\text{and} \quad \text{(iv)} & N_{x\phi} = 0 & \text{or} \quad \dfrac{\partial N_{x\phi}}{\partial x} = 0
\end{array} \qquad (8\text{-}6)
$$

The same boundary conditions exist at the right edge, where $\phi = 2\phi_c$. On account of symmetry, the eight boundary conditions reduce themselves to four. It is therefore enough if the boundary conditions are formulated at the left edge only and the four resulting equations are solved for the four unknown arbitrary constants A_n, B_n, C_n, and D_n. It is to be noted that, in formulating the four boundary conditions, the quantities M_ϕ, Q_ϕ, N_ϕ, and $N_{x\phi}$ to be used are the sum of the membrane and bending effects. Thus, for example, when we say $M_\phi = 0$ at $\phi = 0$, we mean that $[(M_\phi)_m + (M_\phi)_b]$ evaluated at $\phi = 0$ is set to zero. The suffixes m and b, respectively, stand for the membrane and bending effects. Instead of the membrane solution, we may use the particular integral, as already explained. We may attach the suffix p to the contribution of the particular integral. Equations (8-6) may also be written as

$$\text{(i)} \qquad (M_\phi)_b = -(M_\phi)_m = 0$$

noting that the membrane state does not cause any M_ϕ.

Alternatively,

$$(M_\phi)_b = -(M_\phi)_p$$

(ii) $\quad (Q_\phi)_b = -(Q_\phi)_m = 0$

Alternatively,

$$(Q_\phi)_b = -(Q_\phi)_p \qquad (8\text{-}7)$$

(iii) $\quad (N_\phi)_b = -(N_\phi)_m$

(iv) $\quad \left(\dfrac{\partial N_{x\phi}}{\partial x}\right)_b = -\left(\dfrac{\partial N_{x\phi}}{\partial x}\right)_m$

8-6 Single Shell with Edge Beams

Next, consider a symmetrically loaded single shell provided with identical edge beams along both its edges. As before, it is enough if four boundary conditions are formulated at the left edge. The depth of edge beams that are usually provided is very large compared with their width. Such deep and narrow edge beams possess negligible stiffness to resist torsion, or bending in the horizontal plane. They are adequately rigid to bend only in the vertical plane. If this is the case, the following two boundary conditions involve only very little error:

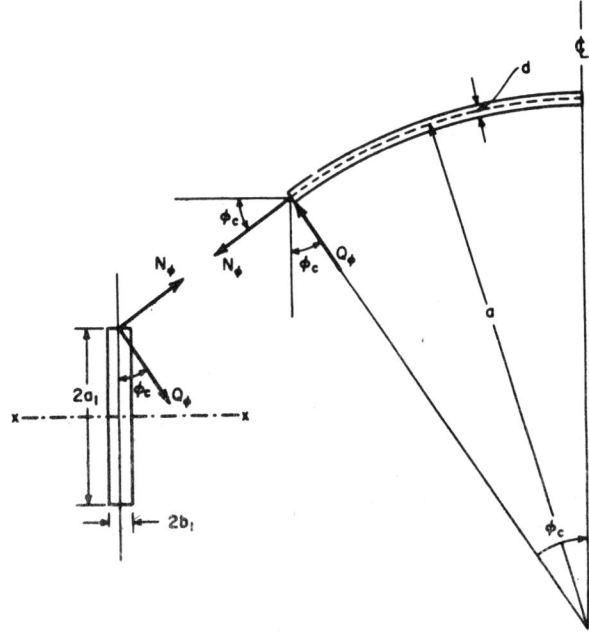

Fig. 8-1a

At $\phi = 0$,

(i) $M_\phi = 0$ (8-8a)

(ii) $H = N_\phi \cos \phi_c + Q_\phi \sin \phi_c = 0$ (8-8b)

The first boundary condition means that the edge beam has no torsional resistance and the second that it cannot receive any horizontal force (Fig. 8-1a). H is the resolved component, in the horizontal direction,

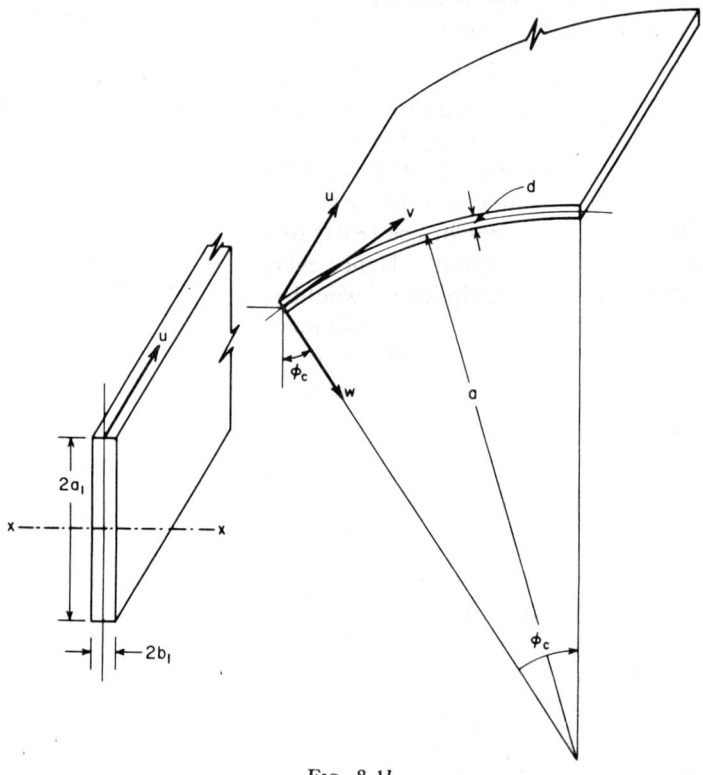

FIG. 8-1b

of the forces transferred to the edge beam. The other two boundary conditions follow from considerations of continuity between the edge of the shell and the edge member, at their junction. They may be stated as follows:

(iii) The longitudinal displacement u at the edge of the shell = the longitudinal displacement of the edge beam at its junction with the shell. This relation must hold good all along the edge.

(iv) The vertical deflection of the shell edge = the vertical deflection of the edge beam at its junction with the shell.   This relation must also be satisfied all along the edge.   To formulate these boundary conditions explicitly, it is necessary to derive expressions for the stress and displacement of the edge beam at its junction with the shell.

Let $2a_1$ and $2b_1$ be the dimensions of the edge beam (Fig. 8-1b).   Let its weight per unit length be W.   Expanding the load in a Fourier series and retaining only the first term, we arrive at $W' = (4/\pi)W$.   The forces acting on the edge beam are

(a) A vertical force of $(N_\phi)_{m+b} \sin \phi_c - (Q_\phi)_{m+b} \cos \phi_c$.

(b) Dead weight of $W' \cos (\pi x/l)$.

(c) A shear $(N_{x\phi})_{m+b}$ acting on the top of the edge beam.   This force has an eccentricity of a_1 about the axis x-x of the beam.   It may therefore be replaced by a central force and a bending moment about x-x.   It is to be remembered that for the quantities N_ϕ, Q_ϕ, and $N_{x\phi}$ we are using a Fourier series in $\cos (\pi x/l)$ or $\sin (\pi x/l)$ of which only the first term is being considered.   Hence every time that these expressions were differentiated a multiplier π/l would result.   Similarly, integration would introduce a multiplier l/π.   We may write down the central force and bending moment to which the effects of $N_{x\phi}$ listed under (c) reduce themselves before we proceed to formulate the boundary conditions.

$$\text{Central force} = -\int_{l/2}^{0} [(N_{x\phi})_{m+b}]_{\phi=0} \sin \frac{\pi x}{l} \, dx = \frac{l}{\pi} [(N_{x\phi})_{m+b}]_{\phi=0}$$

$$\text{Bending moment (hogging) about } x\text{-}x = \frac{l}{\pi} a_1 [(N_{x\phi})_{m+b}]_{\phi=0}$$

$$\begin{aligned}
\text{Stress in the edge beam at} \atop \text{its junction with the shell} \Big\} &= \frac{l}{\pi} \frac{1}{A} (N_{x\phi})_{m+b} + \left(\frac{l}{\pi}\right) \frac{a_1^2}{I} (N_{x\phi})_{m+b} \\
&+ [(N_\phi)_{m+b} \sin \phi_c - (Q_\phi)_{m+b} \cos \phi_c] \left(\frac{l}{\pi}\right)^2 \frac{a_1}{I} \\
&- W' \left(\frac{l}{\pi}\right)^2 \frac{a_1}{I}
\end{aligned}$$

where A is the area of the cross section and I the moment of inertia about the axis x-x of the edge beam.   In this expression, tensile stresses are regarded as positive as usual.   The displacement u of the edge beam, at the junction, may be obtained by integrating the above expression once and dividing the result by E.   Hence the displacement of the edge beam at the junction may be written as

$$\begin{aligned}
u &= \left(\frac{l}{\pi}\right)^2 \frac{1}{AE} (N_{x\phi})_{m+b} + \left(\frac{l}{\pi}\right)^2 \frac{a_1^2}{IE} (N_{x\phi})_{m+b} \\
&+ [(N_{x\phi})_{m+b} \sin \phi_c - (Q_\phi)_{m+b} \cos \phi_c] \left(\frac{l}{\pi}\right)^3 \frac{a_1}{IE} - W' \left(\frac{l}{\pi}\right)^3 \frac{a_1}{IE}
\end{aligned}$$

The vertical deflection of the edge beam may easily be verified to be

$$[(N_\phi)_{m+b} \sin \phi_c - (Q_\phi)_{m+b} \cos \phi_c] \left(\frac{l}{\pi}\right)^4 \frac{1}{EI} + N_{x\phi} \frac{l}{\pi} a_1 \left(\frac{l}{\pi}\right)^2 \frac{1}{EI} - W' \left(\frac{l}{\pi}\right)^4 \frac{1}{EI}$$

In this expression, deflections upward are regarded as positive. The first term represents the vertical deflection caused by the vertical components of N_ϕ and Q_ϕ transferred to the edge beam, the second term the upward deflection due to the hogging moment caused by $N_{x\phi}$, and the third term the downward deflection caused by the dead weight. We are now ready to form the third and fourth boundary conditions, which may be expressed as follows:

At $\phi = 0$,

$(u)_{m+b}$ of the shell edge

$$= \left(\frac{l}{\pi}\right)^2 \frac{1}{AE} (N_{x\phi})_{m+b} + \left(\frac{l}{\pi}\right)^2 \frac{a_1^2}{I} (N_{x\phi})_{m+b}$$

$$+ [(N_\phi)_{m+b} \sin \phi_c - (Q_\phi)_{m+b} \cos \phi_c] \left(\frac{l}{\pi}\right)^3 \frac{a_1}{I} - W' \left(\frac{l}{\pi}\right)^3 \frac{a_1}{I} \qquad (8\text{-}8c)$$

$(v)_{m+b} \sin \phi_c - (w)_{m+b} \cos \phi_c$ of the shell edge

$$= [(N_\phi)_{m+b} \sin \phi_c - (Q_\phi)_{m+b} \cos \phi_c] \left(\frac{l}{\pi}\right)^4 \frac{1}{EI}$$

$$+ N_{x\phi} \left(\frac{l}{\pi}\right)^3 \frac{a_1}{EI} - W' \left(\frac{l}{\pi}\right)^4 \frac{1}{EI} \qquad (8\text{-}8d)$$

The four boundary conditions (8-8) are enough to solve for the four unknown constants A_n, B_n, C_n, and D_n. Fischer [29] has shown that the approximate formulation discussed here is sufficiently accurate in most cases. If a more rigorous solution is desired, we may formulate the boundary conditions as follows:

(i) The longitudinal displacement of the shell edge and the edge member shall be the same at their junction.

(ii) The vertical deflection of the shell edge and the edge beam shall be the same.

(iii) The horizontal deflection of the shell edge and the edge beam shall be the same at their junction.

(iv) The rotation of the shell edge shall be equal to the rotation of the edge beam.

These conditions of continuity provide the four equations necessary to solve for the four arbitrary constants.

8-7 Inner Shell of a Multiple Group of Shells with Featheredge Beams

A design approximation usually made in the treatment of an interior shell of a multiple group is to treat it as a symmetrical problem. The inner shell may be provided with edge beams. Sometimes no edge

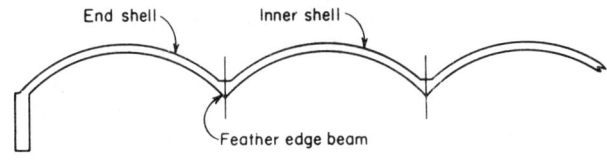

End shell   Inner shell

Feather edge beam

FIG. 8-2

beam is provided and the edges of the adjacent shells form a sharp edge which is sometimes referred to as a featheredge (Fig. 8-2). The boundary conditions to be used are

(i) $\delta_h = 0$; i.e., the horizontal displacement of the edge is zero.

(ii) $V = 0$; i.e., the vertical component of the forces transferred by the shell to the edge beam is zero. This has to be so, because there is no member at the junction to carry the load.

(iii) $\partial N_{x\phi}/\partial x = 0$. This is because there is no member at the junction to carry the tension induced by the shear.

(iv) $\vartheta = 0$, since no rotation can take place at the edge because of the assumed symmetry.

8-8 Inner Shell with Small Edge Members

Instead of providing a featheredge, the junction is sometimes thickened to provide a small edge member (Fig. 8-3) which, because of its negli-

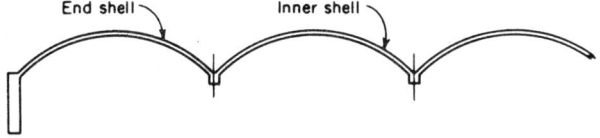

End shell   Inner shell

FIG. 8-3

gible dimensions, functions as a tie but is incapable of offering any resistance to bending. The boundary conditions discussed in the previous article are now modified as

(i) $\delta_h = 0$.

(ii) $V = 0$. In formulating this condition, the dead weight of the edge member is to be considered.

(iii) Stress in the edge member at the junction = stress at the shell edge.

(iv) $\vartheta = 0$.

8-9   Inner Shell with Edge Beams

Sometimes a deep and narrow edge beam is provided at the valley of the inner shell (Fig. 8-4). Three of the four boundary conditions listed in Art. 8-8 remain the same.

(i) $\delta_h = 0$.

(ii) $\vartheta = 0$.

(iii) Stress in the edge beam at the junction = stress in the shell at its edge.

The fourth boundary condition becomes

(iv) Vertical deflection of the shell edge = vertical deflection of the edge beam.

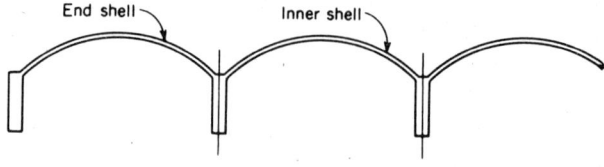

End shell   Inner shell

FIG. 8-4

In these calculations, the width of the edge beam to be considered is one half its actual width, because only one half of the edge beam provided in the valley belongs to one shell.

8-10   Single Shell with Nominal Edge Members Supported on Walls or on Closely Spaced Rows of Flexible Columns

It is assumed that, if the shell is supported on a wall, there is a layer of kraft paper provided to permit free movement of the shell over the wall. Again, if the shell is supported on a closely spaced row of columns, the connection provided between the edge beam and the columns is weak enough to permit horizontal movement of the shell. Under these conditions, the following relations hold:

(i) $H = 0$.

(ii) $M_\phi = 0$.

(iii) $\delta v = 0$; i.e., the deflection of the shell edge in the vertical direction is zero.

(iv) Longitudinal stress in the shell at its edge = longitudinal stress in the edge member at its junction.

8-11 The End Shell

The end shell of a group of multiple shells represents an asymmetric problem in that its support conditions are different at its two edges. But a simplified treatment, often found satisfactory in practice, is to regard the part of the shell extending from the crown to its junction with the inner shell as one half of a symmetrical shell, with the boundary conditions appropriate to an inner shell with edge beams, discussed in Art. 8-9. The other part of the shell extending from the crown to the outer edge is regarded as one half of a single shell with edge beams and treated in the manner described in Art. 8-6. This approximate method is described in the ASCE Manual. A rigorous solution will involve eight arbitrary constants and the formulation of eight boundary conditions, four at each edge.

SELECTION OF SHELL PARAMETERS

8-12 Proper Selection of Parameters

Analysis of shells being a lengthy process, it is essential to start with properly chosen dimensions to avoid a recalculation. Some guiding rules which the designer may find useful are discussed in the paragraphs that follow.

8-13 Selection of Type

The first decision involved in planning shell roofs for covering a given large area is the choice between long and short shells. In general, either multiple long shells or continuous short shells would be the answer (Fig. 8-5). Functional and operational considerations also influence the choice. For covering very large areas, for hangars, warehouses, etc., short shells with closely spaced arch traverses transferring their reactions to the ground are usually economical (Fig. 8-6). Chord widths as large as 400 ft are possible with shells of this type. The span of short shells may be chosen as between one-sixth and one-third of the chord width [30]. A practical limit on the span of long reinforced-concrete shells is about 100 ft. For longer shells, prestressing will prove economical. It is good practice to provide expansion joints at every 150 ft along the length of the shell. Hence the limiting span of long shells may be taken as 150 ft. The need for expansion joints in the transverse direction is not so great as the shell is free to "breathe" and flatten out when temperature changes take place.

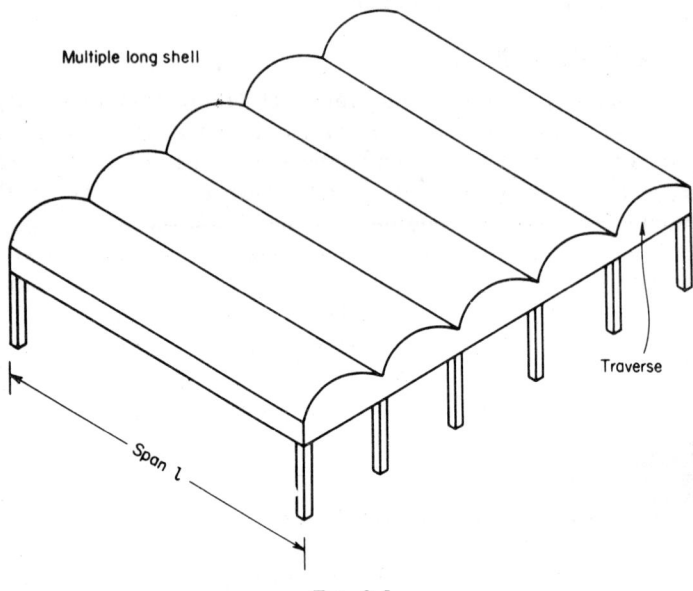

Multiple long shell

Traverse

Span l

FIG. 8-5a

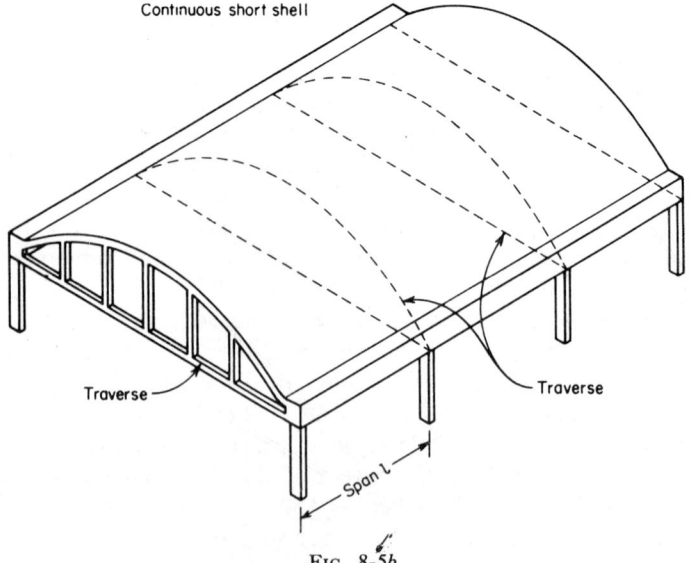

Continuous short shell

Traverse

Traverse

Span l

FIG. 8-5b

PHOTO 8-1. Short shells of 90-ft chord width and continuous over two spans of 35-ft roofing warehouses at New Delhi, India.

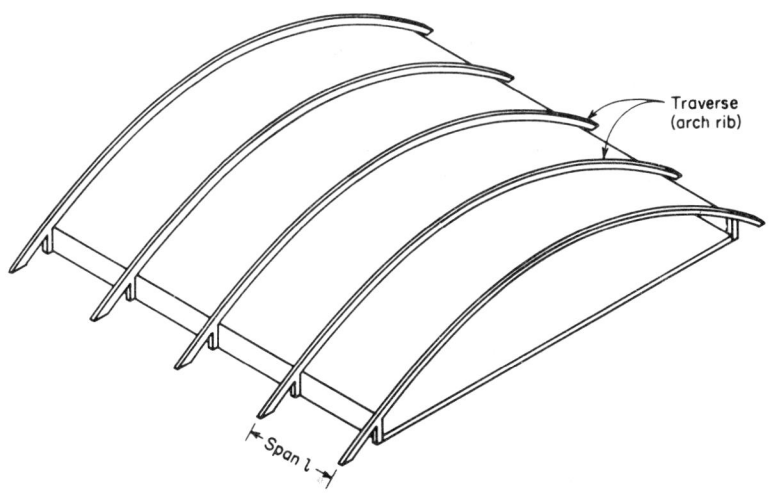

Traverse
(arch rib)

Span l

FIG. 8-6

8-14 Radius

The radius of a cylindrical shell has to be chosen keeping acoustic considerations in view. It is desirable to see that the center of curvature does not lie at the working level.

8-15 Semicentral Angle

The practice is to keep the semicentral angle between 30 and 45°. If the angle exceeds 45°, concreting becomes difficult without the use of top forms. If the angle is below 40°, wind load can be ignored, because it causes only a suction on the shell.[1] Between the limits of 30 and 40°, it is desirable to choose as large a central angle as possible with the object of getting a high structural depth for the shell.

8-16 Thickness

The minimum thickness of reinforced-concrete cylindrical shells is governed by practical considerations such as accommodating reinforcement and providing adequate cover. The thickness of very large shells would depend on considerations of buckling. The type and number of layers of reinforcement provided would also have a bearing on the minimum thickness. It is possible to do with a smaller thickness for precast shells. According to the Dutch Report [31], the usual recommended thickness is between 7 and 8 cm. The minimum thickness specified in the Indian Code is 5 cm. A minimum of 4 cm is recommended by the Institute for Typification of the German Democratic Republic at Berlin [32].

8-17 Structural Depth

In short shells, adequate structural depth would usually be available in the shell itself without the provision of edge beams. The rise of a short shell shall not be less than one-tenth of the chord width. Nominal edge members are sometimes provided to stiffen the edge of the shell. In long shells, it is generally not possible to obtain adequate structural depth without the provision of edge beams. For single shells, the total depth between the crown of the shell and the bottom of the edge beam shall be between one-sixth and one-twelfth of the span. The larger figure is applicable to smaller spans [30].

8-18 Width of Edge Beam

A width of two to three times the thickness of the shell would usually suffice. A minimum of 6 in. is demanded by practical considerations.

[1] Lundgren, H., "Cylindrical Shells," vol I, "Cylindrical Roofs", p. 46, The Danish Technical Press, The Institution of Danish Civil Engineers, Copenhagen, 1960.

DESIGN OF REINFORCEMENT

8-19 Types of Reinforcing Steel

Billet steel intermediate-grade reinforcing bars may be used as for slabs. It is preferable to use deformed bars starting from No. 3. Plain bars may be used in those countries where deformed bars are not manufactured. Welded-wire fabric is also suitable for use in the body of the shell. In the United States, the recommendations given in the CRSI Design Handbook may be followed.

8-20 Size and Spacing of Reinforcement

Plain No. 2 bar is the smallest bar that is usually employed. In the body of the shell, it is good practice to restrict the maximum size of bars as follows:

Max. size	*Shell thickness, in.*
No. 3	$1\frac{1}{2}$–2
No. 4	2–$2\frac{1}{2}$
No. 5	$2\frac{1}{2}$ and above

Bars of larger size may be used in the edge members and the thickened parts of the shell. Irrespective of whether bars or welded fabrics are used, the spacing of bars or wires shall preferably be limited to 40 diameters or five times the thickness of the shell, whichever is less. A further desirable stipulation is that the area of the unreinforced-concrete panel enclosed by the grids shall in no case exceed 15 times the square of the thickness of the shell. Concrete protective cover for reinforcement may be the same as for slab structures.

8-21 Scheme of Reinforcement

The stress resultants may be conveniently grouped as follows for purposes of designing reinforcement:

(i) N_x for the design of longitudinal reinforcement

(ii) N_ϕ and M_ϕ for the design of transverse reinforcement

(iii) N_x, N_ϕ, and $N_{x\phi}$ for the design of reinforcement for principal tension

In a long shell with edge beams, the neutral axis usually lies close to the junction between the shell and the edge member so that the bulk of the longitudinal tensile force caused by N_x is concentrated in the latter. Because the stress in the steel is proportional to its distance from the neutral axis, it is economical to arrange all this steel at the bottom of the edge beam. However, nominal reinforcement shall be provided elsewhere in the edge member. It is often necessary to arrange the steel in the edge member in several rows. If so, the permissible stress is assumed at the centroid of the steel, provided the stress in the lowest layer does not exceed the permissible stress by more than 15%. In a short shell, the distribution of longitudinal stress is far from linear and there may be more than one tensile zone. Hence the longitudinal steel area in the various tensile zones is arrived at by dividing the total tensions by the permissible stress in steel. Compressive forces caused by N_x do not call for anything but nominal reinforcement.

The design for M_ϕ and N_ϕ is carried out in the same way as for a reinforced-concrete section subjected to axial load and bending moment. To facilitate tying together of the longitudinal and transverse reinforcement and to prevent their dislocation during concreting, it is the usual practice to lay the transverse steel touching the longitudinal steel which is arranged along the middle surface of the shell. The transverse steel will be above or below the longitudinal steel depending upon whether the bending moment at the section is hogging or sagging.

From the stress resultants N_x, N_ϕ, and $N_{x\phi}$, the magnitudes and directions of the principal stresses at all points of the shell can be found and two orthogonal families of stress trajectories or isostatics tangential to the principal directions can be sketched in. If the reinforcement is arranged to follow the tension trajectories, the steel consumption will be a minimum. If this scheme of trajectory reinforcement is followed, only nominal steel need be provided along the compression trajectories. Although the trajectory scheme is economical, it is very impractical because of the difficulties involved in fabrication and laying of the reinforcement. In practice, bars inclined at 45° to the axis of the shell are provided near the supports in addition to the longitudinal and transverse reinforcements which form a rectangular grid (Figs. 8-11 and 8-16). Sometimes, a diagonal grid, inclined at 45° to the axis of the shell, is used instead of a rectangular grid and the inclined bars.

8-22 Consumption of Steel

It is difficult to give a single formula for the consumption of steel for the entire range of cylindrical shells used in practice. The following formula proposed by Zalewski and Krzeminski [33] is found to be quite satisfactory for estimating the steel per square meter of plan area:

$$\text{Quantity of steel} = \left[\frac{l(l + B)}{20f} + 6\right] \quad \text{kg/m}^2$$

In the fps system, the formula may be recast as

$$\left[\frac{l(l + B)}{320f} + 1.23\right] \quad \text{psf}$$

where l is the span of the shell, B is the chord width, and f is the rise measured from the springing to the crown.

This formula is applicable if the shell is designed for snow load. In tropical countries where shells are designed only for a nominal live load of about 15 psf and for no snow load, this formula gives too high a value for the shell alone but is a very good approximation to the quantity of total steel in the shell, columns, and foundation footings.

8-23 Step-by-Step Analysis and Design of Cylindrical Shells

A step-by-step procedure for the analysis of cylindrical shells is illustrated by means of two worked examples that follow. The first example relates to a typical single short shell without edge beams and the second to a single long shell with deep and narrow edge beams.

Design Example 8-1

DESIGN OF A SINGLE SHORT CYLINDRICAL
SHELL WITHOUT EDGE BEAMS

1. *DATA*

   Geometry

   Span $l = 35$ ft
   Radius $a = 73.5$ ft
   Thickness $d = 0.25$ ft
   Semicentral angle $\phi_c = 37°48'$

   Loads

   Dead weight $= 37.5$ psf of shell surface
   Live load $\quad = 15.0$ psf of shell surface
   Total load $g = \overline{52.5}$ psf of shell surface

2. *PARAMETERS*

$$\rho = \left(\frac{12\pi^4 a^6}{l^4 d^2}\right)^{1/8} = 14.51$$

$$\kappa = \frac{\pi^2 a^2}{l^2 \rho^2} = 0.20673$$

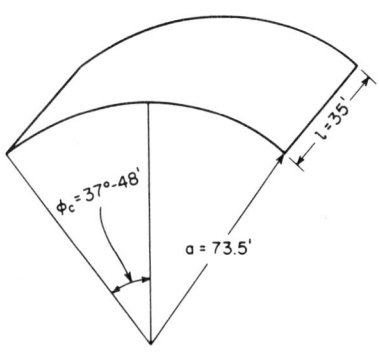

Fig. 8-7

The shell is short. Hence the D-K-J method will be used for the analysis.

$$\phi_c = 0.6597344 \text{ radian}$$
$$2\phi_c = 1.3194688 \text{ radians}$$
$$\sin \phi_c = 0.61291$$
$$\cos \phi_c = 0.79016$$

$$\lambda_n = \frac{\pi a}{l} = 6.5973 \text{ (considering only the first Fourier term)}$$

3. ROOTS OF THE CHARACTERISTIC EQUATION (7-37) AND RELATED QUANTITIES

$$\alpha_1 = \frac{\rho}{8^{1/4}} \{\sqrt{(1 + \kappa \sqrt{2})^2 + 1} + (1 + \kappa \sqrt{2})\}^{1/2} = 14.7570$$

$$\alpha_2 = \frac{\rho}{8^{1/4}} \{\sqrt{(1 - \kappa \sqrt{2})^2 + 1} - (1 - \kappa \sqrt{2})\}^{1/2} = 6.2064$$

$$\beta_1 = \frac{\rho}{8^{1/4}} \{\sqrt{(1 + \kappa \sqrt{2})^2 + 1} - (1 + \kappa \sqrt{2})\}^{1/2} = 5.0854$$

$$\beta_2 = \frac{\rho}{8^{1/4}} \{\sqrt{(1 - \kappa \sqrt{2})^2 + 1} + (1 - \kappa \sqrt{2})\}^{1/2} = 11.9950$$

4. MULTIPLIERS IN THE MATRIX

Quantity	Multiplier	
$N_{x\phi}$	$\dfrac{-D\rho^2}{a^2}\left(\dfrac{\rho^2}{\lambda a}\right)\sin\dfrac{\lambda_n x}{a}$	$-18{,}269\,(0.4342)\sin\dfrac{\lambda_n x}{a}$
Q_ϕ	$\dfrac{-D\rho^2}{a^2}\left(-\dfrac{1}{a\sqrt{2}}\right)\cos\dfrac{\lambda_n x}{a}$	$-18{,}269\,(-0.0096204)\cos\dfrac{\lambda_n x}{a}$
N_ϕ	$\dfrac{-D\rho^2}{a^2}\left(-\dfrac{\rho^2}{a}\right)\cos\dfrac{\lambda_n x}{a}$	$-18{,}269\,(-2.8646)\cos\dfrac{\lambda_n x}{a}$
M_ϕ	$\dfrac{-D\rho^2}{a^2}\left(\dfrac{1}{\sqrt{2}}\right)\cos\dfrac{\lambda_n x}{a}$	$-18{,}269\,(0.70712)\cos\dfrac{\lambda_n x}{a}$

5. TABLE OF B_1, B_2, B_3, AND B_4

Quantity	B_1	B_2	B_3	B_4
$N_{x\phi}$	+5.0854	+14.757	−11.995	−6.2064
Q_ϕ	+5.3576	−18.3558	−20.0158	+9.2956
N_ϕ	0	−1	0	+1
M_ϕ	+1.29235	−1	−0.70765	−1

6. MEMBRANE FORCES AND DISPLACEMENTS (7-73)

(i) N_ϕ

$$N_\phi = -\frac{4}{\pi}ga\cos(\phi_c - \phi)\cos\frac{\pi x}{l} \qquad (7\text{-}73a)$$

$$= -4{,}913.1\cos(\phi_c - \phi)\cos\frac{\pi x}{l}$$

$$N_{\phi(\phi=0)} = -3{,}882.1\cos\frac{\pi x}{l}$$

ϕ	0°	2°21'45"	4°43'30"	7°5'15"	9°27'	12°36'	15°45'	18°54'	23°37'30"	28°21'	37°48'
$e^{-\alpha_1\phi}$	1	0.54418	0.29613	0.16115	0.087691	0.038960	0.017309	0.0076899	0.0022772	0.0006743	0.00005913
$e^{-\alpha_2\phi}$	1	0.77421	0.59941	0.46407	0.35928	0.25542	0.18158	0.12908	0.077373	0.046378	0.016663
$\sin \beta_1\phi$	0	0.20815	0.40719	0.58839	0.74381	0.89938	0.98510	0.99431	0.86478	0.58536	−0.21180
$\sin \beta_2\phi$	0	0.47467	0.83558	0.99621	0.91808	0.48271	−0.15508	−0.72784	−0.23144	−0.34106	0.99823
$\cos \beta_1\phi$	1	0.97809	0.91335	0.80858	0.66839	0.43718	0.17202	−0.10651	−0.50215	−0.81076	−0.97730
$\cos \beta_2\phi$	1	0.88016	0.54937	0.086907	−0.39639	−0.87578	−0.98790	−0.68574	0.97285	0.94004	−0.05951
$e^{-\alpha_1\phi} \sin \beta_1\phi$	0	0.11327	0.12058	0.094817	0.065226	0.035040	0.017051	0.0076461	0.0019692	0.00039473	−0.00001252
$e^{-\alpha_1\phi} \cos \beta_1\phi$	1	0.53226	0.27047	0.13030	0.058612	0.017032	0.0029774	−.00081901	−0.0011435	−0.00054673	−0.00005779
$e^{-\alpha_2\phi} \sin \beta_2\phi$	0	0.36750	0.50085	0.46231	0.32985	0.12329	−0.028158	−0.093953	−0.017908	−0.015818	0.016633
$e^{-\alpha_2\phi} \cos \beta_2\phi$	1	0.68143	0.32930	0.040331	−0.14242	−0.22369	−0.17938	−0.088518	0.075273	0.043597	−0.0009916

(ii) $N_{x\phi}$

$$N_{x\phi} = + \frac{8gl}{\pi^2} \sin (\phi_c - \phi) \sin \frac{\pi x}{l} \tag{7-73b}$$

$$= 1{,}489.4 \sin (\phi_c - \phi) \sin \frac{\pi x}{l}$$

$$N_{x\phi(\phi=0)} = +912.86 \sin \frac{\pi x}{l}$$

(iii) N_x

$$N_x = - \frac{8gl^2}{\pi^3 a} \cos (\phi_c - \phi) \cos \frac{\pi x}{l} \tag{7-73c}$$

$$= - 225.76 \cos (\phi_c - \phi) \cos \frac{\pi x}{l}$$

7. MATRIX FOR STRESS RESULTANTS

$$
\begin{bmatrix} N_{x\phi} \\ \bar{Q}_\phi \\ N_\phi \\ M_\phi \end{bmatrix}
= -18{,}269
\begin{bmatrix}
0.4342 \times \sin \frac{\pi x}{l} & \cdot & \cdot & \cdot \\
\cdot & -0.0096204 \times \cos \frac{\pi x}{l} & \cdot & \cdot \\
\cdot & \cdot & -2.8646 \times \cos \frac{\pi x}{l} & \cdot \\
\cdot & \cdot & \cdot & 0.70712 \times \cos \frac{\pi x}{l}
\end{bmatrix}
$$

$$
\times
\begin{bmatrix}
5.0854 & 14.757 & -11.995 & -6.2064 \\
+5.3576 & -18.3558 & -20.0158 & +9.2956 \\
0 & -1 & 0 & +1 \\
1.29235 & -1 & -0.70765 & -1
\end{bmatrix}
\begin{bmatrix}
1 & \cdot & \cdot & \cdot \\
\cdot & 1 & \cdot & \cdot \\
\cdot & \cdot & 1 & \cdot \\
\cdot & \cdot & \cdot & 1
\end{bmatrix}
\begin{bmatrix} A_n \\ B_n \\ C_n \\ D_n \end{bmatrix}
$$

$$
\begin{bmatrix} N_{x\phi} \\ \bar{Q}_\phi \\ N_\phi \\ M_\phi \end{bmatrix}
= -18{,}269
\begin{bmatrix}
(+2 & 6.4078B_n & -5.2082C_n & -2.7048D_n) \sin \frac{\pi x}{l} \\
(-0.0: & 0.17660B_r & 0.19259C_n & -0.08943D_n) \cos \frac{\pi x}{l} \\
(0 & +2.8646B_n & 0 & -2.8646D_n) \cos \frac{\pi x}{l} \\
(+0.91384A_n & -0.70712B_n & -0.50040C_n & -0.70712D_n) \cos \frac{\pi x}{l}
\end{bmatrix}
$$

8. *BOUNDARY CONDITIONS*

1. The transverse moment M_ϕ at the shell edge is equal to zero.

$$(0.91384A_n - 0.70712B_n - 0.5004C_n - 0.70712D_n) \cos \frac{\pi x}{l} = 0$$

2. The radial shear force $\bar{Q}_\phi$ at the shell edge is equal to zero.

$$(-0.051543A_n + 0.17660B_n + 0.19259C_n - 0.08943D_n) \cos \frac{\pi x}{l} = 0$$

3. The tangential force N_ϕ at the shell edge is equal to zero.

$$(N_\phi)_b + (N_\phi)_m = 0$$

$$-18,269(2.8646B_n - 2.8646D_n) \cos \frac{\pi x}{l} - 3,882.1 \cos \frac{\pi x}{l} = 0$$

4. The shear force $N_{x\phi}$ at the shell edge is equal to zero.

$$(N_{x\phi})_b + (N_{x\phi})_m = 0$$

$$-18,269(2.2181A_n + 6.4078B_n - 5.2082C_n - 2.7048D_n) \sin \frac{\pi x}{l} + 912.86 \sin \frac{\pi x}{l} = 0$$

9. *SOLUTION OF THE EQUATIONS*

The equations are solved for the constants A_n, B_n, C_n, and D_n. Their values are given below

$$A_n = +0.16354236$$
$$B_n = +0.04871276$$
$$C_n = +0.0561665$$
$$D_n = +0.1228930$$

10. *CALCULATION OF STRESS RESULTANTS*

Having obtained the values of A_n, B_n, C_n, and D_n, the values of N_x, N_ϕ, and M_ϕ are calculated at the midspan of the shell and $N_{x\phi}$ at the traverse. (These can be calculated either from the general formula or from the expressions in the final matrix in step 7 of this example.) The membrane values of N_x, N_ϕ, and $N_{x\phi}$ are superimposed over the corresponding bending values to obtain the final stress pattern in the shell. The membrane values of N_x, N_ϕ, and $N_{x\phi}$, the bending values of N_x, N_ϕ, $N_{x\phi}$, and M_ϕ, and their final values are given below.

Membrane Values of N_z, N_ϕ, and $N_{z\phi}$

ϕ	0°	2°21'45"	4°43'30"	7°5'15"	9°27'0"	12°36'0"	15°45'0"	18°54'0"	23°37'30"	28°21'0"	37°48'0"
N_z	−178	−184	−189	−194	−199	−204	−209	−214	−219	−223	−226
N_ϕ	−3,882	−4,003	−4,117	−4,224	−4,324	−4,446	−4,554	−4,648	−4,764	−4,846	−4,913
$N_{z\phi}$	+913	+864	+813	+761	+707	+634	+559	+483	+365	+245	0

Bending Values of N_z, N_ϕ, $N_{z\phi}$, and M_ϕ

ϕ	0°	2°21'45"	4°43'30"	7°5'15"	9°27'0"	12°36'0"	15°45'0"	18°54'0"	23°37'30"	28°21'0"	37°48'0"
N_z	+46,100	+11,346	−5,314	−10,720	−9,920	−5,155	−545	+1,995	+833	+669	−493
N_ϕ	+3,882	+2,915	+988	−620	−1,475	−1,545	−933	−226	+557	+332	−55
$N_{z\phi}$	−913	+6,447	+6,971	+4,615	+1,721	−1,056	−2,040	−1,709	+746	+369	0
M_ϕ	0	+128	−30	−282	−482	−559	−444	−244	+148	+85	+5

Final Values of N_z, N_ϕ, $N_{z\phi}$, and M_ϕ

The final values of the stress resultants (Fig. 8-8) are obtained by adding their membrane and bending values.

ϕ	0°	2°21'45"	4°43'30"	7°5'15"	9°27'0"	12°36'0"	15°45'0"	18°54'0"	23°37'30"	28°21'0"	37°48'0"
N_z	+45,922	+11,162	−5,503	−10,914	−10,119	−5,359	−754	+1,781	+614	+446	−719
N_ϕ	0	−1,088	−3,129	−4,844	−5,799	−5,991	−5,487	−4,874	−4,207	−4,514	−4,968
$N_{z\phi}$	0	+7,311	+7,784	+5,376	+2,428	−422	−1,481	−1,226	+1,111	+615	0
M_ϕ	0	+128	−30	−282	−482	−559	−444	−244	+148	+85	+5

N_z, N_ϕ, and M_ϕ are at midspan and $N_{z\phi}$ at the traverse. N_z, N_ϕ, and $N_{z\phi}$ are in lb/ft and $M_{z\phi}$ in lb ft/ft.

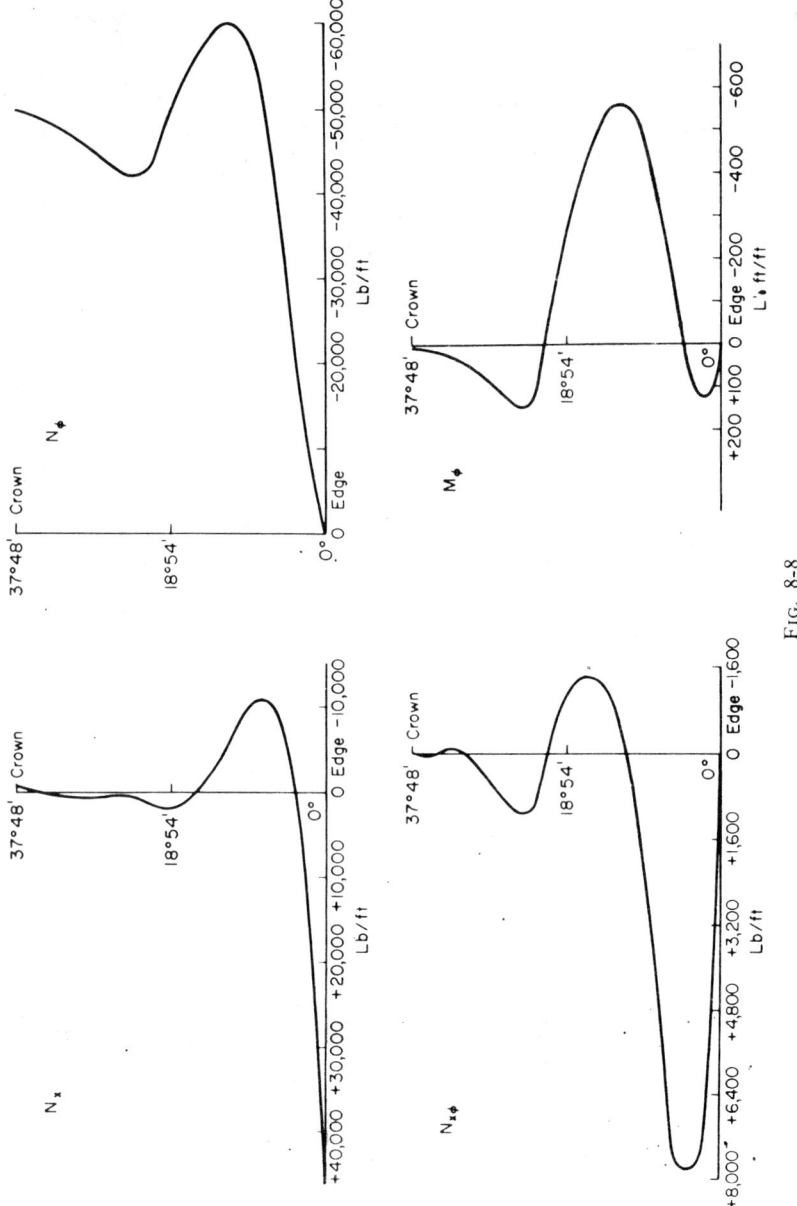

Fig. 8-8

11. CHECK FOR STATICS

Check 1

The resultant longitudinal force on any cross section is zero, i.e., $\Sigma N_x\,ds = 0$.
For calculating the total force, the cross section is divided into 20 parts as shown in
Fig. 8-9. The calculation is shown below in a tabular form.

Point	ϕ	N_x, lb/ft	Distance, ft, from previous point
0	0°	+45,922	0
1	1°10′53″	+25,000	1.51533
2	2°21′45″	+11,162	1.51533
3	3°32′38″	0	1.51533
4	4°43′30″	−5,503	1.51533
5	5°54′23″	−7,500	1.51533
6	7°5′15″	−10,914	1.51533
7	8°16′8″	−11,100	1.51533
8	9°27′0″	−10,119	1.51533
9	11°1′30″	−7,700	2.02044
10	12°36′0″	−5,359	2.02044
11	14°10′30″	−3,000	2.02044
12	15°45′0″	−754	2.02044
13	17°19′30″	+1,350	2.02044
14	18°54′0″	+1,781	2.02044
15	21°15′45″	+1,350	3.03066
16	23°37′30″	+614	3.03066
17	25°59′15″	+550	3.03066
18	28°21′0″	+446	3.03066
19	33°4′30″	−50	6.06131
20	37°48′0″	−719	6.06131

$$\Sigma N_x\,ds = \{\tfrac{1}{3} \times 1.51533[(45,922 - 10,119) + 4(25,000 + 0 - 7,500 - 11,100)$$
$$+ 2(11,162 - 5,503 - 10,914)] + \tfrac{1}{3} \times 2.02044[(-10,119 + 1,781)$$
$$+ 4(-7,700 - 3,000 + 1,350) + 2(-5,359 - 754)]$$
$$+ \tfrac{1}{3} \times 3.03066[(1,781 - 719) + 4(1,350 + 550 - 50) + 2(614 + 446)]\}$$
$$= (23,196 - 5,152 + 50,511 - 15,153 - 22,450 + 11,276 - 5,559 - 11,026$$
$$- 6,815 + 1,200 - 20,750 - 8,075 + 3,770 - 7,218 - 1,016 + 1,799$$
$$- 726 + 5,460 + 2,227 - 202 + 1,241 + 901)$$
$$= 101,581 - 104,142$$
$$= \text{approximately zero}$$

Check 2

The internal resisting moment is equal to the bending moment due to loading, i.e.,
$\Sigma N_x\,ds\,y = M$.

Point	ϕ	$\cos(\phi_c - \phi)$	$\cos(\phi_c - \phi) - \cos\phi_c$	$y = R[\cos(\phi_c - \phi) - \cos\phi_c]$	N_x, lb/ft	ds = effective width of strip	$N_x\,ds$	$N_x\,ds\,y$
1	1°10'53"	0.8026	0.01247	0.9166	+25,000	2.2730	+56,825	+52,083
2	2°21'45"	0.8148	0.02459	1.8074	+11,162	1.51533	+16,914	+30,570
3	3°32'38"	0.8266	0.03641	2.6761	0	1.51533	0	0
4	4°43'30"	0.8380	0.04780	3.5133	−5,503	1.51533	−8,339	−29,297
5	5°54'23"	0.8490	0.05887	4.3270	−7,500	1.51533	−11,365	−49,176
6	7°5'15"	0.8597	0.06958	5.1141	−10,914	1.51533	−16,538	−84,579
7	8°16'8"	0.8701	0.07993	5.8749	−11,100	1.51533	−16,820	−98,800
8	9°27'0"	0.8801	0.08991	6.6084	−10,119	1.51533	−17,889	−118,219
9	11°1'30"	0.8928	0.1026	7.5433	−7,700	1.76789	−15,570	−117,200
10	12°36'0"	0.9048	0.1147	8.4283	−5,359	2.02044	−10,828	−91,257
11	14°10'30"	0.9162	0.1260	9.2632	−3,000	2.02044	−6,061	−56,144
12	15°45'0"	0.9269	0.1367	10.0475	−754	2.02044	−1,523	−15,306
13	17°19'30"	0.9368	0.1467	10.7795	+1,350	2.02044	+2,728	+29,407
14	18°54'0"	0.9461	0.1559	11.4609	+1,781	2.52555	+3,598	+51,551
15	21°15'45"	0.9586	0.1685	12.3826	+1,350	3.03066	+4,091	+50,657
16	23°37'30"	0.9696	0.1794	13.1859	+614	3.03066	+1,861	+24,537
17	25°59'15"	0.9788	0.1887	13.8673	+550	3.03066	+1,667	+23,117
18	28°21'0"	0.9864	0.1963	14.4259	+446	4.54599	+2,028	+29,249
19	33°4'30"	0.9966	0.2064	15.1733	−50	6.06131	−303	−4,598
20	37°48'0"	1	0.2098	15.4232	−719	3.03066	−218	−33,608
							$\Sigma N_x\,ds =$	−406,913 ft lb

For half the shell, i.e., from 0 to 37.8° only.

For the complete shell (i.e., from 0 to 75.6°),

$$\Sigma N_z \, ds \, y = 2 \times 406,913$$

$$= 813,826 \text{ ft lb}$$

$$\text{Actual internal resisting moment} = \frac{\pi^3}{32} \times 813,826$$

$$= 788,000 \text{ ft lb}$$

$$\text{External bending moment due to loads} = \frac{(52.5 \times 73.5 \times 1.3195) \times 35^2}{8}$$

$$= 779,636 \text{ ft lb}$$

The two quantities agree very closely.

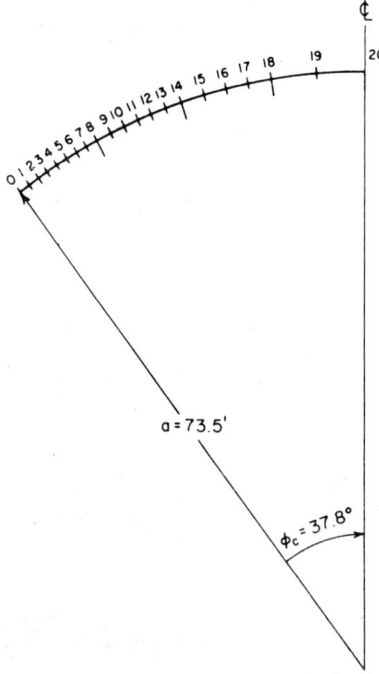

Fɪɢ. 8-9

Check 3

The vertical component of the shear at the traverse is equal to half the total load on the shell, i.e., $\Sigma N_{x\phi} \, ds \, \sin (\phi_c - \phi) = \text{load}/2$.

Point	ϕ	$\sin(\phi_c - \phi)$	$N_{x\phi}$, lb/ft	ds = effective width	$N_{x\phi}\,ds$	$\dfrac{N_{x\phi}\,ds}{\sin(\phi_c - \phi)}$
1	1°10′53″	0.5965	+3,500	2.273	+7,950	+4,750
2	2°21′45″	0.5798	+7,311	1.51533	+11,079	+6,424
3	3°32′38″	0.5628	+7,823	1.51533	+11,854	+6,672
4	4°43′30″	0.5457	+7,784	1.51533	+11,795	+6,437
5	5°54′23″	0.5284	+6,720	1.51533	+10,183	+5,380
6	7°5′15″	0.5107	+5,376	1.51533	+8,146	+4,161
7	8°16′8″	0.4929	+3,990	1.51533	+6,046	+2,980
8	9°27′0″	0.4748	+2,428	1.76789	+4,292	+2,038
9	11°1′30″	0.4505	+910	2.02044	+1,839	+828
10	12°36′0″	0.4258	−422	2.02044	−853	−363
11	14°10′30″	0.4007	−1,200	2.02044	−2,425	−971
12	15°45′0″	0.3754	−1,481	2.02044	−2,992	−1,123
13	17°19′30″	0.3498	−1,550	2.02044	−3,132	−1,095
14	18°54′0″	0.3239	−1,226	2.52555	−3,096	−1,003
15	21°15′45″	0.2847	+100	3.03066	+303	+86
16	23°37′30″	0.2449	+1,130	3.03066	+3,425	+839
17	25°59′15″	0.2047	+1,250	3.03066	+3,788	+775
18	28°21′0″	0.1642	+630	4.54599	+2,864	+470
19	33°4′30″	0.0824	−250	6.06131	−1,515	−125
20	37°48′0″	0	+138	3.03066	+418	0

$$\Sigma\, N_{x\phi}\,ds\,\sin\,(\phi_c - \phi) = \quad +37,130 \text{ lb}$$

For half the shell, i.e., from 0 to 37.8° only.

For the complete shell (i.e., from 0 to 75.6°),

$$\Sigma\, N_{x\phi}\,ds\,\sin\,(\phi_c - \phi) = 2 \times 37,130$$
$$= 74,260 \text{ lb}$$

$$\text{Actual vertical component of the shear} = \frac{\pi^2}{8} \times 74,260$$
$$= 91,615 \text{ lb}$$

$$\text{Half the total load} = \frac{(52.5 \times 73.5 \times 1.3195) \times 35}{2}$$
$$= 89,101 \text{ lb}$$

The two values are nearly equal.

12. DESIGN OF REINFORCEMENT

Longitudinal Reinforcement

The longitudinal tensile force near the edge is equal to 83,800 lb. Using intermediate-grade steel, the area required = 83,800/20,000 = 4.19 in.². Ten No. 6 bars will give an area of 4.4 in.². The length of the tension zone = 4.5 ft. Hence No. 6 bars are provided at spacing varying from $2\frac{1}{2}$ in. at the edge to $8\frac{1}{2}$ in. at a distance of 4.5 ft from the edge. The second tensile zone extends from 21.23 to 42.00 ft from the edge, i.e., over a length of 21 ft. Tensile force = 17,780 lb. Area of steel required = 0.89 in.². No. 2 bars at 9 in. will be provided. Nominal reinforcement in the compression zone can also be No. 2 bars at 9 in.

Transverse Reinforcement

M_ϕ and N_ϕ at the following points are considered:

ϕ	M_ϕ , ft lb/ft	N_ϕ , lb/ft
2°21′45″	+128	−1,088
12°36′0″	−559	−5,991
23°37′30″	+148	−4,207

No. 2 bars at 6 in. are adequate.

Reinforcement for Principal Tension

To carry the principal tension, diagonal reinforcement will be provided near the corners of the shell. The magnitudes and directions of the principal tensile stresses at various points on the shell are shown in Fig. 8-10. The maximum tension of 10,300 lb/ft occurs at quarter span at 3 ft 0 in. from the edge. Using intermediate-grade steel, this force needs No. 4 bars at $4\frac{1}{2}$ in. At the traverse, where the maximum tension is 7,784 lb/ft, No. 4 bars at 6 in. will do. At 12 ft from the edge along the traverse, the tensile stress

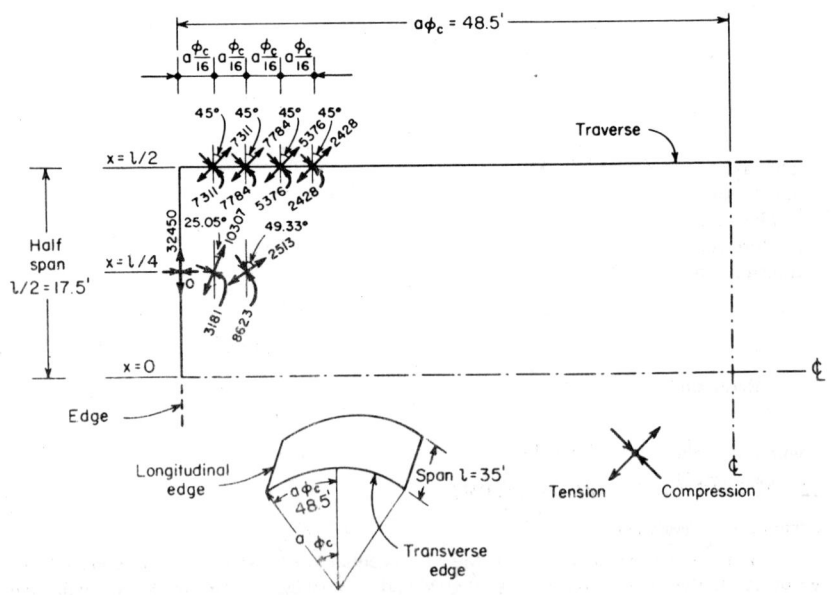

Fig. 8-10

drops to a value of 2,428 lb/ft.   No. 3 bars at 9 in. will be sufficient to carry this tension. At other points on the traverse, the maximum value of $N_{x\phi}$ is 1,550 lb/ft (refer to the table for Static Check 3), resulting in a principal tension of equal magnitude.   This needs a reinforcement of 0.077 in.²/ft.   There is, however, no need to provide these diagonal bars.   The nominal reinforcement of No. 2 bars at 9 in. in the longitudinal direction and No. 2 bars at 6 in. in the transverse direction will give a component of $0.05 \times \frac{12}{9} \cos^2 45° + 0.05 \times \frac{12}{6} \sin^2 45° = 0.083$ in.².

13. *REINFORCEMENT FOR LONGITUDINAL BENDING MOMENT AT THE
JUNCTION OF THE SHELL WITH THE TRAVERSE*

The longitudinal bending moment at the junction is approximately determined by the following formula given in ASCE Manual 31 (pp. 93-97)

$$M_x = \frac{d}{3.56} N_\phi$$

where d is the thickness of the shell and N_ϕ is the transverse stress at the crown of the shell at the traverse.   The value of N_ϕ is zero at the traverse because of the sinusoidal loading assumed.   But as the load is, in fact, uniform, the value of N_ϕ at this point is taken as equal to $\pi/4$ times its maximum value at midspan corresponding to the first term of the Fourier loading.   Hence

$$N_\phi = \frac{\pi}{4}(-4,968) = -3,902$$

The final value of N_ϕ is considered here though it is accurate enough to use the membrane value, as stated in the Manual.

$$M_x = \frac{3}{12 \times 3.56}(-3,902) = -274 \text{ ft lb/ft}$$

This value will, however, get reduced because of the axial compression in the arch. The axial compression at the crown of the arch (from Design Example 20-1) = 3.5883 tons = 8,037 lb.

The cross section of the arch is a rectangle 8 by 24 in. with eight No. 4 bars as reinforcement.   With concrete of 3,000-psi strength, modular ratio is 9.   The transformed area of the arch rib $= 8 \times 24 + (9 - 1) \times 8 \times 0.196 = 204.5$ in.².

Compressive stress σ_ϕ in the rib $= 8037/204.5 = 39.3$ psi

Reduction in the longitudinal moment due to arch compression $= -\dfrac{d^2}{3.56}\sigma_\phi$

where d is the thickness of the shell at the junction with the traverse.
Assuming $d = 3.5$ in. at this point,

Reduction in the moment $= -\dfrac{3.5 \times 3.5}{3.56}(-39.3) = +135$ lb ft/ft

Hence net moment $= -274 + 135 = -139$ ft lb

Longitudinal reinforcement available consists of No. 2 bars at 9 in. Providing additional No. 2 bars at 9 in. for a length of about 3 ft 3 in. near the traverse, total longitudinal reinforcement available at the junction at crown is $0.05 \times 12/4.5 = 0.133$ in.^{2}/ft.

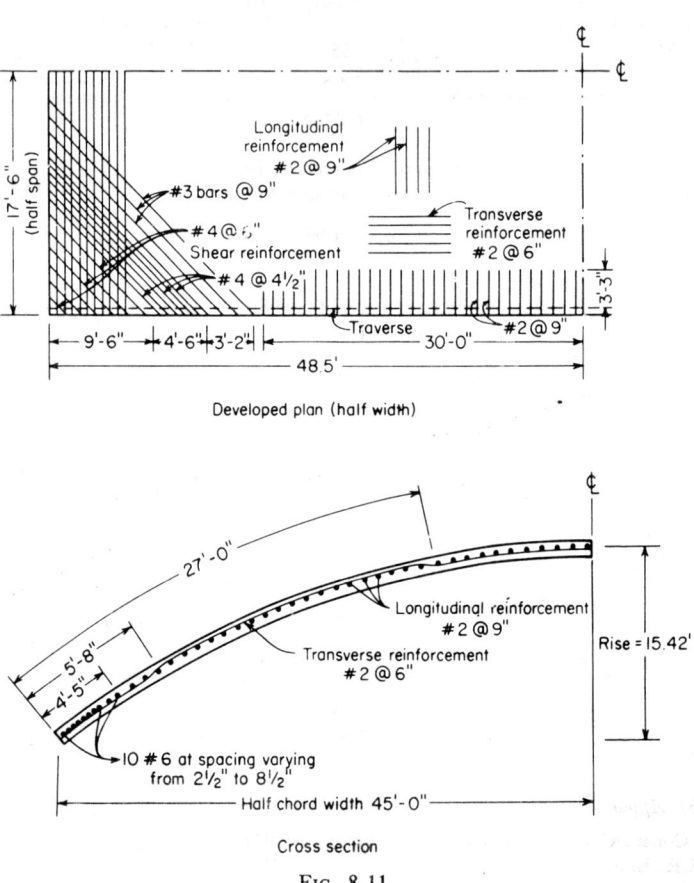

Developed plan (half width)

Cross section

FIG. 8-11

Assuming this steel to be located at the middle of the shell thickness, the shell can develop a resisting moment of 333 ft lb/ft. Hence the reinforcement provided is ample. The scheme of reinforcement is shown in Fig. 8-11.

14. CHECK FOR BUCKLING

The shell is next checked for safety against buckling failure using the procedure given in ACI publication "Concrete Shell Structures—Practice and Commentary."

(a) Apparent Factor of Safety F_1 in the Longitudinal Direction

$$K = \frac{d^2}{12a^2}$$

$$= \frac{0.25 \times 0.25}{12 \times 73.5 \times 73.5}$$

$$= 0.9641 \times 10^{-6}$$

$$l/a = \frac{35}{73.5}$$

$$= 0.4762$$

Knowing K and l/a, the value of q_2 is read off from the graph (Fig. 11) on page 428 of "Stresses in Shells" by W. Flügge, Springer-Verlag, Berlin, 1960.

$$q_2 = 1.975 \times 10^{-3}$$

$$\text{Buckling load per foot length} = q_2 \frac{Ed}{(1 - \nu^2)}$$

$$\therefore \text{Buckling stress} = \frac{q_2 E}{(1 - \nu^2)}$$

$$= \frac{1.975 \times 10^{-3} \times 2.88 \times 10^6}{(1 - 0.0256)}$$

$$= 5,837 \text{ psi}$$

Maximum compressive stress in the shell in the longitudinal direction $= \dfrac{10,914}{12 \times 3}$

$$= 304 \text{ psi}$$

Factor of safety in the longitudinal direction is

$$F_1 = \frac{5,837}{304}$$

$$= 19.2$$

(b) Apparent Factor of Safety F_2 in the Transverse Direction

Considering a cylindrical shell under radial pressure, the critical buckling stress (R.R. Bradshaw, *Journal of the ACI*, March, 1963, pp. 313-328),

$$f_{cr} = \frac{K_\nu \pi^2 E d^2}{12(1 - \nu^2)l^2}$$

where K_ν is to be read from the graph (Fig. 8, p. 322 of Bradshaw's paper) for any known value of Z_L where

$$Z_L = l^2 \sqrt{1 - \nu^2}/ad$$

$$= 60.66$$

From the graph $K_\nu = 10$

∴. Critical buckling stress in the transverse direction

$$f_{cr} = \frac{10 \times \pi^2 \times 2.88 \times 10^6 \times 3^2}{12 \times 0.9744 \times 35 \times 35 \times 144}$$

$$= 1{,}237 \text{ psi}$$

Compressive stress in the shell at the point where maximum longitudinal stress occurs = 4,844 lb/ft

$$= 135 \text{ psi}$$

Factor of safety in the transverse direction $= F_2 = \dfrac{1{,}237}{135}$

$$= 9.2$$

∴. Combined factor of safety $F = \dfrac{19.2 \times 9.2}{19.2 + 9.2}$

$$= 6.22$$

The value obtained is greater than 5 specified in "Practice and Commentary." The check is also applied at two other points,

(i) at $12° - 36'$ from the edge, where N_ϕ is maximum:

$$N_\phi = -5{,}991 \text{ lb/ft}$$

and

$$N_x = -5{,}359 \text{ lb/ft}$$

The combined factor of safety works out to 6.25.

(ii) at $9° - 27'$ from the edge:

$$N_\phi = -5{,}799 \text{ lb/ft}$$
$$N_x = -10{,}119 \text{ lb/ft}$$

The combined factor of safety at this point works out to 5.6.
Hence the shell is safe against buckling.

Design Example 8-2

DESIGN OF A SINGLE LONG CYLINDRICAL SHELL WITH EDGE BEAMS

1. DATA

Geometry

Shell

Span l	= 83.25 ft
Radius a	= 25.00 ft
Thickness d	= 0.25 ft
Semicentral angle ϕ_c	= 35°

Edge Beam

Depth $2a_1$ = 5.00 ft
Width $2b_1$ = 0.75 ft

Loads

Dead weight = 37.5 psf of shell surface
Live load   = 12.5 psf of shell surface

Total load g = 50.0 psf of shell surface

2. GEOMETRICAL PROPERTIES

Shell

$$\rho = \left(\frac{12\pi^4 a^6}{l^4 d^2}\right)^{1/8} = 4.19035$$

$$\kappa = \frac{\pi^2 a^2}{l^2 \rho^2} = 0.05069$$

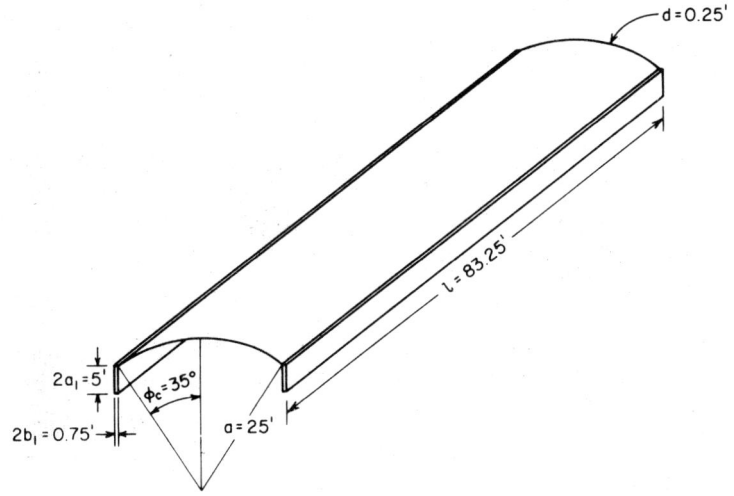

FIG. 8-12

Hence the shell is long. The Schorer method will be used for the analysis.

$$\phi_c = 0.610865 \text{ radian}$$

$$2\phi_c = 1.221730 \text{ radians}$$

$$\sin \phi_c = 0.57358$$

$$\cos \phi_c = 0.81915$$

$$k = \frac{\pi}{l} = 0.03774$$

$$D = \frac{Ed^3}{12} = \frac{360 \times 10^6 (0.25)^3}{12} = 468{,}750$$

$$\lambda_n = \frac{\pi a}{l} = 0.94342 \text{ (considering only the first Fourier term)}$$

Edge Beam

$$EI = 360 \times 10^6 \times \frac{0.75 \times 5^3}{12} = 281.25 \times 10^7$$

$$\frac{1}{(\pi/l)^4 EI} = 1.75325 \times 10^{-4}$$

3. ROOTS OF THE CHARACTERISTIC EQUATION AND RELATED QUANTITIES

$$\alpha_1 = \beta_2 = \rho \, \frac{(\sqrt{2} + 1)^{1/2}}{8^{1/4}} = 3.871379$$

$$\alpha_2 = \beta_1 = \rho \, \frac{(\sqrt{2} - 1)^{1/2}}{8^{1/4}} = 1.603578$$

4. MULTIPLIERS IN THE MATRICES

Quantity	Multiplier
Forces	
M_ϕ	$-\dfrac{D\rho^2}{a^2} \dfrac{1}{\sqrt{2}} \cos \dfrac{\lambda_n x}{a}$ $\qquad -E \times 3.65813 \times 10^{-7} \times 70.71066 \cos \dfrac{\lambda_n x}{a}$
Q_ϕ	$-\dfrac{D\rho^2}{a^2} \dfrac{-1}{a\sqrt{2}} \cos \dfrac{\lambda_n x}{a}$ $\qquad -E \times 3.65813 \times 10^{-7} \times 2.82843 \cos \dfrac{\lambda_n x}{a}$
N_ϕ	$-\dfrac{D\rho^2}{a^2} \dfrac{\rho^2}{a} \cos \dfrac{\lambda_n x}{a}$ $\qquad -E \times 3.65813 \times 10^{-7} \times 70.23613 \cos \dfrac{\lambda_n x}{a}$
$N_{x\phi}$	$-\dfrac{D\rho^2}{a^2} \dfrac{\rho^2}{\lambda_n a} \sin \dfrac{\lambda_n x}{a}$ $\qquad -E \times 3.65813 \times 10^{-7} \times 74.44842 \sin \dfrac{\lambda_n x}{a}$
$\dfrac{\partial N_{x\phi}}{\partial x}$	$\dfrac{-D\rho^2}{a^2} \dfrac{\rho^2}{a^2} \cos \dfrac{\lambda_n x}{a}$ $\qquad -E \times 3.65813 \times 10^{-7} \times 2.80945 \cos \dfrac{\lambda_n x}{a}$
N_x	$-\dfrac{D\rho^2}{a^2} \dfrac{\rho^4}{\sqrt{2}\,\lambda_n^2 a} \cos \dfrac{\lambda_n x}{a}$ $\qquad -E \times 3.65813 \times 10^{-7} \times 9{,}797.96530 \cos \dfrac{\lambda_n x}{a}$
Displacements	
u	$-\dfrac{\lambda_n}{\sqrt{2}\,\rho^2} \sin \dfrac{\lambda_n x}{a}$ $\qquad -37.99176 \times 10^{-3} \sin \dfrac{\lambda_n x}{a}$
$\dfrac{\partial u}{\partial x}$	$-\dfrac{\lambda_n^2}{\sqrt{2}\,\rho^2 a} \cos \dfrac{\lambda_n x}{a}$ $\qquad -1.43369 \times 10^{-3} \cos \dfrac{\lambda_n x}{a}$
v	$+\dfrac{1}{\sqrt{2}\,\rho^2} \cos \dfrac{\lambda_n x}{a}$ $\qquad +40.27025 \times 10^{-3} \cos \dfrac{\lambda_n x}{a}$
w	$+\cos \dfrac{\lambda_n x}{a}$ $\qquad +1{,}000 \times 10^{-3} \cos \dfrac{\lambda_n x}{a}$
ϑ	$+\dfrac{1}{a} \cos \dfrac{\lambda_n x}{a}$ $\qquad +40 \times 10^{-3} \cos \dfrac{\lambda_n x}{a}$

ϕ	0°	10°	20°	30°	35°	40°	50°	60°	70°
$e^{-\alpha_1\phi}$	1	0.5088088	0.2588863	0.1317236	0.09395954	0.06702212	0.03410144	0.01735112	0.008828402
$e^{-\alpha_2\phi}$	1	0.7558766	0.5713495	0.4318697	0.3754725	0.3264402	0.2467485	0.1865114	0.1409796
$\cos\beta_1\phi$	1	0.96109	0.84738	0.66774	0.55738	0.43613	0.17057	-0.10826	-0.37866
$\sin\beta_1\phi$	0	0.27624	0.53098	0.74440	0.83026	0.89989	0.98535	0.99412	0.92554
$\cos\beta_2\phi$	1	0.78028	0.21768	-0.44059	-0.71323	-0.90524	-0.97211	-0.61177	0.01739
$\sin\beta_2\phi$	0	0.62543	0.97602	0.89771	0.70093	0.42491	-0.23462	-0.79104	-0.99985
$e^{-\alpha_1\phi}\cos\beta_1\phi$	1	0.48901105	0.2193751	0.0879571	0.0523712	0.0292304	0.0058167	-0.0018784	-0.003343
$e^{-\alpha_1\phi}\sin\beta_1\phi$	0	0.1405533	0.1374634	0.0980550	0.0780108	0.0603125	0.0336019	0.0172491	0.008171
$e^{-\alpha_2\phi}\cos\beta_2\phi$	1	0.5897954	0.1243714	-0.1902775	-0.2677983	-0.2955067	-0.2398667	-0.1141021	0.002452
$e^{-\alpha_2\phi}\sin\beta_2\phi$	0	0.4722479	0.5576485	0.3876937	0.2631799	0.1387077	-0.0578921	-0.1475380	-0.140959

5. TABLE OF B_1, B_2, B_3, AND B_4

	B_1	B_2	B_3	B_4
M_ϕ	$+1$	-1	-1	-1
Q_ϕ	-2.26780	$+5.47496$	$+5.47496$	-2.26780
N_ϕ	0	$+1$	0	-1
$N_{x\phi}$	$+1.60358$	$+3.87138$	-3.87138	-1.60358
N_x	-1	-1	$+1$	-1
$\dfrac{\partial u}{\partial x}$	-1	-1	$+1$	-1
v	-5.47496	-2.26780	-2.26780	-5.47496
w	$+1$	0	$+1$	0
ϑ	-3.87138	$+1.60358$	-1.60358	$+3.87138$

6. MEMBRANE FORCES AND DISPLACEMENTS

(i) N_ϕ

$$N_\phi = -\frac{4}{\pi} ga \cos (\phi_c - \phi) \cos \frac{\pi x}{l}$$

$$= -1{,}591.54943 \cos (\phi_c - \phi) \cos \frac{\pi x}{l}$$

$$\therefore N_{\phi(\phi=0)} = -1{,}303.72 \cos \frac{\pi x}{l}$$

(ii) $N_{x\phi}$

$$N_{x\phi} = +\frac{8}{\pi^2} gl \sin (\phi_c - \phi) \sin \frac{\pi x}{l}$$

$$= +3{,}373.995 \sin (\phi_c - \phi) \sin \frac{\pi x}{l}$$

$$N_{x\phi(\phi=0)} = +1{,}935.26 \sin \frac{\pi x}{l}$$

(iii) N_x

$$N_x = -\frac{8gl^2}{a\pi^3} \cos (\phi_c - \phi) \cos \frac{\pi x}{l}$$

$$= -3{,}576.34 \cos (\phi_c - \phi) \cos \frac{\pi x}{l}$$

$$N_{x(\phi=0)} = -2{,}929.56 \cos \frac{\pi x}{l}$$

(iv) u

$$u = -\frac{8gl^3}{E\pi^4 ad}\cos(\phi_c - \phi)\sin\frac{\pi x}{l}$$

$$= -\frac{379,082.0}{E}\cos(\phi_c - \phi)\sin\frac{\pi x}{l}$$

$$u_{(\phi=0)} = -\frac{310,525}{E}\sin\frac{\pi x}{l}$$

(v) v

$$v = -\frac{8g}{E\pi d}\left[2\left(\frac{l}{\pi}\right)^2 + \frac{1}{a^2}\left(\frac{l}{\pi}\right)^4\right]\sin(\phi_c - \phi)\cos\frac{\pi x}{l}$$

$$= -\frac{1,117,117.30}{E}\sin(\phi_c - \phi)\cos\frac{\pi x}{l}$$

$$v_{(\phi=0)} = -\frac{640,756.17}{E}\cos\frac{\pi x}{l}$$

(vi) w

$$w = +\frac{8g}{E\pi d}\left[2\left(\frac{l}{\pi}\right)^2 + \frac{1}{a^2}\left(\frac{l}{\pi}\right)^4\right]\cos(\phi_c - \phi)\cos\frac{\pi x}{l}$$

$$= +1,117,084.39\cos(\phi_c - \phi)\cos\frac{\pi x}{l}$$

$$w_{(\phi=0)} = +\frac{915,059.68}{E}\cos\frac{\pi x}{l}$$

7. MATRIX FOR STRESS RESULTANTS

Disturbances from Left Edge

$$\begin{bmatrix}\dfrac{\partial N_{x\phi}}{\partial x}\\ Q_\phi\\ N_\phi\\ M_\phi\end{bmatrix} = -E \times 3.65813 \times 10^{-7}\cos\frac{\pi x}{l}\begin{bmatrix}2.8095 & 0 & 0 & 0\\ 0 & 2.8284 & 0 & 0\\ 0 & 0 & 70.2361 & 0\\ 0 & 0 & 0 & 70.7107\end{bmatrix}$$

$$\times \begin{bmatrix}1.6036 & 3.8714 & -3.8714 & -1.6036\\ -2.2678 & 5.4750 & 5.4750 & -2.2678\\ 0 & +1 & 0 & -1\\ +1 & -1 & -1 & -1\end{bmatrix}\begin{bmatrix}+1 & 0 & 0 & 0\\ 0 & +1 & 0 & 0\\ 0 & 0 & +1 & 0\\ 0 & 0 & 0 & +1\end{bmatrix}\begin{bmatrix}A_n\\ B_n\\ C_n\\ D_n\end{bmatrix}$$

$$\begin{bmatrix}\dfrac{\partial N_{x\phi}}{\partial x}\\ Q_\phi\\ N_\phi\\ M_\phi\end{bmatrix} = -E \times 3.65813 \times 10^{-7}\cos\frac{\pi x}{l}$$

$$\times \begin{bmatrix}4.5053 & 10.8767 & -10.8767 & -4.5053\\ -6.4142 & 15.4855 & 15.4855 & -6.4142\\ 0 & 70.2361 & 0 & -70.2361\\ 70.7107 & -70.7107 & -70.7107 & -70.7107\end{bmatrix}\begin{bmatrix}A_n\\ B_n\\ C_n\\ D_n\end{bmatrix}$$

$$
\begin{bmatrix} \dfrac{\partial N_{x\phi}}{\partial x} \\ Q_\phi \\ N_\phi \\ M_\phi \end{bmatrix} = - E \times 3.65813 \times 10^{-7} \cos \dfrac{\pi x}{l}
$$

$$
\times \begin{bmatrix} (\ +4.5053 A_n & +10.8767 B_n & -10.8767 C_n & -4.5053 D_n) \\ (\ -6.4142 A_n & +15.4855 B_n & +15.4855 C_n & -6.4142 D_n) \\ (\qquad +0 & +70.2361 B_n & +0 & -70.2361 D_n) \\ (+70.7107 A_n & -70.7107 B_n & -70.7107 C_n & -70.7107 D_n) \end{bmatrix}
$$

Disturbances from Right Edge

$$
\begin{bmatrix} \dfrac{\partial N_{x\phi}}{\partial x} \\ Q_\phi \\ N_\phi \\ M_\phi \end{bmatrix} = - E \times 3.65813 \times 10^{-7} \cos \dfrac{\pi x}{l} \begin{bmatrix} 2.8095 & 0 & 0 & 0 \\ 0 & 2.8284 & 0 & 0 \\ 0 & 0 & 70.2361 & 0 \\ 0 & 0 & 0 & 70.7107 \end{bmatrix}
$$

$$
\times \begin{bmatrix} 1.6036 & 3.8714 & -3.8714 & -1.6036 \\ -2.2678 & 5.4750 & 5.4750 & -2.2678 \\ 0 & +1 & 0 & -1 \\ +1 & -1 & -1 & -1 \end{bmatrix}
$$

$$
\times \begin{bmatrix} -0.003343 & 0.008171 & 0 & 0 \\ -0.008171 & -0.003343 & 0 & 0 \\ 0 & 0 & 0.002452 & -0.140959 \\ 0 & 0 & 0.140959 & 0.002452 \end{bmatrix} \begin{bmatrix} A_n \\ B_n \\ C_n \\ D_n \end{bmatrix}
$$

$$
\begin{bmatrix} \dfrac{\partial N_{x\phi}}{\partial x} \\ Q_\phi \\ N_\phi \\ M_\phi \end{bmatrix} = - E \times 3.65813 \times 10^{-7} \cos \dfrac{\pi x}{l} \begin{bmatrix} 4.5053 & 10.8767 & -10.8767 & -4.5053 \\ -6.4142 & 15.4855 & 15.4855 & -6.4142 \\ 0 & 70.2361 & 0 & -70.2361 \\ 70.7107 & -70.7107 & -70.7107 & -70.7107 \end{bmatrix}
$$

$$
\times \begin{bmatrix} (-0.003343 A_n & +0.008171 B_n) \\ (-0.008171 A_n & -0.003343 B_n) \\ (+0.002452 C_n & -0.140959 D_n) \\ (+0.140959 C_n & +0.002452 D_n) \end{bmatrix}
$$

$$
\begin{bmatrix} \dfrac{\partial N_{x\phi}}{\partial x} \\ Q_\phi \\ N_\phi \\ M_\phi \end{bmatrix} = - E \times 3.65813 \times 10^{-7} \cos \dfrac{\pi x}{l}
$$

$$
\times \begin{bmatrix} (-0.1040 A_n & +0.0004 B_n & -0.6618 C_n & +1.5222 D_n) \\ (-0.1051 A_n & -0.1042 B_n & -0.8661 C_n & -2.1985 D_n) \\ (-0.5739 A_n & -0.2348 B_n & -9.9004 C_n & -0.1722 D_n) \\ (+0.3414 A_n & +0.8142 B_n & -10.1407 C_n & +9.7939 D_n) \end{bmatrix}
$$

Sum of the Effects of the Two Edges

$$
\begin{bmatrix} \dfrac{\partial N_{x\phi}}{\partial x} \\ Q_\phi \\ N_\phi \\ M_\phi \end{bmatrix} = -E \times 3.65813 \times 10^{-7} \cos\frac{\pi x}{l}
$$

$$
\times \begin{bmatrix} (\ +4.6093A_n & +10.8763B_n & -10.2149C_n & -6.0275D_n) \\ (\ -6.3091A_n & +15.5897B_n & +16.3516C_n & -4.2157D_n) \\ (\ -0.5739A_n & +70.0013B_n & -9.9004C_n & -70.4083D_n) \\ (+71.0521A_n & -69.8965B_n & -80.8514C_n & -60.9168D_n) \end{bmatrix}
$$

8. MATRIX FOR DISPLACEMENTS

Disturbances from Left Edge

$$
\begin{bmatrix} \dfrac{\partial u}{\partial x} \\ w \\ v \\ \vartheta \end{bmatrix} = 10^{-3}\cos\frac{\pi x}{l} \begin{bmatrix} -1.4337 & 0 & 0 & 0 \\ 0 & 1,000 & 0 & 0 \\ 0 & 0 & 40.2703 & 0 \\ 0 & 0 & 0 & 40 \end{bmatrix}
$$

$$
\times \begin{bmatrix} -1 & -1 & +1 & -1 \\ +1 & 0 & +1 & 0 \\ -5.4750 & -2.2678 & -2.2678 & -5.4750 \\ -3.8714 & 1.6036 & -1.6036 & 3.8714 \end{bmatrix} \begin{bmatrix} +1 & 0 & 0 & 0 \\ 0 & +1 & 0 & 0 \\ 0 & 0 & +1 & 0 \\ 0 & 0 & 0 & +1 \end{bmatrix} \begin{bmatrix} A_n \\ B_n \\ C_n \\ D_n \end{bmatrix}
$$

$$
\begin{bmatrix} \dfrac{\partial u}{\partial x} \\ w \\ v \\ \vartheta \end{bmatrix} = 10^{-3}\cos\frac{\pi x}{l} \begin{bmatrix} 1.4337 & 1.4337 & -1.4337 & 1.4337 \\ 1,000 & 0 & 1,000 & 0 \\ -220.4799 & -91.3250 & -91.3250 & -220.4799 \\ -154.8560 & 64.1440 & -64.1440 & 154.8560 \end{bmatrix} \begin{bmatrix} A_n \\ B_n \\ C_n \\ D_n \end{bmatrix}
$$

$$
\begin{bmatrix} \dfrac{\partial u}{\partial x} \\ w \\ v \\ \vartheta \end{bmatrix} = 10^{-3}\cos\frac{\pi x}{l} \begin{bmatrix} (\ +1.4337A_n & +1.4337B_n & -1.4337C_n & +1.4337D_n) \\ (\ +1,000A_n & +0 & +1,000C_n & +0\ ) \\ (-220.4799A_n & -91.3250B_n & -91.3250C_n & -220.4799D_n) \\ (-154.8560A_n & +64.1440B_n & -64.1440C_n & +154.8560D_n) \end{bmatrix}
$$

Disturbances from Right Edge

$$
\begin{bmatrix} \dfrac{\partial u}{\partial x} \\ w \\ v \\ \vartheta \end{bmatrix}
= 10^{-3} \cos \dfrac{\pi x}{l}
\begin{bmatrix}
-1.4337 & 0 & 0 & 0 \\
0 & 1{,}000 & 0 & 0 \\
0 & 0 & 40.2703 & 0 \\
0 & 0 & 0 & 40
\end{bmatrix}
$$

$$
\times
\begin{bmatrix}
-1 & -1 & +1 & -1 \\
+1 & 0 & +1 & 0 \\
-5.4750 & -2.2678 & -2.2678 & -5.4750 \\
-3.8714 & 1.6036 & -1.6036 & 3.8714
\end{bmatrix}
$$

$$
\times
\begin{bmatrix}
-0.003343 & 0.008171 & 0 & 0 \\
-0.008171 & -0.003343 & 0 & 0 \\
0 & 0 & 0.002452 & -0.140959 \\
0 & 0 & 0.140959 & 0.002452
\end{bmatrix}
\begin{bmatrix} A_n \\ B_n \\ C_n \\ D_n \end{bmatrix}
$$

$$
\begin{bmatrix} \dfrac{\partial u}{\partial x} \\ w \\ v \\ \vartheta \end{bmatrix}
= 10^{-3} \cos \dfrac{\pi x}{l}
\begin{bmatrix}
1.4337 & 1.4337 & -1.4337 & 1.4337 \\
1{,}000 & 0 & 1{,}000 & 0 \\
-220.4799 & -91.3250 & -91.3250 & -220.4799 \\
-154.8560 & 64.1440 & -64.1440 & 154.8560
\end{bmatrix}
$$

$$
\times
\begin{bmatrix}
(-0.003343 A_n & +0.008171 B_n) \\
(-0.008171 A_n & -0.003343 B_n) \\
(+0.002452 C_n & -0.140959 D_n) \\
(+0.140959 C_n & +0.002452 D_n)
\end{bmatrix}
$$

$$
\begin{bmatrix} \dfrac{\partial u}{\partial x} \\ w \\ v \\ \vartheta \end{bmatrix}
= 10^{-3} \cos \dfrac{\pi x}{l}
\begin{bmatrix}
(-0.0165 A_n & +0.0069 B_n & +0.1986 C_n & +0.2056 D_n) \\
(-3.343 A_n & +8.171 B_n & +2.452 C_n & -140.959 D_n) \\
(+1.4833 A_n & -1.4962 B_n & -31.3025 C_n & +12.3325 D_n) \\
(-0.0064 A_n & -1.4797 B_n & +21.6710 C_n & +9.4214 D_n)
\end{bmatrix}
$$

Sum of the Effects of the Two Edges

$$
\begin{bmatrix} \dfrac{\partial u}{\partial x} \\ w \\ v \\ \vartheta \end{bmatrix}
= 10^{-3} \cos \dfrac{\pi x}{l}
\begin{bmatrix}
(+1.4172 A_n & +1.4406 B_n & -1.2351 C_n & +1.6393 D_n) \\
(+996.657 A_n & +8.171 B_n & +1{,}002.452 C_n & -140.959 D_n) \\
(-221.9632 A_n & -89.8268 B_n & -60.0225 C_n & -232.8124 D_n) \\
(-154.8496 A_n & +65.6237 B_n & -85.8150 C_n & +145.4346 D_n)
\end{bmatrix}
$$

9. BOUNDARY CONDITIONS

1. The resultant horizontal force at the shell edge is equal to zero, i.e.,

$$(N_\phi)_b \cos \phi_c + (Q_\phi)_b \sin \phi_c + (N_\phi)_m \cos \phi_c = 0$$

$$(N_\phi)_m = -1{,}303.72 \cos \frac{\pi x}{l}$$

$$(N_\phi)_m \cos \phi_c = -1{,}067.94 \cos \frac{\pi x}{l}$$

$$(N_\phi)_b = -E \times 3.65813 \times 10^{-7} \cos \frac{\pi x}{l} (-0.5739 A_n$$
$$+ 70.0013 B_n - 9.9004 C_n - 70.4083 D_n)$$

$$(N_\phi)_b \cos \phi_c = -E \times 2.99656 \times 10^{-7} \cos \frac{\pi x}{l} (-0.5739 A_n$$
$$+ 70.0013 B_n - 9.9004 C_n - 70.4083 D_n)$$

$$(Q_\phi)_b = -E \times 3.65813 \times 10^{-7} \cos \frac{\pi x}{l} (-6.3091 A_n$$
$$+ 15.5897 B_n + 16.3516 C_n - 4.2157 D_n)$$

$$(Q_\phi)_b \sin \phi_c = -E \times 2.09823 \times 10^{-7} \cos \frac{\pi x}{l} (-6.3091 A_n$$
$$+ 15.5897 B_n + 16.3516 C_n - 4.2157 D_n)$$

$$(N_\phi)_b \cos \phi_c + (Q_\phi)_b \sin \phi_c + (N_\phi)_m \cos \phi_c = -E \times 10^{-7} \cos \frac{\pi x}{l}$$
$$\times (-14.9577 A_n + 242.4739 B_n + 4.6423 C_n - 219.8282 D_n) - 1{,}067.94 \cos \frac{\pi x}{l}$$

Equating this to zero, we get

$$-(360 \times 10^6) 10^{-7} \cos \frac{\pi x}{l} (-14.9577 A_n + 242.4739 B_n + 4.6423 C_n - 219.8282 D_n)$$
$$= 1{,}067.94 \cos \frac{\pi x}{l}$$

Simplifying,

$$+14.9577 A_n - 242.4739 B_n - 4.6423 C_n + 219.8282 D_n = +29.665 \tag{1}$$

2. The transverse moment M_ϕ at the shell edge is equal to zero, i.e.,

$$+71.0521 A_n - 69.8965 B_n - 80.8514 C_n - 60.9168 D_n = 0 \tag{2}$$

3. Vertical deflection of the shell edge is equal to the vertical deflection of the edge beam, i.e.,

$$[-(w)_m \cos \phi_c + (v)_m \sin \phi_c] + [-(w)_b \cos \phi_c + (v)_b \sin \phi_c]$$
$$= \frac{1}{k^4 EI} [(N_\phi)_{m+b} \sin \phi_c - (Q_\phi)_b \cos \phi_c + (N_{x\phi})_{m+b} k a_1 - W']$$

$[-(w)_m \cos \phi_c + (v)_m \sin \phi_c]$

$$= \frac{1}{E}(-915,059.68 \times 0.81915 - 640,756.17 \times 0.57358) \cos \frac{\pi x}{l}$$

$$= -\frac{1,117,096}{E} \cos \frac{\pi x}{l} = -0.003103 \cos \frac{\pi x}{l}$$

$[-(w)_b \cos \phi_c + (v)_b \sin \phi_c]$

$$= 10^{-3} \cos \frac{\pi x}{l} [-0.81915(+996.657A_n + 8.171B_n + 1,002.452C_n - 140.959D_n)$$

$$+ 0.57358(-221.9632A_n - 89.8268B_n - 60.0225C_n - 232.8124D_n)]$$

$$= 10^{-3} \cos \frac{\pi x}{l}(-943.7253A_n - 58.2162B_n - 855.5863C_n - 18.0699D_n)$$

The vertical deflection of the shell edge (measured upward) is therefore equal to

$$(-0.9437A_n - 0.0582B_n - 0.8556C_n - 0.0181D_n - 0.003103) \cos \frac{\pi x}{l}$$

The vertical deflection of the edge beam is calculated below.

$$(N_\phi)_m \sin \phi_c = -747.79 \cos \frac{\pi x}{l}$$

$$(N_\phi)_b \sin \phi_c = -E \times 3.65813 \times 10^{-7} \times 0.57358 \cos \frac{\pi x}{l}$$

$$\times (-0.5739A_n + 70.0013B_n - 9.9004C_n - 70.4083D_n)$$

$$= (+43.3503A_n - 5,287.6392B_n + 747.840C_n + 5,318.3825D_n) \cos \frac{\pi x}{l}$$

$$(N_\phi)_{m+b} \sin \phi_c = (+43.3503A_n - 5,287.6392B_n + 747.840C_n$$

$$+ 5,318.3825D_n - 747.79) \cos \frac{\pi x}{l}$$

$$-(Q_\phi)_b \cos \phi_c = +E \times 3.65813 \times 10^{-7} \times 0.81915 \cos \frac{\pi x}{l}$$

$$\times (-6.3091A_n + 15.5897B_n + 16.3516C_n - 4.2157D_n)$$

$$= (-680.6011A_n + 1,681.7560B_n + 1,763.9468C_n - 454.7733D_n)$$

$$(N_{x\phi})_m ka_1 = +1,935.26 \times 0.03774 \times 2.5$$

$$= +182.59$$

$$(N_{x\phi})_b ka_1 = +\left(\frac{\partial N_{x\phi}}{\partial x}\right)_b a_1$$

$$= -2.5E \times 3.65813 \times 10^{-7}(+4.6093A_n + 10.8763B_n$$

$$-10.2149C_n - 6.0275D_n)$$

$$= -(+1,517.5277A_n + 3,580.8227B_n - 3,363.0689C_n - 1,984.4441D_n)$$

$$+(N_{x\phi})_{m+b} ka_1 = (-1,517.5277A_n - 3,580.8227B_n$$

$$+3,363.0689C_n + 1,984.4441D_n + 182.59)$$

$$-W' = \frac{4}{\pi} \times \text{weight of the edge beam per foot length}$$

$$= -\frac{4}{\pi} \times 562.5$$

$$= -716.20$$

$$\frac{1}{k^4 EI} = 1.75325 \times 10^{-4}$$

$$= \frac{1}{5,703.69}$$

The vertical deflection of the edge beam (measured upward) is therefore equal to

$$\frac{1}{5,703.69} (-2,154.7785A_n - 7,186.7059B_n + 5,874.8557C_n + 6,848.0533D_n - 1,281.40)$$

Therefore, boundary condition 3 is given by

$$(-2,154.7785A_n - 7,186.7059B_n + 5,874.8557C_n + 6,848.0533D_n - 1,281.40)$$
$$= 5,703.69(-0.9437A_n - 0.0582B_n - 0.8556C_n - 0.0181D_n - 0.003103)$$

Simplifying, we get

$$+3,227.7938A_n - 6,854.7511B_n + 10,754.9329C_n + 6,951.2901D_n = 1,263.7 \quad (3)$$

4. Longitudinal displacement of the shell edge is equal to the longitudinal displacement of the edge beam at its junction with the shell, i.e.,

$$(u)_m + (u)_b = \left\{ [(N_\phi)_{m+b} \sin \phi_c - (Q_\phi)_{m+b} \cos \phi_c] a_1 \frac{1}{k^4 EI} k + \left[\left(\frac{N_{x\phi}}{k^2} \right)_{m+b} \frac{1}{AE} \right] \right.$$
$$\left. + \left[(N_{x\phi})_{m+b} a_1^2 k^2 \frac{1}{k^4 EI} \right] - a_1 \frac{1}{k^4 EI} kW' \right\}$$

u for Shell

$$[(u)_m]_{\phi=0} = -\frac{310,525}{E} \sin \frac{\pi x}{l}$$

$$= -0.00086257 \sin \frac{\pi x}{l}$$

$$(u)_b = \frac{1}{k} \left(\frac{\partial u}{\partial x} \right)_0$$

$$= \frac{10^{-3}}{0.03774} (+1.4172A_n + 1.4406B_n - 1.2351C_n + 1.6393D_n) \sin \frac{\pi x}{l}$$

$$= (+0.03755A_n + 0.03817B_n - 0.03273C_n + 0.04344D_n) \sin \frac{\pi x}{l}$$

$$(u)_m + (u)_b = (+0.03755A_n + 0.03817B_n - 0.03273C_n + 0.04344D_n - 0.00086257) \sin \frac{\pi x}{l}$$

u for Edge Beam

$$(N_\phi)_{m+b} \sin \phi_c = (+43.3503A_n - 5,287.6392B_n + 747.840C_n + 5,318.3825D_n - 747.79)$$

$$-(Q_\phi)_{m+b} \cos \phi_c = (-680.6011A_n + 1,681.7560B_n + 1,763.9468C_n - 454.7733D_n)$$

$$[(N_\phi)_{m+b} \sin \phi_c - (Q_\phi)_{m+b} \cos \phi_c]$$
$$= (-637.2508 A_n - 3,605.8832 B_n + 2,511.7868 C_n + 4,863.6092 D_n - 747.79)$$

$$a_1 \frac{1}{k^4 EI} k = 2.5 \times \frac{1}{5,703.69} \times 0.03774$$
$$= 0.00001654$$

$$[(N_\phi)_{m+b} \sin \phi_c - (Q_\phi)_{m+b} \cos \phi_c] a_1 \frac{1}{k^4 EI} k$$
$$= (-0.01054 A_n - 0.05964 B_n + 0.04154 C_n + 0.08044 D_n - 0.01237)$$

$$\left(\frac{N_{z\phi}}{k^2}\right)_m = + \frac{1,935.26}{0.0014243}$$
$$= + 1,358,745$$

$$\left(\frac{N_{z\phi}}{k^2}\right)_b = - \frac{E \times 3.65813 \times 10^{-7}}{k^3}(+ 4.6093 A_n + 10.8763 B_n$$
$$- 10.2149 C_n - 6.0275 D_n)$$
$$= - 10^6(+11.2932 A_n + 26.6480 B_n - 25.0275 C_n - 14.7680 D_n)$$

$$\left(\frac{N_{z\phi}}{k^2}\right)_{m+b} = - 10^6(+11.2932 A_n + 26.6480 B_n - 25.0275 C_n - 14.7680 D_n - 1.358745)$$

$$\left(\frac{N_{z\phi}}{k^2}\right)_{m+b} \frac{1}{AE} = (-0.008365 A_n - 0.019738 B_n + 0.018538 C_n + 0.010939 D_n + 0.001006)$$

$$(N_{z\phi})_m = + 1,935.26$$

$$(N_{z\phi})_b = - \frac{E \times 3.65813 \times 10^{-7}}{k}(+ 4.6093 A_n + 10.8763 B_n$$
$$- 10.2149 C_n - 6.0275 D_n)$$
$$= - (+ 16,084.023 A_n + 37,952.544 B_n$$
$$- 35,644.608 C_n - 21,032.79 D_n)$$

$$(N_{z\phi})_{m+b} = - (+ 16,084.023 A_n + 37,952.544 B_n$$
$$- 35,644.608 C_n - 21,032.79 D_n - 1,935.26)$$

$$(N_{z\phi})_{m+b} a_1{}^2 k^2 \frac{1}{k^4 EI} = (- 0.025102 A_n - 0.059233 B_n + 0.055631 C_n$$
$$+ 0.032826 D_n + 0.003020)$$

$$-a_1 \frac{1}{k^4 EI} kW' = - 2.5 \times \frac{1}{5,703.69} \times 0.03774 \times 716.2$$
$$= - 0.011847$$

$$\therefore \left\{[(N_\phi)_{m+b} \sin \phi_c - (Q_\phi)_{m+b} \cos \phi_c] a_1 \frac{1}{k^4 EI} k \right.$$
$$+ \left[\left(\frac{N_{z\phi}}{k^2}\right)_{m+b} \frac{1}{AE}\right] + \left[(N_{z\phi})_{m+b} a_1{}^2 k^2 \frac{1}{k^4 EI}\right] - \left. a_1 \frac{1}{k^4 EI} kW' \right\}$$
$$= (-0.044007 A_n - 0.138611 B_n + 0.115709 C_n + 0.124205 D_n - 0.020191)$$

Condition 4 is therefore as given below.

$$(+0.03755A_n + 0.03817B_n - 0.03273C_n + 0.04344D_n - 0.00086257)$$
$$= (-0.044007A_n - 0.138611B_n + 0.115709C_n + 0.124205D_n - 0.020191)$$

i.e.,

$$+0.081557A_n + 0.176781B_n - 0.148439C_n - 0.080765D_n = -0.0193284 \qquad (4)$$

The four equations are listed below:

$$14.9577A_n - 242.4739B_n - 4.6423C_n + 219.8282D_n = +29.665$$
$$71.0521A_n - 69.8965B_n - 80.8514C_n - 60.9168D_n = 0$$
$$3,227.7938A_n - 6,854.7511B_n + 10,754.9329C_n + 6,951.2901D_n = +1,263.7$$
$$0.081557A_n + 0.176781B_n - 0.148439C_n - 0.080765D_n = -0.0193284$$

10. SOLUTION OF THE EQUATIONS

The equations are solved for the constants A_n, B_n, C_n, and D_n. Their values are given below.

$$A_n = +0.0208467$$
$$B_n = -0.0662277$$
$$C_n = +0.029562$$
$$D_n = +0.061070$$

11. CALCULATION OF STRESS RESULTANTS

N_x, N_ϕ, and M_ϕ are calculated at the midspan of the shell and $N_{x\phi}$ at the traverse. The general formula given in Art. 8-1 is used to calculate the values. The multipliers are taken from step 4 of this example and the values of the constants B_1, B_2, B_3, and B_4 from step 5. The contributions due to bending are superimposed on the membrane values to obtain the final values of the stress resultants. While calculating the contributions due to bending, it is to be noted that the effect of the farther edge is to be added to that of the nearer edge with appropriate sign as explained in Art. 7-18. For example,

At $\phi = 0°$, $N_x = (N_x)_{\phi=0°} + (N_x)_{\phi=(70°-0°)}$
At $\phi = 10°$, $N_x = (N_x)_{\phi=10°} + (N_x)_{\phi=(70°-10°)}$
... ...
At $\phi = 35°$, $N_x = (N_x)_{\phi=35°} + (N_x)_{\phi=(70°-35°)} = 2(N_x)_{\phi=35°}$

Similarly N_ϕ and M_ϕ are calculated. As $N_{x\phi}$ is an antisymmetric function, the calculation will be as shown below.

At $\phi = 0°$, $N_{x\phi} = (N_{x\phi})_{\phi=0°} - (N_{x\phi})_{\phi=(70°-0°)}$
At $\phi = 10°$, $N_{x\phi} = (N_{x\phi})_{\phi=10°} - (N_{x\phi})_{\phi=(70°-10°)}$
... ...
At $\phi = 35°$, $N_{x\phi} = (N_{x\phi})_{\phi=35°} - (N_{x\phi})_{\phi=(70°-35°)} = 0$

The values of N_x, $N_{x\phi}$, N_ϕ, and M_ϕ are tabulated below and shown in Fig. 8-13.

	$\phi = 0°$	$\phi = 10°$	$\phi = 20°$	$\phi = 30°$	$\phi = 35°$
N_x at midspan, lb/ft	−3,134	−9,736	−13,011	−14,162	−14,278
$N_{x\phi}$ at traverse, lb/ft	+6,452	+5,340	+3,430	+1,172	0
N_ϕ at midspan, lb/ft	−87	−934	−1,568	−1,902	−1,944
M_ϕ at midspan, lb ft/ft	0	+169	−61	−284	−317

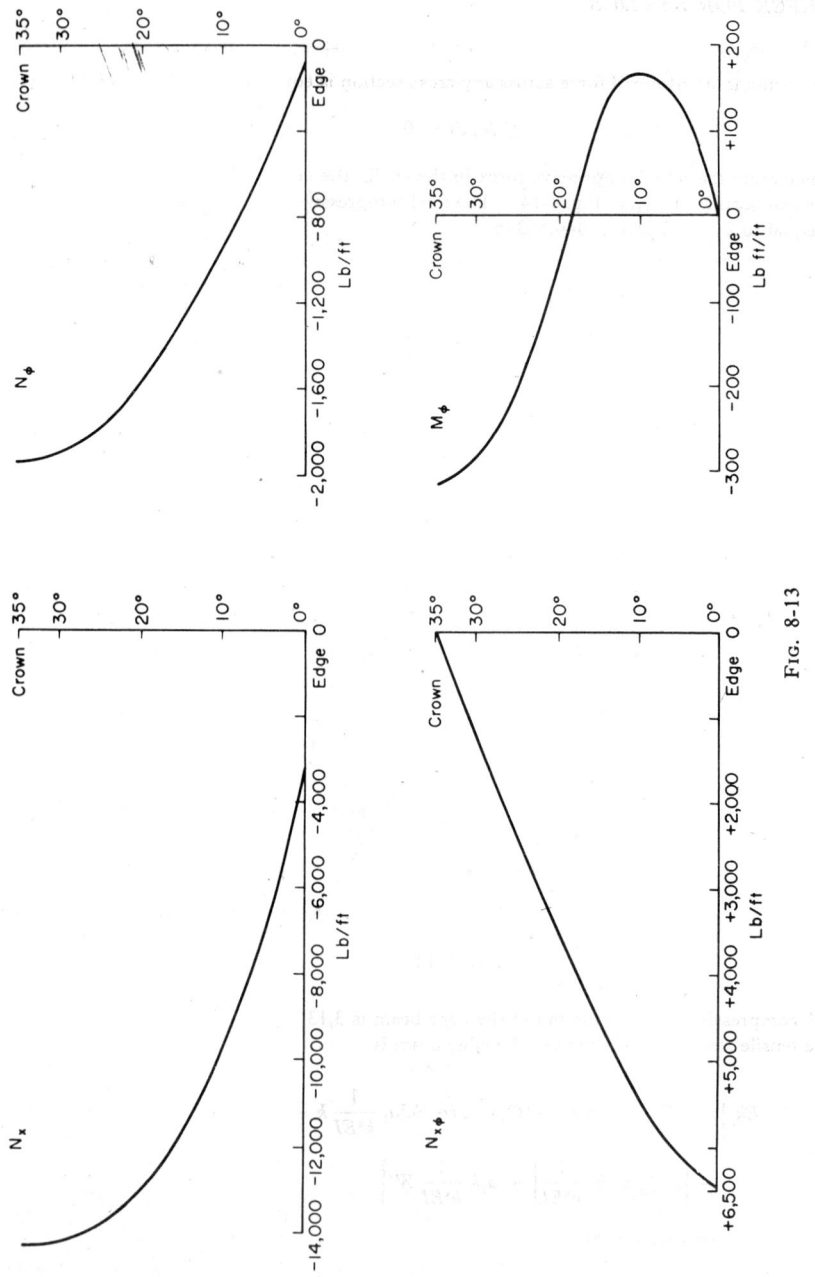

Fig. 8-13

12. CHECK FOR STATICS

Check 1

The resultant longitudinal force across any cross section is equal to zero, i.e.,

$$\Sigma N_z \, ds = 0$$

For calculating the total compressive force in the shell, the cross section is divided into seven equal parts as shown in Fig. 8-14. The total compressive force in the shell is found to be equal to $2 \times 172{,}280 = 344{,}560$ lb.

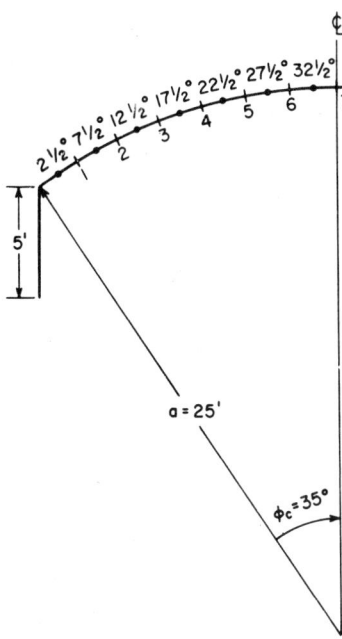

FIG. 8-14

The compressive stress at the top of the edge beam is $3{,}134/0.25 = 12{,}536$ psf. The tensile stress at the bottom of the edge beam is

$$Ek \left\{ - [(N_\phi)_{m+b} \sin \phi_c - (Q_\phi)_{m+b} \cos \phi_c] a_1 \frac{1}{k^4 EI} k + \left[\left(\frac{N_{x\phi}}{k^2} \right) \frac{1}{AE} \right] \right.$$

$$\left. - \left[(N_{x\phi}) a_1{}^2 k^2 \frac{1}{k^4 EI} \right] + a_1 k \frac{1}{k^4 EI} W' \right\}$$

$$= 103{,}700 \text{ psf}$$

It is easily verified that the neutral axis is at a depth of 0.539 ft below the top of the edge beam. Compressive force in each edge beam is $12{,}536/2 \times \frac{9}{12} \times 0.539 = 2{,}535$ lb.

Hence the total compressive force in the shell and the edge beams is $344,560 + 2(2,535) = 349,630$ lb.

$$\text{Tensile force in each edge beam} = \frac{103,700}{2} \times \frac{9}{12} \times 4.461$$

$$= 173,475 \text{ lb}$$

$$\text{Total tensile force in the two edge beams} = 346,950 \text{ lb}$$

$$\text{Resultant longitudinal force} = 346,950 - 349,630$$

This is approximately equal to zero.

Check 2

The internal resisting moment is equal to the bending moment due to the loading. Moments of the N_x forces in the shell are taken about the bottom of the edge beam.

No.	ϕ, deg (1)	$y = a \cos (\phi_c - \phi) - \cos \phi_c$ (2)	$y + 5'$ (3)	Force(compressive) (4)	Moment (3) × (4)
1	$2\frac{1}{2}$	0.6050	5.6050	11,180	62,664
2	$7\frac{1}{2}$	1.6950	6.6950	18,600	124,527
3	$12\frac{1}{2}$	2.6175	7.6175	23,510	179,087
4	$17\frac{1}{2}$	3.3625	8.3625	27,650	231,223
5	$22\frac{1}{2}$	3.9275	8.9275	29,590	264,165
6	$27\frac{1}{2}$	4.3050	9.3050	30,680	285,477
7	$32\frac{1}{2}$	4.4950	9.4950	31,070	295,010
				$\Sigma = 172,280$	1,442,153

$$\Sigma \text{ moment} = 1,442,153 \text{ ft lb}$$

$$\text{Moment due to compressive force (in edge beam)} = 2,535 \left(5 - \frac{0.539}{3} \right) = 12,220 \text{ ft lb}$$

$$\text{Moment due to tensile force (in edge beam)} = 173,475 \times \frac{4.461}{3} = 257,900 \text{ ft lb}$$

$$\text{Net moment} = 2 \, (1,442,153 + 12,220) - (257,900 \times 2)$$

$$= 2,392,946 \text{ ft lb}$$

The values of the longitudinal forces obtained as a result of taking only the first term of the Fourier series are $32/\pi^3$ of their true values. Therefore, the actual values may be obtained by multiplying the above values by $\pi^3/32$.

$$\therefore \text{ Actual value of moment} = 2,392,946 \times \frac{\pi^3}{32}$$

$$= 2,318,000 \text{ ft lb}$$

$$\text{Bending moment due to loads} = \frac{l^2}{8} \times \text{weight per ft length of shell}$$

$$[\text{Arc length of shell} = a\phi = 30.545 \text{ ft}$$

$$\therefore \text{ Weight per ft} = (30.545 \times 50) + (2 \times 5 \times 0.75 \times 150)$$

$$= 2,652.25 \text{ lb}]$$

$$\therefore \text{ Moment} = 2,298,000 \text{ ft lb}$$

These two values agree very closely.

Check 3

The vertical component of the shear at the traverse is equal to half the total load on the shell.

Calculation of the vertical component of the shear forces is shown in a tabular form below.

ϕ, deg	Width of strip	$(\phi_c - \phi)°$	$\sin(\phi_c - \phi)$	Shear force in strip	Vertical component
$2\frac{1}{2}$	2.1817	$32\frac{1}{2}$	0.53730	$6,350 \times 2.1817$	7,443.7
$7\frac{1}{2}$	2.1817	$27\frac{1}{2}$	0.46175	$5,750 \times 2.1817$	5,792.6
$12\frac{1}{2}$	2.1817	$22\frac{1}{2}$	0.38268	$4,930 \times 2.1817$	4,116.0
$17\frac{1}{2}$	2.1817	$17\frac{1}{2}$	0.30071	$3,950 \times 2.1817$	2,591.4
$22\frac{1}{2}$	2.1817	$12\frac{1}{2}$	0.21644	$2,900 \times 2.1817$	1,369.5
$27\frac{1}{2}$	2.1817	$7\frac{1}{2}$	0.13053	$1,750 \times 2.1817$	498.3
$32\frac{1}{2}$	2.1817	$2\frac{1}{2}$	0.04362	620×2.1817	58.9

$$\Sigma = 21,870 \quad N_{x_y}$$

Total vertical component $= 21,870 \times 2 = 43,740$ lb

The vertical shear in the edge beam at the traverse is equal to

$$\frac{1}{k}[(N_\phi)_{m+b} \sin\phi_c - (Q_\phi)_{m+b} \cos\phi_c - W'] = 22,980 \text{ lb}$$

Vertical shear in two edge beams $= 45,960$ lb

$$\therefore \text{ Total vertical component of shear} = 43,740 + 45,960$$

$$= 89,700 \text{ lb}$$

The values of shear obtained by taking only the first term of the Fourier series are only $8/\pi^2$ of their actual values. The actual shears may therefore be obtained by multiplying these values by $\pi^2/8$.

$$\therefore \text{ Corrected value of shear} = \frac{\pi^2}{8} \times 89{,}700$$

$$= 110{,}700 \text{ lb}$$

$$\text{Half the load on the shell} = 2{,}652.25 \times 41.625$$

$$= 110{,}400 \text{ lb}$$

Hence check 3 is satisfied.

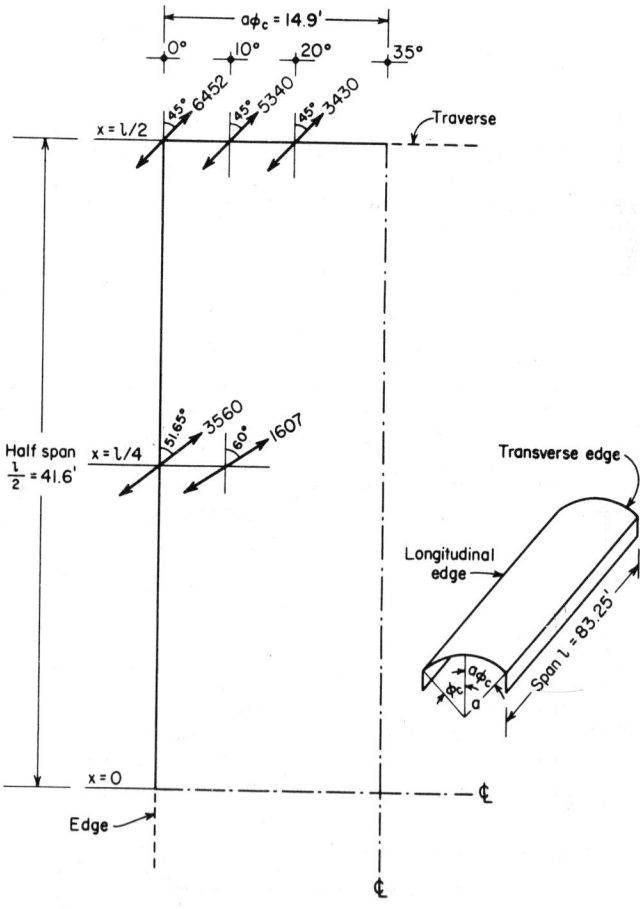

Fig. 8-15

13. *DESIGN OF REINFORCEMENT*

Longitudinal Reinforcement

Total tension in the edge beam = 173,475 lb (refer to Static Check 1).

With a permissible tensile stress of 20,000 psi in steel, an area of 8.67 in.² would be required if the centroid of the steel is located at the center of the tensile force. The theoretical center of the tensile force is at a height of one-third the depth of the tensile zone, i.e., at 4.46/3 = 1.49 ft from the bottom of the edge beam. The center of compression forces is at 1.15 ft below the crown of the shell. (This is easily verified from the calculation made for the second check for statics.) The theoretical lever arm is therefore

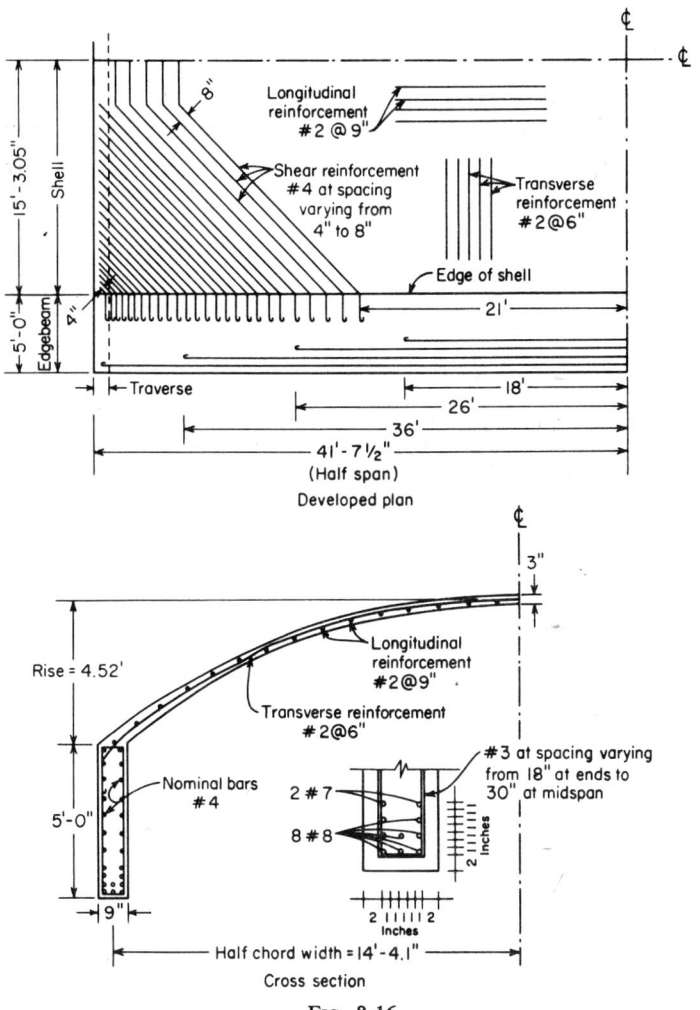

Developed plan

Cross section

Fig. 8-16

equal to $9.52 - 1.15 - 1.49 = 6.88$ ft.   For maximum efficiency, the tensile steel is located as close to the bottom of the edge beam as possible.   Assuming that the centroid of the steel is at 5 in. above the bottom of the edge beam, the actual lever arm is equal to $9.52 - 1.15 - \frac{5}{12} = 7.95$ ft.   Corresponding to this, the tensile force is reduced to $173,475 \times 6.88/7.95 = 150,000$ lb.   Hence the area of steel gets reduced to 7.50 in.^{2}. Eight No. 8 and two No. 7 bars with an area of 7.52 in.^{2} will do.   The centroid is at 4.92 in. from the bottom of the edge beam as against 5 in. assumed.   It may be noted that, in this calculation, any changes in the position of the center of compression forces and the position of neutral axis as a consequence of the change in the position of the tensile steel are ignored.   It can be seen that the stress in the lowest bar exceeds the stress in the centroid of the steel by 4.9 %.

In the body of the shell, which is in compression, only nominal reinforcement is required.   No. 2 bars at 9 in. center to center will be provided.

Transverse Reinforcement

M_ϕ and N_ϕ at the following points are considered:

ϕ	M_ϕ , ft lb/ft	N_ϕ , lb/ft
10°	+169	−934
35°	−317	−1,944

No. 2 bars at 6 in. are adequate.

Reinforcement for Principal Tension

The magnitudes and directions of the principal tension at various points on the shell are shown in Fig. 8-15.   The maximum tension of 6,452 lb/ft, which occurs at the traverse, needs No. 4 bars at 4 in.   At quarter span, edge, the principal tension is only 3,560 lb/ft which can be resisted by concrete.   Hence No. 4 bars at spacing varying from 4 in. at the traverse to 8 in. at midspan will be provided as shown in Fig. 8-16.

CHAPTER 9

BEAM THEORY OF CYLINDRICAL SHELLS

9-1 Advantages of the Beam Method

The average designer unfamiliar with advanced mathematics is bound to ask, "Is there no simple method of designing cylindrical shells based on a strength of materials approach?" Fortunately, it so happens that a large class of cylindrical shells can be analyzed with sufficient accuracy by regarding them as beams of curved cross section. The familiar Mc/I and VQ/Ib formulas may be used to determine the longitudinal stress N_x and the shear stress $N_{x\phi}$, respectively. To determine M_ϕ and N_ϕ one need only be familiar with the principles of arch analysis. Although Finsterwalder [34] and Aas-Jakobsen [35] were the first to suggest the beam approximation, the credit for its detailed development must go to Lundgren [2]. The advantages of the beam theory may be summed up as follows:

(i) It brings shell analysis within the reach of those who are unfamiliar with the techniques of advanced mathematics.

(ii) Unlike the analytical theory, shells with noncircular directrices can be dealt with.

(iii) It can be applied to shells with nonuniform thickness.

(iv) It can handle shells strengthened by ribs in the longitudinal and transverse directions.

(v) It is also claimed by one author[1] that line loads carried by shells can be treated by this method. One has, however, to be careful to see that such loading does not cause local deformations which might violate the assumptions of the beam theory.

(vi) Structural action of the shell is easily visualized.

[1] Chronowicz, A., "The Design of Shells, A Practical Approach," 2d ed., p. 168, Crosby Lockwood & Son, Ltd., London, 1960.

9-2 Assumptions

The following assumptions are implied in the beam theory:

(i) The deformations of the cross section in its plane are negligible.

(ii) M_x—and hence Q_x—and $M_{x\phi}$ may be ignored.

(iii) The strain γ_{xy} caused by the shear force $N_{x\phi}$ and the lateral contraction are neglected.

The first assumption may be replaced by the Bernoulli assumption of plane sections remaining plane after deformation, if the shell beam is subjected to simple bending. For the second assumption to be true, the shell has to be long. Thus the applicability of the beam theory is limited to fairly long shells. The ranges within which it is valid will be examined in the next article. The assumptions on which the beam method is based can never be exactly fulfilled. If these assumptions are sufficiently satisfied, the beam method may be regarded as applicable.

9-3 Range of Validity

The assumptions listed above do not give any idea of the range of shells for which the beam method may be confidently applied at least for purposes of preliminary design.

Studies indicate that the beam method is applicable to the following classes of shells provided they are uniformly loaded:

(i) Single shells without edge beams if $l/a > 5$.

(ii) Long single shells with not too deep edge beams if $l/a > 3$.

(iii) Interior shells with featheredge beams of a group of multiple shells with $l/a > 1.67$.

(iv) Interior shells with edge beams of a multiple group of shells if $l/a > 3$. These limits may provide some guidelines to the designer. The validity of the beam theory for other loading and boundary conditions will need careful examination by the designer before he decides to use it.

Chinn,[1] who studied two interior shells with $l/a = 2$ and 1.10 with featheredge beams by the beam method, reported close agreement with the results of the analytical theory. This would indicate that the beam theory is applicable to uniformly loaded interior shells even when $l/a = 1.10$. A higher value of $l/a = 1.67$ has been suggested in line with the comment made by Parme and Conner[2] on

[1] Chinn, James, "Cylindrical Shell Analysis Simplified by Beam Method," *Journal of the American Concrete Institute*, vol. 55, p. 1183, May, 1959.

[2] Parme and Conner, discussion of James Chinn, "Cylindrical Shell Analysis Simplified by Beam Method," *Journal of the American Concrete Institute*, vol. 55, part 2, p. 1583, December, 1959.

Chinn's paper to take into account the influence of other parameters such as ϕ_c .

9-4 Beam Analysis

There are two distinct steps in the beam method of analysis. In the first, the shell is regarded as a beam of curved cross section (Fig. 9-1) and the familiar Mc/I and VQ/Ib formulas are applied to determine N_x and $N_{x\phi}$. This step may be called *beam analysis*. Referring to Fig. 9-1 and assuming that the shell is carrying only vertical loading, symmetrically distributed over the cross section,

$$N_x = \frac{M_{yy}}{I_{yy}} z \, d \qquad (9\text{-}1)$$

where M_{yy} is the bending moment at any cross section computed as

for a simply supported beam and I_{yy} is the moment of inertia about the axis yy. It is easily verified that

$$I_{yy} = 2d \int_{\phi=0}^{\phi=\phi_c} a \, d\phi \left(a \cos \phi - \frac{a \sin \phi_c}{\phi_c} \right)^2$$

$$= a^3 d \left[\phi_c + \sin \phi_c \left(\cos \phi_c - \frac{2 \sin \phi_c}{\phi_c} \right) \right] \qquad (9\text{-}2)$$

Formulas (9-1) and (9-2) are applicable only to shells of symmetric cross section. If the shell cross section is unsymmetrical, yy and zz are not principal axes of inertia and hence the product of inertia would also be involved. The application of the beam theory to such asymmetric shells will be discussed in the next chapter.

The beam analysis also enables $N_{x\phi}$ to be found by the use of the well-known VQ/Ib formula. For a shell subjected to vertical loading, symmetrically distributed over the cross section, it is easily verified that

$$N_{x\phi} = \frac{VQd}{2I_{yy}d} = \frac{VQ}{2I_{yy}} \qquad (9\text{-}3)$$

where V is the vertical shearing force at the cross section, computed as for a simple beam, and Q is the first static moment of the cross section up to the point under consideration about the axis yy found from the expression

$$Q = a \left(\frac{\sin \phi}{\phi} - \frac{\sin \phi_c}{\phi_c} \right) 2a\phi d = 2a^2 d \left(\sin \phi - \frac{\phi}{\phi_c} \sin \phi_c \right) \qquad (9\text{-}4)$$

9-5   Arch Analysis

The second step in the beam approximation may be described as the *arch analysis*. The object of this analysis is to find M_ϕ, Q_ϕ, and N_ϕ in the shell. Consider a free body in the form of an elementary arch included between two adjacent cross sections of the shell which are dx apart. The equilibrium of the arch is maintained by two sets of forces,

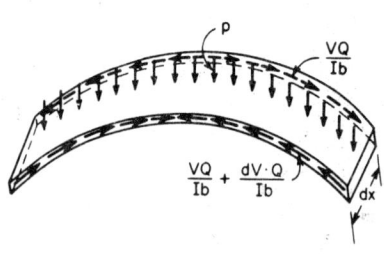

Fig. 9-2

namely, the load acting on the element and the force $\partial N_{x\phi}/\partial x$ (Fig. 9-2). The latter is known as the *specific shear*. The specific shear at any point, acting in the direction of the tangent to the shell arch, may be resolved into horizontal and vertical components. It is clear that the sum of the vertical components of the specific shear would balance the load on the shell arch; the horizontal components of the specific shear which are symmetrically disposed about the crown balance themselves. First, consider a single shell with or without edge beams. It is clear that the elementary shell arch cut out from such a shell will not develop any restraining forces or moments at its ends. Hence we have a statically determinate arch. The transverse moment M_ϕ at any point in the shell arch may therefore be found as the algebraic sum of the moments caused by the loading and the horizontal and vertical components of the specific shear. Next let us consider the elementary arch cut out from an interior shell of a symmetrically loaded group of multiple shells. As the shell arch is restrained at the ends, it would behave as a fixed arch. If the loading on the shell arch is symmetrically distributed across the cross section, one would expect the degree of indeterminacy to be three. But the degree of indeterminacy involved is only two, as no vertical reactions develop at the ends, the vertical loading on the shell being fully balanced by the sum of the vertical components of the specific shears. The elementary shell arch fixed at the ends and acted upon by the load and the components of the specific shear may be analyzed by any method applicable to fixed arches to determine the transverse moments M_ϕ. The column-analogy method[1] is particularly convenient for this purpose. Chinn[2] recommends the elastic-center approach which is analytically identical with column analogy. When once M_ϕ is found, Q_ϕ and N_ϕ may be found rather simply as explained in the example that follows.

9-6 The Modified Beam Method

In long shells with $l/a > 4$, the edges of the shell tend to deflect outward, causing the shell to flatten out. This behavior often alters the value of M_ϕ computed from the beam theory. Sometimes the sign of the moment also gets reversed. The flattening is primarily caused by the direct (membrane) stresses. Although the value of M_ϕ for such long shells would tend to be small, a correction for the flattening due to membrane stresses is called for if the results of the beam theory are to agree with the values of M_ϕ computed from the analytical theory. Parme and Conner[3] draw attention to the need for such a correction in their discussion on Chinn's paper already referred to. The resulting approach is called the modified beam method by them.

9-7 The Longitudinal Variation of M_ϕ

The beam theory would indicate that M_ϕ does not vary along the length of the shell. At the supports of the shell, it is reasonable to assume that M_ϕ is zero as the traverses are very stiff compared with the shell. Analytical theories indicate that the increase of M_ϕ from zero at the support to its maximum value at midspan closely follows a sine curve for shells with $l/r < 2.50$. Even for very long shells, M_ϕ does not build up to its maximum value before the quarter span section is reached. Hence it is rational to assume that M_ϕ is not constant as indicated by the beam theory but varies as a sine curve along the length of the shell.

9-8 Design Example

An example of an interior shell of a multiple group is given to illustrate the procedure employed to analyze a shell by the beam theory. Results obtained are shown compared with those obtained by using the method described in ASCE Manual 31. The problem considered is in fact

---

[1] Cross, Hardy, "Column Analogy," *University of Illinois, Engineering Experiment Station Bulletin* 215, vol. 28, no. 7, Oct. 14, 1930.

[2] Chinn, James, "Cylindrical Shell Analysis Simplified by Beam Method," *Journal of the American Concrete Institute*, vol. 55, p. 1183, May, 1959.

[3] Parme and Conner, discussion of James Chinn, "Cylindrical Shell Analysis Simplified by Beam Method," *Journal of the American Concrete Institute*, vol. 55, part 2, p. 1583, December, 1959.

Example 2 discussed in the Manual. The same problem has also been solved by Chinn by his method.

Design Example 9-1

ANALYSIS OF AN INTERIOR SHELL OF A SYMMETRICALLY LOADED MULTIPLE SHELL ASSEMBLY BY THE BEAM THEORY

1. *DATA*

Geometry

Span l	$= 62$ ft
Radius a	$= 31$ ft
Thickness d	$= 3.75$ in. $= 0.3125$ ft
Semicentral angle ϕ_c	$= 40°$

The shell has no edge beams (Fig. 9-3).

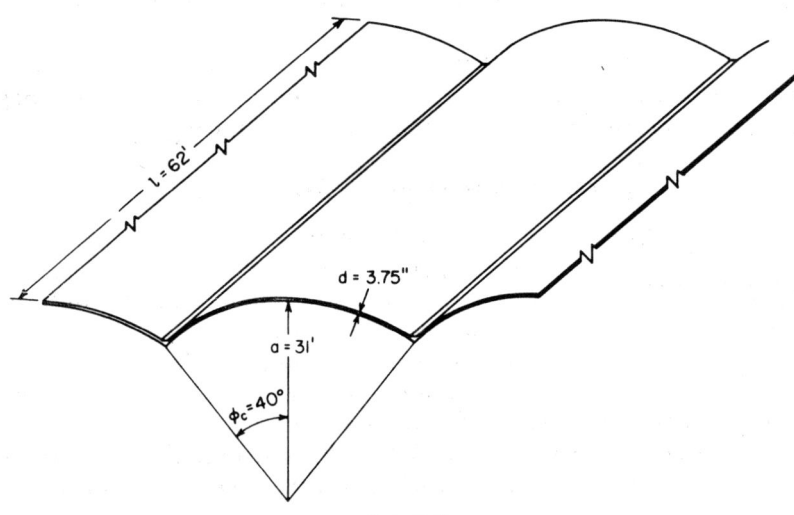

FIG. 9-3

Loading

Dead weight $= 47$ psf of surface area
Snow load $\quad= 25$ psf of horizontal projection

2. *TRIGONOMETRIC QUANTITIES*

$\phi_c \quad = 40° = 0.69813$ radian
$\sin \phi_c = 0.64279$
$\cos \phi_c = 0.76604$

3. *SECTIONAL PROPERTIES*

$\bar{z} =$ depth of neutral axis below crown

$$= a\left(1 - \frac{\sin \phi_c}{\phi_c}\right) = 31\left(1 - \frac{0.64279}{0.69813}\right) = 2.4574 \text{ ft}$$

G is the center of gravity and yy is the neutral axis (Fig. 9-4).

$$I_{yy} = a^3d \left[\phi_c + \sin \phi_c \left(\cos \phi_c - \frac{2 \sin \phi_c}{\phi_c} \right) \right]$$

$$= 31^3 \times 0.3125 \left[0.69813 + 0.64279 \left(0.76604 - \frac{1.28558}{0.69813} \right) \right]$$

$$= 9{,}309.6875 \, (0.69813 - 0.691267) = 63.911 \text{ ft}^4$$

One half of the shell cross section is divided into 8 equal parts 0-1, 1-2,..., 7-8 as

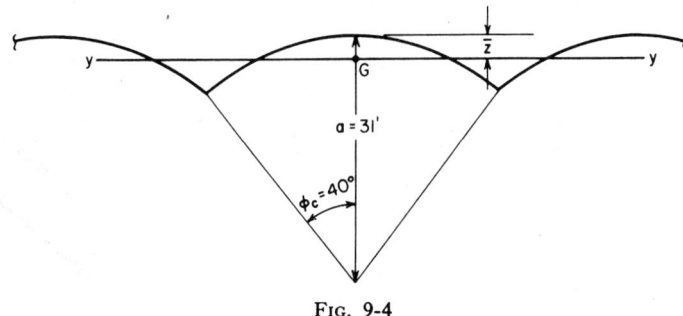

Fɪɢ. 9-4

shown in Fig. 9-5. The first moments of area at points 0, 1, 2,..., 8 about the neutral axis, found from formula (9-4), are given below.

Point	Angle from crown, deg	Q, ft³
0	40	0
1	35	6.6904
2	30	10.7554
3	25	12.5389
4	20	12.3879
5	15	10.6754
6	10	7.7796
7	5	4.0910
8	0	0

4. BEAM CALCULATION

The values of N_x at midspan and $N_{x\phi}$ at the traverses are calculated using formulas (9-1) and (9-3), respectively, namely,

$$N_x = \frac{M_{yy}}{I_{yy}} zd \qquad \text{and} \qquad N_{x\phi} = \frac{VQ}{2I_{yy}}$$

M_{vv} = bending moment at the midspan section due to applied loading

     = load per foot length $\times$ $l^2/8$

     = (47 $\times$ arc length + 25 $\times$ chord length) $(62^2/8)$

     = (47 $\times$ 43.284 + 25 $\times$ 39.853) $(62^2/8)$

     = 3,030.67 $\times$ $(62^2/8)$

     = 1,456,230 lb ft

V = shear force at traverse

   = load per foot length $\times$ $l/2$

   = 3,030.67 $\times$ $\dfrac{62}{2}$ = 93,951 lb

z is the vertical distance of any point from the neutral axis.

$$\therefore\ N_z = \frac{M_{vv}}{I_{vv}}\, dz$$

$$= \frac{1,456,230}{63.911} \times 0.3125\, z$$

$$= 7,120.3\, z$$

$$N_{z\phi} = \frac{V}{2I_{vv}}\, Q$$

$$= \frac{93,951}{2 \times 63.911}\, Q$$

$$= 735.03\, Q$$

Using these expressions, the values of N_z and $N_{z\phi}$ are calculated. They are given in Table (i).

Table (i)

Point	z, ft	Q, ft³	N_z, lb/ft	$N_{z\phi}$, lb/ft
0	+4.795	0	+34,145	0
1	+3.149	6.6904	+22,421	+4,918
2	+1.696	10.7554	+12,074	+7,906
3	+0.447	12.5389	+3,183	+9,216
4	−0.588	12.3879	−4,185	+9,105
5	−1.401	10.6754	−9,977	+7,847
6	−1.987	7.7796	−14,144	+5,718
7	−2.339	4.0910	−16,657	+3,007
8	−2.457	0	−17,497	0

5. ARCH CALCULATION

A slice of the shell, 1 ft in length, is considered as an arch which is in equilibrium under the action of the applied loads and the specific shears $\partial N_{z\phi}/\partial x$. The values of specific shears are obtained by dividing $N_{z\phi}$ by $l/2$. They are entered in column 5 of Table (ii). In column 2, which gives the lengths of the intervals, half of the length of interval 0-1 is entered against point 0, and half the length of interval 7-8 against point 8. Full lengths of intervals are entered against points 1 to 7. Columns 3 and 4 give the horizontal and vertical projections of the lengths given in column 2. Values of $\partial N_{x\phi}/\partial x$ per foot width at points 0, 1, 2,..., 8 are given in column 5. Columns 6 and 7 give the total horizontal and vertical forces due to $\partial N_{x\phi}/\partial x$ at points 0, 1, 2,..., 8. Columns 8 and 9 give the applied loading, due to dead weight and snow load on the shell, at points 0, 1, 2,..., 8. Column 6 gives the horizontal forces at all points. The vertical forces, which are obtained by the summation of columns 7, 8, and 9, are entered in column 10. All horizontal forces are acting inward while the vertical forces are designated $+ve$ if acting downward. The forces are shown in Fig. 9-6. At this stage a check can be applied on the calculations. The sum of loads in column 7 shall be equal to the sum of the loads in columns 8 and 9.

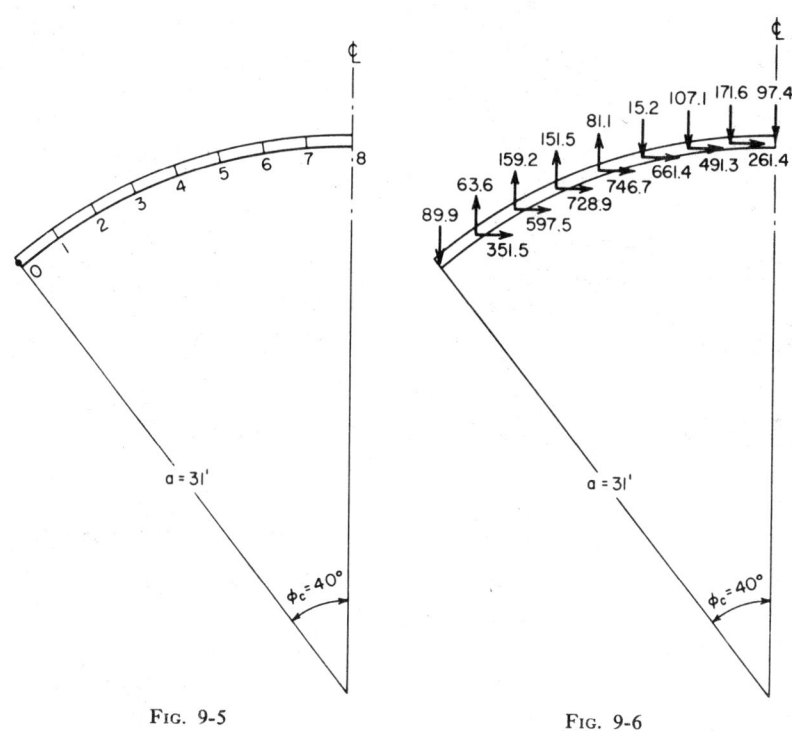

FIG. 9-5   FIG. 9-6

From the forces obtained above, the statically determinate moments at points 0, 1, 2,..., 8 are calculated. Sagging moments are assumed positive. The values are given below.

$$m_0 = 0 \qquad\qquad m_4 = -2,655.8$$
$$m_1 = -193.0 \qquad m_5 = -3,685.3$$
$$m_2 = -763.9 \qquad m_6 = -4,566.8$$
$$m_3 = -1,630.0 \qquad m_7 = -5,176.9$$
$$m_8 = -5,437.0$$

The bending-moment diagram is plotted (Fig. 9-7).

Table (ii)

Point	ds, ft	dy, ft	dz, ft	$\dfrac{\partial N_{x\phi}}{\partial x}$, lb/ft/ft	$\dfrac{\partial N_{x\phi}}{\partial x}dy = H$	$\dfrac{\partial N_{x\phi}}{\partial x}dz$	$g\ ds$	$p\ dy$	$V = \text{col. 7} + \text{col. 8} + \text{col. 9}$
(1)	(2)	(3)	(4)	(5)	(6)	(7)	(8)	(9)	(10)
0	1.353	1.055	0.847	0	0	0	63.5734	26.3725	+89.9459
1	2.705	2.215	1.551	158.6	351.5	−246.1	127.1468	55.3825	−63.6127
2	2.705	2.342	1.352	255.0	597.5	−344.9	127.1468	58.5500	−159.2396
3	2.705	2.451	1.143	297.3	728.9	−339.9	127.1468	61.2800	−151.4741
4	2.705	2.541	0.925	293.7	746.7	−271.8	127.1468	63.5275	−81.0888
5	2.705	2.612	0.700	253.1	661.4	−177.2	127.1468	65.3100	+15.2321
6	2.705	2.663	0.469	184.5	491.3	−86.7	127.1468	66.5800	+107.0755
7	2.705	2.694	0.236	97.0	261.4	−22.9	127.1468	67.3550	+171.6308
8	1.353	1.352	0.029	0	0	0	63.5734	33.8050	+97.3784

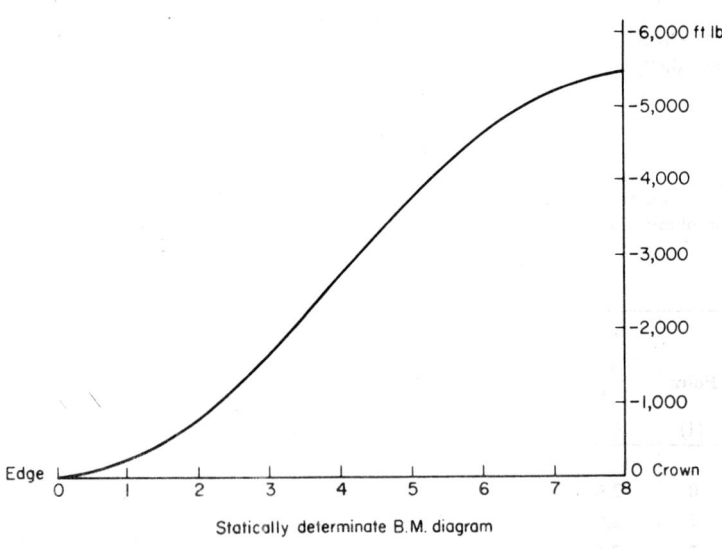

Statically determinate B.M. diagram

Fig. 9-7

Use of the Column-analogy Method

The transverse moments M_ϕ and the transverse stresses N_ϕ in the shell are evaluated using column analogy. The area of the bending-moment diagram and its moment about the yy axis are first calculated.

(i) Area of the bending-moment diagram. The area is found by using Simpson's rule for areas.

Area of the bending-moment diagram (for the entire cross section)

$$= -2 \left\{ \frac{2.705}{3} \left[(0 + 5,437.0) + 4 (193.0 + 1,630.0 + 3,685.3 + 5,176.9) \right. \right.$$
$$\left. \left. + 2 (763.9 + 2,655.8 + 4,566.8) \right] \right\}$$
$$= -1.80333 (5,437 + 42,746 + 1,5973)$$
$$= -1.80333 \times 64,156 = -115,696 \text{ units}$$

(ii) Moment of the area of the bending-moment diagram about the yy axis (from Fig. 9-7)

$$M = -2 [+1.3525 \times 5,437 \times 2.457 + 2.705 (+5,176.9 \times 2.339 + 4,566.8 \times 1.987$$
$$+ 3,685.3 \times 1.401 + 2,655.8 \times 0.588 - {}^\bullet 1,630.0 \times 0.447 - 763.9 \times 1.696$$
$$- 193.0 \times 3.149) - 1.3525 \times 0 \times 4.795]$$
$$= -172,898 \text{ units}$$

Properties of the Analogous Column

Load P on the analogous column $= -115,696/EI$, causing tension throughout. $M = 172,898/EI$, causing tension near the crown and compression at the edge.

$$\text{Area } A \text{ of the analogous column} = \frac{2R\phi_c}{EI} = \frac{43.284}{EI}$$

$$I_{vv} = \frac{63.911}{0.3125EI} = \frac{204.515}{EI}$$

$\therefore$ Stress on the analogous column, which is equal to the indeterminate moment in the shell, is given by

$$\frac{P}{A} \pm \frac{M}{I_{vv}} z = -\frac{115,696}{43.284} \pm \frac{172,898}{204.515} z$$
$$= -2,673 \pm 845.4 \, z$$

The indeterminate moments are subtracted from the determinate moments of Fig. 9-7 to obtain the final moments M_ϕ. The calculations are shown in Table.

Table (iii)

Point	$\dfrac{P}{A}$	$\dfrac{M}{I_{vv}}$	z	$\dfrac{M}{I_{vv}} z$	$m_i = \dfrac{P}{A} + \dfrac{M}{I_{vv}} z$	m_s	$M_\phi = m_s - m_i$
(1)	(2)	(3)	(4)	(5)	(6)	(7)	(8)
0	−2,673	845.4	4.795	4,054.1	+1,381	0	−1,381
1	−2,673	845.4	3.149	2,662.1	−11	−193	−182
2	−2,673	845.4	1.696	1,433.6	−1,239	−764	+475
3	−2,673	845.4	0.447	377.9	−2,295	−1,630	+665
4	−2,673	845.4	−0.588	−496.9	−3,170	−2,656	+514
5	−2,673	845.4	−1.401	−1,184.6	−3,858	−3,685	+173
6	−2,673	845.4	−1.987	−1,679.2	−4,352	−4,567	−215
7	−2,673	845.4	−2.339	−1,977.7	−4,651	−5,177	−526
8	−2,673	845.4	−2.457	−2,077.5	−4,750	−5,437	−687

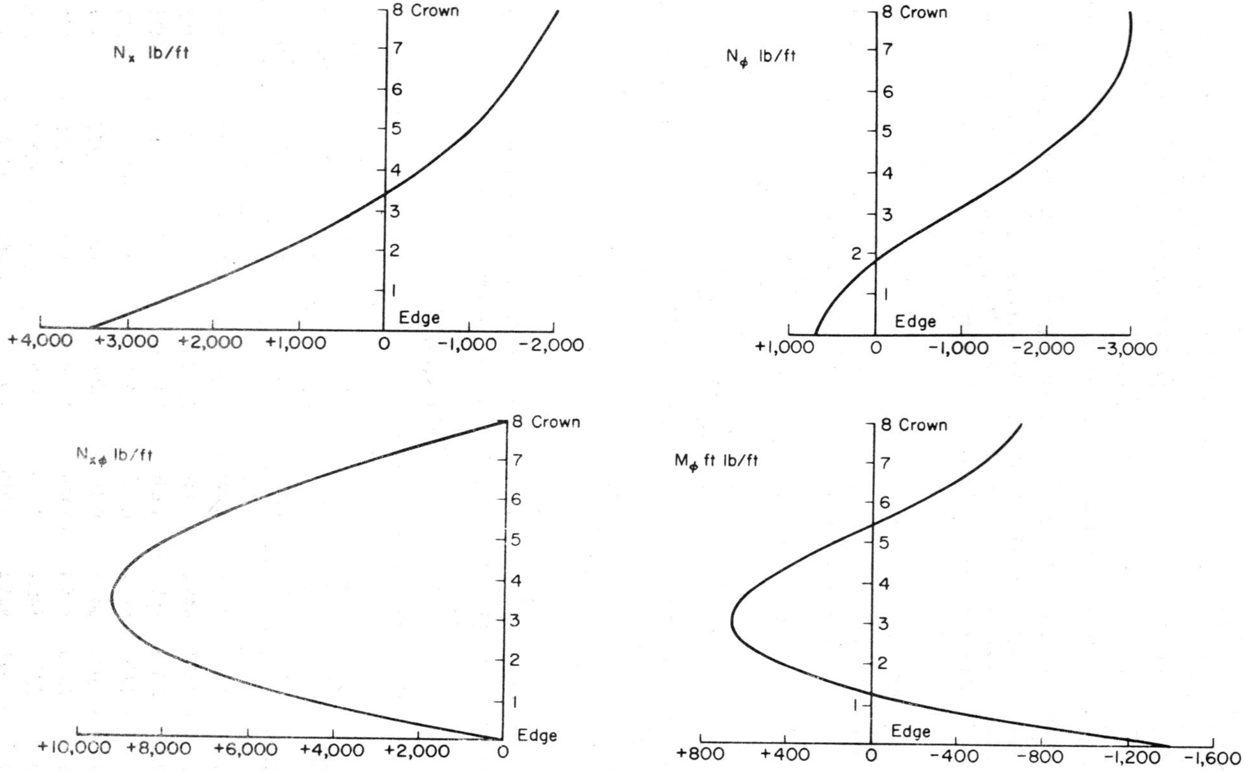

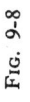

N_ϕ lb/ft

8 Crown
7
6
5
4
3
2
1
Edge

+1,000   0   −1,000   −2,000   −3,000

N_x lb/ft

8 Crown
7
6
5
4
3
2
1
Edge

+4,000   +3,000   +2,000   +1,000   0   −1,000   −2,000

M_ϕ ft lb/ft

8 Crown
7
6
5
4
3
2
1
Edge

+800   +400   0   −400   −800   −1,200   −1,600

$N_{x\phi}$ lb/ft

8 Crown
7
6
5
4
3
2
1
Edge

+10,000   +8,000   +6,000   +4,000   +2,000   0

FIG. 9-8

We shall now proceed to calculate N_ϕ. The horizontal force H_i acting at the centroidal level is given by $M/I_{yy} = 845.4$ lb/ft. This is equal to the horizontal force at the edge of the shell and acts outward as the statically determinate moment (Fig. 9-7) is hogging. This outward force of 845.4 lb at point 0 is in addition to the forces shown in Fig. 9-6. All the vertical and horizontal forces at points 0, 1, 2,..., 8 are thus known. N_ϕ is now calculated as shown in Table (iv). Columns 3 and 4 give the cumulative vertical and horizontal forces at various points. N_ϕ is equal to $(\Sigma H) \cos \phi + (\Sigma V) \sin \phi$. Q_ϕ can be found by adding the radial components of ΣH and ΣV.

Table (iv). Calculation of N_ϕ

Point (1)	ϕ, deg (2)	ΣV (3)	ΣH (4)	$(\Sigma H) \cos \phi$ (5)	$(\Sigma H) \sin \phi$ (6)	$N_\phi =$ col. 5 + col. 6 (7)
0	40	$+90$	$+845.4$	$+647.6$	$+57.9$	$+706$
1	35	$+26.4$	$+493.9$	$+404.6$	$+15.1$	$+420$
2	30	-132.8	-103.6	-89.7	-66.4	-156
3	25	-284.2	-832.5	-754.5	-120.1	-875
4	20	-365.3	$-1,579.2$	$-1,484.0$	-124.9	$-1,709$
5	15	-350.1	$-2,240.6$	$-2,164.2$	-90.6	$-2,255$
6	10	-243.0	$-2,731.9$	$-2,690.4$	-42.2	$-2,733$
7	5	-71.6	$-2,993.3$	$-2,981.9$	-6.2	$-2,988$
8	0	$+25.3$	$-2,993.3$	$-2,993.3$	0	$-2,993$

Note: The vertical and horizontal forces are designated $+ve$ if they are acting downward and outward, respectively. N_ϕ is therefore tensile if $+ve$ and compressive if $-ve$. The values of N_x, N_ϕ, $N_{x\phi}$, and M_ϕ are shown plotted in Fig. 9-8.

CHAPTER 10

NORTHLIGHT CYLINDRICAL SHELLS

10-1 Advantages of Northlight Shell Roofs

Northlight cylindrical shell roofs provide ample and uniform daylighting and large column-free spaces for factory buildings located anywhere in the Northern Hemisphere. They are especially suitable for factories inside which precision operations are to be carried out.

The daylighting performances of a sawtooth roof and a cylindrical northlight roof are shown compared in Fig. 10-1. The daylight reaching the working plane in both cases is composed of two parts, the sky component and the reflected component. The first depends only on the geometry of the opening as seen from the point of observation and the sky luminance. The second depends on the distribution of brightness on the reflecting ceiling surfaces. This brightness for a sawtooth-roof ceiling drops exponentially as one recedes from O to A (Fig. 10-1a). But, for a northlight shell, the reflecting ceiling being curved, its luminance is more or less uniform as we proceed from O to A (Fig. 10-1b). The reflected illumination reaching the working plane in both cases is obtained by averaging the luminance. It is obvious that the averaged luminance is much higher for the northlight shell than for the sawtooth roof of the same average slope. The result is that the working plane receives a higher value of the reflected component of daylight and the illumination is nearly shadow-free with a northlight shell roof. The inclination of the glazing window to the vertical is to be so selected that, while maximum advantage is taken of sky luminance, direct sunlight is excluded and collection of dust is prevented. For countries where the sky is generally overcast, the probable distribution of sky luminance (Fig. 10-2) is given by

$$B_\theta = B_z \frac{1 + 2 \sin \theta}{3}$$

From this point of view, it is an advantage to keep the window inclination to the vertical as small as possible. The need to exclude direct sunlight as far as possible sets the higher limit to this angle.   However, some direct sunlight penetration through northlight openings is unavoidable during early morning and late evening hours in northern latitudes.   The exact duration of such penetration on any day can be calculated with the

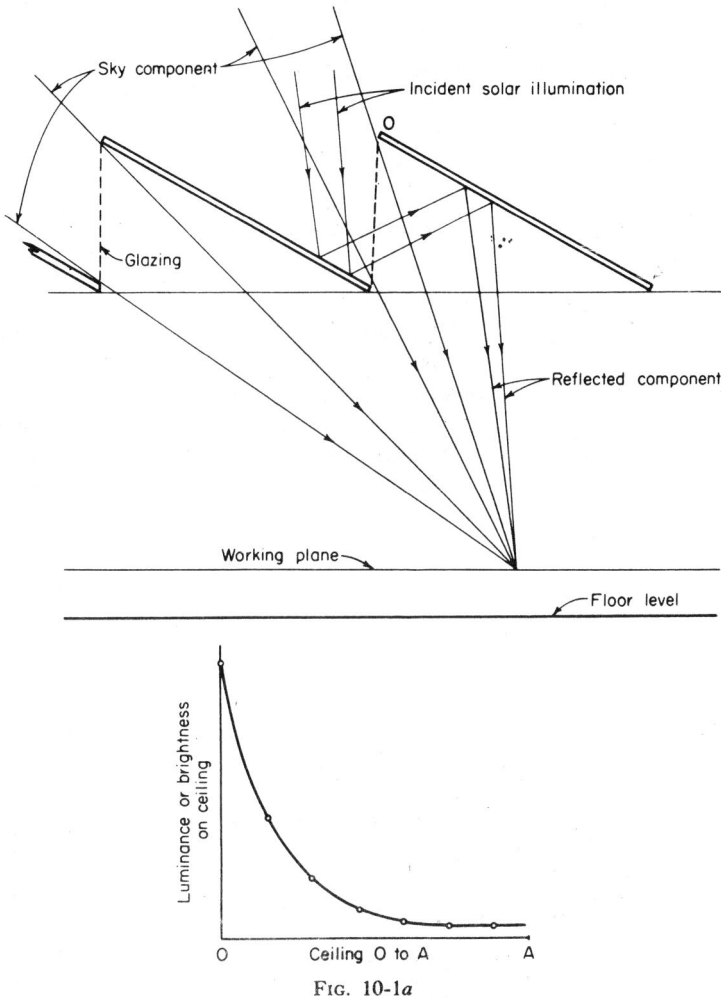

Fig. 10-1a

aid of solar charts available for the purpose.   The greater the inclination of the window to the vertical, the greater is the duration of penetration of direct sunlight.   We can at best hope only to exclude the noon sun.   This can be achieved if the inclination to the vertical is limited to the latitude of the place minus $23\frac{1}{2}°$.   The other consideration influencing

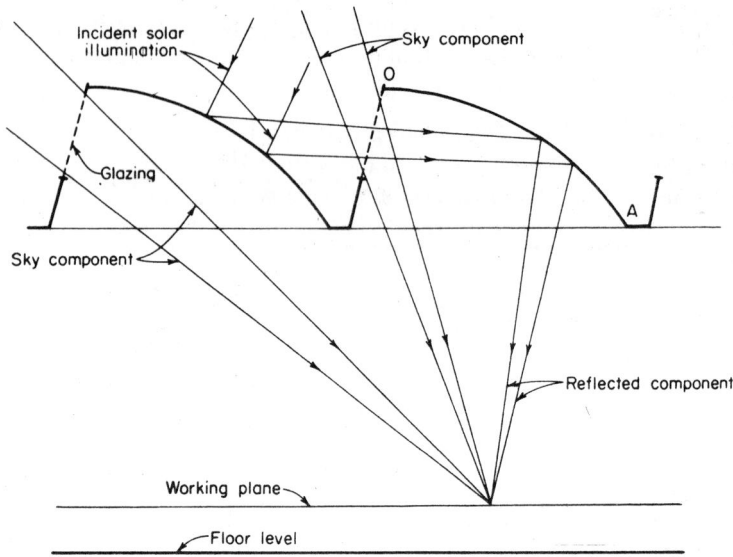

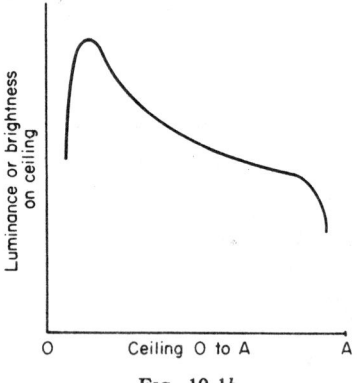

FIG. 10-1b

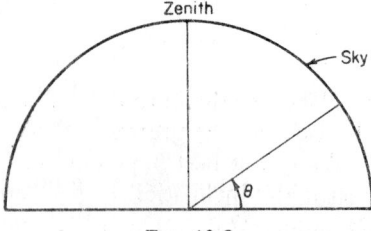

FIG. 10-2

the inclination of the glazing is the need to avoid dust and snow from collecting on it.   Based on these considerations, an angle of 15 to 30° to the vertical is suitable for Canada and the northern United States.   It is desirable to keep the window vertical in the southern United States [36].

10-2 Selection of Shell Parameters

The parts of a northlight shell are shown labeled in Fig. 10-3.   The bottom beam is generally referred to as the *valley* or *gutter beam*.   The top beam is usually known as the *glazing beam*.   The two beams may be connected by mullions carrying glazing of wired glass.

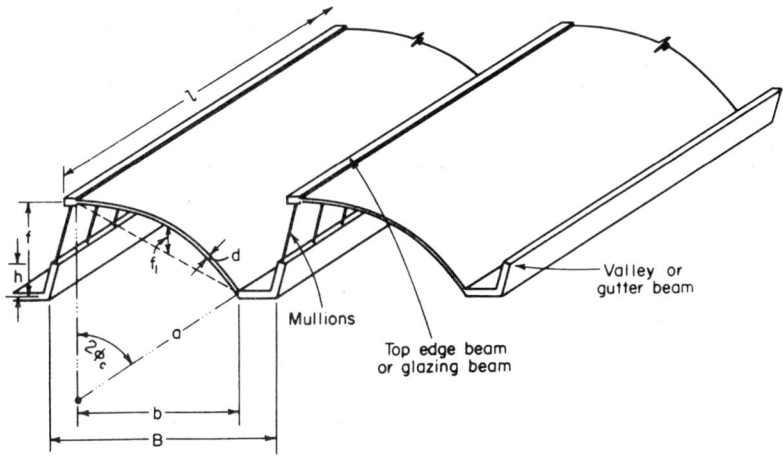

FIG. 10-3

The span of northlight shell, if not prestressed, is limited to about 30 m (100 ft) [37].   The more usual spans lie in the range of 12 to 25 m (40 to 80 ft) [31].   With prestressing, spans up to 150 ft are possible. Where large unobstructed column-free spaces are desired, the spacing of columns at right angles to the span may be made as large as 110 ft by supporting a number of shells on a single traverse (Fig. 10-4a) which may be a deep prestressed girder or a truss whose bottom chord is prestressed.   If a truss is employed, the diagonals may be designed to act as stiffeners for the shells.   Where closer spacing of columns in the transverse direction is not objectionable, each shell may be provided with its own traverse simply supported on columns (Fig. 10-4b). Shells may be designed to be longitudinally continuous over two or more spans.   Radius of northlight shells varies between 20 and 40 ft [36]. The Dutch Report [31] gives the following guiding rules for the selection of shell dimensions: Radius *a* is to be less than 12 m; the normal ratio of width b to span l may be between 1:2 and 1:4; the height f shall not be smaller than $l/6$; the height f_1 of the circular segment shall as a rule be greater than $l/18$; the height h of the gutter edge shall not generally be less than $l/18$.

10-3 Methods of Analysis

The analysis of a northlight shell is more time-consuming than that of a symmetric cylindrical shell. This is because the former is asymmetric and the boundary conditions that exist at the top and bottom edges, being somewhat indeterminate, are not amenable to exact formulation.

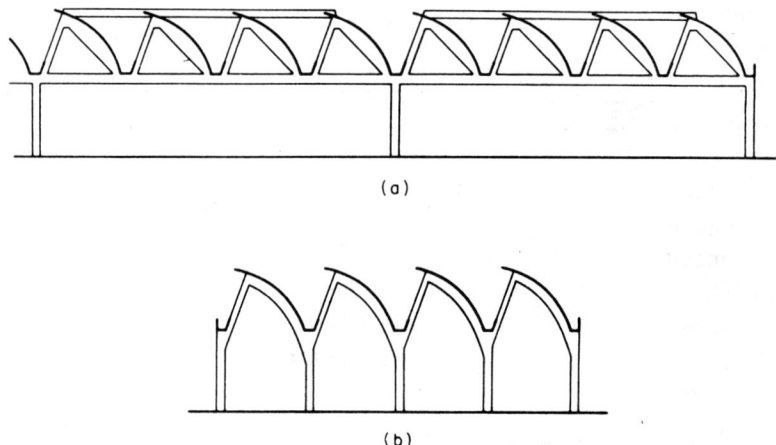

(a)

(b)

Fig. 10-4

Because eight boundary conditions are involved, as against the four for symmetric cylindrical shells, the design time is roughly doubled. If the shell is long, i.e., $l/a \geqslant \pi$, the beam method described in Chapter 9 may be employed to give satisfactory results [38]. If the shell is short, the analytical method involving the formulation of eight boundary conditions is recommended. The labor involved can be materially reduced by using tabulated values available in the book "Circular Cylindrical Shells" by Rüdiger and Urban [39] or the ASCE Manual 31 [3]. Simplifications employed by various authors in applying the analytical method are discussed in detail in a later section.

10-4 The Beam Method

The beam method described in the last chapter needs a small modification before it can be applied to the northlight shell. Instead of using the Mc/I formula, we have now to use the formulas relating to unsymmetrical bending because the shell-beam cross section is asymmetric. Otherwise the procedure remains the same as before. The application of the method is explained in detail, with the aid of the design example included at the end of this chapter. The method described has the advantage that it avoids the use of higher mathematics, but it is exactly equivalent to the Lundgren beam method [2].

10-5 The Analytical Method

For short or intermediate northlight shells, we have to employ one of the analytical methods such as the D-K-J theory and apply eight boundary conditions, four at the top and four at the bottom edge. The boundary conditions given below may be regarded as reasonably rigorous. They relate to an interior shell of a group of shells.

Bottom Edge

(i) The displacements of the shell edge and the gutter beam at their junction along the normal to the shell are equal ($w_{shell} = w_{edge\,beam}$).

(ii) The displacements of the shell edge and the edge beam in the tangential direction are equal ($v_{shell} = v_{edge\,beam}$).

(iii) The displacements of the shell edge and the edge beam are equal along the generator at their junction ($u_{shell} = u_{edge\,beam}$).

(iv) The rotation ϑ of the tangent at the shell edge is equal to the twist of the gutter beam.

Top Edge

(i) The displacements of the shell edge and the edge beam are equal along the generator at their junction.

(ii) The rotation ϑ of the tangent to the shell edge at the junction is equal to the angle of twist of the edge beam.

(iii) The resultant of the vertical and horizontal reactions and of concentrated loads at the shell edge lies in the plane of the mullions. This is a consequence of the inability of the mullions to receive any forces normal to their plane.

(iv) The deflection of the upper shell edge in the direction of the window mullions is equal to the deflection of the gutter beam in the same direction. This implies that the mullions are extremely rigid in their own planes.

If an external shell (Fig. 10-5) on the north side is considered, the boundary condition (iv) corresponding to the top edge alone will change. It is modified as follows:

(iv) The deflection of the top edge of the shell in the direction of the mullions is zero. This is because, in this case, the valley edge beam is expected to be prevented from deflecting by the provision of a wall or a line of columns.

If an external shell on the south side is considered (Fig. 10-6), all boundary conditions except (iv) corresponding to the top edge will remain the same as for the interior shell. Boundary condition (iv) for the top edge is rewritten as

(iv) The deflection of the top edge in direction of the window posts is equal to the deflection of the valley beam of the interior shell in the same direction.

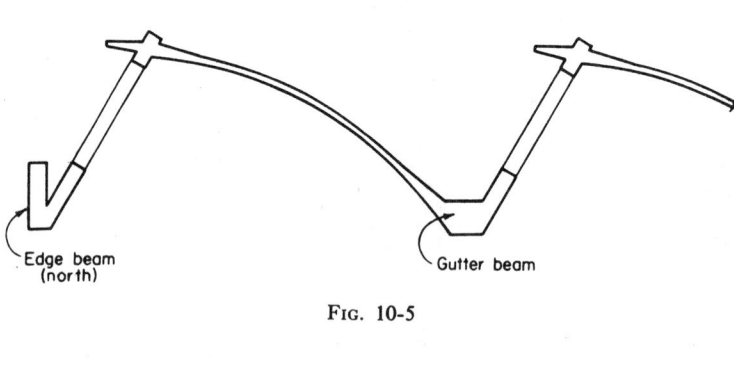

Edge beam (north) Gutter beam

FIG. 10-5

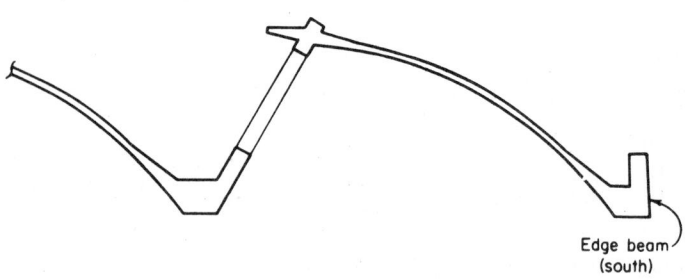

Edge beam (south)

FIG. 10-6

The top edge beam is usually of much smaller proportions than the valley beam. Hence certain approximations are permissible in formulating the boundary conditions at the top edge. Thus boundary condition (ii) corresponding to the top edge may be simplified as $M_\phi = 0$. This implies that the top edge beam has no torsional rigidity. This simplified boundary condition is employed by Mast [36]. If the top edge beam amounts to no more than a local thickening of the shell, we may, following Mast [36] and Wilby [40], assume that the edge beam cannot carry any tension and formulate the first boundary condition at the top edge as $\partial N_{x\phi}/\partial \phi = 0$. But in most northlight shells the top edge beam is not of such negligible dimensions and it would seem desirable to adhere to the more rigorous formulation of this boundary condition. Apart from resorting to simplified boundary conditions, there are also other means employed to shorten the analysis. Mast [36] prefers to solve the shell in the first instance by formulating boundary conditions at the bottom edge, ignoring conditions at the top edge. He next solves for the top edge and returns to the bottom edge, and so on. By this means, an iterative solution is possible. The advantage is that one need solve only for four unknowns at a time. Wilby uses the Tottenham method, which is nothing but the Schorer method with the additional assumption that the edge beams at the top and bottom edges do not rotate. While this assumption can be justified to some extent in dealing with the valley beam, its application to the top edge beam is open to question. The other boundary condition of $\partial N_{x\phi}/\partial\phi = 0$ which he employs at the top would imply a beam of extremely negligible proportions. To assume at the same time that the top edge would not rotate would mean that it has infinite torsional rigidity. These two assumptions are clearly contradictory. Wilby shows, by means of an example, that the stress resultants are insensitive to this error in formulating the boundary conditions at the top. A limitation of the Wilby method is that, like the beam method, it is applicable only to long shells.

10-6 Selection of Shapes

In circular cylindrical northlight shells, because of the pronounced asymmetry of the profile, the center of gravity and the shear center are far apart (Fig. 10-7a). Hence the cross section is subjected to a twisting moment which is countered by the couple formed by the forces that develop in the mullions. This means that the equilibrium of the shell cannot be maintained in the absence of the mullions. This can be a serious disadvantage if moving forms are employed. A more suitable profile from this point of view is that suggested by Mihailescu and Ungureanu [41]. In this S-shaped profile the center of gravity and the shear center are located very close to each other (Fig. 10-7b).

10-7 Preliminary Design

The simple rules given by Chronowicz [42] are often useful for purposes of preliminary design.

Design Example 10-1

ANALYSIS OF A NORTHLIGHT SHELL BY BEAM THEORY

1. *DATA*

Span $l = 60$ ft

Radius $a = 18$ ft

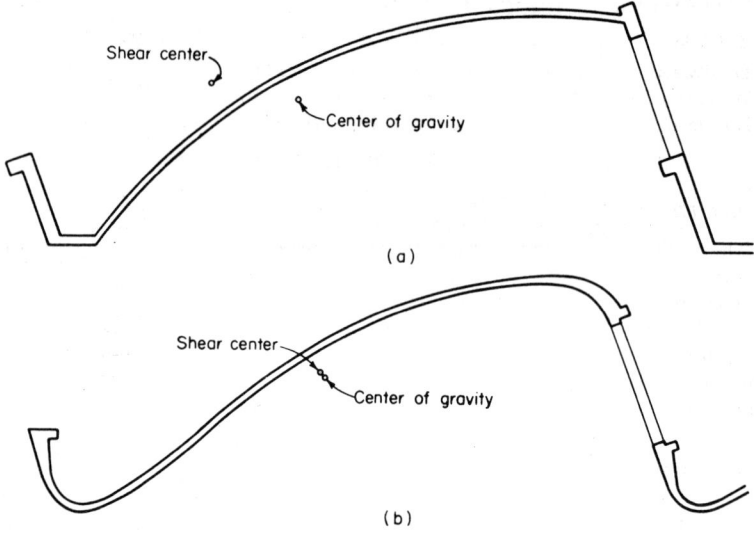

(a)

Shear center

Center of gravity

(b)

FIG. 10-7

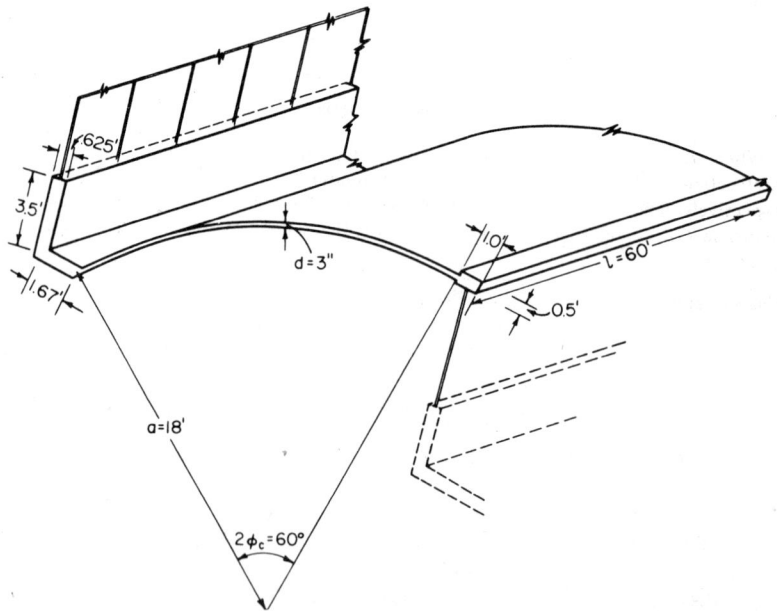

FIG. 10-8

Thickness of the shell $d = 3$ in.
Central angle $2\phi_c = 60°$, (Figure 10-8).

2. LOADS

Dead weight of the shell　　　　$= 37.5$ psf of surface area
Waterproofing, finishing, etc. $= 7.5$ psf of surface area
Live load　　　　　　　　　　$= 15.0$ psf of surface area

　　　　Total　　　　　　$g = 60.0$ psf of shell surface

3. SECTIONAL PROPERTIES

The sectional properties of the shell are calculated from Fig. 10-9. The origin is chosen at O, the midpoint of the 60° arc, the axes Oz and Oy being chosen along the normal and tangent to the shell arc at O. The cross section is divided into 24 equal parts, the reinforcement being replaced by equivalent concrete area. The shell itself is divided into 12 parts, the lower edge beam into 10 parts, and the upper edge beam into 2 parts (Fig. 10-9).

Equivalent area of each element into which the shell and the edge beams are divided

$$= 18 \left(\frac{60}{12} \times \frac{\pi}{180}\right) \times 0.25$$

$$= 0.3925 \text{ ft}^2$$

The coordinates of the points 0 to 24 are entered in columns 2 and 4 of Table (i).

The mean values of the coordinates, which will be approximately equal to the coordinates of the centers of gravity of the elements, are entered in columns 3 and 5. The coordinates of the centroid G of the cross section are given by

$$\left(\frac{\sum z \, \Delta A}{\sum \Delta A}, \frac{\sum y \, \Delta A}{\sum \Delta A}\right) = \left(+\frac{23.553}{24}, -\frac{79.667}{24}\right)$$

$$= (+0.9813, -3.3194)$$

When the position of the centroid is thus determined, the moments and product of inertia about the axes Gz_1 and Gy_1, passing through the centroid and parallel to Oz and Oy respectively, are next calculated. From these values, the orientation of the principal axes of inertia GZ and GY is arrived at and the principal moments of intertia I_{ZZ} and I_{YY} about these axes are calculated. The detailed calculations are given below.

Moments of Inertia

$$I_{z_1 z_1} = \Delta A \left[\sum y^2 - 24 \, (3.3194)^2\right] = \Delta A[1,483.76 - 24(3.3194)^2]$$

$$= 1,219.318 \, \Delta A$$

$$I_{v_1 v_1} = \Delta A \left[\sum z^2 - 24 \, (0.9813)^2\right] = \Delta A[50.715 - 24(0.9813)^2]$$

$$= +27.604 \, \Delta A$$

Product of Inertia

$$I_{v_1 z_1} = \Delta A \left[\sum yz - 24(-3.3194 \times 0.9813)\right] = \Delta A(-32.468 + 24 \times 3.3194 \times 0.9813)$$

$$= +45.684 \, \Delta A$$

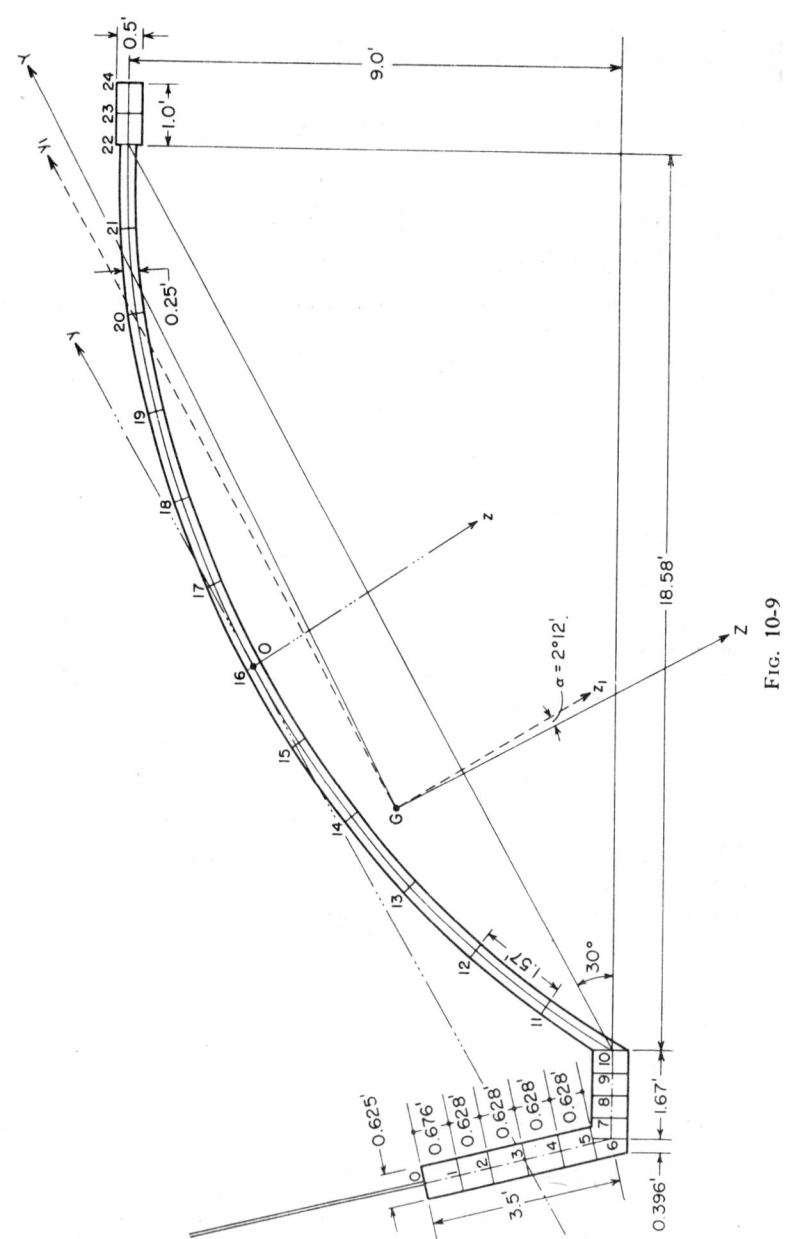

Fig. 10-9

If the principal axes are inclined to Gz_1 and Gy_1 at an angle α (the angle z_1GZ measured in the counterclockwise direction from Gz_1) then

$$\tan 2\alpha = -\frac{2I_{y_1z_1}}{I_{z_1z_1} - I_{y_1v_1}} = -\frac{2 \times 45.684}{1,219.318 - 27.604} = -0.07674$$

$$\therefore \ 2\alpha = -(4°24')$$

$$\therefore \ \alpha = -(2°12')$$

Therefore, the principal moments of inertia are

$$I_{ZZ} = I_{z_1z_1}\cos^2\alpha + I_{y_1v_1}\sin^2\alpha - I_{y_1z_1}\sin 2\alpha$$

$$= [1,219.318(0.99926)^2 + 27.604(0.03839)^2 + 45.684 \times (0.07672)]\,\varDelta A$$

$$= 1,221.0594\,\varDelta A = 1,221.0594 \times 0.3925 = 479.266 \text{ ft}^4$$

$$I_{YY} = I_{y_1v_1}\cos^2\alpha + I_{z_1z_1}\sin^2\alpha + I_{y_1z_1}\sin 2\alpha$$

$$= [1,219.318(0.03839)^2 + 27.604(0.99926)^2 - 45.684 \times 0.07672]\,\varDelta A$$

$$= 25.855\,\varDelta A = 25.855 \times 0.3925 = 10.148 \text{ ft}^4$$

4. BEAM CALCULATION

The coordinates of the points with reference to the principal axes GZ and GY are next calculated using the formulas

$$Z = (z - 0.9813)\cos\alpha - (y + 3.3194)\sin\alpha$$
$$Y = (z - 0.9813)\sin\alpha + (y + 3.3194)\cos\alpha$$

where $\alpha = 2°12'$.

These coordinates are entered in columns 11 and 12 of Table (i). The load is resolved into its components W_Z and W_Y along the two principal axes. W_Z causes bending about the YY axis and W_Y about the ZZ axis. The moments caused by W_Z and W_Y are denoted by M_Z and M_Y, respectively.

Calculation of N_x

The stress N_x/d at any point in the shell is then given by the formula

$$\frac{N_x}{d} = \frac{M_Z}{I_{YY}}Z + \frac{M_Y}{I_{ZZ}}Y$$

where

$$M_Z = \frac{W_Z l^2}{8}$$

$$M_Y = \frac{W_Y l^2}{8}$$

The load per foot length of the shell and its components W_Z and W_Y will now be calculated.

(i) Load from shell $= g \times$ arc length $= 60\left(18 \times 60\,\dfrac{\pi}{180}\right) = 1,131$ lb

(ii) Weight of the two edge beams $\hspace{2cm} = \hspace{0.5cm} 560$ lb

(iii) Weight of window posts (approximately) $\hspace{1cm} = \hspace{0.5cm} \underline{\hspace{0.3cm}250 \text{ lb}}$

$\hspace{10cm} 1,941$ lb

Say, 1,950 lb/ft length of shell

Table (i).

Point	z, ft		y, ft		Σz	Σy	z^2	y^2
1	2	3	4	5	6	7	8	9
0	−1.768		−9.397		0	0		
		−1.44		−9.499			2.0736	90.23
1	−1.123		−9.601		−1.44	−9.499		
		−0.826		−9.695			0.682	93.993
2	−0.53		−9.790		−2.266	−19.194		
		−0.228		−9.884			0.052	97.693
3	+0.07		−9.978		−2.494	−29.078		
		+0.373		−10.072			0.14	101.45
4	+0.673		−10.167		−2.121	−39.150		
		+0.972		−10.261			0.945	105.29
5	+1.272		−10.356		−1.149	−49.411		
		+1.45		−10.364			2.10	107.412
6	+1.618		−10.372		+0.301	−57.775		
		+1.717		−10.201			2.95	104.06
7	+1.816		−10.030		+2.018	−69.976		
		+1.92		−9.857			3.69	97.16
8	+2.014		−9.684		+3.938	−79.833		
		+2.113		−9.513			4.465	90.05
9	+2.212		−9.342		+6.051	−89.346		
		+2.311		−9.171			5.34	84.11
10	+2.41		−9.0		+8.362	−98.517		
		+2.05		−8.31			4.20	69.056
11	+1.69		−7.61		+10.412	−106.827		
		+1.39		−6.89			1.93	47.47
12	+1.09		−6.16		+11.802	−113.719		
		+0.85		−5.41			0.72	29.27
13	+0.61		−4.66		+12.652	−119.127		
		+0.44		−3.89			0.194	15.13
14	+0.27		−3.13		+13.092	−123.017		
		+0.17		−2.35			0.03	5.52
15	+0.07		−1.57		+13.262	−125.367		
		+0.04		−0.78			0	0.61
16	0		0		+13.302	−126.147		
		+0.04		+0.78			0	0.61
17	+0.07		+1.57		+13.342	−125.367		
		+0.17		+2.35			0.03	5.52
18	+0.27		+3.13		+13.512	−123.017		
		+0.44		+3.89			0.194	15.13
19	+0.61		+4.66		+13.952	−119.127		
		+0.85		+5.41			0.72	29.27
20	+1.09		+6.16		+14.802	−113.717		
		+1.39		+6.89			1.93	47.47
21	+1.69		+7.61		+16.192	−106.827		
		+2.05		+8.31			4.20	69.056
22	+2.41		+9.0		+18.242	−98.517		
		+2.53		+9.21			6.40	84.82
23	+2.66		+9.43		+20.772	−89.307		
		+2.78		+9.64			7.73	92.93
24	+2.91		+9.86		+23.553	−79.667		
$\Sigma =$							50.715	1,483.76

Beam Calculation

zy	Z, ft	Y, ft	$\dfrac{N_x}{d}$, psf	Force on each element, lb	$\dfrac{(\text{col. 14})}{450}$	$\dfrac{\partial N_{x\phi}}{\partial x}$	$N_{x\phi}$, lb/ft
10	11	12	13	14	15	16	17
	−2.514	−6.178	−187,026.3			0	0
+13.68				−63,585.1	−141.3		
	−1.862	−6.358	−136,974.2			−141.3	−4,239
+8.01				−44,730.7	−99.40		
	−1.262	−6.524	−90,952.7			−240.70	−7,221.03
+2.25				−26,503.3	−58.9		
	−0.651	−6.687	−44,095.7			−299.60	−8,987.91
−3.76				−8,185.6	−18.19		
	−0.045	−6.855	+2,385.9			−317.79	−9,533.61
−9.974				+10,058.2	+22.35		
	+0.560	−7.020	+48,866.1			−295.44	−8,863.08
−15.03				+24,379.7	+54.18		
	+0.907	−7.023	+75,361.5			−241.26	−7,237.77
−17.52				+32,293.7	+71.76		
	+1.0916	−6.674	+89,192.3			−169.50	−5,084.85
−18.93				+37,719.5	+83.82		
	+1.276	−6.321	+103,008.9			−85.68	−2,570.22
−20.10				+43,595.2	+96.88		
	+1.461	+5.971	+116,839.4			+11.20	+336.12
−21.194				+48,573.9	+107.94		
	+1.646	−5.621	+130,670.8			+119.14	+3,574.38
−17.036				+39,459.0	+87.69		
	+0.873	−4.260	+70,394.8			+206.83	+6,204.99
−9.58				+17,555.4	+39.01		
	+0.218	−2.835	+19,059.4			+245.84	+7,375.35
−4.599				−831.8	−1.85		
	−0.320	−1.354	−23,297.8			+243.99	+7,319.91
−1.71				−15,380.1	−34.18		
	−0.718	+0.162	−55,072.3			+209.81	+6,294.57
−0.40				−25,774.8	−57.28		
	−0.978	+1.713	−76,264.1			+152.53	+4,576.26
−0.03				−32,150.8	−71.45		
	−1.108	+3.279	−87,561.9			+81.08	+2,432.88
+0.03				−34,486.1	−76.64		
	−1.098	+4.851	−88,163.9			+4.44	+133.8
+0.40				−32,765.9	−72.87		
	−0.159	+6.418	−78,796.0			−68.43	−2,050.59
+1.71				−26,967.5	−59.93		
	−0.677	+7.960	−58,618.1			−128.36	−3,848.43
+4.599				−16,926.2	−37.61		
	−0.255	+9.476	−27,630.2			−165.97	−4,976.85
+9.58				−2,927.1	−6.51		
	+0.288	+10.948	+12,715.1			−172.480	−5,172.0
+17.036				+14,752.2	+32.78		
	+0.955	+12.365	+62,455.5			−139.70	−4,188.48
+23.30				+27,942.4	+62.09		
	+1.188	+12.804	+79,926.3			−77.61	−2,325.69
+26.80				+34,799.6	+77.33		
	+1.421	+13.244	+97,396.6			0	0
−32.468				85.9			

$$\therefore \quad M_Z = \frac{(1,950 \times \cos 27°40') \times 60^2}{8} = 776.2185 \text{ kip ft}$$

$$M_Y = \frac{-(1,950 \times \sin 27°48') \times 60^2}{8} = -409.257 \text{ kip ft}$$

$$\therefore \quad \left(\frac{N_x}{d}\right) = \frac{776,218.5 \times Z}{(25.855) \times 0.3925} - \frac{409,257 \times Y}{479.266} = 76,489.845Z - 853.925Y$$

From the above expression the values of N_x/d at all points are calculated and entered in column 13 of Table (i). N_x is zero on the neutral axis. Hence the equation to the neutral axis is given by $76,489.845Z - 853.925Y = 0$ or $89.57Z - Y = 0$.

The longitudinal force on each elemental area is calculated as the average of N_x/d at the two points adjacent to the element, multiplied by the area of the element, namely, 0.3925. The values of the longitudinal force are given in column 14 of Table (i).

Calculation of $N_{x\phi}$

The shear stress is calculated using the principle that in a beam subjected to uniform loading the difference between the specific shears at any two points is equal to the longitudinal force between those two points divided by the bending-moment factor, which is $l^2/8$ for a simply supported beam. Hence

$$\text{Difference between specific shears at points } A \text{ and } B = \frac{\text{longitudinal force on } AB}{l^2/8}$$

$l^2/8 = 450$. The values of the longitudinal force divided by 450 are entered in column 15. The specific shears at all points given in column 16 are obtained by adding the differences of specific shears already obtained. The shear stress $N_{x\phi}$ is equal to the specific shear multiplied by $l/2$. The values of $N_{x\phi}$ are given in column 17.

5. ARCH CALCULATION

A slice of the shell, 1 ft in length, is regarded as an arch subjected to the action of external loads and the specific shears. The calculations are shown in Table (ii). For this calculation, we refer again to the originally chosen axes Oz and Oy.

Columns 2, 3, 4, and 5 of Table (i) giving the values of the coordinates are not repeated in Table (ii). The external loads on all elemental areas are entered in column 2. The components of the load in the Oz and Oy directions are entered in columns 3 and 4.

The specific shears $\partial N_{x\phi}/\partial x$ on all elemental areas are entered in column 5. Columns 6 and 7 give the projections of the distances between the midpoints of the elements in the Oz and Oy directions.

Specific shears on elemental areas in the z and y directions are calculated and entered in columns 8 and 9. The resultant forces q_z and q_v in the two directions are calculated as below and entered in columns 10 and 11.

$$q_z = w_z + \frac{\partial N_{x\phi}}{\partial x} \Delta z$$

$$q_v = w_v + \frac{\partial N_{x\phi}}{\partial x} \Delta y$$

Columns 12 and 13 give the cumulative loads N_z and N_y starting from point 0 up to point 22.

$$\left(N_z = \sum q_z \quad \text{and} \quad N_v = \sum q_v\right)$$

Now starting from point 0 (and assuming that the shell is not connected to the adjacent shells), the increments of the bending moments from point to point of the cross section are calculated as

$$\Delta M_\phi = N_y\, \delta z - N_z\, \delta y$$

and entered in column 18 [Table (ii)]. δz and δy here are the projected lengths of the elements and are entered in columns 14 and 15. They are obtained from columns 2 and 4 of Table (i). The cumulative summation of these increments gives the transverse bending moment $M_{\phi,0}$ at all the points of the cross section. This is entered in column 19. This procedure has resulted in a bending moment of $+6,873.392$ at point 24. But as the only indeterminate loading on the arch is the force in the window posts, this moment must be balanced by the moment caused by the force in the window posts.

This moment (i.e., correcting moment) is a linear function of y and z and may be represented as

$$M_{\phi,1} = Az + By + C$$

The constants can be evaluated by the boundary conditions:

(i) $M_{\phi,1} = 0$ at point 0.
(ii) $M_{\phi,1} = 0$ at point 24.
(iii) $M_{\phi,1}$ at point 16 $(y = 0,\ z = 0)$ = component of the force in window posts perpendicular to the line joining 16 and 24, multiplied by the perpendicular distance between 16 and 24.

The distance between the two rows of window posts is nearly 20 ft.

$$\therefore\ \text{Axial force in window posts} = \frac{-6,873.392}{20}$$
$$= -343.6696\ \text{lb}$$

$$\therefore\quad \text{Moment } M_{\phi,1} \text{ at point 16} = -343.6694 \times 9.64 \times 0.95372$$
$$= -343.6694 \times 9.19386$$
$$= -3,159.6502\ \text{lb ft}$$

From boundary condition (iii),

$$C = -3,159.6502$$
$$\therefore\ M_{\phi,1} = Az + B_y - 3,159.6502$$

Applying boundary conditions 1 and 2 we get

$$0 = -1.768A - 9.397B - 3,159.6502$$
$$-6,873.392 = 2.91A + 9.86B - 3,159.6502$$

Solving the above two equations,

$$A = -377.6864$$
$$B = -265.18044$$

Hence

$$M_{\phi,1} = -377.6864z - 265.18044y - 3,159.6502$$

According to the above expression for $M_{\phi,1}$, the values of $M_{\phi,1}$ for every elemental area have been calculated and entered in column 20 [Table (ii)].

Net transverse moment $M_\phi = M_{\phi,0} + M_{\phi,1}$

The values are entered in column 21 [Table (ii)].

Table (ii). Arch Calculation

Point	W	W_z	W_y	$\dfrac{\partial N_{z\phi}}{\partial x}$	Δz	Δy	$\dfrac{\partial N_z}{\partial x}\,\Delta z$	$\dfrac{\partial N_{z\phi}}{\partial x}\,\Delta y$	q_z	q_y	N_z	N_y	δz
1	2	3	4	5	6	7	8	9	10	11	12	13	14
0	0	0	0	0	+0.328	-0.102	0	0	0	0	0	0	0
1	187.56	162.43	-93.78	-141.30	+0.614	-0.196	-86.76	+27.70	+75.67	-66.09	+75.67	-66.09	+0.645
2	62.56	54.18	-31.28	-240.70	+0.598	-0.189	-143.94	+45.49	-89.76	+14.21	-14.09	-51.87	+0.593
3	62.56	54.18	-31.28	-299.60	+0.601	-0.188	-180.06	+56.32	-125.88	+25.04	-139.97	-26.83	+0.604
4	62.56	54.18	-31.28	-317.79	+0.599	-0.189	-190.35	+60.06	-136.17	+28.78	-276.14	+1.95	+0.599
5	62.56	54.18	-31.28	-295.44	+0.478	-0.103	-141.22	+30.43	-87.04	-0.85	-363.18	+1.10	+0.599
6	36.45	31.57	-18.22	-241.26	+0.267	+0.163	-64.42	-39.33	-32.85	-57.55	-396.03	-56.44	+0.346
7	36.45	31.57	-18.22	-169.50	+0.203	+0.344	-34.41	-58.31	-2.84	-76.53	-398.86	-132.97	+0.198
8	36.45	31.57	-18.22	-85.67	+0.193	+0.344	-16.54	-29.47	+15.04	-47.69	-383.83	-180.66	+0.198
9	36.45	31.57	-18.22	+11.20	+0.198	+0.342	+2.22	+3.49	+33.79	-14.39	-350.04	-195.05	+0.198
10	36.45	31.57	-18.22	+119.15	-0.861	+0.861	-31.10	+102.59	+0.47	+84.37	-349.57	-110.68	+0.198
11	94.2	81.58	-47.1	+206.83	-0.66	+1.42	-136.51	+293.70	-54.93	+246.60	-404.50	+135.92	-0.720
12	94.2	81.58	-47.1	+245.85	-0.54	+1.48	-132.76	+363.85	-51.18	+316.75	-455.67	+452.67	-0.60
13	94.2	81.58	-47.1	+244.00	-0.41	+1.52	-100.04	+370.88	-18.46	+323.78	-474.13	+776.45	-0.48
14	94.2	81.58	-47.1	+209.82	-0.27	+1.53	-56.66	+323.12	+24.92	+276.02	-449.21	+1,052.47	-0.34
15	94.2	81.58	-47.1	+152.54	-0.13	+1.57	+19.83	+239.49	+61.75	+192.39	-387.46	+1,244.86	-0.20
16	94.2	81.58	-47.1	+81.10	0	+1.57	0	+127.32	+81.58	+80.22	-305.88	+1,325.08	-0.07
17	94.2	81.58	-47.1	+4.46	+0.13	+1.57	+0.58	+7.00	+82.16	-40.10	-223.72	+1,284.98	+0.07
18	94.2	81.58	-47.1	-68.35	+0.27	+1.53	-18.46	-104.58	+63.13	-151.68	-160.59	+1,133.30	+0.20
19	94.2	81.58	-47.1	-128.28	+0.41	+1.52	-52.60	-194.99	+28.99	-242.09	-131.61	+891.21	+0.34
20	94.2	81.58	-47.1	-165.90	+0.54	+1.48	-89.58	-245.53	-8.00	-292.63	-139.61	+598.59	+0.48
21	94.2	81.58	-47.1	-172.40	+0.66	+1.42	-113.78	-244.80	-32.20	-291.90	-171.82	+306.69	+0.60
22	94.2	81.58	-47.1	-139.62	+0.48	+0.90	-67.02	-125.65	+14.56	-172.75	-157.25	+133.93	+0.72
23	100	86.603	-50.0	-77.52	+0.25	+0.43	-19.38	-33.34	+67.22	-83.34	-90.03	+50.60	+0.25
24	100	86.603	-50.0	0	+0.13	+0.22	0	0	+86.603	-50.00	0	0	+0.25

Table (ii). Arch Calculation (Continued)

δy	$-N_z \delta y$	$N_v \delta z$	ΔM_ϕ	$M_{\phi,0}$	$M_{\phi,1}$	M_ϕ	θ	$\cos \theta$	$\sin \theta$	$N_v \cos \theta$	$N_z \sin \theta$	N_ϕ (25 + 26)
15	16	17	18	19	20	21	22	23	24	25	26	27
0	0	0	0	0	0	0						
−0.204	0	0	0	0	−189.511	−189.511						
−0.189	+14.302	−39.188	−24.886	−24.886	−363.359	−388.245						
−0.188	−2.649	−31.331	−33.98	−58.866	−541.629	−622.495						
−0.189	+26.454	−16.071	−42.525	−101.391	−717.744	−819.135						
−0.189	−52.191	+1.170	−51.021	−152.412	−893.858	−1,046.27						
−0.016	−5.811	+0.382	−5.429	−157.841	−1,020.295	−1,178.136						
+0.342	+135.441	−11.176	+124.265	−33.576	−1,185.769	−1,219.345						
+0.346	+138.007	−26.328	+111.679	+78.103	−1,352.304	−1,274.201						
+0.342	+131.27	−35.77	+95.50	+173.603	−1,517.776	−1,344.173						
+0.342	+119.714	−38.62	+81.094	+254.697	−1,683.2504	−1,428.553						
+1.390	+485.90	+79.69	+565.59	+820.287	−1,779.917	−959.630	−30°	0.86603	−0.50000	−95.85	+174.78	+78.93
+1.450	+586.666	−81.553	+505.113	+1,325.40	−1,937.817	−612.417	−25°	0.90631	−0.42260	+123.19	+170.940	+294.13
+1.50	+683.51	−217.283	+466.227	+1,791.627	−2,154.298	−362.671	−20°	0.93969	−0.34202	+425.37	+155.85	+581.22
+1.53	+725.422	−263.992	+461.43	+2,253.057	−2,431.610	−178.553	−15°	0.96593	−0.25882	+749.99	+122.71	+872.70
+1.56	+700.766	−210.494	+490.272	+2,743.329	−2,769.755	−26.426	−10°	0.98481	−0.17365	+1,036.48	+78.01	+1,114.49
+1.57	+608.311	−87.14	+521.171	+3,264.50	−3,159.650	+104.85	−5°	0.99619	−0.08716	+1,240.12	+33.77	+1,273.89
+1.57	+480.23	+92.76	+572.99	+3,837.49	−3,602.421	+235.069	0°	1.00000	0	+1,325.08	0	+1,325.08
+1.56	+349.002	+256.996	+605.998	+4,443.488	−4,091.640	+351.848	+5°	0.99619	+0.08716	+1,280.08	−90.50	+1,260.58
+1.53	+245.709	+385.322	+631.031	+5,074.519	−4,625.780	+448.739	+10°	0.98481	+0.17365	+1,116.08	−27.89	+1,088.19
+1.50	+197.414	+427.782	+625.196	+5,699.715	−5,204.839	+494.876	+15°	0.96593	+0.25882	+860.85	−34.06	+826.79
+1.45	+202.437	+359.152	+561.589	+6,261.304	−5,815.963	+445.341	+20°	0.93969	+0.34202	+562.49	−47.75	+514.74
+1.39	+238.824	+220.815	+459.639	+6,720.943	−6,456.498	+264.445	+25°	0.90631	+0.42260	+277.95	−72.61	+205.34
+0.43	+67.602	+33.483	+101.085	+6,822.028	−6,664.748	+157.08	+30°	0.86603	+0.50000	+115.99	−78.30	+37.69
+0.43	+38.714	+12.65	+51.364	+6,873.392	−6,873.396	0						

Force N_ϕ

The transverse force N_ϕ is calculated from the formula $N_\phi = N_x \sin \theta + N_y \cos \theta$ for the shell portion at an interval of 5° and entered in column 27 [Table (ii)]. The values of N_x, N_ϕ, $N_{x\phi}$, and M_ϕ are shown plotted in Fig. 10-10.

6. STATIC CHECKS

Check 1

The resultant longitudinal force on any cross section is zero, i.e., $\Sigma N_x \, ds = 0$. From column 14 of Table (i),

$$\sum N_x \, ds = (331{,}128.93 - 331{,}214.86) \approx 0$$

Check 2

The internal resisting moment is equal to the bending moment due to loading, i.e.,

$$\sum N_x \, ds \, y = M$$

The check is made at the midspan section, and the moments are taken about the Oy axis. The calculations are shown in the table below.

Point	$N_x \, ds$, kip	y, ft	$N_x \, ds \, y$, kip ft
0	−63.5851	−1.44	+91.55966
1	−44.73066	−0.826	+36.94752
2	−26.50326	−0.228	+6.042743
3	−8.185556	+0.373	−3.05321
4	+10.058196	+0.972	+9.77656
5	+24.379658	+1.45	+35.350504
6	+32.293687	+1.717	+55.44826
7	+37.719491	+1.92	+72.42142
8	+43.595227	+2.113	+92.116715
9	+48.573868	+2.311	+112.25421
10	+39.459128	+2.05	+80.89121
11	+17.555401	+1.39	+24.402007
12	−0.831772	+0.85	−0.707006
13	−15.380117	+0.44	−6.76725
14	−25.77475	+0.17	−4.38170
15	−32.15080	+0.04	−1.28603
16	−34.48614	+0.04	−1.37945
17	−32.76587	+0.17	−5.570198
18	−26.967521	+0.44	−11.865709
19	−16.926224	+0.85	−14.38729
20	−2.92709	+0.139	−4.06853
21	+14.752214	+2.05	+30.2420
22	+27.94243	+2.53	+70.69435
23⎫ 24⎭	+34.79963	+2.78	+96.74290

$$\Sigma = 761.4243 \text{ kip ft}$$

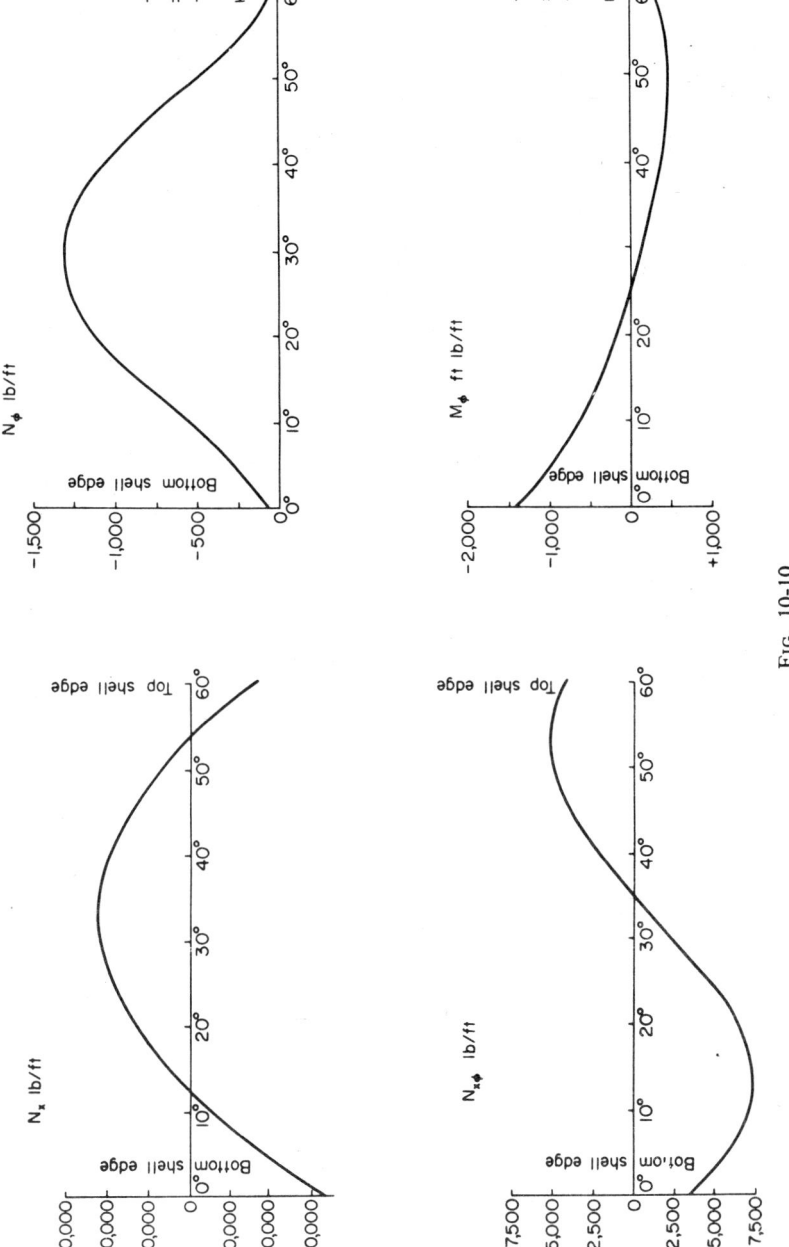

Fig. 10-10

219

Bending moment about axis $Oy = \left(1{,}950 \times \cos 30° \times \dfrac{60^2}{8}\right) \dfrac{1}{1{,}000} = 759.941$ kip ft

Therefore, the values agree very closely.

Check 3

Vertical component of shear = external load in that direction

External load on half span $= \dfrac{Wl}{2} = \dfrac{1{,}950 \times 60}{2} = 58{,}500.0$ lb

The vertical component of shear is calculated in the table below.

Point	Shear force, lb/ft	Width of strip under consideration, ft	Shear force in the strip, lb	Slope of tangent θ, deg	$\sin \theta$	Vertical component acting upward, lb
0	−1,060	0.338	−358.28	77.5	0.9763	+349.78
1	−4,239	0.652	−2,763.83	77.5	0.9763	+2,698.3
2	−7,221	0.628	−4,543.79	77.5	0.9763	+4,427.3
3	−8,988	0.628	−5,644.46	77.5	0.9763	+5,510.6
4	−9,543	0.628	−5,987.35	77.5	0.9763	+5,845.4
5	−8,863	0.512	−4,537.86	77.5	0.9763	+4,430.3
6	−7,238	0.396	−2,866.25	0		
7	−5,085	0.396	−2,013.66	0		
8	−2,570	0.396	−1,017.72	0		
9	+366	0.396	+144.94	0		
10	+3,574	0.983	+3,513.24	60	0.866	+3,042.46
11	+6,025	1.57	+9,459.25	55	0.8192	+7,749.0
12	+7,375	1.57	+11,578.75	50	0.766	+8,869.32
13	+7,920	1.57	+12,434.4	45	0.7071	+8,792.36
14	+6,295	1.57	+9,883.15	40	0.6428	+6,352.89
15	+4,576	1.57	+7,184.32	35	0.5736	+4,120.92
16	+2,433	1.57	+3,819.81	30	0.5000	+1,909.9
17	+134	1.57	+210.38	25	0.4226	+88.91
18	−2,051	1.57	−3,337	20	0.3420	−1,154.93
19	−3,848	1.57	−6,041.36	15	0.2588	−1,563.50
20	−4,977	1.57	−7,813.89	10	0.1736	−1,356.49
21	−5,172	1.57	−8,120	5	0.0872	−708.064
22	−4,188	1.035		0		
23	−2,326	0.5		0		
24		0.25		0		

$\Sigma = 59{,}404.42$

Vertical component of shear = 59,404.42 lb

The values agree very closely.

CHAPTER 11

PRESTRESSED CYLINDRICAL SHELLS

11-1 Objects of Prestressing

As the span of a cylindrical shell becomes large, the tension in the edge beam reaches very high values demanding the provision of very heavy reinforcement. The lengths of the bars involved pose several practical problems. Laps, even if they are staggered, are not fully effective in transmitting tensile forces through the medium of cracked concrete. Hence welding becomes inescapable. With increasing spans, the deflections become excessive and they, in turn, cause heavy transverse moments. A high transverse moment at the junction between the edge beam and the shell can often be taken care of by local thickening of the shell membrane. A heavy transverse moment at the crown cannot be so easily catered to. For these reasons, a design in reinforced concrete often tends to be uneconomical. Prestressing the shell is the answer. The following advantages result if the shell is prestressed:

(i) Steel consumption is considerably reduced. The steel requirement for the shell, columns, and foundation per square foot of covered area for a shell roof of 33 ft chord width and 148 ft 6 in. span is reported to be as low as 3.37 lb [43].

(ii) There is considerable reduction in the transverse moments. In fact, as Marshall [44] has shown, it is sometimes possible to eliminate transverse moments altogether so that the stresses in the shell are reduced to those corresponding to the membrane plus a constant compression.

(iii) The development of cracks due to stresses set up by shrinkage and temperature is inhibited [45].

(iv) Deflections are considerably reduced.

(v) If the prestress is so adjusted that the shell membrane is subjected to compression, the roof can be made watertight.

(vi) From the meager test results now available, it would appear that prestressing has also a favorable influence in providing an adequate load factor against collapse by inelastic buckling [46]. The zones of heavy compression in the shell can be relieved by a judicious application of prestressing.

11·2 Methods of Prestressing Cylindrical Shells

The prestress is most conveniently introduced by poststressing either the edge member or the lower parts of the shell membrane by curved cables or by a combination of both the techniques (Fig. 11-1). Al-

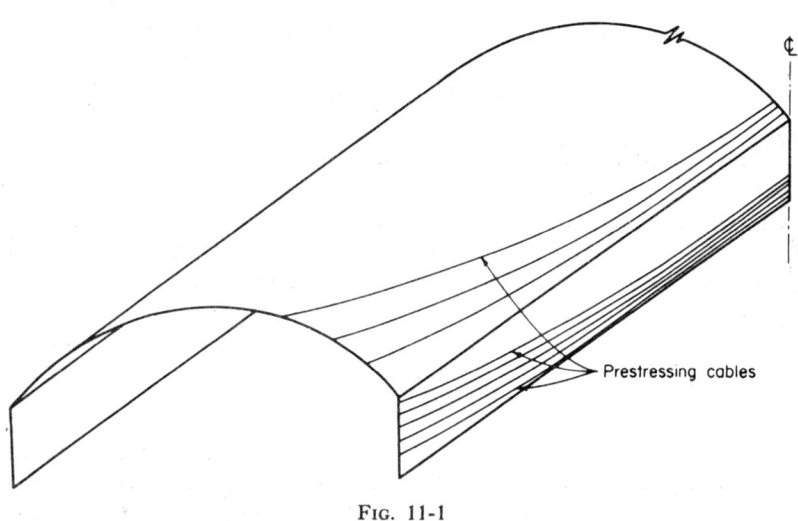

Prestressing cables

FIG. 11-1

though curved cables in the lower parts of the shell close to its junction with the edge beam can be very effective in countering the dead-weight and live-load deflections of the shell, their draping presents practical difficulties as they have to be placed along space curves. Hence the commoner practice is to posttension the edge member only. Some of the earliest examples of such structures were the shell roofs of the tank garages built at Meerut, India, in 1941 [47]. These shells had a span of 120 ft, a chord width of 35 ft, and a thickness of just $2\frac{1}{2}$ in. Another notable structure of this period is the shell roof for the airport hangar at Karachi built in 1942 [48]. Posttensioning of edge members is particularly advantageous for multiple shells.

11-3 Preliminary Analysis

To start with, the approximate prestressing force and the eccentricity at which it is to be applied have to be determined. If the shell to be prestressed is long, as it usually is, it is accurate enough to analyze it by the beam method for purposes of preliminary design. Such a simplified approach is, however, unsuitable for making the final analysis [49].

11-4 Final Analysis

If the prestressed shell treated is long, its analysis may be adequately dealt with by using the Schorer theory developed in Chapter 8. The analysis proceeds in the same manner as for a nonprestressed shell. The only difference is that while formulating boundary conditions for the shell at its junction with the edge member, the effect of the prestressing force has also to be taken into account.

11-5 Formulation of Boundary Conditions for a Prestressed Shell

The calculations are best done by using the Schorer theory in matrix form presented in Chapter 8. Let us, for instance, consider an interior shell of a group of multiple shells. The boundary conditions applicable to this shell at its junction with the edge beam may be stated as follows:

(i) The horizontal displacement of the shell edge is zero. This boundary condition is the same as that applicable to a nonprestressed shell. This condition may be formulated as

$$(v)_{m+b} \cos \phi_c + (w)_{m+b} \sin \phi_c = 0$$

(ii) The rotation of the shell edge is zero. This condition is also the same as that applicable to a nonprestressed shell.

(iii) The vertical deflection of the shell edge is equal to the vertical deflection of the edge beam at their junction. This condition may be formulated as follows:

Referring to Fig. 11-2, the effects of the prestressing force on the edge beam may be considered as composed of three independent effects[1]:

(a) An upward vertical force caused by the curvature of the cable

---

[1] This treatment is based on the paper "Analysis of Prestressed Cylindrical Shell Roofs" by R. Dabrowski, *Journal of the Structural Division, Proceedings of the ASCE*, vol. 89, no. ST5, October, 1963.

(*b*) An end moment caused by the eccentricity of the cable at the ends of the beam

(*c*) An axial compression caused by the prestressing force H

Each of these effects, being uniform along the span, may be expanded by using the same type of Fourier series as already employed

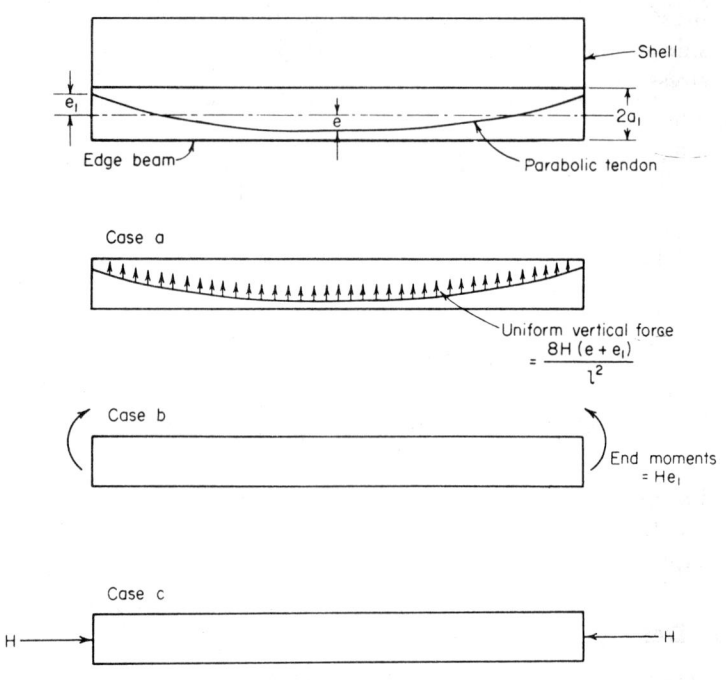

Fig. 11-2. (*Adapted from "Analysis of Prestressed Cylindrical Shell Roofs" by R. Dabrowski, Journal of the Structural Division, Proceedings of the ASCE, vol. 89, no. ST5, October, 1963.*)

for expanding a uniform load. It is now easy to formulate this boundary condition as

$$-(w)_{m+b} \cos \phi_c + (v)_{m+b} \sin \phi_c$$

$$= \frac{1}{k^4 EI} \Big[(N_\phi)_{m+b} \sin \phi_c - (Q_\phi)_b \cos \phi_c$$

$$+ (N_{x\phi})_{m+b} \, ka_1 - W' + \frac{4}{\pi} \frac{8H(e + e_1)}{l^2} - \frac{4}{\pi} He_1 k^2 \Big]$$

where

$$k = \frac{\pi}{l}$$

and

$$W' = \frac{4}{\pi} W$$

(W is the dead weight of the edge member per unit length)
This condition is exactly the same as that applicable for the non-prestressed shell except for the last two terms on the right-hand side which account for the vertical deflection caused by the pre-stress.

(iv) The longitudinal displacement of the shell edge is equal to the longitudinal displacement of the edge beam at its junction with the shell. This condition may be formulated thus:

$$(u)_{m+b} = [(N_\phi)_{m+b} \sin \phi_c - (Q_\phi)_b \cos \phi_c] \frac{a_1}{k^3 EI}$$

$$+ \frac{1}{k^2 AE} (N_{x\phi})_{m+b} + \frac{a_1{}^2}{k^2 EI} (N_{x\phi})_{m+b} - \frac{a_1}{k^3 EI} W'$$

$$+ \frac{4}{\pi} \frac{8H(e+e_1)}{l^2} \frac{a_1}{k^3 EI} - \frac{4}{\pi} He_1 \frac{a_1}{kEI} - \frac{4}{\pi} H \frac{1}{kAE}$$

Except for the last three terms on the right-hand side which account for the longitudinal displacement caused by the prestress, the expression is the same as (8-8a).

11-6 Design Example

In Design Example 11-1, the application of these principles to the analysis of a prestressed interior shell is explained in detail. In Chapter 22, the same example has again been worked out using the powerful matrix-progression technique.

Design Example 11-1

*ANALYSIS OF A LONG INTERIOR CYLINDRICAL SHELL
WITH PRESTRESSED EDGE BEAMS*

1. *DATA*

   Geometry

   Shell

   Span l = 163.75 ft
   Radius a = 26.375 ft
   Thickness d = 0.25 ft
   Semicentral angle ϕ_{e_1} = 39.5°

Edge Beam

Depth $2a_1$ = 7.25 ft
Width $2b_1$ = 0.2917 ft = $3\frac{1}{2}$ in.

Loads

Dead weight   = 37.5 psf
Waterproofing = 5.0 psf
Live load   = 12.5 psf

   Total load g = 55.0 psf of surface area

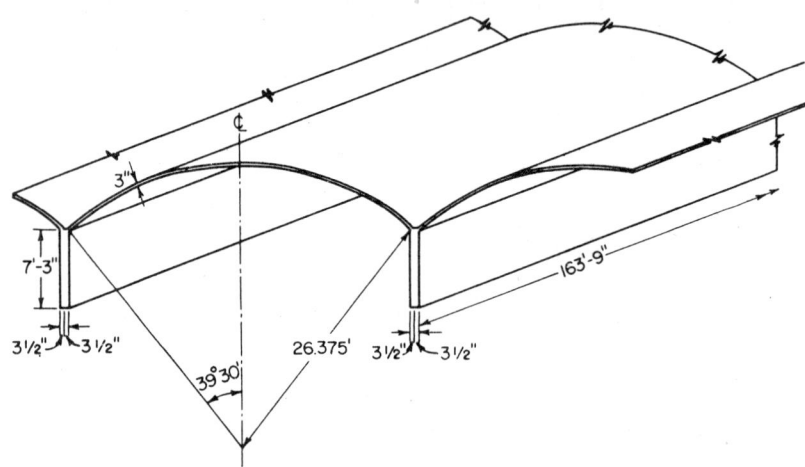

FIG. 11-3

2. GEOMETRICAL PROPERTIES AND OTHER QUANTITIES

Shell

$$\rho = \left(\frac{12\pi^4 a^6}{l^4 d^2} \right)^{1/8} = 3.11022$$

$$\kappa = \left(\frac{\pi a}{l\rho} \right)^2 = 0.02647$$

Since the shell is long, the Schorer method will be used for the analysis.

$$\phi_c = 0.689405 \text{ radian}$$
$$2\phi_c = 1.378810 \text{ radians}$$
$$\sin \phi_c = 0.63608$$
$$\cos \phi_c = 0.77162$$

$$k = \frac{\pi}{l} = 0.019185$$

$$D = \frac{Ed^3}{12} = 468,750$$

$$\lambda = \frac{\pi a}{l} = 0.506012$$

Edge Beam

$$EI = 360 \times 10^6 \left(\frac{0.2917 \times 7.25^3}{12} \right) = 333.44336 \times 10^7$$

$$\frac{1}{k^4 EI} = 2.21362 \times 10^{-3}$$

3. ROOTS OF THE CHARACTERISTIC EQUATION AND RELATED QUANTITIES

$$\alpha_1 = \beta_2 = \rho \frac{(\sqrt{2} + 1)^{1/2}}{8^{1/4}} = 2.873470$$

$$\alpha_2 = \beta_1 = \rho \frac{(\sqrt{2} - 1)^{1/2}}{8^{1/4}} = 1.190231$$

4. MULTIPLIERS IN THE MATRICES

		Multiplier
Forces		
M_ϕ	$-\dfrac{D\rho^2}{a^2} \dfrac{1}{\sqrt{2}}$	$-0.018107E \times 10^{-3} (0.707107)$
Q_ϕ	$-\dfrac{D\rho^2}{a^2} \dfrac{1}{a\sqrt{2}}$	$-0.018107E \times 10^{-3} (2.680973 \times 10^{-2})$
N_ϕ	$-\dfrac{D\rho^2}{a^2} \dfrac{\rho^2}{a}$	$-0.018107E \times 10^{-3} (0.366767)$
$N_{x\phi}$	$-\dfrac{D\rho^2}{a^2} \dfrac{\rho^2}{\lambda a}$	$-0.018107E \times 10^{-3} (0.724818)$
$\dfrac{\partial N_{x\phi}}{\partial x}$	$-\dfrac{D\rho^2}{a^2} \dfrac{\rho^2}{a^2}$	$-0.018107E \times 10^{-3} (0.013906)$
N_x	$-\dfrac{D\rho^2}{a^2} \dfrac{\rho^4}{\sqrt{2}\lambda^2 a}$	$-0.018107E \times 10^{-3} (9.797957)$
Displacements		
u	$-\dfrac{\lambda}{\sqrt{2}\rho^2}$	$-36.988213 \times 10^{-3}$
$\dfrac{\partial u}{\partial x}$	$-\dfrac{\lambda^2}{\sqrt{2}\rho^2 a}$	-0.709630×10^{-3}
v	$+\dfrac{1}{\sqrt{2}\rho^2}$	$+73.097465 \times 10^{-3}$
w	$+1.0$	$+1{,}000 \times 10^{-3}$
ϑ	$+\dfrac{1}{a}$	$+37.914692 \times 10^{-3}$

ϕ	0°	10°	20°	30°	39.5°	50°	60°	70°	79°
$e^{-\alpha_1\phi}$	1	0.60561	0.36677	0.22212	0.13793	0.08147	0.04934	0.02988	0.01903
$e^{-\alpha_2\phi}$	1	0.81242	0.66003	0.53622	0.44019	0.35392	0.28754	0.23360	0.19377
$\cos\beta_1\phi$	1	0.97850	0.91493	0.81201	0.68182	0.50737	0.31876	0.11639	−0.07025
$\sin\beta_1\phi$	0	0.20624	0.40362	0.58364	0.73152	0.86173	0.94784	0.99320	0.99753
$\cos\beta_2\phi$	1	0.87686	0.53775	0.06620	−0.39878	−0.80565	−0.99123	−0.93268	−0.68195
$\sin\beta_2\phi$	0	0.48075	0.84310	0.99781	0.91705	0.59239	0.13218	−0.36070	−0.73140
$e^{-\alpha_1\phi}\cos\beta_1\phi$	1	0.59259	0.33556	0.18036	0.09405	0.04133	0.01573	0.00348	−0.00134
$e^{-\alpha_1\phi}\sin\beta_1\phi$	0	0.12490	0.14803	0.12964	0.10090	0.07020	0.04677	0.02968	0.01898
$e^{-\alpha_2\phi}\cos\beta_2\phi$	1	0.71238	0.35493	0.03550	−0.17554	−0.28514	−0.28502	−0.21788	−0.13214
$e^{-\alpha_2\phi}\sin\beta_2\phi$	0	0.39058	0.55647	0.53505	0.40367	0.20966	0.03801	−0.08426	−0.14172

5. TABLE OF B_1, B_2, B_3, AND B_4

Quantity	B_1	B_2	B_3	B_4
M_ϕ	$+1.0$	-1.0	-1.0	-1.0
Q_ϕ	-1.68324	$+4.06370$	$+4.06370$	-1.68324
N_ϕ	0	$+1.0$	0	-1.0
$N_{x\phi}$	$+1.19023$	$+2.87347$	-2.87347	-1.19023
N_x	-1.0	-1.0	$+1.0$	-1.0
$\dfrac{\partial u}{\partial x}$	-1.0	-1.0	$+1.0$	-1.0
v	-4.06370	-1.68324	-1.68324	-4.06370
w	$+1.0$	0	$+1.0$	0
ϑ	-2.87347	$+1.19023$	-1.19023	$+2.87347$

6. MEMBRANE FORCES AND DISPLACEMENTS

$$N_\phi = -\frac{4}{\pi} ga \cos(\phi_c - \phi) \cos\frac{\pi x}{l}$$

$$= -1{,}846.99 \cos(\phi_c - \phi) \cos\frac{\pi x}{l}$$

$$\therefore (N_\phi)_{\phi=0} = -1{,}425.14 \cos\frac{\pi x}{l}$$

$$N_{x\phi} = +\frac{8}{\pi^2} gl \sin(\phi_c - \phi) \sin\frac{\pi x}{l}$$

$$= +7{,}299.30 \sin(\phi_c - \phi) \sin\frac{\pi x}{l}$$

$$\therefore (N_{x\phi})_{\phi=0} = +4{,}643.08 \sin\frac{\pi x}{l}$$

$$N_x = -\frac{8gl^2}{a\pi^3} \cos(\phi_c - \phi) \cos\frac{\pi x}{l}$$

$$= -14{,}426.91 \cos(\phi_c - \phi) \cos\frac{\pi x}{l}$$

$$\therefore (N_x)_{\phi=0} = -11{,}131.80 \cos\frac{\pi x}{l}$$

$$u = -\frac{8gl^3}{E\pi^4 ad} \cos(\phi_c - \phi) \sin\frac{\pi x}{l}$$

$$= -3{,}007{,}908.59 \cos(\phi_c - \phi) \cos\frac{\pi x}{l}$$

$$\therefore (u)_{\phi=0} = -2{,}320{,}902.27 \sin\frac{\pi x}{l}$$

$$v = -\frac{8g}{E\pi d}\left[2\left(\frac{l}{\pi}\right)^2 + \frac{1}{a^2}\left(\frac{l}{\pi}\right)^4\right]\sin(\phi_c - \phi)\cos\frac{\pi x}{l}$$

$$= -\frac{8{,}988{,}416.65}{E}\sin(\phi_c - \phi)\cos\frac{\pi x}{l}$$

$$\therefore (v)_{\phi=0} = -\frac{5{,}717{,}531.83}{E}\cos\frac{\pi x}{l}$$

$$w = +\frac{8g}{E\pi d}\left[2\left(\frac{l}{\pi}\right)^2 + \frac{1}{a^2}\left(\frac{l}{\pi}\right)^4\right]\cos(\phi_c - \phi)\cos\frac{\pi x}{l}$$

$$= +\frac{8{,}988{,}416.65}{E}\cos(\phi_c - \phi)\cos\frac{\pi x}{l}$$

$$\therefore (w)_{\phi=0} = +\frac{6{,}935{,}462.29}{E}\cos\frac{\pi x}{l}$$

7. MATRIX FOR STRESS RESULTANTS

Disturbances from Left Edge

$$\begin{bmatrix} \dfrac{\partial N_{x\phi}}{\partial x} \\ Q_\phi \\ N_\phi \\ M_\phi \end{bmatrix} = -1.810658E \times 10^{-7} \times \cos\frac{\pi x}{l} \begin{bmatrix} 1.39041 & 0 & 0 & 0 \\ 0 & 2.68097 & 0 & 0 \\ 0 & 0 & 36.67669 & 0 \\ 0 & 0 & 0 & 70.71066 \end{bmatrix}$$

$$\times \begin{bmatrix} 1.19023 & 2.87347 & -2.87347 & -1.19023 \\ -1.68324 & 4.06370 & 4.06370 & -1.68324 \\ 0 & 1.0 & 0 & -1.0 \\ 1.0 & -1.0 & -1.0 & -1.0 \end{bmatrix} \begin{bmatrix} 1 & 0 & 0 & 0 \\ 0 & 1 & 0 & 0 \\ 0 & 0 & 1 & 0 \\ 0 & 0 & 0 & 1 \end{bmatrix} \begin{bmatrix} A_n \\ B_n \\ C_n \\ D_n \end{bmatrix}$$

or

$$\begin{bmatrix} \dfrac{\partial N_{x\phi}}{\partial x} \\ Q_\phi \\ N_\phi \\ M_\phi \end{bmatrix} = -1.810658E \times 10^{-7} \times \cos\frac{\pi x}{l}$$

$$\times \begin{bmatrix} 1.65491 & 3.99529 & -3.99529 & -1.65491 \\ -4.51272 & 10.89467 & 10.89467 & -4.51272 \\ 0 & 36.67669 & 0 & -36.67669 \\ 70.71066 & -70.71066 & -70.71066 & -70.71066 \end{bmatrix} \begin{bmatrix} A_n \\ B_n \\ C_n \\ D_n \end{bmatrix}$$

or

$$
\begin{bmatrix} \dfrac{\partial N_{x\phi}}{\partial x} \\ Q_\phi \\ N_\phi \\ M_\phi \end{bmatrix} = -1.810658E \times 10^{-7} \times \cos\frac{\pi x}{l}
$$

$$
\times
\begin{bmatrix}
(\;\;\;1.65491A_n & +3.99529B_n & -3.99529C_n & -1.65491D_n) \\
(\;-4.51272A_n & +10.89467B_n & +10.89467C_n & -4.51272D_n) \\
(\;\;\;\;\;\;\;0 & +36.67669B_n & +0 & -36.67669D_n) \\
(+70.71066A_n & -70.71066B_n & -70.71066C_n & -70.71066D_n)
\end{bmatrix}
$$

Disturbances from Right Edge

$$
\begin{bmatrix} \dfrac{\partial N_{x\phi}}{\partial x} \\ Q_\phi \\ N_\phi \\ M_\phi \end{bmatrix} = -1.810658E \times 10^{-7} \times \cos\frac{\pi x}{l}
\begin{bmatrix}
1.39041 & 0 & 0 & 0 \\
0 & 2.68097 & 0 & 0 \\
0 & 0 & 36.67669 & 0 \\
0 & 0 & 0 & 70.71066
\end{bmatrix}
$$

$$
\times
\begin{bmatrix}
1.19023 & 2.87347 & -2.87347 & -1.19831 \\
-1.68324 & 4.06370 & 4.06370 & -1.68324 \\
0 & 1.0 & 0 & -1.0 \\
1.0 & -1.0 & -1.0 & -1.0
\end{bmatrix}
$$

$$
\times
\begin{bmatrix}
-0.00134 & +0.01898 & 0 & 0 \\
-0.01898 & -0.00134 & 0 & 0 \\
0 & 0 & -0.13214 & -0.14172 \\
0 & 0 & +0.14172 & -0.13214
\end{bmatrix}
\begin{bmatrix} A_n \\ B_n \\ C_n \\ D_n \end{bmatrix}
$$

or

$$
\begin{bmatrix} \dfrac{\partial N_{x\phi}}{\partial x} \\ Q_\phi \\ N_\phi \\ M_\phi \end{bmatrix} = -1.810658E \times 10^{-7} \times \cos\frac{\pi x}{l}
$$

$$
\times
\begin{bmatrix}
1.65491 & 3.99529 & -3.99529 & -1.65491 \\
-4.51272 & 10.89467 & 10.89467 & -4.51272 \\
0 & 36.67669 & 0 & -36.67669 \\
70.71066 & -70.71066 & -70.71066 & -70.71066
\end{bmatrix}
\begin{bmatrix}
(-0.00134A_n & +0.01898B_n) \\
(-0.01898A_n & -0.00134B_n) \\
(-0.13214C_n & -0.14172D_n) \\
(+0.14172C_n & -0.13214D_n)
\end{bmatrix}
$$

or

$$
\begin{bmatrix} \dfrac{\partial N_{x\phi}}{\partial x} \\ Q_\phi \\ N_\phi \\ M_\phi \end{bmatrix} = -1.810658E \times 10^{-7} \times \cos\dfrac{\pi x}{l}
$$

$$
\times \begin{bmatrix} (-0.07804A_n & +0.02607B_n & +0.29340C_n & +0.78489D_n) \\ (-0.20073A_n & -0.10021B_n & -2.07915C_n & -0.94771D_n) \\ (-0.69607A_n & -0.04902B_n & -5.19787C_n & +4.84639D_n) \\ (+1.24748A_n & +1.43649B_n & -0.67764C_n & +19.36481D_n) \end{bmatrix}
$$

Sum of the Effects of the two Edges

$$
\begin{bmatrix} \dfrac{\partial N_{x\phi}}{\partial x} \\ Q_\phi \\ N_\phi \\ M_\phi \end{bmatrix} = -1.810658E \times 10^{-7} \times \cos\dfrac{\pi x}{l}
$$

$$
\times \begin{bmatrix} (+1.73316A_n & +3.96973B_n & -4.28924C_n & -2.44011D_n) \\ (-4.31199A_n & +10.99488B_n & +12.97382C_n & -3.56502D_n) \\ (-0.69607A_n & +36.62767B_n & -5.19787C_n & -31.83030D_n) \\ (+71.95814A_n & -69.27417B_n & -71.38830C_n & -51.34585D_n) \end{bmatrix}
$$

8. *MATRIX FOR DISPLACEMENTS*

Disturbances from Left Edge

$$
\begin{bmatrix} \dfrac{\partial u}{\partial x} \\ w \\ v \\ \vartheta \end{bmatrix} = 10^{-3} \times \cos\dfrac{\pi x}{l} \begin{bmatrix} -0.70963 & 0 & 0 & 0 \\ 0 & 1{,}000 & 0 & 0 \\ 0 & 0 & 73.09747 & 0 \\ 0 & 0 & 0 & 37.91469 \end{bmatrix}
$$

$$
\times \begin{bmatrix} -1.0 & -1.0 & 1.0 & -1.0 \\ 1.0 & 0 & 1.0 & 0 \\ -4.06370 & -1.68324 & -1.68324 & -4.06370 \\ -2.87347 & 1.19023 & -1.19023 & 2.87347 \end{bmatrix} \begin{bmatrix} 1 & 0 & 0 & 0 \\ 0 & 1 & 0 & 0 \\ 0 & 0 & 1 & 0 \\ 0 & 0 & 0 & 1 \end{bmatrix} \begin{bmatrix} A_n \\ B_n \\ C_n \\ D_n \end{bmatrix}
$$

or

$$
\begin{bmatrix} \dfrac{\partial u}{\partial x} \\ w \\ v \\ \vartheta \end{bmatrix} = 10^{-3} \times \cos\dfrac{\pi x}{l}
$$

$$
\times \begin{bmatrix} 0.70963 & 0.70963 & -0.70963 & 0.70963 \\ 1{,}000 & 0 & 1{,}000 & 0 \\ -297.04620 & -123.04055 & -123.04055 & -297.04620 \\ -108.94670 & +45.12722 & -45.12722 & +108.94670 \end{bmatrix} \begin{bmatrix} A_n \\ B_n \\ C_n \\ D_n \end{bmatrix}
$$

or

$$
\begin{bmatrix} \dfrac{\partial u}{\partial x} \\ w \\ v \\ \vartheta \end{bmatrix} = 10^{-3} \times \cos\dfrac{\pi x}{l}
$$

$$
\times \begin{bmatrix}
(& +0.70963A_n & +0.70963B_n & -0.70963C_n & +0.70963D_n) \\
(& +1{,}000A_n & +0 & +1{,}000C_n & +0 &) \\
(-297.04620A_n & -123.04055B_n & -123.04055C_n & -297.04620D_n) \\
(-108.94670A_n & +45.12722B_n & -45.12722C_n & +108.94670D_n)
\end{bmatrix}
$$

Disturbances from Right Edge

$$
\begin{bmatrix} \dfrac{\partial u}{\partial x} \\ w \\ v \\ \vartheta \end{bmatrix} = 10^{-3} \times \cos\dfrac{\pi x}{l}
\begin{bmatrix}
-0.70963 & 0 & 0 & 0 \\
0 & 1{,}000 & 0 & 0 \\
0 & 0 & 73.09747 & 0 \\
0 & 0 & 0 & 37.91469
\end{bmatrix}
$$

$$
\times \begin{bmatrix}
-1.0 & -1.0 & 1.0 & -1.0 \\
1.0 & 0 & 1.0 & 0 \\
-4.06370 & -1.68324 & -1.68324 & -4.06370 \\
-2.87347 & 1.19023 & -1.19023 & 2.87347
\end{bmatrix}
$$

$$
\times \begin{bmatrix}
-0.00134 & 0.01898 & 0 & 0 \\
-0.01898 & -0.00134 & 0 & 0 \\
0 & 0 & -0.13214 & -0.14172 \\
0 & 0 & 0.14172 & -0.13214
\end{bmatrix}
\begin{bmatrix} A_n \\ B_n \\ C_n \\ D_n \end{bmatrix}
$$

or

$$
\begin{bmatrix} \dfrac{\partial u}{\partial x} \\ w \\ v \\ \vartheta \end{bmatrix} = 10^{-3} \times \cos\dfrac{\pi x}{l}
\begin{bmatrix}
0.70963 & 0.70963 & -0.70963 & 0.70963 \\
1{,}000 & 0 & 1{,}000 & 0 \\
-297.04620 & -123.04055 & -123.04055 & -297.04620 \\
-108.94670 & 45.12722 & -45.12722 & 108.94670
\end{bmatrix}
$$

$$
\times \begin{bmatrix}
(-0.00134A_n & +0.01898B_n) \\
(-0.01898A_n & -0.00134B_n) \\
(-0.13214C_n & -0.14172D_n) \\
(+0.14172C_n & -0.13214D_n)
\end{bmatrix}
$$

or

$$\begin{bmatrix} \dfrac{\partial u}{\partial x} \\ w \\ v \\ \vartheta \end{bmatrix} = 10^{-3} \times \cos\dfrac{\pi x}{l}$$

$$\times \begin{bmatrix} (-0.01442A_n & +0.01252B_n & +0.19434C_n & +0.00680D_n) \\ (-1.33650A_n & +18.97860B_n & -132.13820C_n & -141.7214D_n) \\ (+2.73214A_n & -5.47308B_n & -25.73950C_n & +56.68863D_n) \\ (-0.71084A_n & -2.12797B_n & +21.40308C_n & -8.00053D_n) \end{bmatrix}$$

Sum of the Effects of the Two Edges

$$\begin{bmatrix} \dfrac{\partial u}{\partial x} \\ w \\ v \\ \vartheta \end{bmatrix} = 10^{-3} \times \cos\dfrac{\pi x}{l}$$

$$\times \begin{bmatrix} (\ +0.69521A_n & +0.72215B_n & -0.51529C_n & +0.71643D_n) \\ (\ +998.6635A_n & +18.97860B_n & +867.86180C_n & -141.7214D_n) \\ (-299.77834A_n & -117.56746B_n & -97.30105C_n & -353.73483D_n) \\ (-108.23585A_n & +47.25519B_n & -66.53030C_n & +116.94723D_n) \end{bmatrix}$$

9. PRELIMINARY CALCULATION FOR ESTIMATING THE PRESTRESSING FORCE REQUIRED

The shell is first analyzed by the beam theory, and the tensile stress f_b at the bottom of the edge beam is calculated. A prestressing force H in each edge beam is then applied at the maximum possible eccentricity e and the compressive stress at the bottom of the edge beam is obtained in terms of H as

$$\frac{H}{A} + \frac{He}{I}y_b$$

where

A = area of cross section of the shell and the beam, on one side of the crown
I = moment of inertia of the above area about the neutral axis
y_b = distance of the bottom of edge beam from the neutral axis

The magnitude of the prestressing force H required to cause zero stress at the bottom of the edge beam is obtained by equating f_b and $H/A + (He/I)y_b$. It is found to be equal to 440,000 lb in this case. It is thus estimated that a prestressing force of about 440,000 lb is required to cause zero stress at the bottom of the edge beam.

10. PRESTRESSING FORCE AND ECCENTRICITY

With a view to study the effect of the prestressing force on the stresses in the shell and to arrive at the optimum value of H, various prestressing forces are applied. The magnitudes of the forces and the corresponding eccentricities are given below. The eccentricities given are the distances of the points of application of H below the mid-height of the edge beam.

H, lb	e, ft
200,000	3.292
400,000	3.083
500,000	2.958
600,000	2.875
800,000	2.667
1,000,000	2.458
1,200,000	2.250

At the traverses, the prestressing cables are anchored at a distance of 1.625 ft above the centroid of the edge beams. This value of the end eccentricity e_1 is also calculated from the beam theory. This is to ensure that at the support sections there is no tension at the crown of the shell.

In the present example, detailed calculations are given for a prestressing force of 400,000 lb only. For the other values of H, only the final stresses are given in the form of a graph (Fig. 11-4).

11. BOUNDARY CONDITIONS

(i) The horizontal displacement of the shell edge is equal to zero, i.e.,

$$(v)_{m+b} \cos \phi_c + (w)_{m+b} \sin \phi_c = 0$$

$$\left\{\left[-\frac{5,717,531.83}{E} + 10^{-3}(-299.77834A_n - 117.56746B_n \right.\right.$$

$$\left. -97.30105C_n - 353.73483D_n)\right]0.77162$$

$$+ \left[+\frac{6,935,462.29}{E} + 10^{-3}(998.66350A_n + 18.97860B_n \right.$$

$$\left. \left. + 867.86180C_n - 141.72140D_n)\right]0.63608\right\} = 0$$

or

$$+403.94089A_n - 78.64276B_n + 476.96940C_n - 363.09078D_n = 0 \tag{1}$$

(ii) The rotation of the shell edge is zero, i.e.,

$$\vartheta = 0$$

or

$$-108.23585A_n + 47.25519B_n - 66.53030C_n + 116.94723D_n = 0 \tag{2}$$

(iii) Vertical deflection of the shell edge is equal to the vertical deflection of the edge beam, i.e.,

$$-(w)_{m+b} \cos \phi_c + (v)_{m+b} \sin \phi_c$$

$$= \frac{1}{k^4 EI}[(N_\phi)_{m+b} \sin \phi_c - (Q_\phi)_b \cos \phi_c + (N_{x\phi})_{m+b}ka_1 - W']$$

$$+ \frac{4}{\pi}\frac{8H(e + e_1)}{l^2}\frac{1}{k^4 EI} - \frac{4}{\pi}He_1\frac{1}{k^2 EI}$$

where

$$-(w)_{m+b} \cos \phi_c = \left[-\frac{6,935,462.29}{E} - 10^{-3}(998.66350A_n + 18.97860B_n \right.$$
$$\left. + 867.86180C_n - 141.72140D_n) \right]0.77162$$

$$(v)_{m+b} \sin \phi_c = \left[-\frac{5,717,531.83}{E} + 10^{-3}(-299.77834A_n - 117.56746B_n \right.$$
$$\left. - 97.30105C_n - 353.73483D_n) \right]0.63608$$

Simplifying,

Vertical deflection of shell edge =

$$10^{-3}(- 961.25770A_n - 89.42855B_n - 731.47179C_n$$
$$- 115.65850D_n - 24.96757) \cos \frac{\pi x}{l}$$

$$\frac{-1}{k^4 EI} = 2.21362 \times 10^{-3}$$

$$(N_\phi)_{m+b} \sin \phi_c = [-1,425.14 - 1.810658E \times 10^{-7}(-0.69607A_n + 36.62767B_n$$
$$- 5.19787C_n - 31.83030D_n)]0.63608$$
$$= + 28.86138A_n - 1,518.70489B_n + 215.52112C_n$$
$$+ 1,319.79048D_n - 906.50392$$

$$-(Q_\phi)_b \cos \phi_c = [1.810658E \times 10^{-7}(-4.31199A_n + 10.99488B_n$$
$$+ 12.97382C_n - 3.56502D_n)]0.77162$$
$$= - 216.87452A_n + 552.99532B_n + 652.53776C_n - 179.30509D_n$$

$$(N_{x\phi})_{m+b}ka_1 = (N_{x\phi})_m ka_1 + (N_{x\phi})_b ka_1$$
$$= (N_{x\phi})_m ka_1 + \left(\frac{\partial N_{x\phi}}{\partial x} \right)_b a_1$$
$$= + 4,643.08 \times 0.019185 \times 3.625 - 1.810658E \times 10^{-7}(1.73316A_n$$
$$+ 3.96974B_n - 4.28924C_n - 2.44011D_n)3.625$$
$$= 322.91126 - 409.53078A_n - 938.01181B_n + 1,013.50778C_n$$
$$+ 576.57623D_n$$

$$-W' = -\frac{4}{\pi} \times 0.29167 \times 7.25 \times 150 = -403.83$$

$$\frac{4}{\pi} \frac{8H(e + e_1)}{l^2 k^4 EI} - \frac{4}{\pi} He_1 \frac{1}{k^2 EI} = +0.909366$$

Vertical deflection of the edge beam is therefore equal to

$$10^{-3}(-1,322.74906A_n - 4,214.08854B_n + 4,165.93412C_n$$
$$+ 3,800.91580D_n - 2,185.78465 + 909.366)$$

Equating the deflections of the shell edge and the edge beam, we get

$$361.49136A_n + 4,124.65999B_n - 4,896.56520C_n - 3,916.5743D_n$$
$$= 909.366 - 2,160.81688$$

or

$$361.49136A_n + 4,124.65999B_n - 4,896.56520C_n - 3,916.5730D_n = -1,251.4509 \qquad (3)$$

(iv) Longitudinal displacement of shell edge = longitudinal displacement of edge beam at its junction with the shell, i.e.,

$$u_m + u_b = \bigg\{[(N_\phi)_{m+b} \sin \phi_c - (Q_\phi)_{m+b} \cos \phi_c]a_1 \frac{1}{k^4 EI} k$$
$$+ \left(\frac{N_{x\phi}}{k^2}\right)_{m+b} \frac{1}{AE} + (N_{x\phi})_{m+b} a_1{}^2 k^2 \frac{1}{k^4 EI}$$
$$- \frac{a_1}{k^4 EI} kW' + \frac{4}{\pi} \frac{8H(e + e_1)}{l^2} \frac{a_1}{k^3 EI} - \frac{4}{\pi} He_1 \frac{a_1}{kEI} - \frac{4}{\pi} H \frac{1}{kAE}\bigg\}$$

where

$$u_m = -\frac{2,320,902.27}{E} \sin \frac{\pi x}{l}$$

$$u_b = \frac{1}{k} \frac{\partial u}{\partial x} = 10^{-3}(36.23736A_n + 37.64076B_n - 26.85865C_n + 37.34269D_n)$$

$$u_m + u_b = 10^{-3}(36.23736A_n + 37.64076B_n - 26.85865C_n + 37.34269D_n - 6.44695)$$

$$[(N_\phi)_{m+b} \sin \phi_c - (Q_\phi)_{m+b} \cos \phi_c] \frac{a_1 k}{k^4 EI}$$
$$= 10^{-3}(-28.94352A_n - 148.67027B_n + 133.63887C_n$$
$$+ 175.57719D_n - 139.55627)$$

$$\left(\frac{N_{x\phi}}{k^2}\right)_{m+b} \frac{1}{AE} = 10^{-3}(-21.01578A_n - 48.13570B_n + 52.00990C_n$$
$$+ 29.58800D_n + 16.57074)$$

$$(N_{x\phi})_{m+b} \frac{a_1{}^2 k^2}{k^4 EI} = 10^{-3}(-63.04732A_n - 144.40705B_n + 156.02967C_n$$
$$+ 88.76399D_n + 49.71224)$$

$$\frac{a_1 kW'}{k^4 EI} = -62.17009 \times 10^{-3} + \frac{4}{\pi} \frac{8H(e + e_1)}{l^2} \frac{a_1}{k^3 EI} - \frac{4}{\pi} He_1 \frac{a_1}{kEI}$$
$$- \frac{4}{\pi} H \frac{1}{kAE}$$
$$= + 0.0283716$$

Equating the longitudinal displacement of shell edge and that of edge beam, we get

$$10^{-3}(149.24631A_n + 378.85379B_n - 368.53669C_n - 256.58650D_n)$$
$$= -0.12899644 + 0.11015 - 0.046895 - 0.034867$$

Simplifying,

$$149.24631A_n + 378.85379B_n - 368.53669C_n - 256.58650D_n = -100.626 \qquad (4)$$

Solving the above four equations, we get

$$A_n = +0.0092687$$
$$B_n = -0.1650380$$
$$C_n = +0.0392118$$
$$D_n = +0.0975728$$

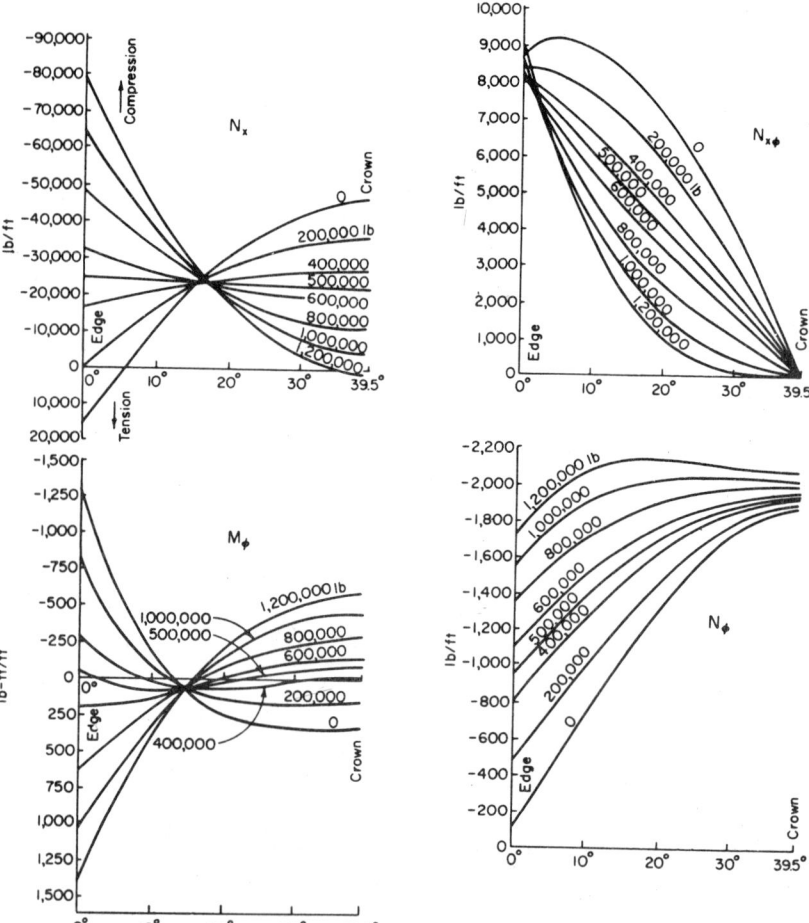

FIG. 11-4

12. *CALCULATION OF STRESS RESULTANTS*

Substituting the values of A_n, B_n, C_n, and D_n in the expressions for the stress resultants arrived at in step 7, the stresses in the shell due to edge disturbances are obtained for various points. Adding the membrane stresses to the stresses due to the edge disturbances, the final stresses are obtained. They are shown in the table below.

Quantity	0°	10°	20°	30°	39.5°
N_x	−16,806	−21,132	−24,319	−26,258	−26,906
$N_{x\phi}$	8,195	6,533	4,542	2,328	0
N_ϕ	−815	−1,280	−1,632	−1,851	−1,926
Q_ϕ	110	31	−6	−11	0
M_ϕ	−280	22	64	17	−10

13. *EFFECT OF PRESTRESSING FORCE ON THE STRESS DISTRIBUTION IN THE SHELL*

The calculations are repeated with the other values of prestressing force, namely, 200,000, 500,000, 600,000,..., 1,200,000 lb and for zero prestressing force. The stress distribution in all cases is shown plotted in Fig. 11-4.

14. *SELECTION OF OPTIMUM PRESTRESSING FORCE*

From a critical examination of Fig. 11-4, it is seen that the optimum prestressing force to be applied is about 500,000 lb. At this prestressing force, the transverse moment is about one-fifteenth of that in a non-prestressed shell. The shell and the edge beam are entirely in compression. Any further increase in the prestressing force is not beneficial. For higher values of the prestressing force, the transverse moment changes sign and registers a steep increase.

The arrangement of prestressing tendons in the edge beam is shown in Fig. 11-5.

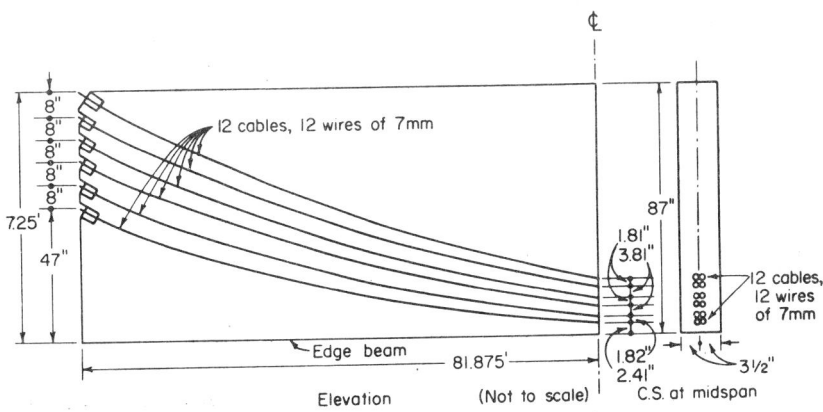

Fɪɢ. 11-5

11-7 Prestressed Northlight Shells

Prestressing can also be applied with advantage to northlight shells. Haas [50] has given details of a project involving a series of northlight shells continuous over two spans each of 40 m.

11-8 Cracking, Buckling, and Failure

There are very few papers dealing with these aspects of the problem. Cretu [51] refers to the Rumanian practice which demands a computation to ensure that cracking does not occur at loads less than 1.20 dead weight or 1.10 (dead weight + snow load). Similarly an other computation is required to ensure that failure does not occur at a load less than 2.5 times the dead weight or 2.2 (dead weight + snow load). Failure of the shell may be brought about either by flexure caused by beam action or by transverse moments. In checking for the former type of failure, the shell may be regarded as a posttensioned beam and the relevant formulas given in the ACI-ASCE Code [52] may be used. To check for safety against transverse moments, the resisting moment of the shell membrane may be computed as for a reinforced-concrete slab.

Buckling computations may be carried out in the same manner as for reinforced-concrete cylindrical shells. This subject is discussed in detail in Chapter 22.

The influence of prestressing on buckling loads has been investigated by Haas [53]. At his request, the late Prof. Torroja carried out tests on a one-tenth reinforced mortar scale model of a northlight shell continuous over two spans of 40 m each. At the midspan section of the prototype, there were 26 Freyssinet cables each of 12 numbers of 5-mm wire. These were simulated in the model at the rate of 7 wires of 2 mm for 10 Freyssinet cables, the loading scale being $\frac{1}{100}$. Plastic tubes of 2.5 mm inner diameter were used to accommodate the model prestressing wires. The test indicated a load factor of 2 against dead weight plus live load. The load factor for the prototype may be inferred to be more than this figure, because the prestressing wires in the model were left ungrouted.

CHAPTER 12

FOLDED PLATES

12-1 Historical Note

Almost akin to shells in structural action are folded plates or hipped plates. They consume a little more material than continuously curved cylindrical shells. But the extra cost on this account is many times offset by the saving effected on forms. Prismatic, prismoidal, pyramidal, or curved in plan, they find application as roofs, coal bunkers, cooling towers, staircases, etc. (Fig. 12-1). Known in German as "Faltwerke,"

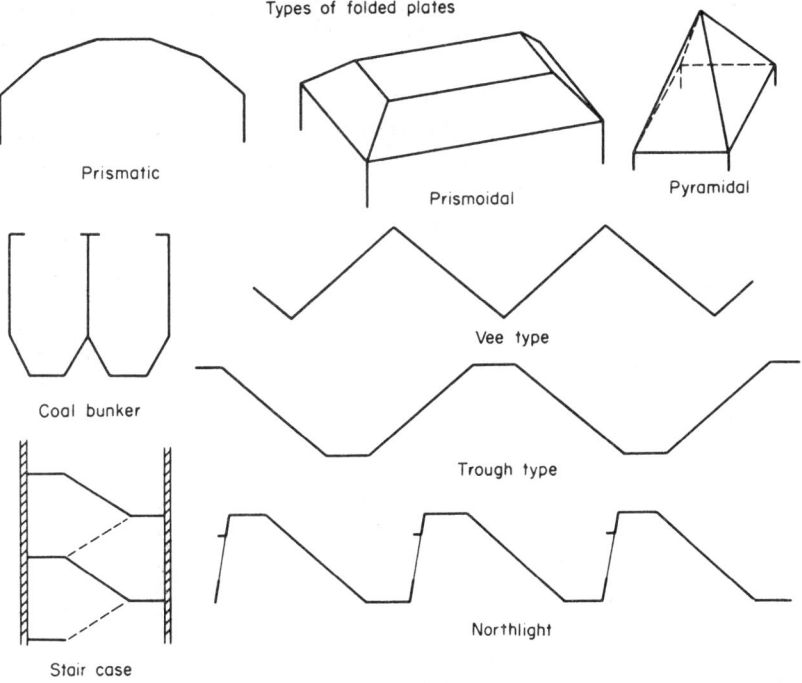

Types of folded plates

Prismatic

Prismoidal

Pyramidal

Coal bunker

Vee type

Stair case

Trough type

Northlight

Fig. 12-1

they were first employed in that country by Ehlers for large coal bunkers in 1924. He published his first paper on their structural analysis in 1930 [53]. In this early simplified treatment, the plates were regarded as hinged to each other along their junctions so that longitudinal sliding between them is prevented. The transverse moments at the joints

PHOTO 12-1. Northlight folded plates continuous over three spans, roofing the Indo-Swiss Training Centre Workshop for Precision Mechanics at Chandigarh, India.

were ignored. Gruber, in 1932, put forward a rigorous theory [54] which took into account the rigidity of the joints and the relative displacements between them. He established that large errors could result if the rigidity of the joints is ignored. In the course of the next few years many European engineers—Craemer, Ohlig, and Girkmann among them—made contributions to the subject. Most of the methods developed in Europe were based on the theory of elasticity and led to differential and algebraic equations. It is only since World War II that engineers in the United States have become interested in this form of construction. Starting from 1947, American engineers have done much to simplify the analysis of folded plates by the development of numerical-distribution procedures which are very suitable for use in the design office. The stress-distribution procedure was first presented in a paper by Winter and Pei [55] in 1947. But this treatment ignored the relative displacement of joints. Gaafar in 1954 [56] published a simple method for analyzing folded plates taking joint displacements into account. The method subsequently developed by Simpson [57] would also commend itself to design engineers because of its extreme simplicity. Another method which has found favor in the United States is Whitney's adaptation [58] of a method originally proposed by Girkmann [59]. Yitzhaki [60] and Vlasov [61] have also written extensively on folded plates. Only the Whitney and Simpson methods are proposed to be developed in detail in this book. These methods are adequate to deal with most of the problems with which we shall be concerned. All the available methods have been discussed in detail in the Phase I Report of the ASCE Task Committee on Folded Plate Construction [62]. The Committee recommends a modified version of the Gaafar method as being more suitable for use in the design office when manual compu-tational methods are employed.

12-2 Scope and Assumptions

Being primarily interested in folded plates as roofs, we shall restrict our treatment to prismatic folded plates consisting of rectangular plates, each plate being of uniform thickness. The following further assump-tions are usually made:

(i) The structure is monolithic and the joints are rigid.

(ii) The material is elastic, homogeneous, and isotropic.

(iii) The length of each plate is more than twice its width.

(iv) In all plates, plane sections remain plane after deformation. (It is, however, to be carefully noted that a plane cross section of the entire structure does not necessarily remain plane after deforma-tion.)

12-3 "Plate" and "Slab" Action

The structural behavior of folded plates is characterized by "slab" and "plate" actions. The loads acting normal to each plate cause it to bend transversely between the ridges as a continuous slab. This is known as slab action. Because of assumption (iii), the plate functions as a one-way slab and longitudinal slab action may be ignored. The plates supported at their ends on traverses bend under the action of loads acting in their plane. This is described as "plate" action. Because of assumption (iii), the bending stresses resulting from plate action may be considered to have a linear distribution across each plate.

12-4   Resolution of Ridge Loads

Let us, in the first instance, regard the plates to be hinged at the joints. The effect of moments at the joints can be superimposed later. Loads such as P_n (Fig. 12-2) applied at the ridges where adjacent plates meet are

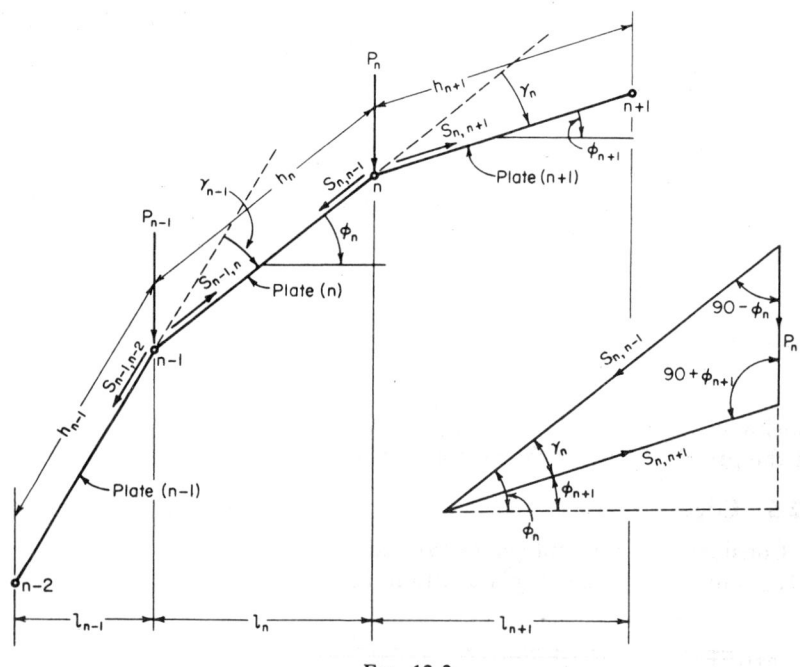

Fig. 12-2

known as ridge loads. Considering a unit length of the folded plate, $P_1 = p_1 h_1 + \frac{1}{2} p_2 h_2$, $P_2 = \frac{1}{2}(p_2 h_2 + p_3 h_3)$, and $P_n = \frac{1}{2}(p_n h_n + p_{n+1} h_{n+1})$. P_n may be resolved into plate loads $S_{n,n-1}$ and $S_{n,n+1}$ lying respectively in the planes of the nth and $(n+1)$th plates by means of the triangle of forces.

In the designation of plate loads such as $S_{n,n-1}$, the first subscript n stands for the joint at which the load acts and the second subscript $n-1$ indicates the joint toward which the plate load is directed. Thus the plate load $S_{n,n-1}$ is directed from n to $n-1$ at ridge n if it is positive. If its value turns out to be negative, the arrow of $S_{n,n-1}$ must be directed toward n at the ridge n. It is clear that the net plate load carried by the nth plate is $(S_{n,n-1} - S_{n-1,n})$.

From the triangle of forces, it is easily verified that

$$S_{n,n-1} = \frac{P_n \cos \phi_{n+1}}{\sin \gamma_n} \tag{12-1}$$

and

$$S_{n,n+1} = \frac{P_n \cos \phi_n}{\sin \gamma_n} \tag{12-2}$$

The following rules are to be observed in measuring the angles: The angle ϕ is to be measured from each plate, pointing toward the higher-order joint, to a horizontal reference line pointing to the right in a

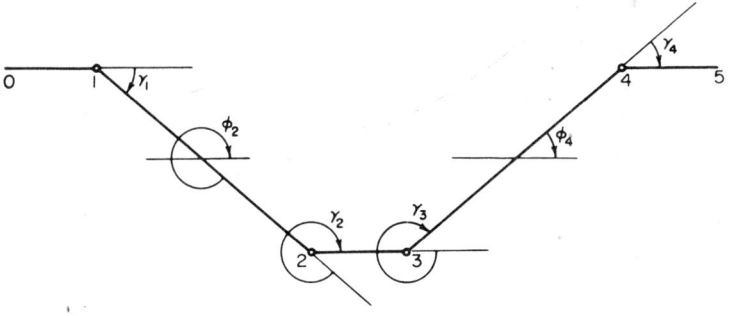

Fig. 12-3

clockwise direction and the angle γ clockwise from the extended direction of the previous plate to the next as indicated in Fig. 12-3.

12-5 Edge Shears

Consider two adjacent plates (n) and $(n + 1)$ submitted to moments $M_{0,n}$ and $M_{0,n+1}$ due to plate action (Fig. 12-4). If the plates were

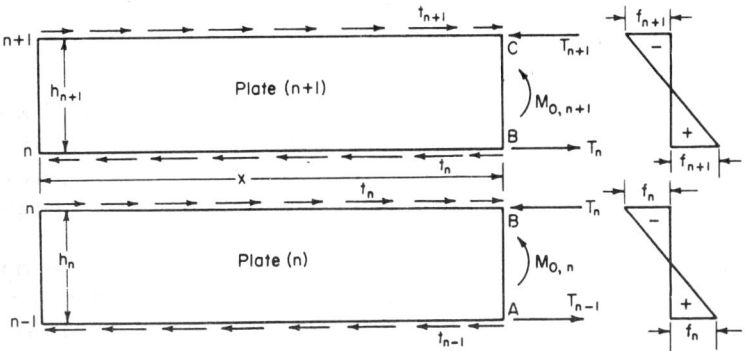

Fig. 12-4

independent, they would develop different fiber stresses f_n and f_{n+1} at the edge n. But the two plates being connected together, they cannot bend independently. Hence shear stresses would develop along their common edge n. The magnitude of these shear stresses would be such that the fiber stresses that develop in plates n and $n + 1$ along their common edge are the same. In Fig. 12-4, the edge shear forces T_{n-1}, T_n, and T_{n+1} are related to the corresponding edge shear stresses as follows:

$$T_{n-1} = - \int_0^x t_{n-1}\, dx \qquad (12\text{-}3a)$$

$$T_n = - \int_0^x t_n\, dx \qquad (12\text{-}3b)$$

$$T_{n+1} = - \int_0^x t_{n+1}\, dx \qquad (12\text{-}3c)$$

Referring to Fig. 12-4, the fiber stress at B of plate $(n + 1)$ is given by

$$\frac{M_{0,n+1}}{Z_{n+1}} + \frac{T_n h_{n+1}}{2Z_{n+1}} + \frac{T_n}{A_{n+1}} - \frac{T_{n+1}}{A_{n+1}} + \frac{T_{n+1} h_{n+1}}{2Z_{n+1}}$$

where Z_{n+1}, A_{n+1}, and h_{n+1} stand, respectively, for the modulus of section, area, and width of plate $(n + 1)$. In like manner, the fiber stress at B of plate (n) may be written as

$$- \frac{M_{0,n}}{Z_n} - \frac{T_{n-1} h_n}{2Z_n} + \frac{T_{n-1}}{A_n} - \frac{T_n}{A_n} - \frac{T_n h_n}{2Z_n}$$

Again, Z_n, A_n, and h_n stand respectively for the modulus of section, area, and width of the plate (n). The moduli of section Z_{n+1} and Z_n may be written as $Z_{n+1} = d_{n+1} h_{n+1}^2/6$ and $Z_n = d_n h_n^2/6$ where d_{n+1} and d_n are the thicknesses of the respective plates. Observing that the fiber stress at B for plates AB and BC has to be the same as the plates are monolithically connected, we may equate the two expressions derived above to obtain

$$\frac{T_{n-1}}{A_n} + 2\left(\frac{T_n}{A_n} + \frac{T_n}{A_{n+1}}\right) + \frac{T_{n+1}}{A_{n+1}} = -\frac{1}{2}\left(\frac{M_{0,n}}{Z_n} + \frac{M_{0,n+1}}{Z_{n+1}}\right) \qquad (12\text{-}4)$$

This relation may be called the theorem of three edge shears. It is similar to the theorem of three moments. This similarity is made use of in deriving a stress-distribution procedure similar to the moment-distribution procedure applicable to continuous beams.

12-6 Stress Distribution

We have just seen that continuity of stresses along the edge n common to both the plates (n) and $(n + 1)$ is secured by the application of edge shears. When plates $(n + 1)$ and (n) are regarded as bending independently, fiber stresses f_{n+1} and f_n develop at the point B of plates AB and BC, respectively. The application of edge shears has the effect of correcting the values of f_n and f_{n+1} so that they become equal. To correct these stresses, it is not necessary to solve for the edge shears using Equation (12-4). The correction can be effected directly by the following stress-distribution and carry-over procedure. Let us first

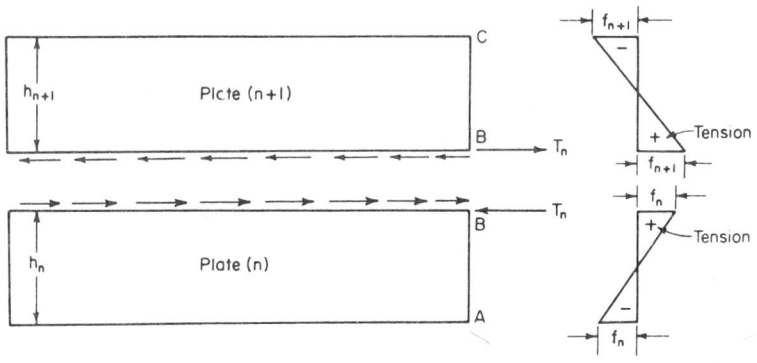

Fig. 12-5

consider the effect of the edge shear T_n only. Referring to Fig. 12-5, we may write

$$\frac{T_n}{A_{n+1}} + \frac{T_n h_{n+1}}{2Z_{n+1}} + f_{n+1} = -\frac{T_n}{A_n} - \frac{T_n h_n}{2Z_n} + f_n \tag{12-5}$$

This relation is nothing but a statement of the continuity of stress at B. Transposing terms and simplifying,

$$4T_n \left(\frac{1}{A_n} + \frac{1}{A_{n+1}} \right) = (f_n - f_{n+1}) \tag{12-6}$$

Referring to Equation (12-5), it is seen that the stress f_{n+1} is to be corrected by the addition of a stress $4T_n/A_{n+1}$ to account for the effect of the edge shear T_n. But from Equation (12-6),

$$\frac{4T_n}{A_{n+1}} = (f_n - f_{n+1}) \frac{A_n}{A_n + A_{n+1}} \tag{12-7}$$

Similarly the stress f_n is to be corrected by the addition of

$$-(f_n - f_{n+1}) \frac{A_{n+1}}{A_n + A_{n+1}}$$

We thus arrive at the distribution theorem. The stresses to be distributed to plates $(n + 1)$ and (n) at B are respectively

$$(f_n - f_{n+1}) \frac{A_n}{A_n + A_{n+1}} \quad \text{and} \quad -(f_n - f_{n+1}) \frac{A_{n+1}}{A_n + A_{n+1}}$$

The application of the edge shear T_n at B will also induce stresses at A and C which are, respectively,

$$\frac{T_n}{A_{n+1}} - \frac{T_n h_{n+1}}{2Z_{n+1}} = -\frac{2T_n}{A_{n+1}} = -\frac{1}{2}(f_n - f_{n+1}) \frac{A_n}{A_n + A_{n+1}}$$

and

$$-\frac{T_n}{A_n} + \frac{T_n h_n}{2Z_n} = +\frac{2T_n}{A_n} = +\frac{1}{2}(f_n - f_{n+1}) \frac{A_{n+1}}{A_n + A_{n+1}}$$

These may be visualized as "carry-overs" of the stresses distributed at B. It is easily verified that the carry-over factor is $-\frac{1}{2}$. The effect of edge shears T_{n+1} and T_{n-1} emanating from the edges A and C is next dealt with in a similar manner. The carry-overs reaching B from A and C will necessitate further distribution and so on as in the moment-distribution procedure. It is to be noted that in applying the distribution process described here, proper signs are to be attached to f_n and f_{n+1} which are regarded as positive if tensile.

12-7   Plate Deflections and Rotations

In Fig. 12-6, v_n and v_{n+1} stand for the plate deflections taking place in the plane of the plates (n) and $(n + 1)$; $w_{n,n-1}$, $w_{n,n+1}$, $w_{n-1,n}$, and $w_{n+1,n}$ are deflections of the plates, at right angles to their planes, occurring at the joints. Thus $w_{n,n+1}$ is the deflection at the joint n of the plate connecting the joints n and $n + 1$. The first subscript indicates the joint at which the deflection is taking place. Referring to Fig. 12-6, we have

$$v_n = (PQ + QR) = w_{n,n-1} \tan \gamma_n + \frac{v_{n+1}}{\cos \gamma_n}$$

or

$$w_{n,n-1} = v_n \cot \gamma_n - \frac{v_{n+1}}{\sin \gamma_n} \tag{12-8}$$

Also

$$w_{n,n+1} = ST = (SQ + QT) = v_{n+1} \tan \gamma_n + \left(v_n - \frac{v_{n+1}}{\cos \gamma_n}\right)\frac{1}{\sin \gamma_n}$$

$$= \frac{v_n}{\sin \gamma_n} - v_{n+1} \cot \gamma_n \tag{12-9}$$

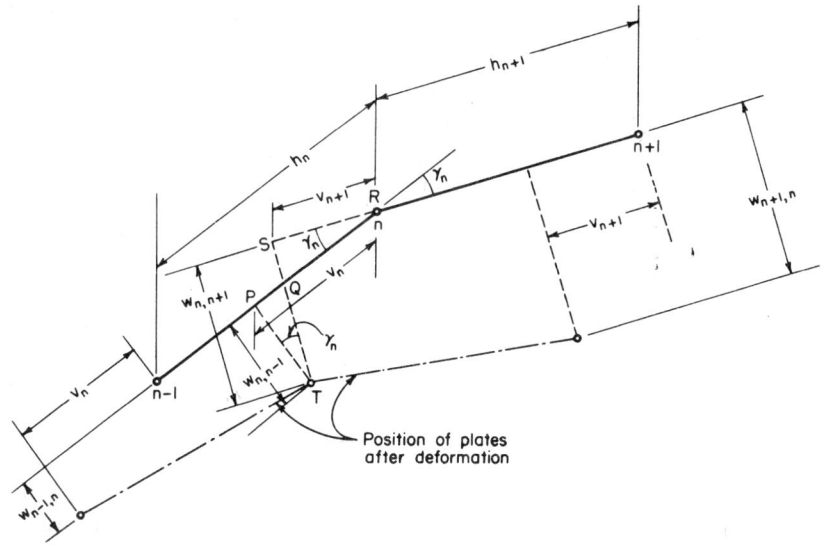

Fig. 12-6

Similar expressions may be written for $w_{n-1,n}$ and $w_{n+1,n}$. Thus

$$w_{n-1,n} = \frac{v_{n-1}}{\sin \gamma_{n-1}} - v_n \cot \gamma_{n-1} \tag{12-10}$$

and

$$w_{n+1,n} = v_{n+1} \cot \gamma_{n+1} - \frac{v_{n+2}}{\sin \gamma_{n+1}} \qquad ,$$

These expressions enable the rotation of plates (n) and $(n + 1)$ to be calculated. Denoting these by ϑ_n and ϑ_{n+1}, we may write

$$\vartheta_n = \frac{1}{h_n}(w_{n,n-1} - w_{n-1,n}) = \frac{1}{h_n}\left[v_n(\cot \gamma_n + \cot \gamma_{n-1}) - \frac{v_{n+1}}{\sin \gamma_n} - \frac{v_{n-1}}{\sin \gamma_{n-1}}\right]$$

and

$$\vartheta_{n+1} = \frac{1}{h_{n+1}}\left[v_{n+1}(\cot \gamma_{n+1} + \cot \gamma_n) - \frac{v_{n+2}}{\sin \gamma_{n+1}} - \frac{v_n}{\sin \gamma_n}\right] \tag{12-11}$$

The reduction of the included angle at the joint as a consequence of the rotations ϑ_n and $\vartheta_{n+1} = (\vartheta_{n+1} - \vartheta_n)$.

12-8   Effects of Joint Moments

Because the plates are rigidly connected together, moments will develop at the joints.   Let us consider them to be sagging moments as indicated in Fig. 12-7.   These moments will result in upward reactions

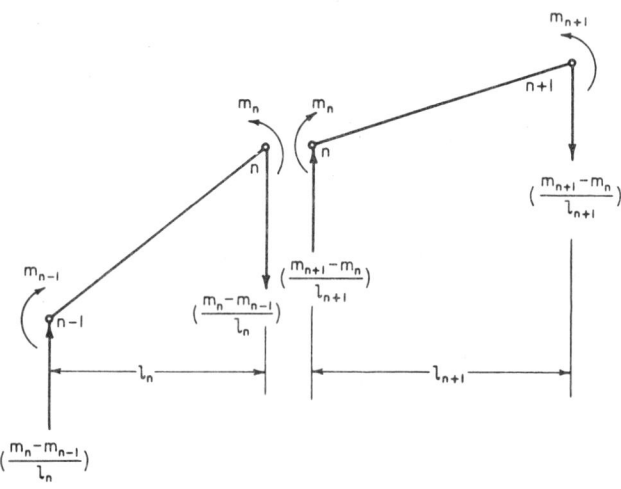

Fɪɢ. 12-7

at the ridges.   Thus, at the ridge n, the moments will cause an upward reaction of

$$\frac{m_{n+1} - m_n}{l_{n+1}} - \frac{m_n - m_{n-1}}{l_n}$$

But we know that such forces do not really exist at the ridges.   To realize this condition, it is necessary to apply downward forces such as $\varDelta P_n$ at joints whose magnitudes are given by the relation

$$\varDelta p_n = \frac{m_{n+1} - m_n}{l_{n+1}} - \frac{m_n - m_{n-1}}{l_n} \tag{12-12}$$

Thus the influence of the joint moments can be accounted for by applying additional loads at the ridges.   These additional ridge loads may be resolved into plate loads in the same manner as the ridge loads P_n.   Thus the ridge load $\varDelta P_n$ applied at the joint n resolves itself into plate loads $\Delta S_{n,n-1}$ and $\Delta S_{n,n+1}$ which are given by the following formulas:

$$\Delta S_{n,n-1} = \Delta P_n \frac{\cos \phi_{n+1}}{\sin \gamma_n} \tag{12-13}$$

$$\Delta S_{n,n+1} = \Delta P_n \frac{\cos \phi_n}{\sin \gamma_n} \tag{12-14}$$

Again the net plate load on plate n is given by

$$(\Delta S_{n,n-1} - \Delta S_{n-1,n})$$

In the calculation of plate moments, plate deflections, edge shears, and plate rotations, the total plate loads caused by the ridge loads P_n and the additional ridge loads ΔP_n have to be used. It is thus seen that the total plate load R_n on plate n is

$$(S_{n,n-1} - S_{n-1,n}) + (\Delta S_{n,n-1} - \Delta S_{n-1,n})$$

12-9 Fourier Representation

It is convenient to represent all quantities such as loads p_n, ridge loads P_n, edge shears T_n, and plate deflections v_n by Fourier series. The advantage of this procedure is that when once the problem is solved for the central section of the folded plate, the values of such quantities as $M_{0,n}$ and T_n are easily determined for other sections. Thus the fiber stress distribution at all other sections can be found without much additional labor. For the sake of convenience, let the quantities p_n, P_n, T_n, $M_{0,n}$, v_n, etc., already introduced, be the amplitudes at the central section of the folded plate of the first term of the corresponding Fourier series. Thus if $\bar{P}_n$ is the actual uniformly distributed ridge load on ridge n, it is related to P_n by $\bar{P}_n = P_n \sin(\pi x/l)$ or $P_n = (4/\pi)\bar{P}_n$. Similar relations apply to T_n, $M_{0,n}$, v_n, and ϑ_n which also follow a sine distribution. Fourier representation is possible only if the folded plate considered is simply supported on the diaphragms. For use of Fourier series to be possible, it is not necessary that the load should be uniformly distributed along the span of the plate. But the distribution for all plates has to be similar. In such cases, the relevant Fourier series can be worked out and adequate number of terms taken.

12-10 The Whitney Method of Analysis

The method presented by Whitney et al. [58] is a modification of a procedure originally proposed by Girkmann [59]. The only difference in the two presentations relates to the treatment of the end plate. Girkmann treats it as a plate supported on three sides and with a free edge on the fourth.   On the two diaphragms, the plate is regarded as simply supported.   The third side, which is monolithically connected to the next plate, is considered as continuous.   This treatment is quite rigorous.   Whitney simplified the treatment considerably by treating the end plates as cantilevers.   The simplification introduced reduces the number of simultaneous equations to be solved by two.   A step-by-step procedure of applying the Whitney method is presented below.

(1) Replace the uniform loads $\bar{p}_n$ by their first Fourier components and arrive at the corresponding ridge loads P_n .

(2) Resolve the ridge loads and arrive at the plate loads at each joint by using the formulas

$$S_{n,n-1} = P_n \frac{\cos \phi_{n+1}}{\sin \gamma_n}$$

and

$$S_{n,n+1} = P_n \frac{\cos \phi_n}{\sin \gamma_n}$$

(3) Apply additional ridge loads

$$\Delta P_n = \frac{m_{n+1} - m_n}{l_{n+1}} - \frac{m_n - m_{n-1}}{l_n}$$

at the joints to account for the joint moments, and resolve them along the plates at each joint to give

$$\Delta S_{n,n-1} = \Delta P_n \frac{\cos \phi_{n+1}}{\sin \gamma_n}$$

and

$$\Delta S_{n,n+1} = \Delta P_n \frac{\cos \phi_n}{\sin \gamma_n}$$

These expressions involve the unknown joint moments m_2 , m_3 , etc.   These are the unknowns in the problem.

(4) Compute the net plate loads R_n on plates.   Thus the plate load on the nth plate is

$$R_n = (S_{n,n-1} - S_{n-1,n}) + (\Delta S_{n,n-1} - \Delta S_{n-1,n})$$

These expressions will also involve the unknown transverse moments at the joints.

(5) Set up edge-shear equations using formulas (12-4) and solve for the edge shears in terms of the unknowns m_2 , m_3 , etc.

(6) Compute plate deflections v_n using the following formula:

$$v_n = \frac{1}{EI_n} \left(\frac{l}{\pi}\right)^4 \left[R_n + \frac{T_n + T_{n-1}}{2} \left(\frac{\pi}{l}\right)^2 h_n\right]$$

where $I_n = d_n h_n^3/12$ is the moment of inertia of the plate. The second term of this expression representing the deflection v_n' caused by the edge shears may need a word of explanation. Re-

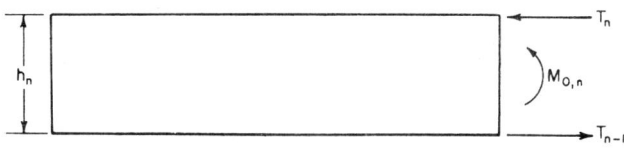

Fig. 12-8

ferring to Fig. 12-8, the moment at the central section of the plate caused by the pair of edge shears T_n and T_{n-1} is equal to

$$\frac{T_n h_n}{2} + \frac{T_{n-1} h_n}{2}$$

Recalling that T_n and T_{n-1} are in fact the central amplitudes of a sine wave, we have

$$-EI_n \frac{d^2 v_n'}{dx^2} = \frac{T_n + T_{n-1}}{2} h_n \sin \frac{\pi x}{l}$$

Integration twice with respect to x gives v_n', the plate deflection due to the edge shears. Hence we may write

$$v_n' = \frac{1}{EI_n} \frac{T_n + T_{n-1}}{2} \left(\frac{l}{\pi}\right)^2 h_n \sin \frac{\pi x}{l}$$

The value of this expression at the center is

$$\frac{1}{EI_n} \frac{T_n + T_{n-1}}{2} \left(\frac{l}{\pi}\right)^2 h_n$$

(7) With the values of the plate deflections v_n calculated in step (6), compute the plate rotations ϑ_n by using formula (12-11).

(8) Determine the change in angle at the joints caused by the slope of the plates due to slab action. The slopes are calculated in two steps.

(a) In the first step the plates are regarded as simply supported at the joints (Fig. 12-9).   Treating the nth plate as simply supported at the joints $n-1$ and n, its slope $\omega_{n,n-1}$ at the joint n may be written down by using the theorem of area moments.   Thus

$$\omega_{n,n-1} = \frac{(p_n \cos \phi_n)\, h_n^3}{24 E J_n}$$

where $J_n = d_n^3/12$ is the moment of inertia of the plate for slab

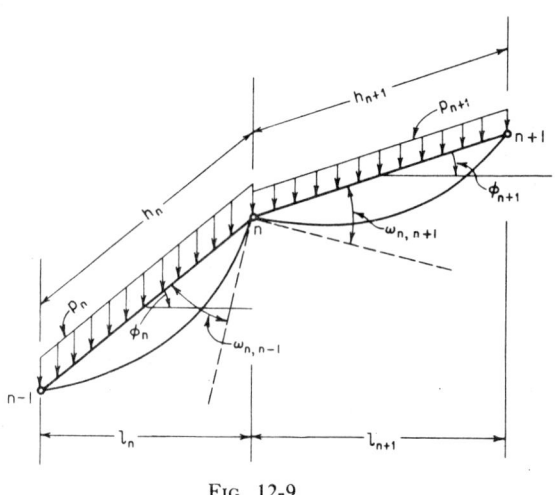

Fɪɢ. 12-9

action.   Noting that $h_n \cos \phi = l_n$, we may rewrite the slope as

$$\omega_{n,n-1} = \frac{p_n\, l_n\, h_n^2}{2E d_n^3}$$

It is to be noted that p_n is the first Fourier component of the actual load $\bar{p}_n$, or $p_n = (4/\pi)\bar{p}_n$.   Similarly,

$$\omega_{n,n+1} = \frac{p_{n+1}\, l_{n+1}\, h_{n+1}^2}{2E d_{n+1}^3}$$

Thus the total angle change at the joint n (tending to reduce the angle) is $(\omega_{n,n-1} + \omega_{n,n+1})$.

(b) Next, the change in angle caused by the joint moments is calculated (Fig. 12-10). Applying the area-moment theorem to plate n, we may write its slope at the joint n as

$$\psi_{n,n-1} = \frac{h_n}{6EJ_n}(2m_n + m_{n-1})$$

It is to be noted that m_{n-1} and m_n are the central amplitudes of the first terms of the Fourier series representing these transverse moments. Similarly,

$$\psi_{n,n+1} = \frac{h_{n+1}}{6EJ_{n+1}}(2m_n + m_{n+1})$$

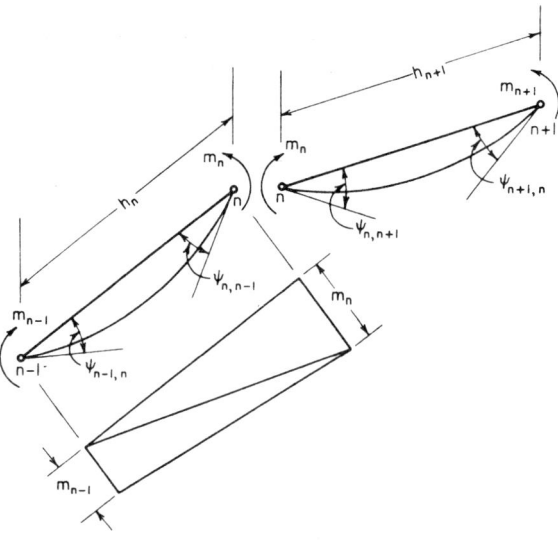

FIG. 12-10

Hence the total change in angle at joint n (tending to reduce the angle) $= (\psi_{n,n-1} + \psi_{n,n+1})$.

(9) Compute the changes in angles ϑ_n and ϑ_{n+1} caused by plate deflections using formula (12-11). The reduction in the included angle at the joint on this account is equal to $(\vartheta_{n+1} - \vartheta_n)$.

(10) At the joints, no angle change can really occur as the adjoining plates are monolithically connected to form rigid joints. Hence the sum of the reductions in angle worked out in steps (8a), (8b), and (9) must be set to zero to give $(\omega_{n,n-1} + \omega_{n,n+1}) + (\psi_{n,n-1} + \psi_{n,n+1}) + (\vartheta_{n+1} - \vartheta_n) = 0$. If a folded plate consists

of n plates, the number of such equations to be formed will be $(n - 3)$ if the problem does not involve symmetry.

(11) Solve the equations formed in step (10) simultaneously and determine the unknown joint moments.

(12) Knowing the joint moments, write down the plate loads R_n and hence the plate moments $M_{0,n}$ and the edge shears T_n. With these values available, compute the fiber stresses in the plate at the central section. From these values arrive at the fiber stresses at the other sections, noting that their distribution along the span follows a sine curve. Correct the values of fiber stresses by multiplying by the factor $\pi^3/32$. This correction is called for as only the first term of the Fourier series has been used.

Design Example 12-1

DESIGN OF A NORTHLIGHT FOLDED PLATE

1. *DATA*

Geometry

Span l = 60 ft
Breadth B = 18 ft

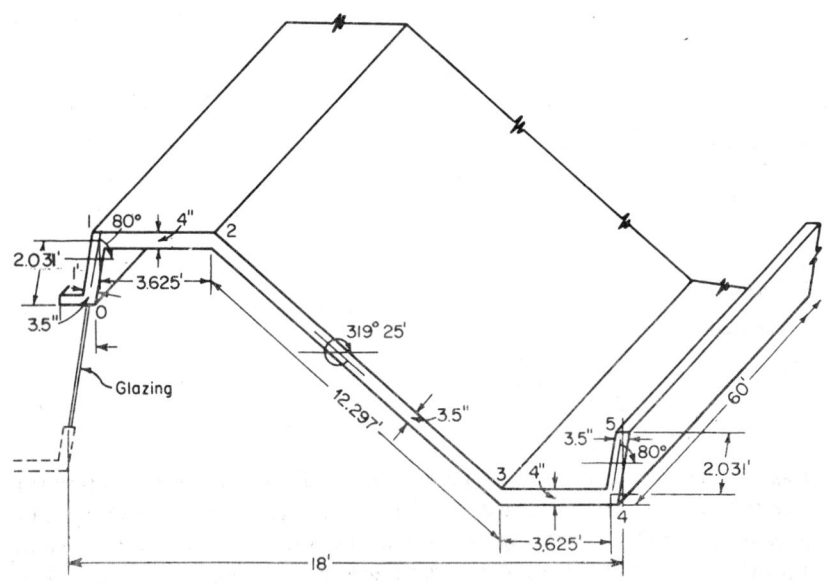

FIG. 12-11

Plate	Plate width h_n, ft	Horizontal projection l_n, ft	Inclination to horizontal ϕ_n	Angle between the plate and the next plate γ_n	Thickness d_n, in.
1	2.031	0.353	80°	80°	3.5
2	3.625	3.625	0°	40°35′	4.0
3	12.297	9.339	(360°–40°35′)	(360°–40°35′)	3.5
4	3.625	3.625	0°	(360°–80°)	4.0
5	2.031	0.353	80°	– – –	3.5

Loads

Dead weight = 12.5 psf/in. thickness of plate
Live load   = 15.0 psf of surface area

2. GEOMETRICAL PROPERTIES AND TRIGONOMETRIC QUANTITIES

Plate	A_n, ft^2	Z_n, ft^3	I_n, ft^4	J_n, ft^4
1	0.5924	0.2005	0.2036	0.0021
2	1.2083	0.7300	1.3232	0.0031
3	3.5866	7.3508	45.1963	0.0021
4	1.2083	0.7300	1.3232	0.0031
5	0.5924	0.2005	0.2036	0.0021

Angle θ	$\sin \theta$	$\cos \theta$	$\tan \theta$	$\cot \theta$	$\sec \theta$	$\operatorname{cosec} \theta$
80°	0.98481	0.17365	5.67128	0.17633	5.75877	1.01543
(360°–80°)	−0.98481	0.17365	−5.67128	−0.17633	5.75877	−1.01543
40°35′	0.65056	0.75947	0.85660	1.16742	1.31672	1.53715
(360°–40°35′)	−0.65056	0.75947	−0.85660	−1.16742	1.31672	−1.53715

3. CALCULATION OF RIDGE LOADS P_n

The loads per foot length of the plates (i.e., $p_n h_n$) are first calculated.

$$\text{Dead weight} = A_n \times 150.0 \text{ lb}$$
$$\text{Live load} = h_n \times 15.0 \text{ lb}$$
$$\text{Total load} = (A_n \times 150.0 + h_n \times 15.0) \text{ lb/ft}$$

Plate 1: $0.5924 \times 150.0 + 2.031 \times 15.0 = 119.325$ lb
Plate 2: $1.2083 \times 150.0 + 3.625 \times 15.0 = 235.620$ lb
Plate 3: $3.5866 \times 150.0 + 12.297 \times 15.0 = 722.445$ lb
Plate 4: $1.2083 \times 150.0 + 3.625 \times 15.0 = 235.620$ lb
Plate 5: $0.5924 \times 150.0 + 2.031 \times 15.0 = 119.325$ lb

Knowing $\bar{p}_n h_n$, the ridge loads P_n are calculated below. Only the first Fourier components are considered. Additional loads of 40 lb/ft at joint 0 and 30 lb/ft at joint 5 are considered due to sunshade and glazing.

$$P_1 = \left(40 + 119.325 + \frac{235.620}{2}\right)\frac{4}{\pi} = 352.859 \text{ lb}$$

$$P_2 = (117.81 + 361.22)\frac{4}{\pi} = 609.923 \text{ lb}$$

$$P_3 = (361.22 + 117.81)\frac{4}{\pi} = 609.923 \text{ lb}$$

$$P_4 = (119.325 + 117.81 + 30.0)\frac{4}{\pi} = 340.127 \text{ lb}$$

4. PLATE LOADS DUE TO P_n

$$S_{1.0} = \frac{352.859}{\sin 80°} = 358.302 \text{ lb}$$

$$S_{1.2} = 352.859 \cot 80° = 62.220 \text{ lb}$$

$$S_{2.1} = 609.923 \cot 40°35' = 712.036 \text{ lb}$$

$$S_{2.3} = \frac{609.923}{\sin 40°35'} = 937.535 \text{ lb}$$

$$S_{3.2} = \frac{609.923}{\sin 40°35'} = 937.535 \text{ lb}$$

$$S_{3.4} = 609.923 \cot 40°35' = 712.036 \text{ lb}$$

$$S_{4.3} = 340.127 \cot 80° = 59.975 \text{ lb}$$

$$S_{4.5} = \frac{340.127}{\sin 80°} = 345.375 \text{ lb}$$

The directions of the forces are shown in Fig. 12-11a.

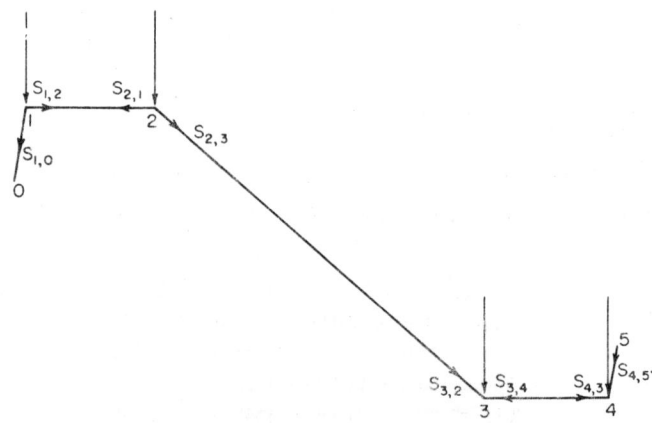

Fɪɢ. 12-11a

5. ADDITIONAL RIDGE LOADS ΔP_n DUE TO TRANSVERSE MOMENTS

(Sagging moments $+ve$)

The known moments m_1 and m_4 are calculated below.

$$m_1 = -(40.0 \times 0.8527 + 119.325 \times 0.1763) = -(55.145 \text{ ft lb}) \times \frac{4}{\pi}$$

$$m_4 = -(30.0 \times 0.3527 + 119.325 \times 0.1763) = -(31.618 \text{ ft lb}) \times \frac{4}{\pi}$$

Note. $m_1 \neq m_4$. But, this difference being small, the torsion caused by this on the cross section is not considered in the analysis.

$$\Delta P_1 = \frac{m_2 - m_1}{3.625} = (0.276m_2 + 19.369)\downarrow$$

$$\Delta P_2 = -\frac{m_2 + 55.145 \times 4/\pi}{3.625} + \frac{m_3 - m_2}{9.339}$$
$$= (-19.369 - 0.383m_2 + 0.107m_3)\downarrow$$

$$\Delta P_3 = \frac{m_2 - m_3}{9.339} + \frac{m_4 - m_3}{3.625}$$
$$= (0.107m_2 - 0.383m_3 - 11.107)\downarrow$$

$$\Delta P_4 = \frac{m_3 - m_4}{3.625} = (0.276m_3 + 11.107)\downarrow$$

6. ADDITIONAL PLATE LOADS DUE TO ΔP_n

$$\Delta S_{1.0} = \frac{0.276m_2 + 19.369}{0.98481} = 0.280m_2 + 19.667$$

$$\Delta S_{1.2} = (0.276m_2 + 19.369) \times 0.176 = 0.049m_2 + 3.415$$

$$\Delta S_{2.1} = \Delta P_2 \cot 40°35' = -22.611 - 0.447m_2 + 0.125m_3$$

$$\Delta S_{2.3} = \frac{\Delta P_2}{\sin 40°35'} = -29.773 - 0.589m_2 + 0.165m_3$$

$$\Delta S_{3.2} = \frac{\Delta P_3}{\sin 40°35'} = 0.165m_2 - 0.589m_3 - 17.073$$

$$\Delta S_{3.4} = \Delta P_3 \cot 40°35' = 0.125m_2 - 0.447m_3 - 12.966$$

$$\Delta S_{4.3} = \Delta P_4 \cot 80° = 0.049m_3 + 1.958$$

$$\Delta S_{4.5} = \frac{\Delta P_4}{\sin 80°} = 0.280m_3 + 11.278$$

7. RESULTANT PLATE LOADS R_n

$$R_n = (S_{n,n-1} - S_{n-1,n}) + (\Delta S_{n,n-1} - \Delta S_{n-1,n})$$
$$R_1 = (358.302 - 0) + (0.280\,m_2 + 19.667 - 0)$$
$$= +377.969 + 0.280\,m_2$$
$$R_2 = +623.790 - 0.496\,m_2 + 0.125\,m_3$$
$$R_3 = -1,828.225 + 0.424\,m_2 + 0.424\,m_3$$
$$R_4 = +637.137 + 0.125\,m_2 - 0.496\,m_3$$
$$R_5 = +356.653 + 0.280\,m_3$$

8. PLATE MOMENTS $M_{0,n}$

$$M_{0,n} = R_n \times \frac{l^2}{\pi^2} = R_n \times 364.756$$

$$M_{0,1} = +137,866.7 + 102.2m_2$$

$$M_{0,2} = +227,531.2 - 180.8m_2 + 45.6m_3$$

$$M_{0,3} = -666,856.6 + 154.7m_2 + 154.7m_3$$

$$M_{0,4} = +232,399.6 + 45.6m_2 - 180.8m_3$$

$$M_{0,5} = +130,091.5 + 102.2m_3$$

9. EDGE-SHEAR EQUATIONS

$$\frac{T_{n-1}}{A_n} + 2T_n\left(\frac{1}{A_n} + \frac{1}{A_{n+1}}\right) + \frac{T_{n+1}}{A_{n+1}} = -\frac{1}{2}\left(\frac{M_{0,n}}{Z_n} + \frac{M_{0,n+1}}{Z_{n+1}}\right)$$

Joint 1

$$4T_1\left(\frac{1}{A_1} + \frac{1}{A_2}\right) + \frac{2T_2}{A_2} = -\left(\frac{M_{0,1}}{Z_1} + \frac{M_{0,2}}{Z_2}\right)$$

$$10.06T_1 + 1.66T_2 = -999,301.07 - 262.04m_2 - 62.46m_3$$

Joint 2

$$\frac{2T_1}{A_2} + 4T_2\left(\frac{1}{A_2} + \frac{1}{A_3}\right) + \frac{2T_3}{A_3} = -\left(\frac{M_{0,2}}{Z_2} + \frac{M_{0,3}}{Z_3}\right)$$

$$1.66T_1 + 4.43T_2 + 0.56T_3 = -220,966.99 + 226.66m_2 - 83.50m_3$$

Joint 3

$$2T_2\frac{1}{A_3} + 4T_3\left(\frac{1}{A_3} + \frac{1}{A_4}\right) + 2T_4\frac{1}{A_4} = -\left(\frac{M_{0,3}}{Z_3} + \frac{M_{0,4}}{Z_4}\right)$$

$$0.56T_2 + 4.43T_3 + 1.66T_4 = -227,635.80 - 83.50m_2 + 226.66m_3$$

Joint 4

$$\frac{2T_3}{A_4} + 4T_4\left(\frac{1}{A_4} + \frac{1}{A_5}\right) = -\left(\frac{M_{0,4}}{Z_4} + \frac{M_{0,5}}{Z_5}\right)$$

$$1.66T_3 + 10.06T_4 = -967,191.03 - 62.46m_2 - 262.04m_3$$

On solving these four equations, T_1, T_2, T_3, and T_4 are evaluated:

$$T_1 = -97,399.31 - 37.32\,m_2 - 1.793\,m_3$$

$$T_2 = -11,617.94 + 68.55\,m_2 - 26.83\,m_3$$

$$T_3 = -14,941.47 - 26.83\,m_2 + 68.55\,m_3$$

$$T_4 = -93,661.51 - 1.79\,m_2 - 37.32\,m_3$$

$$T_1 + T_2 = -109,017.25 + 31.24\,m_2 - 28.63\,m_3$$

$$T_2 + T_3 = -26,559.41 + 41.72\,m_2 + 41.72\,m_3$$

$$T_3 + T_4 = -108,602.98 - 28.63\,m_2 + 31.24\,m_3$$

10. PLATE DEFLECTIONS

$$v_n = \frac{1}{EI_n}\left(\frac{l}{\pi}\right)^4\left[R_n + (T_{n-1} + T_n)\left(\frac{\pi}{l}\right)^2\frac{h_n}{2}\right]\sin\frac{\pi x}{l}$$

$$\left(\frac{l}{\pi}\right)^4 = 10^3 \times 133.047$$

$$\left(\frac{\pi}{l}\right)^2 = 0.0027$$

$$v_1 = \frac{10^3 \times 133.047}{E \times 0.2036}(+377.97 + 0.280m_2 - 271.16 - 0.104m_2 - 0.005m_3)$$

$$= \frac{10^3}{E}(69{,}793.94 + 115.21m_2 - 3.26m_3)$$

$$v_2 = \frac{10^3 \times 133.047}{E \times 1.3232}(+623.79 - 0.49m_2 + 0.13m_3 - 541.72 + 0.16m_2 - 0.14m_3)$$

$$= \frac{10^3}{E}(8{,}252.55 - 34.24m_2 - 1.734m_3)$$

$$v_3 = \frac{10^3 \times 133.047}{E \times 45.1963}(-1{,}828.23 + 0.42m_2 + 0.42m_3 - 447.70 + 0.70m_2 + 0.70m_3)$$

$$= \frac{10^3}{E}(-6{,}699.78 + 3.32m_2 + 3.32m_3)$$

$$v_4 = \frac{10^3 \times 133.047}{E \times 1.3232}(+637.14 + 0.13m_2 - 0.50m_3 - 539.66 - 0.14m_2 + 0.16m_3)$$

$$= \frac{10^3}{E}(+9{,}801.59 - 1.73m_2 - 34.24m_3)$$

$$v_5 = \frac{10^3 \times 133.047}{E \times 0.2036}(+356.65 + 0.28m_3 - 260.76 - 0.104m_3 - 0.005m_2)$$

$$= \frac{10^3}{E}(+62{,}664.54 - 3.26m_2 + 115.21m_3)$$

11. CALCULATION OF ANGLE CHANGES AT THE JOINTS

(1) Angle Changes Due to Plate Deflections

General Formula

$$\vartheta_{n\ n-1} = \frac{1}{h_n}\left[v_n(\cot\gamma_n + \cot\gamma_{n-1}) - \frac{v_{n+1}}{\sin\gamma_n} - \frac{v_{n-1}}{\sin\gamma_{n-1}}\right]$$

Joint 2

$$\vartheta_2 = \vartheta_{2.1} = \frac{1}{h_2}\left[-\frac{v_1}{\sin\gamma_1} + v_2(\cot\gamma_2 + \cot\gamma_1) - \frac{v_3}{\sin\gamma_2}\right]$$

$$= \frac{10^3}{E \times 3.625}(-70{,}870.86 - 116.98m_2 + 3.31m_3 + 11{,}089.37$$

$$- 46.01m_2 - 2.33m_3 + 10{,}298.56 - 5.10m_2 - 5.10m_3)$$

$$= \frac{10^3}{E}(-13{,}650.46 - 46.37m_2 - 1.14m_3)$$

$$\vartheta_3 = \vartheta_{2.3} = \frac{1}{h_3}\frac{v_4 - v_2}{\sin \gamma_2}$$

$$= \frac{10^3}{E \times 12.297 \times 0.651}(+9{,}801.59 - 1.73m_2 - 34.24m_3$$
$$- 8{,}252.55 + 34.24m_2 + 1.73m_3)$$

$$= \frac{10^3}{E}(+193.63 + 4.06m_2 - 4.06m_3)$$

Joint 3

$$\vartheta_{3.2} = \vartheta_{2.3} = \vartheta_3$$

$$\vartheta_4 = \vartheta_{3.4} = \frac{1}{h_4}\left[-\frac{v_3}{\sin \gamma_3} + v_4(\cot \gamma_3 + \cot \gamma_4) - \frac{v_5}{\sin \gamma_4}\right]$$

$$= \frac{10^3}{E \times 3.625}(-10{,}298.56 + 5.10m_2 + 5.10m_3 - 13{,}170.88$$
$$+ 2.33m_2 + 46.01m_3 + 63{,}631.46 - 3.31m_2 + 116.99m_3)$$

$$= \frac{10^3}{E}(+11{,}079.18 + 1.14m_2 + 46.37m_3)$$

(2) *Angle Changes Due to Loads*

$$\omega_n = \frac{4}{\pi}\left(\frac{p_n h_n{}^2 l_n}{2Ed_n{}^3} + \frac{p_{n+1}h_{n+1}^2 l_{n+1}}{2Ed_{n+1}^3}\right)$$

$$\omega_2 = \frac{2}{\pi E}\left(\frac{235.620 \times 3.625^2}{(1/3)^3} + \frac{722.44 \times 12.30 \times 9.339}{(7/24)^3}\right)$$

$$= \frac{10^3}{E}(2{,}181.97)$$

$$\omega_3 = \frac{10^3}{E}(2{,}181.97)$$

(3) *Angle Changes Due to Transverse Moments*
Joint 2

$$\psi_{2.1} = \frac{h_2}{6EJ_2}(2m_2 + m_1) = \frac{1}{E}\frac{3.625}{6 \times 0.0031}(2m_2 + m_1)$$

$$= \frac{10^3}{E}(0.39m_2 - 13.75)$$

$$\psi_{2.3} = \frac{12.297}{6E \times 0.0021}(2m_2 + m_3)$$

$$= \frac{10^3}{E}(1.98m_2 + 0.99m_3)$$

Joint 3

$$\psi_{3.2} = \frac{10^3}{E}(0.99m_2 + 1.98m_3)$$

$$\psi_{3.4} = \frac{3.625}{6E \times 0.0031}(2m_3 + m_4)$$

$$= \frac{10^3}{E}(0.39m_3 - 7.88)$$

Total Change at Each Joint

$$(\omega_{n,n-1} + \omega_{n,n+1}) + (\psi_{n,n-1} + \psi_{n,n+1}) + (\vartheta_{n+1} - \vartheta_n)$$

Joint 2

$$\frac{10^3}{E} (+ 2,181.97 - 13.75 + 0.39m_2 + 1.98m_2 + 0.99m_3$$
$$+ 13,650.46 + 46.37m_2 + 1.14m_3 + 193.63 + 4.06m_2 - 4.06m_3)$$

Simplifying and equating to zero, we get,

$$+16,012.32 + 52.81\, m_2 - 1.94\, m_3 = 0 \tag{1}$$

Joint 3

$$\frac{10^3}{E} (+ 2,181.97 + 0.99m_2 + 1.98m_3 - 7.88 + 0.39m_3$$
$$- 193.63 - 4.06m_2 + 4.06m_3 + 11,079.18 + 1.14m_2 + 46.37m_3)$$

Simplifying and equating to zero,

$$13,059.63 - 1.94\, m_2 + 52.80\, m_3 = 0 \tag{2}$$

On solving the two equations,

$$m_2 = -312.69$$
$$m_3 = -258.75$$

12. CALCULATION OF T_n AND $M_{0,n}$

Substituting the values of m_2 and m_3 in the expressions in steps 8 and 9 for $M_{0,n}$ and T_n, respectively, we get

$$M_{0,1} = +105,908.1 \text{ lb ft}$$
$$M_{0,2} = +272,277.0 \text{ lb ft}$$
$$M_{0,3} = -755,254.1 \text{ lb ft}$$
$$M_{0,4} = +264,931.2 \text{ lb ft}$$
$$M_{0,5} = +103,646.3 \text{ lb ft}$$
$$T_1 = -85,266.3 \text{ lb}$$
$$T_2 = -26,110.5 \text{ lb}$$
$$T_3 = -24,288.5 \text{ lb}$$
$$T_4 = -83,444.9 \text{ lb}$$

Plate 1

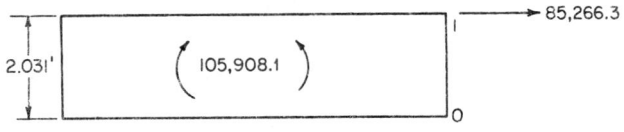

FIG. 12-12

$$\text{Net moment} = +105,908.1 - \left(85,266.3 \times \frac{2.031}{2}\right) = +19,320.2$$

$$f_0 = \frac{19,320.2}{0.2005} + \frac{85,266.3}{0.5924}$$
$$= +\,96,360.1 + 143,933.7$$
$$= +\,240,293.8 \text{ psf} = +1,669 \text{ psi}$$
$$f_1 = -\,96,360.1 + 143,933.7$$
$$= +\,47,573.6 \text{ psf} = +330 \text{ psi}$$

Plate 2

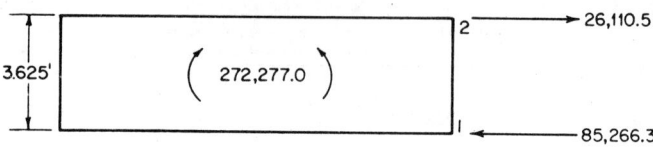

Fɪɢ. 12-13

Net moment $= +\,272,277.0 - 201,870.6 = +70,406.4$

$$f_1 = +\frac{70,406.4}{0.7300} - \frac{59,155.8}{1.2083}$$
$$= 96,447.2 - 48,957.9$$
$$= +\,47,489.3 \text{ psf} = +330 \text{ psi}$$
$$f_2 = -\,145,405.0 \text{ psf} = -1,010 \text{ psi}$$

Plate 3

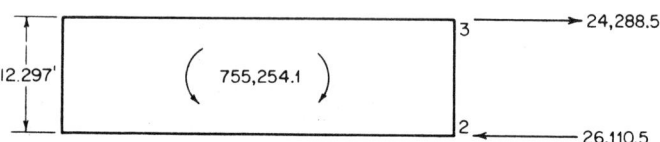

Fɪɢ. 12-14

Net moment $= -\,755,254.1 - 309,878.5 = -1,065,132.6$

$$f_2 = -\frac{1,065,132.6}{7.3508} - \frac{1,822.0}{3.5866}$$
$$= -\,144,900.2 - 508.0$$
$$= -\,145,408.2 \text{ psf} = -1,010 \text{ psi}$$
$$f_3 = +\,144,392.2 \text{ psf} = +1,003 \text{ psi}$$

Plate 4

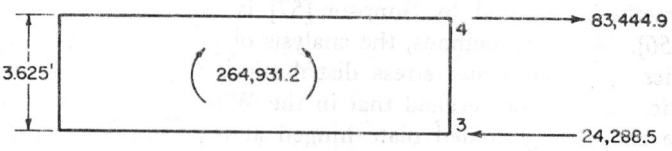

Fɪɢ. 12-15

Net moment $= + 264,931.2 - 195,266.7 = +69,664.5$

$$f_3 = + \frac{69,664.5}{0.7300} + \frac{59,156.4}{1.2083}$$

$$= + 95,430.7 + 48,958.4 = +144,389.1 \text{ psf}$$
$$= +1,003 \text{ psi}$$

$$f_4 = - 46,472.3 \text{ psf} \qquad = -323 \text{ psi}$$

Plate 5

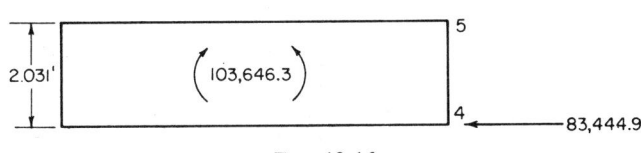

FIG. 12-16

Net moment $= + 103,646.3 - 84,738.3$
$$= + 18,908.0$$

$$f_4 = + \frac{18,908.0}{0.2005} - \frac{83,444.9}{0.5924}$$

$$= + 94,304.3 - 140,859.0$$
$$= - 46,554.7 \text{ psf} = -323 \text{ psi}$$

$$f_5 = - 235,163.3 \text{ psf} = -1,633 \text{ psi}$$

(i)
$$\text{Actual moments} = \left(\frac{\pi}{4}\right) \times \begin{pmatrix} m_2 \\ \text{or} \\ m_3 \end{pmatrix}$$

(ii)
$$\text{Longitudinal stresses} = \frac{\pi^3}{32} \times \begin{pmatrix} f_1 \\ f_2 \\ \cdots \\ f_5 \end{pmatrix}$$

THE SIMPSON METHOD

12-11 Notes on the Simpson and Gaafar Methods

The method proposed by Simpson [57] is very similar to that of Gaafar [56]. In both methods, the analysis of folded plates is reduced to a series of moment and stress distributions involving only simple arithmetic. It may be recalled that in the Whitney method the basic structure chosen is a folded plate hinged at the joints on which the effects of joint moments are subsequently superimposed in the form of additional ridge loads. In the Gaafar and Simpson methods, the basic structure selected is a folded plate with rigid joints. To start with, the ridges are assumed to be nondeflecting. The effects of the ridge displacements are subsequently accounted for. In the Simpson method, this is done by allowing each plate in turn to rotate at midspan by an arbitrary amount. Each of the rotation solutions multiplied by a constant is added to the no-rotation solution to give the final solution.

In the Gaafar method, the relative transverse displacements $\Delta_n = (w_{n,n-1} - w_{n-1,n})$ between two consecutive ridges n and $n-1$ are taken as the unknowns. The method leads to a set of linear simultaneous equations involving them. After the unknowns are solved for, the effects of the differential transverse displacements Δ_n are superimposed on the no-rotation solution corresponding to the folded plate whose ridges are regarded as nondeflecting. The relative transverse displacement Δ_n of the two ends of plate n and its rotation may be related to each other by the equation

$$\psi_n = \frac{\Delta_n}{h_n} \tag{12-15}$$

12-12 Assumptions and Limitations

Apart from the assumptions listed in Art. 12-2, the additional assumption is made that Δ_n and hence ψ_n vary as the ordinates of a sine curve along the span of the folded plate. Or, in other words, denoting the values of Δ_n and ψ_n at midspan by $(\Delta_n)_c$ and $(\psi_n)_c$, we may write

$$\Delta_n = (\Delta_n)_c \sin \frac{\pi x}{l}$$

and

$$\tag{12-16}$$

$$\psi_n = (\psi_n)_c \sin \frac{\pi x}{l}$$

This assumption appears to be reasonable for a simply supported folded plate symmetrically loaded about its midspan. Perhaps it is more accurate to assume that these functions vary as the ordinates of the elastic curve of a simply supported beam loaded in the same manner as the folded plate. Thus, for a uniformly loaded folded plate, the elastic line of the uniformly loaded simply supported beam may offer the best approximation. Gaafar [56] has shown that the sine curve and the elastic line are practically indistinguishable from each other for this condition of loading. Hence the replacement of the elastic line by the sine curve will not make any significant difference, especially as the problem is formulated in terms of the plate deflections v_n which are obtained from Δ_n and ψ_n after four integrations. If, however, the folded plate is continuous over several spans or carries loading which is far from symmetrical about midspan, it is advisable to assume that the variation of Δ_n and ψ_n follows the shape of the elastic line of a beam supported and loaded in the same manner as the folded plate.

12-13 The Simpson Method in Outline

The method is outlined with reference to the folded plate shown in Fig. 12-17.

Step 1

Consider a transverse section of unit length at midspan of the given folded plate. Assuming that the joints do not deflect, arrive at the joint moments by moment distribution. Calculate the reactions at the joints

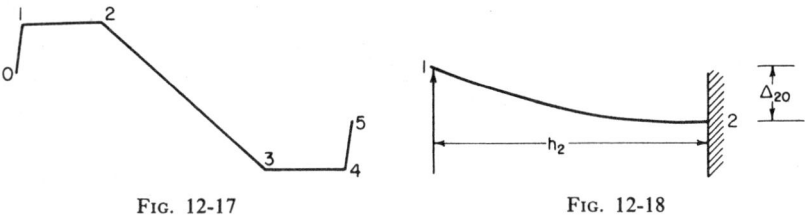

FIG. 12-17   FIG. 12-18

and apply forces equal and opposite to these at the joints. Resolve the loads, thus applied, into plate loads. Calculate the bending stresses caused by the plate loads, assuming each plate to be free to bend independently. These stresses shall be designated as free-edge stresses. Next, establish stress compatibility at the common edges of adjacent plates by stress distribution. The resulting stresses in the plates are those which occur in a folded plate if the joints do not deflect. This solution will be referred to as the no-rotation solution. The solution up to this point is identical with that obtained by the use of the well-known Winter and Pei method which ignores joint displacements.

Step 2

The effects of joint displacements have now to be accounted for by considering the rotations of plates 2, 3, and 4, the first and last plates being regarded as cantilevers. We start with plate 2. Let joint 2 deflect by an arbitrary amount Δ_{20} below the level of joint 1 (Fig. 12-18). The fixing moment induced at 2 as a result is equal to

$3EJ_2\,\Delta_{20}/h_2{}^2 = 3EJ_2\psi_{20}/h_2$, where $\psi_{20} = \Delta_{20}/h_2$. As Δ_{20} is arbitrary, ψ_{20} is an arbitrary rotation of plate 2. Let ψ_{20} be such that the magnitude of the moment induced is 3. Hence $\psi_{20} = h_2/EJ_2$. The arbitrary rotation and the actual rotation of the plate are clearly related by an unknown constant k_2 such that $\psi_2 = k_2\,\psi_{20}$. This arbitrary moment of 3 at joint 2 is next distributed by the moment-distribution procedure. The resulting joint moments and reactions are found. Forces equal and opposite to the reactions are applied at the joints and resolved along the plates as plate loads. The free-edge stresses caused by these loads are next determined. Stress compatibility at the common edges is realized by stress distribution. The resulting stresses shall be referred to as the Case II solution corresponding to an arbitrary rotation of plate 2.

Step 3

Next, consider the effect of an arbitrary rotation of plate 3 (Fig. 12-19). As before, $\psi_3 = k_3\,\psi_{30}$. The moment induced at joints 2 and 3 is

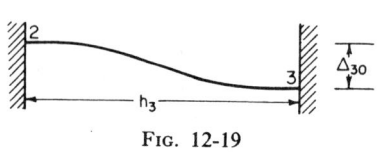

FIG. 12-19

$6EJ_3\,\Delta_{30}/h_3{}^2 = 6EJ_3\,\psi_{30}/h_3$. Let the arbitrary rotation ψ_{30} be such that the magnitude of the moments induced is 6. Hence $\psi_{30} = h_3/EJ_3$. Distribute the moments of 6 units each at joints 2 and 3 by moment distribution. Arrive at the reactions at the joints and apply forces opposite to these at the joints. Resolve these forces into plate loads and compute the free-edge stresses. Correct these by stress distribution to secure stress compatibility at the common edges. The resulting stresses shall be referred to as the Case III solution corresponding to the arbitrary rotation ψ_{30} of plate 3.

Step 4

The Case IV solution corresponding to an arbitrary rotation ψ_{40} of plate 4 is worked in the same manner as the Case III solution. Again, we may note that

$$\psi_4 = k_4\psi_{40} \quad \text{and} \quad \psi_{40} = \frac{h_4}{EJ_4}$$

Step 5

The plate deflections v_n are next worked out. The deflection v_n will consist of the deflection corresponding to the no-rotation solution plus k_2 times the deflection due to the Case II solution plus k_3 times the deflection resulting from the Case III solution plus k_4 times the deflection corresponding to the Case IV solution. To compute these deflections, it is essential to have two formulas—one applicable to the no-rotation solution and the other to the arbitrary rotation solutions.

Formula for Deflection Corresponding to the No-rotation Solution

Let f_t and f_b be the fiber stresses at top and bottom of the plate n at midspan, corresponding to the no-rotation solution (Fig. 12-20). The plate moment at midspan corresponding to the stress distribution shown is $\frac{1}{2}(f_b - f_t)h_n\, d_n(h_n/6)$ or

$$\frac{1}{12}(f_b - f_t)\, d_n h_n{}^2 \qquad (12\text{-}17)$$

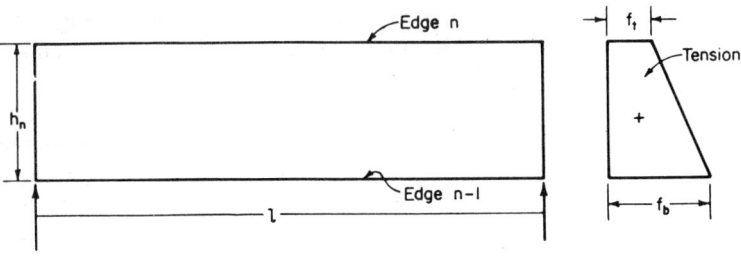

Fig. 12-20

Let the uniformly distributed plate load be p'. Equating the bending and resisting moments at midspan,

$$\frac{p'l^2}{8} = \frac{1}{12}(f_b - f_t)\, d_n h_n{}^2 \qquad (12\text{-}18)$$

The downward plate deflection at midspan is $\frac{5}{384}(p'l^4/EI_n)$. Substituting for p' from (12-18) and noting that $I_n = \frac{1}{12}d_n\, h_n{}^3$, the plate deflection at midspan may be written as

$$\frac{5}{48E}\frac{l^2}{h_n}(f_b - f_t) \qquad (12\text{-}19)$$

Formula for Plate Deflection Corresponding to Arbitrary Rotations

Consider again the same plate n with fiber stresses at midspan of f_b and f_t at its top and bottom fibers caused by an arbitrary rotation of that plate or any other plate. The resisting moment developed at the central section which is equal to the bending moment is again equal to $\frac{1}{12}(f_b - f_t)d_n\, h_n{}^2$. The load on the plate is proportional to ψ_{n0} for which a sine variation along the span has been assumed. Two integrations of this loading will yield the bending moment at the center of the span which will be proportional to $\psi_{n0}(l^2/\pi^2)$. Similarly, the midspan deflection is proportional to $(l^4/\pi^4EI_n)\,\psi_{n0}$. Hence the deflection at midspan may be obtained by multiplying the bending moment at that section by l^2/π^2EI_n. But we have already seen that the bending moment is $\frac{1}{12}(f_b - f_t)\,d_n h_n^2$. The deflection at midspan is therefore

$$\frac{l^2}{\pi^2 Eh_n}(f_b - f_t) \tag{12-20}$$

Step 6

From the plate deflections calculated in step 5, we may now arrive at the transverse joint displacements $w_{n,n-1}$, $w_{n-1,n}$, etc., using formula (12-10). It is to be noted that the plate deflections calculated in step 5 and the transverse joint deflections computed from them in step 6 will involve the unknown constants k_2, k_3, and k_4.

Step 7

From the results of step 6, the plate rotations ψ_2, ψ_3, and ψ_4 may be calculated by using the formulas given below.

$$\psi_n = \frac{1}{h_n}(w_{n,n-1} - w_{n-1,n}) = \frac{1}{h_n}\left[v_n(\cot \gamma_n + \cot \gamma_{n-1}) - \frac{v_{n+1}}{\sin \gamma_n} - \frac{v_{n-1}}{\sin \gamma_{n-1}}\right]$$

and

$$\psi_{n+1} = \frac{1}{h_{n+1}}\left[v_{n+1}(\cot \gamma_{n+1} + \cot \gamma_n) - \frac{v_{n+2}}{\sin \gamma_{n+1}} - \frac{v_n}{\sin \gamma_n}\right]$$

Step 8

Equate ψ_2, ψ_3, and ψ_4 calculated in step 7 to $k_2\,\psi_{20}$, $k_3\,\psi_{30}$, and $k_4\,\psi_{40}$ to obtain a set of three linear simultaneous equations in the unknowns k_2, k_3, and k_4. Solve the equations for k_2, k_3, and k_4.

Step 9

Compute the fiber stresses in the folded plate by combining the stresses of the no-rotation solution with k_2 times the stresses of the Case II solution, k_3 times the stresses of the Case III solution, and k_4 times the stresses of the Case IV solution. For an unsymmetrical problem, the Simpson method leads to $(n - 2)$ simultaneous equations, if n is the number of plates. Even if the cross section of a folded plate is symmetrical, an unsymmetrical problem would result, if it is not symmetrically loaded with respect to its axis of symmetry.

12-14 Symmetrical Problems

For symmetrical problems, the Simpson and Whitney methods will lead to $(n - 2)/2$ equations where n is the number of plates if the number of plates is even and $(n - 3)/2$ if the number of plates is odd. In applying the Simpson method to symmetrical problems, it is observed that a symmetrically oriented pair of plates is to be given equal and symmetrical arbitrary rotations at the same time, if advantage is to be taken of symmetry. In the illustrative example that follows, advantage has been taken of symmetry.

Illustrative Example

The application of the Simpson method to a symmetrical problem is explained in the example that follows.

Design Example 12-2

DESIGN OF A FOLDED PLATE BY THE SIMPSON METHOD

1. *DATA*

   Geometry

   Span $l = 60$ ft

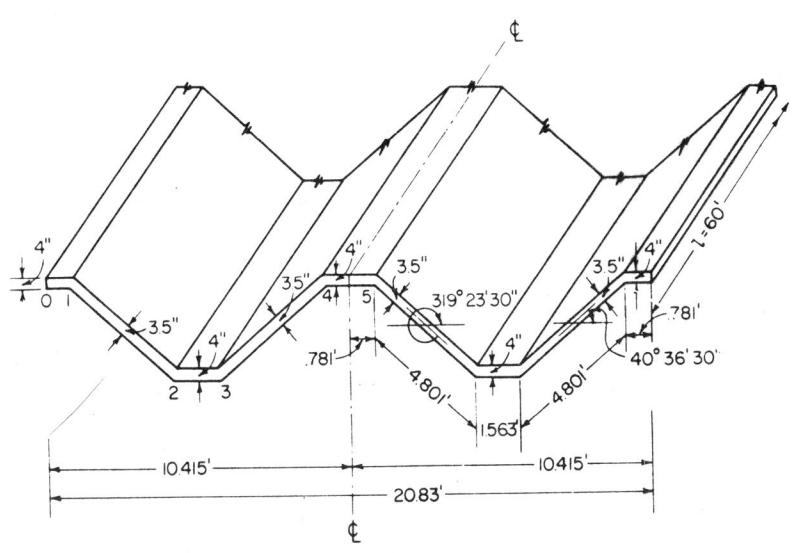

FIG. 12-21

Plate	Plate width h_n, ft	Horizontal projection l_n, ft	Inclination to horizontal ϕ_n	Angle between the plate and the next plate γ_n	Thickness of plate, in.
1	0.781	0.781	0°	40°36′30″	4
2	4.801	3.645	319°23′30″	319°23′30″	3.5
3	1.563	1.563	0°	319°23′30″	4
4	4.801	3.645	40°36′30″	40°36′30″	3.5
5	1.563	1.563	0°	. . .	4

Loads

Dead weight = 12.5 psf of plate/in. thickness
Live load = 15 psf of surface area

2. GEOMETRICAL PROPERTIES AND TRIGONOMETRIC QUANTITIES

Sectional Properties

Plate	A_n, ft²	Z_n, ft³
1	0.2604	0.0339
2	1.4004	1.1206
3	0.5208	0.1356
4	1.4004	1.1206
5	0.5208	0.1356

Trigonometric Quantities

θ	$\sin \theta$	$\cos \theta$	$\tan \theta$	$\cot \theta$	$\mathrm{cosec}\ \theta$	$\sec \theta$
40°36′30″	0.6509	0.7592	0.8574	1.1664	1.5364	1.3172
319°23′30″	−0.6509	0.7592	−0.8574	−1.1664	−1.5364	1.3172

3. LOADS AND FIXED-END MOMENTS ON TRANSVERSE SECTION

Plate	Load, lb	Fixed-end moment, ft lb
1	50.78	19.84
2	282.08	85.68
3	101.56	13.22
4	282.08	85.68
5	101.56	13.22

4. NO-ROTATION SOLUTION

Moment Distribution (No Rotation)

	1	1	2	2	3	3	4	4
Distribution factors	0	1	0.14	0.86	0.82	0.18	0.18	0.82
Initial moments (ft lb)	−19.84	+85.68	−85.68	+13.22	−13.22	+85.68	−85.68	+13.22
		−65.84 →	−32.92					
			+14.81	+90.57 →	+45.29			
				−48.33 ←	−96.67	−21.08 →	−10.54	
			+6.79	+41.54 →	+20.77	+7.43 ←	+14.86	+68.14
				−11.58 ←	−23.15	−5.05 →	−2.52	−34.07
			+1.63	+9.95 →	+4.98	+3.28 ←	+6.55	+30.04
				−3.39 ←	−6.77	−1.48 →	−0.74	−15.02
			+0.48	+2.91 →	+1.46	+1.41 ←	+2.82	+12.94
				−1.18 ←	−2.35	−0.51 →	−0.26	−6.47
			+0.17	+1.01 →	+0.51	+0.60 ←	+1.20	+5.52
				−0.46 ←	−0.91	−0.20 →	−0.10	−2.76
			+0.06	+0.39 →	+0.20	+0.26 ←	+0.51	+2.35
					−0.37	−0.08 →		−1.17
							+0.22	+1.00
								−0.50
							+0.09	+0.41
Final moments	−19.84	+19.84	−94.68	+94.68	−70.26	+70.26	−73.67	+73.67
Reactions due to moments			20.53 ↓	↑ 20.53	15.63 ↑	↓ 15.63	0.94 ↓	↑ 0.94
Reactions due to loads	↑ 50.78	141.04 ↑	↑ 141.04	50.78 ↑	↑ 50.78	141.04 ↑	↑ 141.04	50.78 ↑
Total reactions	171.29 ↑		227.98 ↑		175.26 ↑		192.76 ↑	
Ridge loads P_n (lb)	171.29 ↓		227.98 ↓		175.26 ↓		192.76 ↓	

Calculation of Plate Loads R_n from Ridge Loads P_n

Referring to Fig. 12-22,

$S_{1,0}$ = 171.29 × 1.1664 = 199.79 lb/ft
$S_{1,2}$ = 171.29 × 1.5364 = 263.16 lb/ft
$S_{2,1}$ = 227.98 × 1.5364 = 350.26 lb/ft
$S_{2,3}$ = 227.98 × 1.1664 = 265.91 lb/ft
$S_{3,2}$ = 175.26 × 1.1664 = 204.42 lb/ft
$S_{3,4}$ = 175.26 × 1.5364 = 269.26 lb/ft
$S_{4,3}$ = 192.76 × 1.5364 = 296.14 lb/ft
$S_{4,5}$ = 192.76 × 1.1664 = 224.83 lb/ft
R_1 = +199.79 lb/ft
R_2 = −613.42 lb/ft
R_3 = + 61.49 lb/ft
R_4 = +565.40 lb/ft
R_5 = 0 lb/ft

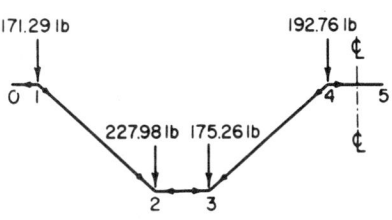

Fɪɢ. 12-22

Longitudinal Stresses Due to Plate Loads

		−ve	Compression
		+ve	Tension

Plate	Load R (kip/ft)	$M = \dfrac{RL^2}{8}$ (kip ft)	$Z(\text{ft}^3)$	$f\,(\text{kip/ft}^2)$	Free edge stresses (kip/ft²)	Stresses after distribution (kip/ft²)
4 – 3	0.565	254.43	1.1206	227.05	227.05 (−) / 227.05 (+)	86.92 (−) / 43.35 (+)
3 – 2	0.061	27.67	0.1356	204.02	204.02 (−) / 204.02 (+)	43.35 / 320.65 (+)
2 – 1	0.613	276.04	1.1206	246.33	246.33 (+) / 246.33 (−)	320.65 (+) / 573.68 (−)
1 – 0	0.200	89.90	0.0339	2651.25	2651.25 (−) / 2651.25 (+)	573.68 (−) / 1612.46 (+)

Stress Distribution (No Rotation)

Distribution factors	0	1		2		3		4	
		0.84	0.16	0.27	0.73	0.73	0.27	0.27	0.73
Free edge stresses (kip/ft²)	+2651.25	−2651.25	−246.33	+246.33	+204.02	−204.02	+227.05	−227.05	
	−1013.67←	+2027.35	−377.57→	+188.79					
			+ 31.31←	− 62.63	+ 168.47→	− 84.24			
	− 13.20←	+26.40	− 4.92→	+ 2.46	− 187.83←	+375.65	−139.65←	+69.82	
			+ 25.78←	−51.57	+ 138.72→	− 69.36	− 21.30←	+42.61	−114.62
	− 10.87←	+ 21.74	− 4.05→	+ 2.02	− 17.52←	+ 35.03	− 13.02→	+ 6.51	+57.31
			+ 2.65←	− 5.30	+ 14.25→	− 7.12	− 6.88←	+13.77	−37.03
	− 1.12←	+ 2.23	− 0.42→	+ 0.21	− 0.09←	+ 0.18	− 0.07→	+0.03	+18.52
			+ 0.04←	− 0.08	+ 0.22→	− 0.11	− 2.50←	+ 5.01	−13.47
	− 0.017←	+0.034	− 0.006→	+0.003	+ 0.87←	− 1.75	+ 0.65→	− 0.33	+ 6.74
			− 0.12←	+ 0.24	− 0.64→	+ 0.32	− 0.96←	+ 1.91	− 5.15
	+ 0.05←	− 0.10	+ 0.02→	− 0.01	+ 0.46←	− 0.93	+ 0.35→	− 0.17	+ 2.57
			− 0.06←	+ 0.13	− 0.35→	+ 0.17	− 0.37←	+ 0.74	− 2.00
					+ 0.20←	− 0.40	+ 0.15→	− 0.07	+ 1.00
	− 0.027←	+0.053	− 0.145→	+0.072	− 0.145←	+0.072	− 0.15→	+ 0.29	− 0.78
	+ 0.038←	−0.077	+ 0.014				− 0.16	+ 0.059	
Stresses after distribution (kip/ft²)	+1612.46	−573.68	−573.68	+320.65	+320.65	+43.35	+43.35	−86.92	−86.92

5. EFFECT OF JOINT DISPLACEMENTS – ARBITRARY ROTATION OF PLATES

Arbitrary Rotation of Plate 2: Case II Solution

	1		2		3		4	
Distribution factors	0	1	0.14	0.86	0.82	0.18	0.18	0.82
Arbitrary moments (kip ft)		+6 −6	+6 → −3					
Initial moments (kip ft)		0	+3					
Moments after distribution (kip ft)	0	0	+2.49	−2.49	−0.27	+0.27	+0.10	−0.10
		0.68↑	↓0.68	1.76↓	↑1.76	0.10↑	↓0.10	
		0.68↑		2.44↓		1.86↑		0.10↓
Ridge loads P_n (kip)		0.68↓		2.44↑		1.86↓		0.10↑

Plate Loads R_n

Knowing the ridge loads P_n, the plate loads R_n are calculated as before. Thus

$$R_1 = +0.80 \text{ kip/ft}$$
$$R_2 = +2.71 \text{ kip/ft}$$
$$R_3 = -5.03 \text{ kip/ft}$$
$$R_4 = +2.71 \text{ kip/ft}$$

Longitudinal Stresses

Plate	Load R (kip/ft)	$M = \dfrac{Rl^2}{\pi^2}$ (kip ft)	Z (ft³)	f(kip/ft²)	Free edge stresses (kip/ft²)	Stresses after distribution (kip/ft²)
4 / 3	2.710	988.45	1.1206	882.07	882.07 / 882.07	1239.57 / 2980.21
3 / 2	5.029	1834.39	0.1356	13524.95	13524.95 / 13524.95	2980.21 / 2756.99
2 / 1	2.710	988.43	1.1206	882.06	882.06 / 882.06	2756.99 / 546.02
1 / 0	0.797	290.66	0.0339	8571.62	8571.62 / 8571.62	546.02 / 4012.80

Stress Distribution: Case II

	1		2		3		4	
0	0.84	0.16	0.27	0.73	0.73	0.27	0.27	0.73
Free edge stresses (kip/ft²)	+8571.62	−8571.62	+882.06	−882.06	−13524.95	+13524.95	+882.07	−882.07
Stresses after distribution (kip/ft²)	+4012.80	+546.02	+546.02	−2756.99	−2756.99	+2980.21	+2980.21	−1239.57	−1239.57

Arbitrary Rotation of Plate 3: Case III Solution

		1		2		3		4	
	0	1	0.14	0.86	0.82	0.18	0.18	0.82
Arbitrary moments (kip ft)				+6	+6			
Moments after distribution (kip ft)	0	0	−0.60	+0.60	+0.70	−0.70	−0.27	+0.27

Ridge arrows:
0.17↓   ↑0.17   0.83↑   ↓0.83   0.27↓   ↑0.27
0.17↓   1.00↑   1.10↓   0.27↑

Ridge loads P_n (kip): 0.17↑   1.00↓   1.10↑   0.27↓

Plate Loads R_n

$$R_1 = -0.19 \text{ kip/ft}$$
$$R_2 = -1.28 \text{ kip/ft}$$
$$R_3 = +2.45 \text{ kip/ft}$$
$$R_4 = -1.28 \text{ kip/ft}$$

Longitudinal Stresses

Plate	Load R (kip/ft)	$M = \dfrac{Rl^2}{\pi^2}$ (kip ft)	$Z(\text{ft}^3)$	$f(\text{kip/ft}^2)$	Free edge stresses (kip/ft²)	Stresses after distribution (kip/ft²)
4 / 3	1.282	467.69	1.1206	417.35	417.35 ... 417.35	602.98 ... 1460.88
3 / 2	2.450	893.50	0.1356	6587.75	6587.75 ... 6587.75	1460.88 ... 1448.59

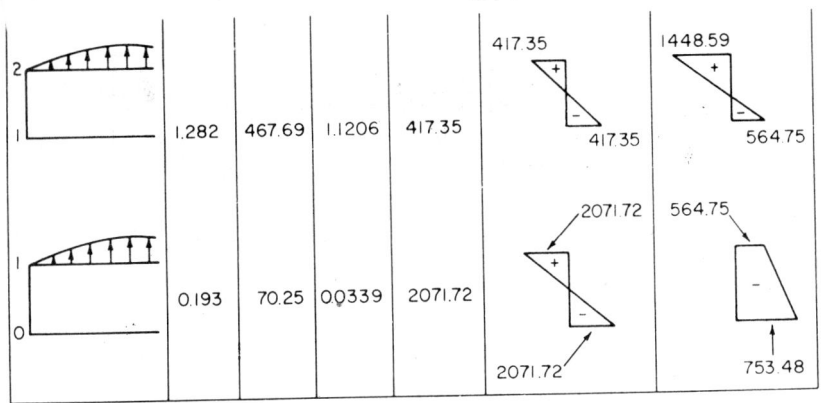

	1.282	467.69	1.1206	417.35
	0.193	70.25	0.0339	2071.72

417.35 417.35 2071.72 2071.72 1448.59 564.75 564.75 753.48

Stress Distribution: Case III

	0	1		2		3		4	
Distribution factors		0.84	0.16	0.27	0.73	0.73	0.27	0.27	0.73
Free edge stresses (kip/ft²)	−2071.72	+2071.72	−417.35	+417.35	+6587.75	−6587.75	−417.35	+417.35	
Stresses after distribution (kip/ft²)	−753.48	−564.75	−564.75	+1448.59	+1448.59	−1460.88	−1460.88	+602.98	+602.98

Arbitrary Rotation of Plate 4: Case IV Solution

	0	1		2		3		4'	
Distribution factors	0	1		0.14	0.86	0.82	0.18	0.18	0.82
Arbitrary moments (kip ft)						+6		+6	
Moments after distribution (kip ft)				+0.36	−0.36	−4.07	+4.07	+3.79	−3.79
		0.10↑		0.10↓	2.84↓	2.84↑	2.16↑	2.16↓	
		0.10↑			2.94↓		5.00↑		2.16↓
Ridge loads P_n (kip)		0.10↓			2.94↑		5.00↓		2.16↑

Plate Loads R_n

$$R_1 = +0.12 \text{ kip/ft}$$
$$R_2 = +4.36 \text{ kip/ft}$$
$$R_3 = -9.24 \text{ kip/ft}$$
$$R_4 = +4.36 \text{ kip/ft}$$

Longitudinal Stresses

Plate	Load R (kip/ft)	$M = \dfrac{Rl^2}{\pi^2}$ (kip ft)	Z (ft³)	f (kip/ft²)	Free edge stresses (kip/ft²)	Stresses after distribution (kip/ft²)
4 — 3	4.355	1588.49	1.1206	1417.53	1417.53 / 1417.53	2208.96 / 5463.80
3 — 2	9.243	3371.57	0.1356	24858.56	24858.56 / 24858.56	5463.80 / 5709.05
2 — 1	4.355	1588.49	1.1206	1417.53	1417.53 / 1417.53	5709.05 / 2973.05
1 — 0	0.117	42.49	0.0339	1252.88	1252.88 / 1252.88	2973.05 / 860.12

Stress Distribution: Case IV

	0		1			2			3			4
		0.84	0.16			0.27	0.73		0.73	0.27		0.27	0.73
Free edge stresses (kip/ft²)	+1252.88	−1252.88	+1417.53		−1417.53	−24858.56	+24858.56		+1417.53	−1417.53		
Stresses after distribution (kip/ft²)	−860.12	+2973.05	+2973.05	−5709.05	−5709.05		+5463.80	+5463.80	−2208.96	−2208.96		

6. PLATE DEFLECTIONS v_n (ft)

All quantities in ft and kip units

Case	Formula	v_1	v_2	v_3	v_4	v_5
I	$v_n = 5\dfrac{(f_b - f_t)l^2}{48Eh_n}$	$+\dfrac{1,049,349}{E}$	$-\dfrac{69,851}{E}$	$+\dfrac{66,552}{E}$	$+\dfrac{10,175}{E}$	0
II	$v_n = \dfrac{(f_b - f_t)l^2}{\pi^2 Eh_n}$	$+\dfrac{1,618,601}{E}$	$+\dfrac{250,931}{E}$	$-\dfrac{1,339,316}{E}$	$+\dfrac{320,578}{E}$	0
III	$v_n = \dfrac{(f_b - f_t)l^2}{\pi^2 Eh_n}$	$-\dfrac{88,115}{E}$	$-\dfrac{152,954}{E}$	$+\dfrac{679,297}{E}$	$-\dfrac{156,792}{E}$	0
IV	$v_n = \dfrac{(f_b - f_t)l^2}{\pi^2 Eh_n}$	$-\dfrac{1,789,657}{E}$	$+\dfrac{659,582}{E}$	$-\dfrac{2,608,235}{E}$	$+\dfrac{582,902}{E}$	0

7. PLATE ROTATIONS ψ_2, ψ_3, AND ψ_4

$$\psi_2 = \frac{1}{h_2}\left[v_2(\cot\gamma_1 + \cot\gamma_2) - \frac{v_1}{\sin\gamma_1} - \frac{v_3}{\sin\gamma_2}\right]$$

$$= \frac{1}{4.8013}\left[v_2 \times 0 - \frac{v_1}{0.6509} + \frac{v_3}{0.6509}\right]$$

$$= \frac{v_3 - v_1}{3.1251}$$

$$\psi_3 = \frac{1}{h_3}\left[v_3(\cot\gamma_2 + \cot\gamma_3) - \frac{v_2}{\sin\gamma_2} - \frac{v_4}{\sin\gamma_3}\right]$$

$$= \frac{1}{1.5625}\left[v_3(-1.1664 - 1.1664) + \frac{v_2}{0.6509} + \frac{v_4}{0.6509}\right]$$

$$= -1.4930 v_3 + \frac{v_2 + v_4}{1.0170}$$

$$\psi_4 = \frac{1}{h_4}\left[v_4(\cot\gamma_3 + \cot\gamma_4) - \frac{v_3}{\sin\gamma_3} - \frac{v_5}{\sin\gamma_4}\right]$$

$$= \frac{1}{4.8013}\left[v_4 \times 0 + \frac{v_3}{0.6509} + 0\right]$$

$$= \frac{v_3}{3.1251}$$

Let

$$\frac{EJ_2}{h_2}\psi_{20} = 1 \text{ kip ft}$$

Then

$$\psi_{20} = \frac{h_2}{EJ_2} = \frac{4.8013}{E[\frac{1}{12} \times 1 \times (\frac{7}{24})^3]} = \frac{2,322}{E}$$

Similarly,

$$\psi_{30} = \frac{h_3}{EJ_3} = \frac{1.5625}{E[\frac{1}{12} \times 1 \times (\frac{1}{3})^3]} = \frac{506}{E}$$

and

$$\psi_{40} = \psi_{20} = \frac{2,322}{E}$$

Now the actual rotations ψ_2, ψ_3, and ψ_4 will be written in terms of the arbitrary rotations ψ_{20}, ψ_{30}, and ψ_{40} and the unknown constants k_2, k_3, and k_4.

$$\psi_2 = \psi_{20} \times k_2 = \frac{2,322}{E} k_2$$

$$\psi_3 = \psi_{30} \times k_3 = \frac{506}{E} k_3$$

$$\psi_4 = \psi_{40} \times k_4 = \frac{2,322}{E} k_4$$

The values of the actual rotations ψ_2, ψ_3, and ψ_4 obtained in the previous step in terms of the arbitrary rotations will be equated to their values obtained in terms of plate deflections. This will result in three simultaneous equations in the unknown constants k_2, k_3, and k_4. Thus,

(i)
$$\psi_2 = \frac{2,322}{E} k_2 = \frac{v_3 - v_1}{3.1251}$$

or

$$\frac{2,322}{E} k_2 = -\frac{314,488}{E} - \frac{946,512}{E} k_2 + \frac{245,534}{E} k_3 - \frac{261,939}{E} k_4$$

or

$$-314,488 - 948,834k_2 + 245,534k_3 - 261,939k_4 = 0 \qquad (1)$$

(ii)
$$\psi_3 = \frac{506}{E} k_3 = -1.4930v_3 + \frac{v_2 + v_4}{1.0170}$$

or

$$\frac{506}{E} k_3 = -\frac{158,038}{E} + \frac{2,561,509}{E} k_2 - \frac{1,318,587}{E} k_3 + \frac{5,115,722}{E} k_4$$

or

$$-158,038 + 2,561,509k_2 - 1,319,093k_3 + 5,115,722k_4 = 0 \qquad (2)$$

(iii)
$$\psi_4 = \frac{2,322}{E} k_4 = \frac{v_3}{3.1251}$$

or

$$\frac{2,322}{E} k_4 = \frac{21,296}{E} - \frac{428,571}{E} k_2 + \frac{217,338}{E} k_3 - \frac{834,616}{E} k_4$$

or

$$21,296 - 428,571k_2 + 217,338k_3 - 836,939k_4 = 0 \qquad (3)$$

Solving the above three simultaneous equations, we get

$$k_2 = -0.72309$$
$$k_3 = -1.50947$$
$$k_4 = +0.00374$$

8. CALCULATION OF LONGITUDINAL STRESSES AND TRANSVERSE MOMENTS

Once the values of the constants k_2, k_3, and k_4 are obtained, the longitudinal stresses at all edges are calculated as the stresses due to the no-rotation solution, plus k_2 times the stresses of the Case II solution, plus k_3 times the stresses of the Case III solutions, plus k_4 times the stresses of the Case IV solution. The transverse moments are also calculated in a similar way.

The longitudinal stresses and the transverse moments—both at midspan—are shown tabulated below.

Longitudinal Stresses at Midspan

Joint	No-rotation solution	Case II	Case III	Case IV	Final stresses kip/ft²	psi
0	+1,613	−2,902	+1,137	−3	−155	−1077
1	−574	−395	+853	+11	−105	−729
2	+321	+1,994	−2,187	−21	+106	+738
3	+43	−2,155	+2,205	+20	+114	+791
4	−87	+896	−910	−8	−109	−757

Transverse Moments at Midspan

Joint	No-rotation solution	Case II	Case III	Case IV	Final moments, ft lb
0	0	0	0	0	0
1	−20	0	0	0,	−20
2	−95	−1,801	+909	+1	−986
3	−70	+192	−1,060	−15	−953
4	−74	−74	+403	+14	+270

The procedure for the calculation of the longitudinal stresses and transverse moments at any other section is illustrated with reference to the quarter span section. The component of the no-rotation solution varies as a parabola for longitudinal stresses and remains constant at all sections for transverse moments. The rotation component of the solution, however, follows a sine variation (assumption of Art. 12-12). The values shown in the following tables are self-explanatory.

Longitudinal Stresses at Quarter Span

Joint	No-rotation solution	Case II	Case III	Case IV	Final stresses	
					kip/ft²	psi
0	+1,209	−2,052	+804	−2	−41	−281
1	−430	−279	+603	+8	−99	−686
2	+241	+1,410	−1,546	−15	+89	+617
3	+33	−1,524	+1,559	+14	+82	+572
4	−65	+634	−644	−6	−81	−561

Transverse Moments at Quarter Span

Joint	No-rotation solution	Case II	Case III	Case IV	Final moments, ft lb
0	0	0	0	0	0
1	−20	0	0	0	−20
2	−95	−1,273	+643	+1	−725
3	−70	+136	−749	−11	−695
4	−74	−52	+285	+10	+169

THE ITERATION METHOD

12-15 The Method in Outline

The iteration method offers a simple means of analyzing certain types of folded plates [63, 64]. The steps involved are the following:

Step 1

Work out the no-rotation solution as in the Simpson method and arrive at the stresses in the plates and transverse moments at the joints.

Step 2

Calculate the plate deflections v_n corresponding to the no-rotation solution. From these, calculate the displacements w_n and the differential ridge displacements Δ_n.

Step 3

Distribute the moments corresponding to Δ_n and compute the shears at the joints. Resolve these into plate loads and determine the free-edge stresses. Correct the free-edge stresses by stress distribution. Compare the stresses thus obtained with the stresses computed in step 1. If the corrections are relatively small, the process may be stopped at this point and the stresses and transverse moments obtained in step 3 may be superimposed over the results in step 1 to give the final stresses and transverse moments.

Step 4

If the corrections are not small, the deflections v_n caused by the correction stresses are found and steps 2 and 3 are repeated as before. The number of cycles of correction required to gain necessary accuracy will depend on each problem.

12-16 Comments on the Iteration Method

The iteration method is not applicable to folded plates of all types. The chief drawback of the method is that it is not possible to determine, a priori, if the method would be satisfactory or not when applied to a given problem. Whether or not the iteration process would converge to an answer would, however, be evident when step 3 is completed. The process does diverge when, for instance, it is applied to a northlight folded plate. For this reason, its general use is not recommended. The Task Committee on Folded Plate Construction of the ASCE[1] in their Phase I report came to the same conclusion. The following extract[2] from the report summarizes the position clearly: "Whether or not the iteration method converges to a solution depends on the relative rigidities of longitudinal plate action and transverse slab action and on the geometry of the structures. The divergence of the method in many cases has been demonstrated. Consequently, it cannot be recommended as a generally effective practical method of folded plate analysis which considers relative joint displacement." In the worked example that follows, the method is shown applied to a symmetrical folded plate of V configuration.

Design Example 12-3

ANALYSIS OF A V-SHAPED FOLDED PLATE BY THE ITERATION METHOD

1. *DATA*

   Geometry

   Span l = 60 ft
   Thickness = 4.5 in. for end plates and 3.5 in. for other plates

---

[1] Phase I Report on Folded Plate Construction, Report of the Task Committee on Folded Plate Construction, Committee on Masonry and Reinforced Concrete, Structural Division, *Journal of the Structural Division, Proceedings of the American Society of Civil Engineers*, vol. 89, ST6, December, 1963, part I.

[2] Extracted by permission of the American Society of Civil Engineers.

Plate	Plate width h_n, ft	Horizontal projection l_n, ft	Inclination to horizontal ϕ_n	Angle between the plate and the next plate γ_n
1	9.6043	7.5	321°20′	282°40′
2	9.6043	7.5	38°40′	77°20′
3	9.6043	7.5	321°20′	282°40′

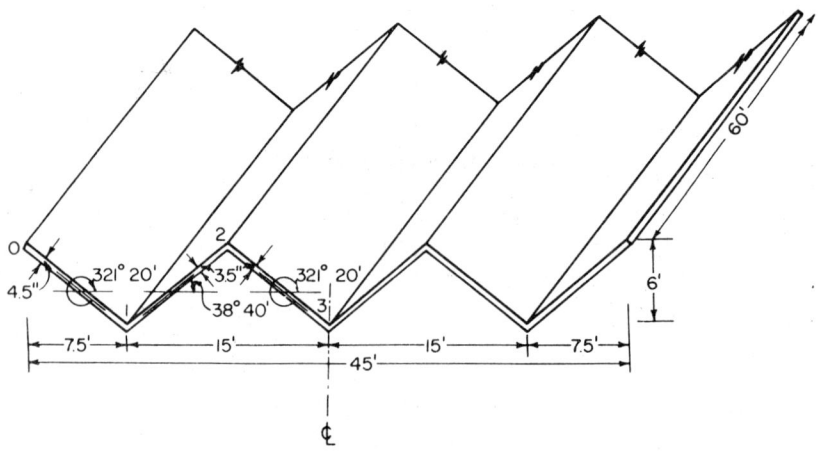

FIG. 12-23

The cross section of the folded plate is symmetrical about the center line through joint 3. Hence the properties of plates 4, 5, and 6 are not tabulated.

Loading

(i) Dead weight = 12.5 psf of surface area/in. thickness
(ii) Live load = 15 psf of surface area

2. *SECTIONAL PROPERTIES*

Plate	A_n, ft²	Z_n, ft³
1	3.602	5.765
2	2.801	4.484
3	2.801	4.484

3. LOADS AND FIXED-END MOMENTS

Plate	Load, lb/ft	Fixed-end moments, lb ft/ft
1	684.306	2,566.149
2	564.253	352.658
3	564.253	352.658

4. NO-ROTATION SOLUTION

Moment Distribution

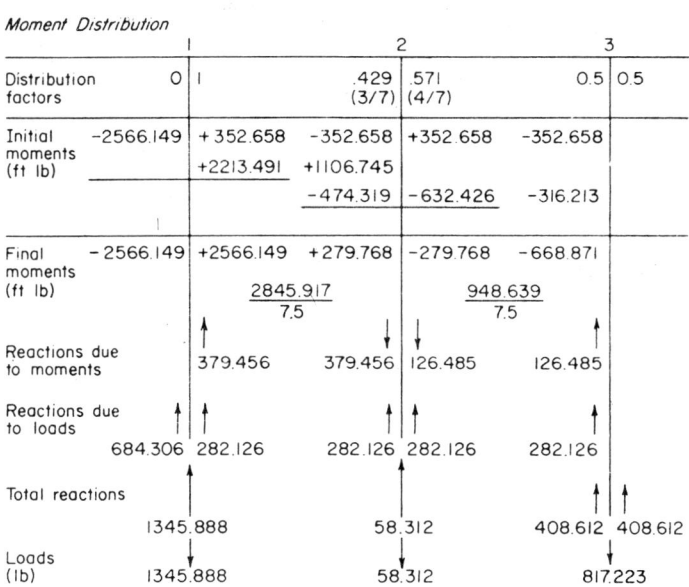

The ridge loads obtained above and the resulting plate loads are shown in Fig. 12-24.

Load on plate $= \dfrac{R}{2 \sin \theta}$, where θ is the inclination of the plate to the horizontal.

Longitudinal Stresses

Tension +ve

Plate	Load R (lb/ft)	M (lb ft)	Z (ft³)	f (lb/ft²)	Free edge stresses (lb/ft²)
3 / 2	700.7	315,333.5	4.484	70,323.9 (488.4 psi)	70323.9 (488.4) (488.4) 70323.9
2 / 1	1123.9	505,738.4	4.484	112,787.0 (783.2 psi)	112787 (783.2) (783.2) 112787
1 / 0	1077.2	484,736.9	5.765	84,080.4 (583.9 psi)	84080.4 (583.9) (583.9) 84080.4

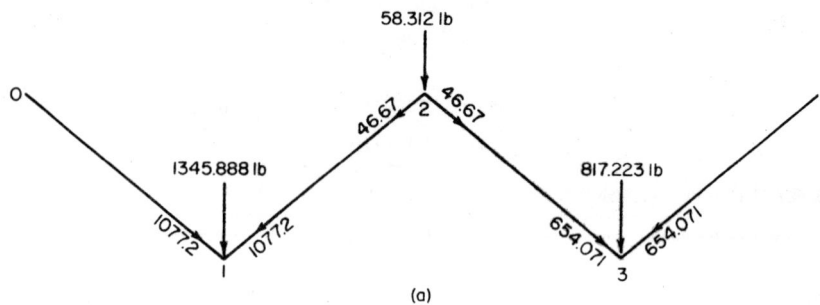

(a)

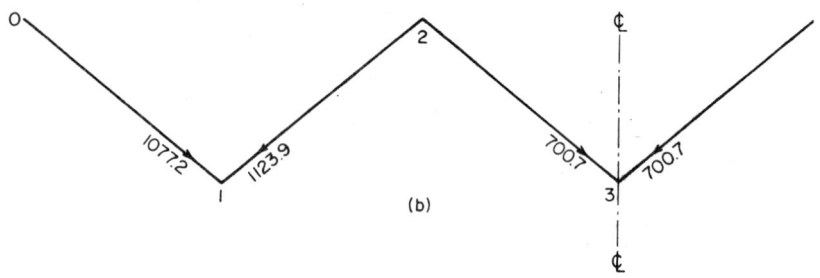

(b)

FIG. 12-24

Stress Distribution (No Rotation)

Distribution factors	0	1		2		3	
		7/16 (.4375)	9/16 (.5625)	1/2 (.5)	1/2 (.5)	1/2 (.5)	1/2 (.5)
Free edge stresses (psi)	−583.891 +583.891		+783.243	−783.243	−488.361 +488.361		
	− 43.608 + 87.217		− 112.136	+ 56.068			
			− 59.704	+ 119.407	− 119.407	+ 59.704	
	+ 13.060 − 26.120		+ 33.583	− 16.791			
			− 4.198	+ 8.396	− 8.396	+ 4.198	
	+ .918 − 1.837		+ 2.361	− 1.181			
			− .295	+ .590	− .590	+ .295	
	+ .0646 − .1291		+ .166				
$f_b - f_t$	− 613.456 +643.022		+643.022	−616.754	− 616.754 +552.558		
	−1256.478 psi		+1259.776 psi		−1169.312 psi		

5. *PLATE DEFLECTIONS DUE TO NO-ROTATION SOLUTION*

$$\text{Deflection} = \frac{5}{48}\frac{(f_b - f_t)l^2}{Eh}$$

$$v_1 = \frac{-7064522.7}{E}\ \text{ft}$$

$$v_2 = \frac{+7083066.2}{E}\ \text{ft}$$

$$v_3 = \frac{-6574434.6}{E}\ \text{ft}$$

6. *RELATIVE DISPLACEMENTS OF JOINTS*

(i) Plate 2

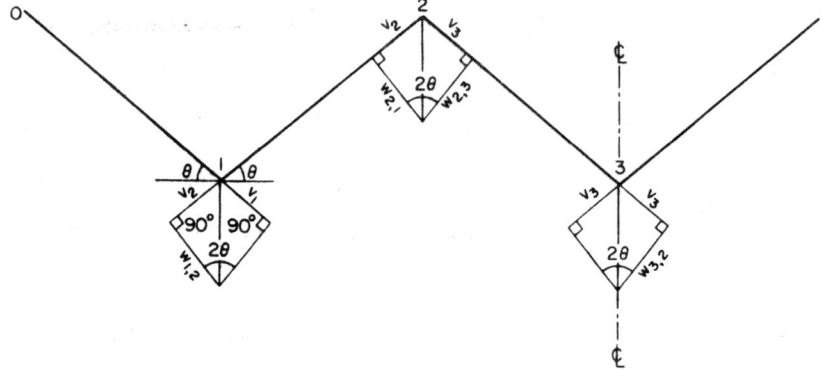

$$\text{F\scriptsize IG.}\ 12\text{-}25$$

Referring to Fig. 12-25,

$$w_{1,2} = v_1 \cosec 2\theta + v_2 \cot 2\theta$$

$$= \frac{7064522.7}{E}\times 1.025 + \frac{7083066.2}{E}\times .225$$

$$= \frac{7241135.8}{E} + \frac{1593689.9}{E} = \frac{8834,825.7}{E}$$

$$w_{2,1} = v_3 \cosec 2\theta + v_2 \cot 2\theta$$

$$= \frac{6574434.6}{E}\times 1.025 + \frac{7083066.2}{E}\times .225$$

$$= \frac{6738795.5}{E} + \frac{1593689.9}{E} = \frac{8332485.4}{E}$$

$$\Delta_2 = (w_{1,2} - w_{2,1}) = \frac{8834825.7}{E} - \frac{8332485.4}{E}$$

$$= \frac{502340.3}{E}$$

$$\therefore\ \text{Transverse moment} = \frac{6EJ_2\Delta_2}{h_2{}^2}$$

$$= \frac{6\times E\times \frac{1}{12}\times 1\left(\frac{7}{24}\right)^3\times \dfrac{502340.3}{E}}{9.6043\times 9.6043} = 67.6\ \text{ft lb}$$

(ii) Plate 3

$$w_{2,3} = v_2 \cosec 2\theta + v_3 \cot 2\theta$$

$$= \frac{7083066.2}{E} \times 1.025 + \frac{6574434.6}{E} \times .225$$

$$= \frac{7260142.9}{E} + \frac{1479247.8}{E} = \frac{8739390.7}{E}$$

$$w_{3,2} = v_3 \cot \theta = \frac{6574434.6}{E} \times 1.25 = \frac{8218043.3}{E}$$

$$\Delta_3 = (w_{2,3} - w_{3,2}) = \frac{521347.4}{E}$$

$$\therefore \text{ Transverse moment} = \frac{6EJ_3\Delta_3}{h_3^2} = \left[\frac{6 \times E \times \frac{1}{12} \times 1 (\frac{7}{24})^3 \times \frac{521347.4}{E}}{92.25} \right]$$

$$= 70.112 \text{ ft lb}$$

7. DISTRIBUTION OF MOMENTS AND STRESSES CAUSED BY JOINT DISPLACEMENTS (FIRST CYCLE OF ITERATION)

Moment Distribution

	1		2			3	
Distribution factors	0	1	3/7	4/7		1/2	1/2
Initial moments (ft lb)	−67.556		−67.556	−70.112		−70.112	
	+67.556		+33.778				
			+44.524	+59.366		+29.683	
Moments (ft lb)	0		+10.746	−10.746		−40.429	
			$\frac{10.746}{7.5}$			$\frac{51.175}{7.5}$	
Reactions due to moments	↑ 1.433			↓↓ 1.433 6.823		↑ 6.823	
Total reactions	↑ 1.433			↓ 8.256		↑ 6.823	6.823
Loads (lb)	↓ 1.433			↑ 8.256		↓ 6.823 x 2 i.e., 13.646	

The plate loads are shown in Fig. 12-26.

Longitudinal Stresses

Tension +ve

Plate	Load R (lb/ft)	$M = \frac{Rl^2}{\pi^2}$ (ft lb)	$Z(ft^3)$	f (lb/ft²)	Free edge stresses (psi)
3 ⟋ 2	4.314	1573.559	4.484	350.926 (2.437 psi)	2.437 / 2.437
2 ⟋ 1	5.461	1991.934	4.484	444.230 (3.085 psi)	3.085 / 3.085
1 ⟋ 0	1.147	418.375	5.765	72.570 (.504 psi)	.504 / .504

Stress Distribution

	O		1		2		3
Distribution factors		.4375	.5625	.5	.5	.5	.5
Free edge stresses (psi)	−.504	+.504	−3.085	+3.085	−2.437	+2.437	
	+.785	−1.570	+2.019	−1.009			
			+1.128	−2.2565	+2.2565	−1.128	
	−.246	+.493	−.635	+.317			
			+.079	−.1585	+.1585	−.079	
	−.017	+.035	−.044	+.022			
			+.005	−.011	+.011	−.005	
	−.001	+.002	−.003	+.0015			
				−.00075	+.00075		
	+.017 psi	−.536	−.536	−.01025	−.01025	+1.225	

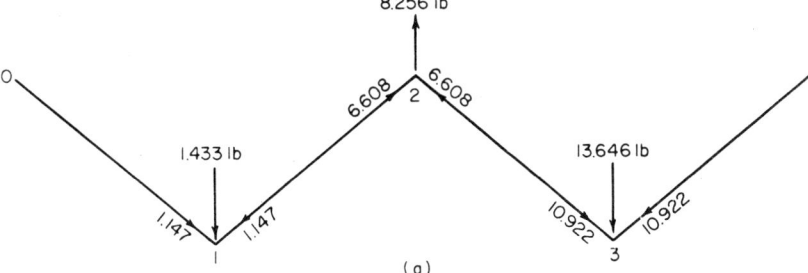

(a)

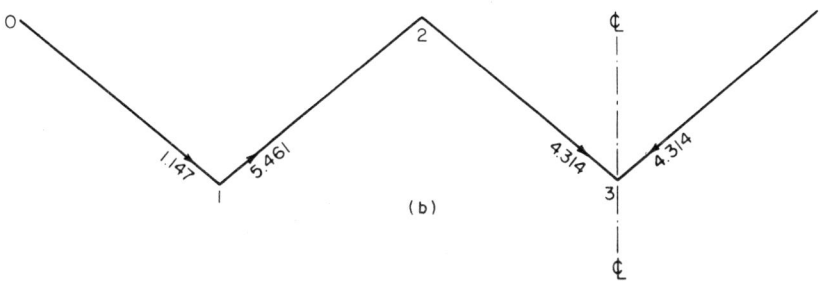

(b)

Fig. 12-26

8. *FINAL STRESSES AND MOMENTS*

It may be noted that these stresses are small compared with those obtained from the no-rotation solution. Hence further cycles of iteration are not necessary.

The final stress resultants are obtained by adding the results due to the no-rotation solution and those due to the first cycle of iteration.

Transverse Moments (lb ft/ft)

Joint	No rotation	First cycle of iteration	Final moments
0	0	0	0
1	−2,566.149	0	−2,566.149
2	+279.768	+10.746	+290.514
3	−668.871	−40.429	−709.300

Longitudinal Stresses (psi)

Joint	No rotation	First cycle of iteration	Final stresses
0	−613.456	+0.0170	−613.439
1	+643.022	−0.5360	+642.486
2	−616.754	−0.01025	−616.765
3	+552.558	+1.2250	+553.783

CONTINUOUS FOLDED PLATES

12-17   Extension of the Simpson Method

The Simpson method described in Art. 12-11 may be easily extended to apply to continuous folded plates. All that we additionally need are relations connecting the fiber stresses and deflection for the nth plate at midspan for the no-rotation and rotation cases. These relations for continuous folded plates correspond to the results (12-19) and (12-20) developed for simply supported folded plates. Consider a uniformly loaded folded plate of two equal spans (Fig. 12-27). Regarding the

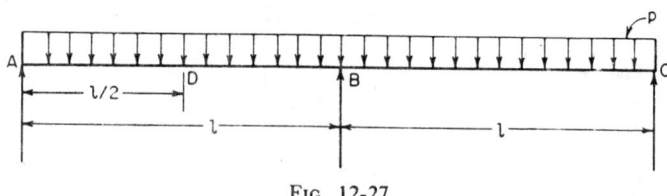

Fɪɢ. 12-27

nth plate as a beam, the deflection and bending moment at D are found to be $\frac{1}{192} pl^4/EI$ and $pl^2/16$, respectively. If f_t and f_b are the fiber stresses at the top and bottom of the plate, we may write the deflection at D as

$$v_D = \frac{1}{12} \frac{f_b - f_t}{Eh_n} l^2 \tag{12-21}$$

Next we proceed to the rotation case. The continuous beam may now be imagined to be loaded with a load proportional to the shape of the elastic line of the deflected continuous beam (Fig. 12-28). For this pattern of loading, the bending moment and deflection at D are $(59/1,120)\,\bar{p}l^2$ and $(227/210 \times 256)\,\bar{p}l^4/EI$, respectively, where $\bar{p}$ is the

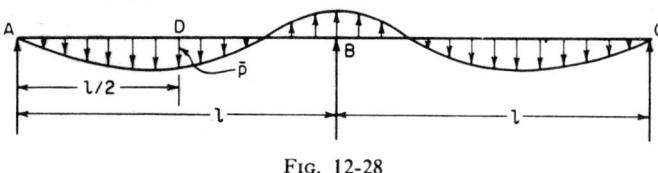

FIG. 12-28

intensity of loading at D. From these two results, the following expression is derived for the deflection at D:

$$v_D = \frac{227}{2,832}\,\frac{(f_b - f_t)}{Eh_n}\,l^2 \tag{12-22}$$

Making use of relations (12-21) and (12-22), we may proceed with the analysis in the same manner as for simply supported folded plates. The details of the analysis are clearly explained in the example that follows.

Design Example 12-4

DESIGN OF A CONTINUOUS NORTHLIGHT FOLDED PLATE

1. *DATA*

Geometry

Span l = 60 ft (Two spans of 60 ft each)
Breadth B = 18 ft

Plate	Plate width h_n, ft	Horizontal projection l_n, ft	Inclination to horizontal ϕ_n	Angle between the plate and the next plate γ_n	Thickness of the plate d_n, in.
1	2.03	0.353	80°	80°	3.5
2	3.625	3.625	0°	40°35′	4.0
3	12.297	9.34	319°25′	319°25′	3.5
4	3.625	3.625	0°	280°	4.0
5	2.03	0.353	80°	. . .	3.5

Loads

Dead weight = 12.5 psf of plate/in. thickness
Live load = 15.0 psf of surface area

2. GEOMETRICAL PROPERTIES AND TRIGONOMETRIC QUANTITIES

Geometrical Properties

Plate	A_n, ft^2	Z_n, ft^3
1	0.5923	0.2005
2	1.2083	0.73
3	3.5867	7.351
4	1.2083	0.73
5	0.5923	0.2005

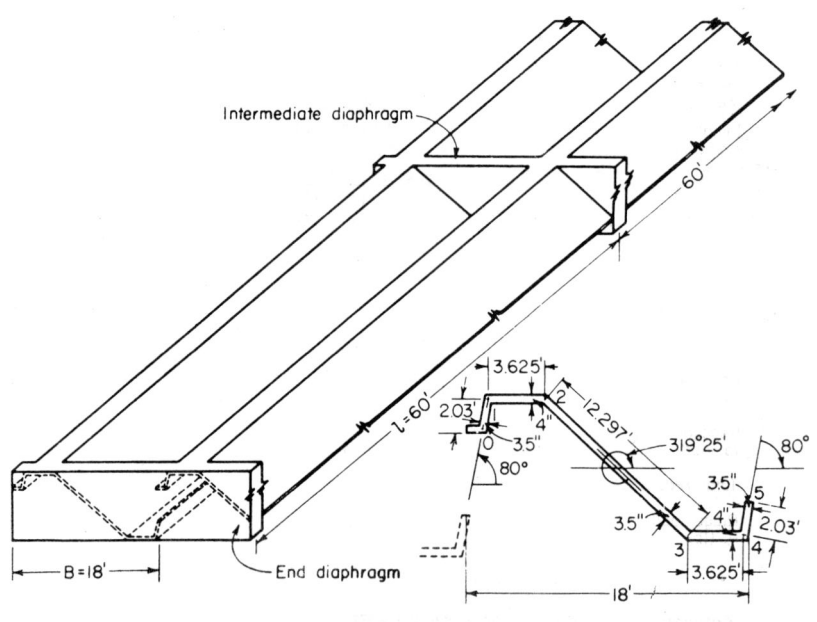

FIG. 12-29

Trigonometric Quantities

θ	$\sin \theta$	$\cos \theta$	$\tan \theta$	$\cot \theta$	$\sec \theta$	$\operatorname{cosec} \theta$
80°	0.98481	0.17365	5.67128	0.17633	5.75877	1.01543
40°35′	0.65056	0.75947	0.85660	1.16742	1.31672	1.53715
319°25′	−0.65056	0.75947	−0.85660	−1.16742	1.31672	−1.53715

3. LOADS AND FIXED-END MOMENTS ON TRANSVERSE SECTION

Joint	Load, lb/ft	Fixed-end moment, ft lb/ft
1	119.31 + 40*	55.14
2	235.63	71.18
3	722.46	562.28
4	235.63	71.18
5	119.31 + 30*	31.62

* The additional loads of 40 lb at joint 0 and 30 lb at joint 5 are due to the sunshade and glazing, respectively.

4. NO-ROTATION SOLUTION

Moment Distribution (No Rotation)

Distribution factors	0	1	.79	.21	.21	.79	1	0
Initial moments (ft lb)	-55.14	+71.18	-71.18	+562.28	-562.28	+ 71.18	-71.18	+31.62
		-16.04 ⟶	- 8.02					
			-381.63	- 101.45 ⟶	-50.72			
				+ 56.89 ⟵	+113.78	+428.04		
			-44.94	- 11.95 ⟶	- 5.97	+ 19.78 ⟵	+39.56	
				- 1.45 ⟵	- 2.90	- 10.91		
			+ 1.14	+ 0.31 ⟶	+0.15			
					-0.03	- 0.12		
Final moments	-55.14	+55.14	-504.63	+504.63	-507.97	+507.97	-31.62	+31.62
Reactions due to moments		124.0↓	↑124.0	0.36↓	↑0.36	131.41↑	↓131.41	
Reactions due to loads	↑159.31	117.82↑	↑117.82	361.23↑	↑361.23	117.82↑	↑117.82	149.31↑
Total reactions	153.13↑		602.69↑		610.82↑		135.72↑	
Ridge loads P_n (lb)	153.13↓		602.69↓		610.82↓		135.72↓	

Calculation of Plate Loads from Ridge Loads

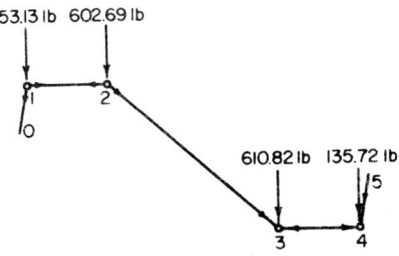

153.13 lb   602.69 lb

610.82 lb   135.72 lb

FIG. 12-30

Referring to Fig. 12-30,

$$S_{1,0} = 153.13 \text{ cosec } 80° = 155.49 \text{ lb}$$
$$S_{1,2} = 153.13 \text{ cot } 80° = 27.00 \text{ lb}$$
$$S_{2,1} = 602.69 \text{ cot } 40° 35' = 703.58 \text{ lb}$$
$$S_{2,3} = 602.69 \text{ cosec } 40° 35' = 926.40 \text{ lb}$$
$$S_{3,2} = 610.82 \text{ cosec } 40° 35' = 938.90 \text{ lb}$$
$$S_{3,4} = 610.82 \text{ cot } 40° 35' = 713.07 \text{ lb}$$
$$S_{4,3} = 135.72 \text{ cot } 80° = 23.93 \text{ lb}$$
$$S_{4,5} = 135.72 \text{ cosec } 80° = 137.81 \text{ lb}$$
$$R_1 = +155.49 \text{ lb/ft}$$
$$R_2 = +676.58 \text{ lb/ft}$$
$$R_3 = -1865.30 \text{ lb/ft}$$
$$R_4 = +689.14 \text{ lb/ft}$$
$$R_5 = +137.81 \text{ lb/ft}$$

Longitudinal Stresses Due to Plate Loads

− ve Compression
+ ve Tension

Plate	Load R (kip/ft)	$M = \dfrac{Rl^2}{16}$ (kip ft)	$Z(ft^3)$	$f(kip/ft^2)$	Free edge stresses (kip/ft²)	Stresses after distribution (kip/ft²)
5 / 4	.138	31.0	.2005	154.66	154.66 / 154.66	43.00 / 68.64
4 / 3	.689	155.06	.73	212.40	212.40 / 212.40	68.64 / 89.84
3 / 2	1.865	419.69	7.351	57.09	57.09 / 57.09	89.84 / 88.88
2 / 1	.677	152.23	.73	208.53	208.53 / 208.53	88.88 / 61.74
1 / 0	.155	34.99	.2005	174.50	174.50 / 174.50	61.74 / 56.38

Stress Distribution (No Rotation)

	0	1		2		3		4		
Distribution factors	O	67	33	75	25	25	75	33	67	
Free edge stresses (kip/ft²)	+174.50	−174.50	+208.53	−208.53	−57.09	+57.09	+212.40	−212.40	+154.66	−154.66
		−128.32 →	+256.63	−126.40 →	+63.20					
				−33.09 →	+66.18	−22.06 →	+11.03			
		+11.09 →	−22.17	+10.92 →	−5.46	−18.04 →	+36.07	−108.21 →	+54.11	
		+4.72 →	−9.43	+3.15 →	−1.58	−51.64 →	+103.27	−209.68 →	+104.84	
		−1.58 →	+3.16	−1.56 →	+0.78	+6.26 →	−12.52	+37.54 →	−18.77	
		−2.06 →	+4.11	−1.37 →	+0.68	−3.10 →	+6.19	−12.58 →	+6.29	
		+0.69 →	−1.38	+0.68 →	−0.34	+0.48 →	−0.95	+2.83 →	−1.42	
			+0.61	−0.21 →	+0.11	−0.24 →	+0.47	−0.95 →	+0.48	
					−0.09	+0.26 →	−0.13			
							+0.04	−0.09 →	+0.05	
Stresses after distribution (kip/ft²)	+56.38	+61.74	+61.74	−88.88	−88.88	+89.84	+89.84	−68.64	−68.64	−43.00

5. EFFECT OF JOINT DISPLACEMENTS – ARBITRARY ROTATION OF PLATES

Arbitrary Rotation of Plate 2: Case II Solution

	1		2		3		4	
Distribution factors	0	1	.79	.21	.21	.79	1	0
Arbitrary moments (kip ft)		+6	+6					
		−6 →	−3					
Initial moments (kip ft)		0	+3					
Moments after distribution (kip ft)	0	0	+.60	−.60	−.25	+.25	0	0
		.166↑	.166↓	.091↓	.091↑	.069↑	.069↓	
		.166↑	.257↓		.160↑		.069↓	
Ridge loads Pₙ (kip)		.166↓	.257↑		.160↓		.069↑	

Plate Loads R_n

Knowing the ridge loads P_n, the plate loads R_n are calculated as before. Thus

$$R_1 = +.168 \text{ kip/ft}$$
$$R_2 = -.329 \text{ kip/ft}$$
$$R_3 = +.148 \text{ kip/ft}$$
$$R_4 = +.199 \text{ kip/ft}$$
$$R_5 = -.07 \text{ kip/ft}$$

Longitudinal Stresses

Plate	Load R (kip/ft)	$M = \dfrac{59}{1120}Rl^2$ (kip ft)	Z (ft^3)	f(kip/ft^2)	Free edge stresses (kip/ft^2)	Stresses after distribution (kip/ft^2)
5 / 4	.07	13.3	.2005	66.31	66.31 / 66.31	51.69 / 37.08
4 / 3	.199	37.7	.73	51.69	51.69 / 51.69	37.08 / .88
3 / 2	.148	28.1	7.351	3.83	3.83 / 3.83	.88 / 23.42
2 / 1	.329	62.3	.73	85.36	85.36 / 85.36	23.42 / 92.39
1 / 0	.168	31.9	.2005	158.96	158.96 / 158.96	92.39 / 125.66

Stress Distribution: Case II

Distribution factors	0	1		2		3		4	
		.67	.33	.75	.25	.25	.75	.33	.67
Free edge stresses (kip/ft^2)	+158.96	−158.96	−85.36	+85.36	+ 3.83	−3.83	+51.69	−51.69	−66.31	+66.31
Stresses after distribution (kip/ft^2)	+125.66	− 92.39	−92.39	+23.42	+23.42	+0.88	+ 0.88	−37.08	− 37.08	+ 51.69

Arbitrary Rotation of Plate 3: Case III Solution

Distribution factors	1		2		3		4	
	0	1	.79	.21	.21	.79	1	0
Arbitrary moments (kip ft)				+6	+6			
Moments after distribution (kip ft)	0	0	−4.29	+4.29	+4.29	−4.29	0	0
		1.183↓	↑1.183	.919↑	↓.919	1.183↓	↑1.183	
	1.183↓		2.102↑		2.102↓		1.183↑	
Ridge loads P_n (kip)	1.183↑		2.102↓		2.102↑		1.183↓	

Plate Loads R_n

$$R_1 = -1.202 \text{ kip/ft}$$
$$R_2 = +2.663 \text{ kip/ft}$$
$$R_3 = 0$$
$$R_4 = -2.663 \text{ kip/ft}$$
$$R_5 = +1.202 \text{ kip/ft}$$

Longitudinal Stresses

Plate	Load R (kip/ft)	$M = \frac{59}{1120} R l^2$ (kip ft)	Z (ft^3)	f(kip/ft^2)	Free edge stresses (kip/ft^2)	Stresses after distribution (kip/ft^2)
5 — 4	1.202	227.9	.2005	1136.68	1136.68 / 1136.68	869.44 / 602.18
4 — 3	2.663	505.0	.73	691.70	691.70 / 691.70	602.18 / 117.63
3 — 2	0	0	7.351	0	0 / 0	117.63 / 117.63
2 — 1	2.663	505.0	.73	691.70	691.70 / 691.70	117.63 / 602.18
1 — 0	1.202	227.9	.2005	1136.68	1136.68 / 1136.68	602.18 / 869.44

Stress Distribution: Case III

Distribution factors	0	1 .67 .33	2 .75 .25	3 .25 .75	4 .33 .67
Free edge stresses (kip/ft^2)	-1136.68 +1136.68	+ 691.70 -691.70	0	0 -691.70 +691.70	+1136.68 -1136.68
Stresses after distribution (kip/ft^2)	-869.44 +602.18	+ 602.18 -117.63	-117.63 -117.63	-117.63 +602.18	+602.18 -869.44

Arbitrary Rotation of Plate 4: Case IV Solution

	1		2		3		4	
Distribution factors	0	1	.79	.21	.21	.79	1	0
Arbitrary moments (kip ft)						+6	+6	
Moments after distribution (kip ft)	0	0	+.25	−.25	−.60	+.60	0	0
		.069↑	↓.069	.091↓	↑.091	.166↑		↓.166
	.069↑		.160↓		.257↑		.166↓	
Ridge loads P_n (kip)	.069↓		.160↑		.257↓		.166↑	

Plate Loads R_n

$$R_1 = +.07 \text{ kip/ft}$$
$$R_2 = -.199 \text{ kip/ft}$$
$$R_3 = -.148 \text{ kip/ft}$$
$$R_4 = +.329 \text{ kip/ft}$$
$$R_5 = -.168 \text{ kip/ft}$$

Longitudinal Stresses

Plate	Load R (kip/ft)	$M = \dfrac{59}{1120} R L^2$ (kip ft)	$Z\,(ft^3)$	$f(kip/ft^2)$	Free edge stresses (kip/ft^2)	Stresses after distribution (kip/ft^2)
5 / 4	.168	31.9	.2005	158.96	158.96 / 158.96	125.66 / 92.39
4 / 3	.329	62.3	.73	85.36	85.36 / 85.36	92.39 / 23.42
3 / 2	.148	28.1	7.351	3.83	3.83 / 3.83	23.42 / .88
2 / 1	.199	37.7	.73	51.69	51.69 / 51.69	.88 / 37.08
1 / 0	.07	13.3	.2005	66.31	66.31 / 66.31	37.08 / 51.69

Stress Distribution: Case IV

Distribution factors	0	1		2		3		4	
		.67	.33	.75	.25	.25	.75	.33	.67
Free edge stresses (kip/ft^2)	+66.31	−66.31	−51.69	+51.69	−3.83	+3.83	+85.36	−85.36	−158.96	+158.96
Stresses after distribution (kip/ft^2)	+51.69	−37.08	−37.08	+0.88	+0.88	+23.42	+23.42	−92.39	−92.39	+125.66

6. PLATE DEFLECTIONS v_n (ft)

All quantities are in ft and kip units

Case	Formula for v_n	v_1	v_2	v_3	v_4	v_5
I	$\dfrac{1}{12}\dfrac{(f_b - f_t)l^2}{Eh_n}$	$-\dfrac{791.8}{E}$	$+\dfrac{12,465.1}{E}$	$-\dfrac{4,360.0}{E}$	$+\dfrac{13,115.6}{E}$	$-\dfrac{3,787.6}{E}$
II	$\dfrac{227}{2,832}\dfrac{(f_b - f_t)l^2}{Eh_n}$	$+\dfrac{30,982.1}{E}$	$-\dfrac{9,218.8}{E}$	$+\dfrac{528.9}{E}$	$+\dfrac{3,021.7}{E}$	$-\dfrac{12,613.1}{E}$
III	$\dfrac{227}{2,832}\dfrac{(f_b - f_t)l^2}{Eh_n}$	$-\dfrac{209,097.7}{E}$	$+\dfrac{57,298.7}{E}$	0	$-\dfrac{57,298.7}{E}$	$+\dfrac{209,097.7}{E}$
IV	$\dfrac{227}{2,832}\dfrac{(f_b - f_t)l^2}{Eh_n}$	$+\dfrac{12,613.1}{E}$	$-\dfrac{3,021.7}{E}$	$-\dfrac{528.9}{E}$	$+\dfrac{9,218.8}{E}$	$-\dfrac{30,982.1}{E}$

7. PLATE ROTATIONS ψ_2, ψ_3, AND ψ_4

$$\psi_2 = \frac{1}{h_2}\left[v_2 \times (\cot \gamma_1 + \cot \gamma_2) - \frac{v_1}{\sin \gamma_1} - \frac{v_3}{\sin \gamma_2}\right]$$

$$= \frac{1}{3.625}\left[v_2 \times (0.17633 + 1.16742) - \frac{v_1}{0.98481} - \frac{v_3}{0.65056}\right]$$

$$= 0.37069v_2 - 0.28012v_1 - 0.42404v_3$$

$$\psi_3 = \frac{1}{h_3}\left[v_3 \times (\cot \gamma_2 + \cot \gamma_3) - \frac{v_2}{\sin \gamma_2} - \frac{v_4}{\sin \gamma_3}\right]$$

$$= \frac{1}{12.2972}\left(v_3 \times 0 - \frac{v_2}{0.65056} + \frac{v_4}{0.65056}\right)$$

$$= 0.125(v_4 - v_2)$$

$$\psi_4 = \frac{1}{h_4}\left[v_4 \times (\cot \gamma_3 + \cot \gamma_4) - \frac{v_3}{\sin \gamma_3} - \frac{v_5}{\sin \gamma_4}\right]$$

$$= \frac{1}{3.625}\left[v_4 \times (-1.16742 - 0.17633) + \frac{v_3}{0.65056} + \frac{v_5}{0.98481}\right]$$

$$= -0.37069v_4 + 0.42404v_3 + 0.28012v_5$$

Let

$$\frac{EJ_2}{h_2} \psi_{20} = 1 \text{ kip ft}$$

Then

$$\psi_{20} = \frac{h_2}{EJ_2} = \frac{3.625}{E \times \frac{1}{12} \times 1 \times (\frac{1}{3})^3} = \frac{1{,}174}{E}$$

Similarly

$$\psi_{30} = \frac{h_3}{EJ_3} = \frac{12.297}{E \times \frac{1}{12} \times 1 \times (\frac{7}{24})^3} = \frac{5{,}947}{E}$$

and

$$\psi_{40} = \psi_{20} = \frac{1{,}174}{E}$$

Now the actual rotations ψ_2, ψ_3, and ψ_4 will be written in terms of the arbitrary rotations ψ_{20}, ψ_{30}, and ψ_{40} and the unknown constants k_2, k_3, and k_4.

$$\psi_2 = \psi_{20} \times k_2 = \frac{1{,}174}{E} k_2$$

$$\psi_3 = \psi_{30} \times k_3 = \frac{5{,}947}{E} k_3$$

$$\psi_4 = \psi_{40} \times k_4 = \frac{1{,}174}{E} k_4$$

The values of the actual rotations ψ_2, ψ_3, and ψ_4 obtained in the previous step in terms of the arbitrary rotations will be equated to their values obtained in terms of plate deflections. This will result in three simultaneous equations in the unknown constants k_2, k_3, and k_4. Thus

(i) $$\psi_2 = \frac{1{,}174}{E} k_2 = 0.37069v_2 - 0.28012v_1 - 0.42404v_3$$

or

$$\frac{1{,}174}{E} k_2 = +\frac{6{,}691}{E} - \frac{12{,}320}{E} k_2 + \frac{79{,}812}{E} k_3 - \frac{4{,}429}{E} k_4$$

or

$$6{,}691 - 13{,}494k_2 + 79{,}812k_3 - 4{,}429k_4 = 0 \tag{1}$$

(ii) $$\psi_3 = \frac{5{,}947}{E} k_3 = 0.125(v_4 - v_2)$$

or

$$\frac{5{,}947}{E} k_3 = +\frac{81}{E} + \frac{1{,}530}{E} k_2 - \frac{14{,}325}{E} k_3 + \frac{1{,}530}{E} k_4$$

or

$$81 + 1{,}530k_2 - 20{,}272k_3 + 1{,}530k_4 = 0 \tag{2}$$

(iii) $\qquad \psi_4 = \dfrac{1,174}{E} k_4 = -0.37069v_4 + 0.42404v_3 + 0.28012v_5$

or

$$\dfrac{1,174}{E} k_4 = -\dfrac{7,772}{E} - \dfrac{4,429}{E} k_2 + \dfrac{79,812}{E} k_3 - \dfrac{12,320}{E} k_4$$

or

$$-7,772 - 4,429k_2 + 79,812k_3 - 13,494k_4 = 0 \qquad (3)$$

Solving the above three simultaneous equations, we get

$$k_2 = +0.76019$$
$$k_3 = -0.00164$$
$$k_4 = -0.83509$$

8. CALCULATION OF LONGITUDINAL STRESSES AND TRANSVERSE MOMENTS

Once the values of the constants k_2, k_3, and k_4 are obtained, the longitudinal stresses at all the edges are calculated as the stresses due to the no-rotation solution, plus k_2 times the stresses of the Case II solution, plus k_3 times the stresses of the Case III solution, plus k_4 times the stresses of the Case IV solution. The transverse moments are also calculated in a similar way.

The longitudinal stresses and the transverse moments—both at midspan—are shown tabulated below.

Longitudinal Stresses at Midspan

Joint	No-rotation solution	Case II	Case III	Case IV	Final stresses kip/ft²	psi
0	+56	+96	+1	−43	+110	+764
1	+62	−70	−1	+31	+22	+153
2	−89	+18	0	−1	−72	−500
3	+90	+1	0	−20	+71	+493
4	−69	−28	−1	+77	−21	−146
5	−43	+39	+1	−105	−108	−750

Transverse Moments at Midspan

Joint	No-rotation solution	Case II	Case III	Case IV	Final moments, lb ft
0	0	0	0	0	0
1	−55	0	0	0	−55
2	−505	+456	+7	−209	−251
3	−508	−190	−7	+501	−204
4	−32	0	0	0	−32
5	0	0	0	0	0

The procedure for calculation of the longitudinal stresses and transverse moments at any other section is illustrated with reference to the support section.

Stresses at the Intermediate Support Section

Stresses at the supports are obtained by multiplying the longitudinal stresses of the no-rotation solution at midspan by the ratio of the support moment to the midspan moment, which is equal to -2 in this case.

Joint	Longitudinal stress, psi
0	-784
1	-857
2	$+1,234$
3	$-1,248$
4	$+953$
5	$+597$

Transverse Moments at the Intermediate Support Section

Transverse moments due to the rotation solution are zero.   Hence the moments at the support section are those due to the no-rotation solution.

Joint	Transverse moment, lb ft
0	0
1	-55
2	-505
3	-508
4	-32
5	0

PRESTRESSED FOLDED PLATES

12-18   Need for Prestressing

With increasing spans, the longitudinal tension assumes very high values demanding the use of large quantities of steel.   Hence the design becomes uneconomical.   Moreover, placing this steel without undue congestion presents practical difficulties.   In long spans, splicing tension bars is also a problem.   Laps are not effective in tension members and welding becomes necessary.   These problems can be overcome by posttensioning the end plates, which results in the following advantages:

(i) The longitudinal tension is considerably reduced or completely eliminated, thus saving steel.

(ii) The transverse moments are also reduced.

(iii) The deflection of the structure can be controlled.

(iv) A crackless structure which offers greater resistance to weathering agents becomes possible.

It is sometimes stated that the object of prestressing is to throw the folded plate back into a pure membrane state. Although this condition cannot be fully realized, the reduction of deflections and transverse moments makes the structure approach a membrane condition. In Example 12-5, the design of a posttensioned folded plate is carried out in detail.

Design Example 12-5

ANALYSIS OF A SIMPLY SUPPORTED PRESTRESSED FOLDED PLATE

1. DATA

Geometry

Span = 83 ft 4 in.

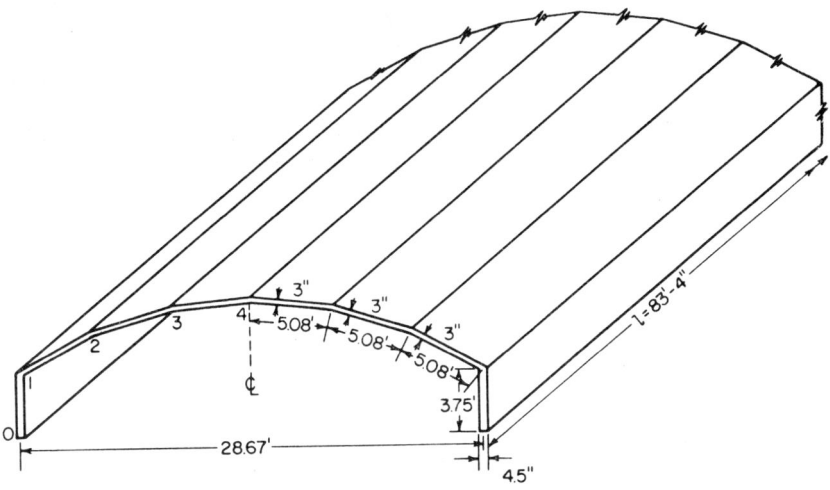

Fig. 12-31

Plate	Plate width h_n , ft	Horizontal projection l_n , ft	Inclination to horizontal ϕ_n	Angle between the plate and the next plate γ_n	Thickness of plate d, in.
1	3.75	0	90°	60°50′	4.5
2	5.08	4.436	29°10′	11°40′	3
3	5.08	4.845	17°30′	11°40′	3
4	5.08	5.054	5°50′	11°40′	3

Loads

Dead weight = 12.5 psf of plate/in. thickness
Live load = 25 psf of horizontal projection

2. GEOMETRICAL PROPERTIES AND TRIGONOMETRIC QUANTITIES

Geometrical Properties

Plate	A_n , ft²	Z_n , ft³
1	1.40625	0.8789
2	1.27	1.075
3	1.27	1.075
4	1.27	1.075

Trigonometric Quantities

Angle θ	sin θ	cos θ	tan θ	cosec θ	sec θ	cot θ
5°50′	0.10164	0.99482	0.10217	9.83980	1.00521	9.78880
11°40′	0.20222	0.97935	0.20648	4.94530	1.02109	4.84309
17°30′	0.30071	0.95372	0.31530	3.32551	1.04853	3.17159
29°10′	0.48735	0.87320	0.55813	2.05202	1.14522	1.79184
60°50′	0.87321	0.48735	1.79169	1.14520	2.05186	0.55811

3. LOADS AND FIXED-END MOMENTS ON TRANSVERSE SECTION

Plate	Load, kip	Fixed-end moments, kip ft
1	0.2109	0
2	0.3014	0.11142
3	0.3116	0.12581
4	0.3168	0.13345

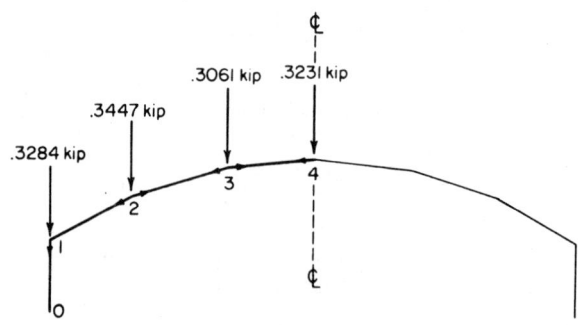

Fɪɢ. 12-32

4. NO-ROTATION SOLUTION

Moment Distribution

Distribution factors	1		2		3		4	
	0	1	3/7	4/7	.5	.5	.5	.5
Initial moments (kip ft)	0	+.11142	−.11142	+.12581	−.12581	+.13345	−.13345	
		−.11142 → −.05571						
			+.01771	+.02361 → +.01180				
				−.00486 ← −.00972	−.00972 → −.00486			
			+.00208	+.00278 → +.00139				
				−.00035 ← −.00070	−.00070 → −.00035			
			+.00015	+.00020 → +.00010				
				−.00005	−.00005 → −.00003			
Final moments (kip ft)	0	0	−.14719	+.14719	−.12298	+.12298	−.13868	+.13868
Reactions due to moments		.03318	.03318	.00499	.00499	.00311	.00311	.00311
Reactions due to loads	.2109	.1507	.1507	.1558	.1558	.15843	.15843	.15843
Total reactions	.3284		.3447		.3061		.3231	
Ridge loads P_n (kip)	.3284		.3447		.3061		.3231	

Calculation of Plate Loads From Ridge Loads

Referring to Fig. 12-32,

$$S_{1,0} = .3284 \text{ kip/ft}$$

$$S_{2,1} = P_2 \frac{\cos\phi_3}{\sin\gamma_2} = .3447\,\frac{\cos 17°\text{-}30'}{\sin 11°\text{-}40'} = 1.6256 \text{ kip/ft}$$

$$S_{2,3} = P_2 \frac{\cos\phi_2}{\sin\gamma_2} = .3447\,\frac{\cos 29°\text{-}10'}{\sin 11°\text{-}40'} = 1.4884 \text{ kip/ft}$$

$$S_{3,2} = P_3 \frac{\cos\phi_4}{\sin\gamma_3} = .3061\,\frac{\cos 5°\text{-}50'}{\sin 11°\text{-}40'} = 1.5060 \text{ kip/ft}$$

$$S_{3,4} = P_3 \frac{\cos\phi_3}{\sin\gamma_3} = .3061\,\frac{\cos 17°\text{-}30'}{\sin 11°\text{-}40'} = 1.4438 \text{ kip/ft}$$

$$S_{4,3} = P_4 \frac{\cos\phi_5}{\sin\gamma_4} = .3231\,\frac{\cos(360°-(5°\text{-}50'))}{\sin 11°\text{-}40'} = 1.5892 \text{ kip/ft}$$

Hence

$$R_1 = +0.3284 \text{ kip/ft}$$
$$R_2 = +1.6256 \text{ kip/ft}$$
$$R_3 = +0.0177 \text{ kip/ft}$$
$$R_4 = +0.1454 \text{ kip/ft}$$

Longitudinal Stresses Due to Plate Loads

−ve Compression
+ve Tension

Plate	Load R (kip/ft)	$M = \dfrac{Rl^2}{8}$ (kip ft)	Z (ft³)	f(kip/ft²)	Free edge stresses (kip/ft²)	Stresses after distribution (kip/ft²)
4 — 3	0.1454	126.22	1.075	117.4	117.4 / 117.4	146 / 174.75
3 — 2	0.0177	15.36	1.075	14.3	14.3 / 14.3	174.75 / 450.65
2 — 1	1.6256	1411.11	1.075	1312.7	1312.7 / 1312.7	450.65 / 329.3
1 — 0	0.3284	285.07	0.8789	324.4	324.4 / 324.4	329.3 / 2.5

Stress Distribution (No-Rotation)

Distribution factors	0	1		2		3		4
		.47	.53	.5	.5	.5	.5	
Free edge stresses (kip/ft²)	+324.4 −324.4	+1312.7 −1312.7		+ 14.3 −14.3		+117.4 −117.4		
	−384.7 ← +769.4	− 867.7 → +433.8						
		− 223.3 ← +446.6		−446.6 → +223.3				
	+ 52.4 ← −104.9	+ 118.4 → −59.2		+ 22.9 ← −45.8		+ 45.8 → −22.9		
		− 20.5 ← +41.05		− 41.05 → +20.5				
	+ 4.8 ← − 9.6	+ 10.9 → − 5.4		+ 5.1 ← −10.25		+ 10.25 → − 5.1		
		− 2.6 ← +5.25		− 5.25 → +2.6				
	+ 0.6 ← −1.2	+ 1.4 → − 0.7		+ 0.6 ← − 1.3		+ 1.3 → −0.6		
		+0.65		− 0.65				
Stresses after distribution (kip/ft²)	− 2.5 +329.3	+ 329.3 −450.65		−450.65 +174.75		+174.75 −146		

5. EFFECT OF JOINT DISPLACEMENTS – ARBITRARY ROTATION OF PLATES

Arbitrary Rotation of Plate 2: Case II Solution

Distribution factors	1		2		3		4
	0	1	3/7	4/7	.5	.5	
Arbitrary moments (kip ft)	+6		+6				
	−6		−3				
Initial moments (kip ft)	0		+3				
Moments after distribution (kip ft)	0	0	+1.628	−1.628	−0.4607	+0.4607	+0.2304	−0.2304
		0.367	0.367	0.431	0.431	0.137	0.137	0.137
Total reactions	0.367		0.798		0.568		0.274
Ridge loads P_n (kip)	0.367		0.798		0.568		0.274

Plate loads R_n

Knowing the ridge loads P_n, the plate loads R_n are calculated as before.

$$R_1 = +0.367 \text{ kip/ft}$$
$$R_2 = -3.764 \text{ kip/ft}$$
$$R_3 = +6.240 \text{ kip/ft}$$
$$R_4 = -4.027 \text{ kip/ft}$$

Longitudinal Stresses

Plate	Load R (kip/ft)	$M = \dfrac{Rl^2}{\pi^2}$ (kip ft)	Z (ft^3)	f (kip/ft^2)	Free edge stresses (kip/ft^2)	Stresses after distribution (kip/ft^2)
4 – 3	4.027	2833.5	1.075	2635.8	2635.8 / 2635.8	2866.4 / 3096.95
3 – 2	6.240	4390.6	1.075	4084.3	4084.3 / 4084.3	3096.95 / 2801.4
2 – 1	3.764	2648.4	1.075	2463.6	2463.6 / 2463.6	2801.4 / 1561
1 – 0	0.367	258.23	0.8789	293.8	293.8 / 293.8	1561 / 927.4

Stress Distribution: Case II

Distribution factors	0	1		2		3		4
		.47	.53	.5	.5	.5	.5	
Free edge stresses (kip/ft^2)	+293.8	−293.8	−2463.6	+2463.6	+4084.3	−4084.3	−2635.8	+2635.8
Stresses after distribution (kip/ft^2)	+927.4	−1561.0	−1561.0	+2801.4	+2801.4	−3096.95	−3096.95	+2866.4

Arbitrary Rotation of Plate 3: Case III Solution

Distribution factors	1		2		3		4
	0	1	3/7	4/7	.5	.5	
Arbitrary moments (kip ft)			+6		+6		
Moments after distribution (kip ft)	0 0		−2.07 +2.07		+2.31 −2.31		−1.145 +1.145
		.467	.467 .904		.904 .682		.682 .682
	0.467		1.371		1.586		1.364
Ridge loads P$_n$ (kip)	0.467		1.371		1.586		1.364

Plate Loads R$_n$

$$R_1 = -0.467 \text{ kip/ft}$$
$$R_2 = +6.466 \text{ kip/ft}$$
$$R_3 = -13.722 \text{ kip/ft}$$
$$R_4 = +14.190 \text{ kip/ft}$$

Longitudinal Stresses

Plate	Load R (kip/ft)	$M = \dfrac{Rl^2}{\pi^2}$ (kip ft)	Z (ft³)	f (kip/ft²)	Free edge stresses (kip/ft²)	Stresses after distribution (kip/ft²)
4 ⟍ 3	14.190	9984.4	1.075	9287.8	9287.8 / 9287.8	8791.8 / 8295.7
3 ⟍ 2	13.722	9655.3	1.075	8981.7	8981.7 / 8981.7	8295.7 / 6121.4
2 ⟍ 1	6.466	4549.6	1.075	4232.2	4232.2 / 4232.2	6121.4 / 2975.8
1 ⟍ 0	0.467	328.6	0.8789	373.9	373.9 / 373.9	2975.8 / 1674.8

Stress Distribution: Case III

	0		1		2		3		4
Distribution factors			.47	.53	.5	.5	.5	.5	
Free edge stresses (kip/ft²)	−373.9	+373.9	+4232.2	−4232.2	−8981.7	+8981.7	+9287.8	−9287.8	
Stresses after distribution (kip/ft²)	−1674.8	+2975.8	+2975.8	−6121.4	−6121.4	+8295.7	+8295.7	−8791.8	

Arbitrary Rotation of Plate 4: Case IV Solution

	1		2		3		4
Distribution factors	0	1	3/7	4/7	.5	.5	
Arbitrary moments (kip ft)					+6		+6
Moments after distribution (kip ft)	0	0	+.687	−.687	−2.770	+2.770	+4.382 / −4.382
		.155	.155 / .714		.714 / 1.415		1.415 / 1.415
		.155	.869		2.129		2.830
Ridge loads P_n (kip)		0.155	0.869		2.129		2.830

Plate Loads R_n

$$R_1 = + 0.155 \text{ kip/ft}$$
$$R_2 = - 4.098 \text{ kip/ft}$$
$$R_3 = + 14.226 \text{ kip/ft}$$
$$R_4 = -23.963 \text{ kip/ft}$$

Longitudinal Stresses

Plate	Load R (kip/ft)	$M = \dfrac{Rl^2}{\pi^2}$ (kip ft)	$Z(ft^3)$	$f(kip/ft^2)$	Free edge stresses (kip/ft^2)	Stresses after distribution (kip/ft^2)
4 ... 3	23.963	16860.6	1.075	15684.3	15684.3 / 15684.3	13452.7 / 11221.0
3 ... 2	14.226	10009.7	1.075	9311.3	9311.3 / 9311.3	11221.0 / 6435.4
2 ... 1	4.098	2883.8	1.075	2682.6	2682.6 / 2682.6	6435.4 / 2526.7
1 ... 0	0.155	109.06	0.8789	124.1	124.1 / 124.1	2526.7 / 1325.2

Stress Distribution: Case IV

Distribution factors	O		1	.47	.53		2	.5	.5		3	.5	.5		4
Free edge stresses (kip/ft^2)	+ 124.1			−124.1	−2682.6		+2682.6		+ 9311.3	−9311.3			−15684.3		+15684.3
Stresses after distribution (kip/ft^2)	+1325.2			−2526.7	−2526.7		+6435.4		+6435.4	−11221.0			−11221.0		+13452.7

6. PLATE DEFLECTIONS v_n (ft)

All quantities are in ft and kip units

Case	Formula for v_n	v_1	v_2	v_3	v_4
I	$5\dfrac{(f_b - f_t)l^2}{48Eh_n}$	$-\dfrac{64,004.4}{E}$	$+\dfrac{111,062.4}{E}$	$-\dfrac{89,055.6}{E}$	$+\dfrac{45,674.4}{E}$
II	$\dfrac{(f_b - f_t)l^2}{\pi^2 Eh_n}$	$+\dfrac{466,903.2}{E}$	$-\dfrac{604,226.4}{E}$	$+\dfrac{816,967.2}{E}$	$-\dfrac{825,969.6}{E}$
III	$\dfrac{(f_b - f_t)l^2}{\pi^2 Eh_n}$	$-\dfrac{872,600.4}{E}$	$+\dfrac{1,260,032.4}{E}$	$-\dfrac{1,996,880.4}{E}$	$+\dfrac{2,366,751.6}{E}$
IV	$\dfrac{(f_b - f_t)l^2}{\pi^2 Eh_n}$	$+\dfrac{722,739.6}{E}$	$-\dfrac{1,241,320.8}{E}$	$+\dfrac{2,445,548.4}{E}$	$-\dfrac{3,417,498.0}{E}$

7. PLATE ROTATIONS ψ_n

$$\psi_2 = \frac{1}{h_2}\left[v_2(\cot \gamma_1 + \cot \gamma_2) - \frac{v_1}{\sin \gamma_1} - \frac{v_3}{\sin \gamma_2}\right]$$

$$= -0.225432v_1 + 1.063224v_2 - 0.973452v_3$$

$$\psi_3 = \frac{1}{h_3}\left[v_3(\cot \gamma_2 + \cot \gamma_3) - \frac{v_2}{\sin \gamma_2} - \frac{v_4}{\sin \gamma_3}\right]$$

$$= -0.973452v_2 + 1.906728v_3 - 0.973452v_4$$

$$\psi_4 = \frac{1}{h_4}\left[v_4(\cot \gamma_3 + \cot \gamma_4) - \frac{v_3}{\sin \gamma_3} - \frac{v_5}{\sin \gamma_4}\right]$$

$$= -0.973452v_3 + 2.880180v_4 \qquad \text{(because } v_5 = -v_4\text{)}$$

Let

$$\frac{EJ_2}{h_2}\psi_{20} = 1 \text{ kip ft}$$

Then

$$\psi_{20} = \frac{h_2}{EJ_2} = \frac{3,901.39}{E}$$

Similarly,

$$\psi_{30} = \psi_{20} = \frac{3,901.39}{E}$$

and

$$\psi_{40} = \psi_{20} = \frac{3,901.39}{E}$$

Now, the actual rotations ψ_2, ψ_3, and ψ_4 will be written in terms of the arbitrary rotations ψ_{20}, ψ_{30}, and ψ_{40} and the unknown constants k_2, k_3, and k_4.

$$\psi_2 = k_2\psi_{20} = \frac{3,901.39}{E} k_2$$

$$\psi_3 = k_3\psi_{30} = \frac{3,901.39}{E} k_3$$

$$\psi_4 = k_4\psi_{40} = \frac{3,901.39}{E} k_4$$

The values of the actual rotations ψ_2, ψ_3, and ψ_4 obtained from the previous step in terms of the arbitrary rotations will be equated to their values obtained in terms of plate deflections. This will result in three simultaneous equations in terms of the unknown constants k_2, k_3, and k_4. Thus,

(i) $\qquad \psi_2 = \dfrac{3,901.39}{E} k_2 = -0.225432v_1 + 1.063224v_2 - 0.973452v_3$

or

$$\frac{3,901.39}{E} k_2 = \frac{219,196.8}{E} - \frac{1,542,945.6}{E} k_2 + \frac{3,480,278.4}{E} k_3 - \frac{3,863,347.2}{E} k_4$$

or

$$219,196.8 - 1,546,847.0k_2 + 3,480,278.4k_3 - 3,863,347.2k_4 = 0 \qquad (i)$$

(ii) $\qquad \psi_3 = \dfrac{3,901.39}{E} k_3 = -0.973452v_2 + 1.906728v_3 - 0.973452v_4$

or

$$\frac{3,901.39}{E} k_3 = -\frac{322,387.2}{E} + \frac{2,949,955.2}{E} k_2 - \frac{7,337,995.2}{E} k_3 + \frac{9,198,100.8}{E} k_4$$

or

$$-322,387.2 + 2,949,955.2k_2 - 7,341,896.6k_3 + 9,198,100.8k_4 = 0 \qquad (ii)$$

(iii) $\qquad \psi_4 = \dfrac{3,901.39}{E} k_4 = 0.973452v_3 + 2.880180v_4$

or

$$\frac{3,901.39}{E} k_4 = +\frac{218,232.0}{E} - \frac{3,174,206.4}{E} k_2 + \frac{8,760,513.6}{E} k_3 - \frac{12,223,598}{E} k_4$$

or

$$218,232.0 - 3,174,206.4k_2 + 8,760,513.6k_3 - 12,227,499k_4 = 0 \qquad (iii)$$

Solving the above three simultaneous equations, we get

$$k_2 = +0.029347$$
$$k_3 = -0.188507$$
$$k_4 = -0.124829$$

8. *CALCULATION OF LONGITUDINAL STRESSES AND TRANSVERSE MOMENTS*

Once the values of the constants k_2, k_3, and k_4 are obtained, the longitudinal stresses at all edges are calculated as the stresses due to the no-rotation solution, plus k_2 times the stresses of the Case II solution, plus k_3 times the stresses of the Case III solution, plus k_4 times the stresses of the Case IV solution. The transverse moments are also calculated in a similar way.

Longitudinal Stresses at Midspan Tension +ve

Joint	No rotation	Case II	Case III	Case IV	Final stresses	
					kip/ft²	psi
0	−2.5	+27.2	+315.7	−165.4	+175	+1,215
1	+329.3	−45.8	−561.0	+315.4	+38	+263
2	−450.6	+82.2	+1,153.9	−803.3	−18	−124
3	+174.7	−90.9	−1,563.8	+1,400.7	−79	−548
4	−146.0	+84.1	+1,657.3	−1,679.3	−84	−583

Transverse Moments Sagging +ve

Joint	No rotation	Case II	Case III	Case IV	Final moments, kip ft
0	0	0	0	0	0
1	0	0	0	0	0
2	−0.1472	+0.0478	+0.3902	−0.0858	+0.2050
3	−0.1230	−0.0135	−0.4355	+0.3458	−0.2262
4	−0.1387	+0.0068	+0.2158	−0.5470	−0.4631

9. *CALCULATION OF THE PRESTRESSING FORCE P*

To arrive at a suitable value of the prestressing force to be applied, an arbitrary value of the prestressing force, say $P = 130$ kip, is applied. The eccentricity of the force is generally governed by the protective cover for the cables, size of the anchorage grips, etc. In the present example an eccentricity of 0.625 ft with respect to the center of gravity of the end plate has been chosen. Incidentally, this happens to be the bottom kern point for the end plate. The free-edge stresses at edges 0 and 1 due to this prestressing force are

$$f_0 = \frac{-130}{1.40625} - \frac{130 \times 0.625 \times 1.875}{1.64795}$$

$$= -184.9 \text{ kip/ft}^2$$

$$f_1 = -\frac{130}{1.40625} + \frac{130 \times 0.625 \times 1.875}{1.64795}$$

$$= 0$$

$$v_1 = \frac{(f_0 - f_1)l^2}{\pi^2 E h_1} = \frac{-34693}{E}$$

(As the curve of the cable is shallow, the parabolic profile has been approximated by a sine curve.) $v_2 = v_3 = v_4 = 0$ in this particular case as f_1 happens to be zero. If this were not so, v_2, v_3, and v_4 would have to be arrived at in the usual manner after carrying out stress distribution.

These deflections are now added to the deflections of the unprestressed folded plate, and the following revised equations are formed:

$$1,576.5 - 10,742.0k_2 + 24,168.6k_3 - 26,828.8k_4 = 0 \qquad \text{(i)}$$

$$-2,238.8 + 20,485.8k_2 - 50,985.4k_3 + 63,875.7k_4 = 0 \qquad \text{(ii)}$$

$$1,515.5 - 22,043.1k_2 + 60,836.9k_3 - 84,913.2k_4 = 0 \qquad \text{(iii)}$$

Solving the above three simultaneous equations, we get

$$k_2 = +0.721671$$
$$k_3 = +0.329168$$
$$k_4 = +0.066340$$

The resulting N_x and M_ϕ values are shown computed in the tables below.

Longitudinal Stresses at Midspan

Joint	No rotation	Due to pre-stressing	Case II	Case III	Case IV	Final stresses	
						kip/ft²	psi
0	−2.5	−184.9	+669.3	−551.3	+87.9	+18.5	+128
1	+329.3	0	−1,126.5	+979.5	−167.6	+14.7	+102
2	−450.6	0	+2,021.7	−2,015.0	+426.9	−17.0	−118
3	+174.7	0	−2,235.0	+2,730.7	−744.4	−74.0	−514
4	−146.0	0	+2,068.6	−2,894.0	+892.5	−78.9	−548

Transverse Moments

Joint	No rotation	Case II	Case III	Case IV	Final moments, kip ft/ft
0	0	0	0	0	0
1	0	0	0	0	0
2	−0.1472	+1.1749	−0.6814	+0.0456	+0.3919
3	−0.1230	−0.3325	+0.7604	−0.1838	+0.1211
4	−0.1387	+0.1663	−0.3769	+0.2907	−0.0586

From a study of the stresses and moments in the unprestressed and prestressed folded plates, we may arrive at the following expressions for the stress resultants in terms of the unknown P.

Case I

Dead Weight plus Full Live Load

Longitudinal Stresses

Joint	Stress, kip/ft^2
0	$+175.0 - 1.203P$
1	$+37.9 - 0.179P$
2	$-17.8 + 0.006P$
3	$-79.3 + 0.0408P$
4	$-83.9 + 0.038P$

Transverse Moments

$$m_2 = +0.205 + 0.0014P$$
$$m_3 = -0.226 + 0.00268P$$
$$m_4 = -0.463 + 0.0031P$$

Case II

Dead Weight and Half Live Load

In practice, the combination of dead weight plus live load may occur only very infrequently, if at all. It is hence more rational to arrive at a prestressing force which will cause a condition of no tension in the folded plate under dead weight plus a realistic fraction, say half, of the live load, provided under dead weight plus full live load the tensile stress is within the permissible limits prescribed in the Tentative Recommendations for Prestressed Concrete reported by the ACI-ASCE Joint Committee.

The stresses and moments in the unprestressed folded plate under dead weight plus half live load are given below.

Joint	Longitudinal stress, kip/ft^2	Transverse moment, kip ft/ft
0	$+159.56$	0
1	$+23.83$	0
2	-18.94	$+0.1005$
3	-63.72	-0.2925
4	-65.14	-0.4959

For any value of P, the stresses and moments are

Joint	Longitudinal stress, kip/ft^2	Transverse moment, kip ft/ft
0	$+159.56 - 1.203P$	0
1	$+23.83 - 0.179P$	0
2	$-18.94 + 0.006P$	$+0.1005 + 0.0014P$
3	$-63.72 + 0.0408P$	$-0.2925 + 0.00268P$
4	$-65.14 + 0.038P$	$-0.4959 + 0.0031P$

We now proceed to choose the prestressing force for which, under dead weight plus half live load, the folded plate is entirely in compression and, with full live load, the residual tension in the end plate is within the permissible limits. Assuming, for instance, that f'_c of the concrete in the edge beams is 5,000 psi, we may permit $3 \sqrt{f'_c} = 212$ psi in tension. Based on these considerations, $P = 133.5$ kip is found to be suitable. The final stress resultants when this prestressing force acts are given below:

(i) *Dead Weight + Half Live Load*

Joint	Longitudinal stress		Transverse moment, ft lb/ft
	kip/ft²	psi	
0	−1.1	−8	0
1	−0.1	−1	0
2	−18.1	−126	+293.0
3	−58.3	−405	+65.0
4	−60.1	−417	−79.5

(ii) *Dead Weight + Full Live Load*

Joint	Longitudinal stress		Transverse moment, ft lb/ft
	kip/ft²	psi	
0	+14.3	+99	0
1	+14.0	+97	0
2	−17.0	−118	+397
3	−73.9	−513	+130
4	−78.8	−547	−47.7

The arrangement of the prestressing cables in the end plate is shown in Fig. 12-33.

12-19 Design of Reinforcement

The reinforcement provided may be classed into two groups.

(*a*) Steel required to take care of the tangential group of forces N_x and $N_{x\phi}$

(*b*) Reinforcement required to resist the transverse moment

Design of steel to resist the tangential stress resultants may be based on principal tensions which develop in the plates.

For finding the principal stresses, it is necessary to work out the shear stresses τ at various points in the folded plates. The shear stress at any point is calculated from the longitudinal shearing force T at that point,

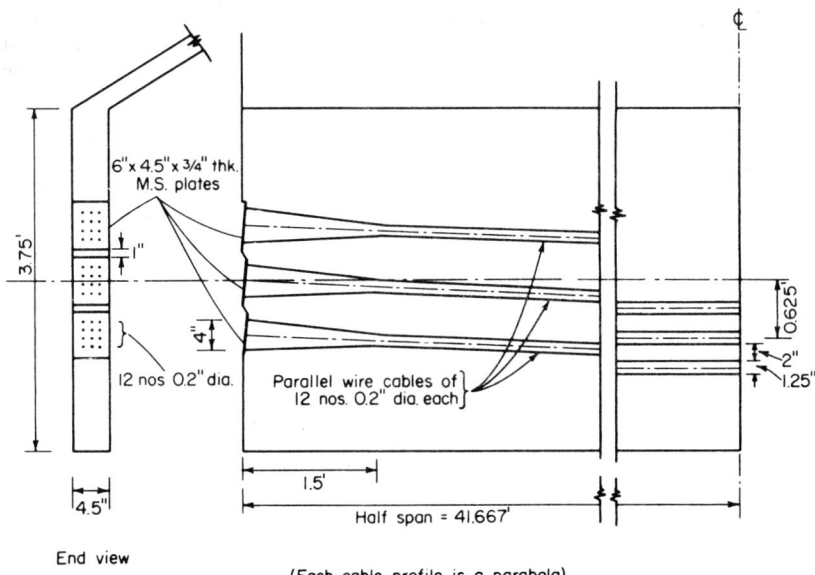

End view

(Each cable profile is a parabola)

FIG. 12-33.   Cable profile in the end plate.

which, in turn, is derived from the longitudinal edge shear forces T_n and the longitudinal stresses f_n .   For example, for plate n (Fig. 12-34),

$$T = T_{n-1} - \frac{1}{2} dy \left(f_{n-1} + f_{n-1} \frac{h_n - y}{h_n} \right) - \frac{1}{2} dy f_n \frac{y}{h_n} \qquad (12\text{-}23)$$

τ is obtained from the relation

$$\tau = \frac{1}{d} \frac{\delta T}{\delta x} \qquad (12\text{-}24)$$

T has the same longitudinal variation as the plate moment $M_{0,n}$.

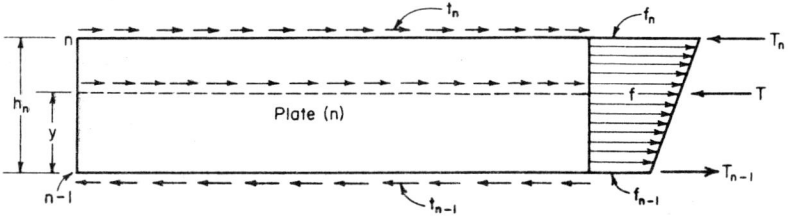

FIG. 12-34

Hence for the no-rotation solution, T at a distance x from the support is related to T_{max} at midspan by

$$T = T_{max} \frac{4x}{l}\left(1 - \frac{x}{l}\right) \tag{12-25}$$

Differentiating and substituting in (12-24), we get

$$\tau = \frac{4T_{max}}{dl}\left(1 - \frac{2x}{l}\right) \tag{12-26}$$

For the correction analysis, i.e., for the rotation solutions,

$$T = T_{max} \sin \frac{\pi x}{l}$$

Hence

$$\tau = \frac{\pi T_{max}}{dl} \cos \frac{\pi x}{l} \tag{12-27}$$

Adding up,

$$\tau = \frac{4T_{1max}}{d}\left(1 - \frac{2x}{l}\right) + \frac{\pi T_{2max}}{dl} \cos \frac{\pi x}{l} \tag{12-28}$$

where T_1 and T_2 are the longitudinal shear forces due to the no-rotation and rotation solutions, respectively.

As the shearing force due to the rotation solution is usually small compared with the value obtained from the no-rotation solution, it is sufficiently accurate for practical purposes to assume a parabolic variation for the shearing forces T_1 as well as T_2 [62]. Thus,

$$\tau = \frac{4T_{max}}{dl}\left(1 - \frac{2x}{l}\right) \tag{12-29}$$

The principal stresses are computed in the usual manner from the formula

$$f_p = \frac{f}{2} \pm \sqrt{\left(\frac{f}{2}\right)^2 + \tau^2} \tag{12-30}$$

where f is the longitudinal stress at the point under consideration.

The following principles underlying the design of reinforcement are extracted from the draft report of the ACI Committee 334 on Concrete Shell Structures.[1]

403 Shell Reinforcement

(a) The stress in the reinforcement may be assumed at the allowable value independently of the strain in the concrete.

[1] Extracted by permission of the American Concrete Institute.

(*b*) Where the tensile stresses vary greatly in magnitude over the shell as in the case of cylindrical shells, the reinforcement capable of resisting the total tension may be concentrated in the region of maximum tensile stress. Where this is done, the percentage of crack control reinforcing in any 12 in. width of shell shall be not less than 0.35 per cent throughout the tensile zone.

(*c*) The principal tensile stresses shall be resisted entirely by reinforcement.

(*d*) Reinforcement to resist the principal tensile stresses, assumed to act at the middle surface of the shell, may be placed either in the general direction of the lines of principal tensile stress (also referred to as parallel to the lines of principal tensile stress), or in two or three directions. In the regions of high tension it is advisable, based on experience, to place the reinforcing in the general direction of the principal stress.

(*e*) The reinforcement may be considered parallel to the lines of principal stress when its direction does not deviate from the direction of the principal stress more than fifteen degrees. Variations in the directions of the principal stress over the cross section of a shell due to moments need not be considered for the determination of the maximum deviation. In areas where the stress in the reinforcing is less than the allowable stress, a deviation greater than fifteen degrees can still be considered parallel placing; a stress decrease of 5 % shall be considered to compensate for each additional degree of deviation. Wherever possible, such reinforcing may run along lines considered most practical for construction, such as straight lines.

(*f*) Where placed in more than one direction, the reinforcement shall resist the components of the principal tension force in each direction.

(*g*) In those areas where the computed principal tensile stress in the concrete exceeds 300 psi, placement in at least one layer of the reinforcing shall be parallel to the principal tensile stress, unless it can be proven that a deviation of the reinforcing from the direction parallel to the lines of principal tensile stress is permissible because of the geometrical characteristics of a particular shell and because for reasons of geometry only insignificant and local cracking could develop.

(*h*) Where the computed principal tensile stress (in psi) in the concrete exceeds numerically the value $2 \sqrt{f'_c}$ (where f'_c is also in psi), the spacing of reinforcement shall not be greater than three times the thickness of the thin shell. When the computed principal tensile stress is numerically less than the value $2 \sqrt{f'_c}$, the reinforcement shall be spaced at not more than five times the thickness of the thin shell, nor more than 18 inches.

(*i*) Minimum reinforcement shall be provided as required in the Building Code (ACI 318-63) even where not required by analysis.

(*j*) The percentage of reinforcement in any 12 in. width of shell shall not exceed $30 f'_c/f_s$. However, the maximum percentage shall not exceed six per cent if $f_s = 20{,}000$ psi, five per cent if $f_s = 25{,}000$ psi, or four per cent if $f_s = 30{,}000$ psi when the latter value is acceptable. If the deviation of the reinforcing from the lines of principal stress is greater than 10 degrees, the maximum percentage shall be only half of the above values.

(*k*) Splices in principal tensile reinforcement shall be kept to a practical minimum. Where necessary they shall be staggered with not more than one third of the bars spliced at any one cross section. Bars shall be lapped only within the same layer. The minimum lap for draped shell reinforcing bars shall be 30 diameters with a 1 ft. 6 in. minimum unless more is required by the Building Code (ACI 318-63), except that the minimum may be 12 in. for reinforcement not required by analysis. The minimum lap for welded wire fabric shall be 8 in. or one mesh, whichever is greater, except that Building Code requirements shall govern where the wire fabric at the splice must carry the full allowable stress.

(*l*) Reinforcement to resist bending moments shall be proportioned and provided in the conventional manner with proper allowance for the direct forces.

These principles are shown applied in the design of reinforcement for the folded plate analyzed in Example 12-2 (Fig. 12-21).

Transverse Reinforcement

At joints 2 and 3, the transverse moment causes tension at top.
Effective depth $= 2 + \frac{7}{16} + \frac{3}{16} = 2.625$ in.

$$A_t = \frac{986 \times 12}{20,000 \times 0.874 \times 2.625} = 0.2585 \text{ in.}^2$$

Provide No. 3 bars 4.5 in. center to center to give 0.294 in.². At joints 4 and 5, the transverse moment is causing tension at bottom

$$A_t = \frac{270 \times 12}{20,000 \times 0.874 \times 2.625} = 0.07 \text{ in.}^2$$

Provide No. 3 bars at 9 in. center to center to give 0.147 in.².

No. 3 bars at 9 in. center to center may be provided in all the plates below the longitudinal steel and touching it. For taking care of hogging moments at joints 3, 2, and 1, No. 3 bars at $4\frac{1}{2}$ in. center to center shall be provided above the longitudinal steel as shown in the drawing (Fig. 12-37).

The same transverse reinforcement may be continued throughout the span. However, if greater economy is desired, the spacing of the bars may be increased from midspan to the support section based on the magnitude of the transverse bending moment.

Longitudinal Reinforcement

Plate 1

The plate is entirely in compression and the stresses are less than what is permissible. Hence, no reinforcement is required. Two No. 3 bars will do as nominal reinforcement.

Plate 2 (Fig. 12-35)

$$A_{t_2} = \frac{2.42 \times 12 \times 3.5 \times \dfrac{738}{2}}{20,000} = 1.875 \text{ in.}^2$$

Provide 3 No. 7 bars and 1 No. 3 bar as shown in Fig. 12-37.

In the remaining part of the tension zone, reinforcement equal to 0.35 % of the area of the unreinforced tensile zone shall be provided. This works out to $(0.35/100) \times 3.5 \times 18 = 0.22$ in.^{2}. Hence 2 No. 3 bars are provided.

Plate 3

$$\text{Total tension} = \frac{738.12 + 791.11}{2} \times 1.563 \times 12 \times 4 = 57,300 \text{ lb}$$

$$A_{t_3} = \frac{57,300}{20,000} = 2.87 \text{ in.}^2$$

∴ Provide 5 No. 7 bars.

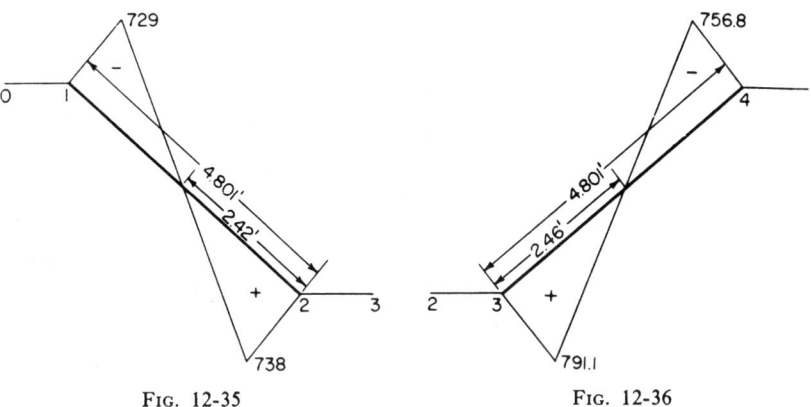

FIG. 12-35   FIG. 12-36

Plate 4 (Fig. 12-36)

$$A_{t_4} = \frac{2.46 \times 12 \times 3.5}{20,000} \times \frac{791.11}{2} = 2.05 \text{ in.}^2$$

3 No. 7 bars and 1 No. 5 bar will suffice (Fig. 12-37).

Plate 5

The plate is entirely in compression and the stress is within the permissible limit. Nominal reinforcement consisting of 3 No. 3 bars is provided.

Shear Reinforcement

Shear reinforcement will be provided to take up the principal tensile stresses. At the support section, the principal tension is inclined at 45° to the longitudinal axis; hence the shear reinforcement will be

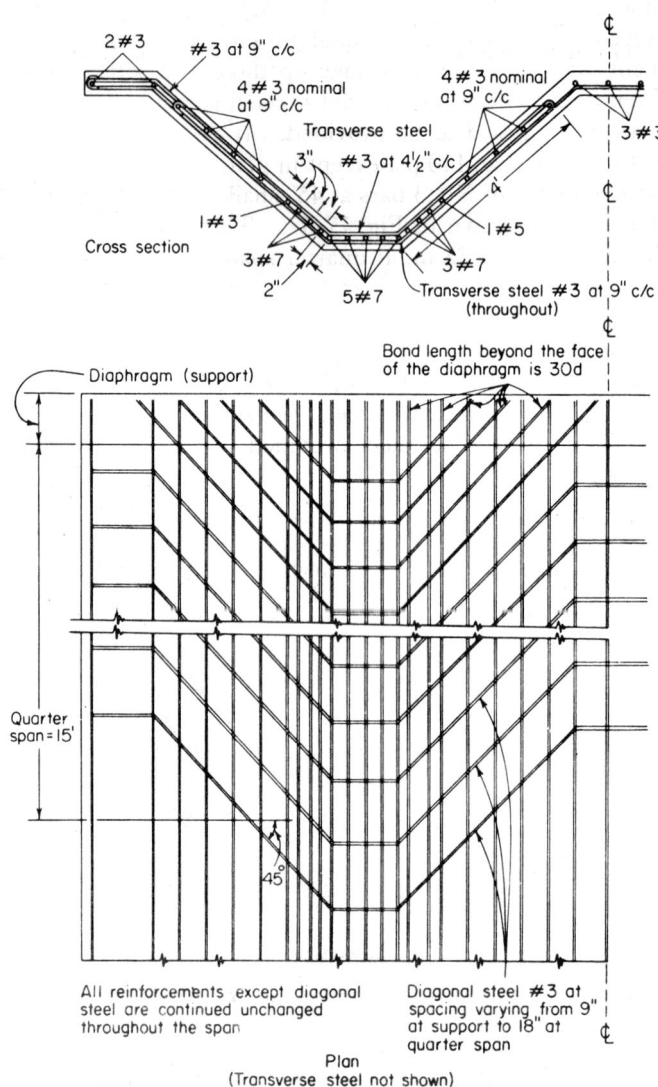

Cross section

2 #3

#3 at 9" c/c

4 #3 nominal at 9" c/c

4 #3 nominal at 9" c/c

3 #3

Transverse steel

3"   #3 at 4½" c/c

1 #3

1 #5

3 #7   3 #7

2"   5 #7   Transverse steel #3 at 9" c/c (throughout)

Bond length beyond the face of the diaphragm is 30d

Diaphragm (support)

Quarter span = 15'

45°

All reinforcements except diagonal steel are continued unchanged throughout the span

Diagonal steel #3 at spacing varying from 9" at support to 18" at quarter span

Plan
(Transverse steel not shown)

Fig. 12-37

provided at 45° as shown in Fig. 12-37. Whenever the principal tensile stress exceeds $2\sqrt{f'_c}$, which is equal to 110 psi in this instance, reinforcement will be designed to take up the entire tension; elsewhere, nominal reinforcement at spacing not exceeding five times the thickness of the plate is proposed to be provided. It is seen that the principal tensile stress is less than 110 psi except in plate 2 in which it is 111.7 psi. Hence nominal steel of No. 3 bars at 45° shall be provided from support to quarter span as shown in Fig. 12-37. The spacing may be varied from 9 in. at support to 18 in. at quarter span.

Check at Quarter Span

The maximum value of the principal tension is 622 psi at joint 2 and is inclined at 2°44′ to the longitudinal axis. Being almost in the same direction as the longitudinal steel, no additional reinforcement is required. The longitudinal steel provided is quite adequate to resist the tensile stress.

CHAPTER 13

CONTINUOUS CYLINDRICAL SHELLS

13-1 Boundary Conditions

The theories developed in Chapter 7 are limited in their application to cylindrical shells simply supported over their traverses. For this particular condition, one of the selected stress resultants, say M_ϕ or w, can be expressed in Fourier series. This is because the prescribed boundary conditions over the traverse, $w = 0$ and $w'' = 0$, can be satisfied by a trigonometric function. The method breaks down if we have to deal with other boundary conditions. A case in point of special practical importance is a shell continuous over the traverses.

13-2 Methods for Dealing with Longitudinally Continuous Cylindrical Shells

None of the available methods [65, 3, 66, 67] is really satisfactory for dealing with both long and short shells. Long shells longitudinally continuous over traverses may be dealt with by a method described by Morice [65], who applied it to a shell without edge beams. This method, suitably modified to take care of edge beams, is presented in this chapter and its use explained in detail by means of examples. Perhaps the method described in the ASCE Manual is the best means that we have at present for investigating the behavior of longitudinally continuous short shells.

13-3 Basic Functions and Their Properties

In the Morice method, basic functions take the place of trigonometric functions. Basic functions, first introduced by Inglis [68], have the interesting property that they reproduce themselves on four successive differentiations. Basic functions may be written in the form

$$F(x) = A_n(\cosh \alpha_n x - \cos \alpha_n x) - (\sinh \alpha_n x - \sin \alpha_n x) \qquad (13-1)$$

Successive differentiations yield

$$\frac{\partial}{\partial x}F(x) = \alpha_n[A_n(\sinh \alpha_n x + \sin \alpha_n x) - (\cosh \alpha_n x - \cos \alpha_n x)]$$
$$= \alpha_n \Gamma(x) \tag{13-2}$$

$$\frac{\partial^2}{\partial x^2}F(x) = \alpha_n^2[A_n(\cosh \alpha_n x + \cos \alpha_n x) - (\sinh \alpha_n x + \sin \alpha_n x)]$$
$$= \alpha_n^2 \Phi(x) \tag{13-3}$$

$$\frac{\partial^3}{\partial x^3}F(x) = \alpha_n^3[A_n(\sinh \alpha_n x - \sin \alpha_n x) - (\cosh \alpha_n x + \cos \alpha_n x)]$$
$$= \alpha_n^3 \pi(x) \tag{13-4}$$

$$\frac{\partial^4}{\partial x^4}F(x) = \alpha_n^4[A_n(\cosh \alpha_n x - \cos \alpha_n x) - (\sinh \alpha_n x - \sin \alpha_n x)]$$
$$= \alpha_n^4 F(x) \tag{13-5}$$

The basic functions possess the property of orthogonality. Hence it is possible to express a given loading on the shell in basic function series.

13-4 Generation of Basic Functions

Appropriate basic functions for shells with different boundary conditions may be found by examining the modes of free vibration of beams of uniform section under the same boundary conditions. It is

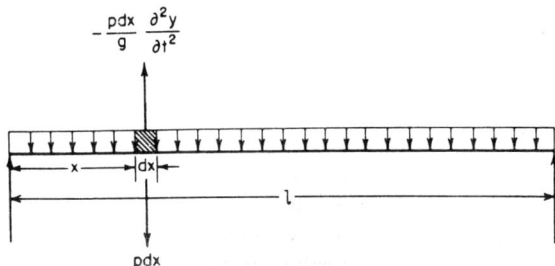

Fig. 13-1

well known that the governing differential equation of a beam of uniform section (Fig. 13-1) undergoing free vibration is given by

$$\frac{\partial^4 y}{\partial x^4} + \frac{p}{gEI}\frac{\partial^2 y}{\partial t^2} = 0 \tag{13-6}$$

where y = dynamic deflection at a point distant x from the origin
p = weight of the beam per unit length
t = time

The other symbols have their usual meanings. Equation (13-6) may be solved by expressing y as

$$y = X(x) T(t) \tag{13-7}$$

where X is a function of x only and T is a function of t only. Substitution of (13-7) in (13-6) leads to

$$X''''T + \frac{p}{gEI} XT^{\cdot\cdot} = 0 \tag{13-8}$$

where the primes and dots denote differentiation with respect to x and t, respectively. Relation (13-8) may now be rewritten as

$$\frac{X''''}{X} = -\frac{p}{gEI} \frac{T^{\cdot\cdot}}{T} = \alpha_n^4 \tag{13-9}$$

Hence the following two differential equations result:

$$X'''' - \alpha_n^4 X = 0 \tag{13-10}$$

$$T^{\cdot\cdot} + \alpha_n^4 \frac{gEI}{p} T = 0 \tag{13-11}$$

We are concerned with the solution of (13-10) only. This solution may be written as

$$X = \sum (C_1 \cos \alpha_n x + C_2 \sin \alpha_n x + C_3 \cosh \alpha_n x + C_4 \sinh \alpha_n x) \tag{13-12}$$

The constants C_1, C_2, C_3, and C_4 are evaluated from boundary conditions. Some of the common boundary conditions that have relevance to shell problems are discussed below.

Case I Simply Supported Ends

Let us consider a beam with simply supported ends. The four boundary conditions are

(i)	$X = 0$	at	$x = 0$
(ii)	$X'' = 0$	at	$x = 0$
(iii)	$X = 0$	at	$x = l$
(iv)	$X'' = 0$	at	$x = l$

Applying the first boundary condition, $C_1 = -C_3$. Applying the second boundary condition, $C_1 = C_3$. Hence $C_1 = C_3 = 0$. We may now write

$$X = C_2 \sin \alpha_n x + C_4 \sinh \alpha_n x$$

Imposing condition (iii),

$$C_2 \sin \alpha_n l + C_4 \sinh \alpha_n l = 0 \tag{13-13}$$

Now, making use of the last boundary condition,

$$-\alpha_n^2 C_2 \sin \alpha_n l + \alpha_n^2 C_4 \sinh \alpha_n l = 0 \tag{13-14}$$

Hence

$$C_4 = 0$$

Also

$$\sin \alpha_n l = 0 \tag{13-15}$$

From (13-15),

$$\alpha_n = \frac{n\pi}{l}$$

where $n = 1, 2, 3, \ldots, \infty$. Hence

$$X = \sum_{n=1}^{\infty} C_2 \sin \frac{n\pi x}{l} \tag{13-16}$$

is the required solution.

In this case, the appropriate basic function is the trigonometric function.

Case II   Both Ends Clamped

As before,

$$X = \sum (C_1 \cos \alpha_n x + C_2 \sin \alpha_n x + C_3 \cosh \alpha_n x + C_4 \sinh \alpha_n x)$$

The boundary conditions are

(i) $X = 0$ at $x = 0$
(ii) $X' = 0$ at $x = 0$
(iii) $X = 0$ at $x = l$
(iv) $X' = 0$ at $x = l$

Making use of the first two boundary conditions,

$$X = \sum [C_1(\cosh \alpha_n x - \cos \alpha_n x) + C_2(\sinh \alpha_n x - \sin \alpha_n x)] \tag{13-17}$$

If now we make use of the last two boundary conditions, the following two relations result:

$$\frac{C_1}{C_2} = -\frac{(\sinh \alpha_n l - \sin \alpha_n l)}{(\cosh \alpha_n l - \cos \alpha_n l)} \tag{13-18}$$

and

$$\begin{vmatrix} \cos \alpha_n l - \cosh \alpha_n l & \sin \alpha_n l - \sinh \alpha_n l \\ -\sin \alpha_n l - \sinh \alpha_n l & \cos \alpha_n l - \cosh \alpha_n l \end{vmatrix} = 0 \tag{13-19}$$

Substituting (13-18) in (13-17), we arrive at the appropriate basic function

$$X = \sum [A_n(\cosh \alpha_n x - \cos \alpha_n x) - (\sinh \alpha_n x - \sin \alpha_n x)] \tag{13-20}$$

where $A_n = -C_1/C_2$. This may be readily recognized as the standard form of the basic function given in (13-1). On expanding (13-19), we arrive at the relation

$$\cosh \alpha_n l \cos \alpha_n l = 1 \tag{13-21}$$

This transcendental equation is most easily solved by iteration. The first three values of $\alpha_n l$ and A_n are

$$\alpha_1 l = 4.730 \qquad A_1 = 1.018$$
$$\alpha_2 l = 7.854 \qquad A_2 = 0.999$$
$$\alpha_3 l = 10.996 \qquad A_3 = 1.000$$

Higher values of $\alpha_n l$ may be approximately found from the relation

$$\alpha_n l = \left(\frac{2n+1}{2}\right)\pi \tag{13-22}$$

Case III   One End Clamped and the Other End Propped

The relevant boundary conditions are

(i) $X = 0$ at $x = 0$
(ii) $X' = 0$ at $x = 0$
(iii) $X = 0$ at $x = l$
(iv) $X'' = 0$ at $x = l$

The first two boundary conditions are the same as for Case II. Hence

$$X = \sum [A_n(\cosh \alpha_n x - \cos \alpha_n x) - (\sinh \alpha_n x - \sin \alpha_n x)]$$

Imposing the third boundary condition, it is easily verified that A_n has the same value as for Case II. From the third and fourth boundary conditions we get

$$\frac{\sinh \alpha_n l - \sin \alpha_n l}{\cosh \alpha_n l - \cos \alpha_n l} = \frac{\sinh \alpha_n l + \sin \alpha_n l}{\cosh \alpha_n l + \cos \alpha_n l}$$

or

$$\tan \alpha_n l = \tanh \alpha_n l \tag{13-23}$$

The first three values of $\alpha_n l$ and A_n are given below

$$\alpha_1 l = 3.927 \qquad A_1 = 1.0$$
$$\alpha_2 l = 7.069 \qquad A_2 = 1.0$$
$$\alpha_3 l = 10.210 \qquad A_3 = 1.0$$

Higher values of $\alpha_n l$ may be approximately computed from the formula

$$\alpha_n l = \left(\frac{4n+1}{4}\right) \pi \tag{13-24}$$

13-5 Application of Basic Functions to Shell Problems

The boundary conditions of an interior shell of a group of longitudinally continuous shells approach those of a clamped shell. For such a shell, the appropriate basic function to be used is that found for Case II. Similarly, the boundary conditions of an end shell of a longitudinally continuous group approach those of a propped cantilever, and hence the basic function found for Case III would apply. Thus, if the shell is not simply supported on the traverses, the load is to be developed in the appropriate basic function series instead of trigonometric series. Hence expressions describing the longitudinal variation of the stress resultants and displacements would involve the basic functions and their derivatives.

13-6 Membrane Stresses and Displacements

By developing the load in the appropriate basic function series, the expressions given in Tables 13-1 and 13-2 for the membrane stresses and displacements are easily derived. The procedure involved is exactly the same as that described in Arts. 6-16 and 7-40 for Fourier loading. The only difference is that the basic function series takes the place of trigonometric series. The values given in Tables 13-1 and 13-2 are quoted[1] from the paper by Morice [65].

---

[1] Used by permission of Prof. P. B. Morice.

Table 13-1. Membrane Stresses

Stress resultant	Support conditions	First term	Second term	Third term
N_ϕ	Clamped ends	$-0.818ag$ $\cos(\phi_c - \phi) F_1(x)$	0	$-0.382ag$ $\cos(\phi_c - \phi) F_3(x)$
	Propped cantilever	$-0.846ag$ $\cos(\phi_c - \phi) F_1(x)$	$-0.829ag$ $\cos(\phi_c - \phi) F_2(x)$	$-0.325ag$ $\cos(\phi_c - \phi) F_3(x)$
$N_{x\phi}$	Clamped ends	$\dfrac{1.626}{\alpha_1} g$ $\sin(\phi_c - \phi) \pi_1(x)$	0	$\dfrac{0.764}{\alpha_3} g$ $\sin(\phi_c - \phi) \pi_3(x)$
	Propped cantilever	$\dfrac{1.692}{\alpha_1} g$ $\sin(\phi_c - \phi) \pi_1(x)$	$\dfrac{1.658}{\alpha_2} g$ $\sin(\phi_c - \phi) \pi_2(x)$	$\dfrac{0.650}{\alpha_3} g$ $\sin(\phi_c - \phi) \pi_3(x)$
N_x	Clamped ends	$\dfrac{1.626}{\alpha_1^2 a} g$ $\cos(\phi_c - \phi) \Phi_1(x)$	0	$\dfrac{0.764}{\alpha_3^2 a} g$ $\cos(\phi_c - \phi) \Phi_3(x)$
	Propped cantilever	$\dfrac{1.692}{\alpha_1^2 a} g$ $\cos(\phi_c - \phi) \Phi_1(x)$	$\dfrac{1.658}{\alpha_2^2 a} g$ $\cos(\phi_c - \phi) \Phi_2(x)$	$\dfrac{0.650}{\alpha_3^2 a} g$ $\cos(\phi_c - \phi) \Phi_3(x)$

13-7 Application of Morice's Method to a Long Shell Continuous over Two Equal Spans without Edge Beams

Referring to Fig. 13-2, it is obvious that because of symmetry, each of the shells will have boundary conditions corresponding to a propped cantilever. Consider one of the shells, say BC, which is clamped at B and propped at C. We shall choose the origin at B and measure x toward C. We shall assume that the shell is acted upon by its dead weight g only.

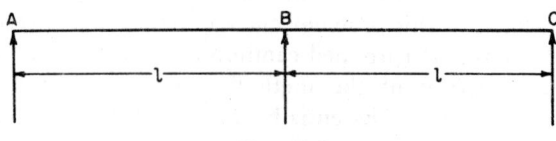

Fig. 13-2

Table 13-2. Membrane Displacements

Displacement	Support conditions	First term	Second term	Third term
u	Clamped	$\dfrac{1.626}{\alpha_1{}^3 E\,da}\,g$ $\cos(\phi_c - \phi)\,\Gamma_1(x)$	0	$\dfrac{0.764}{\alpha_3{}^3 E\,da}\,g$ $\cos(\phi_c - \phi)\,\Gamma_3(x)$
	Propped cantilever	$\dfrac{1.692}{\alpha_1{}^3 E\,da}\,g$ $\cos(\phi_c - \phi)\,\Gamma_1(x)$	$\dfrac{1.658}{\alpha_2{}^3 E\,da}\,g$ $\cos(\phi_c - \phi)\,\Gamma_2(x)$	$\dfrac{0.650}{\alpha_3{}^3 E\,da}\,g$ $\cos(\phi_c - \phi)\,\Gamma_3(x)$
v	Clamped	$-\dfrac{1.626}{\alpha_1{}^4 E\,da^2}\,g$ $\sin(\phi_c - \phi)\,F_1(x)$	0	$-\dfrac{0.764}{\alpha_3{}^4 E\,da^2}\,g$ $\sin(\phi_c - \phi)\,F_3(x)$
	Propped cantilever	$-\dfrac{1.692}{\alpha_1{}^4 E\,da^2}\,g$ $\sin(\phi_c - \phi)\,F_1(x)$	$-\dfrac{1.658}{\alpha_2{}^4 E\,da^2}\,g$ $\sin(\phi_c - \phi)\,F_2(x)$	$-\dfrac{0.650}{\alpha_3{}^4 E\,da^2}\,g$ $\sin(\phi_c - \phi)\,F_3(x)$
w	Clamped	$\dfrac{a^2}{Ed}\dfrac{1.626}{\alpha_1{}^4 a^4}\,g$ $\cos(\phi_c - \phi)\,F_1(x)$	0	$\dfrac{a^2}{Ed}\dfrac{0.764}{\alpha_3{}^4 a^4}\,g$ $\cos(\phi_c - \phi)\,F_3(x)$
	Propped cantilever	$\dfrac{a^2}{Ed}\dfrac{1.692}{\alpha_1{}^4 a^4}\,g$ $\cos(\phi_c - \phi)\,F_1(x)$	$\dfrac{a^2}{Ed}\dfrac{1.658}{\alpha_2{}^4 a^4}\,g$ $\cos(\phi_c - \phi)\,F_2(x)$	$\dfrac{a^2}{Ed}\dfrac{0.650}{\alpha_3{}^4 a^4}\,g$ $\cos(\phi_c - \phi)\,F_3(x)$
ϑ	Clamped	$\dfrac{a}{Ed}\dfrac{1.626}{\alpha_1{}^4 a^4}\,g$ $\sin(\phi_c - \phi)\,F_1(x)$	0	$\dfrac{a}{Ed}\dfrac{0.764}{\alpha_3{}^4 a^4}\,g$ $\sin(\phi_c - \phi)\,F_3(x)$
	Propped cantilever	$\dfrac{a}{Ed}\dfrac{1.692}{\alpha_1{}^4 a^4}\,g$ $\sin(\phi_c - \phi)\,F_1(x)$	$\dfrac{a}{Ed}\dfrac{1.658}{\alpha_2{}^4 a^4}\,g$ $\sin(\phi_c - \phi)\,F_2(x)$	$\dfrac{a}{Ed}\dfrac{0.650}{\alpha_3{}^4 a^4}\,g$ $\sin(\phi_c - \phi)\,F_3(x)$

Morice's approach is based on the Schorer equation with the load developed in terms of the appropriate basic function series instead of Fourier series. For this problem, the appropriate basic function is that developed for Case III (propped cantilever).

In the presentation of the method, we shall employ the matrix-progression technique.[1] Essentially, the process consists in replacing

[1] The reader not familiar with this technique is referred to Arts. 22-3 to 22-11.

the governing eighth-order partial differential equation by eight first-order ordinary differential equations after making the problem unidirectional. The problem is rendered unidirectional by selecting eight functions whose variation in the x direction is defined by the same basic function. This means that, for the time being, we are interested in studying the variation of the selected functions in the ϕ direction only. The eight first-order equations governing the problem may be obtained from the equations of equilibrium and the stress-strain relations of the Schorer theory, which have already been developed in Chapter 7. Let us start by defining the functions to be used as follows:

Let

$$f_1 = w$$

$$f_2 = \vartheta = \frac{1}{a} \frac{\partial w}{\partial \phi} = \frac{1}{a} f_1^{\cdot} \tag{13-25}$$

$$f_3 = M_\phi = -\frac{D}{a^2} w^{\cdot\cdot} = -\frac{D}{a} f_2^{\cdot} \tag{13-26}$$

$$f_4 = Q_\phi = \frac{1}{a} M_\phi^{\cdot} = \frac{1}{a} f_3^{\cdot} \tag{13-27}$$

$$f_5 = N_\phi = -Q_\phi^{\cdot} = -f_4^{\cdot} \tag{13-28}$$

$$f_6 = \frac{\partial N_{x\phi}}{\partial x} = -\frac{1}{a} N_\phi^{\cdot} = -\frac{1}{a} f_5^{\cdot} \tag{13-29}$$

$$f_7 = \int u \, dx \tag{13-30}$$

$$f_8 = v = -\frac{1}{a} f_7^{\cdot} \tag{13-31}$$

$$f_1 = f_8^{\cdot} \tag{13-32}$$

$\int u \, dx$ may be found as follows:

We have the equation of equilibrium

$$\frac{1}{a} \frac{\partial N_{x\phi}}{\partial \phi} + \frac{\partial N_x}{\partial x} = 0$$

Hence

$$\frac{\partial N_x}{\partial x} = -\frac{1}{a} \frac{\partial N_{x\phi}}{\partial \phi} = -\frac{1}{a} \frac{\partial}{\partial \phi} \int \frac{\partial N_{x\phi}}{\partial x} \, dx = -\frac{1}{a\alpha_1} f_6^{\cdot} \pi(x)$$

$$N_x = -\frac{1}{a\alpha_1^2} f_6^{\cdot} \Phi(x)$$

But

$$\frac{N_x}{Ed} = \frac{\partial u}{\partial x}$$

Hence

$$\frac{\partial u}{\partial x} = -\frac{1}{a\alpha_1{}^2 Ed} f_6 \cdot \Phi(x)$$

$$u = -\frac{1}{a\alpha_1{}^3 Ed} f_6 \cdot \Gamma(x)$$

$$\int u \, dx = -\frac{1}{a\alpha_1{}^4 Ed} f_6 \cdot F(x) = f_7 \tag{13-33}$$

Now it will be clear why $\int u \, dx$ was selected instead of $\partial u/\partial x$. It was chosen because its variation along the x direction is defined by the basic function itself and not by one of its derivatives. This is not true of $\partial u/\partial x$. Thus we have succeeded in finding eight quantities the variation of each of which in the x direction is defined by the same basic function. In other words, the problem has been rendered unidirectional. The resulting eight first-order ordinary differential equations may be presented in matrix form as

$$\frac{d}{d\phi}\begin{bmatrix} f_1 \\ f_2 \\ f_3 \\ f_4 \\ f_5 \\ f_6 \\ f_7 \\ f_8 \end{bmatrix} = \begin{bmatrix} 0 & a & 0 & 0 & 0 & 0 & 0 & 0 \\ 0 & 0 & -\dfrac{a}{D} & 0 & 0 & 0 & 0 & 0 \\ 0 & 0 & 0 & a & 0 & 0 & 0 & 0 \\ 0 & 0 & 0 & 0 & -1 & 0 & 0 & 0 \\ 0 & 0 & 0 & 0 & 0 & -a & 0 & 0 \\ 0 & 0 & 0 & 0 & 0 & 0 & -E \, da\alpha_1{}^4 & 0 \\ 0 & 0 & 0 & 0 & 0 & 0 & 0 & -a \\ 1 & 0 & 0 & 0 & 0 & 0 & 0 & 0 \end{bmatrix}\begin{bmatrix} f_1 \\ f_2 \\ f_3 \\ f_4 \\ f_5 \\ f_6 \\ f_7 \\ f_8 \end{bmatrix} \tag{13-34}$$

The eight differential equations comprising (13-34) may be condensed and written as

$$\left\{\frac{dF}{d\phi}\right\} = [A]\{F\} \tag{13-35}$$

The loading-solution matrix $\{P(\phi)\}$ may be constructed with the aid of the membrane stresses and displacements for a propped cantilever given in Tables 13-1 and 13-2. Hence we may write

$$\{P(\phi)\} = \begin{bmatrix} w \\ \vartheta \\ M_\phi \\ Q_\phi \\ N_\phi \\ \dfrac{\partial N_{x\phi}}{\partial x} \\ \int u\,dx \\ v \end{bmatrix} = \begin{bmatrix} +\dfrac{1.692}{E\,d\alpha_1{}^4 a^2}\,g\cos(\phi_c - \phi) \\ +\dfrac{1.692}{E\,d\alpha_1{}^4 a^3}\,g\sin(\phi_c - \phi) \\ 0 \\ 0 \\ -0.846ag\cos(\phi_c - \phi) \\ +1.692g\sin(\phi_c - \phi) \\ +\dfrac{1.692}{E\,d\alpha_1{}^4 a}\,g\cos(\phi_c - \phi) \\ -\dfrac{1.692}{E\,d\alpha_1{}^4 a^2}\,g\sin(\phi_c - \phi) \end{bmatrix}$$ (13-36)

We also need the value of $-\{P(0)\}$. This may be found from $\{P(\phi)\}$ by setting $\phi = 0$ and attaching a minus sign to the result. Thus

$$-\{P(0)\} = \begin{bmatrix} w \\ \vartheta \\ M_\phi \\ Q_\phi \\ N_\phi \\ \dfrac{\partial N_{x\phi}}{\partial x} \\ \int u\,dx \\ v \end{bmatrix} = \begin{bmatrix} -\dfrac{1.692}{E\,d\alpha_1{}^4 a^2}\,g\cos\phi_c \\ -\dfrac{1.692}{E\,d\alpha_1{}^4 a^3}\,g\sin\phi_c \\ 0 \\ 0 \\ +0.846ag\cos\phi_c \\ -1.692g\sin\phi_c \\ -\dfrac{1.692}{E\,d\alpha_1{}^4 a}\,g\cos\phi_c \\ +\dfrac{1.692}{E\,d\alpha_1{}^4 a^2}\,g\sin\phi_c \end{bmatrix}$$ (13-37)

Next we proceed to build up the initial-restraint matrix which expresses all the eight quantities $f_1, f_2, ..., f_8$ in terms of the four quantities $f_1, f_2, f_7,$ and f_8 at the origin $\phi = 0$ (Fig. 13-3).

$$\{F(0)\} = \begin{bmatrix} f_1 \\ f_2 \\ f_3 \\ f_4 \\ f_5 \\ f_6 \\ f_7 \\ f_8 \end{bmatrix} = \begin{bmatrix} 1 & 0 & 0 & 0 \\ 0 & 1 & 0 & 0 \\ 0 & 0 & 0 & 0 \\ 0 & 0 & 0 & 0 \\ 0 & 0 & 0 & 0 \\ 0 & 0 & 0 & 0 \\ 0 & 0 & 1 & 0 \\ 0 & 0 & 0 & 1 \end{bmatrix} \begin{bmatrix} f_1 \\ f_2 \\ f_7 \\ f_8 \end{bmatrix}_{\phi=0}$$ (13-38)

These relations follow from the known boundary conditions at the edge $\phi = 0$ which may be stated as

$$f_3 = 0 \qquad f_4 = 0 \qquad f_5 = 0 \qquad \text{and} \qquad f_6 = 0 \qquad (13\text{-}39)$$

Knowing the initial-restraint matrix and the loading-solution matrix, the value of $\{F(\phi)\}$ at any point ϕ may be written as

$$\{F(\phi)\} = e^{[A]\phi}[\{F(0)\} - \{P(0)\}] + \{P(\phi)\} \qquad (13\text{-}40)$$

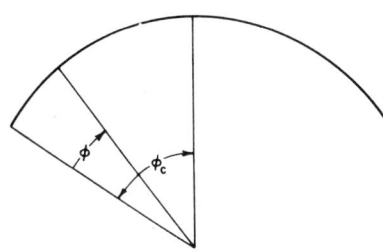

FIG. 13-3

With the aid of (13-40), we may now compute $\{F(\phi)\}$ at the line of symmetry ($\phi = \phi_c$). At this point we have four known boundary conditions:

$$f_2 = f_8 = f_6 = f_4 = 0 \qquad (13\text{-}41)$$

Making use of these conditions, the four initial unknowns f_1, f_2, f_7, and f_8 at the origin ($\phi = 0$) can now be found. The detailed application of the above procedure is described in Design Example 13-1.

Design Example 13-1

ANALYSIS OF A CYLINDRICAL SHELL WITHOUT EDGE BEAMS
CONTINUOUS OVER TWO SPANS
BY THE MATRIX-PROGRESSION METHOD

1. *DATA*

Geometry

Span $l = 75$ ft
(continuous over two spans of 75 ft each)
Radius $a = 25$ ft

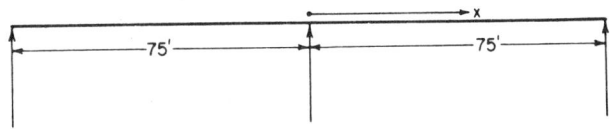

FIG. 13-4

Thickness $d = 0.25$ ft

Semicentral angle $\phi_c = 45°$

Loads

Dead weight = 37.5 psf

Live load   = 15.0 psf

Total load g = 52.5 psf of shell surface

2. TRIGONOMETRIC AND OTHER QUANTITIES

$$\sin \phi_c = 0.70711$$

$$\cos \phi_c = 0.70711$$

$$\alpha_1 = \frac{5}{4}\frac{\pi}{l} = 0.05236$$

3. MATRIX OF INITIAL RELATIONS

The initial-relations matrix equation (13-34) governing the shell actions, i.e., $d\{F\}/d\phi = [A]\{F\}$, is given below.

$$
\frac{d}{d\phi}
\begin{bmatrix}
w \\
\vartheta \\
M_\phi \\
Q_\phi \\
N_\phi \\
\dfrac{\partial N_{x\phi}}{\partial x} \\
\displaystyle\int u\,dx \\
v
\end{bmatrix}
=
\begin{bmatrix}
0 & +25.0 & 0 & 0 & 0 & 0 & 0 & 0 \\
0 & 0 & -0.00005333 & 0 & 0 & 0 & 0 & 0 \\
0 & 0 & 0 & +25.0 & 0 & 0 & 0 & 0 \\
0 & 0 & 0 & 0 & -1.0 & 0 & 0 & 0 \\
0 & 0 & 0 & 0 & 0 & -25.0 & 0 & 0 \\
0 & 0 & 0 & 0 & 0 & 0 & -16{,}911.302 & 0 \\
0 & 0 & 0 & 0 & 0 & 0 & 0 & -25.0 \\
+1.0 & 0 & 0 & 0 & 0 & 0 & 0 & 0
\end{bmatrix}
\begin{bmatrix}
w \\
\vartheta \\
M_\phi \\
Q_\phi \\
N_\phi \\
\dfrac{\partial N_{x\phi}}{\partial x} \\
\displaystyle\int u\,dx \\
v
\end{bmatrix}
$$

4. MEMBRANE ACTIONS AT THE ORIGIN

The membrane actions at the origin, i.e., at $\phi = 0$, are calculated from Equation (13-37) and are given below.

$$\{P(0)\} = \begin{bmatrix} w \\ \vartheta \\ M_\phi \\ Q_\varphi \\ N_\phi \\ \dfrac{\partial N_{x\phi}}{\partial x} \\ \int u \, dx \\ v \end{bmatrix} = \begin{bmatrix} +0.00014857 \\ +0.00000594 \\ 0 \\ 0 \\ -785.15368 \\ +62.812296 \\ +0.00371422 \\ -0.00014857 \end{bmatrix}$$

5. ACTION MATRIX AT THE ORIGIN IN TERMS OF THE RESTRAINT MATRIX

Equation (13-38) for the action matrix at the origin is now written down.

$$\begin{bmatrix} w \\ \vartheta \\ M_\phi \\ Q_\phi \\ N_\phi \\ \dfrac{\partial N_{x\phi}}{\partial x} \\ \int u \, dx \\ v \end{bmatrix} = \begin{bmatrix} 1 & 0 & 0 & 0 \\ 0 & 1 & 0 & 0 \\ 0 & 0 & 0 & 0 \\ 0 & 0 & 0 & 0 \\ 0 & 0 & 0 & 0 \\ 0 & 0 & 0 & 0 \\ 0 & 0 & 1 & 0 \\ 0 & 0 & 0 & 1 \end{bmatrix} \begin{bmatrix} w \\ \vartheta \\ \int u \, dx \\ v \end{bmatrix}$$

i.e.,

$$\{F(0)\} = \begin{bmatrix} w \\ \vartheta \\ 0 \\ 0 \\ 0 \\ 0 \\ \int u \, dx \\ v \end{bmatrix}$$

6. ACTION MATRIX AT ANY ANGLE ϕ

The matrix of shell actions at any angle ϕ from the origin is given by Equation (13-40), i.e.,

$$\{F(\phi)\} = e^{[A]\phi}[\{F(0)\} - \{P(0)\}] + \{P(\phi)\}$$

At the crown of the shell, the actions are given by

$$\{F(\phi_c)\} = e^{[A]\phi_c}[\{F(0)\} - \{P(0)\}] + \{P(\phi_c)\}$$

The calculations are given below.[1]

$$e^{[A]\phi_c} = \begin{bmatrix}
-0.26500337 & +16.875017 & -0.00039967 & -0.00267088 \\
-0.51535806 & -0.16500337 & -0.00003600 & -0.00039967 \\
+86,103.516 & +241,574.09 & -0.26500337 & +16.875017 \\
+26,296.510 & +86,103.516 & -0.51535806 & -0.26500337 \\
-167,145.23 & -657,412.76 & +4.5921875 & +12.883952 \\
+33,876.049 & +167,145.22 & -1.4024806 & -4.5921875 \\
-7.4938587 & -50.079009 & +0.00052713 & +0.00207329 \\
+0.67500070 & +7.4938587 & -0.00010684 & -0.00052713
\end{bmatrix}$$

$$\begin{bmatrix}
+0.00052713 & -0.00207329 & +4.5921875 & -12.883952 \\
+0.00010684 & -0.00052713 & +1.4024806 & -4.5921875 \\
-7.4938587 & +50.079009 & -167,145.23 & +657,412.76 \\
-0.67500070 & +7.4938587 & -33,876.049 & +167,145.23 \\
-0.26500337 & -16.875017 & +126,730.90 & -846,901.22 \\
+0.51535806 & -0.26500337 & -11,415.140 & +126,730.90 \\
-0.00027155 & +0.00076185 & -0.26500337 & -16.87502 \\
+0.00008293 & -0.00027155 & +0.51535806 & -0.26500337
\end{bmatrix}$$

$e^{[A]\phi_c}[\{F(0)\} - \{P(0)\}]$

$$= \begin{bmatrix}
\left(-0.26500337w + 16.875017\vartheta + 4.5921875\int u\,dx - 12.883952v + 0.52507267\right) \\
\left(-0.51535806w - 0.26500337\vartheta + 1.4024806\int u\,dx - 4.5921875v + 0.11117892\right) \\
\left(+86,103.516w + 241,574.09\vartheta - 167,145.23\int u\,dx + 657,412.76v - 8,325.1511\right) \\
\left(+26,296.510w + 86,103.516\vartheta - 33,876.049\int u\,dx + 167,145.23v - 854.44862\right) \\
\left(-167,145.23w - 657,412.76\vartheta + 126,730.90\int u\,dx - 846,901.22v + 284.10007\right) \\
\left(+33,876.049w + 167,145.23\vartheta - 11,415.140\int u\,dx + 126,730.90v + 476.48112\right) \\
\left(-7.4938587w - 50.079009\vartheta - 0.26500337\int u\,dx - 16.875017v - 0.26117063\right) \\
\left(+0.67500070w + 7.4938587\vartheta + 0.51535806\int u\,dx - 0.26500337v + 0.08007207\right)
\end{bmatrix}$$

[1] Comparison with a closed solution for $e^{[A]\phi}$ using eigenvalues shows that 21 terms are necessary for ensuring acceptable accuracy.

$$\{F(\phi_c)\} = e^{[A]\phi_c}[\{F(0)\} - \{P(0)\}] + \{P(\phi_c)\}$$

$$= \begin{bmatrix}
\left(-0.26500337w + 16.875017\vartheta + 4.5921875 \int u\,dx - 12.883952v + 0.52528278\right) \\[2ex]
\left(-0.51535806w - 0.265003370\vartheta + 1.4024806 \int u\,dx - 4.5921875v + 0.11117892\right) \\[2ex]
\left(+86,103.516w + 241,574.09\vartheta - 167,145.23 \int u\,dx + 657,412.76v - 8,325.1511\right) \\[2ex]
\left(+26,296.510w + 86,103.516\vartheta - 33,876.049 \int u\,dx + 167,145.23v - 854.44862\right) \\[2ex]
\left(-167,145.23w - 657,412.76\vartheta + 126,730.90 \int u\,dx - 846,901.22v - 826.27493\right) \\[2ex]
\left(+33,876.049w + 167,145.23\vartheta - 11,415.140 \int u\,dx + 126,730.90v + 476.48112\right) \\[2ex]
\left(-7.4938587w - 50.079009\vartheta - 0.26500337 \int u\,dx - 16.875017v - 0.25591793\right) \\[2ex]
\left(+0.67500070w + 7.4938587\vartheta + 0.51535806 \int u\,dx - 0.26500337v + 0.08007207\right)
\end{bmatrix}$$

In the above matrix, the values of $\{P(\phi_c)\}$ are obtained from Equation (13-36) by substituting $\phi = \phi_c$.

7. BOUNDARY CONDITIONS

$\{F(\phi_c)\}$ gives all the actions at the crown of the shell which is also the line of symmetry. The boundary conditions on this line are

$$\vartheta = 0$$
$$Q_\phi = 0$$
$$\frac{\partial N_{x\phi}}{\partial x} = 0$$
$$v = 0$$

The equations corresponding to the above boundary conditions are written below.

$$-0.51535806w - 0.26500337\vartheta + 1.4024806 \int u\,dx - 4.5921875v = -0.11117892$$

$$+26,296.510w + 86,103.516\vartheta - 33,876.049 \int u\,dx + 167,145.23v = 854.44862$$

$$+33,876.049w + 167,145.23\vartheta - 11,415.140 \int u\,dx + 126,730.90v = -476.48112$$

$$+0.67500070w + 7.4938587\vartheta + 0.51535806 \int u\,dx - 0.26500337v = -0.08007207$$

Solving the above four equations, the values of the four unknowns are obtained as follows:

$$w = +0.35745445$$

$$\vartheta = -0.03168388$$

$$\int u \, dx = -0.20176078$$

$$v = -0.07569537$$

8. STRESSES AT VARIOUS POINTS OF THE SHELL

Once the unknowns are evaluated, the actions at various points on the shell are calculated from

$$\{F(\phi)\} = e^{[A]\phi}[\{F(0)\} - \{P(0)\}] + \{P(\phi)\}$$

The effect of the variation along the span is considered by multiplying the above values by the basic function factors. It is to be noted here that w, ϑ, M_ϕ, Q_ϕ, N_ϕ, and v vary as the basic function (13-1), while N_x and $N_{x\phi}$ do not. N_x is obtained after differentiating $\int u \, dx$ twice with respect to x, and $N_{x\phi}$ after integrating $\partial N_{x\phi}/\partial x$ once with respect to x.

$$N_x = Ed\frac{du}{dx} = Ed\frac{\partial^2}{\partial x^2}\left(\int u \, dx\right)$$

$$= Ed\alpha_1{}^2[A_1(\cosh \alpha_1 x + \cos \alpha_1 x) - (\sinh \alpha_1 x + \sin \alpha_1 x)]\int u \, dx$$

$$N_{x\phi} = \int \frac{\partial N_{x\phi}}{\partial x} \, dx = \frac{1}{\alpha_1}[A_1(\sinh \alpha_1 x - \sin \alpha_1 x) - (\cosh \alpha_1 x + \cos \alpha_1 x)]\frac{\partial N_{x\phi}}{\partial x}$$

The longitudinal variation factors for N_x are worked out below for $x = 0$, $x = l/2$, and $x = l$. The factors for $N_{x\phi}$ and other actions are calculated in a similar manner.

(i) N_x

(a) $x = 0$

Factor $= Ed\alpha_1{}^2[A_1(\cosh \alpha_1 x + \cos \alpha_1 x) - (\sinh \alpha_1 x + \sin \alpha_1 x)]$

$$= 360 \times 10^6 \times 0.25 \times 0.05236^2[1(1 + 1) - 0]$$

$$= 246{,}740 \times 2 = 493{,}480$$

(b) $x = l/2$

Factor $= 246{,}740[1(3.6322765 - 0.38268354) - (3.4919096 + 0.92387947)]$

$$= 246{,}740 \, (-1.16619611)$$

$$= -287{,}747$$

(c) $x = l$

Factor $= 246{,}740 \, [1(25.386865 - 0.70710678) - (25.367163 - 0.70710678)]$

$$\approx 0$$

The stresses N_x, $N_{x\phi}$, N_ϕ, and M_ϕ are calculated for various values of ϕ at $x = 0$ (i.e., midspan), $x = l/2$, and $x = l$. The values are tabulated below.

At $x = 0$ (middle support)

	0°	10°	20°	30°	40°	45°
N_x, lb/ft	−99,565	+4,903	+26,262	+10,997	−5,889	−8,441
$N_{x\phi}$, lb/ft	0	−8,749	−4,078	+491	+866	0
N_ϕ, lb/ft	0	0	0	0	0	0
M_ϕ, lb ft/ft	0	0	0	0	0	0

At $x = l/2$

	0°	10°	20°	30°	40°	45°
N_x	+58,056	−2,859	−15,313	−6,412	+3,434	+4,922
$N_{x\phi}$	0	−2,982	−1,390	+167	+295	0
N_ϕ	0	−869	−1,854	−2,005	−1,790	−1,746
M_ϕ	0	−395	−1,131	−1,618	−1,781	−1,795

At $x = l$ (end support)*

	0°	10°	20°	30°	40°	45°
N_x	−981	+48	+259	+108	−58	−83
$N_{x\phi}$	0	+6,101	+2,843	−342	−604	0
N_ϕ	0	−12	−25	−27	−24	−24
M_ϕ	0	−5	−15	−22	−24	−24

* The values of N_x, N_ϕ, and M_ϕ should be zero at $x = l$. The small values obtained here are due to computational errors.

13-8 Application to a Long Shell with Edge Beams and Continuous over Two Equal Spans

The method described in Art. 13-7 can be extended to a shell with edge beams. All matrices involved, except the initial-restraint matrix, remain the same as before. The initial-restraint matrix applicable to this case may be built up by using the four compatibility relations available at the junction of the shell edge and the edge beam at $\phi = 0$. These conditions are

(i) $$M_\phi = 0 \quad \text{at} \quad \phi = 0 \quad\quad (13\text{-}42)$$

This condition is the same as that stated in Equation (8-8a) for a simply supported shell.

(ii) $$(N_\phi)_{m+b} \cos \phi_c + (Q_\phi)_{m+b} \sin \phi_c = 0 \quad \text{at} \quad \phi = 0 \qquad (13\text{-}43)$$

or

$$(N_\phi)_{m+b} = -(Q_\phi)_{m+b} \tan \phi_c \qquad (13\text{-}44)$$

This again is the same as the boundary condition for a simply supported shell stated in Equation (8-8b).

(iii) The third boundary condition is a statement of the equality of the vertical deflections of the shell edge and the edge beam all along their line of junction. This condition may be stated as

$$(v)_{m+b} \sin \phi_c - (w)_{m+b} \cos \phi_c = [(N_\phi)_{m+b} \sin \phi_c - (Q_\phi)_{m+b} \cos \phi_c] \frac{1}{\alpha_1^4 EI}$$
$$+ \left(\frac{\partial N_{x\phi}}{\partial x}\right)_{m+b} \frac{a_1}{\alpha_1^4 EI} - \frac{W'}{\alpha_1^4} \frac{1}{EI}$$

This condition is identical with the relation (8-8d) for simply supported shells, if (π/l) is replaced by α_1.

Making the substitution,

$$(N_\phi)_{m+b} = -(Q_\phi)_{m+b} \tan \phi_c$$

and simplifying,

$$(w)_{m+b} = (v)_{m+b} \tan \phi_c + (Q_\phi)_{m+b} \frac{1}{\alpha_1^4 EI} \frac{1}{\cos^2 \phi_c}$$
$$- \left(\frac{\partial N_{x\phi}}{\partial x}\right)_{m+b} \frac{a_1}{\alpha_1^4 EI \cos \phi_c} + \frac{W'}{\alpha_1^4} \frac{1}{EI} \frac{1}{\cos \phi_c} \qquad (13\text{-}45)$$

where W' is the first term of W expanded in basic function series.

(iv) The last boundary condition is a statement of the equality of longitudinal stresses all along the line of junction of the shell and the edge beam. This condition may be formulated as

$$E \frac{\partial u}{\partial x} = [(N_\phi)_{m+b} \sin \phi_c - (Q_\phi)_{m+b} \cos \phi_c] \frac{a_1}{\alpha_1^2 I}$$
$$+ \left(\frac{\partial N_{x\phi}}{\partial x}\right)_{m+b} \frac{1}{\alpha_1^2 A} + \left(\frac{\partial N_{x\phi}}{\partial x}\right)_{m+b} \frac{1}{\alpha_1^2} \frac{a_1^2}{I} - \frac{W'}{\alpha_1^2} \frac{a_1}{I} \qquad (13\text{-}46)$$

The variation of N_ϕ, $(\partial N_{x\phi}/\partial x)$, Q_ϕ, and W' in the x direction is defined by the basic function given in (13-1). Because we have integrated these functions twice, the function $\Phi_1(x)$ would appear on the right-hand side. Also the variation of $\partial u/\partial x$ appearing on the left-hand side is defined by the function $\Phi_1(x)$. But we desire that $F_1(x)$ should appear on both sides. This can be achieved by integrating both sides twice with respect to x. No integration constants would appear as $u = v = 0$ at $x = 0$. Hence the fourth boundary condition reduces itself to

$$\left(\int u\,dx\right)_{m+b} = -(Q_\phi)_{m+b}\frac{a_1}{\alpha_1{}^4 EI \cos\phi_c} + \left(\frac{\partial N_{x\phi}}{\partial x}\right)_{m+b}\frac{1}{\alpha_1{}^4 AE}$$
$$+ \left(\frac{\partial N_{x\phi}}{\partial x}\right)_{m+b}\frac{a_1{}^2}{\alpha_1{}^4 EI} - \frac{W'a_1}{\alpha_1{}^4 EI} \qquad (13\text{-}47)$$

With the help of the four boundary conditions formulated above, the initial- or boundary-restraint matrix for a shell with edge beam may be written down as

$$
\begin{bmatrix} f_1 \\ f_2 \\ f_3 \\ f_4 \\ f_5 \\ f_6 \\ f_7 \\ f_8 \end{bmatrix} =
\begin{bmatrix}
0 & \dfrac{1}{\alpha_1{}^4 EI \cos^2\phi_c} & -\dfrac{a_1}{\alpha_1{}^4 EI \cos\phi_c} & \tan\phi_c \\
1 & 0 & 0 & 0 \\
0 & 0 & 0 & 0 \\
0 & 1 & 0 & 0 \\
0 & -\tan\phi_c & 0 & 0 \\
0 & 0 & 1 & 0 \\
0 & -\dfrac{a_1}{\alpha_1{}^4 EI \cos\phi_c} & \left(\dfrac{a_1{}^2}{\alpha_1{}^4 EI} + \dfrac{1}{\alpha_1{}^4 AE}\right) & 0 \\
0 & 0 & 0 & 1
\end{bmatrix}
\begin{bmatrix} f_2 \\ f_4 \\ f_6 \\ f_8 \end{bmatrix}
$$

$$
+ \begin{bmatrix}
\dfrac{W'}{\alpha_1{}^4}\dfrac{1}{EI}\dfrac{1}{\cos\phi_c} \\
0 \\
0 \\
0 \\
0 \\
0 \\
-\dfrac{W'}{EI}\dfrac{a_1}{\alpha_1{}^4} \\
0
\end{bmatrix} \qquad (13\text{-}48)
$$

As before, $\{F(\phi)\}$ may now be written down at $\phi = \phi_c$ where four boundary conditions $f_2 = f_4 = f_6 = f_8 = 0$ are available. The method is shown applied to an actual problem in Design Example 13-2.

Design Example 13-2

ANALYSIS OF A CYLINDRICAL SHELL WITH EDGE BEAMS
AND CONTINUOUS OVER TWO SPANS
BY THE MATRIX-PROGRESSION METHOD

1. *DATA*

   Geometry
   Shell

Span $l = 70$ ft (continuous over two spans of 70 ft each)
Radius $a = 21$ ft
Thickness $d = 0.25$ ft
Semicentral angle $\phi_c = 35° = 0.610865$ radian

Edge Beam
Depth $2a_1 = 3.5$ ft
Width $2b_1 = 0.5$ ft

Loads

Dead weight $= 37.5$ psf
Superimposed load $= 17.5$ psf

Total load $g = 55.0$ psf of shell surface

2. TRIGONOMETRIC AND OTHER QUANTITIES

$$\sin \phi_c = 0.57358$$
$$\cos \phi_c = 0.81915$$
$$\alpha_1 = \frac{5}{4}\frac{\pi}{l} = 0.0560999$$
$$EI \text{ of edge beam} = 360 \times 10^6 \left(\frac{0.5 \times 3.5^3}{12}\right) = 6.43125 \times 10^8$$

3. MATRIX OF INITIAL RELATIONS

The initial-relations matrix equation (13-34) governing the shell actions, i.e., $(d/d\phi)\{F\} = [A]\{F\}$, is given below.

$$\frac{d}{d\phi}\begin{bmatrix} w \\ \vartheta \\ M_\phi \\ Q_\phi \\ N_\phi \\ \partial N_{x\phi} \\ \hline \partial x \\ \int u\, dx \\ v \end{bmatrix}$$

$$= \begin{bmatrix}
0 & +21.0 & 0 & 0 & 0 & 0 & 0 & 0 \\
0 & 0 & -0.00004480 & 0 & 0 & 0 & 0 & 0 \\
0 & 0 & 0 & +21.0 & 0 & 0 & 0 & 0 \\
0 & 0 & 0 & 0 & -1.0 & 0 & 0 & 0 \\
0 & 0 & 0 & 0 & 0 & -21.0 & 0 & 0 \\
0 & 0 & 0 & 0 & 0 & 0 & -18{,}720.145 & 0 \\
0 & 0 & 0 & 0 & 0 & 0 & 0 & -21.0 \\
+1.0 & 0 & 0 & 0 & 0 & 0 & 0 & 0
\end{bmatrix}\begin{bmatrix} w \\ \vartheta \\ M_\phi \\ Q_\phi \\ N_\phi \\ \dfrac{\partial N_{x\phi}}{\partial x} \\ \int u\, dx \\ v \end{bmatrix}$$

4. *MEMBRANE ACTIONS AT THE ORIGIN*

The membrane actions at the origin, i.e., at $\phi = 0$, are calculated from Equations (13-37) and are given below.

$$
\{P(0)\} =
\begin{bmatrix}
w \\
\vartheta \\
M_\phi \\
Q_\phi \\
N_\phi \\
\dfrac{\partial N_{x\phi}}{\partial x} \\
\displaystyle\int u\,dx \\
v
\end{bmatrix}
=
\begin{bmatrix}
+0.00019391 \\
+0.00000647 \\
0 \\
0 \\
-800.41803 \\
+53.377024 \\
+0.00407210 \\
-0.00013578
\end{bmatrix}
$$

5. *ACTION MATRIX AT THE ORIGIN IN TERMS OF THE RESTRAINT MATRIX*

Equation (13-48) for the action matrix at the origin is now written down.

$$
\begin{bmatrix}
w \\
\vartheta \\
M_\phi \\
Q_\phi \\
N_\phi \\
\dfrac{\partial N_{x\phi}}{\partial x} \\
\displaystyle\int u\,dx \\
v
\end{bmatrix}
=
\begin{bmatrix}
0 & +0.00023395 & -0.00033538 & +0.70020755 \\
+1.0 & 0 & 0 & 0 \\
0 & 0 & 0 & 0 \\
0 & +1.0 & 0 & 0 \\
0 & -0.70020755 & 0 & 0 \\
0 & 0 & +1.0 & 0 \\
0 & -0.00033538 & +0.00064102 & 0 \\
0 & 0 & 0 & +1.0
\end{bmatrix}
$$

$$
\times
\begin{bmatrix}
\vartheta \\
Q_\phi \\
\dfrac{\partial N_{x\phi}}{\partial x} \\
v
\end{bmatrix}
+
\begin{bmatrix}
+0.04255909 \\
0 \\
0 \\
0 \\
0 \\
0 \\
-0.06100914 \\
0
\end{bmatrix}
$$

Simplifying, we get

$$\{F(0)\} =$$

$$
\begin{bmatrix} w \\ \vartheta \\ M_\phi \\ Q_\phi \\ N_\phi \\ \dfrac{\partial N_{x\phi}}{\partial x} \\ \displaystyle\int u\,dx \\ v \end{bmatrix}_{\phi=0}
=
\begin{bmatrix}
\left(+0.00023395 Q_\phi \;\; -0.00033538\,\dfrac{\partial N_{x\phi}}{\partial x} \;\; +0.70020755v \;\; +0.04255909 \right) \\[4pt]
\vartheta \\[4pt]
0 \\[4pt]
Q_\phi \\[4pt]
-0.70020755 Q_\phi \\[4pt]
\dfrac{\partial N_{x\phi}}{\partial x} \\[4pt]
\left(-0.00033538 Q_\phi \;\; +0.00064102\,\dfrac{\partial N_{x\phi}}{\partial x} \;\; -0.06100914 \right) \\[4pt]
v
\end{bmatrix}
$$

·6. ACTION MATRIX AT ANY ANGLE ϕ

The matrix of shell actions at any angle ϕ from the origin is given by Equation (13-40), i.e.,

$$\{F(\phi)\} = e^{[A]\phi}[\{F(0)\} - \{P(0)\}] + \{P(\phi)\}$$

At the crown of the shell, the matrix of shell actions is obtained by putting $\phi = \phi_c$. The calculations are given below.[1]

$$
e^{[A]\phi_c} =
\begin{bmatrix}
+0.92156596 & +12.716374 & -0.00017523 & -0.00075023 \\
-0.04891333 & +0.92156596 & -0.00002713 & -0.00017523 \\
+12,511.108 & +22,928.122 & +0.92156596 & +12.716374 \\
+5,851.4996 & +12,511.108 & -0.04891330 & +0.92156596 \\
-47,890.505 & -122,881.49 & +0.56049763 & +1.0271799 \\
+14,928.185 & +47,890.505 & -0.26214718 & -0.56049763 \\
-3.9113124 & -16.746232 & +0.00011461 & +0.00029407 \\
+0.60554161 & +3.9113124 & -0.00003573 & -0.00011461 \\
\end{bmatrix}
$$

$$
\begin{bmatrix}
+0.00011461 & -0.00029407 & +0.56049763 & -1.0271799 \\
+0.00003573 & -0.00011461 & +0.26214718 & -0.56049763 \\
-3.9113124 & +16.746232 & -47,890.505 & +122,881.49 \\
-0.60554161 & +3.9113124 & -14,928.185 & +47,890.505 \\
+0.92156596 & -12.716374 & +73,220.336 & -313,491.89 \\
+0.04891333 & +0.92156596 & -11,335.827 & +73,220.336 \\
-0.00002994 & +0.00005487 & +0.92156596 & -12.716374 \\
+0.00001400 & -0.00002994 & +0.04891333 & +0.92156596 \\
\end{bmatrix}
$$

[1] Comparison with a closed solution for $e^{[A]\phi}$ using eigenvalues shows that 17 terms are necessary for ensuring acceptable accuracy.

$e^{[A]\phi_c}[\{F(0)\} - \{P(0)\}] =$

$$
\begin{bmatrix}
\left(+12.716374\vartheta & -0.00080286Q_\phi & -0.00024385\ \dfrac{\partial N_{x\phi}}{\partial x} & -0.38189244v & +0.10977449\right) \\[2ex]
\left(+0.92156596\vartheta & -0.00029960Q_\phi & +0.00006984\ \dfrac{\partial N_{x\phi}}{\partial x} & -0.59474711v & +0.01549750\right) \\[2ex]
\left(+22,928.122\vartheta & +34.443392Q_\phi & -18.148485\ \dfrac{\partial N_{x\phi}}{\partial x} & +131,641.86v & -361.20392\right) \\[2ex]
\left(+12,511.108\vartheta & +7.7210859Q_\phi & -7.6204098\ \dfrac{\partial N_{x\phi}}{\partial x} & +51,987.769v & +532.40557\right) \\[2ex]
\left(-122,881.49\vartheta & -35.378496Q_\phi & +50.280655\ \dfrac{\partial N_{x\phi}}{\partial x} & -347,025.18v & -5,419.5306\right) \\[2ex]
\left(+47,890.505\vartheta & +6.6994963Q_\phi & -11.351474\ \dfrac{\partial N_{x\phi}}{\partial x} & +83,673.164v & +1,369.7776\right) \\[2ex]
\left(-16.746232\vartheta & -0.00090909Q_\phi & +0.00195737\ \dfrac{\partial N_{x\phi}}{\partial x} & -15.455104v & -0.25419247\right) \\[2ex]
\left(+3.9113124\vartheta & +0.00000085Q_\phi & -0.00020167\ \dfrac{\partial N_{x\phi}}{\partial x} & +1.3455708v & +0.03537717\right)
\end{bmatrix}
$$

$\{F(\phi_c)\} = e^{[A]\phi_c}[\{F(0)\} - \{P(0)\}] + \{P(\phi_c)\} =$

$$
\begin{bmatrix}
\left(+12.716374\,\vartheta & -0.00080286Q_\phi & -0.00024385\ \dfrac{\partial N_{x\phi}}{\partial x} & -0.38189244v & +0.11001121\right) \\[2ex]
\left(+0.92156596\vartheta & -0.00029960Q_\phi & +0.00006984\ \dfrac{\partial N_{x\phi}}{\partial x} & -0.59474711v & +0.01549750\right) \\[2ex]
\left(+22,928.122\vartheta & +34.443392Q_\phi & -18.148485\ \dfrac{\partial N_{x\phi}}{\partial x} & +131,641.86v & -361.20392\right) \\[2ex]
\left(+12,511.108\vartheta & +7.7210859Q_\phi & -7.6204098\ \dfrac{\partial N_{x\phi}}{\partial x} & +51,987.769v & +532.40557\right) \\[2ex]
\left(-122,881.49\vartheta & -35.378496Q_\phi & +50.280655\ \dfrac{\partial N_{x\phi}}{\partial x} & -347,025.18v & -6,396.6606\right) \\[2ex]
\left(+47,890.505\vartheta & +6.6994963Q_\phi & -11.351474\ \dfrac{\partial N_{x\phi}}{\partial x} & +83,673.164v & +1,369.7776\right) \\[2ex]
\left(-16.746232\vartheta & -0.00090909Q_\phi & +0.00195737\ \dfrac{\partial N_{x\phi}}{\partial x} & -15.455104v & -0.24922136\right) \\[2ex]
\left(+3.9113124\vartheta & +0.00000085Q_\phi & -0.00020167\ \dfrac{\partial N_{x\phi}}{\partial x} & +1.3455708v & +0.03537717\right)
\end{bmatrix}
$$

In the above matrix, the values of $\{P(\phi_c)\}$ are obtained from Equation (13-36) by substituting $\phi = \phi_c$.

7. BOUNDARY CONDITIONS

The boundary conditions at the line of symmetry are

$$\vartheta = 0$$
$$Q_\phi = 0$$
$$\frac{\partial N_{x\phi}}{\partial x} = 0$$

and

$$v = 0$$

The equations resulting from the above boundary conditions are written below.

$$+0.92156596\vartheta - 0.00029960Q_\phi + 0.00006984\,\frac{\partial N_{x\phi}}{\partial x} - 0.59474711v = -0.01549750$$

$$+12,511.108\vartheta + 7.7210859Q_\phi - 7.6204098\,\frac{\partial N_{x\phi}}{\partial x} + 51,987.769v = -532.40557$$

$$+47,890.505\vartheta + 6.6994963Q_\phi - 11.351474\,\frac{\partial N_{x\phi}}{\partial x} + 83,673.164v = -1,369.7776$$

$$+3.9113124\vartheta + 0.00000085Q_\phi - 0.00020167\,\frac{\partial N_{x\phi}}{\partial x} + 1.3455708v = -0.03537717$$

Solving the above four equations, the values of the unknown ϑ, Q_ϕ, $\partial N_{x\phi}/\partial x$, and v at $\phi = 0$ are obtained.

$$\vartheta = -0.00072706$$
$$Q_\phi = +94.273232$$
$$\frac{\partial N_{x\phi}}{\partial x} = +51.717050$$
$$v = -0.01648649$$

8. STRESSES AT VARIOUS POINTS IN THE SHELL

When the unknowns have been evaluated, the actions at various points on the shell are calculated from

$$\{F(\phi)\} = e^{[A]\phi}[\{F(0)\} - \{P(0)\}] + \{P(\phi)\}$$

The variation of the actions along the span is considered as explained in step 8 of Design Example 13-1.

The stresses N_x, $N_{x\phi}$, N_ϕ, and M_ϕ at various points on the shell at $x = 0$, $x = l/2$, and $x = l$ are tabulated below.

At $x = 0$

	0°	10°	20°	30°	35°
N_x, lb/ft	−33,692	−5,750	+10,714	+17,997	+188,553
$N_{x\phi}$, lb/ft	−1,844	−5,687	−4,998	−1,909	0
N_ϕ, lb/ft	0	0	0	0	0
M_ϕ, lb ft/ft	0	0	0	0	0

At $x = l/2$

	0°	10°	20°	30°	35°
N_x	+19,646	+3,353	−6,248	−10,494	−10,993
$N_{x\phi}$	−628	−1,938	−1,703	−651	0
N_ϕ	−96	−604	−1,356	−1,845	−1,911
M_ϕ	0	+183	−53	−308	−347

At $x = l^*$

	0°	10°	20°	30°	35°
N_x	−332	−57	+106	+177	+186
$N_{x\phi}$	+1,286	+3,965	+3,485	−1,331	0
N_ϕ	−1	−8	−18	−25	−26
M_ϕ	0	+2	−1	−4	−5

* The values of N_x, N_ϕ, and M_ϕ should be zero. The small values obtained here are due to computational errors.

PART III

SHELLS OF DOUBLE CURVATURE

CHAPTER 14

SURFACES OF REVOLUTION

MEMBRANE THEORY

14-1 Equations of Equilibrium

In the membrane state, equilibrium of an element of the shell (Fig. 14-1) is maintained solely by the in-plane stress resultants N_ϕ, N_θ, $N_{\phi\theta}$, and $N_{\theta\phi}$. Bending moments and transverse shears are absent. Three equations of equilibrium may be written for the element in the directions of the meridian tangent, the tangent to the circle of latitude, and the normal to the element directed inward. The external forces Y, X, and Z per unit area of the element act along these directions.

Let us sum up the forces in the x direction (i.e., the direction of the tangent to the circle of latitude). The contribution of the N_θ forces is $(\partial N_\theta/\partial\theta)\, r_1\, d\theta\, d\phi$. The shear forces $N_{\theta\phi}$ acting on the two sides AB and CD (Fig. 14-2a) contribute a force of $N_{\theta\phi}\, r_1\, d\phi\, d\epsilon$. Observing that $r_0/PA = \cos\phi$ and $d\epsilon = r_0\, d\theta/PA$, $d\epsilon = \cos\phi\, d\theta$. Hence the contribution of the pair of shear forces $N_{\theta\phi}$ in the x direction may be simplified as $N_{\theta\phi}\, r_1 \cos\phi\, d\theta\, d\phi$. The pair of shear forces acting on the sides AD and CB contribute a force of $(\partial/\partial\phi)(N_{\phi\theta}\, r_0)\, d\theta\, d\phi$ in the x direction. We are now ready to write the equation of equilibrium in the x direction as

$$\frac{\partial N_\theta}{\partial\theta}\, r_1\, d\theta\, d\phi + N_{\theta\phi} r_1 \cos\phi\, d\theta\, d\phi + \frac{\partial}{\partial\phi}(N_{\phi\theta} r_0)\, d\theta\, d\phi + X r_1 r_0\, d\theta\, d\phi = 0$$

or

$$\frac{\partial N_\theta}{\partial\theta}\, r_1 + N_{\theta\phi} r_1 \cos\phi + \frac{\partial}{\partial\phi}(N_{\phi\theta} r_0) + X r_1 r_0 = 0 \qquad (14\text{-}1)$$

Next let us deal with the forces acting in the y direction (i.e., the direction of the meridian tangent). The N_ϕ stress resultants contribute a force of $(\partial/\partial\phi)(N_\phi r_0)\, d\theta\, d\phi$ (Fig. 14-2b). Referring to Fig. 14-2c, it is found that the N_θ forces contribute a force of $N_\theta r_1\, d\phi\, d\theta$ directed inward

353

along the radius of the circle of latitude.   The meridian tangent makes
an angle of ϕ with this direction (Fig. 14-2*b*).   Hence the component of
this force along the meridian tangent is $N_\theta \cos \phi \, r_1 \, d\theta \, d\phi$ and it acts
upward.   The shears $N_{\theta\phi}$ acting on the sides *AB* and *CD* provide a

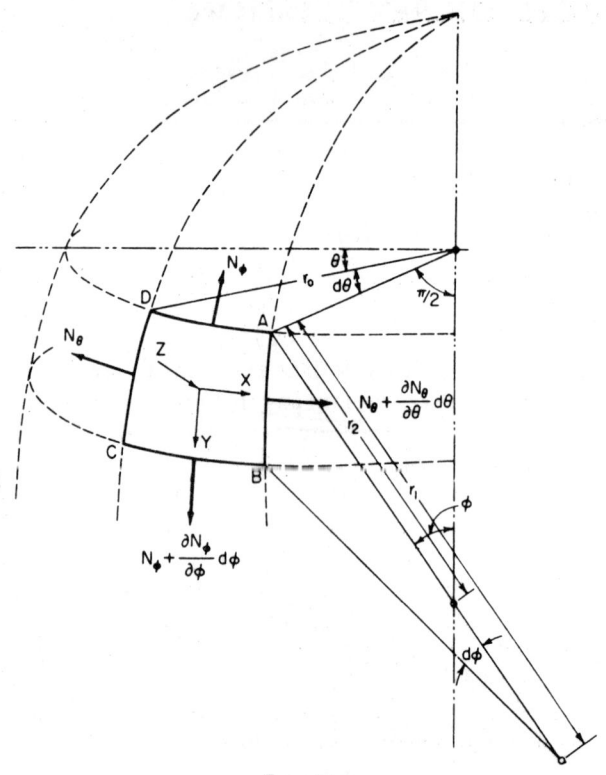

Fɪɢ. 14-1

resultant force of $(\partial N_{\theta\phi}/\partial\theta) \, r_1 \, d\phi \, d\theta$ acting downward in this direction.
We may now write the equation of equilibrium in the y direction as

$$\frac{\partial}{\partial\phi}(N_\phi r_0) \, d\theta \, d\phi - N_\theta \cos \phi \, r_1 \, d\theta \, d\phi + \frac{\partial N_{\theta\phi}}{\partial\theta} r_1 \, d\theta \, d\phi + Y r_1 r_0 \, d\theta \, d\phi = 0$$

or

$$\frac{\partial}{\partial\phi}(N_\phi r_0) - N_\theta r_1 \cos \phi + \frac{\partial N_{\theta\phi}}{\partial\theta} r_1 + Y r_1 r_0 = 0 \qquad (14\text{-}2)$$

Let us now proceed to enumerate the forces in the z direction (i.e.,
the direction of the inward normal).   The contribution of the N_ϕ stress

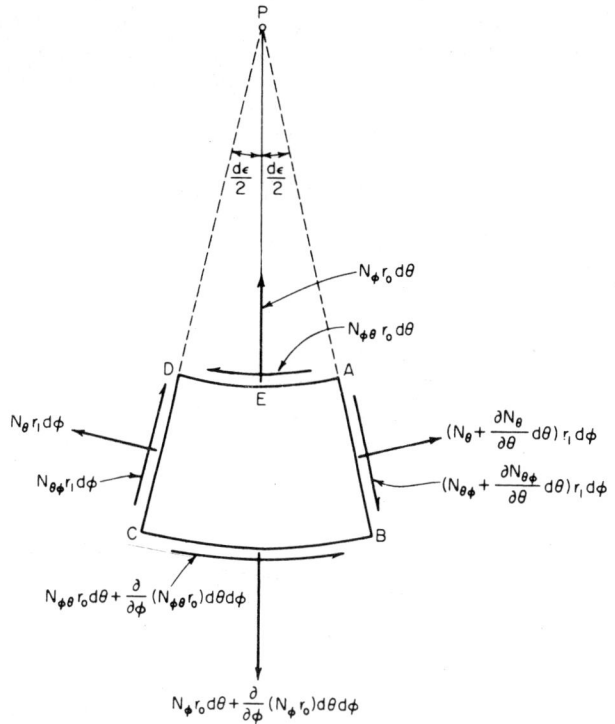

$N_\phi r_o d\theta$

$N_{\phi\theta} r_o d\theta$

D   A

E

$N_\theta r_1 d\phi$

$(N_\theta + \dfrac{\partial N_\theta}{\partial \theta} d\theta) r_1 d\phi$

$N_{\theta\phi} r_1 d\phi$

$(N_{\theta\phi} + \dfrac{\partial N_{\theta\phi}}{\partial \theta} d\theta) r_1 d\phi$

C   B

$N_{\phi\theta} r_o d\theta + \dfrac{\partial}{\partial \phi}(N_{\phi\theta} r_o) d\theta d\phi$

$N_\phi r_o d\theta + \dfrac{\partial}{\partial \phi}(N_\phi r_o) d\theta d\phi$

Fig. 14-2a

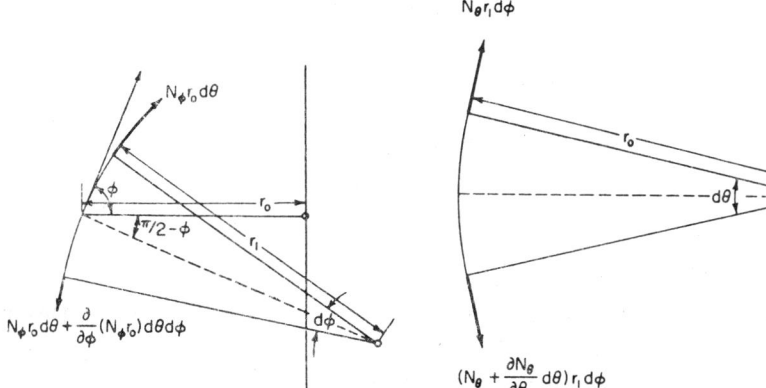

$N_\phi r_o d\theta$

ϕ

$\pi/2 - \phi$

r_o

r_1

$N_\phi r_o d\theta + \dfrac{\partial}{\partial \phi}(N_\phi r_o) d\theta d\phi$

$d\phi$

Fig. 14-2b

$N_\theta r_1 d\phi$

r_o

$d\theta$

$(N_\theta + \dfrac{\partial N_\theta}{\partial \theta} d\theta) r_1 d\phi$

Fig. 14-2c

resultants along the inward-directed normal is $N_\phi r_0 \, d\theta \, d\phi$ (Fig. 14-2b). We have already seen that the N_θ forces produce a resultant force of $N_\theta r_1 \, d\phi \, d\theta$ directed along the radius of the circle of latitude. The component of this force in the normal direction is

$$N_\theta r_1 \, d\phi \, d\theta \sin \phi$$

Hence the equation of equilibrium in the z direction takes the form

$$N_\phi r_0 \, d\theta \, d\phi + N_\theta r_1 \sin \phi \, d\theta \, d\phi + Z r_1 r_0 \, d\theta \, d\phi = 0$$

or

$$N_\phi r_0 + N_\theta r_1 \sin \phi + Z r_1 r_0 = 0 \tag{14-3}$$

14-2 Symmetrically Loaded Shells

Of frequent occurrence in practice are shells of revolution which are symmetrically loaded. In such cases $X = 0$; $N_{\theta\phi} = N_{\phi\theta} = 0$. If these values are inserted in Equations (14-1), (14-2), and (14-3), we arrive at the following equations of equilibrium corresponding to this specialized case:

$$N_\theta r_1 \cos \phi - \frac{d}{d\phi}(N_\phi r_0) = Y r_0 r_1 \tag{14-4}$$

$$\frac{N_\phi}{r_1} + \frac{N_\theta}{r_2} = -Z \tag{14-5}$$

Solving for N_θ from (14-5), we get

$$N_\theta = -r_2 \left(Z + \frac{N_\phi}{r_1} \right)$$

Substituting this value of N_θ in (14-4),

$$-r_2 r_1 \cos \phi \left(Z + \frac{N_\phi}{r_1} \right) - \frac{d}{d\phi}(N_\phi r_0) = Y r_0 r_1$$

Hence

$$-N_\phi r_2 \cos \phi - \frac{d}{d\phi}(N_\phi r_0) = (Y r_0 r_1 + Z r_1 r_2 \cos \phi)$$

Multiplying both sides by $\sin \phi$, this equation may be rewritten as

$$N_\phi r_2 \sin \phi \cos \phi + \frac{d}{d\phi}(N_\phi r_2 \sin \phi) \sin \phi = -r_1 r_2 (Y \sin \phi + Z \cos \phi) \sin \phi$$

or

$$\frac{d}{d\phi}(N_\phi r_2 \sin^2 \phi) = -r_1 r_2 (Y \sin \phi + Z \cos \phi) \sin \phi$$

Integrating both sides and solving for N_ϕ,

$$N_\phi = -\frac{1}{r_2 \sin^2 \phi} \left[\int r_1 r_2 (Y \sin \phi + Z \cos \phi) \sin \phi \, d\phi + C \right]$$

This expression becomes physically more meaningful if it is recast in the form

$$N_\phi = -\frac{1}{2\pi r_2 \sin^2 \phi} \left[\int 2\pi r_1 r_2 (Y \sin \phi + Z \cos \phi) \sin \phi \, d\phi + C \right] \quad (14\text{-}6)$$

A simple physical explanation of (14-6) is possible if we refer to Fig. 14-3. The term $2\pi r_1 r_2 \sin \phi \, d\phi$ stands for the surface area of an

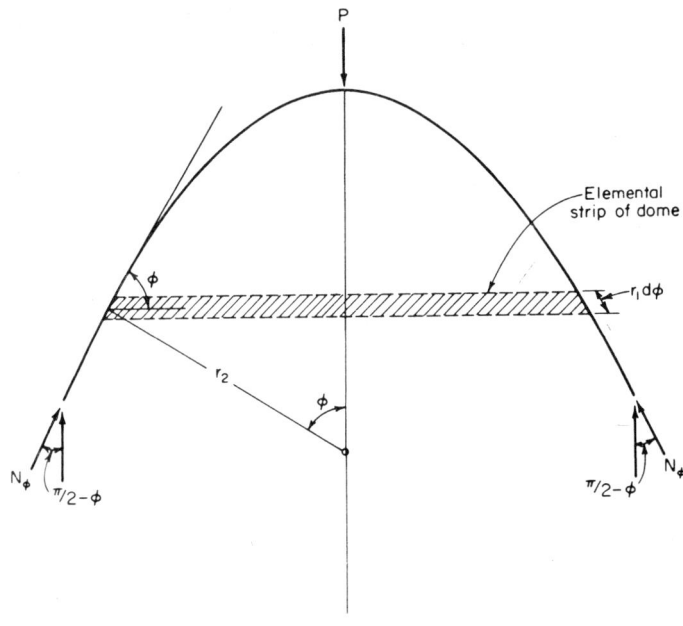

FIG. 14-3

elemental strip of the dome. $(Y \sin \phi + Z \cos \phi)$ stands for the vertical component of the forces per unit area that acts on this elemental strip. Hence $2\pi r_1 r_2 \sin \phi \, (Y \sin \phi + Z \cos \phi)$ stands for the vertical load acting on the strip. The integral $\int 2\pi r_1 r_2 \sin \phi \, (Y \sin \phi + Z \cos \phi) \, d\phi$ represents the vertical load acting on the dome up to the level where the meridional angle is ϕ. The vertical component of the thrust N_ϕ acting around the circle of latitude of ϕ resists this vertical force. Equation

(14-6) is nothing but a mathematical statement of this fact.   Hence we arrive at the following simple rule:

$$N_\phi = -\frac{W}{2\pi r_2 \sin^2 \phi} \tag{14-7}$$

where W is the total vertical load acting on the dome above the level denoted by ϕ.

The constant of integration C can be made use of to account for concentrated loads, if any, applied on the dome above this level.   Thus,

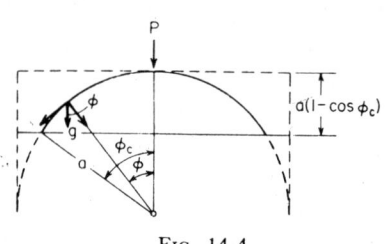

FIG. 14-4

for example, let there be a lantern load P acting at the crown of the shell where $\phi = 0$ (Fig. 14-4).   Obviously, this load also has to be reckoned in computing W.   Its contribution to N_ϕ is clearly seen to be $-(P/2\pi r_2 \sin^2 \phi)$.

Hence the value of C for a dome subjected to the action of a distributed load and a concentrated lantern load at its crown is $P/2\pi$.   If the lantern load is not present, $C = 0$.   We shall now apply these principles to a few specific cases.

THE SPHERICAL SHELL

14-3   Stresses under Own Weight

Let us, as a first example, consider a spherical shell acted upon by its own weight g per unit area of the surface.   It can be shown by integration that the surface area of the dome above the circle of latitude corresponding to ϕ is equal to the surface area of a cylinder whose radius is the same as that of the dome and whose height is equal to the rise of the dome up to ϕ (Fig. 14-4).   Hence the vertical load of the dome up to this level is $2\pi a^2 g(1 - \cos \phi)$.   From (14-7),

$$N_\phi = -\frac{2\pi a^2 g(1 - \cos \phi)}{2\pi a \sin^2 \phi}$$

noting that for a spherical dome $r_2 = a$.   Simplifying,

$$N_\phi = -\frac{ga}{1 + \cos \phi} \tag{14-8}$$

From Fig. 14-4, $Z = g \cos \phi$.   Hence, from (14-5),

$$N_\theta = -a\left(g \cos \phi + \frac{N_\phi}{a}\right) = -ag\left(\cos \phi - \frac{1}{1 + \cos \phi}\right) \tag{14-9}$$

From (14-5) it is evident that N_ϕ will remain compressive for all values of ϕ. The expression for N_θ would change from compression to tension passing through zero, corresponding to the value of ϕ satisfying the relation $\cos\phi + \cos^2\phi - 1 = 0$. This value of $\phi = 51°50'$. For values of ϕ in excess of this angle, N_θ will be tensile. The circle of latitude corresponding to $N_\theta = 0$ is known as the *plane of rupture*.

14-4 Stresses with Concentrated Load *P* at Crown

As already shown,

$$N_\phi = -\frac{P}{2\pi a \sin^2\phi} \tag{14-10}$$

$$\frac{N_\phi}{a} + \frac{N_\theta}{a} = 0 \qquad \text{because} \quad Z = 0$$

Hence

$$N_\theta = -N_\phi = \frac{P}{2\pi a \sin^2\phi} \tag{14-11}$$

14-5 Stresses under Snow Load

Let the intensity of the snow load be p_0 over the horizontal projection of the shell surface (Fig. 14-5). The total vertical load up to the circle of latitude corresponding to $\phi = p_0 \pi a^2 \sin^2\phi$.
Making use of (14-7),

$$N_\phi = -\frac{p_0 \pi a^2 \sin^2\phi}{2\pi a \sin^2\phi} = -\frac{p_0 a}{2} \tag{14-12}$$

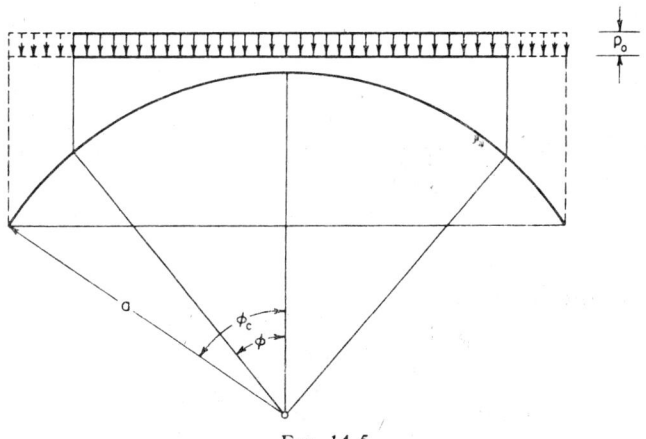

FIG. 14-5

From the relation,

$$N_\theta + N_\phi = -aZ = -p_0 a \cos^2 \phi$$

$$N_\theta = -\frac{p_0 a}{2}(2\cos^2\phi - 1) = -\frac{p_0 a}{2}\cos 2\phi$$

It is clear that for $\phi > \pi/4$, N_θ will turn out to be tensile.

ROTATIONAL HYPERBOLOID OF ONE SHEET

14-6 Stresses under Own Weight

The rotational hyperboloid of one sheet is commonly used for cooling towers of thermal power stations. Referring to Fig. 14-6a, its equation may be written as

$$\frac{r_0^2}{a^2} - \frac{z^2}{b^2} = 1 \qquad (14\text{-}13)$$

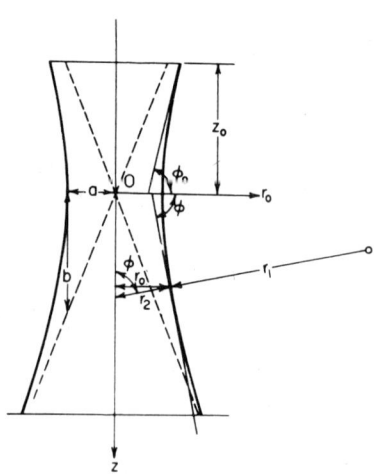

The two principal radii of curvature r_1 and r_2 are found by using formulas (5-42) and (5-43) developed in Chapter 5. Their values are

$$r_2 = a\left[1 + \frac{z^2}{a^2}(\alpha + \alpha^2)\right]^{1/2} \qquad (14\text{-}14)$$

where $\alpha = a^2/b^2$ and

$$r_1 = -a^2 b^2 \left(\frac{r_0^2}{a^4} + \frac{z^2}{b^4}\right)^{3/2} \qquad (14\text{-}15)$$

FIG. 14-6a   The two principal curvatures being , of opposite sign, it is clear that ιe surface is of negative Gauss curvature. The following relation bet ien r_1 and r_2 may be readily verified:

$$r_1 = -\frac{r_2^3}{\alpha a^2} \qquad (14\text{-}16)$$

We may rewrite (14-13) as

$$z = \pm\frac{b}{a}\sqrt{r_0^2 - a^2} \qquad (14\text{-}17)$$

$$\frac{dz}{dr_0} = \tan\phi = \pm\frac{b}{a}\sqrt{\frac{r_0^2}{r_0^2 - a^2}}$$

Hence

$$\tan^2 \phi = \frac{b^2}{a^2} \frac{r_0{}^2}{r_0{}^2 - a^2}$$

$$\cot^2 \phi = \frac{a^2}{b^2} \left(1 - \frac{a^2}{r_0{}^2}\right)$$

$$\frac{b^2}{a^2} \cot^2 \phi = 1 - \frac{a^2}{r_0{}^2}$$

Simplifying,

$$r_0 = \frac{a^2 \sin \phi}{(a^2 \sin^2 \phi - b^2 \cos^2 \phi)^{1/2}} \tag{14-18}$$

Similarly, we may show that

$$z = \frac{b^2 \cos \phi}{(a^2 \sin^2 \phi - b^2 \cos^2 \phi)^{1/2}} \tag{14-19}$$

$$r_2 = \frac{r_0}{\sin \phi} = \frac{a^2}{(a^2 \sin^2 \phi - b^2 \cos^2 \phi)^{1/2}} \tag{14-20}$$

Making use of (14-16),

$$r_1 = \frac{-a^2 b^2}{(a^2 \sin^2 \phi - b^2 \cos^2 \phi)^{3/2}} \tag{14-21}$$

The total vertical load W above the level ϕ may be found by integration. Thus

$$W = g \int 2\pi r_0 r_1 \, d\phi = 2\pi g a^4 b^2 \int_\phi^{\phi_0} \frac{\sin \phi \, d\phi}{(a^2 \sin^2 \phi - b^2 \cos^2 \phi)^2} \tag{14-22}$$

where ϕ and ϕ_0 correspond to z and z_0. To effect the integration of (14-22), let us make the substitution

$$\cos \phi = \frac{a}{\sqrt{a^2 + b^2}} \xi \tag{14-23}$$

Hence

$$\begin{aligned} W &= -\frac{2\pi g a b^2}{\sqrt{a^2 + b^2}} \int_\xi^{\xi_0} \frac{d\xi}{(1 - \xi^2)^2} \\ &= \frac{\pi g}{2} \frac{a b^2}{\sqrt{a^2 + b^2}} \left| \frac{2\xi}{1 - \xi^2} + \ln \frac{1 + \xi}{1 - \xi} \right|_{\xi_0}^{\xi} \end{aligned} \tag{14-24}$$

From (14-7),

$$N_\phi = -\frac{\pi g}{2} \frac{ab^2}{\sqrt{a^2 + b^2}} \frac{1}{2\pi r_0 \sin\phi} [f(\xi) - f(\xi_0)]$$ (14-25)

where

$$f(\xi) = \frac{2\xi}{1 - \xi^2} + \ln\frac{1 + \xi}{1 - \xi}$$

Simplifying,

$$N_\phi = -\frac{g}{4} b^2 \sqrt{a^2 + b^2} \frac{\sqrt{1 - \xi^2}}{(a^2 + b^2 - a^2\xi^2)} [f(\xi) - f(\xi_0)]$$ (14-26)

To find N_θ, we make use of the relation

$$\frac{N_\phi}{r_1} + \frac{N_\theta}{r_2} = -g\cos\phi$$

Simplifying,

$$N_\theta = -\frac{ga^2}{\sqrt{a^2 + b^2}} \frac{\xi}{\sqrt{1 - \xi^2}} + N_\phi \frac{a^2}{b^2}(1 - \xi^2)$$ (14-27)

The analysis of the rotational hyperboloid presented in this article is based on a paper by Soare[1] from which Table 14-1 for calculating $f(\xi)$ is extracted.

Table 14-1*

ξ	$f(\xi)$	ξ	$f(\xi)$	ξ	$f(\xi)$	ξ	$f(\xi)$
0.000	0.000	0.100	0.403	0.275	1.159	0.550	2.814
0.010	0.040	0.110	0.444	0.300	1.278	0.600	3.261
0.020	0.080	0.120	0.485	0.325	1.401	0.650	3.802
0.030	0.120	0.130	0.526	0.350	1.529	0.700	4.480
0.040	0.160	0.140	0.567	0.375	1.661	0.750	5.374
0.050	0.200	0.150	0.609	0.400	1.800	0.800	6.642
0.060	0.241	0.175	0.715	0.425	1.945	0.850	8.638
0.070	0.281	0.200	0.822	0.450	2.111	0.900	12.418
0.080	0.321	0.225	0.932	0.475	2.261	0.950	23.151
0.090	0.362	0.250	1.044	0.500	2.432	1.000	∞

*Used by permission of Mircea Soare.

[1] Soare, Mircea, "Teoria Suprafetelor Subtiri De Rotatie," Studii Şi Cercetări De Mecanică Şi Metalurgie, Tomul III, Nr. 3–4, 1952 (in Rumanian).

Design Example 14-1

ANALYSIS OF A ROTATIONAL HYPERBOLOID

1. *DATA*

Geometry

$a = 30$ ft
$b = 70$ ft
z at base $= +100$ ft
z at top $= z_0 = -30$ ft

Load

Dead weight $g = 150$ lb/ft²

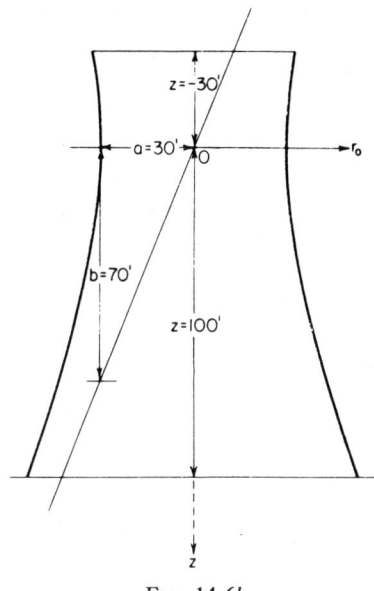

FIG. 14-6*b*

2. *GEOMETRICAL PROPERTIES AND DESIGN FACTORS*
   (i) *At the base*

$$r_0 = a\sqrt{1 + \frac{z^2}{b^2}}$$

$$= 30\sqrt{1 + \frac{100^2}{70^2}} = 30 \times 1.7438 = 52.314 \text{ ft}$$

$$\tan\phi = \frac{b}{a}\sqrt{\frac{r_0^2}{r_0^2 - a^2}}$$

$$= \frac{70}{30}\sqrt{\frac{52.314^2}{52.314^2 - 30^2}} = 2.8482$$

$$\cos\phi = \frac{1}{\sqrt{1 + 2.8482^2}} = 0.33127$$

From Equation (14-23),

$$\cos \phi = \frac{a}{\sqrt{a^2 + b^2}} \, \xi$$

$$\therefore \, \xi = \frac{\cos \phi \, \sqrt{a^2 + b^2}}{a} = \frac{0.33127 \, \sqrt{30^2 + 70^2}}{30} = 0.8409$$

From Table 14-1,

$$f(\xi) = 8.276$$

(ii) *At top*

$$r_0 = 30 \sqrt{1 + \frac{(-30)^2}{70^2}} = 32.6391 \text{ ft}$$

$$\tan \phi = \frac{70}{30} \sqrt{\frac{32.6391^2}{32.6391^2 - 30^2}} = 5.9233$$

$$\cos \phi = \frac{1}{\sqrt{1 + 5.9233^2}} = 0.16647$$

$$\xi_0 = \frac{0.16647 \, \sqrt{30^2 + 70^2}}{30} = 0.4226$$

From Table 14-1,

$$f(\xi_0) = 1.931$$

3. CALCULATION OF STRESSES

(i) *At the base*

Making use of Equation (14-26),

$$N_\phi = -\frac{g}{4} b^2 \sqrt{a^2 + b^2} \frac{\sqrt{1 - \xi^2}}{a^2 + b^2 - a^2 \xi^2} \{f(\xi) - f(\xi_0)\}$$

$$= -\frac{150}{4} \times 70^2 \sqrt{30^2 + 70^2} \frac{\sqrt{1 - 0.8409^2}}{[30^2 + 70^2 - (30^2 \times 0.8409^2)]} \times (8.276 - 1.931)$$

$$= -9,305.387 \text{ lb/ft}$$

$$= -9.305 \text{ kip/ft}$$

Considering Equation (14-27),

$$N_\theta = -\frac{ga^2}{\sqrt{a^2 + b^2}} \frac{\xi}{\sqrt{1 - \xi^2}} + N_\phi \frac{a^2}{b^2} (1 - \xi^2)$$

$$= -\frac{150 \times 30^2}{\sqrt{30^2 + 70^2}} \frac{0.8409}{\sqrt{1 - 0.8409^2}} + \left[-9.305 \frac{30^2}{70^2} (1 - 0.8409^2) \right]$$

$$= -2,754.7 - 500.48 = -3,255.18 \text{ lb/ft}$$

$$= -3.255 \text{ kip/ft}$$

(ii) *At top*

Making use of Equation (14-26),

$$N_\phi = 0 \qquad \text{because} \quad f(\xi_0) - f(\xi_0) = 0$$

Now, by substituting the value of N_ϕ in Equation (14-27), it takes the form

$$N_\theta = -\frac{ga^2}{\sqrt{a^2 + b^2}} \frac{\xi_0}{\sqrt{1 - \xi_0^2}}$$

$$= -\frac{150 \times 30^2}{\sqrt{30^2 + 70^2}} \frac{0.4226}{\sqrt{1 - 0.4226^2}} = -826.548 \text{ lb/ft}$$

$$= -0.827 \text{ kip/ft}$$

4. CHECK FOR STATICS

$$\frac{N_\phi}{r_1} + \frac{N_\theta}{r_2} \quad \text{shall be equal to} \quad -g \cos \phi$$

(i) *At the base*

$$r_2 = \frac{r_0}{\sin \phi}$$

where

$$\sin \phi = \tan \phi \cos \phi = 2.84819 \times 0.33127 = 0.94352$$

Hence

$$r_2 = \frac{r_0}{\sin \phi} = \frac{52.314}{0.94352} = 55.4454$$

$$r_1 = -\frac{r_2^3}{\alpha a^2} \quad \text{where} \quad \alpha = \frac{a^2}{b^2}$$

$$= -\frac{55.4452^3 \times 70^2}{30^4} = -1,031.1168 \text{ ft}$$

Therefore

$$\frac{N_\phi}{r_1} + \frac{N_\theta}{r_2} = \frac{-9,305,387}{-1,031.1168} + \frac{-3,255,18}{55.445}$$

$$= -49.70$$

and

$$-g \cos \phi = -150 \times 0.33127$$

$$= -49.69$$

These two values agree very closely.

(ii) *At top*

$$r_2 = \frac{r_0}{\sin \phi}$$

where

$$\sin \phi = \tan \phi \cos \phi = 5.92331 \times 0.16647$$

$$= 0.98605$$

Hence

$$r_2 = \frac{r_0}{\sin \phi} = \frac{32.6391}{0.98605} = 33.101 \text{ ft}$$

$$r_1 = -\frac{r_2^3}{\alpha a^2} \qquad \left(\text{where } \alpha = \frac{a^2}{b^2}\right)$$

$$= \frac{33.101^3 \times 70^2}{30^4} = -219.398 \text{ ft}$$

Therefore

$$\frac{N_\phi}{r_1} + \frac{N_\theta}{r_2} = 0 + \frac{-826.548}{33.101} = -24.97$$

and

$$-g \cos \phi = -150 \times 0.16647 = -24.97$$

These two values are equal.

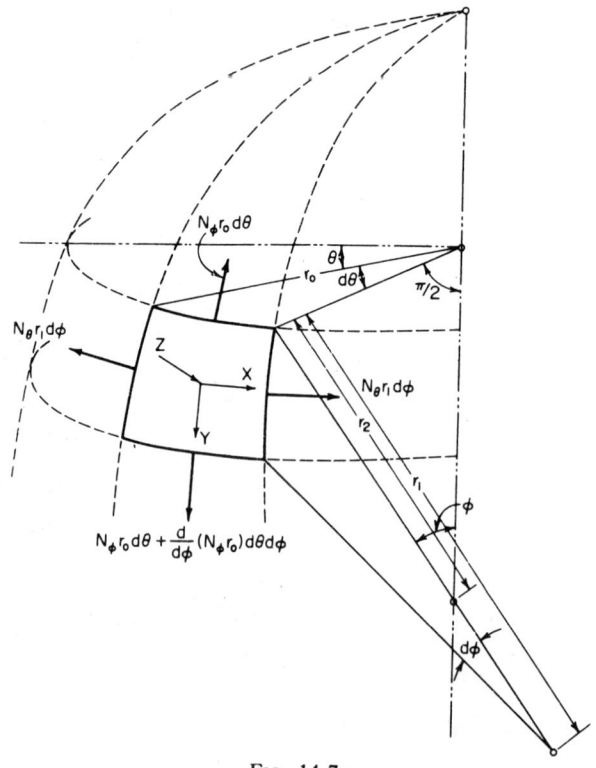

Fig. 14-7a

APPROXIMATE BENDING THEORY OF SHELLS OF REVOLUTION WITH AXISYMMETRIC LOADING

14-7 Equations of Equilibrium

Referring to Figs. 14-7a and b, the following equations of equilibrium are easily established:

$$\frac{d}{d\phi}(N_\phi r_0) - N_\theta r_1 \cos \phi - r_0 Q_\phi + Y r_0 r_1 = 0 \tag{14-28}$$

$$N_\phi r_0 + N_\theta r_1 \sin \phi + \frac{d}{d\phi}(Q_\phi r_0) + Z r_1 r_0 = 0 \tag{14-29}$$

$$\frac{d}{d\phi}(M_\phi r_0) - M_\theta r_1 \cos \phi - Q_\phi r_0 r_1 = 0 \tag{14-30}$$

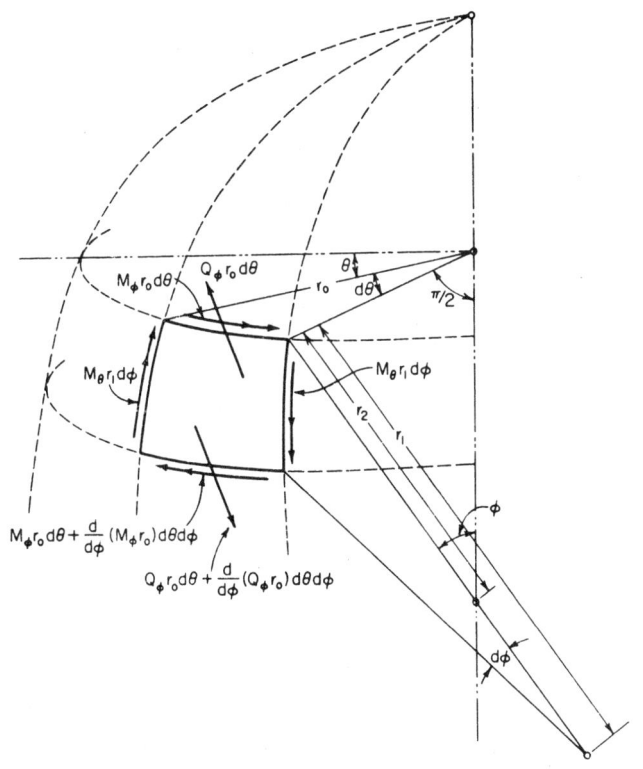

· Fig. 14-7b

14-8 Expressions for Strains, Change of Slope of the Meridian Tangent, and Curvatures

On an analogy with expression (7-5) developed for cylindrical shells,

$$\epsilon_\phi = \frac{1}{r_1} \frac{dv}{d\phi} - \frac{w}{r_1} \tag{14-31}$$

where v is the displacement of a point on the middle surface along the meridian tangent and w its displacement along the normal directed inward. In Fig. 14-8 the displacement of a point P on the middle surface to its new position P' after deformation is sketched. The change suffered by r_0 is

$$\Delta r_0 = v \cos \phi - w \sin \phi \tag{14-32}$$

Hence

$$\text{Strain} = \frac{\Delta r_0}{r_0} = \frac{v \cos \phi}{r_0} - \frac{w \sin \phi}{r_0}$$

Noting that $r_0 = r_2 \sin \phi$, the strain

$$\frac{\Delta r_0}{r_0} = \frac{v}{r_2} \cot \phi - \frac{w}{r_2}$$

Hence the circumferential strain is

$$\epsilon_\theta = \frac{2\pi(r_0 + \Delta r_0) - 2\pi r_0}{2\pi r_0} = \frac{\Delta r_0}{r_0} = \frac{v}{r_2} \cot \phi - \frac{w}{r_2} \tag{14-33}$$

The change in slope ϑ of the meridian tangent may be found in the same way as the rotation of the tangent of a cylindrical shell for which expression (7-7) was developed in Chapter 7. We may thus write

$$\vartheta = \frac{1}{r_1} \left(v + \frac{dw}{d\phi} \right) \tag{14-34}$$

The change in curvature of the meridian curve χ_ϕ may be found as $(1/r_1 \, d\phi)(d\vartheta/d\phi) \, d\phi$. Hence

$$\chi_\phi = \frac{1}{r_1} \frac{d\vartheta}{d\phi} \tag{14-35}$$

The change of curvature in the other principal direction may be found as $(1/r_2' - 1/r_2)$, where $1/r_2'$ is the curvature after deformation. Thus

$$\chi_\theta = \frac{1}{r_2'} - \frac{1}{r_2} = \frac{\sin(\phi + \vartheta)}{r_0 + \Delta r_0} - \frac{\sin \phi}{r_0}$$

$$\approx \frac{\sin \phi + \vartheta \cos \phi}{r_0} - \frac{\sin \phi}{r_0} \approx \frac{\vartheta \cos \phi}{r_0} \approx \frac{\vartheta}{r_2} \cot \phi \tag{14-36}$$

We shall next proceed to find an expression for ϑ in terms of ϵ_ϕ and ϵ_θ.
From Fig. 14-9,

$$\cos\phi = \frac{1}{r_1}\frac{dr_0}{d\phi} \tag{14-37}$$

After deformation, ϕ becomes $(\phi + \vartheta)$, r_1 becomes $r_1(1 + \epsilon_\phi)$, and r_0
changes to $r_0(1 + \epsilon_\theta)$. Hence

$$\cos(\phi + \vartheta) = \frac{1}{r_1(1 + \epsilon_\phi)}\frac{d}{d\phi}r_0(1 + \epsilon_\theta)$$

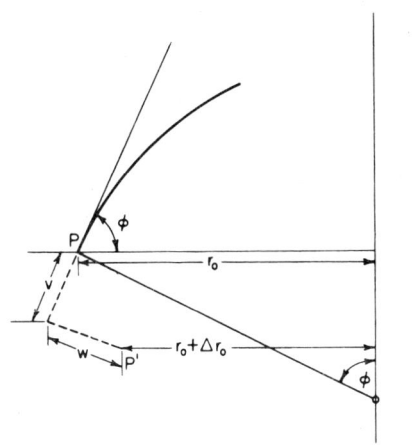

Fig. 14-8   Fig. 14-9

Confining ourselves to small values of ϑ,

$$-\vartheta\sin\phi + \cos\phi(1 + \epsilon_\phi) = \frac{r_0}{r_1}\frac{d\epsilon_\theta}{d\phi} + \frac{1}{r_1}\frac{dr_0}{d\phi} + \frac{\epsilon_\theta}{r_1}\frac{dr_0}{d\phi}$$

or

$$-\vartheta\sin\phi + \epsilon_\phi\cos\phi = \frac{r_0}{r_1}\frac{d\epsilon_\theta}{d\phi} + \frac{\epsilon_\theta}{r_1}\frac{dr_0}{d\phi} = \frac{r_2}{r_1}\sin\phi\frac{d\epsilon_\theta}{d\phi} + \epsilon_\theta\cos\phi$$

Simplifying,

$$\vartheta = (\epsilon_\phi - \epsilon_\theta)\cot\phi - \frac{r_2}{r_1}\frac{d\epsilon_\theta}{d\phi} \tag{14-38}$$

14-9 Stress-Strain and Moment-Curvature Relations

Making use of the expressions for strain and curvature derived in the
last article, we may write down the following stress-strain and moment-
curvature relations:

$$\epsilon_\phi = \frac{1}{Ed}(N_\phi - \nu N_\theta) \tag{14-39}$$

$$\epsilon_\theta = \frac{1}{Ed} \left(N_\theta - \nu N_\phi \right) \tag{14-40}$$

$$M_\phi = -D \left(\frac{1}{r_1} \frac{d\vartheta}{d\phi} + \nu \frac{\vartheta}{r_2} \cot \phi \right) \tag{14-41}$$

and

$$M_\theta = -D \left(\frac{\vartheta}{r_2} \cot \phi + \nu \frac{1}{r_1} \frac{d\vartheta}{d\phi} \right) \tag{14-42}$$

14-10 The Geckeler Approximation

In shells of revolution which are symmetrically loaded the bending stresses are generally confined to a narrow region close to the boundaries; elsewhere, in the interior of the shells, the membrane state of stress remains undisturbed. Moreover, the bending stresses get rapidly damped out as we move away from the boundaries. If the angle $\phi > 30°$, it is sufficiently accurate to superimpose an edge effect on the membrane stresses. It is not necessary to make an elaborate bending analysis. An approximate theory for dealing with the edge effect is that generally attributed to Geckeler [69]. The derivation that follows is based on the development given in a book by Kolykunov [70]. We have already noted that the edge effect is in the nature of a damped wave with a large damping coefficient. Hence successive differentiation of the functions representing this wave results in progressively increasing quantities, each successive result being the damping factor times the immediately preceding result. Hence, in expressions for stresses and deformations where variables and their derivatives occur together, we need retain only the highest derivative and drop the rest. On this basis, we may make use of Equation (14-30) to write

$$Q_\phi = \frac{1}{r_1} \frac{dM_\phi}{d\phi} \tag{14-43}$$

But from (14-41),

$$M_\phi \approx -\frac{D}{r_1} \frac{d\vartheta}{d\phi} \tag{14-44}$$

Or, from (14-43) and (14-44),

$$Q_\phi = -\frac{D}{r_1^2} \frac{d^2\vartheta}{d\phi^2} \tag{14-45}$$

We are now dealing with the unloaded shell. Hence $Y = 0$ and $Z = 0$. Referring to Fig. 14-10, it is easily verified that

$$2\pi r_0 Q_\phi \cos\phi + 2\pi r_0 N_\phi \sin\phi = 0$$

or

$$N_\phi = -Q_\phi \cot\phi$$

Let us denote

$$Q_\phi r_2 = U$$

so that

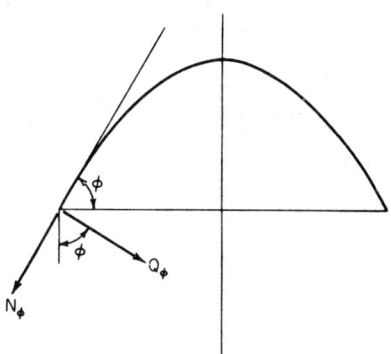

$$N_\phi = -\frac{U}{r_2}\cot\phi \qquad (14\text{-}46)$$

Fig. 14-10

Making use of (14-29) and noting that $Z = 0$,

$$r_1 N_\theta \sin\phi = -N_\phi r_0 - \frac{d}{d\phi}(Q_\phi r_0) = Q_\phi \cot\phi \, r_2 \sin\phi - \frac{d}{d\phi}(Q_\phi r_2 \sin\phi)$$

$$= Q_\phi r_2 \cos\phi - \frac{d}{d\phi}(Q_\phi r_2 \sin\phi) = U\cos\phi - \frac{d}{d\phi}(U\sin\phi)$$

Hence

$$N_\theta = -\frac{1}{r_1}\frac{dU}{d\phi} \qquad (14\text{-}47)$$

From Equation (14-40),

$$\epsilon_\theta = \frac{1}{Ed}(N_\theta - \nu N_\phi) = -\frac{1}{Ed}\left(\frac{1}{r_1}\frac{dU}{d\phi} + \frac{\nu}{r_2}U\cot\phi\right)$$

$$\approx -\frac{1}{Ed}\frac{1}{r_1}\frac{dU}{d\phi} \qquad (14\text{-}48)$$

Retaining only the highest derivative, (14-38) may be rewritten as

$$\vartheta \approx -\frac{r_2}{r_1}\frac{d\epsilon_\theta}{d\phi} \qquad (14\text{-}49)$$

Substituting for the value of U in (14-48) from (14-45), we get

$$\epsilon_\theta \approx \frac{D}{Ed}\frac{r_2}{r_1{}^3}\frac{d^3\vartheta}{d\phi^3} \qquad (14\text{-}50)$$

Substituting for ϵ_θ from (14-50), (14-49) may now be rewritten as

$$\vartheta = -\frac{D}{Ed}\frac{r_2{}^2}{r_1{}^4}\frac{d^4\vartheta}{d\phi^4}$$

or

$$\frac{r_2{}^4}{r_1{}^4}\frac{d^4\vartheta}{d\phi^4} + \frac{Ed}{D}r_2{}^2\vartheta = 0$$

Noting that

$$D = \frac{Ed^3}{12(1-v^2)}$$

This equation may be reduced to the form

$$\left(\frac{r_2}{r_1}\right)^4\frac{d^4\vartheta}{d\phi^4} + 4\lambda^4\vartheta = 0 \tag{14-51}$$

where

$$\lambda = \sqrt[4]{3(1-v^2)\frac{r_2{}^2}{d^2}}$$

If we choose Q_ϕ as the variable, the differential equation will take the form

$$\left(\frac{r_2}{r_1}\right)^4\frac{d^4Q_\phi}{d\phi^4} + 4\lambda^4Q_\phi = 0 \tag{14-52}$$

The fourth-order equations (14-51) and (14-52) are generally known as the Geckeler equations. They have the same form as the equation for the bending of a beam on an elastic foundation. For a spherical shell for which $r_1 = r_2 = a$, Equation (14-52) assumes the form

$$\frac{d^4Q_\phi}{d\phi^4} + 4\lambda^4Q_\phi = 0 \tag{14-53}$$

The solution for Q_ϕ may be written as

$$Q_\phi = C_1e^{\lambda\phi}\cos\lambda\phi + C_2e^{\lambda\phi}\sin\lambda\phi + C_3e^{-\lambda\phi}\cos\lambda\phi + C_4e^{-\lambda\phi}\sin\lambda\phi \tag{14-54}$$

The third and fourth terms of this solution are to be discarded as they imply that the edge disturbance fades away rapidly as we approach the edge at which it, in fact, originates. We may therefore write

$$Q_\phi = C_1e^{\lambda\phi}\cos\lambda\phi + C_2e^{\lambda\phi}\sin\lambda\phi \tag{14-55}$$

It is expedient to measure the variable angle from the edge of the shell rather than from the crown. This change may be effected by introducing a new variable $\alpha = (\phi_c - \phi)$, the angle α being measured from the edge (Fig. 14-11). Making this substitution, Equation (14-55) may be recast as

$$Q_\phi = Ce^{-\lambda\alpha} \sin(\lambda\alpha + \beta) \tag{14-56}$$

In this equation, C and β are the two arbitrary constants to be deter-

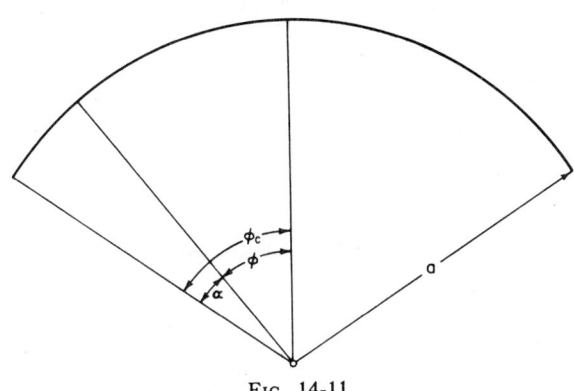

Fig. 14-11

mined from the prescribed boundary conditions. Expression for the other stress resultants are easily derived. Thus

$$N_\phi = -Q_\phi \cot\phi = -C\cot(\phi_c - \alpha)e^{-\lambda\alpha}\sin(\lambda\alpha + \beta) \tag{14-57}$$

$$N_\theta \approx -\frac{dQ_\phi}{d\phi} = -C\lambda\sqrt{2}\,e^{-\lambda\alpha}\sin\left(\lambda\alpha + \beta - \frac{\pi}{4}\right) \tag{14-58}$$

$$\vartheta = \frac{1}{Ed}\frac{d^2Q_\phi}{d\phi^2} = -\frac{2\lambda^2}{Ed}Ce^{-\lambda\alpha}\cos(\lambda\alpha + \beta) \tag{14-59}$$

$$M_\phi = -\frac{D}{a}\frac{d\vartheta}{d\phi} = \frac{2\sqrt{2}\,\lambda^3 DC}{aEd}\sin\left(\lambda\alpha + \beta + \frac{\pi}{4}\right)$$

$$= \frac{a}{\sqrt{2}\,\lambda}Ce^{-\lambda\alpha}\sin\left(\lambda\alpha + \beta + \frac{\pi}{4}\right) \tag{14-60}$$

Another expression that we need is for the horizontal deflection of a point on the shell. Denoting this by δ, we have

$$\delta = \epsilon_\theta\, a\sin\phi = \frac{a\sin\phi}{Ed}(N_\theta - \nu N_\phi) \tag{14-61}$$

Substituting for N_θ and N_ϕ,

$$\delta = -\frac{a\sin\phi}{Ed}\left(\frac{1}{a}\frac{dU}{d\phi} + \nu Q_\phi \cot\phi\right)$$

Hence

$$\delta \approx - \frac{\sin\phi}{Ed} \frac{dU}{d\phi}$$

Substituting for Q_ϕ from (14-56), we may finally write

$$\delta \approx - \frac{a\lambda\sqrt{2}\,C}{Ed} \sin\left(\phi_c - \alpha\right) e^{-\lambda\alpha} \sin\left(\lambda\alpha + \beta - \frac{\pi}{4}\right) \qquad (14\text{-}62)$$

The horizontal deflection at the shell edge may be derived from (14-62) by evaluating that expression at $\alpha = 0$. Hence

$$(\delta)_{\alpha=0} = - \frac{a\lambda\sqrt{2}\,C}{Ed} \sin\phi_c \sin\left(\beta - \frac{\pi}{4}\right) \qquad (14\text{-}63)$$

14-11   Stresses and Displacements Caused by Edge Moment ($-M_0$)

Imagine a spherical shell acted upon by an edge moment $(-M_0)$ applied at $\alpha = 0$ (Fig. 14-12). The boundary conditions corresponding to this case are

$$(M_\phi)_{\alpha=0} = -M_0 \qquad \text{and} \qquad (N_\phi)_{\alpha=0} = 0$$

Making use of the second boundary condition, it is clear from (14-57) that $\beta = 0$. Now, making use of the first boundary condition and Equation (14-60),

$$C = \frac{-2\lambda M_0}{a} \qquad (14\text{-}64)$$

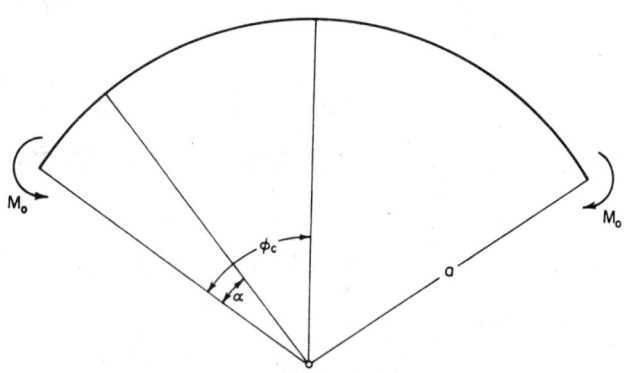

Fig. 14-12

Expressions for the stress resultants and displacements may now be written down. Thus

$$M_\phi = -\sqrt{2}\,M_0 e^{-\lambda\alpha} \sin\left(\lambda\alpha + \frac{\pi}{4}\right) \qquad (14\text{-}65)$$

$$M_\theta = \nu M_\phi = -\nu\,\sqrt{2}\,M_0 e^{-\lambda\alpha} \sin\left(\lambda\alpha + \frac{\pi}{4}\right) \qquad (14\text{-}66)$$

$$N_\phi = +\frac{2\lambda M_0}{a} \cot(\phi_c - \alpha)\, e^{-\lambda\alpha} \sin\lambda\alpha \qquad (14\text{-}67)$$

$$N_\theta = +\frac{2\sqrt{2}\,\lambda^2 M_0}{a}\, e^{-\lambda\alpha} \sin\left(\lambda\alpha - \frac{\pi}{4}\right) \qquad (14\text{-}68)$$

$$Q_\phi = -\frac{2\lambda M_0}{a}\, e^{-\lambda\alpha} \sin\lambda\alpha \qquad (14\text{-}69)$$

$$(\delta)_{\alpha=0} = -\frac{2 M_0 \lambda^2}{Ed} \sin\phi_c \qquad (14\text{-}70)$$

$$(\vartheta)_{\alpha=0} = \frac{4\lambda^3}{E\,da}\, M_0 \qquad (14\text{-}71)$$

14-12 Effects of a Horizontal Load H_0 applied at $\alpha = 0$

Next, consider a spherical shell acted upon by a horizontal edge load H_0 acting outward at $\alpha = 0$ (Fig. 14-13). The boundary conditions relevant to this case are

$$M_\phi = 0 \qquad \text{at} \quad \alpha = 0$$

and

$$H_0 \cos\phi_c = (N_\phi)_{\alpha=0}$$

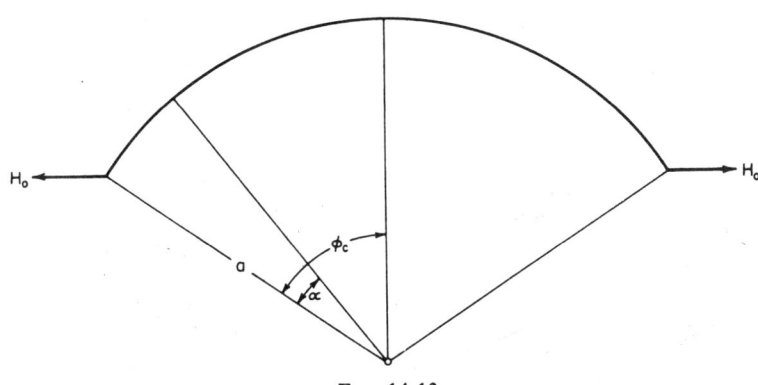

Fig. 14-13

From the first boundary condition and Equation (14-60),

$$\beta = -\frac{\pi}{4}$$

Referring to (14-57) and making use of the second boundary condition,

$$C = H_0 \sqrt{2} \sin \phi_c \tag{14-72}$$

Expressions for stress resultants and displacements easily follow and may now be written down.

$$Q_\phi = \sqrt{2} H_0 \sin \phi_c \, e^{-\lambda\alpha} \sin\left(\lambda\alpha - \frac{\pi}{4}\right) \tag{14-73}$$

$$N_\phi = -\sqrt{2} H_0 \sin \phi_c \cot(\phi_c - \alpha) \, e^{-\lambda\alpha} \sin\left(\lambda\alpha - \frac{\pi}{4}\right) \tag{14-74}$$

$$N_\theta = -2H_0\lambda \sin \phi_c \, e^{-\lambda\alpha} \sin\left(\lambda\alpha - \frac{\pi}{2}\right) \tag{14-75}$$

$$M_\phi = \frac{aH_0}{\lambda} \sin \phi_c \, e^{-\lambda\alpha} \sin \lambda\alpha \tag{14-76}$$

$$M_\theta = \frac{a\nu H_0}{\lambda} \sin \phi_c \, e^{-\lambda\alpha} \sin \lambda\alpha \tag{14-77}$$

$$(\vartheta)_{\alpha=0} = -\frac{2\lambda^2}{Ed} H_0 \sin \phi_c \tag{14-78}$$

$$(\delta)_{\alpha=0} = \frac{2a\lambda H_0 \sin^2 \phi_c}{Ed} \tag{14-79}$$

14-13 Shells of Revolution with Edge Beams

If a membrane state of equilibrium is to be maintained, a shell of revolution needs to be supported as shown in Fig. 14-14. This is seldom practicable, because the wall or column supporting the dome will not be able to resist the large thrusts transferred to it. A ring beam is therefore usually provided. The provision of the ring beam disturbs the membrane equilibrium. The bending stresses brought into play as a result can be approximately computed by applying the Geckeler theory. Let the breadth and depth of edge beam provided be $2b_1$ and $2a_1$.

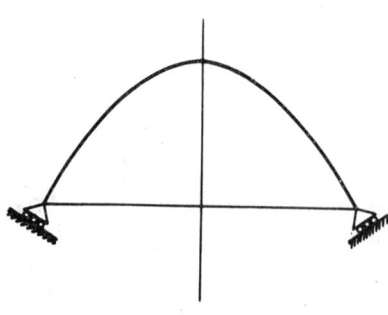

Fig. 14-14

In Fig. 14-15 the ring beam is shown severed from the shell as a free body. It is necessary to evaluate the two redundants X_1 and X_2 before we can compute the bending stresses in the shell.

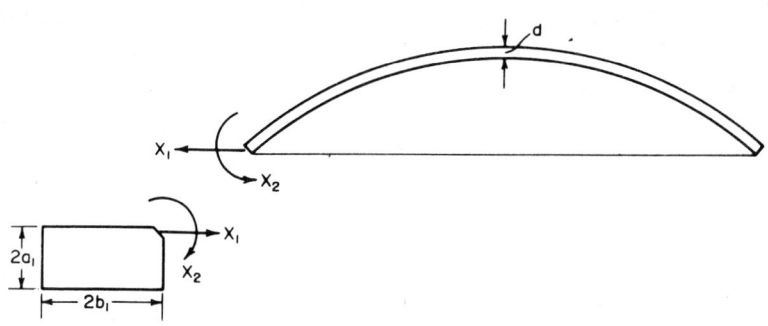

FIG. 14-15

14-14 Membrane Displacements of a Spherical Shell

(a) *Dead Weight*

If the stresses due to dead weight derived in (14-8) and (14-9) are substituted in (14-61) and its value found at $\phi = \phi_c$, we arrive at the expression for δ_{mg}, the horizontal membrane displacement of the shell edge. Thus

$$\delta_{mg} = \frac{a^2 g \sin \phi_c}{Ed} \left(\frac{1 + \nu}{1 + \cos \phi_c} - \cos \phi_c \right) \tag{14-80}$$

The change in slope of the tangent to the shell edge ϑ_{mg} due to dead weight can be found by using (14-38). We may easily verify that

$$\vartheta_{mg} = \frac{1}{Ed} \left\{ \cot \phi_c (1 + \nu)(N_\phi - N_\theta) - ga \sin \phi_c \left[1 + \frac{1 + \nu}{(1 + \cos \phi_c)^2} \right] \right\} \tag{14-81}$$

In Equation (14-81), the values of N_ϕ and N_θ to be substituted are those derived in (14-8) and (14-9) for $\phi = \phi_c$. Expression (14-81) can be further simplified and recast as

$$\vartheta_{mg} = - \frac{ag}{Ed} (2 + \nu) \sin \phi_c \tag{14-82}$$

(b) *Snow Load*

Similar expressions for the horizontal displacement of the shell edge and the change in slope of the tangent to the shell edge due to snow load may be easily derived. They are

$$\delta_{mp_0} = \frac{p_0 a^2 \sin \phi_c}{Ed} \left(\frac{1 + \nu}{2} - \cos^2 \phi_c \right) \tag{14-83}$$

and

$$\vartheta_{mp_0} = -\frac{ap_0}{Ed}(3+\nu)\sin\phi_c\cos\phi_c \tag{14-84}$$

In the above expressions the suffix m stands for membrane, g for dead weight, and p_0 for snow load.

14-15 Membrane Displacements of the Ring Beam

(a) Dead Weight

The horizontal component of the dead-weight thrust transmitted by the shell causes a hoop tension of

$$\frac{ga}{1+\cos\phi_c}\cos\phi_c(a\sin\phi_c)$$

$$\text{Hoop stress in ring beam} = \frac{ga\cos\phi_c(a\sin\phi_c)}{(1+\cos\phi_c)\,A}$$

where $A = 4a_1b_1$ is the area of the ring beam.

$$\text{Hoop strain} = \frac{ga\cos\phi_c(a\sin\phi_c)}{(1+\cos\phi_c)\,AE}$$

Hence

$$\text{Increase in radius of ring} = \frac{ga\cos\phi_c(a\sin\phi_c)^2}{(1+\cos\phi_c)\,AE}$$

This is also the amount δ'_{mg} by which the edge of the ring beam moves away from the edge of the shell. Thus

$$\delta'_{mg} = \frac{ga\cos\phi_c(a\sin\phi_c)^2}{(1+\cos\phi_c)\,AE} \tag{14-85}$$

It is assumed that the thrust of the dome passes through the center of the edge beam. Hence no bending moments are transferred by the shell to the edge beam in the membrane state. Consequently the rotation of the edge beam due to dead weight is zero.

(b) Snow Load

Proceeding as before, we may establish the following expression for the horizontal displacement of the ring beam edge due to snow load:

$$\delta'_{mp_0} = \frac{p_0 a}{2}\cos\phi_c\frac{(a\sin\phi_c)^2}{AE} \tag{14-86}$$

As before, the edge-beam rotation is zero.

14-16 "Gaps" between Shell Edge and Ring Beam due to Dead Weight and Snow Load

(a) Dead Weight

Imagine that the ring beam and shell have been separated from each other. Under the action of dead weight, it is seen from Equation (14-80) that the shell edge will deflect horizontally by an amount

$$\delta_{mg} = \frac{a^2 g}{Ed} \sin \phi_c \left(\frac{1 + \nu}{1 + \cos \phi_c} - \cos \phi_c \right)$$

The expression within the parentheses is negative, indicating that the deflection is inward. The same load acting on the ring beam causes it to move away from the shell edge by δ'_{mg} which, being positive, indicates that the movement is outward. As the movements of the shell edge and the ring beam are in opposite directions, the two deflections add up to give a cumulative gap $\Delta_1 = (\delta_{mg} + \delta'_{mg})$. Similarly, under the action of the dead load, an angular gap of ϑ_{mg} will develop. It is to be noted that the edge beam does not rotate because of the dead-weight reactions transmitted to it. Hence the total angular gap $\Delta_2 = \vartheta_{mg}$ because $\vartheta'_{mg} = 0$.

(b) Snow Load

Proceeding in a similar manner, we may show that the horizontal gap between the shell edge and the ring beam caused by snow load is $(\delta_{mp_0} + \delta'_{mp_0})$. The angular gap is obviously ϑ_{mp_0}.

14-17 Action of Redundants on Shell and Ring Beam

The function of the redundants X_1 and X_2 is to close the gaps caused by loading. Let us study the effect of a unit load $H_0 = 1$ applied to the shell edge as well as the ring beam at their junction in the same direction as X_1. From Equation (14-79), it is seen that the shell edge will move horizontally *outward* to close the gap by $(2a\lambda \sin^2 \phi_c)/Ed$. At the same time, the edge of the ring beam will move *inward* by a certain amount to close the gap. This amount may be found by treating the unit load applied at the top of the edge beam as equal to

(i) A central load of 1

(ii) A twisting moment $T = (1 \times a_1)$ applied about the center of the beam

The first effect causes the edge to move inward by $(a \sin \phi_c)^2 / AE$. The twisting moment $(1 \times a_1)$ may be resolved into bending moments

M (Fig. 14-16). It is clear that $Ta \sin \phi_c = M$. The bending moment will cause a compression in the top fiber of $(6Ta \sin \phi_c)/[2b_1(2a_1)^2]$.

$$\text{Strain in top fiber} = \frac{6Ta \sin \phi_c}{2b_1(2a_1)^2} \frac{1}{E}$$

$$\text{Change in radius at top of edge beam} = \frac{6T(a \sin \phi_c)^2}{2b_1(2a_1)^2 E}$$

$$= \frac{6a_1(a \sin \phi_c)^2}{2b_1(2a_1)^2 E} = \frac{3(a \sin \phi_c)^2}{AE}$$

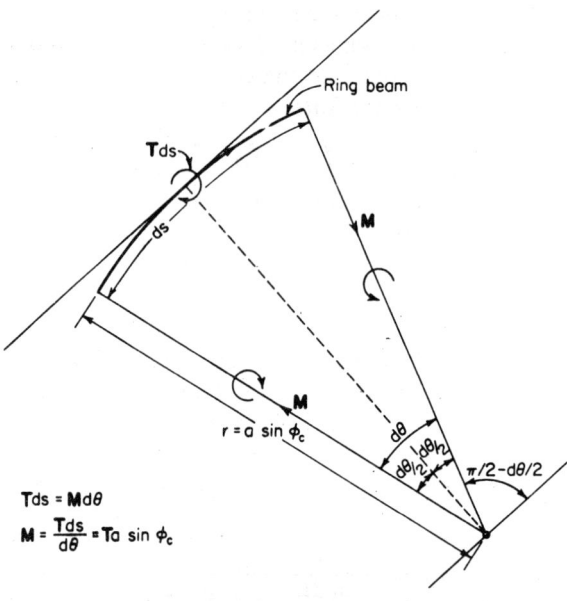

$$Tds = Md\theta$$
$$M = \frac{Tds}{d\theta} = Ta \sin \phi_c$$

Fɪɢ. 14-16

This change in radius also represents the amount by which the edge moves inward at the junction with the shell. Thus the total horizontal "gap" closed is

$$\frac{2a\lambda \sin^2 \phi_c}{Ed} + \frac{4(a \sin \phi_c)^2}{AE}$$

There is one more effect to be taken into account. We already saw that, because of the twisting moment T, the top of the edge beam undergoes compression causing the radius at the top to decrease; at the same time, the bottom fibers of the beam being in tension, the radius at this level increases by the same amount. The result is that the beam rotates by an angle equal to

$$\frac{6a_1(a \sin \phi_c)^2}{2b_1(2a_1)^2 E} \frac{2}{2a_1} = \frac{3(a \sin \phi_c)^2}{a_1 AE}$$

which tends to close the angular gap caused by the loading. The unit horizontal load also causes ϑ, the angle made by the tangent to the shell edge, to change by an amount equal to $-(2\lambda^2/Ed) \sin \phi_c$. Hence the total angular gap closed by the unit force applied in the direction of the redundant X_1 is equal to

$$\frac{3(a \sin \phi_c)^2}{a_1 AE} - \frac{2\lambda^2}{Ed} \sin \phi_c$$

The displacement, linear or angular, of a structural system caused by a unit load applied to it is known as its *flexibility*. The system that we are considering consists of a shell and the edge beam. Its flexibility can be defined by four *influence coefficients*

$$\begin{matrix} f_{11} & f_{12} \\ f_{21} & f_{22} \end{matrix}$$

corresponding to the two redundants X_1 and X_2. *Let us consider the influence coefficient* f_{11}. *It stands for the gap closed in the direction of action* X_1 *due to the application of a unit force to the shell as well as the edge beam in the direction of* X_1. Similarly, f_{21} would mean the gap closed in the direction of X_2 by the application of a unit force in the direction of X_1. It is now easy to see that

$$f_{11} = \frac{2a\lambda \sin^2 \phi_c}{Ed} + \frac{4(a \sin \phi_c)^2}{AE} \tag{14-87}$$

and

$$f_{21} = \frac{3(a \sin \phi_c)^2}{a_1 AE} - \frac{2\lambda^2}{Ed} \sin \phi_c \tag{14-88}$$

The influence coefficients f_{12} and f_{22} are to be found in a similar manner by studying the effect of a unit moment in the direction of X_2 applied to the shell and the ring beam at their junction. These coefficients are

$$f_{12} = -\frac{2\lambda^2}{Ed} \sin \phi_c + \frac{3(a \sin \phi_c)^2}{a_1 AE} \tag{14-89}$$

and

$$f_{22} = \frac{4\lambda^3}{E \, da} + \frac{6(a \sin \phi_c)^2}{2b_1(2a_1)^2} \frac{2}{E} \frac{1}{2a_1} = \frac{4\lambda^3}{E \, da} + \frac{3(a \sin \phi_c)^2}{AEa_1^2} \tag{14-90}$$

14-18 Solution for Redundants

The four influence coefficients just derived may be represented by a matrix $[F]$ thus:

$$[F] = \begin{bmatrix} f_{11} & f_{12} \\ f_{21} & f_{22} \end{bmatrix}$$

It may be noted that $f_{12} = f_{21}$. This is a consequence of the reciprocity theorem of Maxwell. The matrix $[F]$ is known as the flexibility matrix. Let Δ_1 and Δ_2 be the gaps in the directions of X_1 and X_2 due to *any* condition of loading such as dead weight or snow load. We may then formulate the following two equations:

$$f_{11}X_1 + f_{12}X_2 + \Delta_1 = 0$$

$$f_{21}X_1 + f_{22}X_2 + \Delta_2 = 0$$

These equations in matrix notation become

$$\begin{bmatrix} f_{11} & f_{12} \\ f_{21} & f_{22} \end{bmatrix} \begin{bmatrix} X_1 \\ X_2 \end{bmatrix} = \begin{bmatrix} -\Delta_1 \\ -\Delta_2 \end{bmatrix}$$

or, more briefly,

$$[F]\{X\} = \{-\Delta\} \tag{14-91}$$

The redundants $\{X\}$ may now be found as

$$\{X\} = [F]^{-1}\{-\Delta\} \tag{14-92}$$

If now we wish to determine the redundants corresponding to another condition of loading, all that we need change is the column matrix $\{-\Delta\}$ representing the gaps caused by that particular condition of loading. The inverse matrix $[F]^{-1}$, when once computed, can be repeatedly used to investigate several conditions of loading. If we did not use the *flexibility-matrix* method of solution, two simultaneous equations in X_1 and X_2 would have to be solved for each condition of loading. The power and economy of the matrix method will become evident when applied to a shell of revolution which is required to be analyzed for several conditions of loading—dead weight, snow load, prestress, loss of prestress, temperature changes, and various critical combinations of these. The use of this technique for the analysis of a prestressed dome is explained in detail in Design Example 14-2.

14-19 Prestressed Domes

Modern architectural practice favors shallow spherical domes with a rise to span ratio of 1:7 to 1:8. Domes of such shallow rise have also the advantage that ϕ_c can be kept below 45° so that the entire dome may be kept in compression under both dead and snow loads. High values of ϕ_c, say in excess 45°, are also objectionable from the point of view of construction, because such domes will call for top forms. But shallow domes transfer very large thrusts to the walls or columns on which they are supported. As the diameter of the dome increases, we need reinforced-concrete edge beams of unmanageable sizes to absorb the very large thrusts that are transferred. In such cases, prestressing will prove advantageous. The ring beams of domes may conveniently be prestressed by winding wires around them using a special machine such as the one developed by the Preload Corporation [71]. This machine delivers wires at a constant tension of 140,000 psi. The wires used are usually of 0.162 in. diameter. A cover coat of 3 in. is usually provided to protect the wires. It is the usual practice to thicken the dome in the vicinity of its junction with the ring beam to ensure smooth flow of stresses. When shallow domes are used to roof large halls, the ring beam is supported on a number of columns arranged on the periphery. It is accurate enough for purposes of design to ignore the vertical deflection of the ring beam.[1]

The analysis of a prestressed dome is explained in detail in Design Example 14-2.

Design Example 14-2

DESIGN OF A PRESTRESSED DOME

1. *DATA*

    Geometry

    Radius of curvature of dome $a = 173.33$ ft
    Semicentral angle $\phi_c = 28°$
    Thickness $d = 2\frac{1}{2}$ in. = 5/24 ft
    Span of dome $2a \sin \phi_c = 162.75$ ft
    Rise (1/8 × span of dome) = 20.34 ft
    Width of ring beam $2b_1 = 3$ ft
    Depth of ring beam $2a_1 = 1.5$ ft

    Loading

    Dead weight $g = 31.25$ psf of surface area of dome
    Snow load $p_0 = 25$ psf on horizontal projection
    Poisson's ratio ν is assumed to be zero

2. *SECTIONAL PROPERTIES AND TRIGONOMETRIC RATIOS*

    Cross-sectional area of ring beam $A = 4.5$ ft²

$$\lambda = \sqrt[4]{3\,\frac{a^2}{d^2}} = 37.96096$$

---

[1] Timoshenko, S., and S. Woinowsky-Krieger, "Theory of Plates and Shells," 2d ed., p. 555, McGraw-Hill Book Company, New York, 1959.

ϕ_c	$\sin \phi_c$	$\cos \phi_c$	$\tan \phi_c$	$\cot \phi_c$	$\sin^2 \phi_c$	$\cos^2 \phi_c$
28°	0.46947	0.88295	0.53171	1.88073	0.22040	0.77960

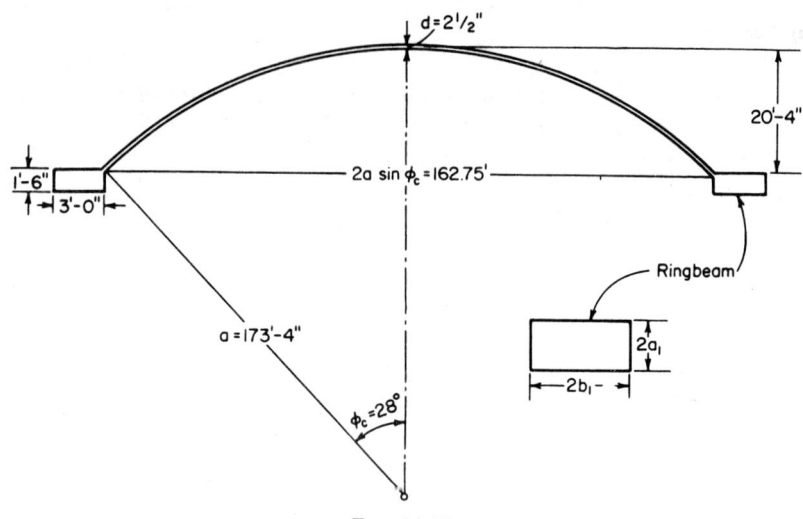

FIG. 14-17

3. MEMBRANE STRESSES AT $\phi_c = 28°$

(a) *Due to Dead Weight*

$$N_\phi = -\frac{ag}{1 + \cos \phi_c} = -\frac{173.33 \times 31.25}{1.88295} = -2,876.64 \text{ lb/ft}$$

$$N_\theta = -ag \left(\cos \phi_c - \frac{1}{1 + \cos \phi_c} \right)$$

$$= -173.33 \times 31.25 \left(0.88295 - \frac{1}{1.88295} \right)$$

$$= -1,905.93 \text{ lb/ft}$$

(b) *Due to Snow Load*

$$N_\phi = -\frac{p_0 a}{2} = -\frac{25 \times 173.33}{2}$$

$$= -2,166.63 \text{ lb/ft}$$

$$N_\theta = -\frac{p_0 a}{2} \cos 2\phi_c = -\frac{25 \times 173.33}{2} \times 0.55919$$

$$= -1,211.56 \text{ lb/ft}$$

(c) *Due to Dead Weight and Snow Load*

$$N_\phi = -2,876.64 - 2,166.63 = -5,043.27 \text{ lb/ft}$$
$$N_\theta = -1,905.93 - 1,211.56 = -3,117.49 \text{ lb/ft}$$

4. MEMBRANE DISPLACEMENTS

(a) *Due to Dead Weight*

$$\delta_{mg} = \frac{ga^2 \sin \phi_c}{Ed} \left(\frac{1 + \nu}{1 + \cos \phi_c} - \cos \phi_c \right)$$

$$= \frac{1}{E} \left[\frac{31.25 \times 30,043.29 \times 0.46947}{0.20833} \left(\frac{1}{1.88295} - 0.88295 \right) \right]$$

$$= -\frac{744,450}{E}$$

$$\delta'_{mg} = -\frac{ga \cos \phi_c (a \sin \phi_c)^2}{(1 + \cos \phi_c)AE}$$

$$= -\frac{1}{E} \left[\frac{31.25 \times 173.33 \times 0.88295 \, (81.375)^2}{1.88295 \times 4.5} \right]$$

$$= -\frac{3,737,577}{E}$$

$$\vartheta_{mg} = \frac{1}{Ed} \left\{ \cot \phi_c (1 + \nu)(N_\phi - N_\theta) - ga \sin \phi_c \left[1 + \frac{1}{(1 + \cos \phi_c)^2} \right] \right\}$$

$$= \frac{1}{E \times 0.20833} \left\{ 1.88073 \times 1 \, (-2,876.64 + 1,905.93) \right.$$

$$\left. - 31.25 \times 81.375 \left[1 + \frac{1}{(1.88295)^2} \right] \right\}$$

$$= -\frac{24,413}{E}$$

Horizontal gap due to dead weight,

$$\Delta_1 = \delta_{mg} + \delta'_{mg}$$

$$= -\frac{744,450}{E} - \frac{3,737,577}{E}$$

$$= -\frac{4,482,027}{E}$$

Angular gap due to dead weight,

$$\Delta_2 = \vartheta_{mg} = -\frac{24,413}{E}$$

(b) *Due to Snow Load*

$$\delta_{mp_0} = \frac{p_0 a^2 \sin \phi_c}{Ed} \left(\frac{1 + \nu}{2} - \cos^2 \phi_c \right)$$

$$= \frac{1}{E} \left[\frac{25 \times 30,043.29 \times 0.46947}{0.20833} (0.5 - 0.77960) \right]$$

$$= -\frac{473,239}{E}$$

$$\delta'_{mp_0} = -\frac{p_0 a}{2} \cos \phi_c \frac{(a \sin \phi_c)^2}{AE}$$

$$= -\frac{1}{E} \left[\frac{25 \times 173.33}{2} \times 0.88295 \frac{(81.375)^2}{4.5} \right]$$

$$= -\frac{2,815,071}{E}$$

$$\vartheta_{mp_0} = -\frac{a p_0}{Ed} (3 + \nu) \sin \phi_c \cos \phi_c$$

$$= -\frac{1}{E} \left(\frac{173.33 \times 25}{0.20833} \times 3 \times 0.46947 \times 0.88295 \right)$$

$$= \frac{25,866}{E}$$

Horizontal gap due to snow load,

$$\Delta_1 = \delta_{mp_0} + \delta'_{mp_0}$$

$$= -\frac{473,239}{E} - \frac{2,815,071}{E}$$

$$= -\frac{3,288,310}{E}$$

Angular gap due to snow load,

$$\Delta_2 = \vartheta_{mp_0} = -\frac{25,866}{E}$$

5. APPROXIMATE PRESTRESSING FORCE

Total N_ϕ due to dead weight and snow load $= -5,043.27$ lb/ft

$\therefore$ Hoop tension in ring beam $=$ total $N_\phi \times \cos \phi_c \times a \sin \phi_c$

$$= -5,043.27 \times 0.88295 \times 81.375$$

$$= -362,359 \text{ lb}$$

Approximate prestressing force estimated at 20% more than the hoop tension

$$= -362,359 \times 1.2$$

$$= -434,831 \text{ lb}$$

150 wires of 0.162 in. diameter stressed initially to 140,000 psi will do, i.e.,

$$\text{Prestressing force } P = 150 \frac{\pi}{4} (0.162)^2 \times 140,000 = 432,852 \text{ lb}$$

6. HORIZONTAL DISPLACEMENT CAUSED BY PRESTRESSING

$$\text{Horizontal displacement caused by prestressing} = \frac{Pa \sin \phi_c}{AE} = \frac{1}{E} \left(\frac{432,852 \times 81.375}{4.5} \right)$$

$$= + \frac{7,827,407}{E}$$

7. LOSS IN PRESTRESS

Loss of stress in wires = 40,000 psi.   Hence

$$\text{Total loss of stress } \Delta P = \frac{40,000}{140,000} \times P$$

$$= 0.2857P$$

$$= 0.2857 \times 432,852$$

$$= 123,666 \text{ lb}$$

8. HORIZONTAL DISPLACEMENT DUE TO LOSS IN PRESTRESS

$$\text{Horizontal displacement due to loss of prestress} = \frac{\Delta Pa \sin \phi_c}{AE}$$

$$= \frac{1}{E} \frac{123,666 \times 81.375}{4.5}$$

$$= + \frac{2,236,294}{E}$$

9. FLEXIBILITY MATRIX

$$f_{11} = \frac{2a\lambda}{Ed} \sin^2 \phi_c + \frac{4a^2 \sin^2 \phi_c}{AE}$$

$$= \frac{2a \sin^2 \phi_c}{E} \left(\frac{\lambda}{d} + \frac{2a}{A} \right)$$

$$= \frac{1}{E} \left[2 \times 173.33 \times 0.22040 \left(\frac{37.96096}{0.20833} + \frac{2 \times 173.33}{4.5} \right) \right]$$

$$= + \frac{19,807.795}{E}$$

$$f_{12} = \left(\frac{6a^2 \sin^2 \phi_c}{E \times 2b_1(2a_1)^2} - \frac{2\lambda^2}{Ed} \sin \phi_c \right)$$

$$= \frac{1}{E} \left(\frac{6 \times 6,621.8906}{3 \times (1.5)^2} - \frac{2 \times 1,441.0351 \times 0.46947}{0.20833} \right)$$

$$= - \frac{608.4906}{E}$$

$$f_{21} = \left(\frac{6a^2 \sin^2 \phi_c}{E \times 2b_1(2a_1)^2} - \frac{2\lambda^2}{Ed} \sin \phi_c \right)$$

$$= -\frac{608.4906}{E}$$

$$f_{22} = \left(\frac{4\lambda^3}{E\,da} + \frac{12a^2 \sin^2 \phi_c}{E2b_1(2a_1)^3} \right)$$

$$= \frac{1}{E} \left(\frac{4 \times 54,703.0758}{0.20833 \times 173.33} + \frac{12 \times 6,621.8906}{3 \times 3.375} \right)$$

$$= + \frac{13,907.804}{E}$$

Hence writing the values in the matrix form,

$$[F] = \begin{bmatrix} f_{11} & f_{12} \\ f_{21} & f_{22} \end{bmatrix} = \frac{1}{E} \begin{bmatrix} +19,807.795 & -608.4906 \\ -608.4906 & +13,907.804 \end{bmatrix}$$

The inverted matrix

$$[F]^{-1} = \begin{bmatrix} +0.0000505537 & +0.00000221183 \\ +0.00000221183 & +0.0000719994 \end{bmatrix}$$

10. CALCULATION OF REDUNDANTS X_1 AND X_2

(a) Due to Dead Weight

$$\begin{bmatrix} X_1 \\ X_2 \end{bmatrix} = [F]^{-1} \begin{bmatrix} -\Delta_1 \\ -\Delta_2 \end{bmatrix}$$

$$= \begin{bmatrix} +0.0000505537 & +0.00000221183 \\ +0.00000221183 & +0.0000719994 \end{bmatrix} \begin{bmatrix} +4,482,027 \\ +24,413 \end{bmatrix}$$

i.e.,

$$\begin{bmatrix} X_1 \\ X_2 \end{bmatrix} = \begin{bmatrix} +226.6371 \\ +11.6712 \end{bmatrix}$$

(b) Due to Snow Load

$$\begin{bmatrix} X_1 \\ X_2 \end{bmatrix} = [F]^{-1} \begin{bmatrix} -\Delta_1 \\ -\Delta_2 \end{bmatrix}$$

$$= \begin{bmatrix} +0.0000505537 & +0.00000221183 \\ +0.00000221183 & +0.0000719994 \end{bmatrix} \begin{bmatrix} +3,288,310 \\ +25,866 \end{bmatrix}$$

i.e.,

$$\begin{bmatrix} X_1 \\ X_2 \end{bmatrix} = \begin{bmatrix} +166.2935 \\ +9.1355 \end{bmatrix}$$

(c) Due to Prestressing Force

$$\begin{bmatrix} X_1 \\ X_2 \end{bmatrix} = [F]^{-1} \begin{bmatrix} -Pa \sin \phi_c \\ A \\ 0 \end{bmatrix}$$

$$= \begin{bmatrix} +0.0000505537 & +0.00000221183 \\ +0.00000221183 & +0.0000719994 \end{bmatrix} \begin{bmatrix} -7,827,407 \\ 0 \end{bmatrix}$$

i.e.,

$$\begin{bmatrix} X_1 \\ X_2 \end{bmatrix} = \begin{bmatrix} -395.7045 \\ -17.3129 \end{bmatrix}$$

(*d*) *Due to Loss in Prestressing Force*

$$\begin{bmatrix} X_1 \\ X_2 \end{bmatrix} = [F]^{-1} \begin{bmatrix} + \dfrac{\Delta Pa \sin \phi_c}{A} \\ 0 \end{bmatrix}$$

$$= \begin{bmatrix} +0.0000505537 & +0.00000221183 \\ +0.00000221183 & +0.0000719994 \end{bmatrix} \begin{bmatrix} +2,236,294 \\ 0 \end{bmatrix}$$

i.e.,

$$\begin{bmatrix} X_1 \\ X_2 \end{bmatrix} = \begin{bmatrix} +113.0584 \\ +4.9465 \end{bmatrix}$$

11. CALCULATION OF STRESSES AT THE EDGE OF THE DOME

Corrective Direct Stresses Due to X_1 and X_2

$$N_\phi \text{ due to } X_1 \text{ (in psi)} = \frac{X_1 \cos \phi_c}{30}$$

$$N_\theta \text{ due to } X_1 \text{ (in psi)} = \left(2\lambda X_1 \sin \phi_c - 2\lambda^2 \frac{X_2}{a}\right)\frac{1}{30}$$

These formulas are used for arriving at the corrective direct stresses given in column 4 of Table 14-2.

Bending Stresses

Bending stresses given in column 6 are arrived at by using the formula

$$\frac{6X_2 \times 24 \times 24}{1 \times 25 \times 144} = \frac{24X_2}{25} \quad \text{psi}$$

Total Direct and Bending Stresses

Total direct and bending stresses tabulated in column 7 are obtained by superposing the results of columns 5 and 6.

12. TABLE OF STRESSES

The edge stresses of the prestressed dome are shown tabulated in Table 14-2.

14-20 Eccentric Concentrated Loads

Exact analytical solutions covering such cases are not available. The stresses induced being of a local nature, Reissner suggests that for purposes of calculation, the eccentric load may be assumed to be acting at the crown. The validity of this assumption has been verified experimentally.[1]

[1] Voss, Walter C., Howard R. Stanley, and G. H. Albert, "Thin-shelled Domes Loaded Eccentrically," *Transactions of the American Society of Civil Engineers,* paper 2336, vol. 113, p. 293, 1948.

All stresses in psi

Table 14-2. Edge Stresses in the Prestressed Dome

Tension $+ve$

1	Load	Membrane stress	Corrective direct stress due to bending	Total direct stress	Bending stress Out-side	Bending stress In-side	Total direct and bending stresses Outside face	Total direct and bending stresses Inside face
	2	3	4	5	6	6	7	7
Meridional								
Before volume changes	g	−96	+7	−89	+11	−11	−78	−100
	p_0	−72	+5	−67	+9	−9	−58	−6
	P	0	−12	−12	−17	+17	−29	+5
	$-\Delta P$	0	+3	+3	+5	−5	+8	−2
After volume changes	$g + P$	−96	−5	−101	−6	+6	−107	−95
	$g + p_0 + P$	−168	0	−168	+3	−3	−165	−171
	$g + P - \Delta P$	−96	−2	−98	−1	+1	−99	−97
	$g + p_0 + P - \Delta P$	−168	+3	−165	+8	−8	−157	−173
Circumferential								
Before volume changes	g	−64	+263	+199	0	0	+199	+199
	p_0	−40	+193	+153	0	0	+153	+153
	P	0	−461	−461	0	0	−461	−461
	$-\Delta P$	0	+132	+132	0	0	+132	+132
After volume changes	$g + P$	−64	−198	−262	0	0	−262	−262
	$g + p_0 + P$	−104	−5	−109	0	0	−109	−109
	$g + P - \Delta P$	−64	−66	−130	0	0	−130	−130
	$g + p_0 + P - \Delta P$	−104	+127	+23	0	0	+23	+23

14-21 Asymmetric Loading

The treatment in this chapter has been limited to shells of revolution which are subject to axisymmetric loading. In such cases, the Geckeler approximation is valid, provided ϕ_c is not too small. If the shell carries asymmetric loading, it is perhaps best dealt with using a numerical technique such as the finite-difference method. A detailed exposition of the method as applied to shells may be found in the book by Soare [72].

14-22 Digital Computation

The general-purpose program for use on the IBM 7090 computer developed by Singhal, Villaveces, and Utku [73] can handle a symmetrically loaded thin shell of revolution with any open meridian curve. The shell may have any arbitrary variations in thickness, material properties, and loading along the meridian curve, and any physically meaningful boundary conditions. The finite-difference method has been employed in this formulation. A large class of problems relating to thin shells of revolution may be studied by using this versatile program.

CHAPTER 15

MEMBRANE THEORY FOR SHELLS OF DOUBLE CURVATURE OTHER THAN SHELLS OF REVOLUTION

15-1 The Membrane State of Stress

A state of stress in which the stresses in the shell are constant over its thickness may be defined as the *membrane state*. A more mathematical approach would be to regard the membrane theory as a particular case of the more exact bending theory. Thus the membrane theory results if certain effects in the bending theory are ignored. Both these approaches lead to the same result. Until very recently, it was the common practice to design shells of double curvature on the membrane hypothesis. Under certain conditions, the results of the membrane theory may be regarded as good enough for purposes of structural design. The designer is still obliged to a large extent to use the membrane theory, as very few solutions of the bending theory of shells of double curvature, which take into account realistic boundary conditions, are available.

15-2 Geometrical Relations

Let the equation of the shell of double curvature be given in the form

$$z = f(x, y)$$

We shall use Monge's notation developed in Chapter 5. The radius vector $\mathbf{r}$ of a point on the surface of the shell may be represented as $\mathbf{r} = x\mathbf{i} + y\mathbf{j} + z\mathbf{k}$. Hence

$$\frac{\partial \mathbf{r}}{\partial x} = \mathbf{i} + \frac{\partial z}{\partial x}\mathbf{k} \qquad (15\text{-}1)$$

and

$$\frac{\partial \mathbf{r}}{\partial y} = \mathbf{j} + \frac{\partial z}{\partial y}\mathbf{k} \qquad (15\text{-}2)$$

Referring to the element of a shell shown in Fig. 15-1, the length AB measured on the shell is

$$\left|\frac{\partial \mathbf{r}}{\partial x}\right| dx = \sqrt{1 + \left(\frac{\partial z}{\partial x}\right)^2}\, dx = \sqrt{1 + p^2}\, dx$$

Similarly, the length AD is   (15-3)

$$\left|\frac{\partial \mathbf{r}}{\partial y}\right| dy = \sqrt{1 + \left(\frac{\partial z}{\partial y}\right)^2}\, dy = \sqrt{1 + q^2}\, dy$$

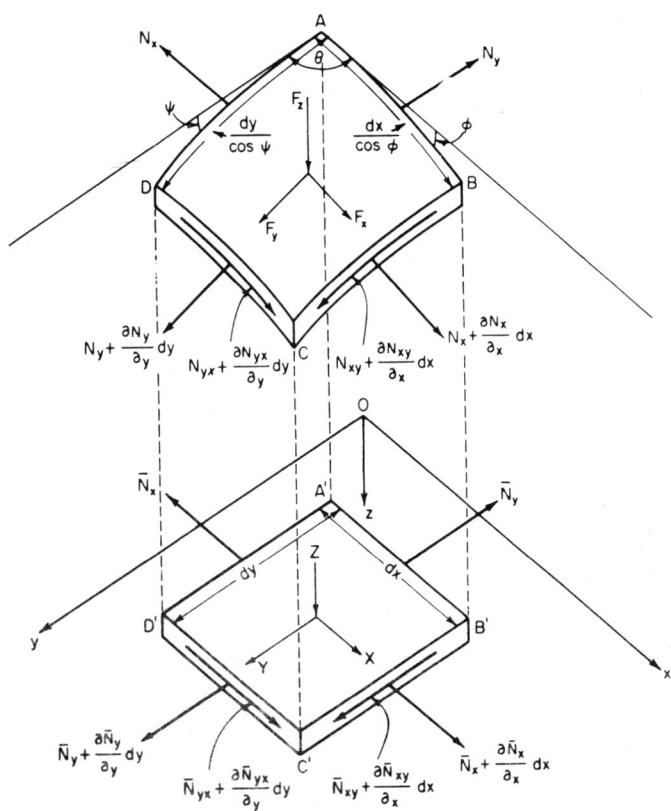

Fig. 15-1

The following relations are easily verified:

$$\cos \phi = \frac{A'B'}{AB} = \frac{1}{\sqrt{1 + p^2}} \tag{15-4}$$

$$\cos \psi = \frac{A'D'}{AD} = \frac{1}{\sqrt{1 + q^2}} \tag{15-5}$$

The angle between AB and AD is found by forming the dot product $(\partial \mathbf{r}/\partial x) \cdot (\partial \mathbf{r}/\partial y)$. From (15-1) and (15-2) this dot product is found to be $(\partial z/\partial x)(\partial z/\partial y) = pq$. But we also know that the dot product is equal to the product of the length of the two vectors and the cosine of the angle θ included between them. Hence

$$\sqrt{1 + p^2} \, \sqrt{1 + q^2} \cos \theta = pq$$

or

$$\cos \theta = \frac{pq}{\sqrt{1 + p^2} \, \sqrt{1 + q^2}} \tag{15-6}$$

The area of the element $ABCD = AB \cdot AD \sin \theta$. We may evaluate this quantity by forming the vector or cross product of $(\partial \mathbf{r}/\partial x) \, dx$ and $(\partial \mathbf{r}/\partial y) \, dy$. The magnitude of the product vector is $\sqrt{1 + p^2 + q^2} \, dx \, dy$. But we know that the magnitude of the vector product must be equal to $AB \cdot AD \sin \theta$. Hence

$$\text{Area of the element} = \sqrt{1 + p^2 + q^2} \, dx \, dy \tag{15-7}$$

15-3 "Pseudo" Stress Resultants

It is expedient to introduce "pseudo" stress resultants $\overline{N}_x$, $\overline{N}_y$, and $\overline{N}_{xy}$ in place of the real stress resultants N_x, N_y, and N_{xy}. Similarly, we shall also introduce pseudo loads X, Y, and Z in the directions x, y, and z in place of the real loads F_x, F_y, and F_z. The pseudo stress resultant $\overline{N}_x$ is such that it exerts the same force in the x direction on the projected side $A'D'$ as the real stress resultant does on the side AD. Hence

$$\overline{N}_x \, dy = N_x \, dy \, \frac{\cos \phi}{\cos \psi}$$

or

$$N_x = \overline{N}_x \sqrt{\frac{1 + p^2}{1 + q^2}} \tag{15-8}$$

Similarly,

$$N_y = \overline{N_y} \sqrt{\frac{1 + q^2}{1 + p^2}} \tag{15-9}$$

$$N_{xy} = \overline{N_{xy}} \tag{15-10}$$

The fictitious loads X, Y, and Z are so defined that

(Real load) $\times$ (area $ABCD$) $=$ (fictitious load) $\times$ (projected area $A'B'C'D'$)

This relationship applies in all the three directions x, y, and z. Thus

$$F_x \sqrt{1 + p^2 + q^2} \, dx \, dy = X \, dx \, dy$$

or

$$X = F_x \sqrt{1 + p^2 + q^2} \tag{15-11}$$

Similarly,

$$Y = F_y \sqrt{1 + p^2 + q^2} \tag{15-12}$$

and

$$Z = F_z \sqrt{1 + p^2 + q^2} \tag{15-13}$$

15-4 Equations of Equilibrium

Referring to the projected element shown in Fig. 15-2, the equations of equilibrium in the x and y directions may be formulated thus:

$$\frac{\partial \overline{N_x}}{\partial x} + \frac{\partial \overline{N_{xy}}}{\partial y} + X = 0 \tag{15-14}$$

$$\frac{\partial \overline{N_y}}{\partial y} + \frac{\partial \overline{N_{xy}}}{\partial x} + Y = 0 \tag{15-15}$$

For formulating the equation of equilibrium in the z direction, we have to go back to the element $ABCD$ of Fig. 15-1.

Vertical component of the normal force acting on face $AD = N_x \sqrt{\frac{1 + q^2}{1 + p^2}} \frac{\partial z}{\partial x} dy$

$$= \overline{N_x} \frac{\partial z}{\partial x} dy$$

This force acts upward.

Vertical component of the normal force acting on $BC = \overline{N_x} \frac{\partial z}{\partial x} dy$

$$+ \frac{\partial}{\partial x} \left(\overline{N_x} \frac{\partial z}{\partial x} \right) dx \, dy$$

Hence the resultant of the vertical forces on the pair of sides AD and BC = $(\partial/\partial x)[\overline{N_x}\,(\partial z/\partial x)]\,dx\,dy$ acting downward. Similarly, the resultant of the vertical forces acting on the pair of sides AB and CD is a downward force of

$$\frac{\partial}{\partial y}\left(\overline{N_y}\,\frac{\partial z}{\partial y}\right)dx\,dy$$

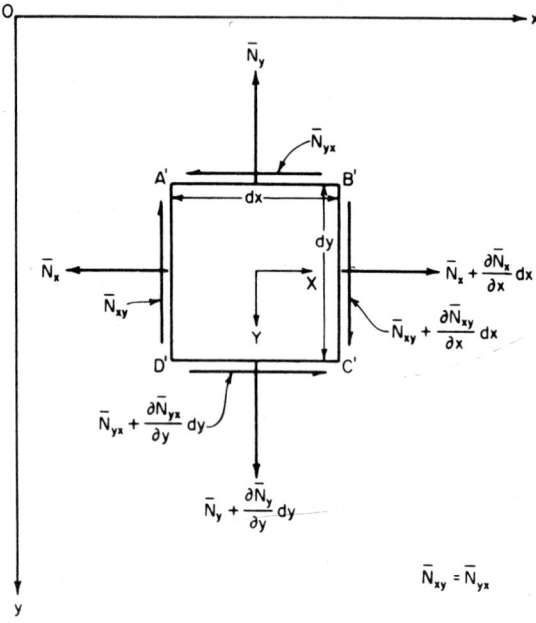

Fig. 15-2

Next, let us consider the contribution of the shears acting on the four sides.

Vertical component of shear force acting on AD = $\overline{N_{xy}}\,dy\,\tan\psi$

$$= \overline{N_{xy}}\,\frac{\partial z}{\partial y}\,dy$$

This force acts upward.

Vertical component of shear force on side BC = $\overline{N_{xy}}\,\frac{\partial z}{\partial y}\,dy$

$$+ \frac{\partial}{\partial x}\left(\overline{N_{xy}}\,\frac{\partial z}{\partial y}\right)dx\,dy$$

This force acts downward.

Summing up the vertical components of the shear forces acting on AD and BC, we get a net downward force of $(\partial/\partial x)[\overline{N_{xy}}(\partial z/\partial y)]\,dx\,dy$. Proceeding in a similar manner, it can be shown that the net downward force contributed by the shear forces acting on the pair of faces AB and CD is $(\partial/\partial y)[\overline{N_{xy}}(\partial z/\partial x)]\,dx\,dy$. The load z contributes a downward force of $Z\,dx\,dy$. Summing up all forces acting on the element in the z direction, we get

$$\frac{\partial}{\partial x}\left(\overline{N_x}\frac{\partial z}{\partial x}\right) + \frac{\partial}{\partial y}\left(\overline{N_y}\frac{\partial z}{\partial y}\right) + \frac{\partial}{\partial x}\left(\overline{N_{xy}}\frac{\partial z}{\partial y}\right) + \frac{\partial}{\partial y}\left(\overline{N_{xy}}\frac{\partial z}{\partial x}\right) + Z = 0$$

This may be expanded as

$$\overline{N_x}\frac{\partial^2 z}{\partial x^2} + 2\overline{N_{xy}}\frac{\partial^2 z}{\partial x\,\partial y} + \overline{N_y}\frac{\partial^2 z}{\partial y^2}$$

$$+ \left(\frac{\partial\overline{N_x}}{\partial x} + \frac{\partial\overline{N_{xy}}}{\partial y}\right)p + \left(\frac{\partial\overline{N_y}}{\partial y} + \frac{\partial\overline{N_{xy}}}{\partial x}\right)q + Z = 0$$

Making use of (15-14) and (15-15) and substituting for the quantities within the brackets, we may recast the above equation as

$$\overline{N_x}\frac{\partial^2 z}{\partial x^2} + 2\overline{N_{xy}}\frac{\partial^2 z}{\partial x\,\partial y} + \overline{N_y}\frac{\partial^2 z}{\partial y^2} = -Z + pX + qY$$

Or, the equation of equilibrium in the z direction takes the form

$$r\overline{N_x} + 2s\overline{N_{xy}} + t\overline{N_y} = -Z + pX + qY \tag{15-16}$$

15-5 Stress Function

We may reduce the three equations of equilibrium (15-14), (15-15), and (15-16) into a single differential equation by introducing a stress function ϕ. This was first suggested by Pucher [74] in 1934. The stress function ϕ is so defined that

$$\overline{N_x} = \frac{\partial^2\phi}{\partial y^2} - \int_{x_0}^{x} X\,dx \tag{15-17}$$

$$\overline{N_y} = \frac{\partial^2\phi}{\partial x^2} - \int_{y_0}^{y} Y\,dy \tag{15-18}$$

and

$$\overline{N_{xy}} = -\frac{\partial^2\phi}{\partial x\,\partial y} \tag{15-19}$$

It may be verified by substitution of these results back into equations (15-14) and (15-15) that the stress function ϕ satisfies the equations of equilibrium in the x and y directions. Substitution in (15-16) leads to the following differential equation:

$$r \frac{\partial^2 \phi}{\partial y^2} - 2s \frac{\partial^2 \phi}{\partial x \, \partial y} + t \frac{\partial^2 \phi}{\partial x^2} = pX + qY - Z + r \int X \, dx + t \int Y \, dy \quad (15\text{-}20)$$

The solution of a problem in the membrane theory of shells therefore reduces itself to finding a function ϕ which satisfies Equation (15-20) and the prescribed boundary conditions, when the shell is acted upon by specified X, Y, and Z loads.

15-6   Type of Differential Equation

Let us denote the right-hand side of Equation (15-20) by L so that it may be rewritten as

$$r \frac{\partial^2 \phi}{\partial y^2} - 2s \frac{\partial^2 \phi}{\partial x \, \partial y} + t \frac{\partial^2 \phi}{\partial x^2} = L \quad (15\text{-}21)$$

This second-order differential equation is elliptic, parabolic, or hyperbolic[1,2] according as

$$rt - s^2 \gtreqless 0 \quad (15\text{-}22)$$

As already shown in Art. 5-30, these are also the conditions for a shell surface to be synclastic, developable, or anticlastic.

15-7   Characteristic Lines

Consider a point P near the boundary of the plan projection of a shell (Fig. 15-3). Let the values of ϕ and $\partial \phi / \partial n$, its derivative in the direction of the normal to the curve, be specified on the boundary. From these values, $\partial \phi / \partial x$ and $\partial \phi / \partial y$ may be found. We may write down the following total differentials:

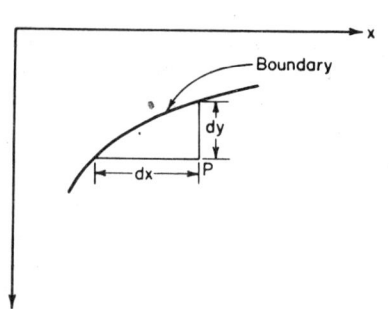

FIG. 15-3

$$d \left(\frac{\partial \phi}{\partial x} \right) = \frac{\partial^2 \phi}{\partial x^2} \, dx + \frac{\partial^2 \phi}{\partial x \, \partial y} \, dy \quad (15\text{-}23)$$

and

$$d \left(\frac{\partial \phi}{\partial y} \right) = \frac{\partial^2 \phi}{\partial x \, \partial y} \, dx + \frac{\partial^2 \phi}{\partial y^2} \, dy \quad (15\text{-}24)$$

[1] For a more detailed discussion of this subject the reader may consult A. Sommerfield, "Partial Differential Equations," Academic Press Inc., New York, 1957.

[2] Sneddon, I. N., "Elements of Partial Differential Equations," McGraw-Hill Book Company, New York, 1957.

Relations (15-21), (15-23), and (15-24) furnished us three simultaneous equations in the three unknowns $\partial^2\phi/\partial x^2$, $\partial^2\phi/\partial x\,\partial y$, and $\partial^2\phi/\partial y^2$. These derivatives in fact represent the stresses at the point P. A unique solution for these three unknowns exists if and only if the determinant of the system of equations is different from zero, i.e.,

$$\begin{vmatrix} r & -2s & t \\ 0 & dy & dx \\ dy & dx & 0 \end{vmatrix} \neq 0$$

i.e.,

$$r\,dx^2 + 2s\,dx\,dy + t\,dy^2 \neq 0 \qquad (15\text{-}25)$$

If this determinant vanishes, we have

$$r\,dx^2 + 2s\,dx\,dy + t\,dy^2 = 0 \qquad (15\text{-}26)$$

or

$$t\left(\frac{dy}{dx}\right)^2 + 2s\frac{dy}{dx} + r = 0$$

or

$$\frac{dy}{dx} = \frac{-s \pm \sqrt{s^2 - rt}}{t} \qquad (15\text{-}27)$$

Let us first consider the case where r, s, and t are constants. Integrating (15-27), we get

$$y = \frac{-s \pm \sqrt{s^2 - rt}}{t}\,x + C \qquad (15\text{-}28)$$

Equation (15-28) defines two families of straight lines on the plan projection of the shell. These are known as the *characteristic lines of the surface*. If r, s, and t are not constants, the characteristic lines are in general two families of curves. It is evident from (15-27) that for the characteristic lines to be real, $s^2 - rt \geqslant 0$. *This means that only shells of negative or zero Gauss curvature can have real characteristic lines.* These lines are imaginary for shells of positive Gauss curvature.

Characteristic lines are in fact the plan projections of the asymptotic lines on the surface. Asymptotic lines may be defined as surface curves whose tangents at every point coincide with the asymptotic directions. Through every point on a surface whose Gauss curvature is negative, there pass two asymptotic lines which are such that their normal curvature is everywhere zero. Their directions may therefore be found by equating the numerator of expression (5-24) for the normal curvature to zero to get

$$r\,dx^2 + 2s\,dx\,dy + t\,dy^2 = 0$$

This is seen to be identical with (15-26) which defines the direction of the characteristic lines. Hence in the literature on shells the terms asymptotic lines and characteristic lines are sometimes loosely used as though they are synonymous and hence interchangeable.

Characteristic lines play a very important role in the membrane theory of anticlastic shells. As we shall see later in our discussion of the membrane theory as applied to hyperbolic paraboloids and conoids, the orientation of the characteristic lines would decide if membrane equilibrium is at all possible, and if possible whether the resulting stresses and displacements would exhibit discontinuities or not. The presence of real characteristic lines for such surfaces also imposes certain restrictions on the type of boundary conditions that are admissible. Synclastic shells do not present similar difficulties, because their characteristic lines are imaginary.

15-8 Open Boundaries

Consider a shell of negative Gauss curvature whose boundaries are characteristic lines (Fig. 15-4). Choosing the characteristic lines $\xi = 0$ and $\eta = 0$, oblique to each other in general, as the coordinate axes, Equation (15-21) may be recast as[1]

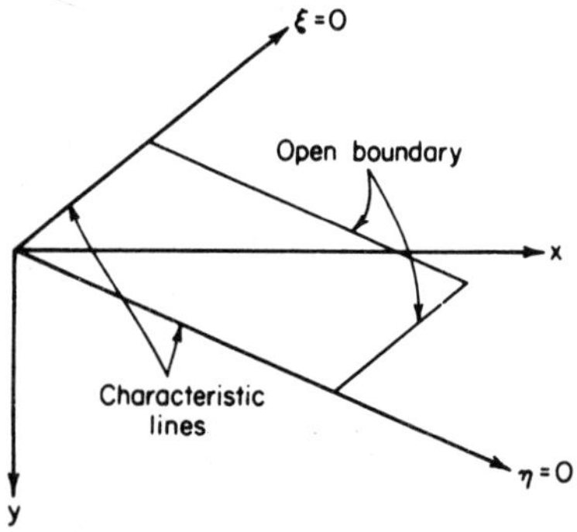

Fig. 15-4

$$\frac{\partial^2 \phi}{\partial \xi\, \partial \eta} = f\left(\frac{\partial \phi}{\partial \xi}, \frac{\partial \phi}{\partial \eta}, \xi, \eta\right) \quad (15\text{-}29)$$

Integration gives

$$\phi = \int\!\!\int f\, d\xi\, d\eta + \psi_1(\xi) + \psi_2(\eta) \quad (15\text{-}30)$$

Hence ϕ may be found, except for first-order terms, at all points if its value is known on $\xi = 0$ and $\eta = 0$. Hence it is appropriate to prescribe *one* boundary condition on each of the adjacent characteristic lines $\xi = 0$ and $\eta = 0$. No boundary conditions can be specified on the two remaining sides, which are known as *open boundaries*. We shall make use of this result in applying the membrane theory to the hyperbolic paraboloid and the conoid.

---

[1] For a detailed discussion of types of partial differential equations the reader may consult A. Sommerfield, "Partial Differential Equations," Academic Press Inc., New York, 1957.

CHAPTER 16

MEMBRANE THEORY OF SYNCLASTIC SHELLS

16-1 The Rotational Paraboloid and the Elliptic Paraboloid

Among synclastic shells, the rotational paraboloid and the elliptic paraboloid are the two surfaces that are most frequently favored for roofing very large column-free areas, square or rectangular in plan. Another form that has sometimes been employed is a shallow spherical calotte on a square or rectangular ground plan [75]. Doganoff [76] reports that in Germany an area as large as 30 by 40 m (98.5 by 131.2 ft) has been roofed over by a spherical calotte shell only 7 cm (2.75 in.) thick. This application bears eloquent testimony to the efficiency of this structural form. If the shell is shallow, the surface of a spherical calotte becomes indistinguishable from that of the corresponding rotational paraboloid and may be so regarded to simplify the analysis.

We have already seen in Chapter 5 that the rotational paraboloid and the elliptic paraboloid are surfaces of translation whose equation may be represented in the form $z = f_1(x) + f_2(y)$. Vertical sections of an elliptic paraboloid are parabolas and its horizontal sections are ellipses. Hence the name elliptic paraboloid. A shell roof in the form of an elliptic paraboloid or rotational paraboloid over a rectangular or square ground plan is usually supported by "shear diaphragms" on all the four edges (Fig. 16-1). The diaphragms are assumed to be stiff enough in their own planes to receive the shear load transferred by the shell; but they cannot carry any load applied normal to their planes.

THE ELLIPTIC PARABOLOID

16-2 Geometry

Referring to Fig. 16-2, the equation of an elliptic paraboloid assumes the form

$$z = \frac{f_x}{a^2} x^2 + \frac{f_y}{b^2} y^2 \qquad (16\text{-}1)$$

The total central rise of the shell $f = (f_x + f_y)$. Expressions for p, q, r, s, and t are found by differentiation. Thus

$$p = \frac{2f_x x}{a^2} \qquad q = \frac{2f_y y}{b^2} \qquad r = \frac{2f_x}{a^2} \qquad s = 0 \qquad \text{and} \qquad t = \frac{2f_y}{b^2} \qquad (16\text{-}2)$$

If R_1 and R_2 are the radii of curvature of the two generating parabolas, it is easily seen that

$$\frac{1}{R_1} = \frac{2f_x}{a^2} \qquad \frac{1}{R_2} = \frac{2f_y}{b^2} \qquad (16\text{-}3)$$

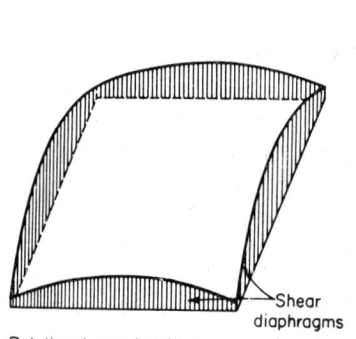

Rotational paraboloid on a square plan

FIG. 16-1

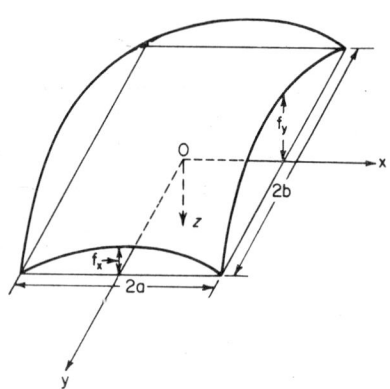

FIG. 16-2

16-3 Membrane Differential Equation

The substitution of r, s, and t, just derived, in Equation (15-20) leads to the following partial differential equation to be satisfied by the Pucher stress function for an elliptic paraboloid

$$\frac{2f_x}{a^2} \frac{\partial^2 \phi}{\partial y^2} + \frac{2f_y}{b^2} \frac{\partial^2 \phi}{\partial x^2} = pX + qY - Z + \int X \, dx + \int Y \, dy \qquad (16\text{-}4)$$

In most practical applications, $X = Y = 0$. In such cases, we may simplify (16-4) as

$$\frac{2f_x}{a^2} \frac{\partial^2 \phi}{\partial y^2} + \frac{2f_y}{b^2} \frac{\partial^2 \phi}{\partial x^2} = -Z \qquad (16\text{-}5)$$

16-4 Stresses Due to Snow Load

When a uniformly distributed snow load of p_0 is acting on the shell, Equation (16-5) takes the form

$$\frac{2f_x}{a^2} \frac{\partial^2 \phi}{\partial y^2} + \frac{2f_y}{b^2} \frac{\partial^2 \phi}{\partial x^2} = -p_0 \qquad (16\text{-}6)$$

The corresponding homogeneous solution may be written as

$$\frac{2f_x}{a^2}\frac{\partial^2 \phi}{\partial y^2} + \frac{2f_y}{b^2}\frac{\partial^2 \phi}{\partial x^2} = 0 \tag{16-7}$$

We may seek a solution in the form

$$\phi = X(x) \cdot Y(y) \tag{16-8}$$

Substitution in (16-7) leads to the equation

$$\frac{2f_x}{a^2} Y''X + \frac{2f_y}{b^2} X''Y = 0$$

where the primes stand for partial differentiation with respect to x or y. We may now write

$$\frac{2f_y a^2}{2f_x b^2}\frac{X''}{X} = -\frac{Y''}{Y} = \lambda^2 \tag{16-9}$$

This leads to two ordinary differential equations of the form

$$Y'' + \lambda^2 Y = 0 \tag{16-10}$$

$$X'' - \frac{f_x}{f_y}\frac{b^2}{a^2}\lambda^2 X = 0 \tag{16-11}$$

Taking note of symmetry, we may try a solution in the form

$$\phi = -\sum_{1,3,5,\ldots}^{\infty} A_n \cosh \beta_n x \cos \lambda_n y \tag{16-12}$$

where

$$\lambda_n = \frac{n\pi}{2b}$$

and

$$\beta_n = \sqrt{\frac{f_x}{f_y}}\frac{b}{a}\lambda_n = \sqrt{\frac{f_x}{f_y}}\frac{n\pi}{2a}$$

A particular integral may be assumed as

$$\phi = -\frac{a^2}{4f_x}p_0 y^2 \tag{16-13}$$

Combining the solution of the homogeneous equation and the particular integral, the complete solution may now be written as

$$\phi = -\sum_{n-1,3,5,\ldots}^{\infty} A_n \cosh \beta_n x \cos \lambda_n y - \frac{a^2}{4f_x}p_0 y^2 \tag{16-14}$$

The expressions for the projected stress resultants become

$$\bar{N}_x = \frac{\partial^2 \phi}{\partial y^2} = \sum_{n=1,3,5,\ldots}^{\infty} A_n \lambda_n^2 \cosh \beta_n x \cos \lambda_n y - \frac{a^2}{2f_x} p_0 \qquad (16\text{-}15)$$

$$\bar{N}_y = \frac{\partial^2 \phi}{\partial x^2} = - \sum_{n=1,3,5,\ldots}^{\infty} A_n \beta_n^2 \cosh \beta_n x \cos \lambda_n y \qquad (16\text{-}16)$$

$$\bar{N}_{xy} = - \frac{\partial^2 \phi}{\partial x \, \partial y} = - \sum_{n=1,3,5,\ldots}^{\infty} A_n \beta_n \lambda_n \sinh \beta_n x \sin \lambda_n y \qquad (16\text{-}17)$$

Because the traverses cannot receive any loads normal to their planes, we have the following boundary conditions:

$$\bar{N}_x = 0 \qquad \text{at} \quad x = \pm a$$

$$\bar{N}_y = 0 \qquad \text{at} \quad y = \pm b$$

The second boundary condition is automatically satisfied because of the form of the expression that we have assumed for the stress function. To apply the first boundary condition, it is necessary to expand the uniform load p_0 in Fourier series in the y direction to get

$$p_0 = \sum_{n=1,3,5,\ldots}^{\infty} \frac{4}{n\pi} (-1)^{(n-1)/2} p_0 \cos \lambda_n y \qquad (16\text{-}18)$$

The application of the first boundary condition now results in the following relation:

$$A_n \lambda_n^2 \cosh \beta_n a \cos \lambda_n y = \frac{a^2}{2f_x} \frac{4}{n\pi} (-1)^{(n-1)/2} p_0 \cos \lambda_n y$$

or

$$A_n = \frac{2 p_0 a^2 (-1)^{(n-1)/2}}{n\pi f_x \lambda_n^2 \cosh \beta_n a} \qquad (16\text{-}19)$$

Knowing A_n, we may now write down the following expressions for the projected stress resultants:

$$\bar{N}_x = \frac{p_0 a^2}{f_x} \left(\frac{2}{\pi} \sum_{n=1,3,5,\ldots}^{\infty} \frac{(-1)^{(n-1)/2} \cosh \beta_n x \cos \lambda_n y}{n \cosh \beta_n a} - \frac{1}{2} \right) \qquad (16\text{-}20)$$

$$\bar{N}_y = - \frac{p_0 b^2}{f_y} \left(\frac{2}{\pi} \sum_{n=1,3,5,\ldots}^{\infty} \frac{(-1)^{(n-1)/2} \cosh \beta_n x \cos \lambda_n y}{n \cosh \beta_n a} \right) \qquad (16\text{-}21)$$

and

$$\bar{N}_{xy} = - \frac{p_0 ab}{\sqrt{f_x f_y}} \left(\frac{2}{\pi} \sum_{n=1,3,5,\ldots}^{\infty} \frac{(-1)^{(n-1)/2} \sinh \beta_n x \sin \lambda_n y}{n \cosh \beta_n a} \right) \qquad (16\text{-}22)$$

The corresponding real stresses may be found by using relations (15-8), (15-9), and (15-10). For values of $f_x/f_y > 1$, convergence of the formulas for the pseudo stress resultants is fairly rapid and hence three or four terms of the series would give satisfactory accuracy. However, expression (16-22) for the shear converges rather slowly at $x = \pm a$. Parme [77] shows how the convergence can be improved by recasting (16-22) in the form

$$\bar{N}_{xy} = -\frac{p_0 ab}{\sqrt{f_x f_y}} \left[\frac{1}{2\pi} \log_e \left(\sec \frac{\pi y}{2b} + \tan \frac{\pi y}{2b} \right)^2 \right.$$
$$\left. -\frac{2}{\pi} \sum_{n=1,3,5,\ldots}^{\infty} (1 - \tanh \beta_n a) \frac{(-1)^{(n-1)/2}}{n} \sin \lambda_n y \right] \qquad (16\text{-}23)$$

He also notes that for values of $f_x/f_y > 1$, $\tanh \beta_n a \approx 1$ and hence the second term in expression (16-23) may be ignored except perhaps for $n = 1$. Tables are also available in Parme's paper for the values of $\bar{N}_x$, $\bar{N}_y$, and $\bar{N}_{xy}$ for a wide range of elliptic paraboloids with f_x/f_y values ranging from 0.20 to 1.00. If the shell is shallow, the dead weight of the shell may be regarded as uniformly distributed without appreciable error. Hence these tables may also be used for finding $\bar{N}_x$, $\bar{N}_y$, and $\bar{N}_{xy}$, when the shell is acted upon by dead weight. We may also use them for computing stresses in shallow spherical calottes.

THE ROTATIONAL PARABOLOID

16-5 Geometry

Because the two parabolas involved are identical for a rotational paraboloid,

$$\frac{f_x}{a^2} = \frac{f_y}{b^2} \qquad (16\text{-}24)$$

or

$$\frac{1}{R_1} = \frac{1}{R_2} = \frac{1}{R} \qquad (16\text{-}25)$$

$$r = t = \frac{1}{R} \quad \text{and} \quad s = 0 \qquad (16\text{-}26)$$

The equation may therefore be written in the form

$$z = \frac{1}{2R}(x^2 + y^2) \qquad (16\text{-}27)$$

Hence

$$p = \frac{x}{R} \quad \text{and} \quad q = \frac{y}{R}$$

16-6   Stresses under Snow Load p_0

The relevant expressions for $\bar{N}_x$, $\bar{N}_y$, and $\bar{N}_{xy}$ may be deduced from (16-20), (16-21), and (16-22). Because $f_x/a^2 = f_y/b^2$, it is seen from (16-12) that $\beta_n = \lambda_n$. Hence

$$\bar{N}_x = \frac{p_0 a^2}{f_x}\left(\frac{2}{\pi}\sum_{n=1,3,5,\dots}^{\infty}\frac{(-1)^{(n-1)/2}\cosh\lambda_n x\cos\lambda_n y}{n\cosh\lambda_n a} - \frac{1}{2}\right) \qquad (16\text{-}28)$$

$$\bar{N}_y = -\frac{p_0 b^2}{f_y}\left(\frac{2}{\pi}\sum_{n=1,3,5,\dots}^{\infty}\frac{(-1)^{(n-1)/2}\cosh\lambda_n x\cos\lambda_n y}{n\cosh\lambda_n a}\right) \qquad (16\text{-}29)$$

$$\bar{N}_{xy} = -\frac{p_0 ab}{\sqrt{f_x f_y}}\left(\frac{2}{\pi}\sum_{n=1,3,5,\dots}^{\infty}\frac{(-1)^{(n-1)/2}\sinh\lambda_n x\sin\lambda_n y}{n\cosh\lambda_n a}\right) \qquad (16\text{-}30)$$

If the shell is shallow, there is very little error in assuming the dead weight of the shell as uniformly distributed and using formulas (16-28), (16-29), and (16-30) to compute the pseudo stress resultants.

16-7   Stresses under Dead Weight g

The method explained in Art. 16-4 may still be used, provided the dead weight g is expanded in double Fourier series as

$$g = \sum\sum A_{mn}\cos\frac{m\pi x}{2a}\cos\frac{n\pi y}{2b}$$

16-8   Alternative Method for Snow Load

Let a rotational paraboloid on a rectangular ground plan be submitted to the action of a snow load p_0. Let us seek a stress function in the form

$$\phi = \sum_{m=1,3,5,\dots}^{\infty}\sum_{n=1,3,5,\dots}^{\infty} A_{mn}\cos\frac{m\pi x}{2a}\cos\frac{n\pi y}{2b} \qquad (16\text{-}31)$$

The uniform load p_0 may be expressed in double Fourier series as

$$p_0 = \sum_{m=1,3,5,\dots}^{\infty}\sum_{n=1,3,5,\dots}^{\infty} C_{mn}\cos\frac{m\pi x}{2a}\cos\frac{n\pi y}{2b} \qquad (16\text{-}32)$$

The membrane differential equation (15-20) assumes the following form for a rotational paraboloid:

$$\frac{\partial^2\phi}{\partial x^2} + \frac{\partial^2\phi}{\partial y^2} = -ZR \qquad (16\text{-}33)$$

When $Z = p_0$, Equation (16-33) takes the form

$$\frac{\partial^2 \phi}{\partial x^2} + \frac{\partial^2 \phi}{\partial y^2} = -p_0 R \tag{16-34}$$

Substituting for the derivatives of ϕ and p_0 from (16-31) and (16-32),

$$-A_{mn}\left[\left(\frac{m\pi}{2a}\right)^2 + \left(\frac{n\pi}{2b}\right)^2\right] = -C_{mn}R$$

or

$$A_{mn} = \frac{R}{\left[\left(\frac{m\pi}{2a}\right)^2 + \left(\frac{n\pi}{2b}\right)^2\right]} C_{mn} \tag{16-35}$$

It may be easily verified that

$$C_{mn} = \frac{16 p_0 (-1)^{[(m+n)/2-1]}}{mn\pi^2} \tag{16-36}$$

Hence

$$A_{mn} = \frac{64(-1)^{[(m+n)/2-1]} p_0 R}{mn\pi^4 \left[\left(\frac{m}{a}\right)^2 + \left(\frac{n}{b}\right)^2\right]} \tag{16-37}$$

Expressions for the stress resultants may be written down by finding the relevant derivatives of ϕ. But the series involved converges very slowly.

16-9 Polynomial Stress Functions

Both the methods discussed in the preceding articles lead to tedious and lengthy calculations if the load acting is not uniform. A simpler approach would be to construct a polynomial stress function. Fischer [78] derives such a stress function for rotational paraboloids on a square base. In the next article, we propose to develop a polynomial stress function applicable to a rotational paraboloid on a rectangular base submitted to the action of the dead weight of the shell. The procedure described can be easily adapted and applied to an elliptic paraboloid as well. Although we have chosen the dead weight as the load acting on the shell for purposes of illustration, the method can take care of any symmetrical loading expressed in polynomial form.

16-10 A Polynomial Stress Function for Dead Weight g

Let us select the stress function in the form

$$\phi = (x^2 - a^2)(y^2 - b^2)(Ax^2 + By^2 + C) \tag{16-38}$$

It automatically satisfies the desired boundary conditions of $\bar{N}_x = 0$ at

$x = \pm a$ and $\bar{N}_y = 0$ at $y = \pm b$. We may work out the derivatives of ϕ from (16-38). Thus

$$\frac{\partial^2 \phi}{\partial y^2} = \bar{N}_x = 2[Ax^4 + 6Bx^2y^2 + (C - Aa^2 - Bb^2)x^2$$
$$- 6Ba^2y^2 + a^2(Bb^2 - C)] \tag{16-39}$$

$$\frac{\partial^2 \phi}{\partial x^2} = \bar{N}_y = 2[By^4 + 6Ax^2y^2 + (C - Aa^2 - Bb^2)y^2$$
$$- 6Ab^2x^2 + b^2(Aa^2 - C)] \tag{16-40}$$

$$-\frac{\partial^2 \phi}{\partial x\,\partial y} = \bar{N}_{xy} = -8(Ax^3y + Bxy^3) + 4xy(Aa^2 + Bb^2 - C) \tag{16-41}$$

When the load acting is the dead weight, Equation (16-33) assumes the form

$$\frac{\partial^2 \phi}{\partial x^2} + \frac{\partial^2 \phi}{\partial y^2} = -gR\sqrt{1 + p^2 + q^2} = -gR\left[1 + \left(\frac{x}{R}\right)^2 + \left(\frac{y}{R}\right)^2\right]^{1/2}$$

Expanding the right-hand side by the binomial theorem and limiting ourselves to the first two terms,

$$\frac{\partial^2 \phi}{\partial x^2} + \frac{\partial^2 \phi}{\partial y^2} = -gR\left[1 + \frac{1}{2R^2}(x^2 + y^2)\right] \tag{16-42}$$

Substituting for the derivatives of ϕ from (16-39) and (16-40) and equating the coefficients of like terms on the left- and right-hand sides of the equation, we arrive at the following three simultaneous equations in the three unknowns A, B, and C:

$$2(C - Aa^2 - Bb^2 - 6Ab^2) = -\frac{g}{2R}$$

$$2(C - Aa^2 - Bb^2 - 6Ba^2) = -\frac{g}{2R} \tag{16-43}$$

and

$$2a^2(Bb^2 - C) + 2b^2(Aa^2 - C) = -gR$$

From the first two equations of the set, we find that

$$Ab^2 = Ba^2$$

Making use of this result, B may be eliminated to give a set of two simultaneous equations

$$2C - 2Aa^2 - 2A\frac{b^4}{a^2} - 12Ab^2 = -\frac{g}{2R}$$

and

$$2b^2(a^2 + b^2)\,A - 2(a^2 + b^2)\,C = -gR$$

or

$$C - Aa^2 - \frac{Ab^4}{a^2} - 6Ab^2 = -\frac{g}{4R}$$

and

$$b^2A - C = -\frac{gR}{2(a^2 + b^2)} \tag{16-44}$$

Knowing A, B, and C, the stresses are easily found. The method gives satisfactory accuracy if the shell is not too deep. If more accuracy is required, higher-power terms can be included in the stress function and more terms retained in the binomial expansion. The calculation procedure is explained in detail in Design Example 16-1.

Design Example 16-1

MEMBRANE ANALYSIS OF A PARABOLOID OF REVOLUTION

1. *DATA*

   Geometry
   $2a = 45$ ft
   $2b = 60$ ft
   Rise at crown $f = f_1 + f_2 = 8$ ft
   Thickness $d = 3$ in.

   Loads
   Total load on surface $= g = 60$ psf

2. *GEOMETRICAL PROPERTIES*

   Choosing the origin at the crown as shown in Fig. 16-3, the equation of the surface is given by

   $$z = \frac{x^2}{2R} + \frac{y^2}{2R}$$

   Noting that $z = f$ at $x = a$ and $y = b$, the radius of curvature of the surface is given by

   $$R = \frac{1}{2f}(a^2 + b^2)$$

   $$= \frac{1}{2 \times 8}(22.5^2 + 30^2) = 87.89 \text{ ft}$$

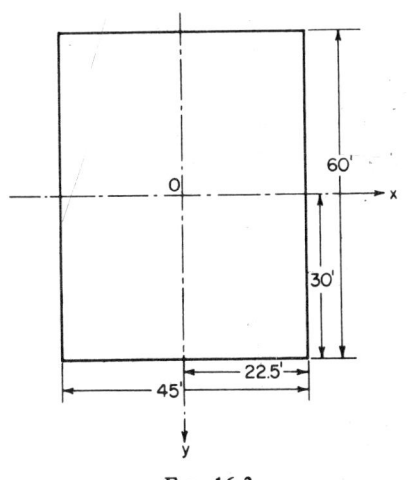

Fig. 16-3

3. STRESS FUNCTION

The stress function given in (16-38), namely, $\phi = (x^2 - a^2)(y^2 - b^2)(Ax^2 + By^2 + C)$, is chosen. Referring to Art. 16-9, the three equations in A, B, and C are written down:

$$2(C - Aa^2 - Bb^2 - 6Ab^2) = -\frac{g}{2R} \tag{i}$$

$$2(C - Aa^2 - Bb^2 - 6Ba^2) = -\frac{g}{2R} \tag{ii}$$

$$2a^2(Bb^2 - C) + 2b^2(Aa^2 - C) = -gR \tag{iii}$$

Substituting the values of a, b, R, and g, the equations are rewritten as

$$5{,}906.25\ A + 900\ B - C = 0.17066672$$
$$506.25\ A + 3{,}937.5\ B - C = 0.17066672$$
$$455{,}625\ A + 455{,}625\ B - 1{,}406.25\ C = -2{,}636.718$$

Solving the equations, we get

$$A = +0.00030966$$
$$B = +0.00055050$$
$$C = +2.15369050$$

4. CALCULATION OF STRESS RESULTANTS

The stresses $\bar{N}_x$, $\bar{N}_y$, and $\bar{N}_{xy}$ at various points are now calculated making use of expressions (16-39), (16-40), and (16-41), respectively. The real stresses N_x and N_y are calculated from $\bar{N}_x$ and $\bar{N}_y$ making use of the relations (15-8) and (15-9). From the real stresses the principal stresses P_1 and P_2 and the inclination of P_1 to the x axis are also calculated. These are shown tabulated below and plotted for some typical sections in Fig. 16-4.

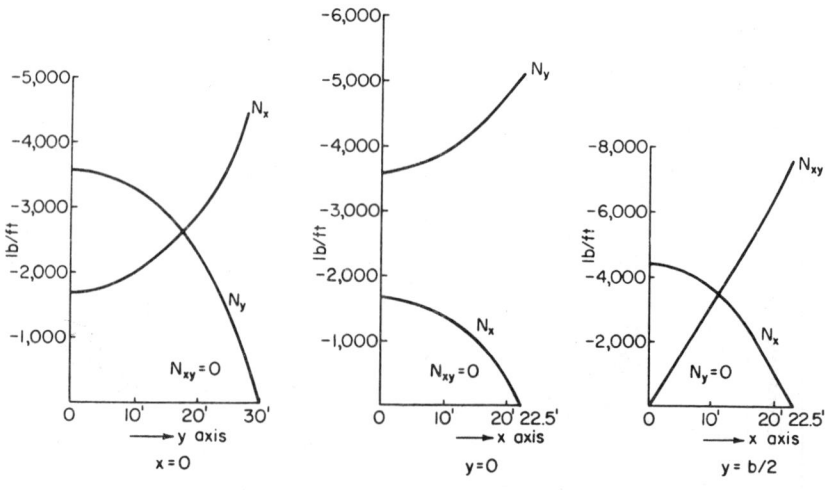

FIG. 16-4

Coordinates		Projected stresses, psf			Real stresses, psf		Principal stresses, psf		Inclination of P_1 to x axis, deg
x	y	$\bar{N}_x$	$\bar{N}_y$	$\bar{N}_{xy}$	N_x	N_y	P_1	P_2	
0	0	−1,679	−3,594	0	−1,679	−3,594	−1,679	−3,594	0
5.625	0	−1,583	−3,700	0	−1,587	−3,693	−1,587	−3,693	0
11.25	0	−1,289	−4,018	0	−1,300	−3,985	−1,300	−3,985	0
16.875	0	−774	−4,547	0	−788	−4,465	−788	−4,465	0
22.5	0	0	−5,288	0	0	−5,122	0	−5,122	0
0	7.5	−1,867	−3,422	0	−1,860	−3,435	−1,860	−3,435	0
5.625	7.5	−1,760	−3,521	−267	−1,757	−3,527	−1,717	−3,566	−8.40
11.25	7.5	−1,430	−3,819	−554	−1,437	−3,802	−1,313	−3,925	−12.55
16.875	7.5	−856	−4,315	−881	−868	−4,253	−653	−4,468	−13.75
22.5	7.5	0	−5,009	−1,267	0	−4,870	310	−5,180	−13.74
0	15.0	−2,431	−2,863	0	−2,397	−2,904	−2,397	−2,904	0
5.625	15.0	−2,289	−2,942	−597	−2,261	−2,979	−1,923	−3,316	−29.49
11.25	15.0	−1,853	−3,181	−1,234	−1,842	−3,200	−1,113	−3,929	−30.58
16.875	15.0	−1,103	−3,577	−1,950	−1,107	−3,564	−31	−4,640	−28.89
22.5	15.0	0	−4,133	−2,785	0	−4,062	1,416	−5,477	−26.95
0	22.5	−3,372	−1,792	0	−3,267	−1,850	−1,850	−3,267	0
5.625	22.5	−3,171	−1,838	−1,052	−3,078	−1,894	−1,278	−3,693	30.32
11.25	22.5	−2,559	−1,977	−2,164	−2,499	−2,024	−85	−4,439	41.87
16.875	22.5	−1,514	−2,209	−3,395	−1,494	−2,239	1,549	−5,282	−41.87
22.5	22.5	0	−2,533	−4,804	0	−2,533	3,702	−6,235	−37.62
0	30.0	−4,689	0	0	−4,437	0	0	−4,437	0
5.625	30.0	−4,405	0	−1,696	−4,177	0	602	−4,779	19.53
11.25	30.0	−3,546	0	−3,471	−3,384	0	2,169	−5,553	32.01
16.875	30.0	−2,090	0	−5,404	−2,014	0	4,490	−6,504	39.72
22.5	30.0	0	0	−7,576	0	0	7,576	−7,576	−45.00

OTHER SYNCLASTIC SHELLS

16-11 Finding Stress Functions

Csonka [79] has developed an inverse method for determining stress functions suitable for a large class of synclastic shells under vertical loading. The method consists in assuming a polynomial stress function which satisfies the boundary conditions and has a number of unknown coefficients.

Let the load at any point corresponding to the assumed stress function be Z^*, the actual load at the point being Z. Now we may stipulate that $(Z - Z^*) = 0$ at a number of selected points. The number of points at which this equality is to be written down will depend on the number of unknown coefficients in the selected stress function. Solving for the unknowns we arrive at the required stress function. Csonka [80] has also described several other methods of finding the stress function. Numerical techniques, such as the finite-difference method, can also be used with advantage to solve the fundamental differential equation of the membrane theory. Detailed information on these two methods is available in the books by Soare [72] and by Beles and Soare [81].

CHAPTER 17

MEMBRANE THEORY OF ANTICLASTIC SHELLS

17-1 Introductory Remarks

The hyperbolic paraboloid, the conoid, and the hyperboloid of one sheet are the most important members of the anticlastic group from the builders' point of view. The membrane theory of the hyperboloid of revolution of one sheet has already been discussed in Chapter 14. Barring its bold use for the assembly building at Chandigarh, India, and for a planetarium at St. Louis, Mo., most of the known applications of the hyperboloid of one sheet as a shell of revolution are as cooling towers of thermal power stations [82]. Parts of one-sheet hyperboloids may be used as roofing units [83]. The units are obtained by cutting

PHOTO 17-1. Shell roof, inspired by the hyperbolic cooling tower, over the Legislative Assembly Hall, Chandigarh, India. [*Courtesy of Gammon India (Private) Ltd., Bombay. Architectural design by Le Corbusier for Capital Project, Chandigarh. Contractors: Gammon India (Private) Ltd.*]

off a strip of a one-sheet hyperboloid by four planes, two of them at right angles to the axis of rotation and the other two parallel to it. They may be cut off from a one-sheet hyperboloid in which the minimum distance from the rotational axis may be between 80 and 500 ft. The width of the units is usually about 8 ft. Such units may be used for spans of up to 60 ft without ties. As these units were first produced

PHOTO 17-2. One-sheet hyperboloid roofing a planetarium at St. Louis, Mo. [*Courtesy of Portland Cement Association. Architects: Hellmuth, Obara & Kassabaum. Engineers: Albert Alper (Structural) and Harold P. Brehm (Mechanical). Contractors: Gamble Construction Co.*]

PHOTO 17-3. Hyperbolic cooling towers at Burnpur steel plant, India. [*Courtesy of Gammon India (Private) Ltd., Bombay. Designers: L. G. Mouchel and Partners, United Kingdom.*]

in Silberkuhl, Germany, they are sometimes called Silberkuhl shells. The units may be pretensioned in a factory, the prestressing wires being arranged along their straight-line generators [84]. All the three members of the anticlastic group share one feature in common: they are *ruled surfaces*. While the conoid is singly ruled, the other two are doubly ruled. This is an advantage from the constructional angle, because such ruled shells can be cast on forms composed of straight planks. The hyperbolic paraboloid appears to have been first used in France by Aimond in 1933. It owes its present vogue in architecture to the Mexican architect-engineer Felix Candela who more or less rediscovered it. He was the first to exploit this exciting and versatile structural form for roofing houses, factories, churches, and pavilions. Its use as a foundation has

PHOTO 17-4. Silberkuhl shells. [*Courtesy of NORMKO, Gesellschaft für Normkonstruktionen und Statik m.b.H. Proprietor: Bernhard Werning, Textile Industrie. Ochtrup/Westf (West Germany). Architect: Hermann Oeinck, Ochtrup/Westf. Producer: A.u.H. Hilgers, Factory for Prefab. Concrete Elements, Mönchengladbach. Erection: A.u.H. Hilgers.*]

also been reported [85]. Corbusier made spectacular use of it in his design for the Philips Pavilion for the Brussels World Fair in 1958 [86, 87].

THE HYPERBOLIC PARABOLOID

17-2 Geometry

The hyperbolic paraboloid may be visualized as a *translational surface* generated by a convex parabola moving over a concave parabola or vice versa (Fig. 17-1). These will be referred to as the *principal* parabolas. The equation of the surface may be written in the form

$$z = \frac{x^2}{2R_1} - \frac{y^2}{2R_2}$$

(17-1)

Vertical sections of the surface are parabolas and horizontal sections are

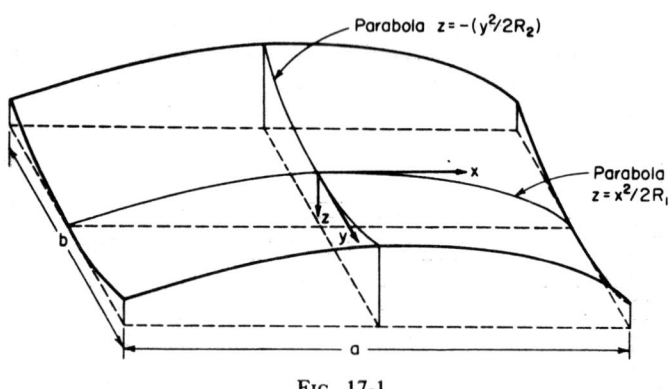

Parabola $z = -(y^2/2R_2)$

Parabola $z = x^2/2R_1$

FIG. 17-1

hyperbolas, hence the name hyperbolic paraboloid. Setting $z = 0$ in (17-1), we get

$$\frac{x^2}{2R_1} - \frac{y^2}{2R_2} = 0$$

or

$$\left(\frac{x}{\sqrt{2R_1}} + \frac{y}{\sqrt{2R_2}}\right)\left(\frac{x}{\sqrt{2R_1}} - \frac{y}{\sqrt{2R_2}}\right) = 0$$

(17-2)

Equation (17-2) represents a pair of straight lines lying entirely on the surface. It may be verified that these straight lines form the asymptotes of all hyperbolas obtained by horizontal sections of the surface. Hence they constitute the *asymptotes of the surface*. We may note that their inclination to the x axis is given by γ where

$$\tan \gamma = \sqrt{\frac{R_2}{R_1}}$$

(17-3)

In general, the asymptotes will be oblique to each other. If $R_2 = R_1$, the asymptotes become orthogonal and the resulting hyperbolic paraboloid is known as a *rectangular hyperbolic paraboloid*. It is easily seen that for such a surface the two parabolas involved in generating the surface are identical.

In many applications, it is convenient to express the equation of the surface with the asymptotes chosen as the axes of coordinates. The desired coordinate transformation is shown in Fig. 17-2. Let x' and y' be the coordinates in the new system. They are related to the coordinates x and y by the relations

$$x = (x' + y') \cos \gamma \qquad \text{and} \qquad y = (x' - y') \sin \gamma \qquad (17\text{-}4)$$

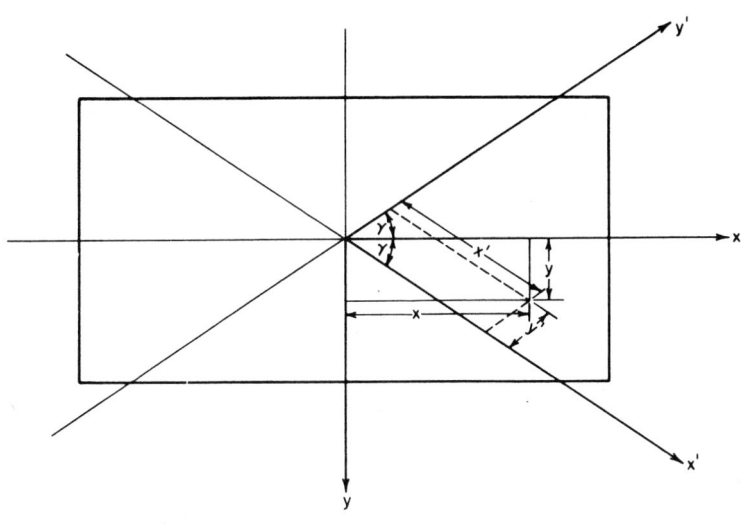

FIG. 17-2

Substituting these relations in (17-1) and noting that

$$\tan \gamma = \sqrt{\frac{R_2}{R_1}}$$

we get

$$z = \frac{2 \sin^2 \gamma}{R_2} x'y' \qquad (17\text{-}5)$$

Let us now consider the special case where the asymptotes are orthogonal. Choosing the asymptotes as axes of coordinates and dropping the primes for the sake of simplicity, we may represent the equation of the surface as

$$z = \frac{xy}{c} \qquad (17\text{-}6)$$

This surface may also be visualized as being formed by two sets of straight-line generators (Fig. 17-3). It is easily seen that the z coordinate of any point is fxy/ab. Or, the equation to the surface is

$$z = \frac{f}{ab}xy \qquad (17\text{-}7)$$

Comparing (17-7) with (17-6),

$$c = \frac{ab}{f} \qquad (17\text{-}8)$$

A simple way of visualizing a hyperbolic paraboloid is to regard it as a warped surface formed by elevating or depressing one corner of a rectangle by a certain amount while the other three corners remain at their original level. The surface shown in Fig. 17-3 is formed by two

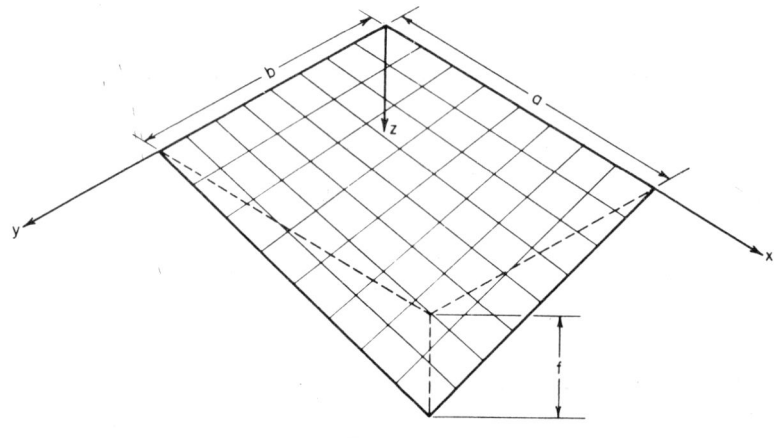

Fig. 17-3

sets of straight-line generators lying entirely on the surface. Members of each system are skew to the other members of the system and to one of the asymptotes. Similarly the members of the other family are skew to one another and to the other asymptote. Thus it is easy to see that the hyperbolic paraboloid is a *doubly ruled surface*. The plan projections on the x-y plane of these generators constitute two families of parallel lines which are the characteristic lines of the surface. They form rectangles if the shell is a rectangular hyperbolic paraboloid. From (15-28), it is clear that the characteristic lines are inclined at an angle of $\tan^{-1}\sqrt{R_2/R_1}$ to the x axis.

The hyperbolic paraboloid may be used in many different ways to form a roof. Some of these possibilities are illustrated in Fig. 17-4a–i.

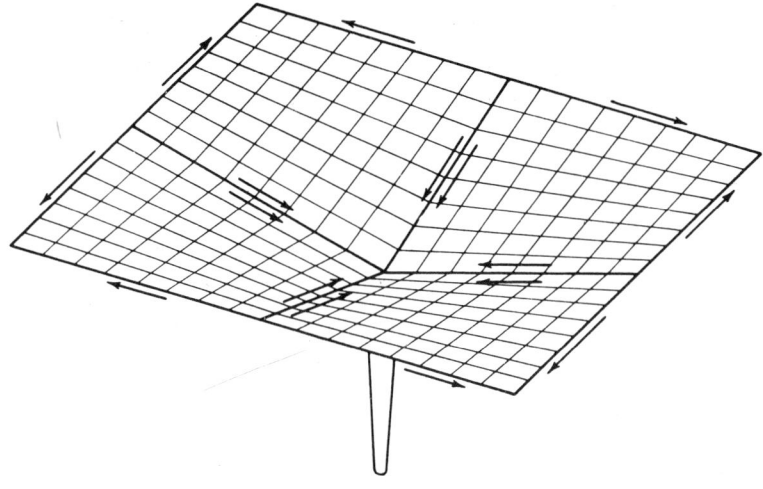

Fig. 17-4a

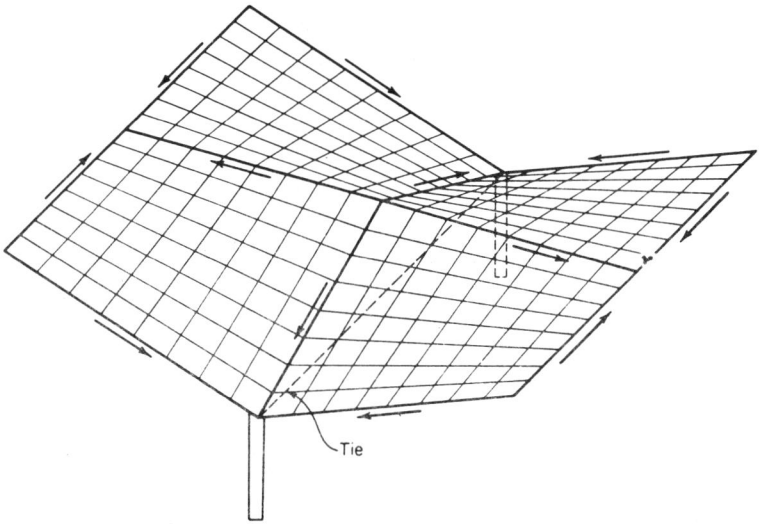

Fig. 17-4b

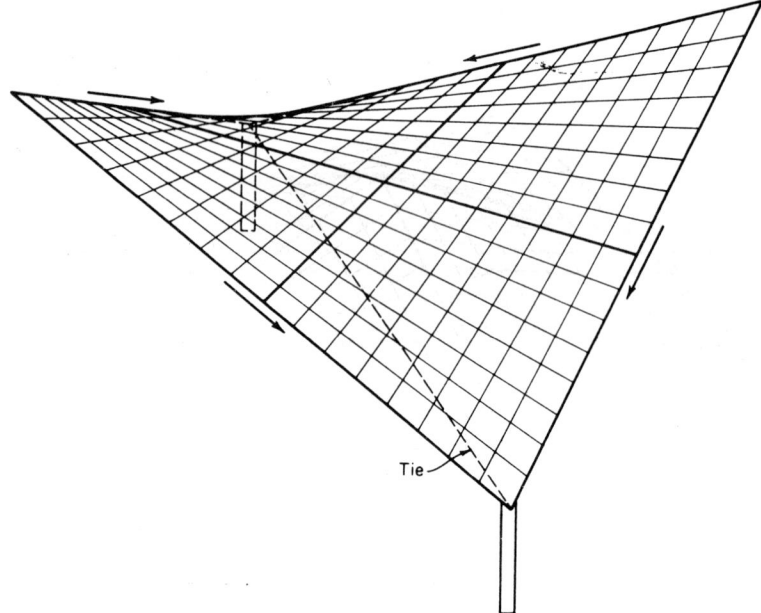

Tie

FIG. 17-4c

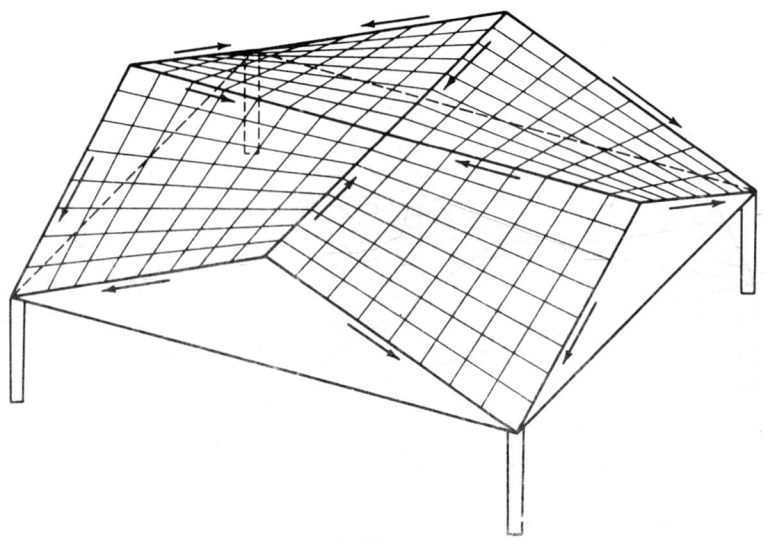

FIG. 17-4d

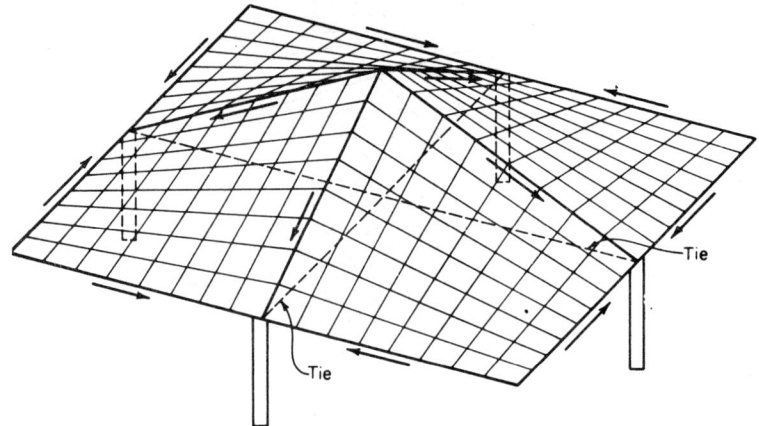

FIG. 17-4*e*

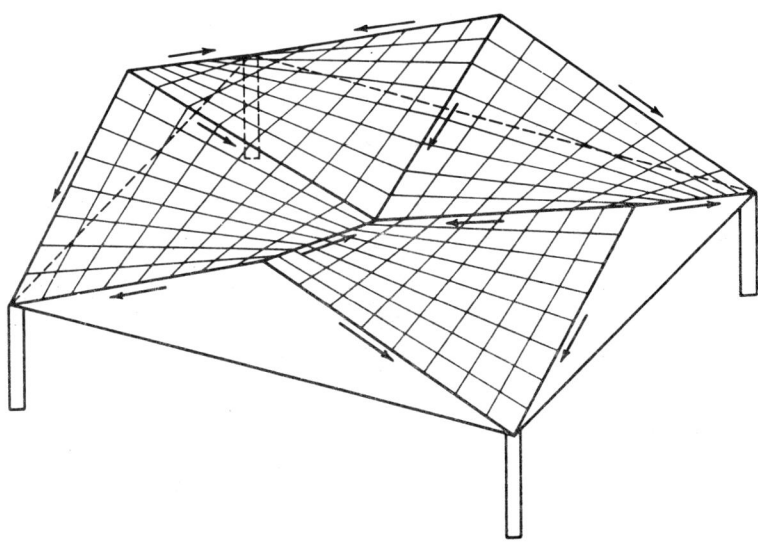

FIG. 17-4*f*

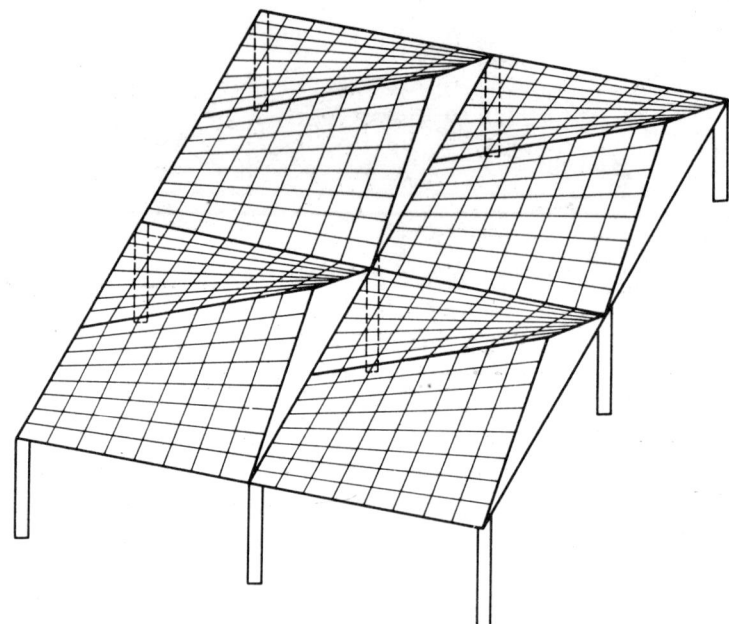

FIG. 17-4*g*

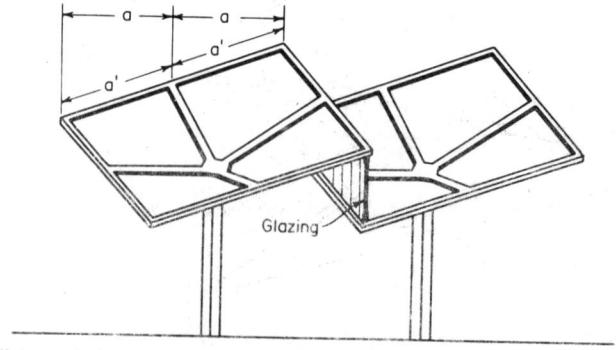

Tilted hyperbolic paraboloids with
northlight glazing

FIG. 17-4*h*

17-3 Membrane Theory of Rectangular Hyperbolic Paraboloids with Straight-line Generators as Boundaries

Reverting to Equation (17-6) for a rectangular hyperbolic paraboloid, we may find p, q, r, s, and t of the surface by differentiation. Thus

$$p = \frac{\partial z}{\partial x} = \frac{y}{c} \qquad q = \frac{x}{c} \qquad r = 0 \qquad s = \frac{1}{c} \qquad t = 0$$

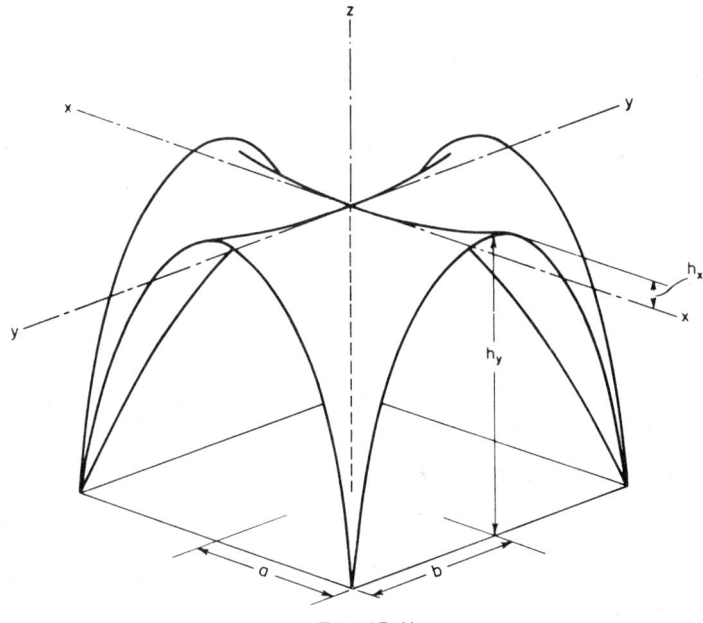

FIG. 17-4i

Inserting these values in Equation (15-16), we get

$$\frac{2}{c} \overline{N_{xy}} = -Z + pX + qY$$

Denoting the right-hand side of this equation by L,

$$\overline{N_{xy}} = \frac{cL}{2} \tag{17-9}$$

Inserting this value in Equation (15-14) and integrating with respect to x,

$$\overline{N_x} = -\int \left(\frac{c}{2}\frac{\partial L}{\partial y} + X\right) dx + f_1(y) \tag{17-10}$$

Similarly, making use of Equation (15-15),

$$\overline{N}_y = -\int \left(\frac{c}{2} \frac{\partial L}{\partial x} + Y\right) dy + f_2(x) \tag{17-11}$$

Consider a rectangular hyperbolic paraboloid under the action of the dead weight g of the shell. The value of L corresponding to this case is

$$-Z = -g \sqrt{1 + p^2 + q^2} = -\frac{g}{c} \sqrt{c^2 + x^2 + y^2}$$

Hence, from (17-9),

$$\overline{N}_{xy} = -\frac{g}{2} \sqrt{c^2 + x^2 + y^2} \tag{17-12}$$

From (17-10),

$$\overline{N}_x = \frac{g}{2} \int \frac{y \, dx}{\sqrt{c^2 + x^2 + y^2}} + f_1(y)$$

$$= \frac{gy}{2} \log_e \left(x + \sqrt{x^2 + y^2 + c^2}\right) + f_1(y) \tag{17-13}$$

Similarly,

$$\overline{N}_y = \frac{gx}{2} \log_e \left(y + \sqrt{x^2 + y^2 + c^2}\right) + f_2(x) \tag{17-14}$$

$f_1(y)$ and $f_2(x)$ are to be evaluated from the boundary conditions which depend upon the manner in which the shell is supported.

17-4 The Umbrella Roof

Consider the arrangement of the umbrella roof formed by four abutting hyperbolic paraboloids resting on four trusses (Fig. 17-5) along their edges. Taking any one of the hyperbolic paraboloids, say $OACB$, it abuts against the adjacent hyperbolic paraboloids along the edges OA and OB. Along the two remaining edges AC and BC, it is supported on trusses which are stiff in their own planes but are incapable of resisting any loads applied normal to their planes. Choosing the origin at O, we may formulate the boundary conditions as follows:

$$\overline{N}_x = 0 \quad \text{at} \quad x = a$$

$$\overline{N}_y = 0 \quad \text{at} \quad y = b$$

These two enable the arbitrary functions $f_1(y)$ and $f_2(x)$ appearing in (17-13) and (17-14) to be evaluated. The following expressions for the stresses result:

$$\overline{N}_x = \frac{gy}{2} \log_e \left(\frac{x + \sqrt{x^2 + y^2 + c^2}}{a + \sqrt{a^2 + y^2 + c^2}}\right) \tag{17-15}$$

and

$$\overline{N}_y = \frac{gx}{2} \log_e \left(\frac{y + \sqrt{x^2 + y^2 + c^2}}{b + \sqrt{x^2 + b^2 + c^2}} \right) \qquad (17\text{-}16)$$

These, together with Equation (17-12), define the state of stress in the shell. We may note, in passing, that no boundary conditions were applied along the edges OA and OB which are open boundaries. Along these edges, one should expect both a normal stress as well as a shear.

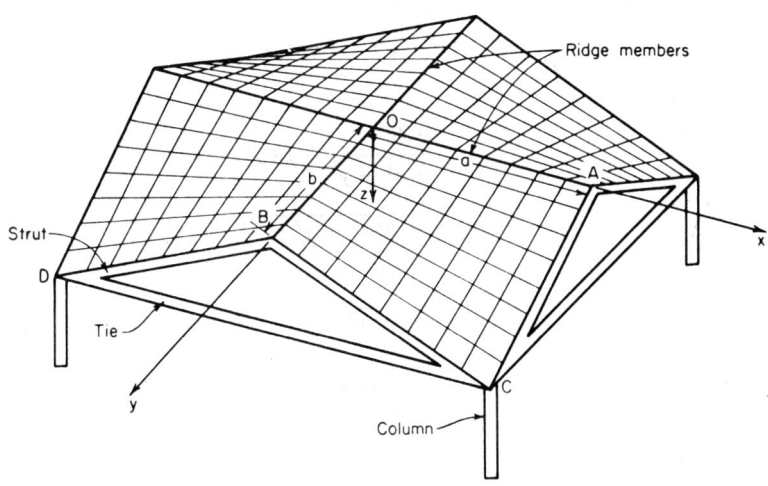

Fig. 17-5

Let us examine the case when the shell considered is shallow. For a shallow shell p^2 and q^2 may be neglected in comparison with 1 and we may write

$$-Z \approx -g$$

Hence from (17-12),

$$\overline{N}_{xy} = -\frac{gc}{2} \qquad (17\text{-}17)$$

Consequently,

$$\overline{N}_x = 0 \quad \text{and} \quad \overline{N}_y = 0$$

Thus we arrive at the important conclusion:

A shallow hyperbolic paraboloid submitted to the action of dead weight develops a state of pure shear unaccompanied by normal stresses.

The stresses that develop in the shell and the manner in which they are transferred to the edge members are shown in Fig. 17-6.   It may be noted that the horizontal ridge members at the junction of adjacent hyperbolic paraboloids and the edge members forming the sloping struts of the trusses carry compression.   The compression carried by these members may be found by summing up the shears transferred to them by the shell.   The axial compression in the strut *BC* varies from zero at *B* to its maximum value at *C*; similarly, the compressive force in the ridge *BO* increases from zero at *B* to its maximum value at *O*.   The

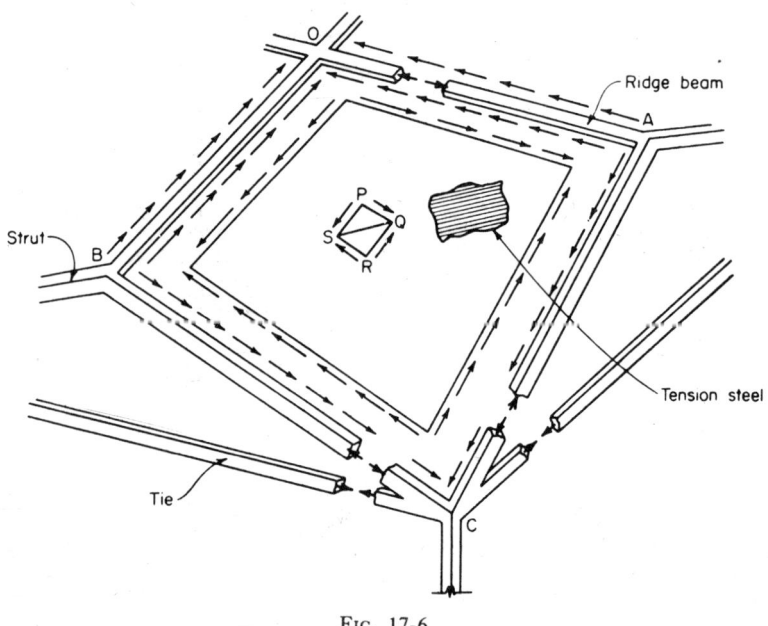

Fɪɢ. 17-6

member *DC* has to be designed as a tie to receive the horizontal component of the force transferred by the member *BC* to the column at *C*. The provision of the tie is necessary to ensure that the column is called upon to carry only a vertical load.   The state of pure shear that develops in a shallow hyperbolic paraboloid under the action of its dead weight is capable of a simple physical explanation.   Referring to Fig. 17-7, the hyperbolic paraboloid may be visualized as being made up of "arches" and "suspension cables" placed at right angles to each other and carrying loads in pure compression and tension, respectively.   At any point on the edge, the normal components of the tension and thrust of the cable and the arch, being equal, cancel each other; the tangential components add up and give rise to a shear. This interpretation also helps us in deciding on the direction in which the shell is to be reinforced. Obviously, tensile steel is to be arranged in the direction of the "cables." This is also clear from the shear stresses shown on the element of the shell (Fig. 17-6). The element which is in pure shear develops diagonal tension along SQ. Hence the steel needs to be arranged parallel to this direction.

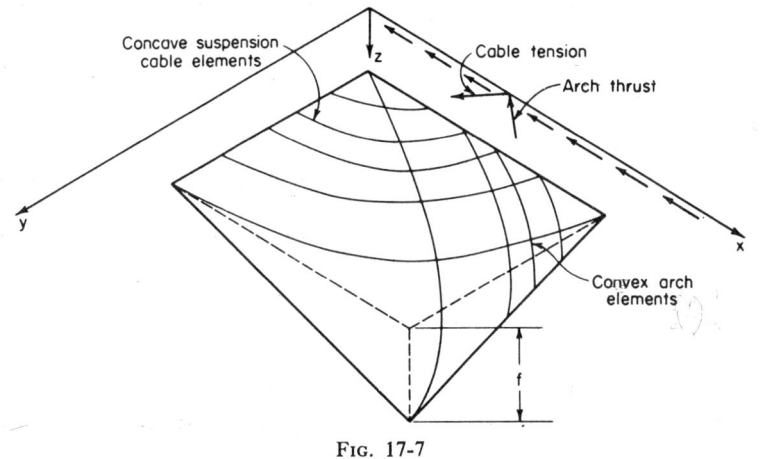

Concave suspension cable elements

Cable tension

Arch thrust

Convex arch elements

FIG. 17-7

17-5 The Inverted Umbrella Roof

Consider the assembly of four hyperbolic paraboloids shown in Fig. 17-8 forming an inverted umbrella roof. Let us study one of the hyperbolic paraboloids, say $OACB$, when acted upon by its dead weight g. Choosing the origin at C and coordinate axes as shown, we formulate the following boundary conditions:

$$\bar{N}_x = 0 \quad \text{at} \quad x = 0$$

and

$$\bar{N}_y = 0 \quad \text{at} \quad y = 0$$

Inserting these boundary conditions in (17-13) and (17-14), we arrive at the following expressions for the projected stresses:

$$\bar{N}_x = \frac{gy}{2} \log_e \left(\frac{x + \sqrt{x^2 + y^2 + c^2}}{\sqrt{y^2 + c^2}} \right) \tag{17-18}$$

and

$$\bar{N}_y = \frac{gx}{2} \log_e \left(\frac{y + \sqrt{x^2 + y^2 + c^2}}{\sqrt{x^2 + c^2}} \right) \tag{17-19}$$

The shear stress $\overline{N_{xy}}$ is given by (17-12).

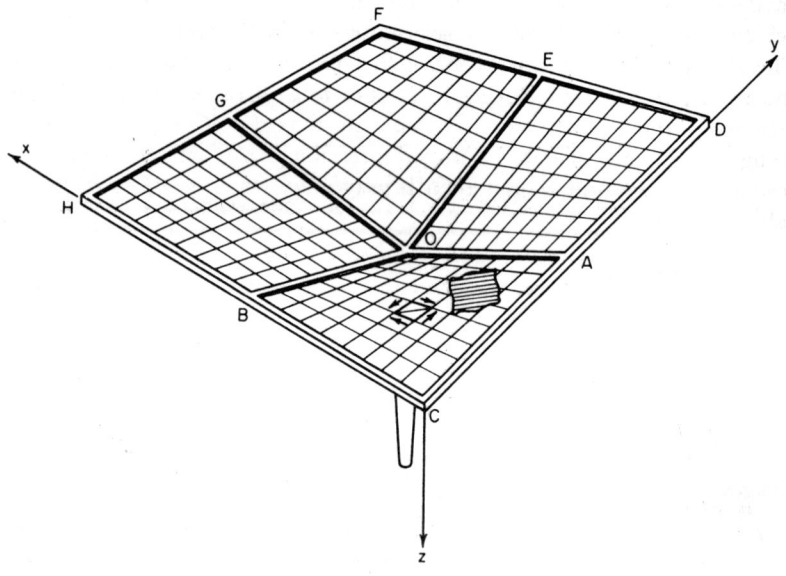

FIG. 17-8

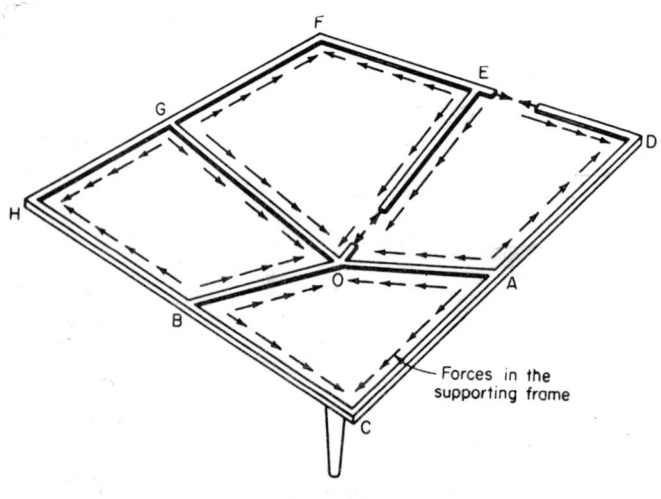

Forces in the supporting frame

FIG. 17-9

If the shell is shallow, the projected stresses $\overline{N_x}$ and $\overline{N_y}$ vanish and we have a state of pure shear. The stresses in the shell and the forces that develop in the supporting members are shown in Figs. 17-8 and 17-9. The sloping ridge members OA, OB, OG, and OE are in compression and the edge members CD, DF, FH, and HC are in tension. The axial tension in the edge members attains its maximum value at midspan, being zero at the corners. The compression in the sloping members is zero at A, B, G, and E, and then it gradually increases to its maximum value at O.

Photo 17-5. Inverted umbrella shells roofing the race track at Scioto Downs, Columbus, Ohio. *(Courtesy of Portland Cement Association. Architects: Kellam & Foley. Engineers: Gensert, Williams & Associates. Contractors: Sheaf Construction Co.)*

17-6 Shallow and Deep H.P. Shells

Most hyperbolic paraboloids used as roofs tend to be shallow. Hence it is of interest to know when a h.p. shell may be so regarded. Expanding (17-12) by the binomial theorem, we get

$$\overline{N_{xv}} = -\frac{gc}{2}\left(1 + \frac{x^2 + y^2}{2c^2} - \frac{1}{8}\frac{(x^2 + y^2)^2}{c^4} + \cdots\right) \qquad (17\text{-}20)$$

The second term within the brackets of this expression attains its maximum value of $(a^2 + b^2)/2c^2$ when $x = a$ and $y = b$. Let $a \leqslant b$.

$$\frac{a^2 + b^2}{2c^2} = \left(\frac{\dfrac{a^2}{b^2} + 1}{\dfrac{2a^2}{f^2}}\right)$$

Because $a/b \leqslant 1$,

$$\left(\frac{\dfrac{a^2}{b^2} + 1}{\dfrac{2a^2}{f^2}}\right) \leqslant \frac{f^2}{a^2}$$

Thus, if $f/a \leqslant \frac{1}{10}$, $(a^2 + b^2)/2c^2 \leqslant \frac{1}{100}$ or $\leqslant 1\%$ of the first term. Again, it is easily verified that the second term will be only 4% of the first term, if $f/a = \frac{1}{5}$.

Hence we may conclude that *for all practical purposes a hyperbolic paraboloid may be considered as shallow if the ratio of f to the shorter side is less than or equal to* 1:5.

17-7 Hyperbolic Paraboloid Supported on Principal Parabolas

There are two possibilities:

Case 1

A hyperbolic paraboloid may be supported on shear diaphragms along two edges with the other two edges free (Fig. 17-10).

Case 2

A hyperbolic paraboloid may be supported along all its four edges on shear diaphragms (Fig. 17-11).

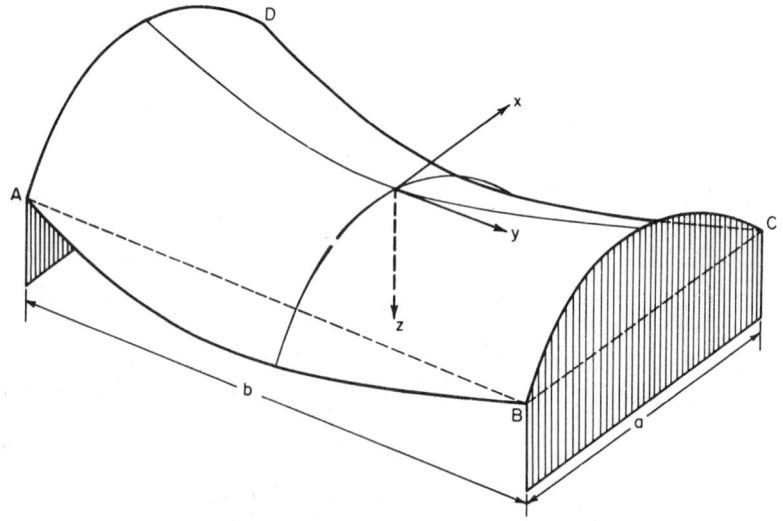

FIG. 17-10

Let us first consider Case 1. We have already seen that the equation of the shell may be written as

$$z = \frac{x^2}{2R_1} - \frac{y^2}{2R_2}$$

The membrane differential equation (15-16) now takes the form

$$\frac{1}{R_1}\frac{\partial^2\phi}{\partial y^2} - \frac{1}{R_2}\frac{\partial^2\phi}{\partial x^2} = -Z \qquad (17\text{-}21)$$

We may represent the load in Fourier series as

$$Z = \sum_{1,3,5,\dots}^{\infty} C_n \cos \alpha_n y \qquad (17\text{-}22)$$

where $\alpha_n = n\pi/b$. We may seek a particular integral of the form $\phi_0 = \phi_n \cos \alpha_n y$, where ϕ_n is a constant. Substitution in (17-21) results in the equation

$$-\phi_n \alpha_n^2 \frac{1}{R_1} = -C_n$$

or

$$\phi_n = \frac{C_n R_1}{\alpha_n^2}$$

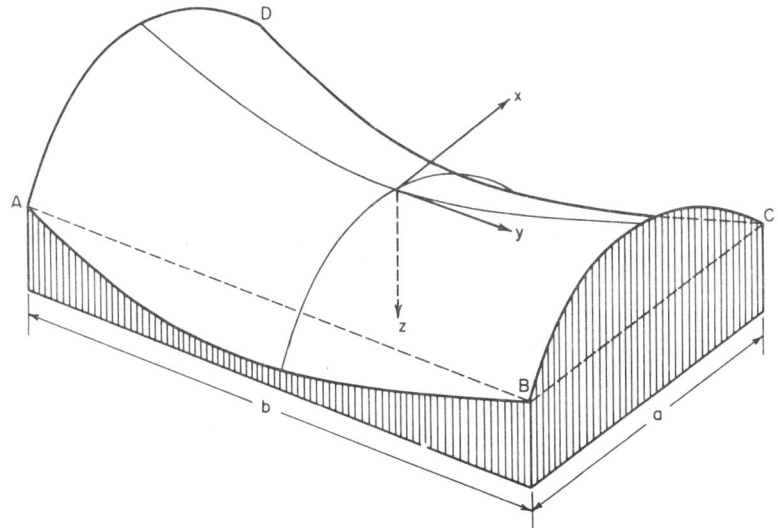

Fɪɢ. 17-11

Hence

$$\phi_0 = \sum_{n=1,3,5,\ldots}^{\infty} \frac{C_n R_1}{\alpha_n^2} \cos \alpha_n y \qquad (17\text{-}23)$$

The particular integral gives the following stresses:

$$\frac{\partial^2 \phi}{\partial y^2} = \bar{N}_x = -C_n R_1 \cos \alpha_n y$$

$$\frac{\partial^2 \phi}{\partial x^2} = \bar{N}_y = 0$$

$$\frac{\partial^2 \phi}{\partial x \, \partial y} = \bar{N}_{xy} = 0$$

The physical meaning of this solution is that the shell behaves as an arch in the x direction and is subjected to compression.

Next, let us solve the homogeneous equation

$$\frac{1}{R_1} \frac{\partial^2 \phi}{\partial y^2} - \frac{1}{R_2} \frac{\partial^2 \phi}{\partial x^2} = 0 \qquad (17\text{-}24)$$

Let us look for a solution in the form

$$\phi_1 = F_1(x) \cos \alpha_n y$$

Substitution in (17-24) leads to the ordinary differential equation

$$\frac{1}{R_2} F_1'' + \frac{\alpha_n^2}{R_1} F_1 = 0$$

or

$$F_1'' + \frac{R_2}{R_1} \alpha_n^2 F_1 = 0 \qquad (17\text{-}25)$$

where the primes denote differentiation with respect to x. The solution of (17-25) is easily seen to be

$$\phi_1 = \sum_{n=1,3,5,\ldots}^{\infty} \left(A_n \cos \sqrt{\frac{R_2}{R_1}} \alpha_n x + B_n \sin \sqrt{\frac{R_2}{R_1}} \alpha_n x \right) \cos \alpha_n y$$

Because we are dealing with a symmetrical problem, only the cosine terms may find a place in the solution. Hence the complementary function becomes

$$\phi_1 = A_n \cos \sqrt{\frac{R_2}{R_1}} \alpha_n x \cos \alpha_n y$$

Combining the particular integral and the complementary functions,

$$\phi = \sum_{1,3,5,\ldots}^{\infty} \left(\frac{C_n R_1}{\alpha_n^2} + A_n \cos \sqrt{\frac{R_2}{R_1}} \alpha_n x \right) \cos \alpha_n y \qquad (17\text{-}26)$$

$$\overline{N_x} = \frac{\partial^2 \phi}{\partial y^2} = -C_n R_1 \cos \alpha_n y - \alpha_n^2 A_n \cos \sqrt{\frac{R_2}{R_1}} \alpha_n x \cos \alpha_n y \qquad (17\text{-}27)$$

$$\overline{N_{xy}} = -\frac{\partial^2 \phi}{\partial x\, \partial y} = -A_n \alpha_n^2 \sqrt{\frac{R_2}{R_1}} \sin \sqrt{\frac{R_2}{R_1}} \alpha_n x \sin \alpha_n y \qquad (17\text{-}28)$$

The boundary conditions are

$$\overline{N_x} = 0 \qquad \text{at} \quad x = \pm \frac{a}{2}$$

and

$$\overline{N_{xy}} = 0 \qquad \text{at} \quad x = \pm \frac{a}{2}$$

Applying the first boundary condition and making use of (17-27), A_n may be found. From the second boundary condition, we get

$$\sin \sqrt{\frac{R_2}{R_1}} \alpha_n \frac{a}{2} = 0$$

or

$$\frac{a}{2} \alpha_n \tan \gamma = k\pi$$

Substituting $\alpha_n = n\pi/b$, and noting that n is odd, the condition to be satisfied is

$$\frac{a}{2} \frac{\pi}{b} \tan \gamma = k\pi$$

or

$$a \tan \gamma = 2bk \qquad (17\text{-}29)$$

where $k = 1, 2, 3, \ldots$. For membrane equilibrium to be possible condition (17-29) must be satisfied. If it is not satisfied, the edges AB and CD will not be free from shear. This is evident from (17-28).

Let us consider the case corresponding to $k = 1$. For this case, we have

$$\tan \gamma = \frac{2b}{a} \qquad (17\text{-}30)$$

Noting that γ is the angle made by the characteristic lines with the x axis, relation (17-30) may be geometrically depicted as shown in Fig. 17-12. Such a representation has been suggested by Bouma [88]. The case shown in Fig. 17-12 will now be studied in greater detail. A characteristic line starting from the corner A will, in general, intersect the opposite edge BC. Similarly, a characteristic line emanating from B will intersect the edge AD. The sides AD and BC are usually called the principal sides and the other two the secondary sides. The hyperbolic paraboloid and the one-sheet hyperboloid possess a unique property. If a force is applied at a point E on any edge, say AB, along the direction of one of the characteristic lines, say EF, it will stress only the strip on which it is applied and would need an equal and opposite force P at the other terminal point F for maintaining equilibrium. We may make use of this property to eliminate unwanted stresses from a given edge. This artifice appears to have been first employed by Aimond [89]. Thus, for instance, in the case under study, we desire that the edges AB and CD shall be free of normal and shear stresses. Let us see if this is possible with the geometrical proportions noted in Fig. 17-12. The particular integral ϕ_0 was found to give a normal compressive stress $\overline{N}_x$ on the edges AB and CD. Our effort now is to eliminate these stresses by applying suitable loads on the boundary AB. Mathematically, this process is equivalent to finding a suitable complementary function which, when combined with the particular integral, leaves the edges AB and CD free of all stresses. Let us start at a point E on the edge AB.

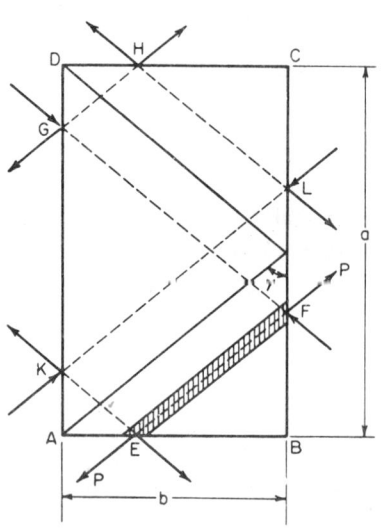

Orientation of characteristic lines for membrane equilibrium to be possible with shear diaphragms at AD and BC and edges AB and CD free

Fig. 17-12

We noted that the particular integral gave a normal compressive stress $\overline{N}_x$ on this edge. Over the length ds we thus have a compressive force of $\overline{N}_x\,ds$ acting on this edge at E. To free this edge of this force, we apply an equal and opposite force and resolve it along the directions of the two characteristic lines EF and EK emanating from this point. At the point F, we introduce a force along the characteristic line FG which is such that its normal component opposes the normal component of the force acting along FE. The tangential components of the two forces would add up and give rise to a shear. But the edge will remain free of normal forces. This is exactly what we

·wish to achieve. At G, the same step is repeated. Consequently at H, we end up with a tension acting along HG equal in magnitude to the one applied at E along EF. Following through the other characteristic line EK in a similar manner, we again end up with a tension at H applied in the direction HL and equal in magnitude to the tension initially applied at E in the direction EK. To sum up, we have, in one stroke, succeeded in wiping out the unwanted normal stresses at E and H of the edges AB and CD. By repeating this process for every point on the edges AB and CD, it is clear that we can ultimately free these edges of all stresses. Shears would appear on the principal sides as a result of this operation. Thus it is seen that a membrane equilibrium is possible with the prescribed boundary conditions. This conclusion confirms what we have already seen with the aid of mathematics. Although we have shown that a membrane equilibrium is possible, the resulting stress distribution is *discontinuous*. Why this is so has been explained lucidly by Flügge [20]. The stress distribution will be different in each of the zones I, II, and III marked on Fig. 17-13.

Now, considering a hyperbolic paraboloid of Case 2 the boundary conditions corresponding to this case are

$$\overline{N_x} = 0 \quad \text{at} \quad x = \pm \frac{a}{2}$$

and

$$\overline{N_y} = 0 \quad \text{at} \quad y = \pm \frac{b}{2}$$

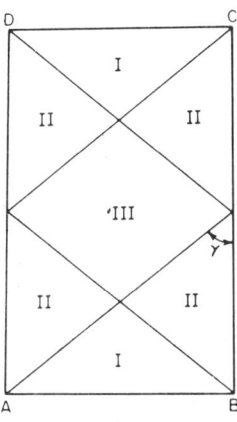

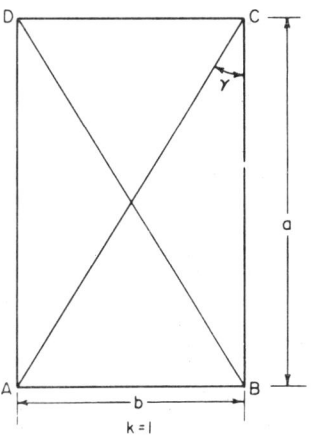

Orientation of characteristic lines for which membrane equilibrium with four shear diaphragms is not possible

Fig. 17-13

Fig. 17-14

Using the first boundary condition, A_n can be found from (17-27) provided

$$\cos \sqrt{\frac{R_2}{R_1}} \alpha_n \frac{a}{2} \neq 0$$

Or, a membrane solution is physically possible *except when*

$$\sqrt{\frac{R_2}{R_1}} \frac{\pi}{b} \frac{a}{2} = k \frac{\pi}{2} \qquad \text{with} \quad k = 1, 3, 5, \ldots$$

i.e.,

$$a \tan \gamma = bk \qquad \text{with} \quad k = 1, 3, 5, \ldots . \tag{17-31}$$

This condition is shown in Fig. 17-14. Figure 17-15 shows the orientation of characteristic lines when membrane equilibrium with

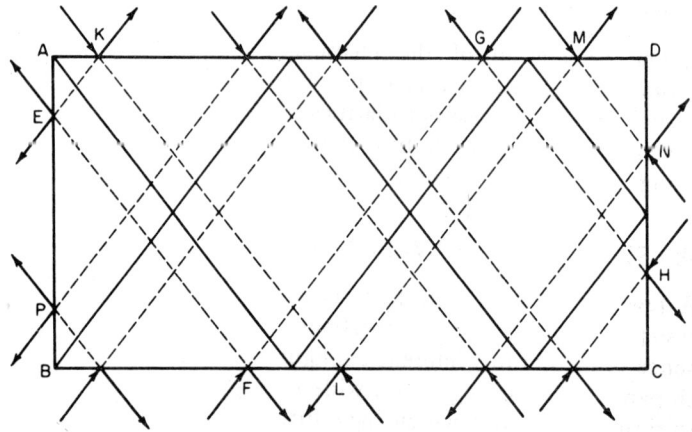

Orientation of characteristic lines for membrane equilibrium
with only shears on the four edges

FIG. 17-15.   *(After Flügge, "Stresses in Shells," Springer Verlag, Berlin.   Used with their permission.)*

four shear diapnragms is possible.   Let us study this case by using the technique of applying loads on the boundary to achieve the boundary conditions desired (Fig. 17-15).   To free the edge AB of normal stresses, we apply a force at E and resolve it along the characteristic lines EK and EF.   Following the chain of generator lines we find that they end up at two different points N and H on the boundary CD.   Similarly the forces applied at another point P on edge AB also terminate at the points N and H.   The normal components of the forces that appear at N and H cancel each other. The tangential components add up and give rise to a shear. Thus the thrusts on AB may be exchanged for shears on CD. Similarly thrusts on CD may be exchanged for shears on AB. In carrying out this process of transformation, we have introduced only shear stresses on the principal edges AD and BC. Thus, on all the four edges AB, CD, AD, and BC, only shear stresses remain. Hence these edges may be supported on shear diaphragms. This is true only if the loading is symmetrical.

The two cases studied in detail emphasize the following features involved in the membrane stress distribution of hyperbolic paraboloids supported on principal parabolas:

(i) A membrane state of equilibrium is not always possible.

(ii) Only certain special types of boundary conditions can be prescribed.

(iii) Even when a membrane equilibrium is found to be physically possible, the stress distribution may exhibit severe discontinuities.

(iv) The orientation of the characteristic lines and the parameter $(a/b)\sqrt{R_2/R_1}$ play a very important role in the membrane theory of such shells. These peculiarities of the hyperbolic paraboloid shell stem from the hyperbolic nature of the differential equation governing its membrane theory.

17-8 Stresses under Wind Loads

A few wind-tunnel tests have been carried out to determine wind-pressure distribution on hyperbolic paraboloid shells [90]. In the absence of results for different configurations of hyperbolic paraboloids with parameters varying over a wide range, one is obliged to depend on empirical assumptions which have little or no rational basis. The so-called cosine and cosine-square laws have sometimes been employed to arrive at stresses caused by wind load [91].

17-9 Lateral Loads

H. P. shells may be analyzed for lateral loads due to earthquake or wind as indicated by Meng and Laushey [92]. These authors came to the conclusion that the stresses induced in the shell by such lateral loads are probably insignificant. But these horizontal forces may cause significant bending in the edge beams and columns. These conclusions are in line with the usual practice followed in the design of rigid frames where the horizontal loads are considered in the design of beams and columns but ignored in the design of the slabs.

17-10   Unsymmetrical Loading

The possibility of unsymmetrical loading on hyperbolic paraboloid shells cannot be ruled out.   Thus it is quite possible that in an inverted umbrella assembly two of the adjacent shells are loaded, the other two being unloaded.   Such a condition of loading is critical for the sloping ribs and the central column.   The membrane theory cannot be used to carry out the stress analysis of such a shell.   We need a bending theory. Unfortunately, however, we do not yet have solutions of the bending theory available to cover such cases which arise frequently in practice. The available test results are also not very conclusive.   They can only, at best, provide qualitative guidance.   Based on the tests carried out at the Portland Cement Association on a 24- by 24-ft inverted umbrella roof, Yu and Kriz [93] come to the conclusion that the only dependable procedure in our present state of knowledge is to regard the rib commc to the loaded shells as a cantilever and design the column for the fuⅡ cantilever moment.   This approach is, of course, somewhat conservative.

17-11   Concentrated Loads

Only a bending theory can take care of concentrated loads on shells. Tests carried out at the Portland Cement Association to which reference h..s already been made also involved the application of concentrated loads. Measurements taken indicate that the "beam-strip" design method, usually employed-for plates, may be adequate in assessing the effect of concentrated loads on shallow h.p. shells.

17-12   Northlight Arrangements

Two possibilities of providing northlight roofs for factory buildings using h.p. shells are illustrated in Fig. 17-4g and h.   In the first, the glazing is accommodated in the triangular gables.   In the second, the h.p. shell is tilted as indicated in Fig. 17-4h.   The shells forming the northlight in Fig. 17-4g may be analyzed as indicated by Rosati [94]. The stresses in the tilted shallow shells in Fig. 17-4h may be found by using formula (17-17).   It is, however, to be noted that in formula (17-8) for c, the quantity a to be used is the projection of the actual sloping length a' of the edge beam.   This method is explained in detail by Parme [95].

17-13   The Groined Vaults

The groined vault shown in Fig. 17-4i may be analyzed by using tables developed by Parme [95].

17-14 Ties and Buttresses

Consider the arrangement of hyperbolic paraboloids proposed for the roof of an assembly hall (Fig. 17-16a). The points marked H represent the high points of the shell which are all at the same level. Similarly the points marked L are at the same level and represent the low points where the roof is to be supported on columns. It is uneconomical to design the columns for the very large horizontal thrusts transferred by the shell. We have therefore to provide ties to relieve the columns of horizontal thrusts. Alternatively, one may provide buttresses to receive the inclined thrust. Some arrangements of ties and buttresses

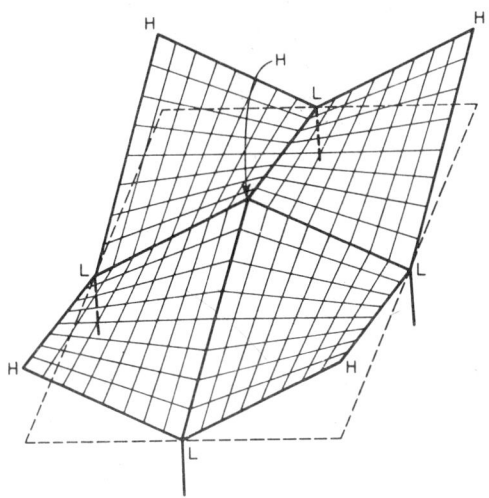

Fig. 17-16a

are shown in Fig. 17-16b. Because the thrusts transmitted by the shell are very nearly horizontal, buttresses are practicable only if the points L are only a few feet above the ground level. The arrangement of ties shown in Fig. 17-16c has the merit that it makes each hyperbolic paraboloid independent. Such an arrangement would permit the same forms to be reused four times. Because the cost of forms constitutes a sizable portion of the total cost of shells, ensuring reuse of forms as many times as possible is a consideration that the designer must always keep in mind. Shells are sensitive to any variation of the distance between their low-level supports. Hence the ties provided should not extend under the action of the heavy tensions carried by them. It is advisable to provide a pretensioned tie in which the tension becomes

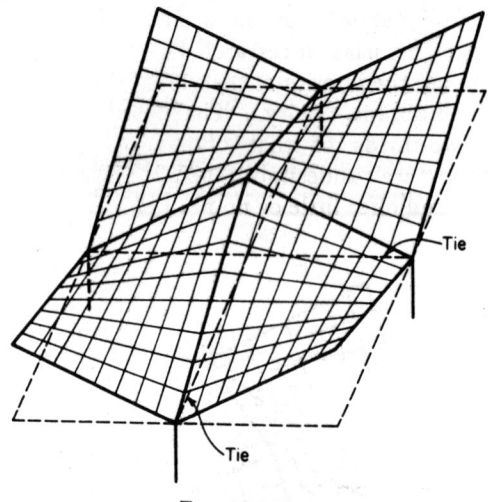

FIG. 17-16*b*(i)

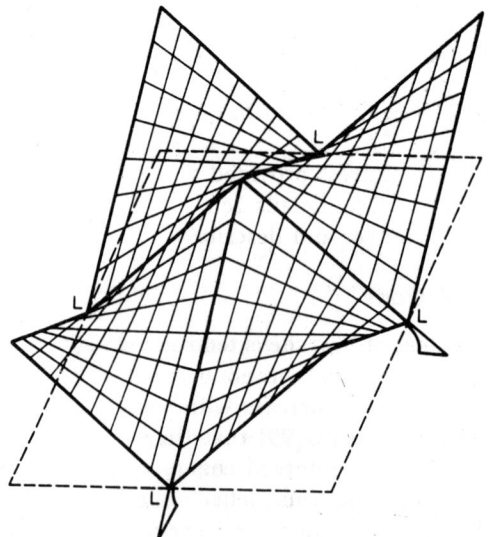

FIG. 17-16*b*(ii)

zero under the action of its own dead weight and the axial tension. This solution was employed by Hajnal-Konyi [96] in his design for a garage at Lincoln. The tests carried out on a model of this shell at the Cement and Concrete Association research laboratory [97] showed that the measured stresses in the interior of the shell closely agreed with those predicted by the membrane theory. The measured force in the tie was only about two-thirds of that obtained from the membrane theory. Exposed ties are undesirable because they are sensitive to temperature stresses and offer little or no fire resistance.

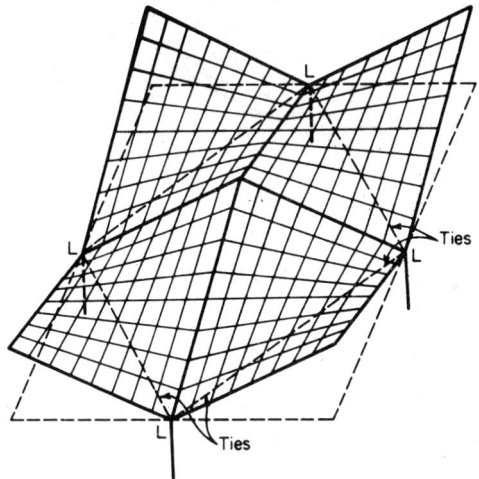

Fig. 17-16c

An example of a steel buttress provided to receive the inclined thrust is shown in Photo 17-6. The details relate to a wide-span h.p. shell, 112 by 132 ft, built at Denver and described by Tedesko [98].

17-15 Creep Deflections

The unsupported edges of h.p. shells tend to deflect on the removal of forms. This deflection is most pronounced at the corners. These deflections have a tendency to increase with time because of creep. It has been reported by De Cossio [99] that the deflection at the end of 60 days on an experimental reinforced-concrete h.p. shell was as much as four to six times the instantaneous deflection on the removal of forms. This confirms our own observations and experience. Prestressing the edge beams of inverted umbrella shells has been found to be helpful. The optimum prestressing force to be applied should be just adequate

PHOTO 17-6. Details of the steel buttress provided for the shell roof at Denver, Colo. *(Courtesy of Anton Tedesko. Architects: I. M. Pei and Partners. Engineers: Anton Tedesko and Roberts and Schaefer Company. Contractors: Webb and Knapp Construction Company.)*

PHOTO 17-7. Hyperbolic paraboloid shell roof at Denver, Colo. *(Courtesy of Anton Tedesko. Architects: I. M. Pei and Partners. Engineers: Anton Tedesko and Roberts and Schaefer Company. Contractors: Webb and Knapp Construction Company.)*

to overcome the axial tension in the edge beam due to the full dead weight plus a nominal part of the live load, say 5 psf.

17-16 Design Example

The details involved in the design of an inverted umbrella roof are explained in Design Example 17-1.

Design Example 17-1

DESIGN OF A HYPERBOLIC PARABOLOID SHELL OF
THE TILTED INVERTED UMBRELLA TYPE

1. *DATA*

   Geometry

   Length $2a$ = 60 ft
   Width $2b$ = 40 ft
   Depth f = 6 ft
   Thickness d = $2\frac{1}{2}$ in.

   Loads

Dead weight	= 31.0 psf of shell surface
Waterproofing, weight of ribs, etc.	= 9.0 psf of shell surface
Live load	= 15.0 psf of shell surface
Total load g	= 55.0 psf of shell surface

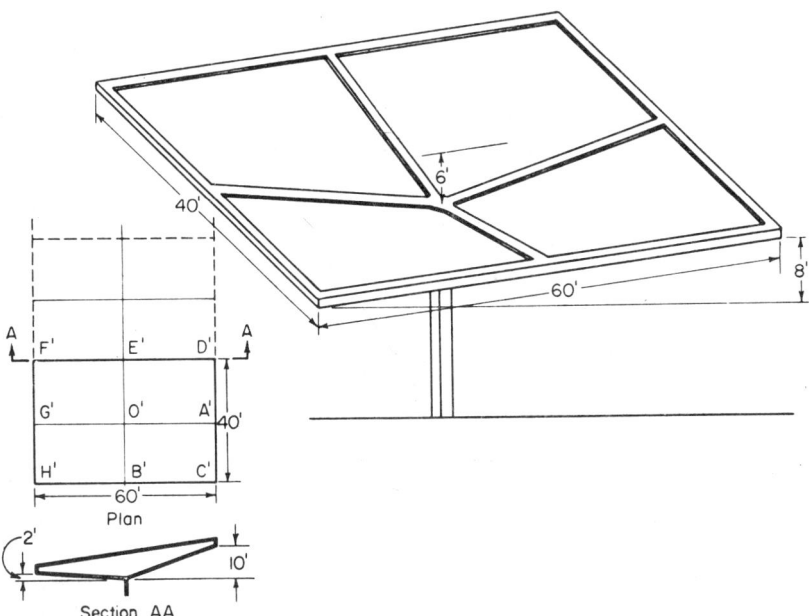

Fig. 17-17

2. *MEMBRANE STRESSES AND FORCES IN EDGE MEMBERS*

$$\text{Membrane shear} = \frac{gab}{2f} = \frac{55 \times 30 \times 20}{2 \times 6}$$

$$= 2{,}750 \text{ lb/ft}$$

The forces in the edge members due to uniform loading over the entire shell surface are calculated below (Fig. 17-18).

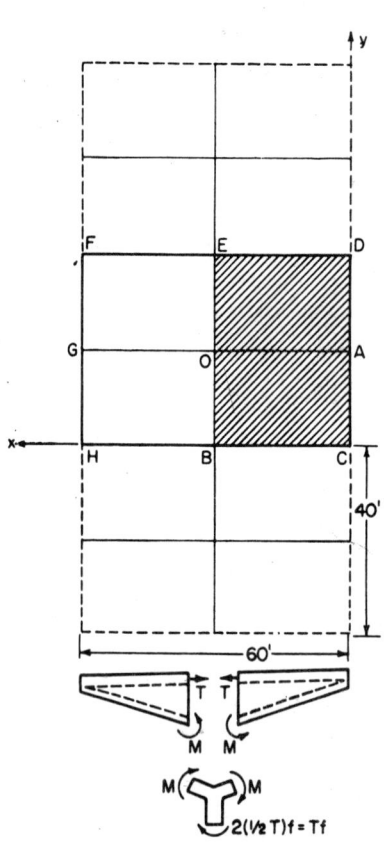

Max. tension in edge beams *FED* and *HBC*

$$= 2{,}750 \times \text{length } B'C' \text{ (Fig. 17-17)}$$

$$= 2{,}750 \times \sqrt{30^2 + 4^2} = 83{,}230 \text{ lb}$$

Max. tension in edge beams *FGH* and *DAC*

$$= 2{,}750 \times \text{length } A'C' \text{ (Fig. 17-17)}$$

$$= 2{,}750 \times 20 = 55{,}000 \text{ lb}$$

Max. compression in rib *GOA*

$$= 2 \times 2{,}750 \times \sqrt{30^2 + 10^2} = 173{,}925 \text{ lb}$$

Max. compression in rib *EOB*

$$= 2 \times 2{,}750 \times \sqrt{20^2 + 6^2} = 114{,}840 \text{ lb}$$

3. *MOMENTS IN COMPRESSION RIBS DUE TO ASYMMETRICAL LIVE LOAD ON SHELL*

The compression ribs are also checked for bending moments caused by live loads acting on only one half of the shell. For example, the rib *GOA* is designed to be safe not only against the direct compression but also against bending moment induced in it because of live load acting on the area *EBCD*.

The bending moment induced in the rib because of live load on only one half of the shell is calculated as shown below. The unbalanced force in each of the edge beams *ED* and *BC* = $(15/55) \times 83.23 = 22.7$ kip.

Assuming that half of this force is taken up by the adjoining shells, shown dotted in Fig. 17-18, the unbalanced force in each edge beam of the shell under consideration is equal to 11.35 kip. The moment to be resisted [95] by the ribs *OA* and *OG* = $\frac{1}{2}(2 \times 11.35)6 = 68.1$ kip ft. Similarly, the moment to be resisted by the ribs *OB* and *OE* due to live load acting only on the area *GACH* works out to 45 kip ft.

4. DESIGN OF REINFORCEMENT

(i) Shell

The shell is subjected to a shear stress of 2,750 lb/ft. This causes a tensile stress of 2,750 lb/ft and a compressive stress of equal magnitude.

Area of steel required = 2,750/20,000 = 0.1375 in.[2] No. 2 bars at 4 in. center to center would suffice. The same reinforcement is provided in the two directions at right angles to each other (Fig. 17-19).

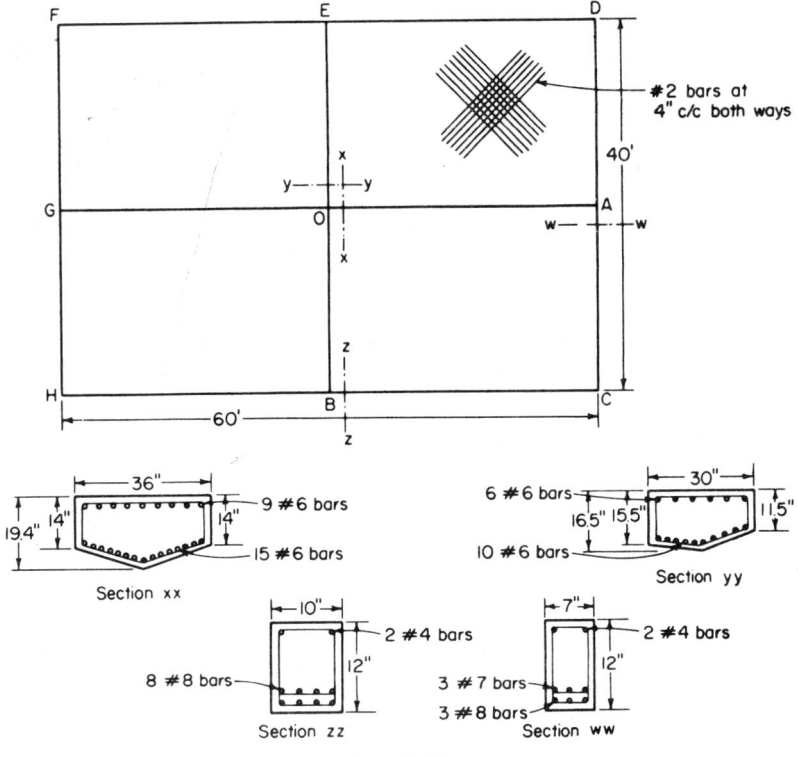

FIG. 17-19

(ii) Edge Beams ED, EF, BC, and BH

Tensile force in the edge beam = 83.23 kip

As the shell is meeting the edge beam at bottom, there is an eccentricity equal to half the depth of the edge beam. The edge beam is assumed to be 10 in. wide and 12 in. deep with a reinforcement of 2 No. 4 bars at top and 8 No. 8 bars at bottom as shown in Fig. 17-19. The cross-sectional dimensions and the reinforcement are arrived at after a number of trials. The section is checked for an axial tension of 83.23 kip and a moment of $83.23 \times \frac{12}{2} = 499.38$ kip in. The stresses in concrete and steel are found to be as given below.

Max. compressive stress in concrete = 1,154 psi

Max. tensile stress in steel = 19,785 psi

(iii) *Edge Beams AD, AC, GF, and GH*

Tensile force = 55 kip, acting at bottom

Assuming the section 7 in. wide and 12 in. deep, the moment due to eccentricity = 55 × 12/2 = 330 kip in.

Reinforcement consisting of 2 No. 4 bars at top and 3 No. 7 and 3 No. 8 bars at bottom is assumed (Fig. 17-19). The stresses are found to be as follows:

Max. compressive stress in concrete = 1,023 psi

Max. tensile stress in steel = 18,945 psi

(iv) *Ribs OA and OG*

Two cases are considered

(a) *Live Load Acting on All the Four Quadrants*

Force = 173.925 kip (compression) acting at the bottom of the rib. The width of the rib is assumed as 36 in. The clear outstand of the rib is taken as 14 in. The average depth of the rib works out to 16.7 in.

Moment due to eccentricity of the compressive force = 173.925 × 16.7/2 = 1,452.27 kip in.

With a reinforcement of 9 No. 6 bars at top and 15 No. 6 bars at bottom, the stresses are as follows:

Max. compressive stress in concrete = 1,088 psi

Max. tensile stress in steel = 8,403 psi

(b) *Live Load Acting on Only Two Quadrants OACB and OADE*

An additional bending moment of 68.1 kip ft is introduced because of the unbalanced live load. The stresses in the rib are checked for an axial compressive force of 173.925 kip and a moment = 1,452.27 + (68.1 × 12) = 2,269.47 kip in. The stresses are found to be as follows:

Max. compressive stress in concrete = 1,702 psi

Max. tensile stress in steel = 21,811 psi

The stresses in concrete and steel are higher than the permissible stresses of 1,350 and 20,000 psi. However, the unbalanced live load assumed here does not normally occur and the design based on this assumption is too conservative. Hence the permissible stresses may be increased by about $33\frac{1}{3}\%$. The actual stresses calculated above are within the increased permissible values.

(v) *Ribs OE and OB*

(a) *Live Load Acting on All the Four Quadrants*

Compressive force = 114.84 kip acting at the bottom. Moment due to eccentricity = 114.84 × 15/2 = 861.3 kip in., where 15 in. is the average depth assumed.

With a reinforcement of 6 No. 6 bars at top and 10 No. 6 bars at bottom, the maximum stresses in concrete and steel work out to 1,021 and 8,330 psi, respectively.

(b) *Live Load Acting on Only Two Quadrants OBHG and OACB*

Additional moment = 45 kip ft. The section is therefore checked for an axial force of 114.84 kip and a moment of 861.3 + (45 × 12) = 1,401.3 kip in. The maximum stresses in concrete and steel work out to 1,689 and 23,525 psi, respectively, which are within the increased permissible values.

THE CONOID

17-17 Definition of a Conoid

A conoid is generated by a variable straight line moving parallel to a plane—known as the *director plane*—with one of its ends on a plane curve and the other on a straight line.   The plane curve and the straight line are known as *directrices* (Fig. 17-20).   It is assumed that both the

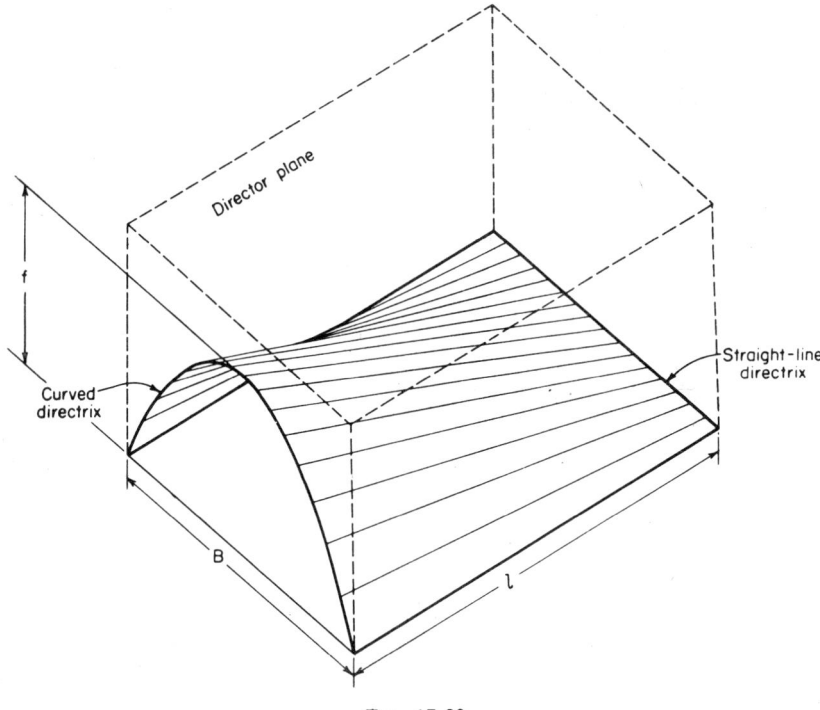

Fɪɢ. 17-20

straight-line directrix and the plane containing the curved directrix are at right angles to the director plane, the curved directrix being moreover symmetrical about its vertical axis.   A part of a conoid, known as a *truncated conoid,* is sometimes employed in preference to a full conoid (Fig. 17-21).

17-18 Geometry of a Conoid

Depending upon the curve used as the directrix, conoids are described as parabolic, circular, or catenary conoids.   Of these, the parabolic conoid is by far the commonest.

Referring to Fig. 17-22, the equation to any conoid may be written in the form

$$z = f(y)\frac{x}{l} \tag{17-32}$$

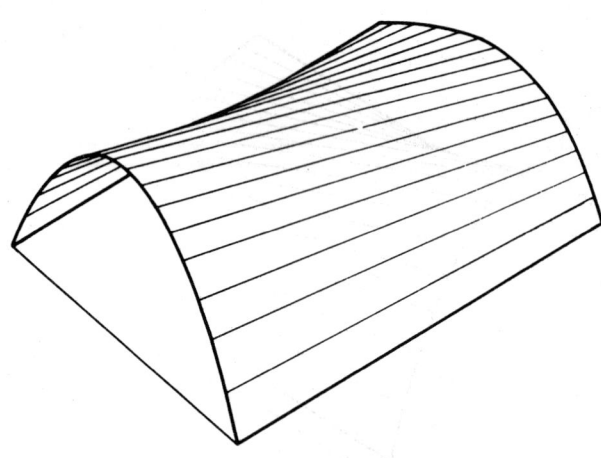

Fig. 17-21

Differentiation leads to the following derivatives:

$$p = \frac{\partial z}{\partial x} = \frac{1}{l}f(y) \tag{17-33}$$

$$q = \frac{\partial z}{\partial y} = f'(y)\frac{x}{l} \tag{17-34}$$

$$r = \frac{\partial^2 z}{\partial x^2} = 0 \tag{17-35}$$

$$s = \frac{\partial^2 z}{\partial x\,\partial y} = \frac{1}{l}f'(y) \tag{17-36}$$

$$t = \frac{\partial^2 z}{\partial y^2} = f''(y)\frac{x}{l} \tag{17-37}$$

where the primes denote differentiation with respect to y.   Knowing the values of r, s, and t, the two principal curvatures at any point on the shell may be written as

$$\frac{1}{R_1} = \frac{r+t}{2} + \sqrt{\left(\frac{r-t}{2}\right)^2 + s^2}$$

and $\tag{17-38}$

$$\frac{1}{R_2} = \frac{r+t}{2} - \sqrt{\left(\frac{r-t}{2}\right)^2 + s^2}$$

For a conoid, r being zero, $rt - s^2 < 0$.   Hence the surface is *anticlastic* and *its Gauss curvature is negative.*   It is seen from (17-38) that at any point on the shell there are, in general, two values of the principal

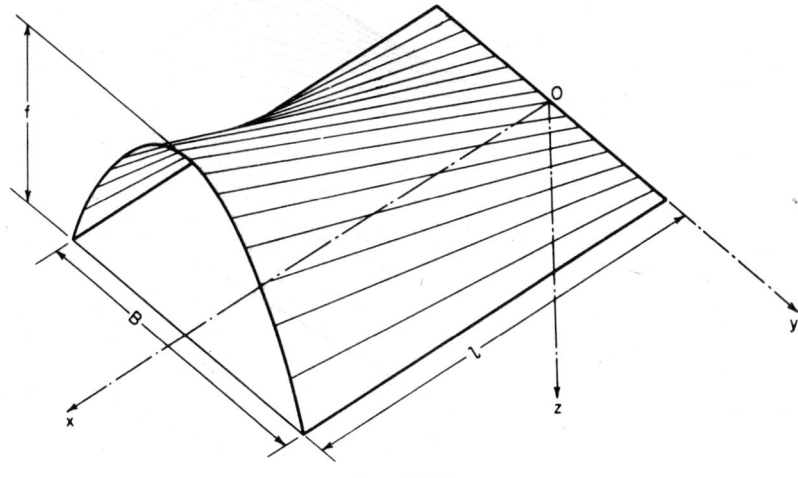

Fig. 17-22

curvature which are different from zero.   Hence the conoid is *doubly curved.*   It is obvious from the manner in which the surface is generated that the conoid is a *ruled surface.*   Being doubly curved, the surface is also *nondevelopable.*

17-19   Types of Conoids

Two kinds of conoids which we shall designate as type I and type II are in general use.   In the first type, illustrated in Fig. 17-23, the glazing is vertical.   In type II the glazing windows are inclined to the vertical. The type II conoid may be obtained by rotating a conoid until the straight-line directrix, originally at the same level as the crown of the curved directrix, is brought down to the springing level of the latter (Fig. 17-24). We shall confine our discussion to parabolic conoids.

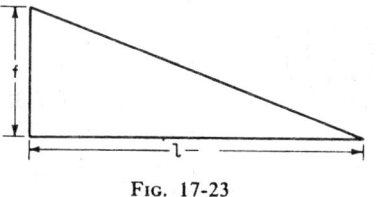

Fig. 17-23

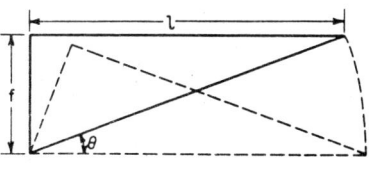

Fig. 17-24

PHOTO 17-8.   Conoid shells (Type I) 40 ft wide and 50 ft span, roofing a railroad workshop in India. [*Courtesy of Indian Railways.   Design and construction: Gammon India (Private) Ltd., Bombay, India.*]

PHOTO 17-9.   Interior view of shells in Photo 17-8.   [*Courtesy of Indian Railways. Design and construction: Gammon India (Private) Ltd., Bombay, India.*]

MEMBRANE THEORY OF THE PARABOLIC CONOID TYPE I

17-20 Introduction

M. Soare [100] has presented a comprehensive treatment of conoids with various directrices. An engineer not well equipped in advanced mathematics would find his treatment somewhat difficult to follow. The membrane theory has also been discussed by G. S. Rao [101] and Bhise and Apte [102]. In what follows, a simplified approach due to the author [103] is presented. The method, which employs a polynomial stress function, leads to results which are in close agreement with Soare's theory if the shell considered is not too deep.

17-21 Geometry

The equation of the type I conoid may be written in the form

$$z = -\frac{4f}{B^2}\left(\frac{B^2}{4} - y^2\right)\frac{x}{l} \tag{17-39}$$

The derivatives are

$$p = -\frac{4f}{lB^2}\left(\frac{B^2}{4} - y^2\right) \tag{17-40}$$

$$q = \frac{8fxy}{lB^2} \tag{17-41}$$

$$r = 0 \tag{17-42}$$

$$s = \frac{8fy}{lB^2} \tag{17-43}$$

and

$$t = \frac{8fx}{lB^2} \tag{17-44}$$

17-22 Stress Function for Snow Load

The snow load p_0, uniform over the horizontal projection of the shell, constitutes a *funicular load* for the shell because all cross sections of the shell are parabolas. It is well known that, under funicular loading, a shell degenerates into a series of *independent arches* so that $\overline{N_x} = \overline{N_{xy}} = 0$. Hence Equation (15-16) governing the membrane equilibrium of the shell assumes the simple form

$$t\overline{N_y} = -p_0$$

or

$$\overline{N_y} = -\frac{p_0}{t} \tag{17-45}$$

Substituting for t from relation (17-45), we get

$$\overline{N_y} = -\frac{p_0}{ax} \qquad (17\text{-}46)$$

where

$$a = \frac{8f}{lB^2}$$

An appropriate stress function is

$$\phi_0 = -\frac{p_0}{a} x (\log_e x - 1) \qquad (17\text{-}47)$$

17-23 Stress Function for Dead Weight g

For this condition of loading, we have

$$X = 0$$
$$Y = 0$$
$$Z = g \sqrt{1 + p^2 + q^2} \qquad (17\text{-}48)$$

Substituting for p and q in (17-48) and expanding by the binomial theorem and restricting ourselves to two terms,

$$-Z = -g \left[1 + \frac{1}{2} \left(\frac{f}{l} \right)^2 \right] - ga^2 \left(\frac{y^4}{8} - \frac{1}{16} B^2 y^2 + \frac{x^2 y^2}{2} \right) \qquad (17\text{-}49)$$

Substituting the values of X, Y, and Z in Equation (15-16),

$$2ay \, \overline{N_{xy}} + ax \, \overline{N_y} = -g \left[1 + \frac{1}{2} \left(\frac{f}{l} \right)^2 \right]$$
$$- ga^2 \left(\frac{y^4}{8} - \frac{1}{16} B^2 y^2 + \frac{x^2 y^2}{2} \right) \qquad (17\text{-}50)$$

Introducing a stress function ϕ, we may rewrite (17-50) as

$$-2ay \frac{\partial^2 \phi}{\partial x \, \partial y} + ax \frac{\partial^2 \phi}{\partial x^2} = -g \left[1 + \frac{1}{2} \left(\frac{f}{l} \right)^2 \right]$$
$$- ga^2 \left(\frac{y^4}{8} - \frac{1}{16} B^2 y^2 + \frac{x^2 y^2}{2} \right) \qquad (17\text{-}51)$$

An appropriate stress function may be found by splitting the load term on the right-hand side into two parts. The first part which is independent of x and y may be regarded as a snow load, and the stress function

$\phi_0(x,y)$ corresponding to it may be found by making use of (17-47). Thus

$$\phi_0 = -\frac{g}{a}\left[1 + \frac{1}{2}\left(\frac{f}{l}\right)^2\right] x\,(\log x - 1) \qquad (17\text{-}52)$$

We are therefore left with the problem of finding a stress function $\phi_1(x,y)$ satisfying the relation

$$-2y\,\frac{\partial^2\phi_1}{\partial x\,\partial y} + x\,\frac{\partial^2\phi_1}{\partial x^2} = -ga\left(\frac{y^4}{8} - \frac{1}{16}B^2 y^2 + \frac{x^2 y^2}{2}\right) \qquad (17\text{-}53)$$

It is easily verified that this equation is satisfied by a stress function of the form

$$\phi_1 = \sum A_{mn} x^m y^n$$

or

$$\phi_1 = \sum A_{mn}(x^m - l^m)\,y^n \qquad (17\text{-}54)$$

The latter form will be chosen as it can be made to satisfy the boundary condition at $x = l$. Substituting the chosen stress function (17-54) in (17-53), we get

$$\sum m(m - 1 - 2n)\,A_{mn} x^{m-1} y^n = -ga\left(\frac{y^4}{8} - \frac{1}{16}B^2 y^2 + \frac{x^2 y^2}{2}\right)$$

The values of A_{mn} corresponding to each term on the right-hand side can be found, term by term, as follows:

Let us, for example, consider the term $-(gay^4/8)$. Equating indices of like powers on both sides, we get $n = 4$ and $m = 1$ so that $1(1 - 1 - 8)A_{mn} = -ga/8$. Hence $A_{mn} = ga/64$. The term in the stress function corresponding to the load term $-(gay^4/8)$ is thus found to be $(ga/64)(x - l)y^4$. Treating the other load term in a similar manner, the stress function may be built up as

$$\phi_1 = \frac{ga}{64}(x - l)\,y^4 + \frac{gaB^2}{64}(l - x)\,y^2 - \frac{ga}{12}(l^3 - x^3)\,y^2$$

Combining this with ϕ_0 already found,

$$\phi = \phi_0 + \phi_1 = -ga\left[\frac{(l - x)\,y^4}{64} - \frac{B^2}{64}(l - x)\,y^2 + \frac{1}{12}(l^3 - x^3)\,y^2\right]$$

$$-\frac{g}{a}\left[1 + \frac{1}{2}\left(\frac{f}{l}\right)^2\right] x\,(\log x - 1) \qquad (17\text{-}55)$$

The expression for the projected stresses may now be written down. Thus

$$\overline{N}_x = \frac{\partial^2 \phi}{\partial y^2} = -ga \left[\frac{3}{16}(l - x)y^2 - \frac{B^2}{32}(l - x) + \frac{1}{6}(l^3 - x^3) \right] \qquad (17\text{-}56)$$

$$\overline{N}_y = \frac{\partial^2 \phi}{\partial x^2} = \frac{ga}{2} xy^2 - \frac{g}{ax} \left[1 + \frac{1}{2}\left(\frac{f}{l}\right)^2 \right] \qquad (17\text{-}57)$$

and

$$\overline{N}_{xy} = -\frac{\partial^2 \phi}{\partial x\, \partial y} = -ga \left(\frac{y^3}{16} - \frac{yB^2}{32} + \frac{x^2 y}{2} \right) \qquad (17\text{-}58)$$

It is easily verified that the stress function chosen satisfies the boundary conditions. From symmetry, the first boundary condition to be imposed is

(i) $$\overline{N}_{xy} = 0 \qquad \text{along} \quad y = 0$$

The second boundary condition is

(ii) $$\overline{N}_x = 0 \qquad \text{at} \quad x = l$$

This is because the traverse at the end $x = l$ is assumed incapable of receiving any forces at right angles to its plane. Because the governing differential equation of the conoid is of the hyperbolic type, no boundary condition may be prescribed on the other two edges, which are "open boundaries."

MEMBRANE THEORY OF CONOID TYPE II

17-24 Rotation of the Load

The stress function and the expressions for the stresses will be derived in two stages. In the first stage, the conoid shall be in the position indicated by the firm lines in Fig. 17-24. Instead of rotating the conoid to assume the position indicated by the dotted lines, it is easier to imagine that it remains stationary and rotate the vertical load g through an angle θ given by $\theta = \tan^{-1}(f/l)$ in a direction opposite to the direction of rotation of the conoid. The use of this artifice suggested by Rosati [94] considerably simplifies the algebra involved.

17-25 Geometry

The equation of the surface may be written as

$$z = hxy^2$$

The derivatives are

$$p = \frac{\partial z}{\partial x} = hy^2 \tag{17-59}$$

$$q = \frac{\partial z}{\partial y} = 2hxy \tag{17-60}$$

$$r = \frac{\partial^2 z}{\partial x^2} = 0 \tag{17-61}$$

$$s = \frac{\partial^2 z}{\partial x \, \partial y} = 2hy \tag{17-62}$$

$$t = \frac{\partial^2 z}{\partial y^2} = 2hx \tag{17-63}$$

17-26 Stress Function for Dead Weight g

The load components are

$$X = -g \sin \theta \sqrt{1 + p^2 + q^2} \tag{17-64}$$
$$Y = 0 \tag{17-65}$$
$$Z = g \cos \theta \sqrt{1 + p^2 + q^2} \tag{17-66}$$

Hence the three equations of equilibrium (15-14), (15-15), and (15-16) assume the following form:

$$\frac{\partial \overline{N_x}}{\partial x} + \frac{\partial \overline{N_{xy}}}{\partial y} - g \sin \theta \sqrt{1 + p^2 + q^2} = 0 \tag{17-67}$$

$$\frac{\partial \overline{N_y}}{\partial y} + \frac{\partial \overline{N_{xy}}}{\partial x} = 0 \tag{17-68}$$

and

$$2y \, \overline{N_{xy}} + x \, \overline{N_y} = -\frac{g}{2h} (hy^2 \sin \theta + \cos \theta) \sqrt{1 + h^2y^4 + 4h^2x^2y^2} \tag{17-69}$$

Expanding the quantity in radicals and restricting ourselves to the first three terms,

$$2y \, \overline{N_{xy}} + x \, \overline{N_y} = -\frac{g}{2h} \cos \theta - g \cos \theta \left(\frac{1}{4} hy^4 + hx^2y^2 - \frac{1}{16} h^3y^8 \right.$$
$$\left. - h^3x^4y^4 - \frac{h^3}{2} x^2y^6 \right) - g \sin \theta \left(\frac{y^2}{2} + \frac{h^2y^6}{4} \right.$$
$$\left. + h^2x^2y^4 - \frac{1}{16} h^4y^{10} - h^4x^4y^6 - \frac{1}{2} h^4x^2y^8 \right) \tag{17-70}$$

The first term $-(g/2h) \cos \theta$, being a constant, may be treated as a "snow load" which is a "funicular load" for the shell. For this loading, $\overline{N_x} = 0$ and $\overline{N_{xy}} = 0$. Hence

$$\overline{N_y} = -\frac{g}{2hx} \cos \theta \tag{17-71}$$

The corresponding term in the stress function is

$$\phi_0 = -\frac{g}{2h} \cos\theta \, x \,(\log_e x - 1) \tag{17-72}$$

The problem therefore reduces itself to finding a stress function $\phi_1(x,y)$ satisfying the relation

$$2y \, \overline{N_{xy}} + x \, \overline{N_y} = -g \cos\theta \left(\frac{1}{4} hy^4 + hx^2y^2 - \frac{1}{16} h^3y^8 - h^3x^4y^4 - \frac{h^3}{2} x^2y^6\right)$$

$$- g \sin\theta \left(\frac{y^2}{2} + \frac{h^2y^6}{4} + h^2x^2y^4 - \frac{1}{16} h^4y^{10}\right)$$

$$- h^4x^4y^6 - \frac{1}{2}h^4x^2y^8\biggr)$$

A solution $\phi = \Sigma A_{mn}(x^m - l^m) \, y^n$ will satisfy the equation, the boundary conditions being the same as for a type I conoid. The coefficients A_{mn} are found in the same manner as for type I. Combining the stress function so found with ϕ_0 already found in (17-72), the complete stress function may be written as

$$\phi = -\frac{g}{2h} \cos\theta \, x \,(\log_e x - 1) + gh \cos\theta \left[(x-l)\frac{y^4}{32} + \frac{1}{6}(x^3 - l^3) \, y^2 \right.$$

$$\left. - \frac{h^0}{256}(x-l) \, y^8 - \frac{h^2}{20}(x^5 - l^5) \, y^4 - \frac{h^2}{60}(x^3 - l^3) \, y^6\right]$$

$$+ g \sin\theta \left[\frac{1}{8}(x-l) \, y^2 + \frac{h^2}{48}(x-l) \, y^6 + \frac{h^2}{18}(x^3 - l^3) \, y^4 \right.$$

$$\left. - \frac{h^4}{320}(x-l) \, y^{10} - \frac{h^4}{40}(x^5 - l^5) \, y^6 - \frac{h^4}{84}(x^3 - l^3) \, y^8\right] \tag{17-73}$$

The expressions for the projected stresses are

$$\overline{N_x} = \frac{\partial^2\phi}{\partial y^2} - \int_l^x X \, dx = gh \cos\theta \left[\frac{12}{32}(x-l) \, y^2 + \frac{1}{3}(x^3 - l^3) \right.$$

$$\left. - \frac{56}{256} h^2(x-l) \, y^6 - \frac{12h^2}{20}(x^5 - l^5) \, y^2 - \frac{30}{60} h^2(x^3 - l^3) \, y^4\right]$$

$$+ g \sin\theta \left[\frac{x-l}{4} + \frac{30}{48}(x-l) \, h^2y^4 + \frac{12}{18}(x^3 - l^3) \, h^2y^2 \right.$$

$$\left. - \frac{90}{320} h^4(x-l) \, y^8 - \frac{30}{40} h^4(x^5 - l^5) \, y^4 - \frac{56}{84} h^4(x^3 - l^3) \, y^6\right]$$

$$+ g \sin\theta \left[(x-l) + \frac{1}{2} h^2y^4(x-l) + \frac{2}{3} h^2(x^3 - l^3) \, y^2 \right.$$

$$\left. - \frac{1}{8} h^4(x-l) \, y^8 - \frac{2}{5}(x^5 - l^5) \, h^4y^4 - \frac{h^4}{3}(x^3 - l^3) \, y^6\right]$$

Simplifying,

$$\overline{N_x} = gh \cos \theta \left[\frac{3}{8} (x - l) y^2 + \frac{1}{3} (x^3 - l^3) - \frac{7}{32} h^2 (x - l) y^6 \right.$$

$$\left. - \frac{3}{5} h^2 (x^5 - l^5) y^2 - \frac{1}{2} h^2 (x^3 - l^3) y^4 \right]$$

$$+ g \sin \theta \left[\frac{5}{4} (x - l) + \frac{9}{8} (x - l) h^2 y^4 \right.$$

$$+ \frac{4}{3} (x^3 - l^3) h^2 y^2 - \frac{13}{32} (x - l) h^4 y^8$$

$$\left. - \frac{23}{20} (x^5 - l^5) h^4 y^4 - (x^3 - l^3) h^4 y^6 \right] \qquad (17\text{-}74)$$

$$\overline{N_{xy}} = - \frac{\partial^2 \phi}{\partial x \, \partial y} = -gh \cos \theta \left[\frac{y^3}{8} + x^2 y - \frac{h^2}{32} y^7 \right.$$

$$\left. - h^2 x^4 y^3 - \frac{3}{10} h^2 x^2 y^5 \right]$$

$$- g \sin \theta \left[\frac{y}{4} + \frac{1}{8} h^2 y^5 + \frac{2}{3} h^2 x^2 y^3 \right.$$

$$\left. - \frac{1}{32} h^4 y^9 - \frac{3}{4} h^4 x^4 y^5 - \frac{2}{7} h^4 x^2 y^7 \right] \qquad (17\text{-}75)$$

$$\overline{N_y} = \frac{\partial^2 \phi}{\partial x^2} = - \frac{g}{2hx} \cos \theta + gh \cos \theta \left(xy^2 - h^2 x^3 y^4 - \frac{1}{10} h^2 xy^6 \right)$$

$$+ g \sin \theta \left(\frac{1}{3} h^2 xy^4 - \frac{1}{2} h^4 x^3 y^6 - \frac{1}{14} h^4 xy^8 \right) \qquad (17\text{-}76)$$

17-27 Design Example

In Design Example 17-2, the membrane analysis of a conoid of type I is given. The results obtained by the author's method are shown compared with stresses calculated by Soare's theory.

Design Example 17-2

DESIGN OF A PARABOLIC CONOID

1. *DATA*

Geometry

    Span L = 30 ft
    Chord width B = 60 ft
    Rise at high end f = 12 ft
    Rise at low end   = 3 ft
    Thickhess d = 3 in.

This is a type I conoid.

Loads

Dead weight	= 37.5 psf
Waterproofing, etc.	= 7.5 psf
Live load	= 15.0 psf
Total load g	= 60.0 psf of shell surface

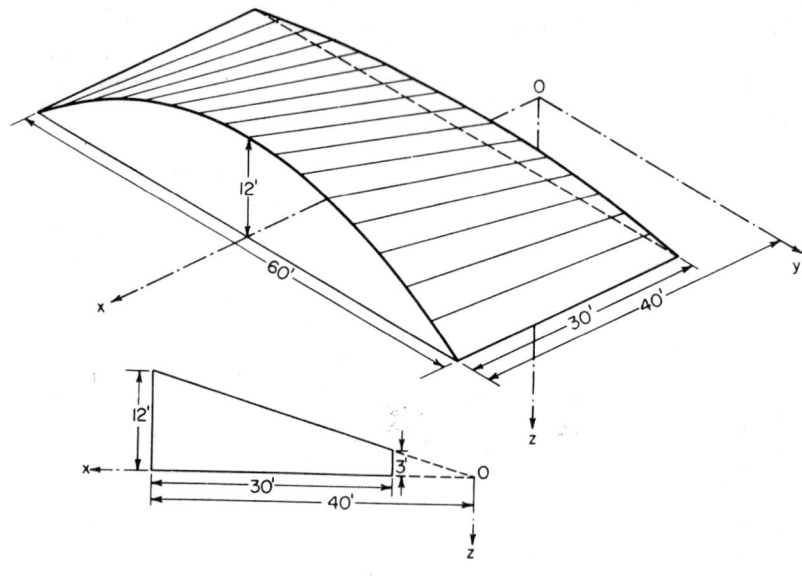

FIG. 17-25

2. GEOMETRICAL PROPERTIES

Referring to Fig. 17-25, the theoretical span l is calculated from the relation $l/12 = (l - L)/3$. Hence

$$l = 40 \text{ ft}$$

$$a = \frac{8f}{lB^2} = \frac{8 \times 12}{40 \times 60^2} = 0.0006667$$

Max. slope of tangent at high end $= \tan^{-1}\left(\dfrac{2f}{B/2}\right)$

$$= \tan^{-1} 0.8 = 38°40'$$

3. CALCULATION OF STRESS RESULTANTS

The projected stresses $\bar{N}_x$, $\bar{N}_y$, and $\bar{N}_{xy}$ at various points on the surface of the shell are calculated by using formulas (17-56), (17-57), and (17-58), respectively. The real stresses N_x and N_y are calculated from $\bar{N}_x$ and $\bar{N}_y$ making use of the relations (15-8) and (15-9). From the real stresses the principal stresses P_1 and P_2 and their directions are also calculated. The results are shown tabulated below.

The stresses N_x, N_y, and N_{xy} are also calculated using the formulas given by Soare [100] and are tabulated below. The stresses obtained by both the methods are shown plotted in Fig. 17-26 for a few typical sections.

Analysis of the Conoid by Author's Method

Coordinates		Projected stresses, psf			Real stresses, psf		Principal stresses, psf		Inclination of P_1 to x axis, deg
x	y	$\bar{N}_x$	$\bar{N}_v$	$\bar{N}_{xv}$	N_x	N_v	P_1	P_2	
10.0	0	−285	−9,405	0	−298	−9,008	−298	−9,008	0
17.5	0	−290	−5,374	0	−302	−5,148	−302	−5,148	0
25.0	0	−255	−3,762	0	−266	−3,603	−266	−3,603	0
32.5	0	−164	−2,894	0	−171	−2,772	−171	−2,772	0
40.0	0	0	−2,351	0	0	−2,252	0	−2,252	0
10.0	7.5	−298	−9,394	18	−309	−9,054	−309	−9,054	0.12
17.5	7.5	−299	−5,355	−13	−310	−5,174	−310	−5,174	−0.16
25.0	7.5	−261	−3,734	−61	−269	−3,622	−268	−3,624	−1.04
32.5	7.5	−167	−2,857	−126	−171	−2,787	−165	−2,793	−2.75
40.0	7.5	0	−2,306	−207	0	−2,264	19	−2,283	−5.19
10.0	15.0	−336	−9,360	29	−342	−9,177	−342	−9,177	0.19
17.5	15.0	−328	−5,296	−33	−331	−5,245	−331	−5,245	−0.38
25.0	15.0	−280	−3,649	−128	−279	−3,670	−274	−3,675	−2.17
32.5	15.0	−177	−2,748	−258	−172	−2,819	−147	−2,843	−5.51
40.0	15.0	0	−2,171	−421	0	−2,281	75	−2,357	−10.13
10.0	22.5	−399	−9,304	28	−398	−9,328	−398	−9,328	0.18
17.5	22.5	−375	−5,197	−65	−366	−5,327	−365	−5,328	−0.75
25.0	22.5	−312	−3,509	−208	−295	−3,716	−282	−3,728	−3.47
32.5	22.5	−193	−2,565	−403	−175	−2,829	−115	−2,889	−8.44
40.0	22.5	0	−1,946	−647	0	−2,250	173	−2,423	−14.95
10.0	30.0	−488	−9,225	8	−478	−9,408	−478	−9,408	0.05
17.5	30.0	−442	−5,059	−116	−417	−5,360	−414	−5,363	−1.35
25.0	30.0	−356	−3,312	−308	−319	−3,703	−291	−3,731	−5.15
32.5	30.0	−215	−2,309	−566	−180	−2,754	−61	−2,873	−11.88
40.0	30.0	0	−1,631	−893	0	−2,089	329	−2,418	−20.26

Analysis of the Conoid by Soare's Method

Coordinates		Projected stresses, psf			Real stresses psf	
x	y	$\bar{N}_x$	$\bar{N}_y$	$\bar{N}_{xy}$	N_x	N_y
10.0	0	−420	−9,000	0	−438	−8,620
17.5	0	−391	−5,143	0	−408	−4,926
25.0	0	−323	−3,600	0	−337	−3,448
32.5	0	−198	−2,769	0	−207	−2,652
40.0	0	0	−2,250	0	0	−2,155
10.0	7.5	−412	−8,989	−15	−428	−8,664
17.5	7.5	−383	−5,123	−46	−397	−4,951
25.0	7.5	−316	−3,572	−93	−325	−3,465
32.5	7.5	−193	−2,733	−157	−198	−2,665
40.0	7.5	0	−2,205	−238	0	−2,165
10.0	15.0	−392	−8,955	−30	−400	−8,780
17.5	15.0	−363	−5,065	−91	−367	−5,016
25.0	15.0	−297	−3,489	−185	−296	−3,509
32.5	15.0	−180	−2,627	−309	−175	−2,695
40.0	15.0	0	−2,077	−462	0	−2,182
10.0	22.5	−363	−8,899	−45	−362	−8,922
17.5	22.5	−335	−4,969	−136	−327	−5,093
25.0	22.5	−271	−3,355	−272	−256	−3,553
32.5	22.5	−162	−2,458	−450	−147	−2,711
40.0	22.5	0	−1,876	−665	0	−2,169
10.0	30.0	−330	−8,822	−59	−324	−8,996
17.5	30.0	−303	−4,837	−178	−286	−5,125
25.0	30.0	−243	−3,175	−354	−217	−3,550
32.5	30.0	−143	−2,236	−578	−120	−2,666
40.0	30.0	0	−1,619	−842	0	−2,073

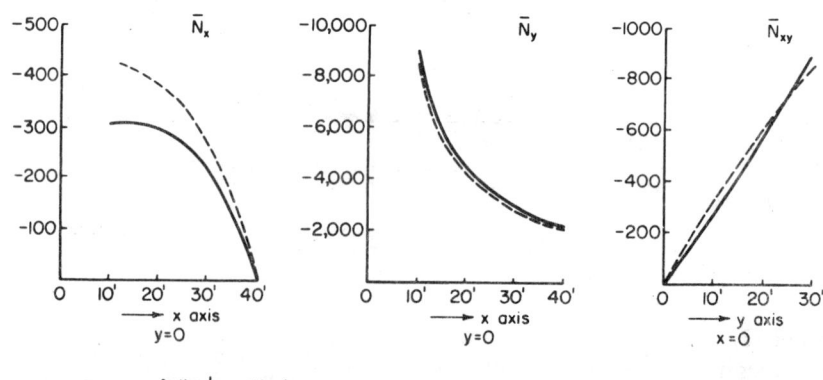

Author's method
Soare's method

Fig. 17-26

17-28   Need for a Bending Theory of Conoids

The membrane theory offers only an inadequate basis for the stress analysis of conoids. If a full conoid is considered, the surface loses all curvature close to the straight-line directrix. A part of the shell close to the straight-line directrix will function as a plate and the membrane theory developed in the previous articles will break down in this neighborhood. The transfer of the load to the traverse at this edge will take place primarily through radial shear as opposed to in-plane shear. Moreover, disturbances emanating from the edge beams will penetrate far into the interior of the shell. This is because the straight edges of the shell carrying the edge beams are characteristic lines. We shall see the full significance of this in a later article. Hence the need for a bending theory.

17-29   Characteristic Lines

Let us consider a parabolic conoid for which

$$r = 0$$

$$s = \frac{8fy}{lB^2}$$

$$t = \frac{8fx}{lB^2}$$

Referring to (15-27), the equation for the characteristic lines may be written as

$$\frac{dy}{dx} = \frac{-s \pm \sqrt{s^2 - rt}}{t}$$

Noting that $r = 0$,

$$\frac{dy}{dx} = \frac{-s \pm s}{t} \tag{17-77}$$

i.e., one set of characteristic lines is given by

$$\frac{dy}{dx} = 0 \quad \text{or} \quad y = c_1 \tag{17-78}$$

where c_1 is an arbitrary constant. It is easily seen that this set consists of straight lines parallel to the x axis. The other set is defined by the equation

$$\frac{dy}{dx} = -\frac{2s}{t} \tag{17-79}$$

Inserting the values of s and t for a conoid,

$$\frac{dy}{dx} = -\frac{2y}{x}$$

Integrating,

$$x^2 y = \text{constant} \tag{17-80}$$

Introducing nondimensional coordinates

$$\xi = \frac{x}{a} \quad \text{and} \quad \eta = \frac{y}{a}$$

The equation of this set of characteristic lines may be written in the form

$$\xi^2 \eta = c_2 \tag{17-81}$$

where c_2 is a constant.   The two sets are shown plotted in Fig. 17-27. Characteristics of a conoid being real, they play an important role in both the membrane and bending theories of such shells.

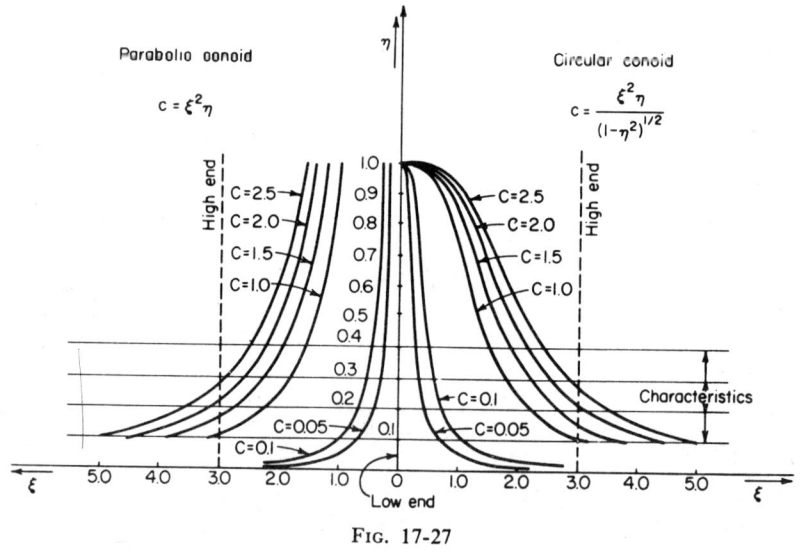

Parabolic conoid

$$c = \xi^2 \eta$$

Circular conoid

$$c = \frac{\xi^2 \eta}{(1-\eta^2)^{1/2}}$$

Fig. 17-27

17-30 Critical Comments on the Membrane Theory

We have come to the end of the membrane theory as applied to shells of double curvature.   It may therefore be opportune at this stage to assess its merits and recognize its limitations.

In many practically useful cases, it gives a fairly accurate picture of the stresses and deformations in the shell far away from certain lines known as *lines of distortion.* In many other cases, it is quite inadequate and resort to a bending theory becomes inescapable. The conditions under which the membrane theory would yield acceptable results have been summarized by Gol'denveizer [104].

Some examples of lines of distortion are listed below:

(i) The edges of the shell

(ii) Lines along which the components of the load or its derivatives are discontinuous

(iii) Lines along which the middle surface of the shell has a break or the curvature of the middle surface changes abruptly

(iv) Lines along which the rigidity or thickness of the shell changes

According to Gol'denveizer two main conditions are to be satisfied if the membrane theory is to give useful results. These are

(i) Lines of distortion shall not be too closely spaced on the shell.

(ii) No line of distortion may touch an asymptotic line of the middle surface.

The second condition may also be paraphrased as follows: *The normal curvature of the surface shall not vanish anywhere on any line of distortion.*

The membrane theory usually gives satisfactory results when applied to synclastic shells. Its application to anticlastic shells has always to be tempered with caution. The reason for this is twofold:

(i) Because the asymptotic lines of synclastic shells are imaginary, there is no possibility of a line of distortion touching an asymptotic line. Hence one of the conditions for the validity of the membrane theory is automatically satisfied.

(ii) Any local disturbance of the membrane state of stress on an asymptotic line is transmitted over a considerable length. This possibility does not exist in synclastic shells whose asymptotic lines are imaginary.

In the light of these observations it is easy to see why the membrane theory is not very suitable for the stress analysis of conoids and hyperbolic paraboloids supported on their straight-line generators.

CHAPTER 18

NEW FORMS OF SHELLS

18-1 Optimum Shapes of Shells

In the design of concrete shell structures, the usual practice is to select the shape of the shell first and then make a stress analysis. In selecting the geometry of the shell, no deliberate effort is made to ensure a desirable state of stress in the material. It is, perhaps, more logical to reverse this process. We may choose a desired state of stress and proceed to find the corresponding shape of the shell. Ideally, a concrete shell, in its membrane state, must carry loads by pure compression unaccompanied by shear stresses so that no tensile stresses develop and reinforcement becomes unnecessary except for taking care of secondary effects due to bending, temperature, and shrinkage. We do this in the design of arches. A parabolic arch is usually preferred because, under uniform loads, it carries loads by pure thrusts unaccompanied by bending. There are advantages in extending this line of reasoning to shells.

18-2 The Catenary Shell

In most shell roofs, the predominant load is the dead weight. It is therefore an advantage to choose the shape of the shell in such a manner that, under this condition of loading, the shell is subjected to pure compression unaccompanied by bending. This may be done by shaping the shell in the form of a catenary which is the *funicular shape* corresponding to dead weight. It is to be noted that this desired state of pure compression can be achieved only for one particular condition of loading. Under other conditions of loading, bending moments would of course develop. Shells shaped in the form of catenaries owe their inspiration to the Ctesiphon (pronounced Tessifon)—a great masonry catenary arch of 83 ft span and 112 ft rise believed to have been built in A.D. 550 and surviving to this day because of its shape on the banks of the Tigris not far from Baghdad.

18-3   Corrugated Catenary Shells

The resistance of catenary shells to lateral loads can be very much improved by corrugating them (Photo 18-1). We already saw that the shell can be maintained in a state of pure compression only under its own dead weight. Under other conditions of loading, bending stresses develop. Hence corrugations are necessary to provide adequate moment of resistance. An ingenious method of casting corrugated shells on flexible molds has been developed and patented by Messrs Barchild Constructions Co. Ltd., London. The method employs hessian (jute fabric) as a flexible mold. In this form of construction, two strip foundations are first laid along the two springings of the catenary shell. A light prefabricated tubular steel rib falsework, shaped to the form of a

PHOTO 18-1. A view of a corrugated catenary shell. (*Courtesy of Central Building Research Institute.*)

catenary, is erected at intervals indicated in Table 18-1, which is extracted from the Bulletin of the Central Building Research Institute, India [105].

Table 18-1*

Span, ft	Spacing of ribs, ft center to center
20	3–4
30	4
40	6
60	6–8
80	8

* Extracted by permission of the Director, Central Building Research Institute, Roorkee, India.

These ribs are backed by timber pieces. Hessian fabric is stretched across these ribs to form a flexible mold. A rendering of 1:3 cement mortar is next applied to the hessian in several layers. The fabric sags under the weight of the rendering to form corrugations. The recommended depths of corrugations at the crown for different spans are given in Table 18-2.

Table 18-2*

Span, ft	Depth of corrugation, in.
Up to 30	8
Up to 40	14
Up to 60	18
Up to 80	24

* Extracted by permission of the Director, Central Building Research Institute, Roorkee, India.

It is reported [105] that during World War II shells of up to 40 ft span and 100 ft length were built of a thickness of only $1\frac{1}{4}$ in. without any steel reinforcement. It is, however, good practice to make the shells $2\frac{1}{2}$ in. thick. The first layer of mortar may be $\frac{3}{4}$ in. thick. The steel is placed in the second layer 1 in. thick. The third layer $\frac{3}{4}$ in. thick provides adequate concrete cover for the reinforcement. Two days after the last coat is applied, the steel ribs may be removed and reerected to cast another length of the shell. This form of construction may prove very economical for warehouses and workshops. The analysis of these shells is carried out by regarding strips of the structure as two-hinged or fixed arches [105]. The usual height to span ratio for these structures is 1:2. This ratio may be reduced to 1:3 for large-span buildings such as hangars. More details are available in a number of papers published on this subject [106, 107].

18-4 Funicular Shells of Double Curvature

The catenary shell may be regarded as a singly curved *funicular shell*. We may now extend this reasoning to shells of double curvature and seek the shapes of shells which carry loads in their membrane state by pure compression unaccompanied by shear. We may find the equation of such a shell by setting $\overline{N}_x = \overline{N}_y = -N$ and $\overline{N}_{xy} = 0$ in Equation (15-16), where N is the desired compressive stress. Let us consider the shell acted upon by its dead weight. For this condition of loading, we get

$$\frac{\partial^2 z}{\partial x^2} + \frac{\partial^2 z}{\partial y^2} = +\frac{g}{N}\sqrt{1 + p^2 + q^2} \tag{18-1}$$

Most shell roofs in practice are shallow. Hence p^2 and q^2 may be ignored in comparison with unity. Hence the equation of a shallow funicular shell of double curvature under the action of dead weight may be written in the form

$$\frac{\partial^2 z}{\partial x^2} + \frac{\partial^2 z}{\partial y^2} = + \frac{g}{N} \qquad (18\text{-}2)$$

This result, in vector notation, may be rewritten as

$$\nabla^2 z = + \frac{g}{N} \qquad (18\text{-}3)$$

This is readily recognized as the equation of the Prandtl membrane occurring in the theory of torsion.[1] The calculation of the heights of the funicular shell at various points involves the solution of Poisson's equation (18-3). Many results already available in the theory of torsion of shafts of different geometrical cross sections may be drawn on to find the shapes of funicular shells with simple geometrical figures as ground plans. Table 18-3 gives the equations of funicular shells over rectangular, square, elliptical, circular, and triangular ground plans.

18-5 Generating Funicular Shapes

A simple experimental technique is available for finding the funicular shape corresponding to a given ground plan. Make a wooden frame whose plan is geometrically similar to that of the shell roof to be built. Stretch a piece of hessian or other flexible fabric across the mold. Load the fabric with a uniform layer of wet concrete or plaster of paris and let it sag. After the concrete has set, invert the surface so formed. Measure its ordinates and multiply them by a scale factor to find the heights of the shell to be built. Photo 18-2 shows a funicular shell roof over an oval ground plan generated by this technique. The shell was designed to roof an assembly hall, oval in plan, with a major axis of 70 ft and a minor axis of 55 ft 6 in. This experimental technique has also been employed by Ramaswamy and Chetty [108] for precasting small funicular-shell-roof units.

18-6 Funicular Shells over Rectangular Ground Plans

Shells of rectangular and square ground plans are of very frequent occurrence in practice. The equations of these shells given in Table 18-3 are rather clumsy to use as they involve series. Referring to

---

[1] Timoshenko, S., and J. N. Goodier, "Theory of Elasticity," 2d ed., p. 268, McGraw-Hill Book Company, New York, 1951.

Fig. 18-1, we may use the following approximate equation to describe the surface of a funicular shell over a rectangular ground plan [109]:

$$z = \frac{5}{8} \frac{g}{N} \frac{1}{a^2 + b^2} (a^2 - x^2)(b^2 - y^2) \qquad (18\text{-}4)$$

Table 18-3*

No.	Ground plan	Equation to the surface
1		$z = \dfrac{16ga^2}{N\pi^3} \displaystyle\sum_{n=1,3,5\ldots}^{\infty} \dfrac{1}{n^3}(-1)^{\frac{n-1}{2}}$ $\left(1 - \dfrac{\cosh\frac{n\pi y}{2a}}{\cosh\frac{n\pi b}{2a}}\right) \cos\dfrac{n\pi x}{2a}$
2		$z = \dfrac{16ga^2}{N\pi^3} \displaystyle\sum_{n=1,3,5\ldots}^{\infty} \dfrac{1}{n^3}(-1)^{\frac{n-1}{2}}$ $\left(1 - \dfrac{\cosh\frac{n\pi y}{2a}}{\cosh\frac{n\pi}{2}}\right) \cos\dfrac{n\pi x}{2a}$
3		$z = \dfrac{g}{2N} \dfrac{a^2 b^2}{a^2 + b^2} \left(\dfrac{x^2}{a^2} + \dfrac{y^2}{b^2} - 1\right)$
4		$z = \dfrac{g}{4N}(x^2 + y^2 - a^2)$
5		$z = \dfrac{g}{2N}\left[\dfrac{1}{2}(x^2 + y^2) - \dfrac{1}{2a}(x^3 - 3xy^2) - \dfrac{2}{27}a^2\right]$

* From "The Theory of a New Shell of Double Curvature," by G. S. Ramaswamy, *Indian Concrete Journal*, Vol. 34, No. 9, September, 1960. Reproduced by permission of the Concrete Association of India, Bombay.

PHOTO 18-2. A funicular shell roof over an oval ground plan roofing an assembly hall at Kanpur, India.

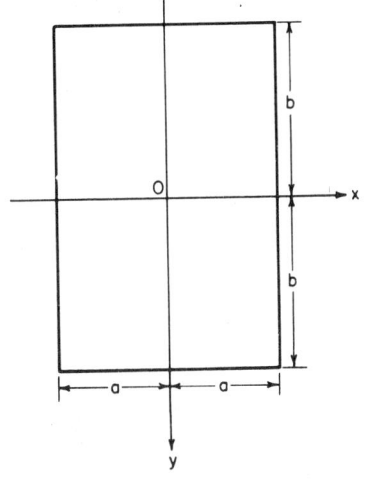

FIG. 18-1

18-7 Bending Theory of Funicular Shells

The membrane theory is inadequate for the analysis of large funicular shells. Resort to the bending theory, to be described in the next chapter, becomes necessary.

PHOTO 18-3.   A view of the casting yard where precast funicular shells are stacked. (*Courtesy of Port Trust, Madras, India.*)

PHOTO 18-4.   A view of the heavy-duty platform built of precast funicular shells for unloading cargo from the ships.   (*Courtesy of Port Trust, Madras, India.*)

18-8 Investigations on Funicular Shells

A number of writers appear to have arrived at the concept of a funicular shell as a result of independent investigations [110, 111]. The author and his colleagues have designed and built quite a number of them in India. Tests carried out in the author's laboratory have repeatedly demonstrated the remarkable load-carrying capacity that these shells possess. Photos 18-3 and 18-4 show prefabricated funicular shells under construction at the Madras Port. These have been designed for a superimposed load of 450 psf.

18-9 Search for New Shapes

The quest for new shapes is a continuing process. The experiments of Horacio Caminos [112] and investigations reported by Saether [113] are efforts in this direction. Such investigations are worthwhile if they result in shapes which are structurally more efficient in carrying loads applied on them.

CHAPTER 19

APPROXIMATE BENDING THEORY OF SHALLOW SHELLS OF DOUBLE CURVATURE

19-1 Shallow Shells

According to Vlasov [114] a shell may be regarded as shallow if the ratio of the rise to the shorter side of a shell of rectangular ground plan is less than 1:5.

19-2 Historical Notes

Marguerre [115], Vlasov [116], and Reissner [117] have independently developed bending theories applicable to shallow shells. Our treatment is based on the work of Vlasov.

19-3 Assumptions

The following assumptions, besides the basic assumptions of the classical theory of shells, are inherent in the Vlasov formulation.

(i) The squares and products of the surface derivatives $p = \partial z/\partial x$ and $q = \partial z/\partial y$ are negligible in comparison with unity.

(ii) The Gauss curvature of the undeformed middle surface cf the shell is very small and can be assumed to be equal to zero.

(iii) The derivatives of r, s, and t may be neglected. This, in effect, amounts to assuming that the principal curvatures of the shell remain constant.

The meaning of the first assumption would become clear if we examine expressions (5-10) for the first quadratic form of a surface. In this expression E and G become unity and F vanishes so that $ds^2 = dx^2 + dy^2$. Hence the distance between two points on the surface is approximated by its projected length.

VLASOV BENDING THEORY

19-4 Shell as a Combination of Disk, Plate, and Membrane

The Vlasov theory is most easily derived if we regard the shallow doubly curved shell as a combination of a disk, plate, and doubly curved membrane. We used this artifice before in Art. 7-20 to derive the D-K-J theory for cylindrical shells. This means that in the equations of equilibrium and the stress-strain relations all terms not figuring in the corresponding relations for the disk, plate, and membrane will be systematically dropped.

19-5 Equations of Equilibrium

Let the equation of the doubly curved shell surface be $z = z(x,y)$. We shall, as usual, denote

$$\frac{\partial^2 z}{\partial x^2} = r$$

$$\frac{\partial^2 z}{\partial x\,\partial y} = s$$

and

$$\frac{\partial^2 z}{\partial y^2} = t$$

Referring to the tangential group of forces acting on the element shown in Fig. 19-1, the following equations of equilibrium are easily written down:

$$\frac{\partial N_x}{\partial x} + \frac{\partial N_{xy}}{\partial y} + X = 0 \tag{19-1}$$

$$\frac{\partial N_y}{\partial y} + \frac{\partial N_{xy}}{\partial x} + Y = 0 \tag{19-2}$$

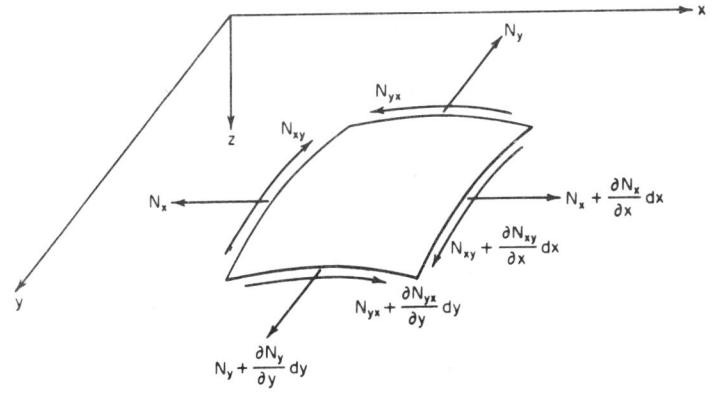

Fɪɢ. 19-1

The normal group of forces is shown in Fig. 19-2. For the sake of convenience, we shall denote $-M_{yx} = M_{xy} = H$. Accordingly, we may modify the forces acting on the element and depict them as in Fig. 19-3. The following equations of equilibrium follow from Figs. 19-2 and 19-3:

$$\frac{\partial M_y}{\partial y} - \frac{\partial H}{\partial x} - Q_v = 0 \tag{19-3}$$

$$\frac{\partial M_x}{\partial x} - \frac{\partial H}{\partial y} - Q_x = 0 \tag{19-4}$$

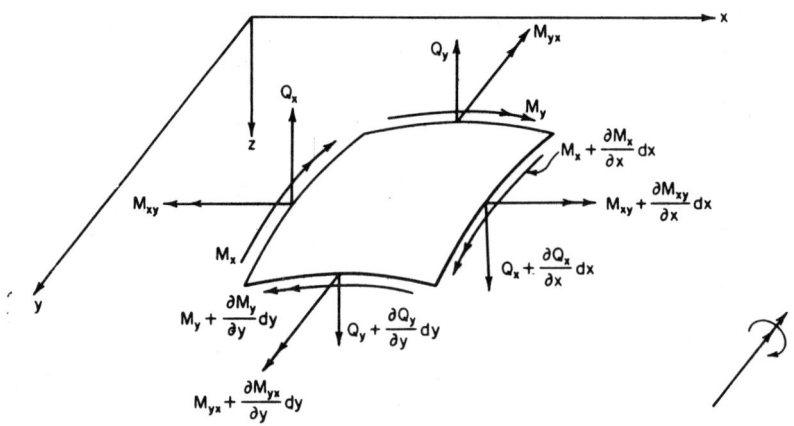

FIG. 19-2

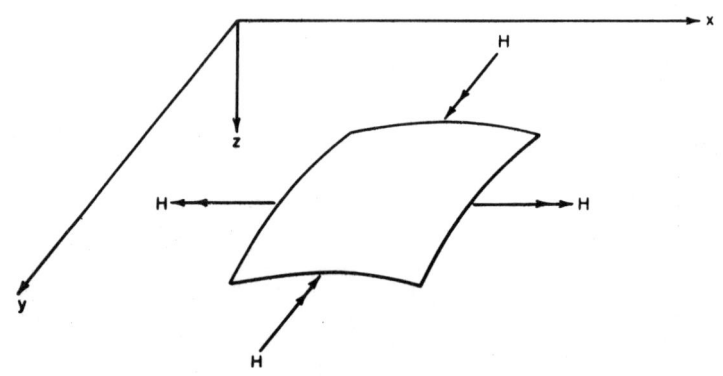

FIG. 19-3

Resolving forces in the z direction,

$$\frac{\partial Q_x}{\partial x} + \frac{\partial Q_y}{\partial y} + N_x \frac{\partial^2 z}{\partial x^2} + 2N_{xy} \frac{\partial^2 z}{\partial x \, \partial y} + N_y \frac{\partial^2 z}{\partial y^2} + Z = 0 \qquad (19\text{-}5)$$

From (19-3) and (19-4),

$$Q_y = \frac{\partial M_y}{\partial y} - \frac{\partial H}{\partial x} \qquad (19\text{-}6)$$

and

$$Q_x = \frac{\partial M_x}{\partial x} - \frac{\partial H}{\partial y} \qquad (19\text{-}7)$$

Substituting these values of Q_y and Q_x in (19-5), we arrive at the following relation:

$$\frac{\partial^2 M_x}{\partial x^2} - 2\frac{\partial^2 H}{\partial x \, \partial y} + \frac{\partial^2 M_y}{\partial y^2} + N_x \frac{\partial^2 z}{\partial x^2} + 2N_{xy} \frac{\partial^2 z}{\partial x \, \partial y} + N_y \frac{\partial^2 z}{\partial y^2} + Z = 0 \qquad (19\text{-}8)$$

19-6   Stress-Strain and Moment-Curvature Relations

The following relations are deduced in the same manner as expressions (7-4) to (7-6) for cylindrical shells by noting that the surface now under consideration has curvature in both x and y directions.   Thus

$$\epsilon_x = \frac{\partial u}{\partial x} - rw \qquad (19\text{-}9)$$

$$\epsilon_y = \frac{\partial v}{\partial y} - tw \qquad (19\text{-}10)$$

The expression for the shear strain assumes the form

$$\gamma_{xy} = \frac{\partial u}{\partial y} + \frac{\partial v}{\partial x} - 2sw \qquad (19\text{-}11)$$

This is the same as expression (7-6) for cylindrical shells except for the last term which results from the twist of the surface.   The expressions for curvature are those figuring in the plate theory.   These are

$$\chi_x = \frac{\partial^2 w}{\partial x^2} \qquad (19\text{-}12)$$

$$\chi_y = \frac{\partial^2 w}{\partial y^2} \qquad (19\text{-}13)$$

and

$$\tau = \frac{\partial^2 w}{\partial x \, \partial y} \qquad (19\text{-}14)$$

The stress-strain and moment-curvature relations may now be written down.

$$N_x = \frac{Ed}{1 - \nu^2} (\epsilon_x + \nu\epsilon_y) \tag{19-15}$$

$$N_y = \frac{Ed}{1 - \nu^2} (\epsilon_y + \nu\epsilon_x) \tag{19-16}$$

$$N_{xy} = N_{yx} = \frac{Ed}{2(1 + \nu)} \gamma_{xy} = \frac{Ed}{2} \frac{1 - \nu}{1 - \nu^2} \gamma_{xy} \tag{19-17}$$

$$M_x = -D(\chi_x + \nu\chi_y) \tag{19-18}$$

$$M_y = -D(\chi_y + \nu\chi_x) \tag{19-19}$$

$$H = D(1 - \nu)\tau \tag{19-20}$$

19-7 Derivation of the Vlasov Equations

From (19-9) and (19-10), we have

$$\frac{\partial^2 \epsilon_x}{\partial y^2} + \frac{\partial^2 \epsilon_y}{\partial x^2} = \frac{\partial^3 u}{\partial x\, \partial y^2} + \frac{\partial^3 v}{\partial y\, \partial x^2} - r\frac{\partial^2 w}{\partial y^2} - t\frac{\partial^2 w}{\partial x^2} \tag{19-21}$$

Again from (19-11),

$$\frac{\partial^2 \gamma_{xy}}{\partial x\, \partial y} = \frac{\partial^3 u}{\partial x\, \partial y^2} + \frac{\partial^3 v}{\partial y\, \partial x^2} - 2s\frac{\partial^2 w}{\partial x\, \partial y} \tag{19-22}$$

From (19-21) and (19-22),

$$\frac{\partial^2 \epsilon_x}{\partial y^2} + \frac{\partial^2 \epsilon_y}{\partial x^2} = \frac{\partial^2 \gamma_{xy}}{\partial x\, \partial y} - r\frac{\partial^2 w}{\partial y^2} + 2s\frac{\partial^2 w}{\partial x\, \partial y} - t\frac{\partial^2 w}{\partial x^2} \tag{19-23}$$

Let us consider the influence of vertical loads only so that $X = Y = 0$. It is expedient at this stage to introduce a stress function ϕ which is such that

$$\frac{\partial^2 \phi}{\partial y^2} = N_x \qquad \frac{\partial^2 \phi}{\partial x^2} = N_y \qquad \text{and} \qquad -\frac{\partial^2 \phi}{\partial x\, \partial y} = N_{xy}$$

Rearranging terms in (19-23), we arrive at the following relation:

$$\left(\frac{\partial^2 \epsilon_x}{\partial y^2} - \frac{\partial^2 \gamma_{xy}}{\partial x\, \partial y} + \frac{\partial^2 \epsilon_y}{\partial x^2}\right) + \left(r\frac{\partial^2 w}{\partial y^2} - 2s\frac{\partial^2 w}{\partial x\, \partial y} + t\frac{\partial^2 w}{\partial x^2}\right) = 0 \tag{19-24}$$

Substituting for the first three terms from the stress-strain relations (19-15), (19-16), and (19-17) and noting the properties of the stress function ϕ, Equation (19-24) takes the form

$$\nabla^4 \phi + Ed\nabla_k^2 w = 0 \tag{19-25}$$

where the operator

$$\nabla_k^2 = \left(r \frac{\partial^2}{\partial y^2} - 2s \frac{\partial^2}{\partial x \, \partial y} + t \frac{\partial^2}{\partial x^2} \right)$$

Now, let us turn our attention to Equation (19-8). Substituting in terms of w for M_x, M_y, and M_{xy} from relations (19-18), (19-19), and (19-20) and writing the stresses N_x, N_{xy}, and N_y in terms of the stress function ϕ, Equation (19-8) may be rewritten as

$$-D\nabla^4 w + \nabla_k^2 \phi + Z = 0$$

or

$$D\nabla^4 w - \nabla_k^2 \phi = Z \tag{19-26}$$

Relations (19-25) and (19-26) constitute the well-known equations of Vlasov for shallow shells. The first of these relations is a statement of compatibility of deformations and the second that of equilibrium. It is interesting to note that (19-26) assumes the form $D\nabla^4 w = Z$, when the membrane contribution $\nabla_k^2 \phi = 0$. This is the well-known equation of plates. Again, if $D = 0$, we have a pure membrane with no bending and $-\nabla_k^2 \phi = Z$, which is easily recognized as the governing differential equation of the membrane theory of doubly curved shells.

19-8 Shells of Variable Curvature

Although Equations (19-25) and (19-26) are strictly applicable only to shells of constant curvature such as translational surfaces of the second degree for which r, s, and t are constants, they may also be employed without appreciable error for the analysis of shallow shells of variable curvature. Thus, for instance, Keshava Rao and Sharma [118] and Hadid [119] have made use of these equations for the analysis of conoids.

19-9 Differential Equations in Terms of Displacements

For application to shells whose boundary conditions are specified in terms of displacements, it is convenient to develop the bending theory for shallow shells in terms of u, v, and w. Substituting for the stresses N_x and N_{xy} in Equation (19-1) in terms of u, v, and w, we arrive at the relation

$$\frac{\partial^2 u}{\partial x^2} + \frac{1-\nu}{2} \frac{\partial^2 u}{\partial y^2} + \frac{1+\nu}{2} \frac{\partial^2 v}{\partial x \, \partial y} - \frac{\partial}{\partial x}[(r + \nu t)w]$$

$$- \frac{\partial}{\partial y}[(1-\nu)sw] + \frac{X}{Ed}(1 - \nu^2) = 0 \tag{19-27}$$

In a similar manner, the equations of equilibrium (19-2) and (19-8) assume the form

$$\frac{\partial^2 v}{\partial y^2} + \frac{1-v}{2}\frac{\partial^2 v}{\partial x^2} + \frac{1+v}{2}\frac{\partial^2 u}{\partial x\,\partial y} - \frac{\partial}{\partial y}[('t + vr)w]$$

$$- \frac{\partial}{\partial x}[(1-v)sw] + \frac{Y}{Ed}(1-v^2) = 0 \qquad (19\text{-}28)$$

$$(r+tv)\frac{\partial u}{\partial x} + (rv+t)\frac{\partial v}{\partial y} - [r^2 + 2vrt + t^2 + 2s^2(1-v)]w$$

$$+ s(1-v)\left(\frac{\partial u}{\partial y} + \frac{\partial v}{\partial x}\right) - \frac{d^2}{12}\nabla^4 w + \frac{Z}{Ed}(1-v^2) = 0 \qquad (19\text{-}29)$$

Equations (19-27), (19-28), and (19-29) are also applicable to shells of variable curvature. For translational shells such as the rotational paraboloid and the elliptic paraboloid, these three equations may be simplified and rewritten as

$$\frac{\partial^2 u}{\partial x^2} + \frac{1-v}{2}\frac{\partial^2 u}{\partial y^2} + \frac{1+v}{2}\frac{\partial^2 v}{\partial x\,\partial y} - \frac{\partial w}{\partial x}(r+vt) + \frac{X}{Ed}(1-v^2) = 0 \qquad (19\text{-}30)$$

$$\frac{\partial^2 v}{\partial y^2} + \frac{1-v}{2}\frac{\partial^2 v}{\partial x^2} + \frac{1+v}{2}\frac{\partial^2 u}{\partial x\,\partial y} - \frac{\partial w}{\partial y}(t+vr) + \frac{Y}{Ed}(1-v^2) = 0 \qquad (19\text{-}31)$$

and

$$(r+vt)\frac{\partial u}{\partial x} + (rv+t)\frac{\partial v}{\partial y} - (r^2 + 2vrt + t^2)w - \frac{d^2}{12}\nabla^4 w + \frac{Z}{Ed}(1-v^2) = 0 \qquad (19\text{-}32)$$

Now, to simplify the equations further, let us put $v = 0$ to get

$$\frac{\partial^2 u}{\partial x^2} + \frac{1}{2}\frac{\partial^2 u}{\partial y^2} + \frac{1}{2}\frac{\partial^2 v}{\partial x\,\partial y} - r\frac{\partial w}{\partial x} + \frac{X}{Ed} = 0 \qquad (19\text{-}33)$$

$$\frac{\partial^2 v}{\partial y^2} + \frac{1}{2}\frac{\partial^2 v}{\partial x^2} + \frac{1}{2}\frac{\partial^2 u}{\partial x\,\partial y} - t\frac{\partial w}{\partial y} + \frac{Y}{Ed} = 0 \qquad (19\text{-}34)$$

$$r\frac{\partial u}{\partial x} + t\frac{\partial v}{\partial y} - (r^2 + t^2)w - \frac{d^2}{12}\nabla^4 w + \frac{Z}{Ed} = 0 \qquad (19\text{-}35)$$

19-10   Eighth-order Equation for Shells of Constant Curvature

The two fourth-order partial differential equations (19-25) and (19-26) can be combined into a single partial differential equation of the eighth order if the shell is of constant curvature, i.e., if r, s, and t are constants. This is the case for second-degree translational surfaces such as the elliptic paraboloid and the hyperbolic paraboloid. The reduction into a single partial differential equation of the eighth order is effected by the introduction of a stress-displacement function $W(x,y)$ which is such that

$$\nabla^4 W = w \tag{19-36}$$

From (19-25) it is evident that

$$\phi = -Ed\,\nabla_k{}^2\,W \tag{19-37}$$

Substituting relations (19-36) and (19-37) in (19-26), we get

$$D\nabla^8 W + Ed\nabla_k{}^2\nabla_k{}^2 W = Z$$

or

$$\nabla^8 W + \frac{12}{d^2}\,\nabla_k{}^2\nabla_k{}^2 W = \frac{Z}{D} \tag{19-38}$$

It is interesting to note that (19-38) would reduce itself to the D-K-J equation for cylindrical shells.

19-11 Expressions for the Shell Actions

In terms of the stress-displacement function, one may write down expressions for the stresses, moments, and displacements as follows:

$$N_x = Ed\,\nabla_k{}^2\,\frac{\partial^2 W}{\partial y^2}$$

$$N_y = Ed\,\nabla_k{}^2\,\frac{\partial^2 W}{\partial x^2}$$

$$N_{xy} = -Ed\,\nabla_k{}^2\,\frac{\partial^2 W}{\partial x\,\partial y}$$

$$M_x = -D\,\nabla^4\left(\frac{\partial^2 W}{\partial x^2} + \nu\,\frac{\partial^2 W}{\partial y^2}\right)$$

$$M_y = -D\,\nabla^4\left(\frac{\partial^2 W}{\partial y^2} + \nu\,\frac{\partial^2 W}{\partial x^2}\right)$$

$$H = D(1-\nu)\,\nabla^4\,\frac{\partial^2 W}{\partial x\,\partial y} \tag{19-39}$$

$$Q_x = -D\,\nabla^6\,\frac{\partial W}{\partial x}$$

$$Q_y = -D\,\nabla^6\,\frac{\partial W}{\partial y}$$

$$Q_x^* = -D\,\nabla^6\,\frac{\partial W}{\partial x} + D(1-\nu)\,\nabla^4\,\frac{\partial^3 W}{\partial x\,\partial y^2}$$

$$Q_y^* = -D\,\nabla^6\,\frac{\partial W}{\partial y} - D(1-\nu)\,\nabla^4\,\frac{\partial^3 W}{\partial x^2\,\partial y}$$

$$w = \nabla^4\,W$$

19-12 The Levy Solution

If two opposite edges at $x = \pm a/2$ (Fig. 19-4) of the shell are simply supported on shear diaphragms on the analogy of plates, we may seek a solution of the Levy type.

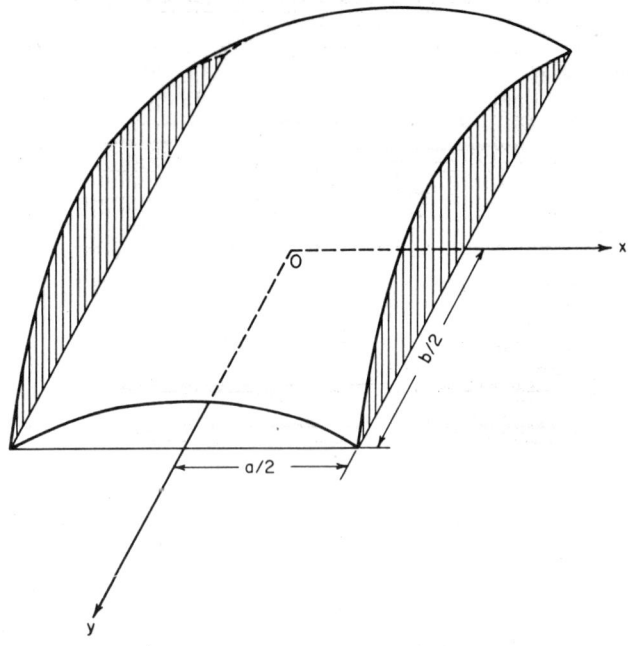

Fig. 19-4

The boundary conditions at the simply supported edges may be formulated as follows: At $x = \pm a/2$,

$$M_x = 0$$

$$w = 0$$

$$N_x = 0$$

$$\epsilon_y = \frac{1}{Ed}\,(N_y - \nu N_x) = 0 \qquad (19\text{-}40)$$

In terms of W, these boundary conditions may be expressed as

$$\nabla^4 W = \nabla^4 \left(\frac{\partial^2 W}{\partial x^2} + \nu\,\frac{\partial^2 W}{\partial y^2} \right) = \nabla_k^2\,\frac{\partial^2 W}{\partial x^2} = \nabla_k^2\,\frac{\partial^2 W}{\partial y^2} = 0 \qquad (19\text{-}41)$$

Let us consider the unloaded shell. It is easily verified that a solution of the form

$$W = \sum_{n=1}^{\infty} C_n e^{\rho \lambda_n y} \cos \lambda_n x \qquad (19\text{-}42)$$

where $\lambda_n = n\pi/a$, would satisfy the prescribed boundary conditions at the edges $x = \pm a/2$. The substitution of (19-42) in (19-38) leads to the following characteristic equation:

$$(-\lambda_n{}^2 + \rho^2 \lambda_n{}^2)^4 + \frac{12}{d^2}(r\rho^2 \lambda_n{}^2 - t\lambda_n{}^2)^2 = 0$$

The eight roots $\rho_1, \rho_2, \rho_3, ..., \rho_8$ of this polynomial equation may be found explicitly. These are given by Apeland and Popov [120]. We may thus write W in terms of eight arbitrary constants as follows:

$$\begin{aligned} W = \sum_{n=1}^{\infty} (&C_1 e^{\rho_1 \lambda_n y} + C_2 e^{\rho_2 \lambda_n y} + C_3 e^{\rho_3 \lambda_n y} \\ &+ C_4 e^{\rho_4 \lambda_n y} + C_5 e^{\rho_5 \lambda_n y} + C_6 e^{\rho_6 \lambda_n y} \\ &+ C_7 e^{\rho_7 \lambda_n y} + C_8 e^{\rho_8 \lambda_n y}) \cos \lambda_n x \qquad (19\text{-}43) \end{aligned}$$

The arbitrary constants $C_1, C_2, C_3, ..., C_8$ are now evaluated by taking into account the boundary conditions at the edges $y = \pm b/2$. Appropriate particular integrals may be found by expanding the load in trigonometric series. Thus, if the load g is constant in the y direction, we may represent it as

$$g = \sum_{n=1}^{\infty} g_n \cos \lambda_n x$$

The particular integral for this case is easily seen to be

$$W = \sum_{n=1}^{\infty} \left(\lambda_n{}^8 + \frac{12}{d^2} t^2 \lambda_n{}^4 \right)^{-1} \frac{g_n}{D} \cos \lambda_n x \qquad (19\text{-}44)$$

This method forms the basis of design tables for translational shells under publication by Apeland and Popov.

19-13 The Navier Solution for Shells Simply Supported at All the Four Edges

If a translational shell is simply supported at all its four edges by shear diaphragms, we may find a solution for ϕ and w in the form of a double trigonometric series. Consider a translational shell of sides a and b

(Fig. 19-5), simply supported on shear diaphragms along its edges $x = 0$, $x = a$, and $y = 0$, $y = b$. It is easily seen that the following functions for ϕ and w would satisfy the prescribed boundary conditions:

$$\phi = \sum A_{mn} \sin \frac{m\pi x}{a} \sin \frac{n\pi y}{b}$$

and

$$w = \sum B_{mn} \sin \frac{m\pi x}{a} \sin \frac{n\pi y}{b}$$

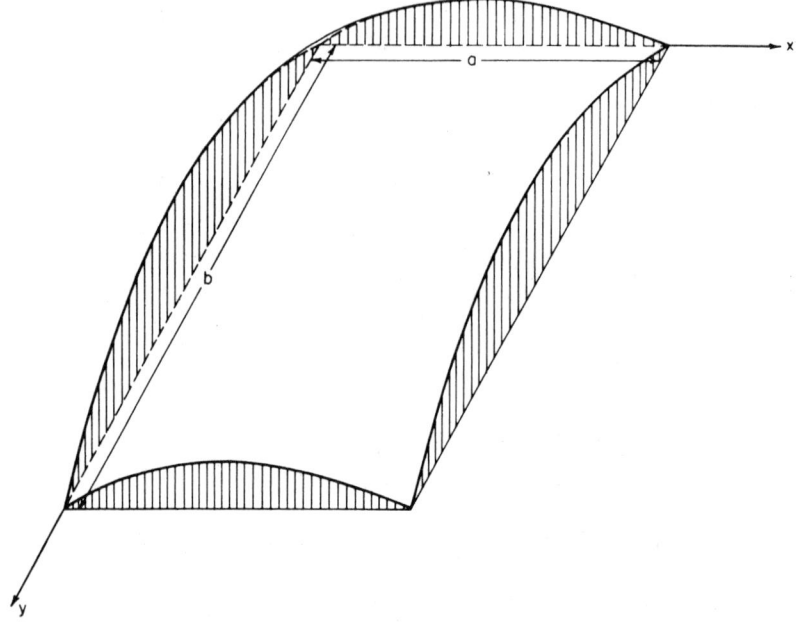

FIG. 19-5

We may also develop the load Z in double series as

$$Z = \sum C_{mn} \sin \frac{m\pi x}{a} \sin \frac{n\pi y}{b}$$

Being a translational surface, $s = 0$. As usual, $\partial^2 z / \partial x^2 = r$ and $\partial^2 z / \partial y^2 = t$. It will be assumed that the surface is of the second degree so that r and t are constants. Let $r = k_1$ and $t = k_2$. One may easily verify the following relations:

$$\nabla^4 \phi = A_{mn}(\alpha_m^2 + \beta_n^2)^2 \sin \alpha_m x \sin \beta_n y$$

where

$$\alpha_m = \frac{m\pi}{a} \quad \text{and} \quad \beta_n = \frac{n\pi}{b}$$

$$\nabla_k^2 = \left(k_2 \frac{\partial^2}{\partial x^2} + k_1 \frac{\partial^2}{\partial y^2} \right)$$

$$\nabla_k^2 w = -B_{mn}(k_2 \alpha_m^2 + k_1 \beta_n^2) \sin \alpha_m x \sin \beta_n y$$

Substitution of these values of $\nabla^4 \phi$ and $\nabla_k^2 w$ in (19-25) leads to the relation

$$A_{mn}(\alpha_m^2 + \beta_n^2)^2 = Ed \, B_{mn}(k_2 \alpha_m^2 + k_1 \beta_n^2)$$

Hence

$$B_{mn} = \frac{A_{mn}(\alpha_m^2 + \beta_n^2)^2}{Ed(k_2 \alpha_m^2 + k_1 \beta_n^2)} \tag{19-45}$$

We may next calculate $D\nabla^4 w$ and $\nabla_k^2 \phi$. These are found as

$$D\nabla^4 w = DB_{mn}(\alpha_m^2 + \beta_n^2)^2 \sin \alpha_m x \sin \beta_n y$$

and

$$\nabla_k^2 \phi = -A_{mn}(k_2 \alpha_m^2 + k_1 \beta_n^2) \sin \alpha_m x \sin \beta_n y$$

Substituting these in (19-26) and noting that $Z = C_{mn} \sin \alpha_m x \sin \beta_n y$, we arrive at A_{mn}. Thus

$$A_{mn} = \frac{C_{mn}(k_2 \alpha_m^2 + k_1 \beta_n^2)}{(k_2 \alpha_m^2 + k_1 \beta_n^2)^2 + \dfrac{d^2}{12}(\alpha_m^2 + \beta_n^2)^4} \tag{19-46}$$

From (19-45), it is now possible to write down B_{mn}. Thus

$$B_{mn} = \frac{C_{mn}(\alpha_m^2 + \beta_n^2)^2}{Ed \left[(k_2 \alpha_m^2 + k_1 \beta_n^2)^2 + \dfrac{d^2}{12}(\alpha_m^2 + \beta_n^2)^4 \right]} \tag{19-47}$$

Some special cases that arise may be noted.

Case 1

If $k_1 = k_2 = 1/R$, the surface becomes a paraboloid or a shallow spherical calotte. The coefficients A_{mn} and B_{mn} for this special case may be simplified as

$$A_{mn} = \frac{RC_{mn}}{(\alpha_m^2 + \beta_n^2)\left[1 + \dfrac{R^2 d^2}{12}(\alpha_m^2 + \beta_n^2)^2 \right]} \tag{19-48}$$

and

$$B_{mn} = \frac{R^2 C_{mn}}{Ed \left[1 + \dfrac{R^2 d^2}{12}(\alpha_m^2 + \beta_n^2)^2 \right]} \tag{19-49}$$

Case 2

If $k_1 = k_2 = 0$, the shell degenerates into a plate. For this case it is easily verified that

$$A_{mn} = 0 \tag{19-50}$$

and

$$B_{mn} = \frac{C_{mn}(\alpha_m^2 + \beta_n^2)^2}{\dfrac{Ed^3}{12}(\alpha_m^2 + \beta_n^2)^4}$$

or

$$B_{mn} = \frac{C_{mn}}{D(\alpha_m^2 + \beta_n^2)^2} \tag{19-51}$$

The expressions for C_{mn} corresponding to uniformly distributed and concentrated loads are readily available in books on plates [121]. Thus, if the uniformly distributed load is p_0, $C_{mn} = 16p_0/\pi^2 mn$. Again if a concentrated load P is applied on the shell at a point whose coordinates are x' and y', $C_{mn} = (4P/ab) \sin(m\pi x'/a) \sin(n\pi y'/b)$. In terms of ϕ and w, expressions for the stresses, moments, and shears may be written as follows, assuming $\nu = 0$ for simplicity.

$$\overline{N_x} = \frac{\partial^2 \phi}{\partial y^2} = -A_{mn}\beta_n^2 \sin \alpha_m x \sin \beta_n y$$

$$\overline{N_y} = \frac{\partial^2 \phi}{\partial x^2} = -A_{mn}\alpha_m^2 \sin \alpha_m x \sin \beta_n y$$

$$\overline{N_{xy}} = -\frac{\partial^2 \phi}{\partial x \partial y} = -A_{mn}\alpha_m\beta_n \cos \alpha_m x \cos \beta_n y$$

$$M_x = -D\frac{\partial^2 w}{\partial x^2} = DB_{mn}\alpha_m^2 \sin \alpha_m x \sin \beta_n y$$

$$M_y = -D\frac{\partial^2 w}{\partial y^2} = DB_{mn}\beta_n^2 \sin \alpha_m x \sin \beta_n y$$

$$H = DB_{mn}\alpha_m\beta_n \cos \alpha_m x \cos \beta_n y$$

$$Q_x = -D\frac{\partial}{\partial x}\left(\frac{\partial^2 w}{\partial x^2} + \frac{\partial^2 w}{\partial y^2}\right)$$
$$= DB_{mn}\alpha_m(\alpha_m^2 + \beta_n^2) \cos \alpha_m x \sin \beta_n y$$

$$Q_y = -D\frac{\partial}{\partial y}\left(\frac{\partial^2 w}{\partial x^2} + \frac{\partial^2 w}{\partial y^2}\right)$$
$$= DB_{mn}\beta_n(\alpha_m^2 + \beta_n^2) \sin \alpha_m x \cos \beta_n y$$

$$Q_x^* = DB_{mn}(\alpha_m^3 + 2\alpha_m\beta_n^2) \cos \alpha_m x \sin \beta_n y$$

$$Q_y^* = DB_{mn}(\beta_n^3 + 2\alpha_m^2\beta_n) \sin \alpha_m x \cos \beta_n y$$

$$(19\text{-}52)$$

19-14 Separation of Plate and Curvature Effects

The convergence of the Navier solution for translational shells is extremely slow. Doganoff [76], for instance, had to take as many as 48 terms to get the required degree of accuracy. This led to suggest that convergence could be very much accelerated by separating out the plate and curvature effects. The curvature effect is rapidly convergent. The slowly convergent plate part may be replaced by some known rapidly convergent plate solution such as the Levy solution. By this means, the Navier solution can be rendered useful for the stress analysis of translational shells. The same artifice has also been employed by Flügge and Conrad [122]. The separation of the plate and curvature effects is possible if B_{mn} is rewritten as

$$B_{mn} = C_{mn} \frac{12}{Ed^3} \left\{ \frac{1}{(\alpha_m{}^2 + \beta_n{}^2)^2} - \frac{\dfrac{12}{d^2}(k_2\alpha_m{}^2 + k_1\beta_n{}^2)^2}{(\alpha_m{}^2 + \beta_n{}^2)^2 \left[(\alpha_m{}^2 + \beta_n{}^2)^4 + \dfrac{12}{d^2}(k_2\alpha_m{}^2 + k_1\beta_n{}^2)^2\right]} \right\} \quad (19\text{-}53)$$

The first and second terms of this expression are respectively identified as the plate and curvature effects. The plate and curvature effects may be superimposed with appropriate signs to yield the complete solution.

19-15 Application to a Rotational Paraboloid

Let us apply the technique explained in the last article to the rotational paraboloid over a rectangular ground plan (Fig. 19-5). The equation of the surface takes the form

$$z = \frac{1}{2R}(x^2 + y^2 - ax - by) \quad (19\text{-}54)$$

Hence

$$\frac{\partial^2 z}{\partial x^2} = k_1 = \frac{1}{R} \qquad \frac{\partial^2 z}{\partial y^2} = k_2 = \frac{1}{R} \qquad \text{and} \qquad \frac{\partial^2 z}{\partial x \, \partial y} = 0$$

$$A_{mn} = \frac{C_{mn}R}{(\alpha_m{}^2 + \beta_n{}^2)\left[1 + \dfrac{R^2 d^2}{12}(\alpha_m{}^2 + \beta_n{}^2)^2\right]}$$

Let us consider a uniform load g due to the dead weight of the shell.

$$A_{mn} = \frac{16gR}{\pi^2 mn(\alpha_m{}^2 + \beta_n{}^2)\left[1 + \dfrac{R^2 d^2}{12}(\alpha_m{}^2 + \beta_n{}^2)^2\right]}$$

The stress function is

$$\phi = \sum_{m=1}^{\infty} \sum_{n=1}^{\infty} A_{mn} \sin \alpha_m x \sin \beta_n y$$

The expressions for the in-plane stresses follow from (19-52). Next we proceed to construct an expression for w. As noted before, it is advantageous to separate the curvature and plate effects. We need only consider here the curvature effect in detail. For the plate part we may make use of the Levy solution already available in the book by Timoshenko.[1] After some simplification, we may write

$$B_{mn}(\text{curvature}) = B'_{mn}$$

$$= \frac{16g}{\pi^2 mnD} \left\{ \frac{1}{(\alpha_m{}^2 + \beta_n{}^2)^2 \left[\dfrac{R^2 d^2}{12} (\alpha_m{}^2 + \beta_n{}^2)^2 + 1 \right]} \right\} \quad (19\text{-}55)$$

Because of the curvature only, $w = B'_{mn} \sin \alpha_m x \sin \beta_n y$. Expressions for moments and shears follow from (19-52).

The application of the method to a rotational paraboloid is explained in detail in Design Example 19-1.

Design Example 19-1

BENDING ANALYSIS OF A PARABOLOID OF REVOLUTION BY SEPARATION OF PLATE AND CURVATURE EFFECTS

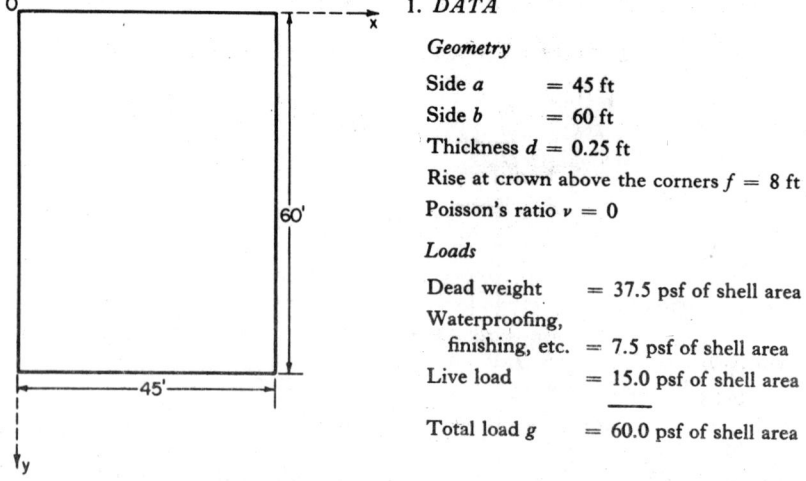

1. DATA

Geometry

Side a	= 45 ft
Side b	= 60 ft
Thickness d	= 0.25 ft
Rise at crown above the corners f	= 8 ft
Poisson's ratio ν	= 0

Loads

Dead weight	= 37.5 psf of shell area
Waterproofing, finishing, etc.	= 7.5 psf of shell area
Live load	= 15.0 psf of shell area
Total load g	= 60.0 psf of shell area

Fig. 19-6

[1] Timoshenko, S., and S. Woinowsky-Krieger, "Theory of Plates and Shells," 2d ed., p. 113, McGraw-Hill Book Company, New York, 1959.

2. GEOMETRICAL PROPERTIES AND OTHER FACTORS

Equation of the surface is given by $z = \dfrac{1}{2R}(x^2 + y^2 - ax - by)$

$$\text{Radius of the surface } R = \frac{1}{2(-f)}\left[\left(\frac{a}{2}\right)^2 + \left(\frac{b}{2}\right)^2 - \frac{a^2}{2} - \frac{b^2}{2}\right] = 87.89 \text{ ft}$$

$$\left.\begin{aligned}\alpha_m &= \frac{m\pi}{a} \\[2mm] \beta_n &= \frac{n\pi}{b}\end{aligned}\right\} \quad \begin{aligned} m &= 1,\ 3,\ 5,\dots \\ n &= 1,\ 3,\ 5,\dots \end{aligned}$$

$$D = \frac{Ed^3}{12} = \frac{288 \times 10^6 \times 0.25^3}{12} = 0.375 \times 10^6$$

3. SOLUTION FOR PLATE

As explained in Art. 19-15, the plate effect is calculated using the Levy solution. The details of calculation are not given, as this solution is well known. The moments M_x, M_y, and M_{xy} at various points on the shell are shown tabulated below.

x	y	M_x	M_y	M_{xy}
0	0	0	0	7,023
5.63	0	0	0	6,188
11.25	0	0	0	4,467
16.88	0	0	0	2,317
22.5	0	0	0	0
0	7.5	0	0	5,947
5.63	7.5	1,694	1,179	5,383
11.25	7.5	2,514	2,052	3,962
16.88	7.5	2,905	2,566	2,080
22.5	7.5	3,022	2,736	0
0	15.0	0	0	4,070
5.63	15.0	2,840	1,452	3,729
11.25	15.0	4,431	2,639	2,802
16.88	15.0	5,228	3,398	1,491
22.5	15.0	5,470	3,657	0
0	22.5	0	0	2,037
5.63	22.5	3,491	1,489	1,874
11.25	22.5	5,582	2,738	1,420
16.88	22.5	6,674	3,561	762
22.5	22.5	7,010	3,847	0
0	30.0	0	0	0
5.63	30.0	3,703	1,485	0
11.25	30.0	5,964	2,737	0
16.88	30.0	7,159	3,567	0
22.5	30.0	7,530	3,856	0

4. SOLUTION FOR CURVATURE

The expressions for A_{mn} and B_{mn}, already derived in Art. 19-15, are reproduced below for ready reference.

$$A_{mn} = \frac{16gR}{\pi^4 mn(\alpha_m{}^2 + \beta_n{}^2)\left[\dfrac{R^2 d^2}{12}(\alpha_m{}^2 + \beta_n{}^2)^2 + 1\right]}$$

$$B_{mn} = \frac{16g}{\pi^2 mn D(\alpha_m{}^2 + \beta_n{}^2)^2\left[\dfrac{R^2 d^2}{12}(\alpha_m{}^2 + \beta_n{}^2)^2 + 1\right]}$$

The expressions for the direct stresses N_x, N_y, and N_{xy} and the moments M_x, M_y, and M_{xy} are given in Equation (19-52) in terms of A_{mn}.

The calculations are carried out using three terms of the Fourier series, i.e., $m = 1, 3$, and 5 and $n = 1, 3$, and 5.

The calculations have also been done with four terms, and it is found that the results do not differ much from those obtained with three terms, indicating the rapid convergence of the curvature solution. The results of curvature solution for various points of the shell are shown below.

x	y	N_x	N_y	N_{xy}	M_x	M_y	M_{xy}
0	0	0	0	−6,986	0	0	6,824
5.63	0	0	0	−5,374	0	0	6,143
11.25	0	0	0	−2,790	0	0	4,458
16.88	0	0	0	−1,307	0	0	2,322
22.5	0	0	0	0	0	0	0
0	7.5	0	0	−4,846	0	0	5,952
5.63	7.5	−2,318	−3,084	−3,918	1,614	1,137	5,395
11.25	7.5	−3,497	−2,430	−2,286	2,517	2,004	3,967
16.88	7.5	−3,922	−1,531	−1,098	2,899	2,518	2,082
22.5	7.5	−4,076	−1,753	0	3,032	2,688	0
0	15.0	0	0	−1,941	0	0	4,058
5.63	15.0	−1,252	−4,204	−1,799	2,760	1,468	3,722
11.25	15.0	−2,122	−3,728	−1,350	4,440	2,664	2,800
16.88	15.0	−2,587	−2,713	−687	5,221	3,425	1,490
22.5	15.0	−2,737	−2,958	0	5,482	3,684	0
0	22.5	0	0	−1,013	0	0	2,044
5.63	22.5	−489	−4,479	−888	3,406	1,469	1,879
11.25	22.5	−1,002	−4,271	−621	5,588	2,708	1,422
16.88	22.5	−1,372	−3,376	−328	6,666	3,528	762
22.5	22.5	−1,493	−3,627	0	7,021	3,813	0
0	30.0	0	0	0	0	0	0
5.63	30.0	−907	−4,692	0	3,625	1,508	0
11.25	30.0	−1,526	−4,480	0	5,973	2,770	0
16.88	30.0	−1,879	−3,574	0	7,152	3,603	0
22.5	30.0	−2,008	−3,859	0	7,542	3,893	0

5. *FINAL STRESSES*

The final stresses and moments in the shell are obtained by combining the plate and curvature solutions and are shown tabulated below.   The values are plotted in Fig. 19-7.

x	y	N_x	N_y	N_{xy}	M_x	M_y	M_{xy}
0	0	0	0	−6,986	0	0	199
5.63	0	0	0	−5,374	0	0	45
11.25	0	0	0	−2,790	0	0	9
16.88	0	0	0	−1,307	0	0	−5
22.5	0	0	0	0	0	0	0
0	7.5	0	0	−4,846	0	0	−5
5.63	7.5	−2,318	−3,084	−3,918	80	42	−12
11.25	7.5	−3,497	−2,430	−2,286	−3	48	−5
16.88	7.5	−3,922	−1,531	−1,098	6	48	−2
22.5	7.5	−4,076	−1,753	0	−10	48	0
0	15.0	0	0	−1,941	0	0	12
5.63	15.0	−1,252	−4,204	−1,799	80	−16	7
11.25	15.0	−2,122	−3,728	−1,350	−9	−25	2
16.88	15.0	−2,587	−2,713	−687	7	−27	1
22.5	15.0	−2,737	−2,958	0	−12	−27	0
0	22.5	0	0	−1,013	0	0	−7
5.63	22.5	−489	−4,479	−888	85	20	−5
11.25	22.5	−1,002	−4,271	−621	−6	30	−2
16.88	22.5	−1,372	−3,376	−328	8	33	0
22.5	22.5	−1,493	−3,627	0	−11	34	0
0	30.0	0	0	0	0	0	0
5.63	30.0	−907	−4,692	0	78	−23	0
11.25	30.0	−1,526	−4,480	0	−9	−33	0
16.88	30.0	−1,879	−3,574	0	7	−36	0
22.5	30.0	−2,008	−3,859	0	−12	−37	0

The variation of M_x and M_y along $y = b/2$ and $x = a/2$ is compared with the results obtained by Doganoff [76] in Fig. 19-8.

19-16   Variational Methods

The Navier and Levy methods discussed in the preceding articles can be applied only to shallow shells with relatively simple boundary conditions.   If the boundary conditions involved are complex, it becomes necessary to take recourse to approximate procedures such as those associated with variational methods due to Galerkin or Kantorovich.[1] The Galerkin method will be discussed here in some detail.

[1] For more details, see L. V. Kantorovich and V. I. Krylev, "Approximate Methods of Higher Analysis," Erven P. Noordhoff, NV, Groningen, Netherlands, 1958, translated by Curtis D. Benster.

19-17   The Galerkin Method

Consider, for instance, a shallow shell of double curvature carrying only vertical loads.   For such a shell, the differential equations in the three displacements u, v, and w may be formed by making use of (19-27), (19-28), and (19-29) by setting $X = Y = 0$.   These three equations may be written as

$$L_1(u) + L_2(v) + L_3(w) = 0 \tag{19-56}$$

$$L_4(u) + L_5(v) + L_6(w) = 0 \tag{19-57}$$

$$L_7(u) + L_8(v) + L_9(w) + \frac{Z}{Ed}(1 - \nu^2) = 0 \tag{19-58}$$

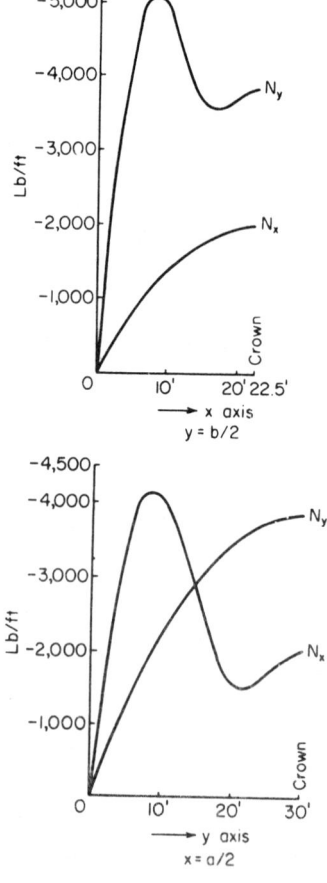

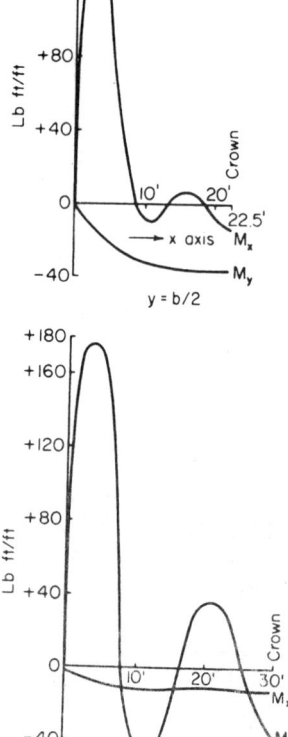

Fig. 19-7a

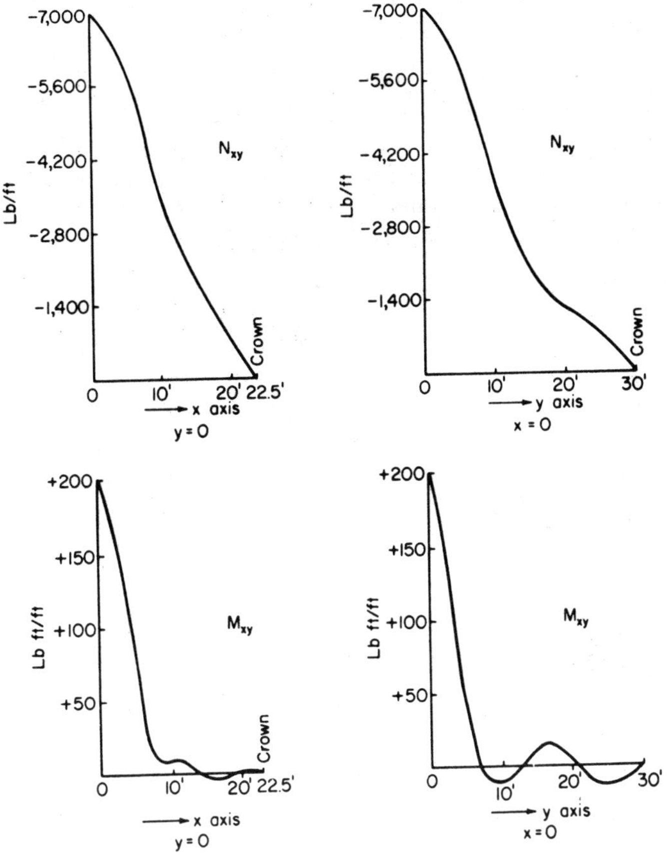

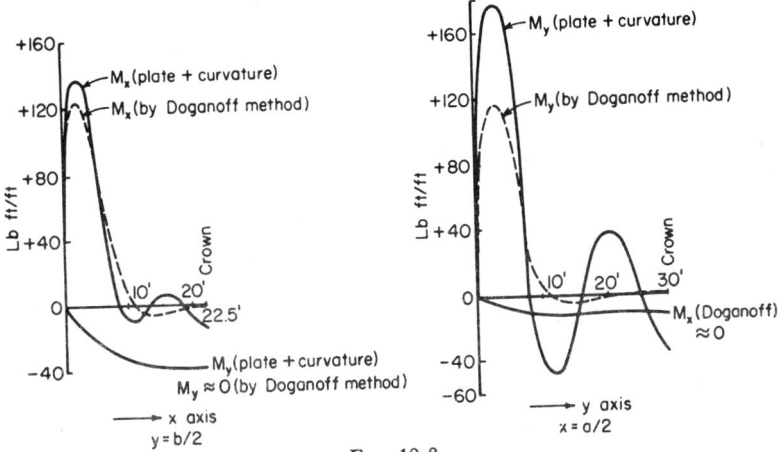

FIG. 19-7b

FIG. 19-8

where L_1, L_2, L_3,..., L_9 are differential operators and u, v, and w are functions of both x and y. Let us assume that homogeneous boundary conditions are prescribed for u, v, and w at the boundaries. To apply the Galerkin procedure to this problem, we pick approximate functions in x and y to represent u, v, and w. These functions are such that they satisfy the prescribed boundary conditions. Thus, let

$$\bar{u}(x, y) = a_1\bar{u}_1(x, y) + a_2\bar{u}_2(x, y) + a_3\bar{u}_3(x, y)$$
$$+ \cdots + a_n\bar{u}_n(x, y) \tag{19-59}$$

$$\bar{v}(x, y) = b_1\bar{v}_1(x, y) + b_2\bar{v}_2(x, y) + b_3\bar{v}_3(x, y)$$
$$+ \cdots + b_n\bar{v}_n(x, y) \tag{19-60}$$

and

$$\bar{w}(x, y) = c_1\bar{w}_1(x, y) + c_2\bar{w}_2(x, y) + c_3\bar{w}_3(x, y)$$
$$+ \cdots + c_n\bar{w}_n(x, y) \tag{19-61}$$

The bars in $\bar{u}$, $\bar{v}$, and $\bar{w}$ denote that they are approximate solutions for u, v, and w. Next, the assumed approximate solutions are substituted in Equations (19-56), (19-57), and (19-58). But since the solutions assumed are not exact, $L_1(\bar{u}) + L_2(\bar{v}) + L_3(\bar{w}) \neq 0$. There will be certain error, say $\varepsilon_1(x,y)$, so that

$$L_1(\bar{u}) + L_2(\bar{v}) + L_3(\bar{w}) = \varepsilon_1(x, y) \tag{19-62}$$

Similarly, let

$$L_4(\bar{u}) + L_5(\bar{v}) + L_6(\bar{w}) = \varepsilon_2(x, y) \tag{19-63}$$

and

$$L_7(\bar{u}) + L_8(\bar{v}) + L_9(\bar{w}) + \frac{Z}{Ed}(1 - \nu^2) = \varepsilon_3(x, y) \tag{19-64}$$

It can now be shown that

$$\iint \varepsilon_1(x, y)\bar{u}_n(x, y) = 0 \tag{19-65}$$

where n takes on values 1, 2,..., n. This results in n equations. The double integration involved is carried out to extend all over the shell. Similarly,

$$\iint \varepsilon_2(x, y)\overline{v_n}(x, y) = 0 \tag{19-66}$$

and

$$\iint \varepsilon_3(x, y)\overline{w_n}(x, y) = 0 \tag{19-67}$$

Thus we have $3n$ equations to determine the $3n$ unknown coefficients $a_1, a_2, a_3, ..., a_n$, $b_1, b_2, b_3, ..., b_n$, and $c_1, c_2, c_3, ..., c_n$. Equations (19-65), (19-66), and (19-67) take on a simple physical meaning if the errors ε_1, ε_2, and ε_3 are regarded as forces and the assumed functions as virtual displacements. Corresponding to n equilibrium configurations, we get n equations for each of the three displacements. The selection of suitable functions and the evaluation of the unknown coefficients are described in detail in Design Example 19-2, wherein the Galerkin method is applied to a clamped paraboloid of revolution.

Design Example 19-2

DESIGN OF A PARABOLOID OF
REVOLUTION WITH
CLAMPED BOUNDARIES

1. *DATA*

Geometry
Side $a = 45$ ft
Side $b = 60$ ft
Thickness $d = 3$ in. (0.25 ft)
Rise at crown above the corners $f = 8$ ft
Poisson's ratio $v = 0$

Loads
Dead weight   = 37.5 psf of shell area
Waterproofing,
  finishing, etc. = 7.5 psf of shell area
Live load   = 15.0 psf of shell area

Total load   = 60.0 psf of shell area

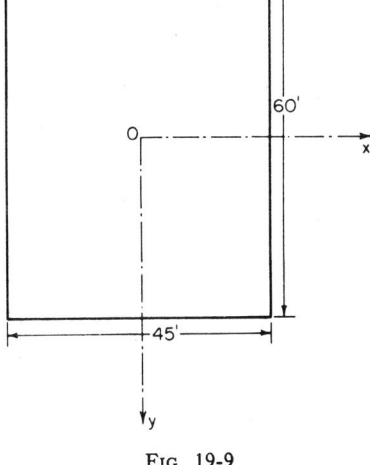

FIG. 19-9

2. *GEOMETRICAL PROPERTIES*

Equation of the surface is given by $z = \dfrac{x^2 + y^2}{2R}$

Radius of the surface $R = \dfrac{(a/2)^2 + (b/2)^2}{2f}$

$$= \frac{a^2 + b^2}{8f} = \frac{45^2 + 60^2}{8 \times 8}$$

$$= 87.89 \text{ ft}$$

Curvature $r = \dfrac{\partial^2 z}{\partial x^2} = \dfrac{1}{R} = 0.011378$

Curvature $t = \dfrac{\partial^2 z}{\partial y^2} = \dfrac{1}{R} = 0.011378$

Curvature $s = \dfrac{\partial^2 z}{\partial x \, \partial y} = 0$

3. *DIFFERENTIAL EQUATIONS EMPLOYED*

The three equations in u, v, and w, namely, (19-33), (19-34), and (19-35), are used in the analysis.

The equations are nondimensionalized by making the following substitutions:

$$a = \lambda b$$

$$x = \tfrac{1}{2}a\bar{x} \qquad y = \tfrac{1}{2}b\bar{y} \qquad z = f\bar{z}$$

$$u = \tfrac{1}{2}a\bar{u} \qquad v = \tfrac{1}{2}b\bar{v} \qquad w = f\bar{w}$$

The equations will now take the following form:

Equation of the surface:

$$\bar{z} = \frac{1}{8Rf}(a^2\bar{x}^2 + b^2\bar{y}^2)$$

Differential equations:

$$\frac{\partial^2\bar{u}}{\partial\bar{x}^2} + \frac{\lambda^2}{2}\frac{\partial^2\bar{u}}{\partial\bar{y}^2} + \frac{1}{2}\frac{\partial^2\bar{v}}{\partial\bar{x}\,\partial\bar{y}} - \frac{f}{R}\frac{\partial\bar{w}}{\partial\bar{x}} = 0$$

$$\frac{1}{2}\frac{\partial^2\bar{u}}{\partial\bar{x}\,\partial\bar{y}} + \frac{\partial^2\bar{v}}{\partial\bar{y}^2} + \frac{1}{2\lambda^2}\frac{\partial^2\bar{v}}{\partial\bar{x}^2} - \frac{f}{R}\frac{\partial\bar{w}}{\partial\bar{y}} = 0$$

$$\frac{\partial\bar{u}}{\partial\bar{x}} + \frac{\partial\bar{v}}{\partial\bar{y}} - \frac{2f}{R}\bar{w} - \frac{4d^2Rf}{3a^4}\left(\frac{\partial^4\bar{w}}{\partial\bar{x}^4} + 2\lambda^2\frac{\partial^4\bar{w}}{\partial\bar{x}^2\,\partial\bar{y}^2} + \lambda^4\frac{\partial^4\bar{w}}{\partial\bar{y}^4}\right) + \frac{R}{Ed}Z = 0$$

4. *FUNCTIONS ASSUMED*

Functions representing the displacements u, v, and w are so chosen that they satisfy the boundary conditions along all the edges. These are given below.

$$\bar{u} = \sum A_{mn}(\bar{x}^{2m-1} - \bar{x}^{2m+1})(\bar{y}^{2n-2} - \bar{y}^{2n})$$

$$\bar{v} = \sum B_{mn}(\bar{x}^{2m-2} - \bar{x}^{2m})(\bar{y}^{2n-1} - \bar{y}^{2n+1})$$

$$\bar{w} = \sum C_{mn}(\bar{x}^{2m-2} - 2\bar{x}^{2m} + \bar{x}^{2m+2})(\bar{y}^{2n-2} - 2\bar{y}^{2n} + \bar{y}^{2n+2})$$

The equations are general, and any desired number of terms can be taken by putting $m = 1, 2, 3,...$ and $n = 1, 2, 3,...$. In this example three terms are taken.

5. *METHOD OF ANALYSIS*

Galerkin's variational method, described in Art. 19-17, is used for the analysis. Equations (19-65), (19-66), and (19-67) are set up and solved for the 27 unknowns, A_{11}, A_{12}, A_{13}, A_{21}, A_{22}, A_{23}, A_{31}, A_{32}, A_{33}, B_{11}, B_{12},..., B_{33}, C_{11}, C_{12},..., C_{33}. Each of these three equations gives rise to k^2 simultaneous equations where k is the number of terms taken. Thus, in this example, each equation results in 3^2, i.e., 9 simultaneous equations, hence 27 simultaneous equations with 27 unknowns. The first differential equation is given below.

$$\int_0^1\int_0^1 \left(\frac{\partial^2\bar{u}}{\partial\bar{x}^2} + \frac{\lambda^2}{2}\frac{\partial^2\bar{u}}{\partial\bar{y}^2} + \frac{1}{2}\frac{\partial^2\bar{v}}{\partial\bar{x}\,\partial\bar{y}} - \frac{f}{R}\frac{\partial\bar{w}}{\partial\bar{x}}\right)(\bar{x}^{2k-1} - \bar{x}^{2k+1})(\bar{y}^{2L-2} - \bar{y}^{2L}) = 0$$

or

$$\int_0^1 \int_0^1 \left\{ \sum_{m=1}^{m=3} \sum_{n=1}^{n=3} A_{mn}[(2m-1)(2m-2)\bar{x}^{2m-3} - (2m+1)(2m)\bar{x}^{2m-1}](\bar{y}^{2n-2} - \bar{y}^{2n}) \right.$$

$$+ \frac{\lambda^2}{2} \sum_{m=1}^{m=3} \sum_{n=1}^{n=3} A_{mn}(\bar{x}^{2m-1} - \bar{x}^{2m+1})[(2n-2)(2n-3)\bar{y}^{2n-4} - (2n)(2n-1)\bar{y}^{2n-2}]$$

$$+ \frac{1}{2} \sum_{m=1}^{m=3} \sum_{n=1}^{n=3} B_{mn}[(2m-2)\bar{x}^{2m-3} - 2m\bar{x}^{2m-1}][(2n-1)\bar{y}^{2n-2} - (2n+1)\bar{y}^{2n}]$$

$$- \frac{f}{R} \sum_{m=1}^{m=3} \sum_{n=1}^{n=3} C_{mn}[(2m-2)\bar{x}^{2m-3} - 2(2m)\bar{x}^{2m-1} + (2m+2)\bar{x}^{2m+1}](\bar{y}^{2n-2} - 2\bar{y}^{2n} + \bar{y}^{2n+2}) \right\}$$

$$\times (\bar{x}^{2k-1} - \bar{x}^{2k+1})_{k=1}^{k=3} (\bar{y}^{2L-2} - \bar{y}^{2L})_{L=1}^{L=3} = 0$$

It will result in the following nine simultaneous equations:

$$\int_0^1 \int_0^1 \left\{ (-6\bar{x})[A_{11}(1 - \bar{y}^2) + A_{12}(\bar{y}^2 - \bar{y}^4) + A_{13}(\bar{y}^4 - \bar{y}^6)] \right.$$

$$+ (6\bar{x} - 20\bar{x}^3)[A_{21}(1 - \bar{y}^2) + A_{22}(\bar{y}^2 - \bar{y}^4) + A_{23}(\bar{y}^4 - \bar{y}^6)]$$

$$+ (20\bar{x}^3 - 42\bar{x}^5)[A_{31}(1 - \bar{y}^2) + A_{32}(\bar{y}^2 - \bar{y}^4) + A_{33}(\bar{y}^4 - \bar{y}^6)]$$

$$+ \frac{\lambda^2}{2}(\bar{x} - \bar{x}^3)[A_{11}(-2) + A_{12}(2 - 12\bar{y}^2) + A_{13}(12\bar{y}^2 - 30\bar{y}^4)]$$

$$+ \frac{\lambda^2}{2}(\bar{x}^3 - \bar{x}^5)[A_{21}(-2) + A_{22}(2 - 12\bar{y}^2) + A_{23}(12\bar{y}^2 - 30\bar{y}^4)]$$

$$+ \frac{\lambda^2}{2}(\bar{x}^5 - \bar{x}^7)[A_{31}(-2) + A_{32}(2 - 12\bar{y}^2) + A_{33}(12\bar{y}^2 - 30\bar{y}^4)]$$

$$+ \tfrac{1}{2}(-2\bar{x})[B_{11}(1 - 3\bar{y}^2) + B_{12}(3\bar{y}^2 - 5\bar{y}^4) + B_{13}(5\bar{y}^4 - 7\bar{y}^6)]$$

$$+ \tfrac{1}{2}(2\bar{x} - 4\bar{x}^3)[B_{21}(1 - 3\bar{y}^2) + B_{22}(3\bar{y}^2 - 5\bar{y}^4) + B_{23}(5\bar{y}^4 - 7\bar{y}^6)]$$

$$+ \tfrac{1}{2}(4\bar{x}^3 - 6\bar{x}^5)[B_{31}(1 - 3\bar{y}^2) + B_{32}(3\bar{y}^2 - 5\bar{y}^4) + B_{33}(5\bar{y}^4 - 7\bar{y}^6)]$$

$$- \frac{f}{R}(-4\bar{x} + 4\bar{x}^3)[C_{11}(1 - 2\bar{y}^2 + \bar{y}^4) + C_{12}(\bar{y}^2 - 2\bar{y}^4 + \bar{y}^6) + C_{13}(\bar{y}^4 - 2\bar{y}^6 + \bar{y}^8)]$$

$$- \frac{f}{R}(2\bar{x} - 8\bar{x}^3 + 6\bar{x}^5)[C_{21}(1 - 2\bar{y}^2 + \bar{y}^4) + C_{22}(\bar{y}^2 - 2\bar{y}^4 + \bar{y}^6) + C_{23}(\bar{y}^4 - 2\bar{y}^6 + \bar{y}^8)]$$

$$- \frac{f}{R}(4\bar{x}^3 - 12\bar{x}^5 + 8\bar{x}^7)[C_{31}(1 - 2\bar{y}^2 + \bar{y}^4) + C_{32}(\bar{y}^2 - 2\bar{y}^4 + \bar{y}^6) + C_{33}(\bar{y}^4 - 2\bar{y}^6 + \bar{y}^8)] \Big\}$$

$$\times \ (\bar{x} - \bar{x}^3)(1 - \bar{y}^2) = 0 \tag{1}$$

$$\int_0^1 \int_0^1 \{ \qquad -do- \qquad \} \ (\bar{x} - \bar{x}^3)(\bar{y}^2 - \bar{y}^4) = 0 \tag{2}$$

$$\int_0^1 \int_0^1 \{ \qquad -do- \qquad \} \ (\bar{x} - \bar{x}^3)(\bar{y}^4 - \bar{y}^6) = 0 \tag{3}$$

$$\int_0^1 \int_0^1 \{ \qquad -do- \qquad \} \ (\bar{x}^3 - \bar{x}^5)(1 - \bar{y}^2) = 0 \tag{4}$$

$$\int_0^1 \int_0^1 \{ \qquad -do- \qquad \}(\bar{x}^3 - \bar{x}^5)(\bar{y}^2 - \bar{y}^4) = 0 \tag{5}$$

$$\int_0^1 \int_0^1 \{ \qquad -do- \qquad \}(\bar{x}^3 - \bar{x}^5)(\bar{y}^4 - \bar{y}^6) = 0 \tag{6}$$

$$\int_0^1 \int_0^1 \{ \qquad -do- \qquad \} \ (\bar{x}^5 - \bar{x}^7)(1 - \bar{y}^2) = 0 \tag{7}$$

$$\int_0^1 \int_0^1 \{ \qquad -do- \qquad \}(\bar{x}^5 - \bar{x}^7)(\bar{y}^2 - \bar{y}^4) = 0 \tag{8}$$

$$\int_0^1 \int_0^1 \{ \qquad -do- \qquad \}(\bar{x}^5 - \bar{x}^7)(\bar{y}^4 - \bar{y}^6) = 0 \tag{9}$$

The second and third differential equations also produce nine simultaneous equations each like that of the first differential equation as shown above. The second differential equation is

$$\int_0^1 \int_0^1 \left(\frac{1}{2} \frac{\partial^2 \bar{u}}{\partial \bar{x} \, \partial \bar{y}} + \frac{\partial^2 \bar{v}}{\partial \bar{y}^2} + \frac{1}{2\lambda^2} \frac{\partial^2 \bar{v}}{\partial \bar{x}^2} - \frac{f}{R} \frac{\partial \bar{w}}{\partial \bar{y}} \right) (\bar{x}^{2k-2} - \bar{x}^{2k})(\bar{y}^{2L-1} - \bar{y}^{2L+1}) = 0$$

or

$$\int_0^1 \int_0^1 \Bigg\{ \frac{1}{2} \sum_{m=1}^{m-3} \sum_{n=1}^{n-3} A_{mn}[(2m - 1)\bar{x}^{2m-2} - (2m + 1)\bar{x}^{2m}][(2n - 2)\bar{y}^{2n-3} - 2n\bar{y}^{2n-1}]$$

$$+ \sum_{m=1}^{m-3} \sum_{n=1}^{n-3} B_{mn}(\bar{x}^{2m-2} - \bar{x}^{2m})[(2n - 1)(2n - 2)\bar{y}^{2n-3} - (2n + 1)(2n)\bar{y}^{2n-1}]$$

$$+ \frac{1}{2\lambda^2} \sum_{m=1}^{m-3} \sum_{n=1}^{n-3} B_{mn}[(2m - 2)(2m - 3)\bar{x}^{2m-4} - 2m(2m - 1)\bar{x}^{2m-2}](\bar{y}^{2n-1} - \bar{y}^{2n+1})$$

$$- \frac{f}{R} \sum_{m=1}^{m-3} \sum_{n=1}^{n-3} C_{mn}(\bar{x}^{2m-2} - 2\bar{x}^{2m} + \bar{x}^{2m+2})[(2n - 2)\bar{y}^{2n-3} - 2(2n)\bar{y}^{2n-1} + (2n + 2)\bar{y}^{2n+1}] \Bigg\}$$

$$\times \ (\bar{x}^{2k-2} - \bar{x}^{2k})(\bar{y}^{2L-1} - \bar{y}^{2L+1}) = 0$$

The third differential equation is

$$\int_0^1 \int_0^1 \left[\frac{\partial \bar{u}}{\partial \bar{x}} + \frac{\partial \bar{v}}{\partial \bar{y}} - \frac{2f}{R} \bar{w} - \frac{4}{3} \frac{d^2 R f}{a^4} \left(\frac{\partial^4 \bar{w}}{\partial \bar{x}^4} + 2\lambda^2 \frac{\partial^4 \bar{w}}{\partial \bar{x}^2 \, \partial \bar{y}^2} + \lambda^4 \frac{\partial^4 \bar{w}}{\partial \bar{y}^4} \right) + \frac{R}{Ed} Z \right]$$

$$\times \ (\bar{x}^{2k-2} - 2\bar{x}^{2k} + \bar{x}^{2k+2})(\bar{y}^{2L-2} - 2\bar{y}^{2L} + \bar{y}^{2L+2}) = 0$$

or

$$
\int_0^1 \int_0^1 \left\{ \sum_{m=1}^{m=3} \sum_{n=1}^{n=3} A_{mn}[(2m-1)\bar{x}^{2m-2} - (2m+1)\bar{x}^{2m}](\bar{y}^{2n-2} - \bar{y}^{2n}) \right.
$$

$$
+ \sum_{m=1}^{m=3} \sum_{n=1}^{n=3} B_{mn}(\bar{x}^{2m-2} - \bar{x}^{2m})[(2n-1)\bar{y}^{2n-2} - (2n+1)\bar{y}^{2n}]
$$

$$
- \frac{2f}{R} \sum_{m=1}^{m=3} \sum_{n=1}^{n=3} C_{mn}(\bar{x}^{2m-2} - 2\bar{x}^{2m} + \bar{x}^{2m+2})(\bar{y}^{2n-2} - 2\bar{y}^{2n} + \bar{y}^{2n+2})
$$

$$
- \frac{4d^2 Rf}{3a^4} \sum_{m=1}^{m=3} \sum_{n=1}^{n=3} C_{mn}[(2m-2)(2m-3)(2m-4)(2m-5)\bar{x}^{2m-6}
$$

$$
- 2(2m)(2m-1)(2m-2)(2m-3)\bar{x}^{2m-4}
$$

$$
+ (2m+2)(2m+1)(2m)(2m-1)\bar{x}^{2m-2}](\bar{y}^{2n-2} - 2\bar{y}^{2n} + \bar{y}^{2n+2})
$$

$$
- \frac{8d^2 Rf\lambda^2}{3a^4} \sum_{m=1}^{m=3} \sum_{n=1}^{n=3} C_{mn}[(2m-2)(2m-3)\bar{x}^{2m-4}
$$

$$
- 2(2m)(2m-1)\bar{x}^{2m-2} + (2m+2)(2m+1)\bar{x}^{2m}]
$$

$$
\times [(2n-2)(2n-3)\bar{y}^{2n-4} - 2(2n)(2n-1)\bar{y}^{2n-2} + (2n+2)(2n+1)\bar{y}^{2n}]
$$

$$
- \frac{4d^2 Rf\lambda^4}{3a^4} \sum_{m=1}^{m=3} \sum_{n=1}^{n=3} C_{mn}(\bar{x}^{2m-2} - 2\bar{x}^{2m} + \bar{x}^{2m+2})[(2n-2)(2n-3)(2n-4)(2n-5)\bar{y}^{2n-6}
$$

$$
- 2(2n)(2n-1)(2n-2)(2n-3)\bar{y}^{2n-4} + (2n+2)(2n+1)(2n)(2n-1)\bar{y}^{2n-2}] + \frac{R}{Ed} Z \right\}
$$

$$
\times (\bar{x}^{2k-2} - 2\bar{x}^{2k} + \bar{x}^{2k+2})(\bar{y}^{2L-2} - 2\bar{y}^{2L} + \bar{y}^{2L+2}) = 0
$$

After solving these 27 simultaneous equations the values of the unknowns A_{11}, A_{12}, A_{13}, A_{21}, A_{22}, A_{23}, A_{31}, A_{32}, A_{33}, B_{11}, B_{12},..., B_{33}, C_{11}, C_{12},..., C_{33} are as follows:

$(A_{11}, \ldots, A_{33})E$	$(B_{11}, \ldots, B_{33})E$	$(C_{11}, \ldots, C_{33})E$
1,373.0633	641.47233	135,810.98
100.9436	3,660.5062	44,820.886
170.37141	1,796.1904	1,554,320.9
5,622.8133	−444.01045	229,448.78
12,106.572	2,293.7857	581,363.43
−5,950.7001	6,165.9499	1,476,490.0
−1,489.2076	578.34496	934,703.35
−6,016.7927	−5,129.9772	−1,657,545.4
12,109.159	4,643.9626	16,763,767.0

The functions u, v, and w become as follows:

$$\bar{u} = \frac{1}{E}\{(\bar{x} - \bar{x}^3)[1{,}373.0633(1 - \bar{y}^2) + 100.9436(\bar{y}^2 - \bar{y}^4) + 170.37141(\bar{y}^4 - \bar{y}^6)]$$

$$+ (\bar{x}^3 - \bar{x}^5)[5{,}622.8133(1 - \bar{y}^2) + 12{,}106.572(\bar{y}^2 - \bar{y}^4) - 5{,}950.7001(\bar{y}^4 - \bar{y}^6)]$$

$$+ (\bar{x}^5 - \bar{x}^7)[-1{,}489.2076(1 - \bar{y}^2) - 6{,}016.7927(\bar{y}^2 - \bar{y}^4) + 12{,}109.159(\bar{y}^4 - \bar{y}^6)]\}$$

$$\bar{v} = \frac{1}{E}\{(1 - \bar{x}^2)[641.47233(\bar{y} - \bar{y}^3) + 3{,}660.5062(\bar{y}^3 - \bar{y}^5) + 1{,}796.1904(\bar{y}^5 - \bar{y}^7)]$$

$$+ (\bar{x}^2 - \bar{x}^4)[-444.01045(\bar{y} - \bar{y}^3) + 2{,}293.7857(\bar{y}^3 - \bar{y}^5) + 6{,}165.9499(\bar{y}^5 - \bar{y}^7)]$$

$$+ (\bar{x}^4 - \bar{x}^6)[578.34496(\bar{y} - \bar{y}^3) - 5{,}129.9772(\bar{y}^3 - \bar{y}^5) + 4{,}643.9626(\bar{y}^5 - \bar{y}^7)]\}$$

$$\bar{w} = \frac{1}{E}\{(1 - 2\bar{x}^2 + \bar{x}^4)[135{,}810.98(1 - 2\bar{y}^2 + \bar{y}^4) + 44{,}820.886(\bar{y}^2 - 2\bar{y}^4 + \bar{y}^6)$$

$$+ 1{,}554{,}320.9(\bar{y}^4 - 2\bar{y}^6 + \bar{y}^8)]$$

$$+ (\bar{x}^2 - 2\bar{x}^4 + \bar{x}^6)[229{,}448.78(1 - 2\bar{y}^2 + \bar{y}^4) + 581{,}363.43(\bar{y}^2 - 2\bar{y}^4 + \bar{y}^6)$$

$$+ 1{,}476{,}490.0(\bar{y}^4 - 2\bar{y}^6 + \bar{y}^8)]$$

$$+ (\bar{x}^4 - 2\bar{x}^6 + \bar{x}^8)[934{,}703.35(1 - 2\bar{y}^2 + \bar{y}^4) - 1{,}657{,}545.4(\bar{y}^2 - 2\bar{y}^4 + \bar{y}^6)$$

$$+ 16{,}763{,}767.0(\bar{y}^4 - 2\bar{y}^6 + \bar{y}^8)]\}$$

With the help of the above functions of $\bar{u}$, $\bar{v}$, and $\bar{w}$, the stress resultants are found out as per formulas given below.

$$N_x = Ed\left(\frac{\partial u}{\partial x} - \frac{1}{R}w\right) = Ed\left(\frac{\partial \bar{u}}{\partial \bar{x}} - \frac{f}{R}\bar{w}\right)$$

$$N_y = Ed\left(\frac{\partial v}{\partial y} - \frac{1}{R}w\right) = Ed\left(\frac{\partial \bar{v}}{\partial \bar{y}} - \frac{f}{R}\bar{w}\right)$$

$$N_{xy} = \frac{Ed}{2}\left(\frac{\partial u}{\partial y} + \frac{\partial v}{\partial x}\right) = \frac{Ed}{2}\left(\lambda\frac{\partial \bar{u}}{\partial \bar{y}} + \frac{1}{\lambda}\frac{\partial \bar{v}}{\partial \bar{x}}\right)$$

$$M_x = -\frac{Ed^3}{12}\frac{\partial^2 w}{\partial x^2} = -\frac{E\,d^3 f}{3a^2}\frac{\partial^2 \bar{w}}{\partial \bar{x}^2}$$

$$M_y = -\frac{Ed^3}{12}\frac{\partial^2 w}{\partial y^2} = -\frac{E\,d^3 f\lambda^2}{3a^2}\frac{\partial^2 \bar{w}}{\partial \bar{y}^2}$$

$$M_{xy} = \frac{Ed^3}{12}\frac{\partial^2 w}{\partial x\,\partial y} = \frac{E\,d^3 f\lambda}{3a^2}\frac{\partial^2 \bar{w}}{\partial \bar{x}\,\partial \bar{y}}$$

$$w = f(\bar{w})$$

The values of the stress resultants are given in the table below and are shown plotted in Fig. 19-10. The results obtained by taking two and four terms are also shown in Fig. 19-11 for comparison.

Coordinates of point		N_x, lb/ft	N_y, lb/ft	N_{xy}, lb/ft	M_x, lb ft/ft	M_y, lb ft/ft	M_{xy}, lb ft/ft
$\bar{y}$	$\bar{x}$						
0.0	0.0	−2,747.18	−2,930.08	0.000	+1.735	+5.250	0.000
	0.2	−2,615.30	−2,922.08	0.000	−2.754	+4.854	0.000
	0.4	−2,517.95	−3,031.38	0.000	−0.923	+4.859	0.000
	0.6	−2,704.69	−3,075.78	0.000	+28.316	+6.066	0.000
	0.8	−2,537.70	−1,909.19	0.000	+37.395	+4.779	0.000
	1.0	−2,753.33	0.00	0.000	−213.985	0.000	0.000
0.2	0.0	−2,607.34	−2,690.90	0.000	+0.952	−2.003	0.000
	0.2	−2,480.75	−2,696.17	−21.489	−2.491	−2.040	+0.508
	0.4	−2,369.77	−2,811.62	−30.598	+0.297	−2.226	−0.542
	0.6	−2,514.93	−2,836.31	−27.037	+26.776	−2.361	−0.908
	0.8	−2,399.92	−1,738.27	−27.111	+33.026	−1.503	+2.859
	1.0	−2,766.92	0.00	−53.052	−195.755	0.000	0.000
0.4	0.0	−2,642.00	−2,484.54	0.000	+0.153	−7.208	0.000
	0.2	−2,515.47	−2,500.70	−44.569	−2.532	−6.733	+0.129
	0.4	−2,387.32	−2,629.95	−82.839	+1.342	−6.946	+0.733
	0.6	−2,487.38	−2,656.50	−109.826	+26.995	−8.835	−0.065
	0.8	−2,360.69	−1,604.49	−124.846	+31.187	−6.962	−3.941
	1.0	−2,796.86	0.00	−137.806	−190.678	0.000	0.000
0.6	0.0	−3,064.76	−2,766.96	0.000	+1.292	+16.094	0.000
	0.2	−2,922.49	−2,752.22	−60.917	−3.494	+16.173	−0.628
	0.4	−2,805.16	−2,870.34	−149.853	−1.090	+16.917	+1.379
	0.6	−2,946.31	−2,962.83	−259.587	+31.503	+17.021	+2.471
	0.8	−2,626.59	−1,821.62	−340.101	+41.619	+10.445	−3.235
	1.0	−2,737.77	0.00	−346.407	−235.582	0.000	0.000
0.8	0.0	−2,227.00	−2,531.70	0.000	+2.112	+34.635	0.000
	0.2	−2,108.55	−2,486.02	−60.547	−2.946	+33.766	−0.231
	0.4	−2,014.26	−2,515.80	−198.081	−3.132	+35.100	−2.594
	0.6	−2,144.28	−2,550.35	−417.428	+23.325	+39.011	+3.966
	0.8	−1,914.53	−1,594.01	−615.810	+35.416	+27.035	+27.738
	1.0	−2,170.98	0.00	−620.815	−186.618	0.000	0.000
1.0	0.0	0.00	−3,049.08	0.000	0.000	−160.643	0.000
	0.2	0.00	−3,081.09	−76.424	0.000	−158.046	0.000
	0.4	0.00	−3,100.88	−229.748	0.000	−164.089	0.000
	0.6	0.00	−2,878.65	−466.651	0.000	−175.873	0.000
	0.8	0.00	−2,027.88	−597.682	0.000	−117.230	0.000
	1.0	0.00	0.00	0.000	0.000	0.000	0.000

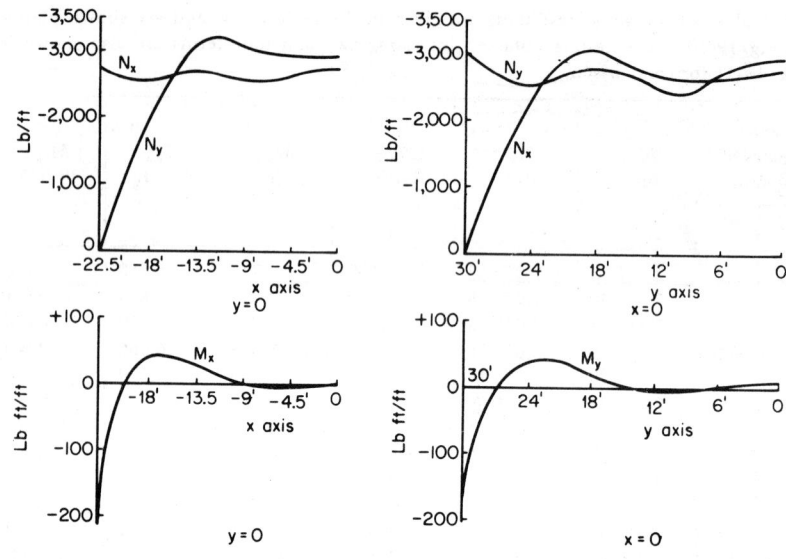

Fig. 19-10a

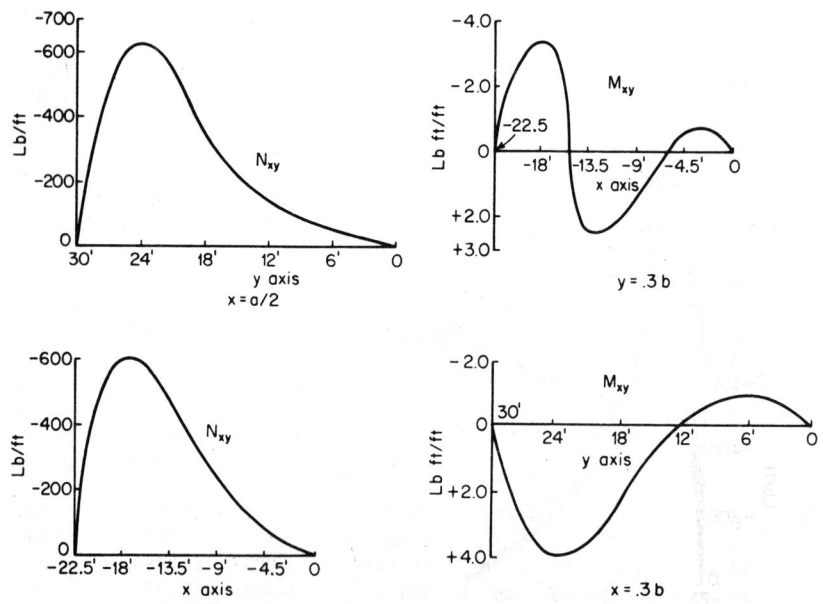

Fig. 19-10b

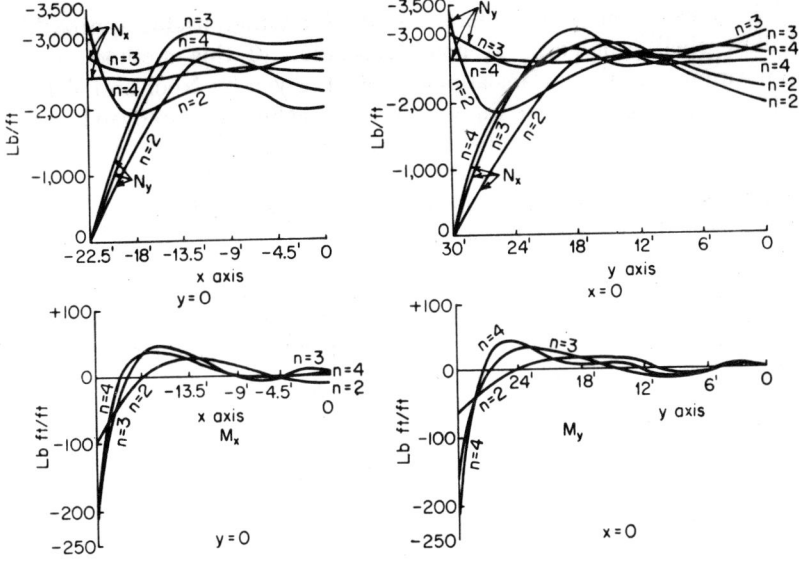

FIG. 19-11a

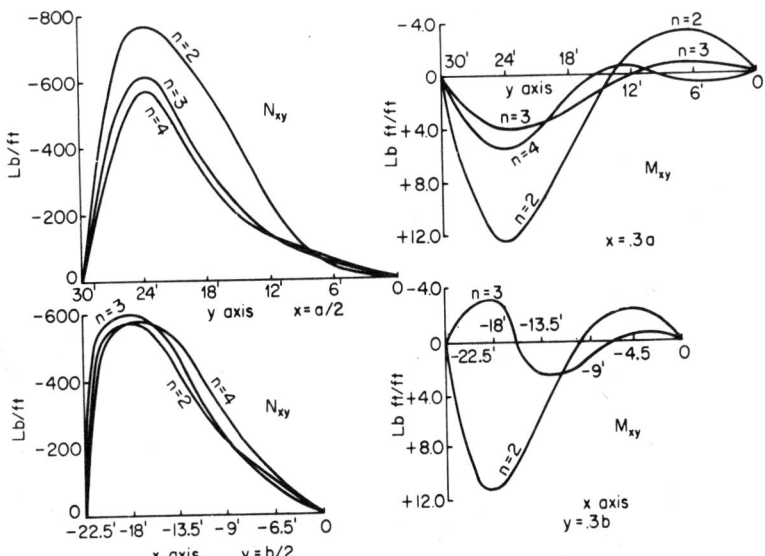

FIG. 19-11b

19-18 The Method of Finite Differences

The method of finite differences offers another useful approach to shallow shell problems. A good exposition of this method and its application to shallow translational shells is given by Noor and Veletsos [123]. The higher-order finite-difference technique employed by them is reported to have given satisfactory accuracy.

19-19 Hyperbolic Paraboloid Shell Supported on Straight-line Characteristics

In many practical applications, hyperbolic paraboloid shells are supported along their straight-line characteristics with support conditions corresponding to those of simply supported or clamped boundaries.

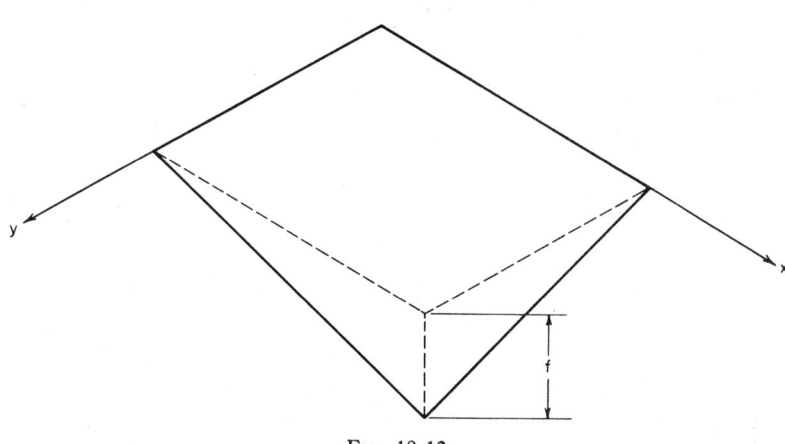

FIG. 19-12

It is not possible, in these instances, to find simple expressions in the form of Fourier series for ϕ and w. Apeland and Popov's [120] solution has the merit of simplicity but it results in unrealistic boundary conditions. At two of the opposite edges $x = 0$ and $x = a$ (Fig. 19-12), the solution satisfies the boundary conditions

$$w = M_x = N_{xy} = u = 0$$

Hence the solution is of limited practical interest. Loof [124] has studied a square hyperbolic paraboloid shell (Fig. 19-13) with "clamped" boundaries. He has worked out two alternative solutions, corresponding to the following sets of boundary conditions:

Case I

At $x = 0$, $\quad u = 0$, $v = 0$, $w = 0$, and $\partial w/\partial x = 0$.

Similar boundary conditions are prescribed at the other three edges.

Case II

At $x = 0$, $\quad v = 0$, $N_x = 0$, $w = 0$, and $\partial w/\partial x = 0$.

Similar boundary conditions are imposed on the other three edges. By using the collocation procedure, he satisfies the boundary conditions at six points on each edge, the resulting equations being solved on a digital

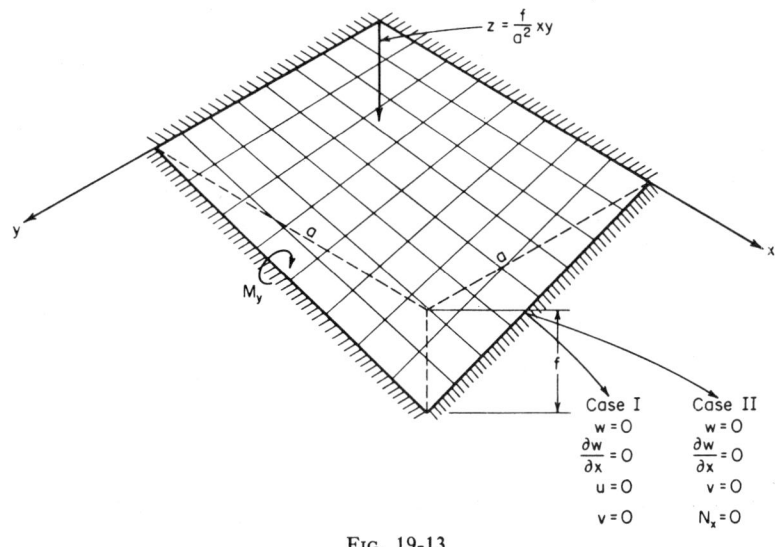

Fig. 19-13

computer. Based on this study, he proposes the following formulas for the clamping moment and radial shear at the midpoint of a clamped edge.

$$M_y = - \, 0.511ga^2 \left(\frac{f}{d}\right)^{-4/3} \tag{19-68}$$

$$Q_y = + \, 1.732ga \left(\frac{f}{d}\right)^{-1} \tag{19-69}$$

where g is the intensity of vertical load. From Equations (19-68) and (19-69), we may draw the qualitative conclusion that the moments tend to be large when f is small, i.e., when the shell is shallow; or when d is large, i.e., when the shell is thick. Loof notes that the results are not materially different for the two alternative sets of boundary conditions studied. A similar calculation made by him for a "hinged" shell with $u = 0$ and $w = 0$ at the boundaries led to the following formulas for the maximum bending moment occurring at a distance of $0.55\,(f/d)^{-1/3}a$ from an edge:

$$M_y = +\,0.149ga^2 \left(\frac{f}{d}\right)^{-4/3} \tag{19-70}$$

The radial shear at the edge is given by

$$Q_y = +\,0.577ga \left(\frac{f}{d}\right)^{-1} \tag{19-71}$$

The formulas proposed by Loof are simple enough for use in the design office to assess the magnitudes of the bending moments in hyperbolic paraboloid shells clamped or simply supported along their straight-line characteristics.

19-20 Vreedenburgh's Approximate Method for Hyperbolic Paraboloids Whose Edges Are Characteristic Lines

An approximate method which is even simpler than that of Loof is that due to Vreedenburgh [86]. Referring to Fig. 19-12, the equation of the hyperbolic paraboloid may be written as

$$z = \frac{f}{ab}\,xy \tag{19-72}$$

It is easily seen that $r = \partial^2 z/\partial x^2 = 0$, $t = \partial^2 z/\partial y^2 = 0$, and $s = f/ab$. Hence the two principal curvatures $1/R_1$ and $1/R_2$ at any point may be written as

$$\frac{1}{R_1} = \frac{1}{R_2} = \frac{f}{ab} \tag{19-73}$$

The shell may approximately be regarded as a *plate on an elastic foundation*. Hence its equation of equilibrium may be written as

$$D\nabla^4 w + Ed\,w\left(\frac{1}{R_1^2} + \frac{1}{R_2^2}\right) = Z \tag{19-74}$$

If the vertical load is g, we may set $Z = g$. Substituting for $1/R_1^2$ and $1/R_2^2$ from Equation (19-73), we arrive at the relation

$$\nabla^4 w + \frac{24f^2}{a^2b^2d^2}\,w = \frac{g}{D} \tag{19-75}$$

The quantity $24f^2/a^2b^2d^2$ represents the modulus of the foundation. Let us suppose that we are studying the edge disturbances emanating from the edge AB. We may now make a radical approximation and treat w as a function of x only so that equation (19-75) may reduce itself to

$$\frac{d^4w}{dx^4} + \frac{24f^2}{a^2b^2d^2}\, w = \frac{g}{D} \tag{19-76}$$

This is strictly permissible only if $b \gg a$. Making the substitution

$$\beta^2 = \sqrt{6}\,\frac{f}{abd}$$

Equation (19-76) may be recast and rewritten as

$$\frac{d^4w}{dx^4} + 4\beta^4 w = \frac{g}{D} \tag{19-77}$$

This relation is at once recognized as the equation governing the structural action of a *beam on an elastic foundation*. The particular integral of Equation (19-77) is easily verified to be

$$w_0 = \frac{g}{4D\beta^4} \tag{19-78}$$

The solution of the homogeneous equation may be written as

$$w = e^{\beta x}(C_1 \cos \beta x + C_2 \sin \beta x) + e^{-\beta x}(C_3 \cos \beta x + C_4 \sin \beta x) \tag{19-79}$$

The first two terms are inadmissible as they increase with increasing x. Ignoring these and combining the homogeneous solution and the particular integral, we arrive at the complete solution, which may be written as

$$w = e^{-\beta x}(C_3 \cos \beta x + C_4 \sin \beta x) + \frac{g}{4D\beta^4} \tag{19-80}$$

Let us study the problem for clamped and simply supported boundary conditions.

Case I Clamped Boundary

The boundary conditions are

(i) $$w = 0 \quad \text{at } x = 0$$

(ii) $$\frac{dw}{dx} = 0 \quad \text{at } x = 0$$

From the first condition,

$$C_3 = -\frac{g}{4D\beta^4} \tag{19-81}$$

and from the second,

$$C_3 = C_4$$

Hence

$$w = \frac{g}{4D\beta^4}[1 - e^{-\beta x}(\cos \beta x + \sin \beta x)] \tag{19-82}$$

Differentiating (19-82), the bending moment is obtained as

$$M_x = -Dw'' = -\frac{g}{2\beta^2} e^{-\beta x}(\cos \beta x - \sin \beta x) \tag{19-83}$$

At the boundary where $x = 0$,

$$M_x = -\frac{g}{2\beta^2} \tag{19-84}$$

The first point of inflection occurs at $x = \pi/4\beta$. At a distance of $3.5/\beta$, the disturbance practically dies down. Differentiating (19-84), we get

$$\frac{dM}{dx} = \frac{g}{\beta} e^{-\beta x} \cos \beta x \tag{19-85}$$

Hence the radial shear per unit length transmitted to the support at $x = 0$ is g/β.

Case II Simply Supported Edge

The relevant boundary conditions are

(i) $\qquad\qquad\qquad w = 0 \qquad$ at $\qquad x = 0$

(ii) $\qquad\qquad\qquad \dfrac{d^2w}{dx^2} = 0 \qquad$ at $\qquad x = 0$

As before

$$C_3 = -\frac{g}{4D\beta^4}$$

$$C_4 = 0$$

Hence

$$w = \frac{g}{4D\beta^4}(1 - e^{-\beta x} \sin \beta x) \tag{19-86}$$

and

$$M_x = -Dw'' = +\frac{g}{2\beta^2} e^{-\beta x} \sin \beta x \tag{19-87}$$

The bending moment M_x attains its maximum value of

$$\frac{g}{2\beta^2 \sqrt{2}} e^{-\pi/4} = 0.16 \frac{g}{\beta^2}$$

at $x = \pi/4\beta$. Differentiating (19-87), we get

$$\frac{dM_x}{dx} = \frac{g}{2\beta} e^{-\beta x}(\cos \beta x - \sin \beta x) \tag{19-88}$$

Hence at $x = 0$

$$\frac{dM_x}{dx} = \frac{g}{2\beta} \tag{19-89}$$

This expression may be made use of to determine the radial shear per unit length transmitted to the support.

19-21· Doganoff's Approximate Method for Shallow Spherical Calottes on a Rectangular Ground Plan

Doganoff [76] has given approximate expressions for w for shallow spherical shells on a rectangular ground plan with both simply supported and clamped boundaries.

Case I Simply Supported Edges

Referring to Fig. 19-5, the following approximate expression is proposed for w in the region $x < a/2$ and $y < b/2$:

$$w = w_0(1 - e^{-\alpha x} \cos \alpha x)(1 - e^{-\alpha y} \cos \alpha y) \tag{19-90}$$

where

$$w_0 = \frac{gR^2}{Ed} \quad \text{and} \quad \alpha = \frac{(3)^{1/4}}{(Rd)^{1/2}}$$

g, R, and d are, as usual, the intensity of vertical load, the radius of curvature, and the thickness of the shell. From expression (19-90) it is possible to derive expressions for the bending moments, twisting moments, and radial shears. The bending moment M_x attains its maximum value of $0.093\ Rdg\ (1 - e^{-\alpha y} \cos \alpha y)$ at $x = \pi/4\alpha$.

Case II Clamped Edges

The expression for w now takes the form

$$w = w_0[1 - e^{-\alpha x}(\sin \alpha x + \cos \alpha x)]\ [1 - e^{\alpha y}(\sin \alpha y + \cos \alpha y)] \tag{19-91}$$

As before, it is possible to derive expressions for the bending and twisting moments in the shell by differentiating the expression for w. The maximum positive moment occurs at $x = \pi/2\alpha$ and its value is

$$(M_x)_{x=\pi/2\alpha} = 0.060 \, Rgd[1 - e^{-\alpha y}(\cos \alpha y + \sin \alpha y)] \qquad (19\text{-}92)$$

The maximum negative moment occurs at $x = 0$ and its value is

$$(M_x)_{x=0} = -0.2885 \, Rdg \, [1 - e^{-\alpha y}(\cos \alpha y + \sin \alpha y)] \qquad (19\text{-}93)$$

When the shell is shallow, a spherical calotte is not very different from a rotational paraboloid. Hence these results may be applied to assess the bending and twisting moments in rotational paraboloids as well. In Design Example 19-1, these results are shown compared with the moments obtained by a more rigorous method.

PART IV

OTHER ASPECTS

CHAPTER 20

DESIGN OF TRAVERSES

20-1 The Function of a Traverse

The supports provided along the curved edges of a shell are known as *traverses*. They are meant to preserve the assumed geometry of the shell. A developable surface, such as a cylinder, will flatten out under loads in the absence of a traverse. Hence it is an indispensable part of a cylindrical shell. They are also necessary for shells of double curvature. But in such structures their role is rather local and they serve the same purpose as an edge member.

20-2 Assumptions in the Design of Traverses

(i) Ideally, a traverse must be absolutely *rigid and undeformable*. Such a condition is difficult, if not impossible, to realize in practice. All that can be done is to ensure that the traverse deflects only by a negligible amount, when loads are applied. If, in some cases, this condition cannot be met, the influence of the deformation of the traverse on the stresses in the shell must be examined and allowed for as a secondary correction. In such cases, the analysis will proceed as follows: In the first instance, the traverse may be regarded as rigid and the stresses in the shell may be computed. The reactions transmitted to the traverse are arrived at. Under the action of these reactions, the deformation of the traverse is computed. Making use of these deformations, the shell stresses may be corrected.

(ii) The second assumption usually made is that the traverse is capable of carrying forces only in its own plane. At right angles to its plane, it is considered to be flexible and incapable of receiving any load. This condition is also never fully realized in practice.

20-3 Types of Traverses

Traverses may be solid diaphragms, arch ribs, tied arches, open-web tied arches (bowstring girders), trusses, or portal frames (Fig. 20-1).

The solid diaphragm is not economical for shells of large sizes. The arched traverses without ties are usually employed for short shells of large chord widths. In such cases, the arches transmit their thrusts to A-frames or buttresses (Fig. 20-1*b*). Where arch-rib traverses are provided, they may be upstand, downstand, or partly upstand and partly downstand with the shell meeting it at its middepth. It is an advantage to have them upstand if movable formwork is proposed to be employed to cast the shells.

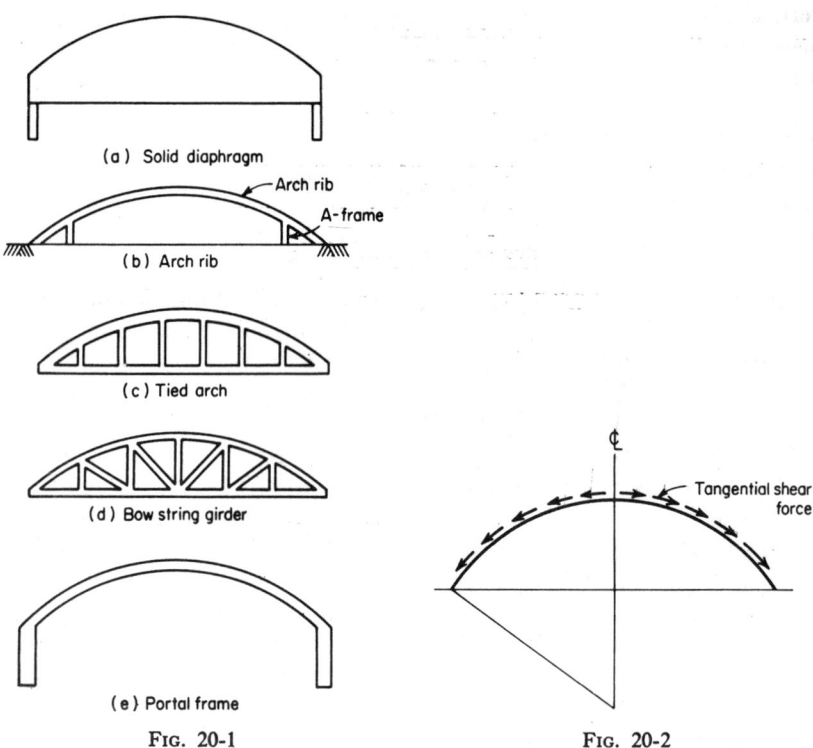

(a) Solid diaphragm

(b) Arch rib

Arch rib

A-frame

(c) Tied arch

(d) Bow string girder

(e) Portal frame

Fig. 20-1

Tangential shear force

Fig. 20-2

20-4 Loads on Traverses

The load acting on the shell is transmitted to the traverses through the medium of tangential shears (Fig. 20-2). In addition, the traverses have to be designed to carry wind and seismic loads acting directly on them.

20-5 Analysis and Design

It is not generally realized that the analysis and design of traverses are as important as those of the shell itself. The author knows of at least one instance where a collapse occurred because of the insufficient attention paid to the detailing of the traverse.

The analysis may be done in two phases—a preliminary analysis to arrive at the sizes of members and a final analysis based on the sizes so assumed. In the preliminary analysis, it is accurate enough to assume that the shell transfers a uniformly distributed vertical reaction on to the traverse. Thus, in a simply supported cylindrical shell, one may assume that the shell transmits half the load on to each traverse as a vertical reaction. The structure is analyzed for this load in the routine manner. Many tables and charts are available for the analysis of portal frames and arches. *Advanced Engineering Bulletins* 7, 8, and 9 published by the Portland Cement Association, and two papers by James Michalos in *Civil Engineering*, namely, "Direct Design of Two-hinged Arches of Constant Section" (vol. 26, no. 1, January, 1956) and "Direct Design of Hingeless Arches of Constant Section" (vol. 26, no. 7, July, 1956), may be found to be specially useful in carrying out the analysis. Zalewski's simplified method [125] may also be employed with advantage for purposes of preliminary design.

In the final analysis, the tangential shear transferred by the shell is resolved into horizontal and vertical components and concentrated at a number of points on the traverse. The structure is analyzed for these loads. Wind or seismic loads assumed to be acting are also concentrated at the same points. In a short shell it is necessary to resolve and concentrate the shell reactions on the traverse at close intervals as the shear may change sign a number of times, and concentrating the reactions only at a few points may result in errors and inaccuracies.

20-6 Support Conditions

The traverses may be hinged to the columns except where the traverses and columns are designed as monolithic units to form portal frames.

20-7 Design of Tied-arch Traverses

A few special considerations need to be taken into account in the design of tied-arch or open-web tied-arch (bowstring-girder) traverses. The influence of the elastic extension of the tie and the effects of temperature changes need consideration in the analysis. The tie being primarily a tension member, special care is called for in making joints in the reinforcing bars provided in it, should such joints become necessary. The bars may be welded or may be provided with threaded-sleeve couplings at the joints. It is not good practice to rely on concrete to transmit the tension at the laps through the medium of bond. If, however, facilities for welding or the provision of threaded sleeves are not available, the provisions of Indian Standard Criteria for the Design of Reinforced Concrete Shell Structures and Folded Plates [30] may be followed. These call for the staggering of laps and a minimum lap length given by

$$\frac{\text{Bar diameter} \times \text{actual tensile stress}}{4 \times \text{permissible average bond stress}}$$

or 30 times the diameter of the bar whichever is greater.

In addition, the composite tension on the transformed area is limited to $0.10 f'_c$. Where the tension exceeds this figure, the code stipulates the provision of helical binders of 6 mm diameter at a pitch not exceeding 7.5 cm. Prestressing the tie member would avoid these difficulties.

20-8 Design Example

The preliminary and final designs of a bowstring-girder traverse are explained in detail in Design Example 20-1.

Design Example 20-1

DESIGN OF A TIED-ARCH TRAVERSE FOR A SHORT CYLINDRICAL SHELL

1. DATA

In this example, a tied-arch traverse (Fig. 20-3) is designed for the short cylindrical shell analyzed in Design Example 8-1. The shell is assumed to meet the arch at the midheight of its cross section. The shear forces on the traverse are taken from step 10 of that example. They are shown plotted in Fig. 20-4 for ready reference.

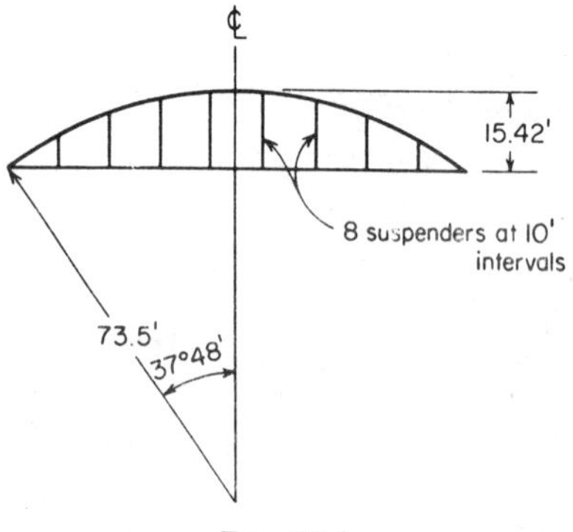

2. PRELIMINARY CALCULATIONS

The tension in the tie of the arch is calculated approximately by the following procedure. The loads from the shell and the traverse are assumed to be uniformly distributed on a span equal to the span of the arch. The total vertical load from the shell = 35/2 (73.5 × 2 × 0.6597) 52.5 = 89,100 lb = 39.777 tons. Assuming that the weight of the arch is 200 lb/ft length,

$$\text{Total weight of the arch} = 200 \times 97$$
$$= 19,400 \text{ lb}$$
$$= 8.661 \text{ tons}$$

Fig. 20-3

The weight due to the suspenders and the tie is assumed to be about 220 lb/ft of arch span. Hence

$$\text{Total weight} = 220 \times 90$$
$$= 19,800 \text{ lb} = 8.839 \text{ tons}$$

Hence

$$\text{Total load on the traverse} = 39.777 + 8.661 + 8.839 = 57.277 \text{ tons}$$

$$\text{Bending moment at midspan} = 57.277 \times \frac{90}{8} = 644.366 \text{ ft tons}$$

$$\text{Approximate tension in tie} = \frac{644.366}{\text{rise of arch at crown}}$$

$$= \frac{644.366}{15.42} = 41.788 \text{ tons}$$

Hence

$$\text{Reinforcement required in tie} = \frac{41.788 \times 2{,}240}{20{,}000} = 4.68 \text{ in.}^2$$

We may, however, provide eight No. 8 bars, giving an area of 6.28 in.^{2}. To accommodate this reinforcement, the section of the tie is chosen as 8 by 10 in.

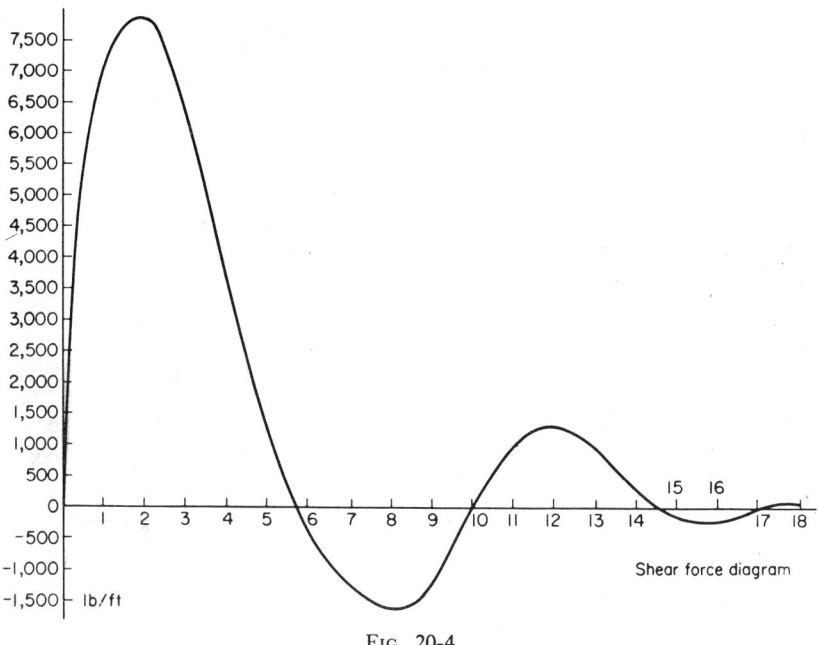

Fig. 20-4

The cross section of the arch is assumed to be rectangular, 8 in. wide and 24 in. deep. The dimension is arrived at by trial. The tie is assumed to be connected to the arch by eight suspenders at 10-ft intervals (Fig. 20-3).

3. FORCES ON THE TRAVERSE

(i) *Forces from the Shell*

The shear forces $N_{x\phi}$ from the shell act on the traverse tangential to the shell arc. These forces are concentrated at 19 points (0 to 18) on the traverse as shown in Fig. 20-5.

The force at each point is resolved into its horizontal and vertical components, as shown in the table below.

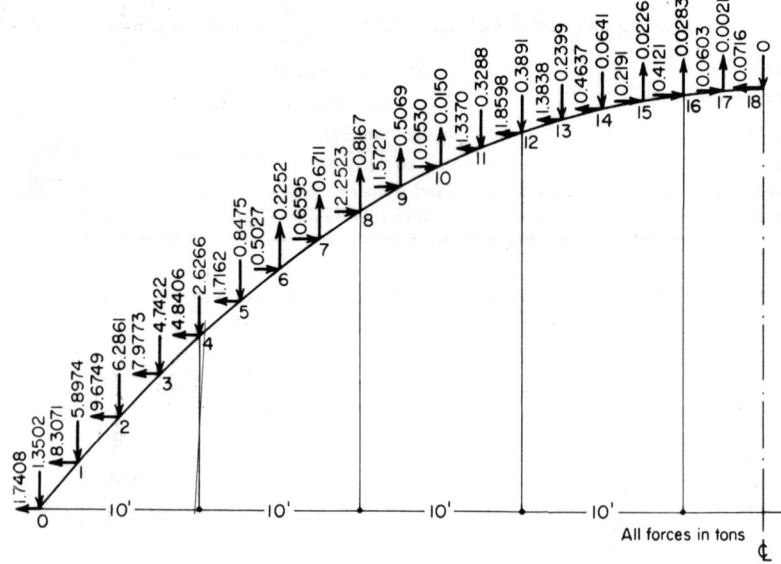

FIG. 20-5

Point	Load, lb, computed from graph	Load × $\pi^2/8$, tons	Vertical component, tons	Horizontal component, tons
0	4,000	2.2030	1.3502	1.7408
1	18,500	10.1890	5.8974	8.3071
2	20,950	11.5384	6.2861	9.6749
3	16,850	9.2803	4.7422	7.9773
4	10,000	5.5076	2.6266	4.8406
5	3,475	1.9139	0.8475	1.7162
6	−1,000	−0.5508	−0.2252	−0.5027
7	−3,250	−1.7900	−0.6711	−1.6595
8	−4,350	−2.3958	−0.8167	−2.2523
9	−3,000	−1.6523	−0.5069	−1.5727
10	−100	−0.0551	−0.0150	−0.0530
11	2,500	1.3769	0.3288	1.3370
12	3,450	1.9001	0.3891	1.8598
13	2,550	1.4044	0.2399	1.3838
14	850	0.4681	0.0641	0.4637
15	−400	−0.2203	−0.0226	−0.2191
16	−750	−0.4131	−0.0283	−0.4121
17	−110	−0.0606	−0.0021	−0.0603
18	130	0.0716	0	0.0716
			+20.484	+32.6411

(ii) *Loads Due to the Dead Weight of the Arch*

$$\text{Weight per foot length of the arch} = 24 \times 8 \times \frac{150}{144} \times \frac{1}{2,240}$$

$$= 0.0893 \text{ ton}$$

This load is uniform on the curved length of the arch as the cross section is assumed to be uniform.

(iii) *Loads Due to the Dead Weight of the Ties and Suspenders*

The cross section of the tie is assumed to be 8 by 10 in. As the reinforcement in it is heavy, the weight of the tie is calculated by considering the actual weights of concrete and steel present in it and is found to be about 95 lb or 0.04241 ton/ft length.

The suspenders are assumed to be 8 by 6 in. in cross section. Hence

$$\text{Their dead weight} = 8 \times 6 \times \frac{150}{144} \times \frac{1}{2,240}$$

$$= 50 \text{ lb} = 0.02232 \text{ ton/ft length}$$

The loads due to the dead weight of the tie and suspenders are assumed to be acting on the arch at the suspender points (i.e., at points 4, 8, 12, and 16). Their values are calculated below.

Point	Load, tons	
4	$10 \times 0.04241 + 5.020 \times 0.02232$	0.5362
8	$10 \times 0.04241 + 9.525 \times 0.02232$	0.6367
12	$10 \times 0.04241 + 12.362 \times 0.02232$	0.7000
16	$10 \times 0.04241 + 13.751 \times 0.02232$	0.7310

4. ANALYSIS OF THE ARCH

(i) *Due to Forces from the Shell*

The coefficients for calculating the horizontal reactions due to the vertical and horizontal loads are worked out from the formulas given by Pippard.[1] The horizontal reactions due to horizontal and vertical loads from the shell are tabulated on page 518.

The vertical reaction due to loads from the shell is equal to the algebraic sum of the vertical loads from points 0 to 18 and is equal to 20.484 tons.

(ii) *Due to Dead Weight of the Arch*

The horizontal reaction due to the dead weight of the arch is given by Pippard[2] as

$$2Wa \left[\frac{\sin 3\phi_c}{6} - \frac{\sin \phi_c}{2} (3 + 6 \cos 2\phi_c + 4\phi_c \sin 2\phi_c) \right.$$

$$\left. + \cos \phi_c (3 \cos \phi_c \sin \phi_c - \phi_c \sin^2 \phi_c - \phi_c \cos^2 \phi_c + 2\phi_c) \right]$$

$$\div [2(4\phi_c - 3 \sin 2\phi_c + 2\phi_c \cos 2\phi_c)]$$

[1] Pippard, A. J. S., "Studies in Elastic Structures," pp. 255, 258, Edward Arnold (Publishers) Ltd., London, 1952.

[2] Pippard, A. J. S., "Studies in Elastic Structures," p. 259, Edward Arnold (Publishers) Ltd., London, 1952.

Point	Horizontal load	Coeffi-cient	Horizontal reaction	Vertical load	Coeffi-cient	Horizontal reaction
0	+1.7408	1.0	+1.7408	+1.3502	0	0
1	+8.3071	0.84750	+7.0403	+5.8974	0.19766	+1.1657
2	+9.6749	0.71874	+6.9537	+6.2861	0.39608	+2.4898
3	+7.9773	0.59435	+4.7413	+4.7422	0.59646	+2.8285
4	+4.8406	0.48188	+2.3326	+2.6266	0.78359	+2.0582
5	+1.7162	0.38263	+0.6567	+0.8475	0.97361	+0.8251
6	−0.5027	0.31009	−0.1559	−0.2252	1.13969	−0.2567
7	−1.6595	0.23363	−0.3877	−0.6711	1.30373	−0.8749
8	−2.2523	0.18043	−0.4064	−0.8167	1.45917	−1.1917
9	−1.5727	0.13144	−0.2067	−0.5069	1.60104	−0.8116
10	−0.0530	0.08758	−0.0046	−0.0150	1.72364	−0.0259
11	+1.3370	0.05948	+0.0795	+0.3288	1.83660	+0.6039
12	+1.8598	0.03588	+0.0667	+0.3891	1.94184	+0.7556
13	+1.3838	0.01551	+0.0215	+0.2399	2.02596	+0.4860
14	+0.4637	0.00914	+0.0042	+0.0641	2.09091	+0.1340
15	−0.2191	0.00294	−0.0006	−0.0226	2.15231	−0.0486
16	−0.4121	0.00082	−0.0003	−0.0283	2.19324	−0.0621
17	−0.0603	0.00049	0	−0.0021	2.22083	−0.0047
18	+0.0716	0	0	0	0	0
			+22.4751			+8.0706

where

W = dead weight of arch per ft length = 0.0893 ton

a = radius of center line of arch which is equal to radius of shell = 73.5 ft

ϕ_c = semicentral angle of arch, equal to that of shell = 37.8° = 0.6597 radian

∴ Horizontal reaction = 2 × 0.0893 × 73.5 × 0.441488 = 5.7945 tons

Vertical reaction = 0.0893 × 73.5 × 0.6597 = 4.3295 tons

(iii) *Due to Dead Weight of Tie and Suspenders*

As already seen, the dead weight of the tie and suspenders acts as vertical loads at points 4, 8, 12, and 16. The horizontal reaction due to these loads is calculated using the appropriate coefficients. The calculation is shown in tabular form.

Point	Load, tons	Coefficient	Horizontal reaction, tons
4	0.5362	0.78359	0.4202
8	0.6367	1.45917	0.9291
12	0.7000	1.94184	1.3593
16	0.7310	2.19324	1.6033
			4.3119

Vertical reaction = 0.5362 + 0.6367 + 0.7000 + 0.7310

= 2.6039 tons

The total horizontal and vertical reactions due to all the above forces are as follows:

$$\text{Total horizontal reaction} = (8.0706 + 22.4751) + 5.7945 + 4.3119$$
$$= 40.6521 \text{ tons}$$
$$\text{Total vertical reaction} = 20.484 + 4.3295 + 2.6039$$
$$= 27.4174 \text{ tons}$$

5. REDUCTION IN HORIZONTAL REACTION DUE TO ELASTIC EXTENSION OF THE TIE

Under the action of axial tension, the reinforcement in the tie undergoes an elongation δ equal to HL/AE.

$$\text{i.e., } \delta = \frac{40.6521(90 \times 12)}{6.2832 \times 13,500} = 0.5176 \text{ in.}$$

The reduction H' in the horizontal reaction due to this yielding of the tie is given by Pippard.[1]

$$H' = \frac{\delta}{(a^3/2EI)[2\phi_c - \sin 2\phi_c + 4 \cos \phi_c(\phi_c \cos \phi_c - \sin \phi_c)]}$$

Substituting $I = 1/12 \times 8 \times 24^3 = 9,216$ in.[4] and $E = 2 \times 10^6$ psi,

$$\text{Reduction } H' \text{ in horizontal reaction} = \frac{0.5176}{2.55696}$$
$$= 0.2024 \text{ ton}$$

Hence

$$\text{Net horizontal reaction} = 40.6521 - 0.2024$$
$$= 40.4497 \text{ tons}$$

6. CALCULATION OF MOMENTS AND THRUSTS AT VARIOUS POINTS ON THE ARCH

Knowing all the forces and reactions, the moments and thrusts at various points of the arch are calculated and shown in tables below.

Moments (in ft tons) (Sagging Positive)

Point	Due to loads from shell	Due to dead weight of arch	Due to weight of tie and suspenders	Due to extension of tie	Total
4	−5.2098	+0.0403	−2.0746	+1.3196	−5.9245
8	−1.0930	+1.3320	−0.8227	+2.2315	+1.6478
12	+2.0169	+2.5799	+1.2544	+2.8057	+8.6569
16	+5.1358	+2.7973	+2.5580	+3.0876	+13.5787
18	+5.4804	+2.3772	+1.8333	+3.1218	+12.8127

[1] Pippard, A. J. S., "Studies in Elastic Structures," Table 10.3, p. 263, Edward Arnold (Publishers) Ltd., London, 1952.

Horizontal Force (in Tons)

Point	Due to loads from shell	Due to dead weight of arch	Due to weight of tie and suspenders	Due to extension of tie	Total
0	+28.8049	+5.7945	+4.3119	−0.2024	+38.7089
4	+2.8456	+5.7945	+4.3119	−0.2024	+12.7496
8	+0.7033	+5.7945	+2.9626	−0.2024	+9.2580
12	+0.9920	+5.7945	+2.9626	−0.2024	+9.5467
16	−2.4962	+5.7945	+1.6033	−0.2024	+4.6992
18	−2.0038	+5.7945	0	−0.2024	+3.5883

Vertical Force (in Tons)

Point	Due to loads from shell	Due to dead weight of arch	Due to weight of tie and suspenders	Due to extension of tie	Total
0	+19.1338	+4.3295	+2.6039	0	+26.0672
4	+2.2081	+3.2625	+2.6039	0	+8.0745
8	+0.4470	+2.2831	+1.4310	0	+4.1611
12	+0.6401	+1.3536	+1.4310	0	+3.4247
16	−0.0304	+0.4509	+0.7310	0	+1.1515
18	+0.4420	0	+0.7310	0	+1.1730

It may be noted here that the deflections of the arch rib have been ignored in the above analysis. These may be considered, if so desired, by following the procedure outlined below. The changes in the lengths of the suspenders on account of the axial loads due to the weight of the suspenders and tie are neglected. The deflections of the arch rib at the suspender points are calculated, and those are assumed to be the same for the tie also. The tie member thus becomes a continuous beam on elastic supports. The secondary moments in the tie due to this uneven settlement of supports give rise to additional reactions which are in turn considered as additional loads on the arch rib at the suspender points. The deflections of the arch due to these loads are calculated and the process repeated to any degree of accuracy required.

7. DESIGN OF REINFORCEMENT

(i) Arch Rib

Concrete for the arch is assumed to have the strength of 3,000 psi. Modulus of elasticity ratio is 9. The arch section is checked for safety against moments and thrusts at two points 0 and 18 (namely, springing and crown). After a few preliminary trials, the reinforcement is assumed to consist of eight No. 4 bars, four at top and four at bottom (Fig. 20-6). The values of the moments and of the thrusts and the stresses developed are shown below.

Point 0 (*Springing*)

$$\text{Moment} = 0$$
$$\text{Horizontal thrust} = 38.7089 \text{ tons}$$
$$\text{Vertical force} = 26.0672 \text{ tons}$$

Resolving the forces along the tangent and the normal,

$$\text{Axial thrust} = (38.7089 \times \cos 37°48') + (26.0672 \times \sin 37°48')$$
$$= 46.5627 \text{ tons}$$

$$\text{Radial shear} = (38.7089 \times \sin 37°48') - (26.0681 \times \cos 37°48')$$
$$= 3.1281 \text{ tons}$$

$$\text{Compressive stress in concrete} = \frac{46.5627 \times 2,240}{24 \times 8 + [(9-1)8 \times 0.196]}$$
$$= 510 \text{ psi}$$

$$\text{Shear stress} = \frac{3.1281 \times 2,240}{8 \times 24} = 36.5 \text{ psi}$$

Point 18 (*Crown*)

$$\text{Moment} = 12.8127 \text{ ft tons}$$
$$\text{Axial thrust} = 3.5883 \text{ tons}$$
$$\text{Radial shear} = -1.1730 \text{ tons}$$

$$\text{Max compressive stress in concrete} = 731 \text{ psi}$$
$$\text{Max tensile stress in steel} = 17,289 \text{ psi}$$
$$\text{Shear stress} = 13.7 \text{ psi}$$

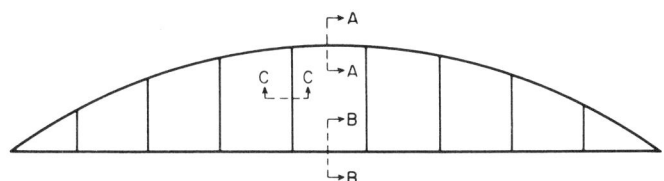

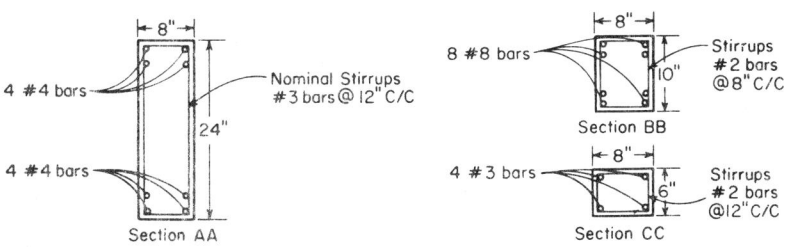

Fig. 20-6

Thus it is seen that the section is safe at all points. As the shear stress is also within the permissible limits, nominal stirrups of No. 3 bars at an average spacing of 12 in. center to center are provided.

(ii) *Tie*

$$\text{Tension in the tie} = 40.4497 \text{ tons}$$

$$\text{Area of steel provided} = 6.28 \text{ in.}^2$$

$$\text{Stress in steel} = \frac{40.4497}{6.28} \times 2,240$$

$$= 14,428 \text{ psi}$$

This is within limits and needs no revision. Stirrups of No. 2 bars are provided at 8-in. centers as shown in Fig. 20-6.

(iii) *Suspenders*

The force in the suspenders being very small, nominal reinforcement consisting of four No. 3 bars, one at each corner, is provided. Stirrups are of No. 2 bars at 12 in. center to center.

20-9 Effective Width of Flange

Where the arch rib of a traverse is downstand, a part of the shell slab tends to act as a compression flange of the traverse. In cylindrical shells, where the shell slab is continuous over the traverse, this width may be taken as $0.76 \sqrt{ad}$ on either side of the traverse. This value would also apply to the shell slab adjacent to an end traverse provided it is rigid enough to restrain rotation of the slab. The equivalent width shall, however, be estimated as $0.38 \sqrt{ad}$ for the shell slab adjacent to the end traverse, which is flexible enough to permit rotation of the slab. Where the shell slab meets the rib at its neutral plane, no effective flange width need be reckoned in the analysis. The analysis for determining the effective width is given in Appendix II of the ASCE Manual 31 [3].

CHAPTER 21

PRACTICAL ASPECTS OF
SHELL CONSTRUCTION

21-1 Design Influenced by Method of Construction

Perhaps in no other form of concrete construction is design so much influenced by the method of construction proposed to be employed. If precast, the size of the shell is limited by the capacity of the lifting appliances available. If poured in place, the size of the shells for a given application has to be judiciously chosen so as to ensure the maximum reuse of the forms. This is important because the cost of formwork constitutes a sizable proportion of the total cost of a shell roof. The economy of shell structures stems from their low consumption of materials—cement and steel—for covering one square foot of area. But this economy will be largely nullified if the cost of forming the shells proves to be an expensive operation. An appreciation of construction problems is therefore indispensable to the designer of shell roofs. The important factors contributing to the cost of a concrete structure are the quantities of materials and formwork involved. These will now be discussed.

21-2 Materials

For shell roofs of moderate size, a concrete mix of 1:2:4 by weight is generally suitable. This will roughly develop a cylinder strength of about 2,500 psi at 28 days with a water-cement ratio of about 0.50. For shells of large proportions, a richer mix of $1:1\frac{1}{2}:3$ by weight is desirable because it would develop a higher modulus of elasticity. In a large shell, elastic stability is usually a problem and a higher modulus of elasticity would offer increased resistance against buckling. In a concrete shell roof, the dead weight is the dominant load. Considerable reduction in dead weight is possible if lightweight structural-grade concrete is employed. Several shells built in lightweight concrete have been described by Pauw and Jenny [126].

Mild-steel deformed bars, preferably of small diameters, are usually preferred for placement in the body of the shell. But bars of such small diameters get easily displaced during construction. Medium-tensile bars have the advantage that they spring back when stepped on. But they are difficult to place on doubly curved surfaces. For singly curved shells, a welded fabric is preferable to bars tied to form a grid. But the fabric is not so suitable for shells of double curvature. In edge beams and traverses, high-strength deformed bars may also be used.

21-3 Formwork

A cubic foot of concrete to be placed on a poured-in-place shell roof calls for 4 to 5 ft² of forms. If the roof is steep, top forms become necessary and this area gets doubled. Top forms become inescapable if the slope of the roof exceeds 45°. They are best avoided as the placing and compaction of the concrete between the two forms presents several practical difficulties.

The centering required also tends to be more voluminous compared with that required for a flat-slab structure of the same column height because of the added rise of the shell. The present trend is to use tubular scaffolding with vertical steel tubes of approximately 2 in. external diameter spaced 5 to 6 ft apart. They are braced in both the horizontal directions at intervals of 5 to 6 ft along their height.

The forms are usually made up of timber battens lined with steel sheeting or plywood. Plastic forms may be used if special surface textures are desired. Steel sheets, if employed as lining material, may be reused up to 200 times [127]. The selection of the lining material and the forms would depend on the number of reuses possible on a given project.

Based on the experience gained on a number of projects, the requirements of tubular scaffolding and accessories needed for different types of shells have been worked out by Coppel and Coulon [128]. This information is given in Table 21-1.

There are two ways of effecting economy in the cost of formwork. If the shells are cast in place, cost of this item can be appreciably reduced by resorting to movable formwork. The other course open is to precast the shells.

CAST-IN-PLACE SHELLS

21-4 Saving in Time

If the number of reuses involved is four or more, the use of movable formwork can be economically justified. A saving in time of about 7 to 10 days per shell would result if this technique is employed.

Table 21-1 *

Type of form	Tubes 40/49 (for 1,000 m³ of volume enclosed) m	Connecting accessories for 1,000 m length of tube			Fixing accessories for 100 m² of area covered		
		Right-angle couplers Nos.	Swivel-type couplers Nos.	End-to-end couplers Nos.	Base plates Nos.	Forks Nos.	Wheels Nos.
Fixed	1,100–1,200	450–550	5–10	60–120	25–35	25–40	
Mobile	800–1,300	450–550	15–30	50–120	15–30	30–50	10–15

* Compiled from "Echafaudages Tubulaires, Théorie et Pratique" by Th. Coppel and J. J. Coulon, Dunod, Paris, 1963. Used by courtesy of Dunod Editeur, Paris.

21-5 Northlight Shells

Northlight shells without glazing mullions may be built using movable formwork employing one of the schemes illustrated in Fig. 21-1.

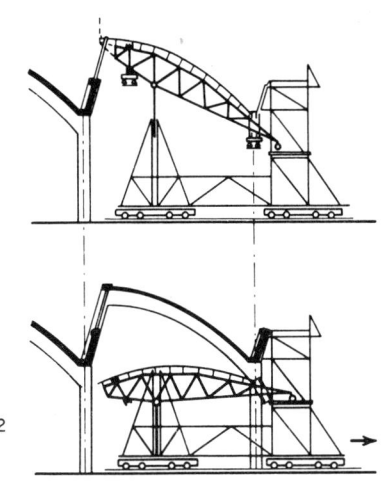

Fig. 21-1a. *(From "Northlight Shell Roofs—New Constructional Methods," by I. Hruban and K. Hruban, Indian Concrete Journal, December, 1959. Used by permission of the Concrete Association of India, Bombay.)*

References 129, 130, and 131 may be consulted for more details. If mullions are to be dispensed with, a suitable profile has to be selected for the northlight shell so that its center of gravity and shear center coincide.

If the shell is to have glazing mullions, it is better to make use of stationary formwork. One of several possible schemes [132] is illustrated in Fig. 21-2a. Some of the other arrangements described in the published literature [133, 134] are shown in Fig. 21-2b and c. Essentially, the formwork consists of a series of timber arches 4 to 6 ft apart sheathed over by planks or panels in small widths to form the surface of the shells. The trusses may be supported only at their ends or propped at intermediate points as well.

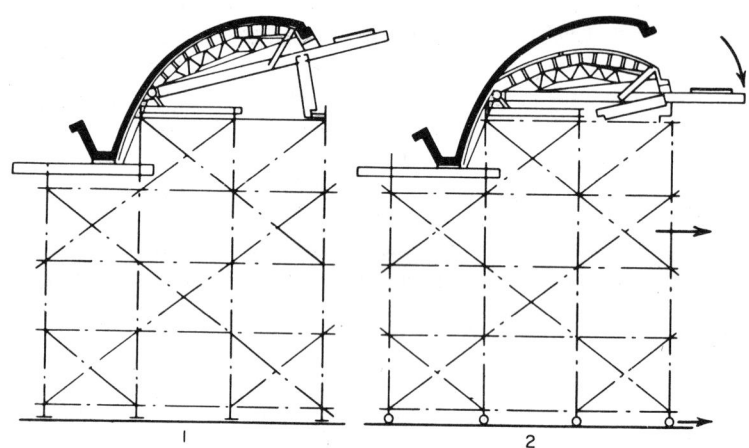

FIG. 21-1b. *(From "Usine Dunlop à Amiens Sheds en Béton Armé," by M. Hahn, IASS Bulletin, no. 1. Used by permission of the International Association for Shell Structures, Madrid.)*

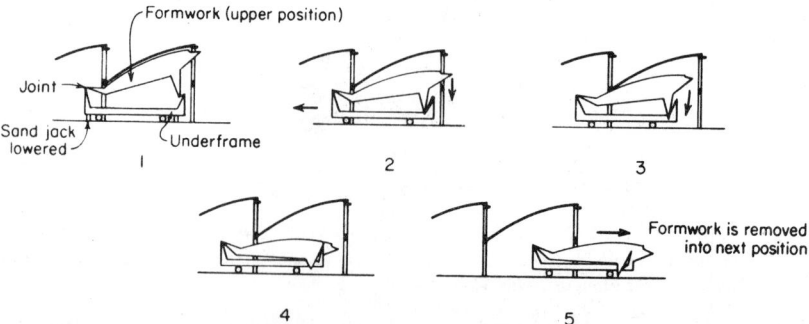

FIG. 21-1c. *(From "Proceedings of the Second Symposium on Concrete Shell Roof Construction", Oslo, July 1957. Used by permission of Dr. A. A. Gvozdev, USSR Academy of Building and Architecture, Moscow, and Teknisk Ukeblad, Norway.)*

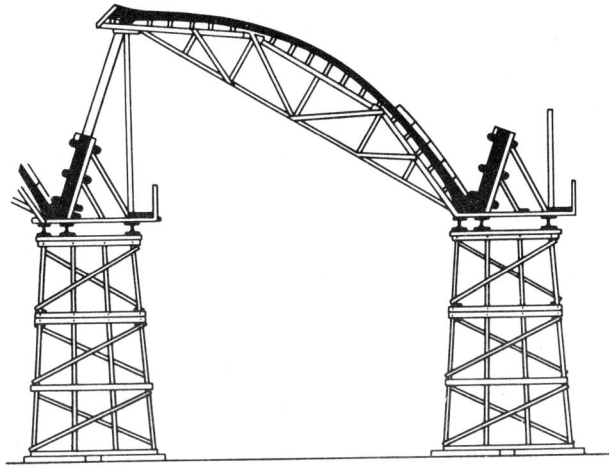

Fig. 21-2a. *(From "Reinforced or Prestressed Concrete (Béton Armé et Précontraint)"*
by A. M. Haas, IASS Bulletin, no. 1. Used by permission of the International Association
for Shell Structures, Madrid.)

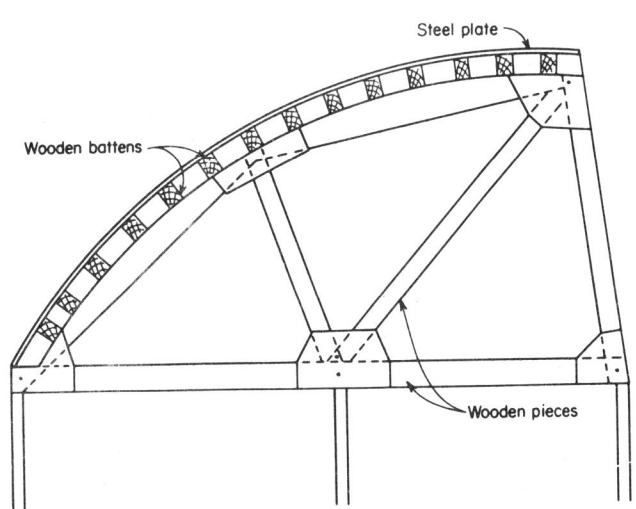

Fig. 21-2b. *(From "Northlight Shell Roofs—Practical Aspects of Construction" by*
N. V. Shastri, Indian Concrete Journal, December, 1959. Used by permission of the
Concrete Association of India, Bombay.)

21-6 Cylindrical Shells

The choice of a fixed or mobile scheme of formwork would depend on whether the shell is short or long, whether it is provided with deep or featheredge beams, and whether the arch ribs of the traverses are upstand or downstand. It would also depend upon whether the type of traverse provided is an openweb truss, trussed girder, or solid diaphragm.

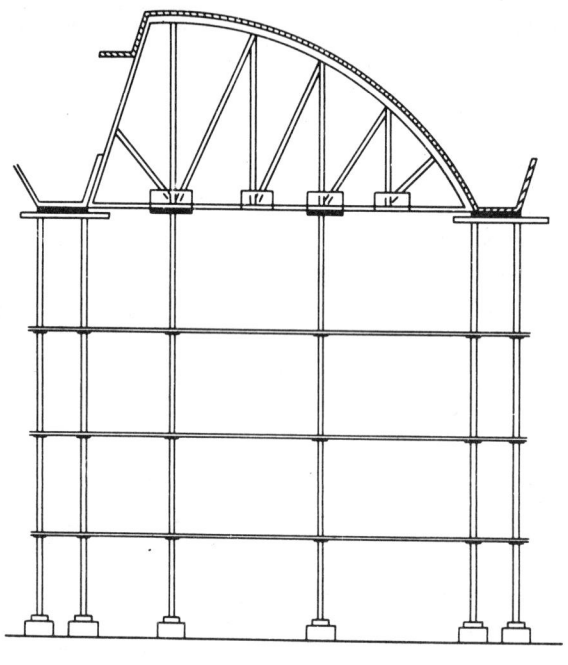

Shuttering of a northlight shell with
wooden trusses and scaffolding

Fig. 21-2c. *(From "Economics of Shell Roof Construction in India" by B. K. Chatterjee, Indian Concrete Journal, December, 1959. Used by permission of the Concrete Association of India, Bombay.)*

If the shells are short, a bowstring girder or a tied arch with the tie supported by suspenders may prove a suitable traverse. Such short shells are usually provided with shallow edge beams. The forms of these edge beams would obstruct the movement of the shell formwork. This problem may be solved by concreting them in advance along with a part of the shell. One of the two schemes described by Esquillan [127] and shown in Fig. 21-3a and b may be adopted to meet this situation. In the latter scheme, a swiveling form is used for casting this part of the shell. A traverse in the form of an arch rib would come in the way of traveling formwork if it is downstand. Hence short shells are usually provided with upstand arch-rib traverses if use of mobile formwork is contemplated. A simple method of forming upstand arch ribs is shown in Fig. 21-4. The movement of formwork is very much facilitated if short shells are supported on A-frame buttresses or if the ties of the arches forming the traverses are placed below the ground level. In

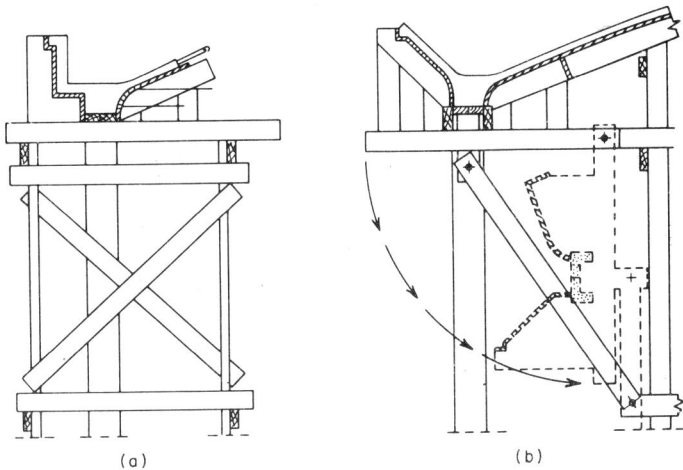

(a)   (b)

FIG. 21-3. *(From "General Report," by N. Esquillan, IASS Bulletin, no. 1. Used by permission of the International Assocation for Shell Structures, Madrid.)*

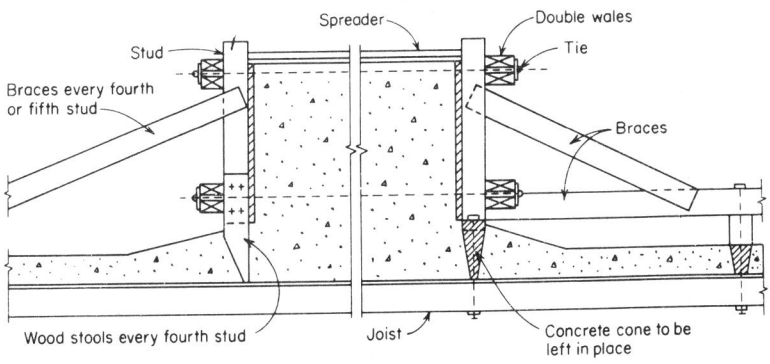

FIG. 21-4. *(From "Formwork for Concrete" by M. K. Hurd. Used by permission of the American Concrete Institute.)*

such arrangements, the traveling formwork is most easily moved forward. The forms need only to be lowered by 2 or 3 in.   Typical mobile formwork suitable for such applications is shown in Fig. 21-5.   However, most short shell roofs are supported on columns with a bowstring girder or a tied arch forming the traverse.   Ingenious methods of overcoming the obstruction caused by these members to the movement of formwork have been given by Coppel and Coulon [135].   This arrangement is shown in Fig. 21-6.   Wherever possible, the tie may be supported independent of the shell formwork.   The tie can be connected to the arch rib after the shell formwork is moved forward to the next bay.

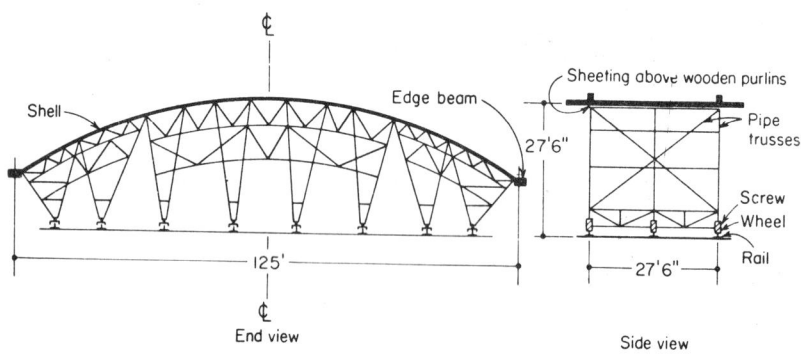

Fig. 21-5

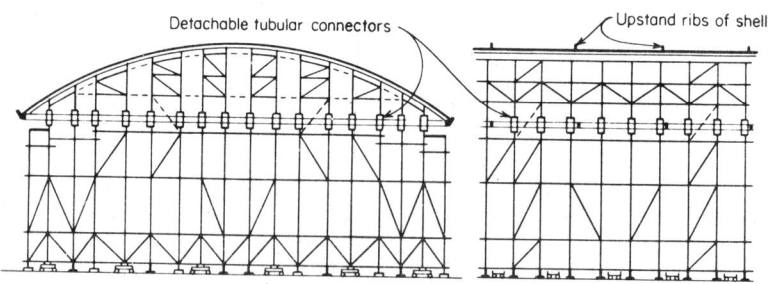

For decentering:
1. Lower the tubes with the help of screws at bottom.
2. Move the scaffold over the wheels till the arch ties touch the vertical tubes.
3. Detach the tubular connectors along the ties.
4. Continue movement till the next line of verticals touches the ties.
5. Replace the connectors, already removed.
6. Detach the connectors now touching the tie.
7. Repeat operations 4 to 6.

Fig. 21-6.   *(From "Echafaudages Tubulaires, Théorie et Pratique" by Th. Coppel and J. J. Coulon, Dunod, Paris, 1963.   Used by permission of Dunod Editeur, Paris.)*

The availability of uninterrupted space between the tie and the arch rib is essential to facilitate easy movement of the traveler.

Movable-formwork schemes are feasible for long multiple cylindrical shells only if the rise of the shell is relatively small. Otherwise each reuse involving the lowering and raising of the forms clear off deep edge beams presents several practical problems. Two possibilities described at the Madrid Colloquium [127] are shown in Fig. 21-7.

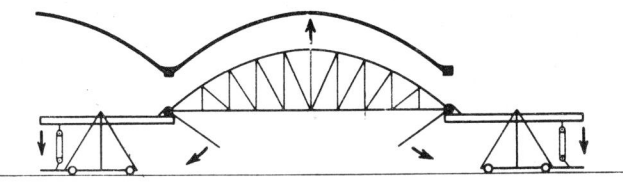

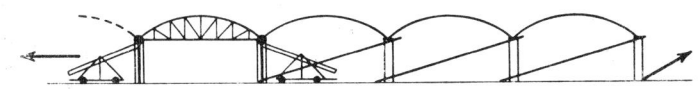

Fig. 21-7a. *(From "General Report" by N. Esquillan, IASS Bulletin, no. 1. Used by permission of the International Association for Shell Structures, Madrid.)*

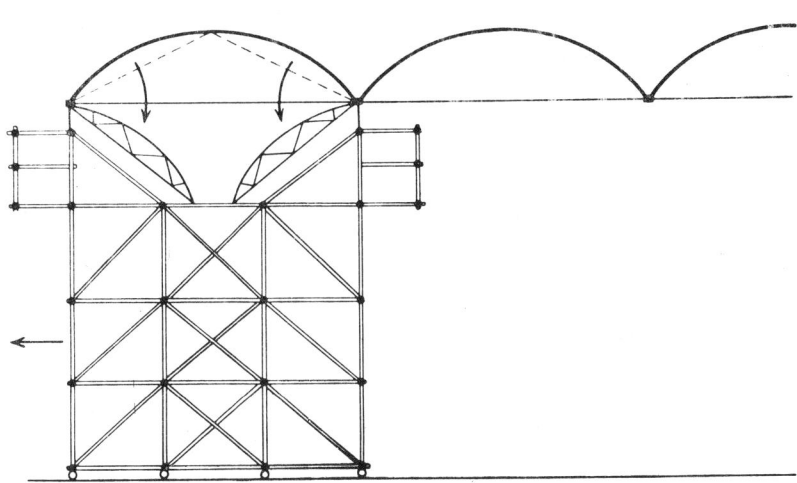

Fig. 21-7b. *(From "General Report" by N. Esquillan, IASS Bulletin, no. 1. Used by permission of the International Association for Shell Structures, Madrid.)*

21-7   Folded Plates

Schemes of movable formwork required for folded-plate construction are in many respects similar to those discussed for long cylindrical shells. The design of formwork for folded plates, however, presents fewer difficulties because no edge beams or curved arches are involved. The forms involve only straight members and the cost of formwork is therefore appreciably less. On this account, folded plates, though they consume a little more cement and steel, often prove to be more economical than or competitive with shells. Typical schemes of movable formwork for folded plates are described in a number of publications. The details given in Fig. 21-8 are extracted from references 136 and 137.

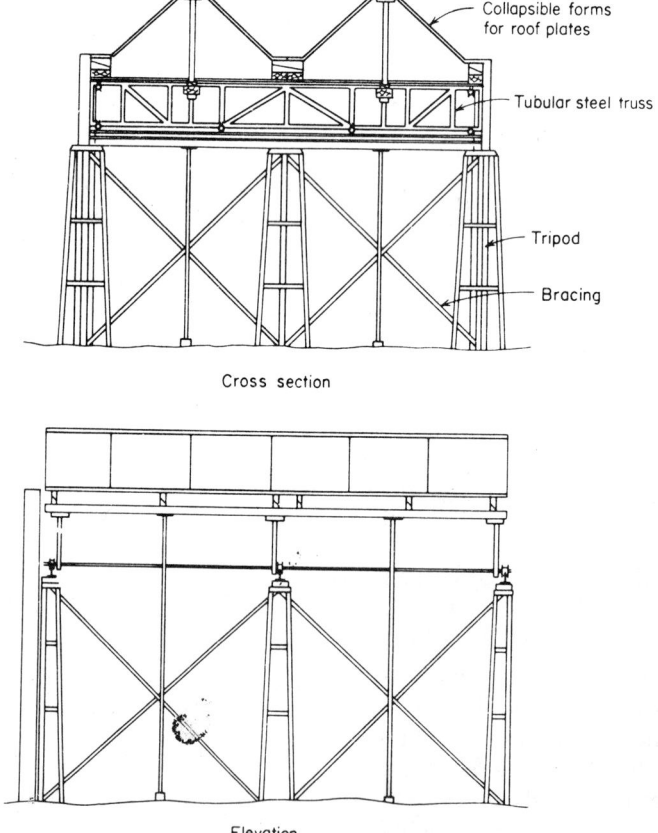

Cross section

Elevation

Fig. 21-8a. *(From "Movable Shuttering for Folded Plate Roof of Museum Building at C.B.R.I., Roorkee" by G. S. Ramaswamy et al., Indian Concrete Journal, July, 1961. Used by permission of the Concrete Association of India, Bombay.)*

21-8 Doubly Curved Shells

Although they are structurally more efficient than singly curved shells, doubly curved surfaces present more difficult forming problems. An exception is the hyperbolic paraboloid with straight boundaries. Being a ruled surface, it can be cast on straight forms. Conoids which are also ruled surfaces may be cast on straight forms supported on a series of arches of varying height. They may be built using mobile formwork. One of the possible schemes of movable formwork is that described in

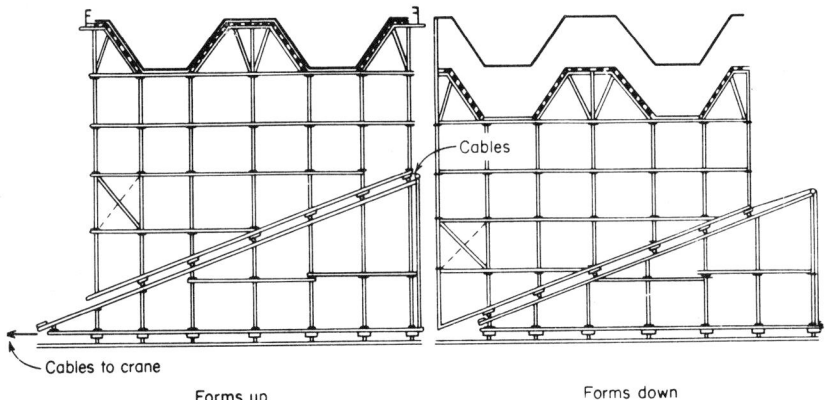

Forms up   Forms down

Fig. 21-8*b*. *(From Engineering News-Record, 1958. Used by permission of Engineering News-Record and McGraw-Hill Book Company.)*

reference 138 and shown in Fig. 21-9. It may be noted that the shape of traverse provided will influence the arrangement to a great extent. Schemes of movable formwork suitable for shallow paraboloids of revolution are described in references 139 and 140. Details of one such scheme are shown in Fig. 21-10. This arrangement is suitable only if the headroom measured from the floor level to the lowest tie is more than the rise of the shell. Similar schemes of mobile formwork are also suitable for building funicular shells. Doubly curved shells other than those already mentioned are usually roofs of monumental structures. As only one structure of its kind is involved, the formwork employed has to be of the fixed type. If the roof to be built is a spherical dome, one may use the ingenious arrangement devised for building the Dallas auditorium [141]. By breaking up the dome into pie-shaped cantilever elements, a large number of reuses of the formwork become possible. While building intricate forms for casting shells of complex shapes, a model is often helpful in explaining the sequence of work and difficulties involved to the construction crew.

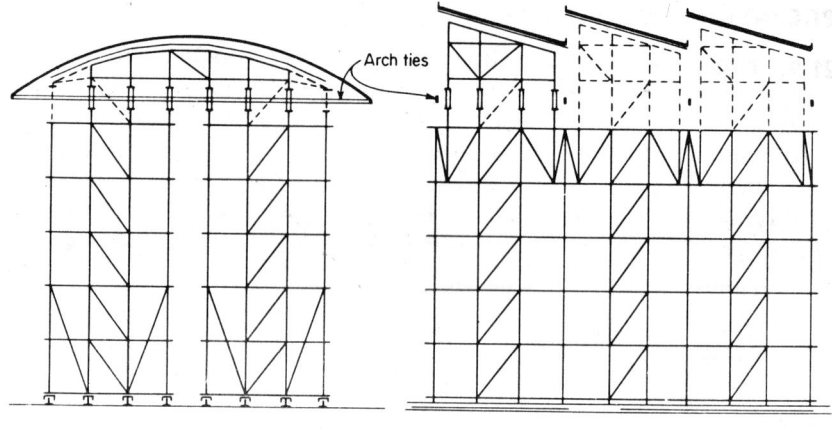

For decentering:
  1. Lower the tubes with the help of screws at bottom.
  2. Move the scaffold over the wheels till the arch ties touch the vertical tubes.
  3. Detach the tubular connectors along the ties.
  4. Continue movement till the next line of verticals touches the ties.
  5. Replace the connectors already removed.
  6. Detach the connectors now touching the tie.
  7. Repeat operations 4 to 6.

FIG. 21-9. *(From "Echafaudages Tubulaires, Théorie et Pratique" by Th. Coppel and J. J. Coulon, Dunod, Paris, 1963. Used by permission of Dunod Editeur, Paris.)*

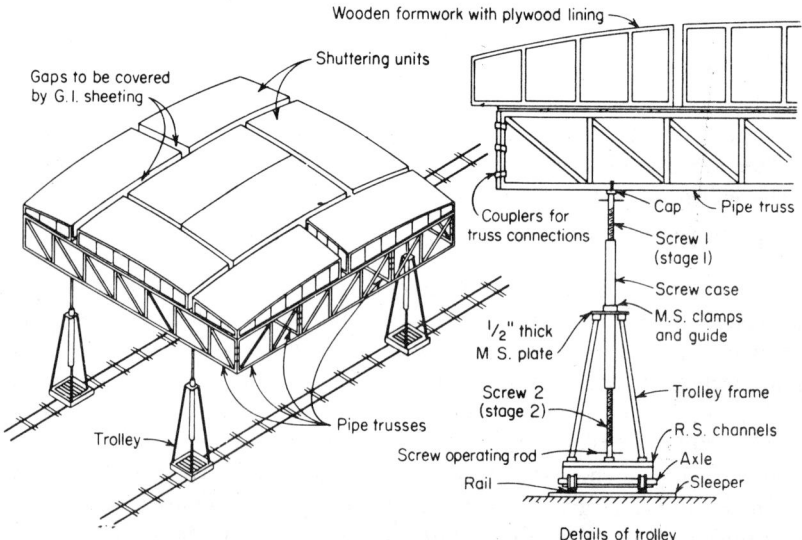

FIG. 21-10. Movable formwork for a paraboloid shell. *(From "Economical Roofs for Industrial and Storage Structures" by Prof. G. S. Ramaswamy, Cement and Concrete, October–December, 1962. Used by permission of Sahu Cement Service, New Delhi.)*

PREFABRICATED SHELLS

21-9 Advantages of Prefabrication

The advantages of prefabrication may be summed up as follows:

(i) Assembly-line techniques can be employed. This results in higher productivity and closer quality control.

(ii) Work in the open is reduced to a minimum.

(iii) Prefabricated shells need only a small fraction of the scaffolding required by cast-in-place shells. Scaffolding is required only for supporting the precast members when they are being joined together.

(iv) Large number of reuses are possible for the forms employed.

(v) Construction is speeded up as there are no interruptions due to bad weather. The precasting can be done under cover. Construction need not be seasonal.

(vi) Shrinkage cracks are minimized as the structure is cast in small elements.

(vii) Site labor, especially skilled labor such as carpenters, can be reduced.

(viii) Shapes which are difficult or expensive to cast in place may be handled.

However, prefabrication presupposes the availability of handling and hoisting equipment. For its large-scale application, a certain degree of standardization of roofs is a prerequisite. Such standardization has been attempted in the Soviet Union and several Eastern European countries.

The applicability of prefabrication techniques to various types of shells will now be discussed.

21-10 Northlight Shells

If crane facilities permit, full-length northlight shells may be prefabricated in the manner described by Haas [132]. Alternatively, the shells may be cast in pieces and assembled over supporting beams [129]. The method is shown in Photo 21-1.

21-11 Cylindrical Shells

Short cylindrical shells are not specially suitable for prefabrication.

Long cylindrical shells may be prefabricated in the same manner as northlight shells. A typical scheme is that employed for shells built on the Dywidag system [142]. The details involved are schematically shown in Fig. 21-11. Pretensioning of cylindrical-shell units on the long-line method is also possible [143].

PHOTO 21-1. Erection of northlight shells precast in pieces. *(From "Northlight Shell Roofs—New Constructional Methods," by I. Hruban and K. Hruban, Indian Concrete Journal, December, 1959. Used by permission of the Concrete Association of India, Bombay.)*

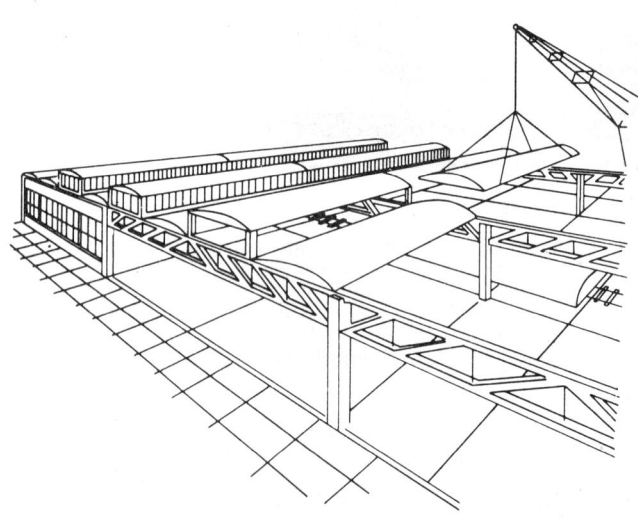

FIG. 21-11. *(Reproduced from "Prefabricated Concrete for Industrial and Public Structures" by L. Mokk. Courtesy of the Publishing House of the Hungarian Academy of Sciences, Budapest.)*

21-12 Folded Plates

Folded-plate roofs or individual plates forming them may be precast by employing a variety of methods. They may also be pretensioned. Details of several such applications are given by Harry [144].

21-13 Doubly Curved Shells

Typical of precast pretensioned doubly curved shells are the Silberkuhl shells developed in Germany. These find many applications as roofing elements [145]. Corrugated shells also lend themselves to prefabrica-

PHOTO 21-2. A view of a prestressed storage silo built at Neyveli, India. *(Courtesy of Gammon India (Private) Ltd., Bombay. Proprietors: The Neyveli Lignite Corporation Ltd., Neyveli, India. Design and Construction: Gammon India (Private) Ltd., Bombay.)*

tion. They may be cast in stacks one over the other and later on assembled in the field to form a corrugated shell roof [146]. Prestressing in the longitudinal direction is sometimes helpful in connecting the units. A shell built in India to serve as a storage silo by this technique is shown in Photo 21-2.

21-14 Lift Shells

Costly shoring and curved formwork required for casting shells may sometimes be avoided by casting the shell on the ground on forms or on a mold of shaped compacted earth covered by a thin layer of mortar with a bond breaker. The shell is then jacked up into its final position. Perhaps some of the largest shells built employing this technique were those designed by Esquillan for the aircraft hangar at Marignane, France [147]. Six shells, each of 101.50 m (333 ft) chord width and 9.80 m (32 ft) span, forming a principal bay covering an area of 101.50 by 58.80 m (333 by 193 ft) were cast on the ground and jacked up as one piece to rest on columns 19 m (62 ft) high. The weight involved was 4,200 tons. Views of the shells before and after lifting are shown in Photo 21-3. A similar technique was employed for building the shell roof over the prefabrication plant at Leningrad [131]. Each shell measuring 40 by

PHOTO 21-3. Shell roof of the aircraft hanger at Marignane, France. *(Courtesy of N. Esquillan.)*

40 m (131 by 131 ft) was built up of prefabricated units on light tubular scaffolding erected on the ground. The shell was then jacked up to 11 m (36 ft) by means of jacks placed at the four corners. The method employed is very similar to the familiar lift-slab technique [148]. The arrangement employed is schematically shown in Fig. 21-12. A notable United States example is the Warner auditorium roof in Indiana. The dome, 268 ft in diameter, was cast on rammed earth serving as formwork. Lifting jacks placed over the 36 columns were used to hoist the roof into its final position[1].

[1] For more details about this dome, reference may be made to "Lightweight Concrete Domes in USA," *Concrete and Constructional Engineering*, vol. LVIII, no. 4, pp. 181–182, April, 1963.

21-15 Placing and Compaction of Concrete

Concrete for placement on shell roofs shall preferably be of wet-earth consistency. If it is too fluid, it will tend to slump down the steep slopes. If too dry, proper compaction by vibration becomes difficult. Needle and form vribrators may be used for compaction only if they are suitably adapted by means of special attachments. Otherwise the resulting energy dissipation is too large to be absorbed by the relatively

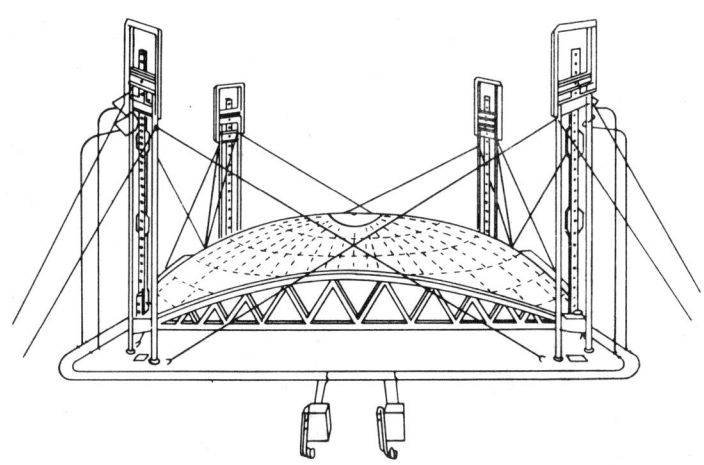

Fig. 21-12. *(From Civil Engineering, October, 1958. Used by permission of the American Society of Civil Engineers.)*

small volume of concrete involved in a thin shell. A suitable arrangement using a pneumatic hammer is shown in Fig. 21-13. Alternatively we may use a pneumatic hammer which is attached to a rod welded to a plate [149]. Top forms, if used, shall be tied down to resist the uplift pressure exerted by the wet concrete as schematically indicated in Fig. 21-14. Needle vibrators may be used to compact concrete placed inside the top forms.

The reinforcement placed on the sloping surface needs to be tied properly. Fastening of cover blocks to the reinforcement may be done by means of binding wire inserted in the blocks during casting. If the exposure of the flat cover blocks on the finished surface of the shell is not desirable, circular snap-on cover blocks may be employed instead. Different types of cover blocks are shown in Fig. 21-15.

The side forms of slender edge beams may be braced against bulging by the provision of concrete blocks with centered holes, through which

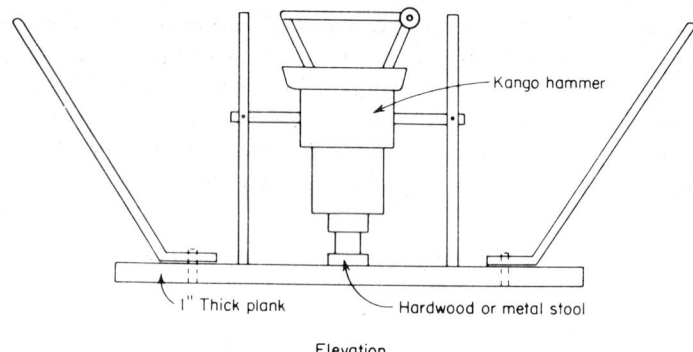

1" Thick plank   Hardwood or metal stool

Elevation

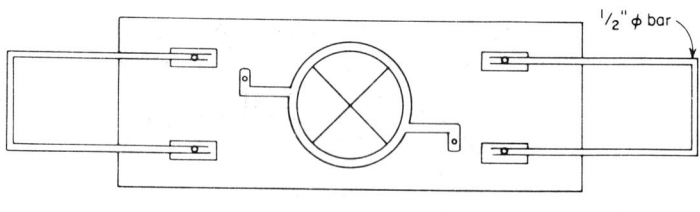

1/2" φ bar

Plan

FIG. 21-13

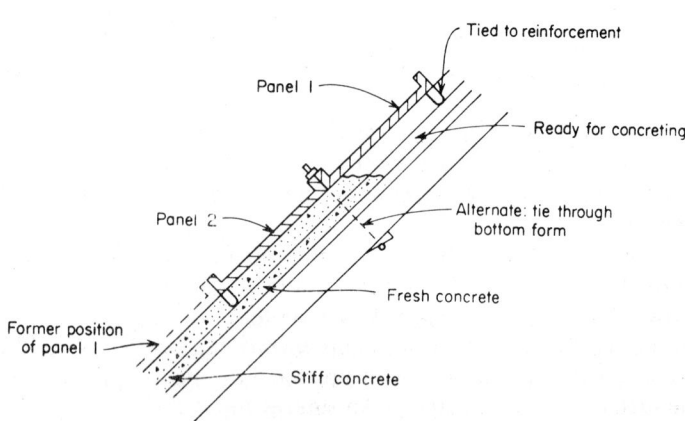

Tied to reinforcement

Panel 1

Ready for concreting

Panel 2

Alternate: tie through
bottom form

Fresh concrete

Former position
of panel 1

Stiff concrete

FIG. 21-14. *(From "Formwork for Concrete" by M. K. Hurd.   Used by permission of the American Concrete Institute.)*

a bolt can be passed (Fig. 21-16).   If this arrangement is used, the mortar
block will appear on the finished concrete.   If this is objectionable, a
coil tie with a cone spreader may be used (Fig. 21-17).

Placement of concrete on shell roofs can also be effected by the shot-
crete or guniting process [150].   Top forms can be eliminated even on
steep slopes if this method is used.   A wet or dry process may be
employed.   In the dry process, sand and cement are mixed dry and
pumped through a hose to a nozzle where water is added under pressure

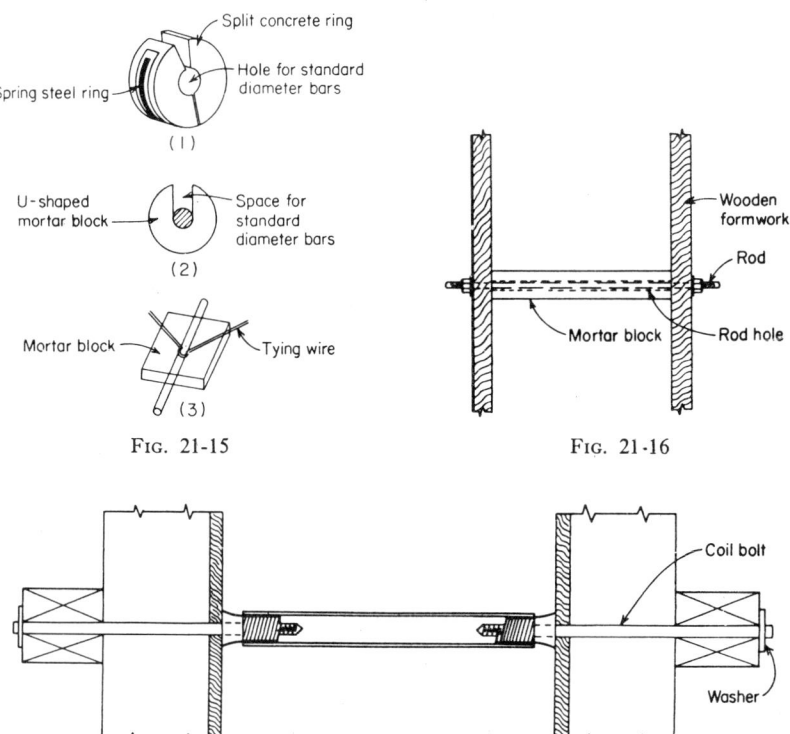

Fig. 21-15

Fig. 21-16

Fig. 21-17.   *(From "Formwork for Concrete" by M. K. Hurd.   Used by permission*
of the American Concrete Institute.)

from another hose.   The mixture jets out of the nozzle in the form of a
thick mist.   The jet is directed on to the surface to be concreted.   In the
wet process, sand, cement, and water are mixed in a pressurized chamber
and fed into the delivery hose.   Compressed air is used to jet the mixture
on to the surface to be concreted.   An interesting application of guniting
to cast shell roofs is described by Flint and Low [151].   Guniting tends
to be somewhat expensive, unless the shell surface is steep.

It is good practice to start concreting from the low points and move upward toward the high points. It is desirable to place the concrete in a symmetrical pattern, as otherwise the shoring will have to be braced for the effect of unsymmetrical loads. Construction joints which are preferably located in zones of compressive stress are to be indicated in detail on the design drawings. If a shell cannot be poured in one day, it may be sectioned into two or more placements with an interval of two or three days between successive pours to minimize shrinkage stresses [152].

21-16 Curing

Thin shells develop shrinkage cracks if not properly cured. In moderate weather (40 to 70°F), use of ordinary methods such as the use of membrane curing compounds or wet curing are usually satisfactory, although wet curing usually produces better results. In hot weather, water or wet-burlap curing is advisable. Use of retarders may be recommended. In cold weather, precautions against freezing are usually called for. The use of accelerators would generally be necessary.

21-17 Decentering

The decentering scheme and sequences must invariably be worked out by the designer for all shell structures of large size or unusual proportions. The decentering sequence should be such that the shell is not called upon to carry concentrated or unsymmetrical loads for which it is not designed. The guiding principle is that decentering should start from the point of maximum deflection and proceed symmetrically toward points of minimum deflection, the decentering of the edge members proceeding simultaneously with the adjoining shell. The designer should indicate the spacing of the shoring if edge members are to be reshored. Traverses are to be decentered after the forms under the shell are stripped. The designer should indicate the strength to be attained before decentering is to commence. For shells of moderate size and normal proportions, forms may be stripped as soon as the concrete attains 80 % of its 28-day cylinder strength. If the shell is sensitive to buckling or where large dead-weight deflections are to be avoided, it may be desirable to specify that the concrete should attain a minimum modulus of elasticity. Where such a stipulation is made, the modulus of elasticity may be found by using lightly reinforced beam specimens 4 in. by 6 in. by 6 ft long [152]. It is essential to ensure that forms do not stick to the shell. This will result in large loads being applied to the green concrete. Use of form joints that can be loosened or small panels that can be removed are some of the precautions necessary.

Enough wedges and screw jacks need to be provided at appropriate points to facilitate easing of the forms. The decentering of mobile forms calls for special care. It is essential that the forms are gradually lowered so that the form as a whole does not adhere to the shell surface.

The dead weight being the principal load for which shells are designed, a major part of the deflections to be expected occurs when the structure becomes self-supporting on stripping the forms. Wherever excessive deflections are expected, the designer shall prescribe suitable precambering of the forms to counter them.

Typical sequences for stripping the forms for a cylindrical shell, a northlight shell, and a hyperbolic paraboloid are shown in Figs. 21-18, 21-19, and 21-20.

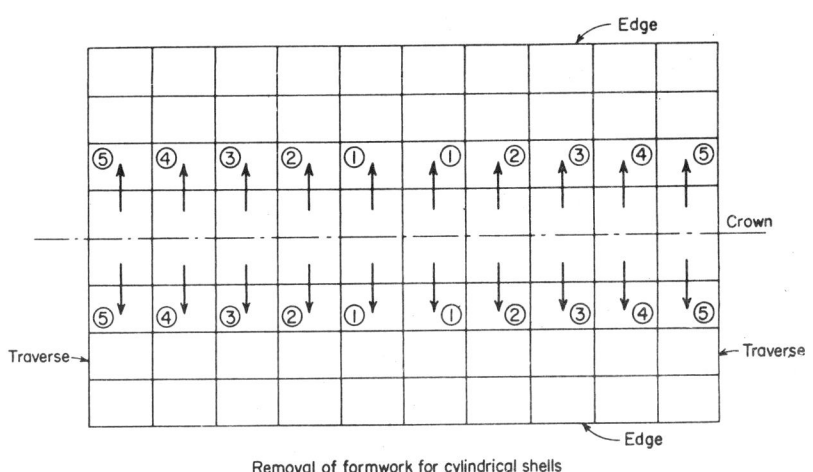

Removal of formwork for cylindrical shells
(numbers indicate the sequence)

Fig. 21-18

21-18 Fixtures and Service Installations

Fixtures for supporting any hanging elements are to be embedded in the concrete at the time of placement. Wherever concentrated loads such as ceiling fans are to be hung from the shell, a mesh of reinforcement may be provided to distribute the load. Electrical conduits can generally be accommodated inside the shell without any additional thickening. They are preferably laid at the center of the shell thickness parallel to the main reinforcement. A minimum concrete cover of $\frac{3}{4}$ in. is desirable. Heavier conduits are preferably located in edge beams, traverses, or columns.

21-19 Thermal Insulation and Waterproofing

Whether the insulation is to be provided on the top or bottom surface of the shell would depend on climatic conditions. In a tropical climate, where the object is to exclude heat, the insulation is most effective if it is laid on the top of the roof. In temperate and cold climates, keeping

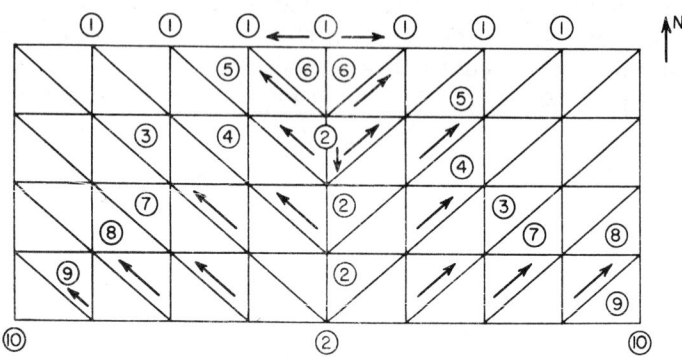

Figures in circles indicate the sequence
of the removal of formwork

FIG. 21-19. *(From "Northlight Shell Roofs—Practical Aspects of Construction" by N. V. Shastri, Indian Concrete Journal, December, 1959. Used by permission of the Concrete Association of India, Bombay.)*

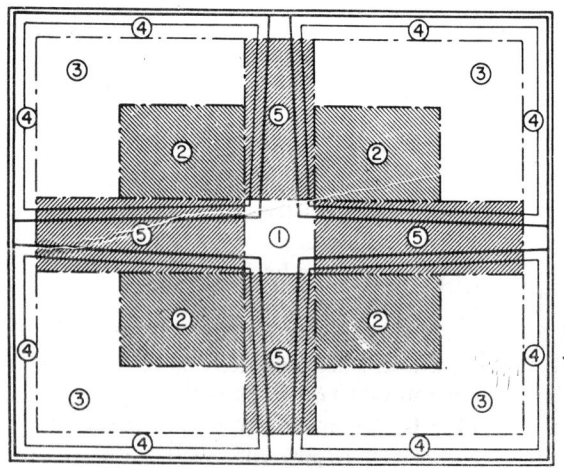

Figures in circles indicate the sequence
of the removal of formwork

FIG. 21-20. *(From "Formwork for Concrete" by M. K. Hurd. Used by permission of the American Concrete Institute.)*

the heat in is more important than keeping it out, and hence the provision of insulation on the bottom surface of the shell would serve the purpose better. Moreover, some of the insulating materials may also be employed to act as acoustic ceilings or even as forms. Hence there are several advantages of providing the insulation on the inside surface of a shell roof. Some of the commonly used materials for insulation may be grouped into insulation boards, fibrous materials, and lightweight foam concrete. Insulation boards may be made of fiber glass, wastewood shavings, or foamed plastics such as polystyrene. If glass wool is used, it is preferably laid on the top of the roof as it is subject to blister and disintegration when exposed to condensation inside the building. Foam concrete with a density of 30 lb/ft³ has a thermal conductivity of 0.66 Btu/ft²/hr/°F/in. thickness. A $2\frac{1}{2}$-in.-thick layer of this material provides about the same insulation as 1-in.-thick polystyrene. The foaming agents are proprietary products, and the amount to be used to get a desired density will be furnished by the manufacturer. Portland cement, water, and the foaming agent are mixed in the prescribed proportions to produce foamed concrete of 30 lb/ft³ density. The foamed concrete may be laid either in the form of precast blocks fixed to the shell surface on a bed of mortar or laid in situ. The first method is more common. The foam-concrete slurry may be made in a mixer with a churner attached to it.

If precast blocks of foam concrete are to be laid, it is convenient to make the blocks of a size ranging between 1 by 1 ft and 2 by 2 ft. The smaller blocks are mostly laid to conform to the curved surface of the shell. The mortar to be used for laying them on the roof should be of a lean mix. A proportion 1 of cement, 3 of lime, and 10 of sand has been found to be suitable. Rich mortars are unsuitable as the blocks laid on them develop cracks. Alternatively, foam concrete can be laid in situ on the shell roof. This usually proves to be more economical. The foam-concrete slurry cannot be used as it is, because it will tend to flow down the slopes. The following procedure is therefore followed. Precast blocks of foam concrete are broken into small sizes to serve as an aggregate. This material is next blended with freshly mixed slurry in the ratio of 60:100. If this mix is poured on a roof which is not too steep, it will stay in place and its thickness can be controlled.

Any insulation material laid on top of a shell roof must necessarily be covered by a waterproofing membrane such as bituminous felt or aluminum foil. One layer of felt is normally adequate on the shell surface. Two layers are necessary in the valleys.

Rigid boards, if used as insulation, may also be employed to act as forms for casting the shell. This method of construction has been described in references 153 and 154.

CHAPTER 22

SHELL RESEARCH

22-1 General Trends

During the past ten years, architects have shown considerable interest in exploiting thin shells as roofs for a variety of structures such as assembly halls, auditoriums, factories, warehouses, and grandstands. This has, in turn, stimulated structural engineers to intensify research efforts in this field to fill up the many gaps that still exist in our knowledge relating to the behavior of such structures. Much of the current research in progress all over the world is directed to finding timesaving short cuts which would considerably reduce the labor involved in carrying out the rather lengthy and tedious calculations associated with the design of shells. This is now possible by programming problems of stress analysis relating to shells for solution on digital computers. Another area of interest to research workers has been the development of experimental methods of design involving the use of models. Considerable progress has been made in this direction during the past ten years. With the trend toward larger and thinner shells, stability problems are becoming increasingly important. This is unfortunately a neglected aspect of shell research. The designer has, at present, very few guidelines available to him when he is called upon to check the stability of a large shell roof, especially if it is thin and shallow. Intensive research needs to be carried out to gather more reliable information on the buckling characteristics of reinforced-concrete shells. In what follows, these three aspects of shell research are discussed in some detail.

DIGITAL COMPUTATION

22-2 Matrix Methods of Structural Analysis

The advent of the digital computer has brought about a near revolution in our approach to structural analysis. Relieved of the drudgery of lengthy calculations, the structural engineer can now address himself to more creative tasks. But to match the speed of the computer, he needs an equally powerful and generalized means of formulating his problems. Matrix methods are the answer. In what follows, two powerful matrix methods which are specially suitable for application to cylindrical shells and folded plates are described. They are the matrix-progression method and the closely allied transfer-matrix technique.

22-3 Introduction to Matrix Progression

Matrix progression is a technique developed by Tottenham [155] for the solution of *unidirectional* problems in structural engineering. For the technique to be applicable, the structure has to be unidirectional in the *physical* and *mathematical* sense, or in the mathematical sense only. Thus a beam problem is unidirectional *both* physically and mathematically. It is mathematically unidirectional, because its behavior is governed by an *ordinary differential* equation. A problem, such as that of the cylindrical shell, is neither physically nor mathematically unidirectional but may be rendered mathematically unidirectional by resorting to the device of expanding the load in Fourier series. By this means, the problem is reduced to the solution of an ordinary differential equation. Again, the bending theory of shells of double curvature is governed by three simultaneous partial differential equations in u, v, and w. Solution by matrix progression is possible if these equations are reduced to ordinary differential equations in the direction of one of the coordinates by using variational techniques in the other direction.

22-4 Basic Concepts and Definitions

The basic concepts underlying the method and some of the definitions required for its exposition are best understood by applying them to the beam problem. As is well known, the beam problem is governed by the ordinary differential equation

$$\frac{d^4w}{dx^4} = \frac{p}{EI} \tag{22-1}$$

where w is the deflection at a distance x from the origin O (Fig. 22-1), p the intensity of uniformly distributed load, and EI the flexural rigidity of the beam. Let us first find the homogeneous solution. It is easily verified that it takes the form

$$w = C_1 + C_2 x + C_3 x^2 + C_4 x^3$$

Successive differentiation results in

$$\theta = w' = C_2 + 2C_3x + 3C_4x^2$$

$$-\frac{M}{EI} = w'' = 2C_3 + 6C_4x$$

$$-\frac{Q}{EI} = w''' = 6C_4 \qquad (22\text{-}2)$$

where the primes denote differentiation with respect to x. The quanti-

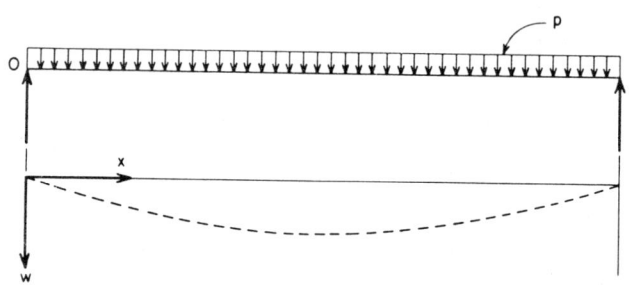

Fig. 22-1

ties w, θ, M, and Q are collectively known as the "beam actions." They may be conveniently represented by a column matrix thus:

$$\{Z\} = \begin{bmatrix} w \\ \theta \\ M \\ Q \end{bmatrix} \qquad (22\text{-}3)$$

$\{Z\}$ is known as the *action matrix*. If we denote the actions at the origin O, i.e., at $x = 0$, by w_0, θ_0, M_0, and Q_0, they may be collectively represented by the matrix

$$\{Z_0\} = \begin{bmatrix} w_0 \\ \theta_0 \\ M_0 \\ Q_0 \end{bmatrix} \qquad (22\text{-}4)$$

It may be noted that the pair M and θ represents corresponding actions as one is a force and the other the displacement caused by it, the terms force and displacement being used in their generalized sense to include moments and rotations as well. Similarly Q and w are corresponding actions. In a beam problem, *two of the four actions will be known or two of the actions would be capable of being expressed in terms of the other two.* For instance, if the beam is simply supported at O, $w_0 = 0$ and $M_0 = 0$. The other two actions Q_0 and θ_0 are therefore the two unknowns. This statement may now be expressed in matrix form as

$$\{Z_0\} = \begin{bmatrix} w_0 \\ \theta_0 \\ M_0 \\ Q_0 \end{bmatrix} = \begin{bmatrix} 0 & 0 \\ 1 & 0 \\ 0 & 0 \\ 0 & 1 \end{bmatrix} \begin{bmatrix} \theta_0 \\ Q_0 \end{bmatrix} \qquad (22\text{-}5)$$

or

$$\{Z_0\} = [K_0]\{\overline{Z_0}\} \qquad (22\text{-}6)$$

$[K_0]$ is known as the *boundary-restraint matrix* and $\{\overline{Z_0}\}$ as the *reduced-action matrix*. It may be noted that the *reduced-action matrix* contains only *half as many terms* as the action matrix at the origin. Let us next consider a beam fixed at the end O. Now the actions $w_0 = \theta_0 = 0$. Hence we are left with the unknowns M_0 and Q_0. One general principle may now be noted. *If one of the actions of a corresponding pair is zero, the other will appear as an unknown.* Both the actions of a corresponding pair cannot be zero. The principles developed so far for beams may at once be extended to cylindrical shells. In this case, we are interested in eight actions. It would therefore follow that four of them have to be known at the origin or four of them must be capable of being expressed in terms of the other four.

Let us now proceed to find the constants C_1, C_2, C_3, and C_4 of the beam problem in terms of the actions at the origin O. This leads to the following relations:

$$w_0 = C_1 \qquad\qquad \theta_0 = C_2$$
$$M_0 = -2EI\,C_3 \qquad Q_0 = -6EI\,C_4 \qquad (22\text{-}7)$$

These relations enable the actions at a point x to be written down in terms of the actions at the origin. Thus

$$\begin{bmatrix} w \\ \theta \\ M \\ Q \end{bmatrix} = \begin{bmatrix} 1 & x & -\dfrac{x^2}{2EI} & -\dfrac{x^3}{6EI} \\ 0 & 1 & -\dfrac{x}{EI} & -\dfrac{x^2}{2EI} \\ 0 & 0 & 1 & x \\ 0 & 0 & 0 & 1 \end{bmatrix} \begin{bmatrix} w_0 \\ \theta_0 \\ M_0 \\ Q_0 \end{bmatrix} \qquad (22\text{-}8)$$

or

$$\{Z\} = [G(x)]\{Z_0\} \qquad (22\text{-}9)$$

$[G(x)]$ is known as the *action-distribution matrix*. Equation $(22\text{-}9)$ is the essence of the matrix-progression technique. The effort always is

to express the actions at any point in terms of the actions at the origin.
Making use of Equations (22-6) and (22-9), we may write

$$\{Z\} = [G(x)][K_0]\{\overline{Z_0}\} \tag{22-10}$$

Let us next proceed to find the influence of the loads. This may be
done by considering a particular integral $w = px^4/24\ EI$ which satisfies
the differential equation (22-1). The following relations are derived by
successive differentiation:

$$\begin{bmatrix} w \\ \theta \\ M \\ Q \end{bmatrix} = \frac{p}{24EI} \begin{bmatrix} x^4 \\ 4x^3 \\ -12EIx^2 \\ -24EIx \end{bmatrix} = \{\check{Z}(x)\} \tag{22-11}$$

$\{\check{Z}(x)\}$ is known as the *loading-solution matrix*.

Now combining the homogeneous solution and the particular integral,
the complete solution may be written as

$$\begin{bmatrix} w \\ \theta \\ M \\ Q \end{bmatrix} = \begin{bmatrix} 1 & x & -\dfrac{x^2}{2EI} & -\dfrac{x^3}{6EI} \\ 0 & 1 & -\dfrac{x}{EI} & -\dfrac{x^2}{2EI} \\ 0 & 0 & 1 & x \\ 0 & 0 & 0 & 1 \end{bmatrix} \begin{bmatrix} w_0 \\ \theta_0 \\ M_0 \\ Q_0 \end{bmatrix} + \frac{p}{24EI} \begin{bmatrix} x^4 \\ 4x^3 \\ -12EIx^2 \\ -24EIx \end{bmatrix} \tag{22-12}$$

or

$$\{Z\} = [G(x)]\{Z_0\} + \{\check{Z}(x)\} \tag{22-13}$$

It may be noted that the distribution matrix must become the unit
matrix at the origin; similarly, the loading-solution matrix must become
the null matrix at this point.

Concentrated loads and reactions acting on the beam may be dealt
with as discontinuities in the shear Q. The other three actions remain
unchanged. Thus the load P acting at C (Fig. 22-2) brings about a

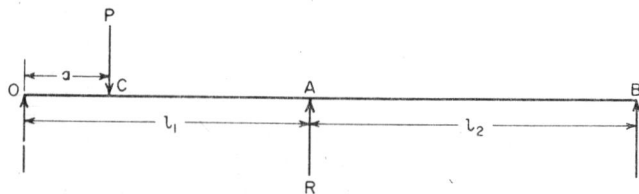

change in the action matrix $\{Z\}$. Infinitesimally to the left of C, the action matrix is $\{Z(a)\}$. Infinitesimally to the right of C, the action matrix is $\{Z(a)\} + \{\Delta Z_c\}$ where

$$\{\Delta Z_c\} = \begin{bmatrix} 0 \\ 0 \\ 0 \\ -P \end{bmatrix} \tag{22-14}$$

Similarly, infinitesimally to the left of A, the action matrix is $\{Z(l_1)\}$. Infinitesimally to the right of A, the action matrix is $\{Z(l_1)\} + \{\Delta Z_A\}$ where

$$\{\Delta Z_A\} = \begin{bmatrix} 0 \\ 0 \\ 0 \\ R \end{bmatrix} \tag{22-15}$$

22-5 General Method for Finding the Distribution Matrix $[G(x)]$

In the beam problem, we were able to construct the distribution matrix in a straightforward manner. This will not generally be the case when we are called upon to deal with complex problems relating to plates and shells. For this purpose, it is necessary to establish a general approach. The procedure consists in reducing the governing differential equation into a number of first-order linear differential equations. In what follows, we shall confine our discussion to cases where it is possible to replace the governing differential equation by a set of first-order linear differential equations with constant coefficients. This set of equations may be represented by the matrix equation

$$\left\{ \frac{dZ}{dx} \right\} = [A]\{Z\} + \{B\} \tag{22-16}$$

where $\{dZ/dx\}$, $\{Z\}$, and $\{B\}$ are column matrices and $[A]$ is a square matrix. If the order of the matrices $\{dZ/dx\}$, $\{Z\}$, and $\{B\}$ is $(n \times 1)$, the order of the matrix $[A]$ is $(n \times n)$. It is now possible to show [156] that the solution of the set of equations comprising (22-16) is given by

$$\{Z(x)\} = e^{[A]x}\{Z_0\} - ([I] - e^{[A]x})[A]^{-1}\{B\} \tag{22-17}$$

The exponential with matrix exponent $[A]$ may be expanded as

$$e^{[A]x} = [I] + [A]\frac{x}{1!} + [A]^2 \frac{x^2}{2!} + [A]^3 \frac{x^3}{3!} + \cdots \tag{22-18}$$

where $[I]$ is a unit matrix. In Equation (22-17), $e^{[A]x}$ is really the distribution matrix $[G(x)]$ and $-([I] - e^{[A]x})[A]^{-1}\{B\}$ is the loading-solution matrix $\{\bar{Z}(x)\}$.

By way of illustration, let us proceed to apply the general method to beams. The governing differential equation (22-1) is reduced to a set of first-order equations defining new functions as follows:

$$w = f_1$$

$$w' = f_2 = f_1'$$

$$w'' = f_3 = f_2'$$

$$w''' = f_4 = f_3'$$

Hence the governing differential equation is equivalent to the following first-order equations:

$$\frac{d}{dx}\begin{bmatrix} f_1 \\ f_2 \\ f_3 \\ f_4 \end{bmatrix} = \begin{bmatrix} 0 & 1 & 0 & 0 \\ 0 & 0 & 1 & 0 \\ 0 & 0 & 0 & 1 \\ 0 & 0 & 0 & 0 \end{bmatrix}\begin{bmatrix} f_1 \\ f_2 \\ f_3 \\ f_4 \end{bmatrix} + \begin{bmatrix} 0 \\ 0 \\ 0 \\ \dfrac{p}{EI} \end{bmatrix} \qquad (22\text{-}19)$$

With some rearrangement, we may write (22-19) in the form

$$\frac{d}{dx}\begin{bmatrix} w \\ \theta \\ M \\ Q \end{bmatrix} = \begin{bmatrix} 0 & 1 & 0 & 0 \\ 0 & 0 & -\dfrac{1}{EI} & 0 \\ 0 & 0 & 0 & 1 \\ 0 & 0 & 0 & 0 \end{bmatrix}\begin{bmatrix} w \\ \theta \\ M \\ Q \end{bmatrix} + \begin{bmatrix} 0 \\ 0 \\ 0 \\ p \end{bmatrix} \qquad (22\text{-}20)$$

This is now in the standard form

$$\left\{ \frac{dZ}{dx} \right\} = [A]\{Z\} + \{B\}$$

Hence the solution may be written in the form

$$\{Z(x)\} = e^{[A]x}\{Z_0\} - ([I] - e^{[A]x})[A]^{-1}\{B\}$$

To proceed further, we need the expansion of the matrix series for $e^{[A]x}$. Let us proceed to find the powers $[A]^2$ and $[A]^3$ of the matrix $[A]$. These are

$$[A]^2 = \begin{bmatrix} 0 & 0 & -\dfrac{1}{EI} & 0 \\ 0 & 0 & 0 & -\dfrac{1}{EI} \\ 0 & 0 & 0 & 0 \\ 0 & 0 & 0 & 0 \end{bmatrix} \qquad (22\text{-}21)$$

and

$$[A]^3 = \begin{bmatrix} 0 & 0 & 0 & -\dfrac{1}{EI} \\ 0 & 0 & 0 & 0 \\ 0 & 0 & 0 & 0 \\ 0 & 0 & 0 & 0 \end{bmatrix}$$ (22-22)

It is easily seen that all higher powers of the matrix $[A]$ are zero in this instance so that we need take only four terms. Thus

$$e^{[A]x} = [I] + [A]\frac{x}{1!} + [A]^2\frac{x^2}{2!} + [A]^3\frac{x^3}{3!}$$ (22-23)

As shown by Pestel and Leckie [157] this may be foreseen by making use of the Cayley-Hamilton theorem. In some problems, the number of terms to be taken is not easy to find. The matrix-series method is not suitable if a digital computer is not available. If a desk calculator is to be used, $[G(x)]$ and $\{Z(x)\}$ will have to be determined analytically.

Substituting this polynomial series of four terms in solution (22-17) for $\{Z\}$ and inserting the values of $[A]^2$ and $[A]^3$ already found, it may be verified that the solution found by applying the general method is exactly equivalent to the solution found in (22-12).

22-6 Some Useful Relations

In many applications, it may be necessary to find the distribution matrix at a point other than the origin. Reverting to the problem of the beam, let us suppose that we need the distribution matrix $[G_a(x)]$ at a point $x = a$ which will enable us to write the actions at x in terms of the actions at a. Let $[G_0(x)]$ be the distribution matrix at the origin. It is easily verified that

$$[G_a(x)] = [G_0(x - a)]$$ (22-24)

The proof is quite simple.

$$\{Z(x)\} = e^{[A]x}\{Z_0\}$$

$$\{Z(a)\} = e^{[A]a}\{Z_0\}$$

On division,

$$\frac{\{Z(x)\}}{\{Z(a)\}} = e^{[A](x-a)}$$

or

$$\{Z(x)\} = e^{[A](x-a)}\{Z(a)\}$$

or

$$\{Z(x)\} = [G_a(x)]\{Z(a)\}$$

Similar relations hold good for the loading-solution matrix as well. Hence

$$\{\bar{Z}_a(x)\} = \{\bar{Z}_0(x - a)\}$$

22-7   Application to the Beam Problem

The concepts developed so far will now be applied to the specific beam problem shown in Fig. 22-3. At the point O, the origin, the action matrix may be written as

$$\begin{bmatrix} w_0 \\ \theta_0 \\ M_0 \\ Q_0 \end{bmatrix} = \begin{bmatrix} 0 & 0 \\ 1 & 0 \\ 0 & 0 \\ 0 & 1 \end{bmatrix} \begin{bmatrix} \theta_0 \\ Q_0 \end{bmatrix}$$

i.e.,

$$\{Z_0\} = [K_0]\{\bar{Z}_0\}$$

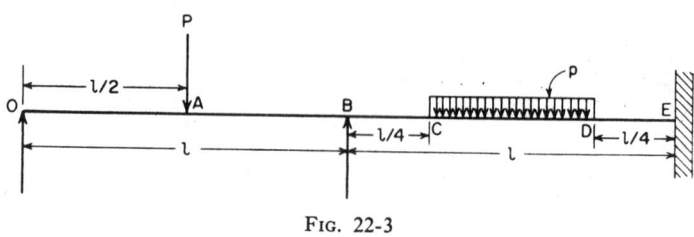

FIG. 22-3

This is because at the support we know that $w_0 = 0$ and $M_0 = 0$. In the region O to A, we may write the action matrix as

$$\{Z(x)\} = [G_0(x)]\{Z_0\} \tag{22-25}$$

where $[G_0(x)]$ is the distribution matrix at O. Making use of formula (22-25), the action matrix infinitesimally to the left of A is given by

$$\left\{Z\left(\frac{l}{2}\right)\right\} = \left[G_0\left(\frac{l}{2}\right)\right]\{Z_0\} \tag{22-26}$$

At A, a concentrated load P acts, bringing about a change of $\{\varDelta Z_A\}$ in the action. This is given by

$$\{\varDelta Z_A\} = \begin{bmatrix} 0 \\ 0 \\ 0 \\ -P \end{bmatrix} \tag{22-27}$$

The action matrix infinitesimally to the right of A is $\{Z(l/2)\} + \{\Delta Z_A\}$. We may now write the action matrix at any point in the region A to B in terms of the action matrix infinitesimally to the right of A. Or, in other words, for the region AB we may regard A as the origin. Hence the action matrix for this zone may be rewritten as

$$\{Z(x)\} = [G_a(x)] \left(\left[G_0 \left(\frac{l}{2} \right) \right] \{Z_0\} + \{\Delta Z_A\} \right)$$

Noting that $[G_a(x)] = [G_0(x - l/2)]$,

$$\{Z(x)\} = \left[G_0 \left(x - \frac{l}{2} \right) \right] \left(\left[G_0 \left(\frac{l}{2} \right) \right] \{Z_0\} + \{\Delta Z_A\} \right)$$

$$= [G_0(x)]\{Z_0\} + \left[G_0 \left(x - \frac{l}{2} \right) \right] \{\Delta Z_A\} \qquad \frac{l}{2} < x < l \qquad (22\text{-}28)$$

Using this formula,

$$\{Z(l)\} = [G_0(l)]\{Z_0\} + \left[G_0 \left(\frac{l}{2} \right) \right] \{\Delta Z_A\} \qquad (22\text{-}29)$$

At B a concentrated load acts in the form of the reaction R_B bringing about a change of $\{\Delta Z_B\}$ in the action matrix. Hence, infinitesimally to the right of B, the action matrix takes the form

$$\{Z(l)\} = [G_0(l)]\{Z_0\} + \left[G_0 \left(\frac{l}{2} \right) \right] \{\Delta Z_A\} + \{\Delta Z_B\} \qquad (22\text{-}30)$$

In the zone BC, the action matrix may be written as

$$\{Z(x)\} = [G_b(x)] \left([G_0(l)]\{Z_0\} + \left[G_0 \left(\frac{l}{2} \right) \right] \{\Delta Z_A\} + \{\Delta Z_B\} \right) \qquad (22\text{-}31)$$

Making use of this relation, the action matrix at C is

$$\left\{ Z \left(\frac{5l}{4} \right) \right\} = \left[G_0 \left(\frac{5l}{4} \right) \right] \{Z_0\} + \left[G_0 \left(\frac{3l}{4} \right) \right] \{\Delta Z_A\} + \left[G_0 \left(\frac{l}{4} \right) \right] \{\Delta Z_B\} \qquad (22\text{-}32)$$

Starting from C the beam is carrying a uniformly distributed load. Hence the loading solution has to be introduced in writing the action matrix in the zone CD. At any point in this region,

$$\{Z(x)\} = [G_c(x)] \left(\left[G_0 \left(\frac{5l}{4} \right) \right] \{Z_0\} + \left[G_0 \left(\frac{3l}{4} \right) \right] \{\Delta Z_A\} \right.$$

$$\left. + \left[G_0 \left(\frac{l}{4} \right) \right] \{\Delta Z_B\} \right) + \{\bar{Z}_c(x)\} \qquad (22\text{-}33)$$

Noting that

$$\{\bar{Z}_c(x)\} = \left\{\bar{Z}_0\left(x - \frac{5l}{4}\right)\right\}$$

$$\{Z(x)\} = [G_0(x)]\{Z_0\} + \left[G_0\left(x - \frac{l}{2}\right)\right]\{\varDelta Z_A\} + [G_0(x - l)]\{\varDelta Z_B\}$$

$$+ \left\{\bar{Z}_0\left(x - \frac{5l}{4}\right)\right\} \qquad \frac{5l}{4} < x < \frac{7l}{4} \qquad (22\text{-}34)$$

Hence the action matrix at D is

$$\left\{Z\left(\frac{7l}{4}\right)\right\} = \left[G_0\left(\frac{7l}{4}\right)\right]\{Z_0\} + \left[G_0\left(\frac{5l}{4}\right)\right]\{\varDelta Z_A\}$$

$$+ \left[G_0\left(\frac{3l}{4}\right)\right]\{\varDelta Z_B\} + \left\{\bar{Z}_0\left(\frac{l}{2}\right)\right\} \qquad (22\text{-}35)$$

In the zone between D and E, the action matrix may be written as

$$\{Z(x)\} = \left[G_0\left(x - \frac{7l}{4}\right)\right]\left(\left[G_0\left(\frac{7l}{4}\right)\right]\{Z_0\} + \left[G_0\left(\frac{5l}{4}\right)\right]\{\varDelta Z_A\}\right.$$

$$+ \left[G_0\left(\frac{3l}{4}\right)\right]\{\varDelta Z_B\} + \left\{\bar{Z}_0\left(\frac{l}{2}\right)\right\}\right)$$

or

$$\{Z(x)\} = [G_0(x)]\{Z_0\} + \left[G_0\left(x - \frac{l}{2}\right)\right]\{\varDelta Z_A\} + [G_0(x - l)]\{\varDelta Z_B\}$$

$$+ \left[G_0\left(x - \frac{7l}{4}\right)\right]\left\{\bar{Z}_0\left(\frac{l}{2}\right)\right\} \qquad \frac{7l}{4} < x < 2l \qquad (22\text{-}36)$$

Inserting $x = 2l$ in Equation (22-36), we find the action matrix at E as

$$\{Z(2l)\} = [G_0(2l)]\{Z_0\} + \left[G_0\left(\frac{3l}{2}\right)\right]\{\varDelta Z_A\} + [G_0(l)]\{\varDelta Z_B\}$$

$$+ \left[G_0\left(\frac{l}{4}\right)\right]\left\{\bar{Z}_0\left(\frac{l}{2}\right)\right\} \qquad (22\text{-}37)$$

To determine the three unknowns Q_0, θ_0, and R_B we have three conditions. They are

$$w = 0 \qquad \text{at} \qquad x = l$$

$$\begin{matrix} w = 0 \\ w' = 0 \end{matrix} \Big\} \qquad \text{at} \qquad x = 2l$$

This means that, to formulate the condition on w at B, i.e., at $x = l$, we have to separate it from the rest of the actions in the action matrix $\{Z(l)\}$. This may be done by premultiplying it by the *isolation matrix*

$$[I_B] = [1 \quad 0 \quad 0 \quad 0] \tag{22-38}$$

Hence the condition at $x = l$ may be formulated as

$$[I_B]\{Z(l)\} = \{0\} \tag{22-39}$$

Similarly at the point E ($x = 2l$), we introduce the isolation matrix $[I_E]$ defined as

$$[I_E] = \begin{bmatrix} 1 & 0 & 0 & 0 \\ 0 & 1 & 0 & 0 \end{bmatrix} \tag{22-40}$$

so that w and w' can be isolated from the rest of the actions contained in the action matrix $\{Z(2l)\}$. The conditions at E ($x = 2l$) are thus formulated as

$$[I_E]\{Z(2l)\} = \{0\} \tag{22-41}$$

The relation (22-40) results in one equation and the relation (22-41) in two equations so that three equations are available to solve for the three unknowns. In the computation of $\{Z(l)\}$ and $\{Z(2l)\}$ we need the quantities given below. They are found by substituting the respective arguments, such as $2l$ and l, in the formulas for $[G(x)]$ and $\{\tilde{Z}(x)\}$ given in relations (22-8) and (22-11).

$$[G_0(2l)] = \begin{bmatrix} 1 & 2l & -\dfrac{2l^2}{EI} & -\dfrac{4}{3}\dfrac{l^3}{EI} \\ 0 & 1 & -\dfrac{2l}{EI} & -\dfrac{2l^2}{EI} \\ 0 & 0 & 1 & 2l \\ 0 & 0 & 0 & 1 \end{bmatrix}$$

$$[G_0(\tfrac{3}{2}l)] = \begin{bmatrix} 1 & \dfrac{3l}{2} & -\dfrac{9}{8}\dfrac{l^2}{EI} & -\dfrac{9}{16}\dfrac{l^3}{EI} \\ 0 & 1 & -\dfrac{3}{2}\dfrac{l}{EI} & -\dfrac{9}{8}\dfrac{l^2}{EI} \\ 0 & 0 & 1 & \dfrac{3l}{2} \\ 0 & 0 & 0 & 1 \end{bmatrix}$$

$$[G_0(l)] = \begin{bmatrix} 1 & l & -\dfrac{l^2}{2EI} & -\dfrac{l^3}{6EI} \\ 0 & 1 & -\dfrac{l}{EI} & -\dfrac{l^2}{2EI} \\ 0 & 0 & 1 & l \\ 0 & 0 & 0 & 1 \end{bmatrix}$$

$$\left[G_0\left(\dfrac{l}{2}\right)\right] = \begin{bmatrix} 1 & \dfrac{l}{2} & -\dfrac{l^2}{8EI} & -\dfrac{l^3}{48EI} \\ 0 & 1 & -\dfrac{l}{2EI} & -\dfrac{l^2}{8EI} \\ 0 & 0 & 1 & \dfrac{l}{2} \\ 0 & 0 & 0 & 1 \end{bmatrix}$$

$$\left[G_0\left(\dfrac{l}{4}\right)\right] = \begin{bmatrix} 1 & \dfrac{l}{4} & -\dfrac{l^2}{32EI} & -\dfrac{l^3}{384EI} \\ 0 & 1 & -\dfrac{l}{4EI} & -\dfrac{l^2}{32EI} \\ 0 & 0 & 1 & \dfrac{l}{4} \\ 0 & 0 & 0 & 1 \end{bmatrix}$$

$$\left\{Z_0\left(\dfrac{l}{2}\right)\right\} = \dfrac{p}{24EI} \begin{bmatrix} \dfrac{l^4}{16} \\ \dfrac{l^3}{2} \\ -3EI\, l^2 \\ -12EI\, l \end{bmatrix}$$

$$[G_0(l)]\{Z_0\} = \begin{bmatrix} \left(l\theta_0 - \dfrac{Q_0 l^3}{6EI}\right) \\ \left(\theta_0 - \dfrac{Q_0 l^2}{2EI}\right) \\ Q_0 l \\ Q_0 \end{bmatrix}, \qquad \left[G_0\left(\dfrac{l}{2}\right)\right]\{\varDelta Z_A\} = \begin{bmatrix} \dfrac{Pl^3}{48EI} \\ \dfrac{Pl^2}{8EI} \\ -\dfrac{Pl}{2} \\ -P \end{bmatrix}$$

$$[G_0(2l)]\{Z_0\} = \begin{bmatrix} \left(2l\theta_0 - \dfrac{4}{3}\dfrac{Q_0 l^3}{EI}\right) \\[2mm] \left(\theta_0 - \dfrac{2Q_0 l^2}{EI}\right) \\[2mm] 2Q_0 l \\[2mm] Q_0 \end{bmatrix}, \qquad \left[G_0\left(\dfrac{3l}{2}\right)\right]\{\varDelta Z_A\} = \begin{bmatrix} \dfrac{9}{16}\dfrac{Pl^3}{EI} \\[2mm] \dfrac{9}{8}\dfrac{Pl^2}{EI} \\[2mm] -\dfrac{3}{2}Pl \\[2mm] -P \end{bmatrix}$$

$$[G_0(l)]\{\varDelta Z_B\} = \begin{bmatrix} -\dfrac{R_B l^3}{6EI} \\[2mm] -\dfrac{R_B l^2}{2EI} \\[2mm] R_B l \\[2mm] R_B \end{bmatrix}, \quad \text{and} \quad \left[G_0\left(\dfrac{l}{4}\right)\right]\left\{\dot{Z}\left(\dfrac{l}{2}\right)\right\} = \dfrac{p}{24EI}\begin{bmatrix} \dfrac{5l^4}{16} \\[2mm] \dfrac{13l^3}{8} \\[2mm] -6EI\,l^2 \\[2mm] -12EI\,l \end{bmatrix}$$

We may now form the three simultaneous equations as

$$l\,\theta_0 - \frac{Q_0 l^3}{6EI} + \frac{Pl^3}{48EI} = 0 \qquad\qquad (22\text{-}42)$$

$$2l\,\theta_0 - \frac{4}{3}\frac{Q_0 l^3}{EI} + \frac{9}{16}\frac{Pl^3}{EI} - \frac{R_B l^3}{6EI} + \frac{5}{384}\frac{pl^4}{EI} = 0 \qquad (22\text{-}43)$$

and

$$\theta_0 - \frac{2Q_0 l^2}{EI} + \frac{9}{8}\frac{Pl^2}{EI} - \frac{R_B l^2}{2EI} + \frac{13}{192}\frac{pl^3}{EI} = 0 \qquad (22\text{-}44)$$

Solving,

$$Q_0 = \frac{11}{28}\left(P - \frac{pl}{16}\right)$$

$$\theta_0 = \frac{1}{112}\frac{l^2}{EI}\left(5P - \frac{11}{24}pl\right)$$

and

$$R_B = \frac{1}{56}\left(43P + \frac{101}{8}pl\right)$$

22-8 Application to a Single Cylindrical Shell

For purposes of illustration, let us consider the Schorer formulation. Making use of the various relations derived in Chapter 7, the eighth-order Schorer differential equation may be replaced by eight first-order equations. Let us first consider the unloaded shell. The factor $\cos(\pi x/l)$ which is a common multiplying factor for all the stress resultants is omitted in all the relations as we are just now concerned with the variation in the ϕ direction only. The problem is thus rendered unidirectional. Referring to Art. 7-32, and introducing functions $f_1, f_2, ..., f_8$ defined below, the following relations are easily established:

$$f_1 = w \tag{22-45}$$

$$f_2 = \vartheta = \frac{1}{a}\frac{\partial w}{\partial \phi} = \frac{1}{a}f_1^{\cdot} \tag{22-46}$$

$$f_3 = M_\phi = -\frac{D}{a^2}w^{\cdot\cdot} = -\frac{D}{a}f_2^{\cdot} \tag{22-47}$$

$$f_4 = Q_\phi = \frac{1}{a}M_\phi^{\cdot} = \frac{1}{a}f_3^{\cdot} \tag{22-48}$$

$$f_5 = N_\phi = -Q_\phi^{\cdot} = -f_4^{\cdot} \tag{22-49}$$

$$f_6 = \frac{\partial N_{x\phi}}{\partial x} = -\frac{1}{a}N_\phi^{\cdot} = -\frac{1}{a}f_5^{\cdot} \tag{22-50}$$

$$f_7 = \frac{\partial u}{\partial x} = \frac{1}{Ed}\frac{a}{\lambda_n^2}f_6^{\cdot} \tag{22-51}$$

$$f_8 = v = \frac{a}{\lambda_n^2}f_7^{\cdot} \tag{22-52}$$

and

$$f_1 = f_8^{\cdot} \tag{22-53}$$

These relations may be arranged in matrix form to give eight first-order equations.

$$\frac{d}{d\phi}\begin{bmatrix} f_1 \\ f_2 \\ f_3 \\ f_4 \\ f_5 \\ f_6 \\ f_7 \\ f_8 \end{bmatrix} = \begin{bmatrix} 0 & a & 0 & 0 & 0 & 0 & 0 & 0 \\ 0 & 0 & -\dfrac{a}{D} & 0 & 0 & 0 & 0 & 0 \\ 0 & 0 & 0 & a & 0 & 0 & 0 & 0 \\ 0 & 0 & 0 & 0 & -1 & 0 & 0 & 0 \\ 0 & 0 & 0 & 0 & 0 & -a & 0 & 0 \\ 0 & 0 & 0 & 0 & 0 & 0 & \dfrac{Ed\lambda_n^2}{a} & 0 \\ 0 & 0 & 0 & 0 & 0 & 0 & 0 & \dfrac{\lambda_n^2}{a} \\ 1 & 0 & 0 & 0 & 0 & 0 & 0 & 0 \end{bmatrix}\begin{bmatrix} f_1 \\ f_2 \\ f_3 \\ f_4 \\ f_5 \\ f_6 \\ f_7 \\ f_8 \end{bmatrix} \tag{22-54}$$

The eight equations may be condensed and written as

$$\left\{\frac{dF}{d\phi}\right\} = [A]\{F\} \tag{22-55}$$

Making use of (22-17), we may write the solution for $\{F\}$ as

$$\{F\} = e^{[A]\phi}\{F(0)\} \tag{22-56}$$

where $\{F(0)\}$ is the value of $\{F\}$ at the edge O (Fig. 22-4) of the shell which is the origin for ϕ.

The formula $-([I] - e^{[A]x})[A]^{-1}\{B\}$ is not very convenient for use in practice for finding the loading-solution matrix. It is often convenient to derive the loading-solution matrix from either the membrane solution or a particular integral. We shall make use of the membrane solution to construct the loading-solution matrix. Let $\{P(\phi)\}$ be the action matrix corresponding to the membrane solution. Its value at

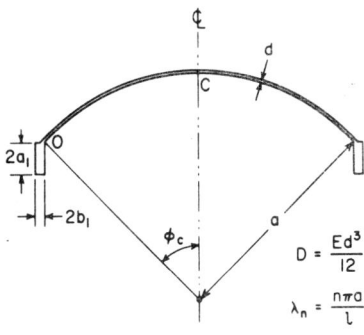

Fig. 22-4

$\phi = 0$ is therefore $\{P(0)\}$. This may not be zero. But we have already seen that the loading-solution matrix must become a null matrix at the origin. Hence the membrane solution, as it is, cannot be used as the loading-solution matrix. So we resort to the artifice of introducing an action of $-\{P(0)\}$ at O. The effect of this will be to reduce the membrane actions to zero at O. However, the introduction of an action $-\{P(0)\}$ at O will produce an action of $-e^{[A]\phi}\{P(0)\}$ at a section at an angle of ϕ from O. Hence

the total action at a section ϕ due to load $= -e^{A[\phi]}\{P(0)\} + \{P(\phi)\}$ (22-57)

We now combine the homogeneous solution (22-56) with the loading solution (22-57) to get the action matrix of a loaded shell as

$$\{F(\phi)\} = e^{[A]\phi}(\{F(0)\} - \{P(0)\}) + \{P(\phi)\} \tag{22-58}$$

$\{P(\phi)\}$ and $\{P(0)\}$ may be written down as follows using the results of the membrane theory:

$$\{P(\phi)\} = \begin{bmatrix} w \\ \\ \vartheta \\ M_\phi \\ Q_\phi \\ N_\phi \\ \dfrac{\partial N_{x\phi}}{\partial x} \\ \dfrac{\partial u}{\partial x} \\ v \end{bmatrix} = \begin{bmatrix} \dfrac{8g}{\pi a \left[Eda \left(\dfrac{\pi}{l} \right)^4 + \dfrac{Ed^3}{12a^5} \right]} \cos(\phi_c - \phi) \\ 0 \\ 0 \\ 0 \\ -\dfrac{4ag}{\pi} \cos(\phi_c - \phi) \\ \dfrac{8g}{\pi} \sin(\phi_c - \phi) \\ -\dfrac{8gl^2}{a\pi^3 Ed} \cos(\phi_c - \phi) \\ -\dfrac{8g}{\pi a \left[Eda \left(\dfrac{\pi}{l} \right)^4 + \dfrac{Ed^3}{12a^5} \right]} \sin(\phi_c - \phi) \end{bmatrix} \tag{22-59}$$

$$\{P(0)\} = \begin{bmatrix} w \\ \\ \vartheta \\ M_\phi \\ Q_\phi \\ N_\phi \\ \dfrac{\partial N_{x\phi}}{\partial x} \\ \dfrac{\partial u}{\partial x} \\ v \end{bmatrix} = \begin{bmatrix} \dfrac{8g}{\pi a \left[Eda \left(\dfrac{\pi}{l} \right)^4 + \dfrac{Ed^3}{12a^5} \right]} \cos \phi_c \\ 0 \\ 0 \\ 0 \\ -\dfrac{4ag}{\pi} \cos \phi_c \\ \dfrac{8g}{\pi} \sin \phi_c \\ -\dfrac{8gl^2}{a\pi^3 Ed} \cos \phi_c \\ -\dfrac{8g}{\pi a \left[Eda \left(\dfrac{\pi}{l} \right)^4 + \dfrac{Ed^3}{12a^5} \right]} \sin \phi_c \end{bmatrix} \tag{22-60}$$

Next we proceed to build up the initial-restraint matrix $[K_0]$ making use of the boundary conditions prescribed at O (Fig. 22-4). We have already seen in Chapter 8 that the relevant boundary conditions are the following:

(i) $$(M_\phi)_{\phi=0} = 0$$

(ii) $$(N_\phi)_{\phi=0} \cos \phi_c + (Q_\phi)_{\phi=0} \sin \phi_c = 0$$

(iii) $E \left(\dfrac{\partial u}{\partial x}\right)_{\phi=0} = $ stress in edge beam at its junction with the shell

(iv) $(v)_{\phi=0} \sin \phi_c - (w)_{\phi=0} \cos \phi_c = $ vertical deflection of edge beam at its junction with the shell

Stress in edge beam at its junction with shell

$$= \left(\frac{l}{\pi}\right)^2 \frac{1}{A} \left(\frac{\partial N_{\phi x}}{\partial x}\right)_{\phi=0} + \left(\frac{l}{\pi}\right)^2 \frac{a_1^2}{I} \left(\frac{\partial N_{x\phi}}{\partial x}\right)_{\phi=0}$$

$$+ \left[(N_\phi)_{\phi=0} \sin \phi_c - (Q_\phi)_{\phi=0} \cos \phi_c\right] \left(\frac{l}{\pi}\right)^2 \frac{a_1}{I} - \frac{4}{\pi} W \left(\frac{l}{\pi}\right)^2 \frac{a_1}{I} \quad (22\text{-}61)$$

But from boundary condition (ii),

$$(N_\phi)_{\phi=0} \cos \phi_c + (Q_\phi)_{\phi=0} \sin \phi_c = 0$$

or

$$(N_\phi)_{\phi=0} = -(Q_\phi)_{\phi=0} \tan \phi_c$$

Making this substitution in (22-61), boundary condition (iii) may be written as

$$\left(\frac{\partial u}{\partial x}\right)_{\phi=0} = \left(\frac{l}{\pi}\right)^2 \frac{1}{AE} \left(\frac{\partial N_{x\phi}}{\partial x}\right)_{\phi=0} + \left(\frac{l}{\pi}\right)^2 \frac{a_1^2}{EI} \left(\frac{\partial N_{x\phi}}{\partial x}\right)_{\phi=0}$$

$$-(Q_\phi)_{\phi=0} \left(\frac{l}{\pi}\right)^2 \frac{a_1}{EI} \frac{1}{\cos \phi_c} - \frac{4}{\pi} W \left(\frac{l}{\pi}\right)^2 \frac{a_1}{EI} \quad (22\text{-}62)$$

where W is the weight per unit length of the edge beam. Next, we proceed to formulate boundary condition (iv).

Vertical deflection of edge beam at its junction with shell

$$= -(Q_\phi)_{\phi=0} \frac{1}{\cos \phi_c} \left(\frac{l}{\pi}\right)^4 \frac{1}{EI} + \left(\frac{l}{\pi}\right)^4 \left(\frac{\partial N_{x\phi}}{\partial x}\right)_{\phi=0} \frac{a_1}{EI} - \frac{4}{\pi} W \left(\frac{l}{\pi}\right)^4 \frac{1}{EI}$$

Vertical deflection of shell at junction with edge beam

$$= (v)_{\phi=0} \sin \phi_c - (w)_{\phi=0} \cos \phi_c$$

Hence, according to boundary condition (iv),

$$(v)_{\phi=0} \sin \phi_c - (w)_{\phi=0} \cos \phi_c = -(Q_\phi)_{\phi=0} \frac{1}{\cos \phi_c} \left(\frac{l}{\pi}\right)^4 \frac{1}{EI}$$

$$+ \left(\frac{l}{\pi}\right)^4 \left(\frac{\partial N_{x\phi}}{\partial x}\right)_{\phi=0} \frac{a_1}{EI} - \frac{4}{\pi} W \left(\frac{l}{\pi}\right)^4 \frac{1}{EI}$$

Simplifying,

$$(w)_{\phi=0} = (v)_{\phi=0} \tan \phi_c + (Q_\phi)_{\phi=0} \left(\frac{l}{\pi}\right)^4 \frac{1}{EI} \frac{1}{\cos^2 \phi_c}$$

$$- \left(\frac{l}{\pi}\right)^4 \left(\frac{\partial N_{x\phi}}{\partial x}\right)_{\phi=0} \frac{a_1}{EI \cos \phi_c} + \frac{4}{\pi} W \left(\frac{l}{\pi}\right)^4 \frac{1}{EI \cos \phi_c} \qquad (22\text{-}63)$$

We are now ready to write down the action matrix $\{F(0)\}$ at O in terms of the initial-restraint matrix and the reduced-action matrix which will contain only four out of the eight actions. Thus

$$\begin{bmatrix} w \\ \vartheta \\ M_\phi \\ Q_\phi \\ N_\phi \\ \dfrac{\partial N_{x\phi}}{\partial x} \\ \dfrac{\partial u}{\partial x} \\ v \end{bmatrix}_{\phi=0} = \begin{bmatrix} \tan \phi_c & 0 & \left(\dfrac{l}{\pi}\right)^4 \dfrac{1}{EI \cos^2 \phi_c} & -\left(\dfrac{l}{\pi}\right)^4 \dfrac{a_1}{EI \cos \phi_c} \\ 0 & 1 & 0 & 0 \\ 0 & 0 & 0 & 0 \\ 0 & 0 & 1 & 0 \\ 0 & 0 & -\tan \phi_c & 0 \\ 0 & 0 & 0 & 1 \\ 0 & 0 & -\left(\dfrac{l}{\pi}\right)^2 \dfrac{a_1}{EI \cos \phi_c} & \left(\dfrac{l}{\pi}\right)^2 \left(\dfrac{1}{AE} + \dfrac{a_1^2}{EI}\right) \\ 1 & 0 & 0 & 0 \end{bmatrix}$$

$$\times \begin{bmatrix} v \\ \vartheta \\ Q_\phi \\ \dfrac{\partial N_{x\phi}}{\partial x} \end{bmatrix}_{\phi=0} + \begin{bmatrix} \dfrac{4W}{\pi} \left(\dfrac{l}{\pi}\right)^4 \dfrac{1}{EI \cos \phi_c} \\ 0 \\ 0 \\ 0 \\ 0 \\ -\dfrac{4W}{\pi} \left(\dfrac{l}{\pi}\right)^2 \dfrac{a_1}{EI} \\ 0 \end{bmatrix}$$

Hence we may write

$$\{F(0)\} = [K_0]\{F(0)\} + \{E\} \qquad (22\text{-}64)$$

Substituting $\{F(0)\}$ from (22-64) in (22-58), we arrive at

$$\{F(\phi)\} = e^{[A]\phi}([K_0]\{F(0)\} + \{E\} - \{P(0)\}) + \{P(\phi)\} \qquad (22\text{-}65)$$

$e^{[A]\phi}$ can be expressed as an infinite matrix series. Thus,

$$e^{[A]\phi} = [I] + \frac{[A]\phi}{1!} + \frac{[A]^2\phi^2}{2!} + \frac{[A]^3\phi^3}{3!} + \cdots$$

The number of terms necessary for any desired degree of accuracy is not always easy to find. Moreover, truncation errors are involved in limiting the series to a definite number of terms. To obviate these difficulties, an alternative procedure will be to express $e^{[A]\phi}$ in closed form containing only eight terms, corresponding to the eighth-order matrix $[A]\phi$ associated with cylindrical shells. This may be done by finding the eight eigenvalues of $[A]\phi$ and solving for the eight constants $C_0, C_1, ..., C_7$ by using the Cayley-Hamilton theorem. Thus we have

$$e^{[A]\phi} = C_0[I] + C_1[A]\phi + C_2[A]^2\phi^2 + \cdots + C_7[A]^7\phi^7$$

This procedure eliminates truncation errors and there is no uncertainty as to the number of terms to be taken.

We may now write the action matrix at the line of symmetry by setting $\phi = \phi_c$ in (22-65).

At the line of symmetry, we have the following four boundary conditions:

(i) $$(v)_{\phi=\phi_c} = 0$$

(ii) $$(\vartheta)_{\phi=\phi_c} = 0$$

(iii) $$(Q_\phi)_{\phi=\phi_c} = 0$$

(iv) $$\left(\frac{\partial N_{x\phi}}{\partial x}\right)_{\phi=\phi_c} = 0 \qquad (22\text{-}66)$$

An isolation matrix $[I_c]$ is introduced so that the four actions listed above are isolated.

$$[I_c] = \begin{bmatrix} 0 & 0 & 0 & 0 & 0 & 0 & 0 & 0 \\ 0 & 1 & 0 & 0 & 0 & 0 & 0 & 0 \\ 0 & 0 & 0 & 0 & 0 & 0 & 0 & 0 \\ 0 & 0 & 0 & 1 & 0 & 0 & 0 & 0 \\ 0 & 0 & 0 & 0 & 0 & 0 & 0 & 0 \\ 0 & 0 & 0 & 0 & 0 & 1 & 0 & 0 \\ 0 & 0 & 0 & 0 & 0 & 0 & 0 & 0 \\ 0 & 0 & 0 & 0 & 0 & 0 & 0 & 1 \end{bmatrix} \qquad (22\text{-}67)$$

We may thus formulate the boundary conditions on the line of symmetry as

$$[I_c]\{F(\phi_c)\} = \{0\} \tag{22-68}$$

This relation enables the four unknowns, namely, $(v)_{\phi=0}$, $(\vartheta)_{\phi=0}$, $(Q_\phi)_{\phi=0}$, and $(\partial N_{x\phi}/\partial x)_{\phi=0}$ to be determined. When once these are known, $\{F(\phi)\}$ at any value of ϕ may be found by using (22-65). Thus the actions at all points in the shell may be found. This method is ideally suitable for being programmed on the digital computer. Design Example 22-1 was worked out by writing such a program in the Fortran language. The computations were carried out on an IBM 1620 machine.

Design Example 22-1

ANALYSIS OF A SINGLE CYLINDRICAL SHELL
BY MATRIX-PROGRESSION METHOD USING THE SCHORER THEORY

1. *DATA*

Geometry

Shell

Span l = 83.25 ft
Radius a = 25 ft
Thickness d = 0.25 ft
Semicentral angle $\phi_c = 35° = 0.610865$ radian

Edge Beam

Depth $2a_1$ = 5 ft
Width $2b_1$ = 0.75 ft

Loads

Dead weight = 37.5 psf
Live load   = 12.5 psf

Total load g = 50.0 psf of shell surface

2. *GEOMETRICAL PROPERTIES AND OTHER QUANTITIES*

$$\sin \phi_c = 0.57358$$

$$\cos \phi_c = 0.81915$$

$$\frac{l}{\pi} = 26.49930$$

$$EI \text{ of edge beam} = 360 \times 10^6 \times \frac{(0.75 \times 5^3)}{12} = 281.25 \times 10^7$$

3. *MATRIX OF INITIAL RELATIONS*

The initial-relations matrix equation (22-54) governing the shell actions, i.e., $(d/d\phi)\{F\} = [A]\{F\}$, is first written down

$$\frac{d}{d\phi}\begin{bmatrix} w \\ \vartheta \\ M_\phi \\ Q_\phi \\ N_\phi \\ \dfrac{\partial N_{x\phi}}{\partial x} \\ \dfrac{\partial u}{\partial x} \\ v \end{bmatrix}$$

$$=\begin{bmatrix} 0 & 25.0 & 0 & 0 & 0 & 0 & 0 & 0 \\ 0 & 0 & -0.00005333 & 0 & 0 & 0 & 0 & 0 \\ 0 & 0 & 0 & 25.0 & 0 & 0 & 0 & 0 \\ 0 & 0 & 0 & 0 & -1.0 & 0 & 0 & 0 \\ 0 & 0 & 0 & 0 & 0 & -25.0 & 0 & 0 \\ 0 & 0 & 0 & 0 & 0 & 0 & 3{,}204{,}157.0 & 0 \\ 0 & 0 & 0 & 0 & 0 & 0 & 0 & 0.03560174 \\ 1.0 & 0 & 0 & 0 & 0 & 0 & 0 & 0 \end{bmatrix}\begin{bmatrix} w \\ \vartheta \\ M_\phi \\ Q_\phi \\ N_\phi \\ \dfrac{\partial N_{x\phi}}{\partial x} \\ \dfrac{\partial u}{\partial x} \\ v \end{bmatrix}$$

4. LOADING-SOLUTION MATRIX

The loading-solution matrix (22-59) giving the membrane stresses is given below.

$$\{P(\phi)\} = \begin{bmatrix} w \\ \vartheta \\ M_\phi \\ Q_\phi \\ N_\phi \\ \dfrac{\partial N_{x\phi}}{\partial x} \\ \dfrac{\partial u}{\partial x} \\ v \end{bmatrix} = \begin{bmatrix} \dfrac{8g}{\pi a[Eda(\pi/l)^4 + Ed^3/12a^5]}\cos(\phi_c - \phi) \\ 0 \\ 0 \\ 0 \\ -\dfrac{4ag}{\pi}\cos(\phi_c - \phi) \\ \dfrac{8g}{\pi}\sin(\phi_c - \phi) \\ -\dfrac{8gl^2}{a\pi^3 Ed}\cos(\phi_c - \phi) \\ -\dfrac{8g}{\pi a[Eda(\pi/l)^4 + Ed^3/12a^5]}\sin(\phi_c - \phi) \end{bmatrix}$$

The membrane actions (22-60) at the origin, i.e., at $\phi = 0$, are as follows:

$$\{P(0)\} = \begin{bmatrix} w \\ \vartheta \\ M_\phi \\ Q_\phi \\ N_\phi \\ \dfrac{\partial N_{x\phi}}{\partial x} \\ \dfrac{\partial u}{\partial x} \\ v \end{bmatrix} = \begin{bmatrix} 0.00091429 \\ 0 \\ 0 \\ 0 \\ -1{,}303.7209 \\ 73.030021 \\ -0.00003255 \\ -0.00064019 \end{bmatrix}$$

5. ACTION MATRIX AT THE ORIGIN IN TERMS OF THE RESTRAINT MATRIX

Equation (22-64) for the action matrix at the origin is now written down.

$$\begin{bmatrix} w \\ \vartheta \\ M_\phi \\ Q_\phi \\ N_\phi \\ \dfrac{\partial N_{x\phi}}{\partial x} \\ \dfrac{\partial u}{\partial x} \\ v \end{bmatrix} = \begin{bmatrix} 0.70020757 & 0 & 0.00026129 & -0.00053508 \\ 0 & 1.0 & 0 & 0 \\ 0 & 0 & 0 & 0 \\ 0 & 0 & 1.0 & 0 \\ 0 & 0 & -0.70020757 & 0 \\ 0 & 0 & 0 & 1.0 \\ 0 & 0 & -0.00000076 & 0.00000208 \\ 1.0 & 0 & 0 & 0 \end{bmatrix} \begin{bmatrix} v \\ \vartheta \\ Q_\phi \\ \dfrac{\partial N_{x\phi}}{\partial x} \end{bmatrix} + \begin{bmatrix} 0.14715813 \\ 0 \\ 0 \\ 0 \\ 0 \\ 0 \\ -0.00042916 \\ 0 \end{bmatrix}$$

$$[K_0]\{F(0)\} = \begin{bmatrix} \left(0.70020757v + 0.00026129Q_\phi - 0.00053508\,\dfrac{\partial N_{x\phi}}{\partial x}\right) \\ \vartheta \\ 0 \\ Q_\phi \\ -0.70020757Q_\phi \\ \dfrac{\partial N_{x\phi}}{\partial x} \\ \left(-0.00000076Q_\phi + 0.00000208\,\dfrac{\partial N_{x\phi}}{\partial x}\right) \\ v \end{bmatrix}$$

$$[K_0]\{\bar{F}(0)\} + \{E\} = \begin{bmatrix} \left(0.70020757v + 0.00026129Q_\phi - 0.00053508\,\dfrac{\partial N_{x\phi}}{\partial x} + 0.14715813\right) \\ \vartheta \\ 0 \\ Q_\phi \\ -0.70020757Q_\phi \\ \dfrac{\partial N_{x\phi}}{\partial x} \\ \left(-0.00000076Q_\phi + 0.00000208\,\dfrac{\partial N_{x\phi}}{\partial x} - 0.00042916\right) \\ v \end{bmatrix}$$

6. ACTION MATRIX AT ANY ANGLE ϕ

The action matrix at any point at an angle ϕ from the origin is given by Equation (22-65), i.e.,

$$\{F(\phi)\} = e^{[A]\phi}([K_0]\{\bar{F}(0)\} + \{E\} - \{P(0)\}) + \{P(\phi)\}$$

where $e^{[A]\phi}$ is calculated as the sum of the series

$$[I] + \frac{[A]\phi}{1!} + \frac{[A]^2\phi^2}{2!} + \frac{[A]^3\phi^3}{3!} + \cdots + \frac{[A]^7\phi^7}{7!}$$

Only eight terms of the series are retained.[1]

At the crown of the shell, which is also the line of symmetry,

$$e^{[A]\phi_c} = [I] + \frac{[A]\phi_c}{1!} + \frac{[A]^2\phi_c{}^2}{2!} + \frac{[A]^3\phi_c{}^3}{3!} + \cdots + \frac{[A]^7\phi_c{}^7}{7!}$$

The terms of the series are calculated and summed up. The sum is given below.

$$e^{[A]\phi_c} = \begin{bmatrix}
+1.0 & +15.271632 & -0.00024877 & -0.00126638 \\
-0.02394701 & +1.0 & -0.00003258 & -0.00024877 \\
+5,145.2341 & +11,225.160 & +1.0 & +15.271632 \\
+2,021.4870 & +5,145.2341 & -0.02394701 & +1.0 \\
-16,546.096 & -50,537.175 & +0.27441248 & +0.59867520 \\
+4,333.8124 & +16,546.096 & -0.10781265 & -0.27441250 \\
+0.00664251 & +0.03381398 & -0.00000028 & -0.00000084 \\
+0.61086526 & +4.6644548 & -0.00005066 & -0.00019340
\end{bmatrix}$$

$$\begin{bmatrix}
+0.00019340 & -0.00059070 & -192.69596 & -0.59867520 \\
+0.00005066 & -0.00019340 & -75.707413 & -0.27441248 \\
-4.6644548 & +23.744611 & +11,618,880 & +50,537.175 \\
-0.61086526 & +4.6644548 & +3,043,258.4 & +16,546.096 \\
+1.0 & -15.271632 & -14,945,645 & -108,345.31 \\
+0.02394701 & +1.0 & +1,957,308.2 & +21,283.642 \\
+0.00000009 & -0.00000019 & +1.0 & +0.02174787 \\
+0.00002363 & -0.00006014 & -16.815895 & +1.0
\end{bmatrix}$$

[1] Comparison with a closed solution for $e^{[A]\phi}$ using eigenvalues shows that 19 terms are necessary for ensuring acceptable accuracy.

7. SHELL ACTIONS AT THE LINE OF SYMMETRY

Knowing $e^{[A]\phi_c}$, $[K_0]\{\bar{F}(0)\}$ in terms of the unknowns and the values of $\{E\}$, $\{P(\phi_c)\}$, and $\{P(0)\}$, the shell actions at the line of symmetry are obtained from $\{F(\phi_c)\} \doteq e^{[A]\phi_c}$ $([K_0]\{\bar{F}(0)\} + \{E\} - \{P(0)\}) + \{P(\phi_c)\}$ in which

$[K_0]\{\bar{F}(0)\} + \{E\} - \{P(0)\}$

$$
= \begin{bmatrix}
\left(0.70020757v + 0.00026129Q_\phi - 0.00053508 \dfrac{\partial N_{x\phi}}{\partial x} + 0.14624384\right) \\[2mm]
\vartheta \\[2mm]
0 \\[2mm]
Q_\phi \\[2mm]
(-0.70020757Q_\phi + 1{,}303.7209) \\[2mm]
\left(\dfrac{\partial N_{x\phi}}{\partial x} - 73.030021\right) \\[2mm]
\left(-0.00000076Q_\phi + 0.00000208 \dfrac{\partial N_{x\phi}}{\partial x} - 0.00039661\right) \\[2mm]
(v + 0.00064019)
\end{bmatrix}
$$

$e^{[A]\phi_c}([K_0]\{\bar{F}(0)\} + \{E\} - \{P(0)\}) =$

$$
\begin{bmatrix}
\left(+0.10153237v +15.271632\vartheta -0.00099368Q_\phi -0.00152671 \dfrac{\partial N_{x\phi}}{\partial x} +0.51755974\right) \\[2mm]
\left(-0.29118035v +1.0\vartheta -0.000023281Q_\phi -0.00033810 \dfrac{\partial N_{x\phi}}{\partial x} +0.10651252\right) \\[2mm]
\left(+54{,}139.906v +11{,}225.160\vartheta +11.028575Q_\phi +45.166085 \dfrac{\partial N_{x\phi}}{\partial x} -11{,}638.570\right) \\[2mm]
\left(+17{,}961.556v +5{,}145.2341\vartheta -0.36302710Q_\phi +9.9146895 \dfrac{\partial N_{x\phi}}{\partial x} -2{,}037.8073\right) \\[2mm]
\left(-119{,}931.01v -50{,}537.175\vartheta +6.9637030Q_\phi -37.514477 \dfrac{\partial N_{x\phi}}{\partial x} +5{,}857.4758\right) \\[2mm]
\left(+24{,}318.210v +16{,}546.096\vartheta -0.65027410Q_\phi +2.7534896 \dfrac{\partial N_{x\phi}}{\partial x} -170.67900\right) \\[2mm]
\left(+0.02639900v +0.033813398\vartheta +0.00000007Q_\phi -0.00000166 \dfrac{\partial N_{x\phi}}{\partial x} +0.00071404\right) \\[2mm]
\left(+1.4277324v +4.6644548\vartheta -0.000037520Q_\phi -0.00042199 \dfrac{\partial N_{x\phi}}{\partial x} +0.13184094\right)
\end{bmatrix}
$$

Hence

$\{F(\phi_c)\} =$

$$
\begin{bmatrix}
\left(+0.10153237v \ +15.271632\vartheta \ -0.00099368Q_\phi \ -0.00152671\dfrac{\partial N_{x\phi}}{\partial x} \ +0.51867588 \right) \\[2ex]
\left(-0.29118035v \ +1.0\vartheta \ -0.00023281Q_\phi \ -0.00033810\dfrac{\partial N_{x\phi}}{\partial x} \ +0.10651252 \right) \\[2ex]
\left(+54,139.906v \ +11,225.160\vartheta \ +11.028575Q_\phi \ +45.166085\dfrac{\partial N_{x\phi}}{\partial x} \ -11,638.570 \right) \\[2ex]
\left(+17,961.556v \ +5,145.2341\vartheta \ -0.36302710Q_\phi \ +9.9146895\dfrac{\partial N_{x\phi}}{\partial x} \ -2,037.8073 \right) \\[2ex]
\left(-119,931.01v \ -50,537.175\vartheta \ +6.9637030Q_\phi \ -37.514477\dfrac{\partial N_{x\phi}}{\partial x} \ +4,265.9264 \right) \\[2ex]
\left(+24,318.210v \ +16,546.096\vartheta \ -0.65027410Q_\phi \ +2.7534896\dfrac{\partial N_{x\phi}}{\partial x} \ -170.67900 \right) \\[2ex]
\left(+0.02639900v \ +0.03381398\vartheta \ +0.00000007Q_\phi \ -0.00000166\dfrac{\partial N_{x\phi}}{\partial x} \ +0.00067430 \right) \\[2ex]
\left(+1.4277324v \ +4.6644548\vartheta \ -0.00003752Q_\phi \ -0.00042199\dfrac{\partial N_{x\phi}}{\partial x} \ +0.13184094 \right)
\end{bmatrix}
$$

8. BOUNDARY CONDITIONS

The boundary conditions at the line of symmetry are

$$\vartheta = 0$$

$$Q_\phi = 0$$

$$\frac{\partial N_{x\phi}}{\partial x} = 0$$

$$v = 0$$

$$-0.29118035v \ +1.0\vartheta \quad -0.00023281Q_\phi \ -0.00033810\frac{\partial N_{x\phi}}{\partial x} = -0.10651252 \tag{1}$$

$$17,961.556v \ +5,145.2341\vartheta \quad -0.36302710Q_\phi \ +9.9146895\frac{\partial N_{x\phi}}{\partial x} = 2,037.8073 \tag{2}$$

$$24,318.210v \ +16,546.096\vartheta \quad -0.65027410Q_\phi \ +2.7534896\frac{\partial N_{x\phi}}{\partial x} = 170.67900 \tag{3}$$

$$1.4277324v \ +4.6644548\vartheta \quad -0.00003752Q_\phi \ -0.00042199\frac{\partial N_{x\phi}}{\partial x} = -0.13184094 \tag{4}$$

Solving the above four equations, the values of the four unknowns v, ϑ, Q_ϕ, and $\partial N_{x\phi}/\partial x$ at the origin are obtained.

$$v = -0.01646998$$

$$\vartheta = -0.00046387$$

$$Q_\phi = +127.18159$$

$$\frac{\partial N_{x\phi}}{\partial x} = +240.26880$$

9. STRESSES AT VARIOUS POINTS IN THE SHELL

The values of the unknowns are the elements of $\{F(0)\}$. Once $\{F(0)\}$ is evaluated, the stresses at various points of the shell are calculated using the action matrix $\{F(\phi)\}$ given in step 6 above. The quantities $\partial N_{x\phi}/\partial x$ and $\partial u/\partial x$ are multiplied by l/π and Ed, respectively, to obtain $N_{x\phi}$ and N_x. The stresses are shown tabulated below.

	0°	10°	20°	30°	35°
N_x, lb/ft	−2,355	−9,644	−13,477	−14,836	−14,934
$N_{x\phi}$, lb/ft	+6,367	+5,327	+3,384	+1,084	0
N_ϕ, lb/ft	−89	−930	−1,560	−1,865	−1,863
Q_ϕ, lb/ft	+127	−22	−62	−29	0
M_ϕ, lb ft/ft	0	+184	−33	−249	−281

22-9   Application to Cylindrical Shells with Prestressed Edge Beams

Let us suppose that the edge beam of an interior cylindrical shell of a multiple group is prestressed by a prestressing force H applied at an eccentricity e at midspan (Fig. 22-5). The eccentricity of the tendon is

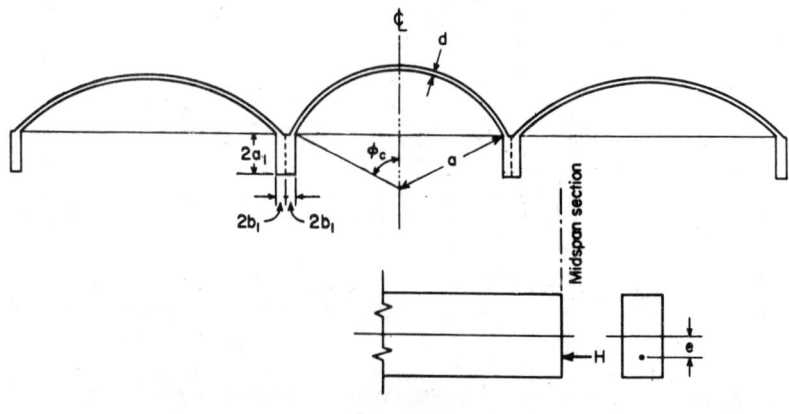

Fɪɢ. 22-5

assumed to vary parabolically, and becomes e_1 above the centroid of the edge beam at the supports of the shell. We have already seen in Chapter 11 that the relevant boundary conditions for such a shell are:

(i) The horizontal deflection at its junction with the edge beam is zero, i.e., $(v)_{\phi=0} \cos \phi_c + (w)_{\phi=0} \sin \phi_c = 0$.

(ii) The change of slope of the tangent is zero, i.e., $(\vartheta)_{\phi=0} = 0$.

(iii) $E(\partial u/\partial x)_{\phi=0}$ = stress in the edge beam at its junction with the shell.

(iv) $(v)_{\phi=0} \sin \phi_c - (w)_{\phi=0} \cos \phi_c$ = vertical deflection of the edge beam at its junction with the shell.

In formulating boundary conditions (iii) and (iv), the influence of the prestressing force is to be reckoned. Boundary conditions (iii) and (iv) take the following forms after some simplification:

Boundary condition (iii):

$$\left(\frac{\partial u}{\partial x}\right)_{\phi=0} = \left(\frac{l}{\pi}\right)^2 \frac{1}{AE} \left(\frac{\partial N_{x\phi}}{\partial x}\right)_{\phi=0} + \left(\frac{l}{\pi}\right)^2 \left(\frac{\partial N_{x\phi}}{\partial x}\right)_{\phi=0} \frac{a_1^2}{EI}$$

$$+ [(N_\phi)_{\phi=0} \sin \phi_c - (Q_\phi)_{\phi=0} \cos \phi_c] \left(\frac{l}{\pi}\right)^2 \frac{a_1}{EI}$$

$$+ \frac{4}{\pi} \frac{8H(e+e_1)}{l^2} \left(\frac{l}{\pi}\right)^2 \frac{a_1}{EI}$$

$$- \frac{4}{\pi} He_1 \frac{a_1}{EI} - \frac{4}{\pi} H \frac{1}{AE} - \frac{4}{\pi} W \left(\frac{l}{\pi}\right)^2 \frac{a_1}{EI} \tag{22-69}$$

Boundary condition (iv) gives:

$$(w)_{\phi=0} = - [(N_\phi)_{\phi=0} \sin \phi_c - (Q_\phi)_{\phi=0} \cos \phi_c] \left(\frac{l}{\pi}\right)^4 \cos \phi_c$$

$$- \left(\frac{l}{\pi}\right)^4 \left(\frac{\partial N_{x\phi}}{\partial x}\right)_{\phi=0} \frac{a_1}{EI} \cos \phi_c$$

$$+ \frac{4}{\pi} W \left(\frac{l}{\pi}\right)^4 \frac{1}{EI} \cos \phi_c - \frac{4}{\pi} \frac{8H(e+e_1)}{l^2} \left(\frac{l}{\pi}\right)^4 \frac{1}{EI} \cos \phi_c$$

$$+ \frac{4}{\pi} He_1 \left(\frac{l}{\pi}\right)^2 \frac{1}{EI} \cos \phi_c \tag{22-70}$$

It is to be noted in this formulation that $2b_1$ must be taken as *half* the actual width of the edge beam because this member is common to two adjacent shells. The action matrix at the origin may now be written as

$$
\begin{bmatrix} w \\ \vartheta \\ M_\phi \\ Q_\phi \\ N_\phi \\ \dfrac{\partial N_{x\phi}}{\partial x} \\ \dfrac{\partial u}{\partial x} \\ v \end{bmatrix}
=
\begin{bmatrix}
\left[-\sin\phi_c\cos\phi_c\left(\frac{l}{\pi}\right)^4\frac{1}{EI}\right] & \left[\cos^2\phi_c\left(\frac{l}{\pi}\right)^4\frac{1}{EI}\right] & \left[-\left(\frac{l}{\pi}\right)^4\frac{a_1}{EI}\cos\phi_c\right] & 0 \\
0 & 0 & 0 & 0 \\
0 & 0 & 0 & 1 \\
0 & 1 & 0 & 0 \\
1 & 0 & 0 & 0 \\
0 & 0 & 1 & 0 \\
\left[\left(\frac{l}{\pi}\right)^2\frac{a_1}{EI}\sin\phi_c\right] & \left[-\left(\frac{l}{\pi}\right)^2\frac{a_1}{EI}\cos\phi_c\right] & \left[\left(\frac{l}{\pi}\right)^2\left(\frac{1}{AE}+\frac{a_1^2}{EI}\right)\right] & 0 \\
\left[\sin^2\phi_c\left(\frac{l}{\pi}\right)^4\frac{1}{EI}\right] & \left[-\sin\phi_c\cos\phi_c\left(\frac{l}{\pi}\right)^4\frac{1}{EI}\right] & \left[\left(\frac{l}{\pi}\right)^4\frac{a_1}{EI}\sin\phi_c\right] & 0
\end{bmatrix}
\begin{bmatrix} N_\phi \\ Q_\phi \\ \dfrac{\partial N_{x\phi}}{\partial x} \\ M_\phi \end{bmatrix}
$$

$$
+
\begin{bmatrix}
\left\{\frac{4}{\pi}W\left(\frac{l}{\pi}\right)^4\frac{1}{EI}\cos\phi_c - \frac{4}{\pi}\left(\frac{l}{\pi}\right)^2\frac{H}{EI}\left[\frac{8(e+e_1)}{\pi^2}-e_1\right]\cos\phi_c\right\} \\
0 \\
0 \\
0 \\
0 \\
0 \\
\left\{-\frac{4}{\pi}W\left(\frac{l}{\pi}\right)^2\frac{a_1}{EI}+\frac{4}{\pi}\frac{H}{E}\left[\frac{8(e+e_1)}{\pi^2}\frac{a_1}{I}-\frac{e_1 a_1}{I}-\frac{1}{A}\right]\right\} \\
\left\{-\frac{4}{\pi}W\left(\frac{l}{\pi}\right)^4\frac{1}{EI}\sin\phi_c+\frac{4}{\pi}\left(\frac{l}{\pi}\right)^2\frac{H}{EI}\left[\frac{8(e+e_1)}{\pi^2}-e_1\right]\sin\phi_c\right\}
\end{bmatrix}
\tag{22-71}
$$

i.e.,
$$\{F(0)\} = [K_0]\{\bar{F}(0)\} + \{E\} \tag{22-72}$$

The action matrix at any point may be found using (22-65). The boundary conditions at the line of symmetry are the same as those listed in (22-66). The rest of the solution proceeds in the same way as for the unprestressed shell. In Design Example 22-2, these principles are applied to the same prestressed shell which we discussed in Chapter 11. The computations were carried out on an IBM 1620 machine.

Design Example 22-2

ANALYSIS OF A PRESTRESSED INTERIOR SHELL
BY MATRIX-PROGRESSION METHOD USING THE SCHORER THEORY

1. DATA

Geometry

Shell

Span l = 163.75 ft
Radius a = 26.375 ft
Thickness d = 0.25 ft
Semicentral angle ϕ_c = 39°30′ = 0.689405 radian

Edge Beam

Depth $2a_1$ = 7.25 ft
Width $2b_1$ = 0.2917 ft = $3\frac{1}{2}$ in.

Loads

Dead weight = 37.5 psf
Waterproofing = 5.0 psf
Live load = 12.5 psf

Total load g = 55.0 psf of shell surface

2. GEOMETRICAL PROPERTIES AND OTHER QUANTITIES

$$\sin \phi_c = 0.63608$$

$$\cos \phi_c = 0.77162$$

$$EI \text{ of edge beam} = 360 \times 10^6 \frac{(0.2917 \times 7.25^3)}{12}$$

$$= 333.44336 \times 10^7$$

3. MATRIX OF INITIAL RELATIONS

The initial-relations matrix equation (22-54) governing the shell actions is first written down.

$$\frac{d}{d\phi}\begin{bmatrix} w \\ \vartheta \\ M_\phi \\ Q_\phi \\ N_\phi \\ \dfrac{\partial N_{x\phi}}{\partial x} \\ \dfrac{\partial u}{\partial x} \\ v \end{bmatrix} = \begin{bmatrix} 0 & 26.375 & 0 & 0 & 0 & 0 & 0 & 0 \\ 0 & 0 & -0.00005627 & 0 & 0 & 0 & 0 & 0 \\ 0 & 0 & 0 & 26.375 & 0 & 0 & 0 & 0 \\ 0 & 0 & 0 & 0 & -1.0 & 0 & 0 & 0 \\ 0 & 0 & 0 & 0 & 0 & -26.375 & 0 & 0 \\ 0 & 0 & 0 & 0 & 0 & 0 & 873{,}719.68 & 0 \\ 0 & 0 & 0 & 0 & 0 & 0 & 0 & 0.00970800 \\ 1.0 & 0 & 0 & 0 & 0 & 0 & 0 & 0 \end{bmatrix}\begin{bmatrix} w \\ \vartheta \\ M_\phi \\ Q_\phi \\ N_\phi \\ \dfrac{\partial N_{x\phi}}{\partial x} \\ \dfrac{\partial u}{\partial x} \\ v \end{bmatrix}$$

4. LOADING-SOLUTION MATRIX

Using the loading-solution matrix (22-59) giving the membrane stresses, the membrane actions at the origin, i.e., at $\phi = 0$, are given below.

$$\{P(0)\} = \begin{bmatrix} w \\ \vartheta \\ M_\phi \\ Q_\phi \\ N_\phi \\ \dfrac{\partial N_{x\phi}}{\partial x} \\ \dfrac{\partial u}{\partial x} \\ v \end{bmatrix} = \begin{bmatrix} 0.01273965 \\ 0 \\ 0 \\ 0 \\ -1{,}425.1853 \\ 89.086794 \\ -0.00012369 \\ -0.01050176 \end{bmatrix}$$

5. PRESTRESSING FORCE

The analysis will be carried out for a prestressing force of 500,000 lb. The effect of the variation of prestressing force on the stresses in the shell and the criterion for selecting the optimum prestressing force have been discussed in detail in Design Example 11-1.

$H =$ prestressing force = 500,000 lb

$e =$ eccentricity at midspan below the centroid of the edge beam = 2.958 ft

$e_1 =$ eccentricity at support sections above the centroid of the edge beam = 1.625 ft

6. ACTION MATRIX AT THE ORIGIN IN TERMS OF THE RESTRAINT MATRIX

Equation (22-71) for the action matrix at the origin is now written down.

$$
\begin{bmatrix} w \\ \vartheta \\ M_\phi \\ Q_\phi \\ N_\phi \\ \dfrac{\partial N_{x\phi}}{\partial x} \\ \dfrac{\partial u}{\partial x} \\ v \end{bmatrix} =
\begin{bmatrix}
-0.00108635 & 0.00131785 & -0.00619110 & 0 \\
0 & 0 & 0 & 0 \\
0 & 0 & 0 & 1 \\
0 & 1 & 0 & 0 \\
1 & 0 & 0 & 0 \\
0 & 0 & 1 & 0 \\
0.00000188 & -0.00000228 & 0.00001427 & 0 \\
0.00089552 & -0.00108635 & 0.00510355 & 0
\end{bmatrix}
$$

$$
\times \begin{bmatrix} N_\phi \\ Q_\phi \\ \dfrac{\partial N_{x\phi}}{\partial x} \\ M_\phi \end{bmatrix} +
\begin{bmatrix} -0.17412700 \\ 0 \\ 0 \\ 0 \\ 0 \\ 0 \\ -0.00053509 \\ 0.14353922 \end{bmatrix}
$$

$$
[K_0]\{\bar{F}(0)\} =
\begin{bmatrix}
\left(-0.00108635 N_\phi +0.00131785 Q_\phi -0.00619110 \dfrac{\partial N_{x\phi}}{\partial x} \right) \\
0 \\
M_\phi \\
Q_\phi \\
N_\phi \\
\dfrac{\partial N_{x\phi}}{\partial x} \\
\left(+0.00000188 N_\phi -0.00000228 Q_\phi +0.00001427 \dfrac{\partial N_{x\phi}}{\partial x} \right) \\
\left(+0.00089552 N_\phi -0.00108635 Q_\phi +0.00510355 \dfrac{\partial N_{x\phi}}{\partial x} \right)
\end{bmatrix}
$$

$$[K_0]\{F(0)\} + \{E\}$$

$$
=
\begin{bmatrix}
\left(-0.00108635N_\phi +0.00131785Q_\phi -0.00619110\dfrac{\partial N_{x\phi}}{\partial x} -0.17412700\right) \\
0 \\
M_\phi \\
Q_\phi \\
N_\phi \\
\dfrac{\partial N_{x\phi}}{\partial x} \\
\left(+0.00000188N_\phi -0.00000288Q_\phi +0.00001427\dfrac{\partial N_{x\phi}}{\partial x} -0.00053509\right) \\
\left(+0.00089552N_\phi -0.00108635Q_\phi +0.00510355\dfrac{\partial N_{x\phi}}{\partial x} +0.14353922\right)
\end{bmatrix}
$$

7. ACTION MATRIX AT ANY ANGLE ϕ

The action matrix at any angle ϕ from the origin is given by Equation (22-65). I.e.,

$$\{F(\phi)\} = e^{[A]\phi}([K_0]\{F(0)\} + \{E\} - \{P(0)\}) + \{P(\phi)\}$$

At the crown of the shell, which is also the line of symmetry, $e^{[A]\phi_c}$ is calculated and given below.[1]

$$
e^{[A]\phi_c} =
\begin{bmatrix}
+0.98891833 & +18.160670 & -0.00035258 & -0.00213737 \\
-0.00487560 & +0.98891833 & -0.00003874 & -0.00035258 \\
+879.83125 & +2,285.4364 & +0.98891833 & +18.160670 \\
+290.32308 & +879.83125 & -0.00487560 & +0.98891833 \\
-2,105.5769 & -7,657.2712 & +0.04950517 & +0.12859389 \\
+463.17458 & +2,105.5769 & -0.01633551 & -0.04950517 \\
+0.00230644 & +0.01398186 & -0.00000014 & -0.00000049 \\
+0.68855620 & +6.2662028 & -0.00008104 & -0.00036839
\end{bmatrix}
$$

$$
\begin{bmatrix}
+0.00036839 & -0.00133973 & -134.49726 & -0.12859389 \\
+0.00008104 & -0.00036839 & -44.380849 & -0.04950517 \\
-6.2662028 & +37.986382 & +5,720,499.7 & +7,657.2712 \\
-0.68855620 & +6.2662028 & +1,258,367.7 & +2,105.5769 \\
+0.98891833 & -18.160670 & -5,474,904.8 & -12,216.230 \\
+0.00487560 & +0.98891833 & +601,605.11 & +2,015.1794 \\
+0.00000006 & -0.00000015 & +0.98891833 & +0.00668450 \\
+0.00005080 & -0.00015394 & -13.246182 & +0.98891833
\end{bmatrix}
$$

[1] Comparison with a closed solution for $e^{[A]\phi}$ using eigenvalues shows that 16 terms are necessary for ensuring acceptable accuracy.

8. SHELL ACTIONS AT THE LINE OF SYMMETRY

Knowing $e^{[A]\phi_c}$, $[K_0]\{F(0)\}$, and $\{E\}$ the shell actions $\{F(\phi_c)\}$ at the line of symmetry are obtained from

$$\{F(\phi_c)\} = e^{[A]\phi_c}([K_0]\{F(0)\} + \{E\} - \{P(0)\}) + \{P(\phi_c)\}$$

in which

$[K_0]\{F(0)\} + \{E\} - \{P(0)\}$

$$= \begin{bmatrix} \left(-0.00108635N_\phi +0.00131785Q_\phi -0.00619110\dfrac{\partial N_{x\phi}}{\partial x} -0.18686665\right) \\ 0 \\ M_\phi \\ Q_\phi \\ (N_\phi +1{,}425.1853) \\ \left(\dfrac{\partial N_{x\phi}}{\partial x} -89.086797\right) \\ \left(+0.00000188N_\phi -0.00000228Q_\phi +0.00001427\dfrac{\partial N_{x\phi}}{\partial x} -0.00041140\right) \\ \left(+0.00089552N_\phi -0.00108635Q_\phi +0.00510355\dfrac{\partial N_{x\phi}}{\partial x} +0.15404098\right) \end{bmatrix}$$

$\therefore \ e^{[A]\phi_c}([K_0]\{F(0)\} + \{E\} - \{P(0)\}) =$

$$\begin{bmatrix} \left(-0.00107373N_\phi -0.00038793Q_\phi -0.01003832\dfrac{\partial N_{x\phi}}{\partial x} -0.00035258M_\phi +0.49510974\right) \\ \left(-0.00004137N_\phi -0.00020409Q_\phi -0.00122435\dfrac{\partial N_{x\phi}}{\partial x} -0.00003874M_\phi +0.15985622\right) \\ \left(+10.381148N_\phi -2.0341816Q_\phi +153.27299\dfrac{\partial N_{x\phi}}{\partial x} +0.98891833M_\phi -13{,}652.875\right) \\ \left(+3.2454717N_\phi -3.7834390Q_\phi +33.176654\dfrac{\partial N_{x\phi}}{\partial x} -0.00487560M_\phi -1{,}787.1549\right) \\ \left(-17.948129N_\phi +23.101069Q_\phi -145.61982\dfrac{\partial N_{x\phi}}{\partial x} +0.04950517M_\phi +3{,}791.3045\right) \\ \left(+2.4364488N_\phi -2.9992392Q_\phi +16.993245\dfrac{\partial N_{x\phi}}{\partial x} -0.01633551M_\phi -104.78282\right) \\ \left(+0.00000539N_\phi -0.00000697Q_\phi +0.00003380\dfrac{\partial N_{x\phi}}{\partial x} -0.00000014M_\phi +0.00028571\right) \\ \left(+0.00016349N_\phi -0.00050511Q_\phi +0.00044106\dfrac{\partial N_{x\phi}}{\partial x} -0.00008104M_\phi +0.11522164\right) \end{bmatrix}$$

Hence

$\{F(\phi_e)\} =$

$$
\begin{bmatrix}
\left(-0.00107373N_\phi \quad -0.00038793Q_\phi \quad -0.01003832\dfrac{\partial N_{x\phi}}{\partial x} \quad -0.00035258M_\phi \quad +0.51161991\right) \\[2ex]
\left(-0.00004137N_\phi \quad -0.00020409Q_\phi \quad -0.00122435\dfrac{\partial N_{x\phi}}{\partial x} \quad -0.00003874M_\phi \quad +0.15985622\right) \\[2ex]
\left(+10.381148N_\phi \quad -2.0341816Q_\phi \quad +153.27299\dfrac{\partial N_{x\phi}}{\partial x} \quad +0.98891833M_\phi \quad -13,652.875\right) \\[2ex]
\left(+3.2454717N_\phi \quad -3.7834390Q_\phi \quad +33.176654\dfrac{\partial N_{x\phi}}{\partial x} \quad -0.00487560M_\phi \quad -1,787.1549\right) \\[2ex]
\left(-17.948129N_\phi \quad +23.101069Q_\phi \quad -145.61982\dfrac{\partial N_{x\phi}}{\partial x} \quad +0.04950517M_\phi \quad +1,944.3114\right) \\[2ex]
\left(+2.4364488N_\phi \quad -2.9992392Q_\phi \quad +16.993245\dfrac{\partial N_{x\phi}}{\partial x} \quad -0.01633551M_\phi \quad -104.78282\right) \\[2ex]
\left(+0.00000539N_\phi \quad -0.00000697Q_\phi \quad +0.00003380\dfrac{\partial N_{x\phi}}{\partial x} \quad -0.00000014M_\phi \quad +0.00012541\right) \\[2ex]
\left(+0.00016349N_\phi \quad -0.00050511Q_\phi \quad +0.00044106\dfrac{\partial N_{x\phi}}{\partial x} \quad -0.00008104M_\phi \quad +0.11522164\right)
\end{bmatrix}
$$

9. BOUNDARY CONDITIONS

The boundary conditions at the line of symmetry are, as in Design Example 22-1,

$$\vartheta = 0$$

$$Q_\phi = 0$$

$$\frac{\partial N_{x\phi}}{\partial x} = 0$$

$$v = 0$$

i.e.,

$$-0.00004137N_\phi \quad -0.00020409Q_\phi \quad -0.00122435\frac{\partial N_{x\phi}}{\partial x} \quad -0.00003874M_\phi = -0.15985622 \tag{1}$$

$$3.2454717N_\phi \quad -3.7834390Q_\phi \quad +33.176654\frac{\partial N_{x\phi}}{\partial x} \quad -0.00487560M_\phi = 1,787.1549 \tag{2}$$

$$2.4364488N_\phi \quad -2.9992392Q_\phi \quad +16.993245\frac{\partial N_{x\phi}}{\partial x} \quad -0.01633551M_\phi = 104.78282 \tag{3}$$

$$0.00016349N_\phi \quad -0.00050511Q_\phi \quad +0.00044106\frac{\partial N_{x\phi}}{\partial x} \quad -0.00008104M_\phi = -0.11522164 \tag{4}$$

Solving the above four equations, the values of the tour unknowns N_ϕ, Q_ϕ, $\partial N_{x\phi}/\partial x$, and M_ϕ at the origin are obtained.

$$N_\phi = -969.37607$$

$$Q_\phi = +57.335553$$

$$\frac{\partial N_{x\phi}}{\partial x} = +155.22773$$

$$M_\phi = -46.410623$$

10. STRESSES AT VARIOUS POINTS IN THE SHELL

The values f the unknowns are the elements of $\{\bar{F}(0)\}$. Having evaluated $\{\bar{F}(0)\}$, the stresses at various points of the shell are calculated using the action matrix $\{F(\phi)\}$ given in step 7 above. They are tabulated below.

	0°	10°	20°	30°	39.5°
N_x, lb/ft	−24,389	−23,649	−23,056	−22,652	−22,515
$N_{x\phi}$, lb/ft	+8,091	+5,971	+3,910	+1,893	0
N_ϕ, lb/ft	−969	−1,407	−1,710	−1,885	−1,939
Q_ϕ, lb/ft	+57	+1.5	−18	−14	0
M_ϕ, lb ft/ft	−46	+72	+24	−55	−86

22-10 Application to Multiple Shells

As an example, consider the multiple-shell assembly shown in Fig. 22-6. Let us suppose that we are required to determine the influence of partial loading g per unit area over the part BD. Shell 1 and shell 3 have identical geometrical properties. Shell 2 has geometrical dimensions which are different from those of shells 1 and 3.

Proceeding in the same manner as for the single shell already discussed, we start from O and set up the action matrix for any point on the first shell. This matrix may be written as

$$\{F_1(\phi)\} = [G_0(\phi)]\{F_1(0)\} \tag{22-73}$$

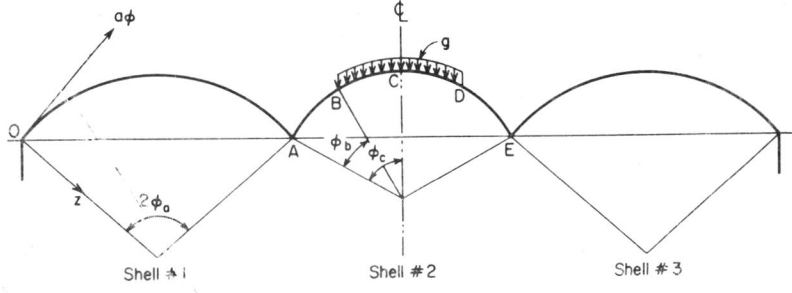

FIG. 22-6

where $[G_0(\phi)]$ is the distribution matrix at O. We already know that $[G_0(\phi)] = e^{[A]\phi}$, where $[A]$ is the matrix already defined in (22-54).

The suffix 1 assigned to $\{F\}$ denotes that the action matrix relates to shell 1. Because there is no loading on shell 1, there is no loading-solution matrix. The action matrix at the point A may be written as

$$\{F_1(2\phi_a)\} = [G_0(2\phi_a)]\{F_1(0)\} \qquad (22\text{-}74)$$

Now, before we proceed to shell 2, the actions at A of shell 1 have to be expressed in the coordinates of shell 2. The rotation of coordinates that takes place at A is illustrated in Fig. 22-7. It is easily seen that the

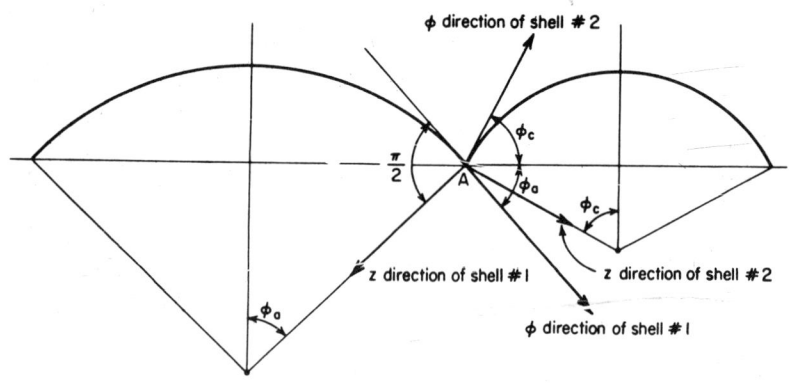

Fig. 22-7

ϕ-Z coordinate system of shell 1 rotates counterclockwise through an angle of $(\phi_a + \phi_c)$ where ϕ_a and ϕ_c, respectively, are the semicentral angles of shells 1 and 2. The *rotation matrix* relating new and old coordinates when an orthogonal system of coordinates is rotated through an angle θ in the counterclockwise direction (Fig. 22-8) is given by

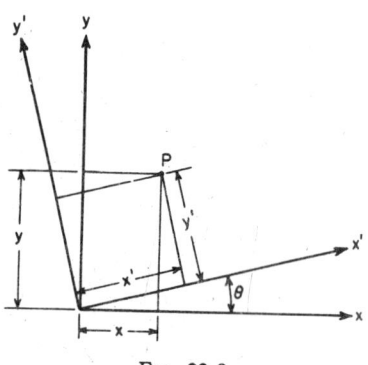

Fig. 22-8

$$\begin{bmatrix} x' \\ y' \end{bmatrix} = \begin{bmatrix} \cos\theta & \sin\theta \\ -\sin\theta & \cos\theta \end{bmatrix} \begin{bmatrix} x \\ y \end{bmatrix}$$

$$(22\text{-}75)$$

where x', y' are the new and x, y the old coordinates. On account of the rotation of the coordinate system, the shell actions transform as shown in matrix (22-76).

$$
\begin{bmatrix}
w \\
\vartheta \\
M_\phi \\
Q_\phi \\
N_\phi \\
\dfrac{\partial N_{x\phi}}{\partial x} \\[4pt]
\dfrac{\partial u}{\partial x} \\[4pt]
v
\end{bmatrix}
=
\begin{bmatrix}
\cos(\phi_a+\phi_c) & 0 & 0 & 0 & 0 & 0 & 0 & \sin(\phi_a+\phi_c) \\
0 & 1 & 0 & 0 & 0 & 0 & 0 & 0 \\
0 & 0 & 1 & 0 & 0 & 0 & 0 & 0 \\
0 & 0 & 0 & \cos(\phi_a+\phi_c) & \sin(\phi_a+\phi_c) & 0 & 0 & 0 \\
0 & 0 & 0 & -\sin(\phi_a+\phi_c) & \cos(\phi_a+\phi_c) & 0 & 0 & 0 \\
0 & 0 & 0 & 0 & 0 & 1 & 0 & 0 \\
0 & 0 & 0 & 0 & 0 & 0 & 1 & 0 \\
-\sin(\phi_a+\phi_c) & 0 & 0 & 0 & 0 & 0 & 0 & \cos(\phi_a+\phi_c)
\end{bmatrix}
\begin{bmatrix}
w \\
\vartheta \\
M_\phi \\
Q_\phi \\
N_\phi \\
\dfrac{\partial N_{x\phi}}{\partial x} \\[4pt]
\dfrac{\partial u}{\partial x} \\[4pt]
v
\end{bmatrix}
$$

$$[O_A]$$

Action matrix of shell 2 at A = Action matrix of shell 1 at A

(22-76)

Hence action matrix $\{F_2(0)\}$ at the origin A of shell 2 may be written as

$$\{F_2(0)\} = [O_A]\{F_1(2\phi_a)\} \qquad (22\text{-}77)$$

Now the action matrix $\{F_2(\phi)\}$ of shell 2 in terms of the action matrix at its origin A may be written as

$$\{F_2(\phi)\} = [G_a(\phi)]\{F_2(0), \qquad 0 < \phi < \phi_b \qquad (22\text{-}78)$$

where $[G_a(\phi)]$ is the distribution matrix at A for shell 2, the angle ϕ now being measured from A. The action matrix at B may therefore be written as

$$\{F_2(\phi_b)\} = [G_a(\phi_b)]\{F_2(0)\} \qquad (22\text{-}79)$$

Starting from B, the load begins to act and hence the loading-solution matrix needs to be introduced. Hence the action matrix in the region B to D assumes the form

$$\{F_2(\phi)\} = [G_b(\phi)]([G_a(\phi_b)]\{F_2(0)\} - \{P(\phi_b)\}) + \{P(\phi)\} \qquad (22\text{-}80)$$

where $\{P(\phi)\}$ is the membrane solution corresponding to the load. This expression may be simplified as

$$\{F(\phi)\} = [G_a(\phi)]\{F_2(0)\} - [G_a(\phi - \phi_b)]\{P(\phi_b)\} + \{P(\phi)\} \qquad (22\text{-}81)$$

The action matrix at the line of symmetry C may now be written as

$$\{F(\phi_c)\} = [G_a(\phi_c)]\{F_2(0)\} - [G_a(\phi_c - \phi_b)]\{P(\phi_b)\} + \{P(\phi_c)\} \qquad (22\text{-}82)$$

Four equations in the four unknown actions may be formed by pre-multiplying (22-82) by the isolation matrix $[I_c]$ defined in Equation (22-67). Thus

$$[I_c]\{F(\phi_c)\} = \{0\} \qquad (22\text{-}83)$$

Once the unknowns are determined, the action matrix at any point on the three shells may be readily written down and the actions found.

22-11   Formulation in Terms of Other Theories

In the above presentation, the formulation was in terms of the Schorer theory. It is equally easy to apply the principles of matrix progression to the D-K-J theory.

APPLICATION OF TRANSFER MATRICES TO FOLDED PLATES

22-12 The Ivar Holand Method

A procedure that is particularly suitable for the analysis of folded plates on a digital computer is that described by Ivar Holand [158]. The merit of this technique will become evident if it is compared with our usual procedures. For instance, if the Whitney method is used, we have already seen that $(n - 3)$ simultaneous equations need to be solved to analyze a folded-plate structure consisting of n plates. Thus, if there are 10 plates, we are called upon to solve 7 equations. But a more careful examination would reveal that there should be no more than four equations in four unknowns, irrespective of the number of plates involved. The transfer-matrix method is built around this idea. The identity of the matrix-progression and the transfer-matrix procedures will become immediately evident if it is recalled that we had also dealt with only four unknowns in analyzing cylindrical shells by matrix progression. The transfer matrix is nothing but the distribution matrix. Except for some minor modifications in notation and some additional explanations, the presentation given below closely follows that of Ivar Holand.

Let us consider the equilibrium of one of the plates of a simply supported folded-plate structure of span l (Fig. 22-9). Free-body diagrams (Figs. 22-10 and 22-11) show the actions of the normal and tangential groups of forces acting on the plate under consideration. As usual, the plate will be assumed to behave as a beam under both "slab" and "plate" actions. It may be noted that each junction in the structure where two plates meet is designated by two numbers. Because the plate functions as a beam under "slab action," the actions at $(n + 1)$ may be written down in the terms of the actions at n by means of the distribution matrix developed in (22-9). Thus

$$
\begin{bmatrix} w \\ \vartheta \\ M \\ Q \end{bmatrix}_{n+1} =
\begin{bmatrix}
1 & h & -\dfrac{h^2}{2EI} & -\dfrac{h^3}{6EI} \\
0 & 1 & -\dfrac{h}{EI} & -\dfrac{h^2}{2EI} \\
0 & 0 & 1 & h \\
0 & 0 & 0 & 1
\end{bmatrix}
\begin{bmatrix} w \\ \vartheta \\ M \\ Q \end{bmatrix}_{n}
\tag{22-84}
$$

This is known as the *transfer matrix* which enables the actions at the forward end of the plate to be written in terms of the actions at its tail end. We may now proceed to build up a similar transfer matrix for the actions of the tangential group. The actions in which we are interested are N_x, N_{xy}, N_y, and v.

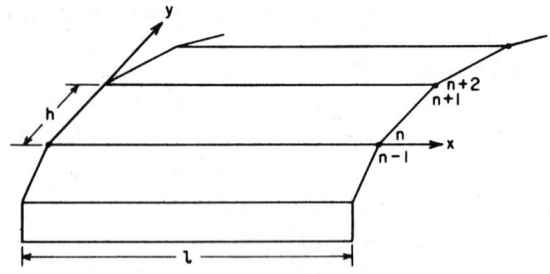

FIG. 22-9

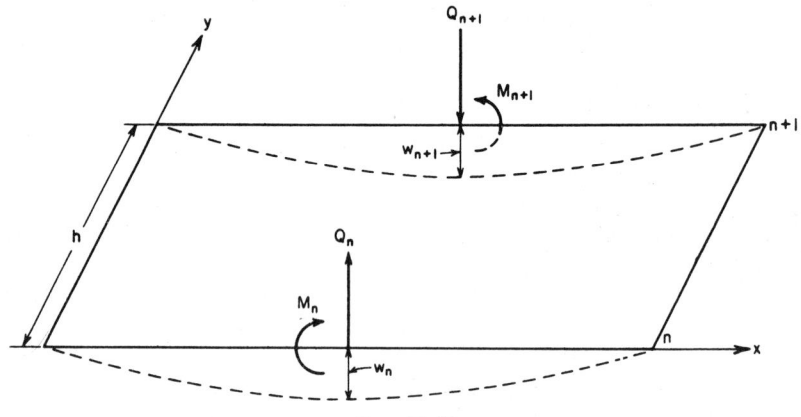

FIG. 22-10

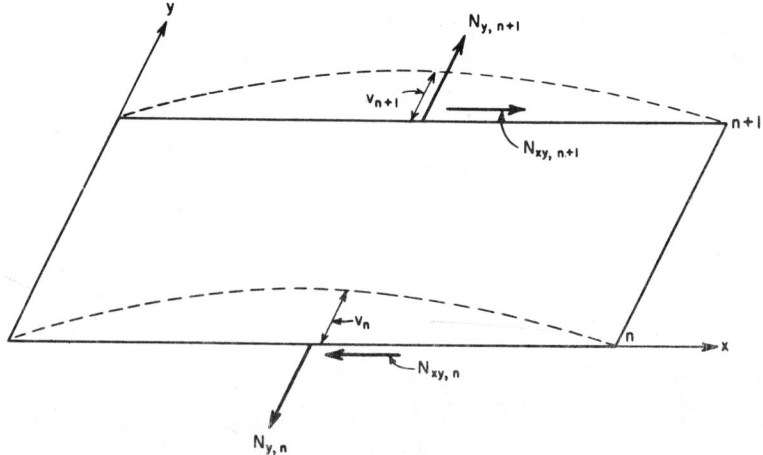

FIG. 22-11

If the plate loads are constant in the x direction, which is generally the case, the actions vary as noted below.

$$N_y = \overline{N_y} \tag{22-85}$$

$$N_{xy} = \overline{N_{xy}} \left(1 - 2\frac{x}{l}\right) \tag{22-86}$$

$$N_x = \overline{N_x} \, 4\frac{x}{l}\left(1 - \frac{x}{l}\right) \tag{22-87}$$

$$v = \bar{v}\,\frac{16}{5}\frac{x}{l}\left(1 - 2\frac{x^2}{l^2} + \frac{x^3}{l^3}\right) \tag{22-88}$$

These relations are the same as those for a simply supported beam which is uniformly loaded. The quantities with the bar denote the maximum values of the actions. It is to be noted that while N_x and v attain their maxima at $x = l/2$, N_{xy} is maximum at the supports. As beam action is assumed, we may write down the relation

$$N_x = N_{x,n} - EI\,\frac{d^2v}{dx^2} - \frac{yd}{I} \tag{22-89}$$

where N_x is the longitudinal stress at a distance y from the edge n (Fig. 22-9) and d is the thickness of the plate.

Substituting for v from relation (22-88) and carrying out two differentiations with respect to x,

$$N_x = N_{x,n} - \frac{192}{5}\frac{Eyd}{l}\left(\frac{x^2}{l^3} - \frac{x}{l^2}\right)\overline{v_n} \tag{22-90}$$

At the midspan section where $x = l/2$, this relation takes the form

$$\overline{N_x} = \overline{N_{x,n}} + \frac{48}{5}\frac{Edh}{l^2}\,\overline{v_n} \tag{22-91}$$

We may now make use of the equilibrium equation

$$\frac{\partial N_x}{\partial x} + \frac{\partial N_{xy}}{\partial y} = 0$$

or

$$\frac{\partial N_{xy}}{\partial y} = -\frac{\partial N_x}{\partial x}$$

From Equation (22-90), we may write N_x as

$$N_x = \overline{N_{x,n}}\frac{4x}{l}\left(1 - \frac{x}{l}\right) - \frac{192}{5}\frac{Eyd}{l}\left(\frac{x^2}{l^3} - \frac{x}{l^2}\right)\overline{v_n} \tag{22-92}$$

Hence

$$-\frac{\partial N_x}{\partial x} = \frac{\partial N_{xy}}{\partial y} = -\overline{N_{x,n}}\left(\frac{4}{l} - \frac{8x}{l^2}\right) + \frac{192}{5}\frac{Eyd}{l}\left(\frac{2x}{l^3} - \frac{1}{l^2}\right)\overline{v_n} \tag{22-93}$$

$$\left(\frac{\partial N_{xy}}{\partial y}\right)_{x=0} = -\frac{4}{l}\,\overline{N_{x,n}} - \frac{192}{5}\frac{Eyd}{l^3}\,\overline{v_n} \tag{22-94}$$

This relation enables us to find $\overline{N_{xy,n+1}}$. Thus

$$N_{xy,n+1} = \overline{N_{xy,n}} + \int_0^h \left(-\frac{4}{l}\,\overline{N_{x,n}} - \frac{192}{5}\,\frac{E\overline{v_n}d}{l^3}\,y\,dy\right)$$

or

$$\overline{N_{xy,n+1}} = \overline{N_{xy,n}} - \frac{4h}{l}\,N_{x,n} - \frac{96}{5}\,\frac{E\,dh^2}{l^3}\,\overline{v_n} \tag{22-95}$$

From relation (22-93) we may write the expression for N_{xy} at any point at a distance y from the edge n. Thus

$$N_{xy} = N_{xy,n} - \overline{N_{x,n}}\left(\frac{4}{l} - \frac{8x}{l^2}\right)y + \frac{96}{5}\,\frac{Ed}{l}\left(\frac{2x}{l^3} - \frac{1}{l^2}\right)y^2\,\overline{v_n}$$

$$= \overline{N_{xy,n}}\left(1 - \frac{2x}{l}\right) - \overline{N_{x,n}}\left(\frac{4}{l} - \frac{8x}{l^2}\right)y$$

$$+ \frac{96}{5}\,\frac{Ed}{l}\left(\frac{2x}{l^3} - \frac{1}{l^2}\right)y^2\overline{v_n} \tag{22-96}$$

Hence

$$-\frac{\partial N_{xy}}{\partial x} = \frac{\partial N_y}{\partial y} = -\frac{8}{l^2}\,\overline{N_{x,n}}\,y + \frac{2}{l}\,\overline{N_{xy,n}} - \frac{192}{5}\,\frac{Ed}{l^4}\,y^2\,\overline{v_n}$$

and

$$\left(\frac{\partial N_y}{\partial y}\right)_{x=l/2} = -\frac{8}{l^2}\,\overline{N_{x,n}}\,y + \frac{2}{l}\,\overline{N_{xy,n}} - \frac{192}{5}\,\frac{Ed}{l^4}\,y^2\,\overline{v_n} \tag{22-97}$$

We may therefore write

$$\overline{N_{y,n+1}} = \overline{N_{y,n}} - \frac{4h^2}{l^2}\,\overline{N_{x,n}} + \frac{2h}{l}\,\overline{N_{xy,n}} - \frac{64}{5}\,\frac{E\,dh^3}{l^4}\,\overline{v_n} \tag{22-98}$$

Making use of the above results, we may now express the actions at $n + 1$ in terms the actions at n by means of a transfer matrix thus:

$$
\begin{bmatrix} \overline{N_y} \\[6pt] \overline{N_{xy}} \\[6pt] \overline{N_x} \\[6pt] \bar{v} \end{bmatrix}_{n+1}
=
\begin{bmatrix}
1 & +\dfrac{2h}{l} & -\dfrac{4h^2}{l^2} & -\dfrac{64}{5}\dfrac{Edh^3}{l^4} \\[8pt]
0 & 1 & -\dfrac{4h}{l} & -\dfrac{96}{5}\dfrac{Edh^2}{l^3} \\[8pt]
0 & 0 & 1 & +\dfrac{48}{5}\dfrac{Edh}{l^2} \\[8pt]
0 & 0 & 0 & 1
\end{bmatrix}
\begin{bmatrix} \overline{N_y} \\[6pt] \overline{N_{xy}} \\[6pt] \overline{N_x} \\[6pt] \bar{v} \end{bmatrix}_n
\tag{22-99}
$$

It is now possible to assemble matrices (22-84) and (22-99) into a single transfer matrix. This is strictly permissible only if the quantities of the normal and tangential groups thus combined vary in the same manner along the fold. Although this condition is not satisfied in this case, the inconsistency involved is not of serious consequence. The values of

M and Q calculated in this manner are fairly accurate except near the ends. In organizing the combined matrix, it is an advantage to separate out certain coefficients. The actions containing these coefficients are denoted by the symbol $\widehat{\ }$ placed on their top. d_k and h_k are arbitrarily chosen dimensions. If the structure is made up of several equal plates, d_k and h_k may be chosen as the dimensions of this plate. With these explanations, it is now possible to write down the relation between $\{Z\}_n$ and $\{Z\}_{n+1}$ where $\{Z\}_n$ and $\{Z\}_{n+1}$ are the action matrices at n and $n+1$.

$$
\begin{bmatrix} \hat{N}_y \\ \hat{N}_{xy} \\ \hat{N}_x \\ \hat{v} \\ \hat{w} \\ \hat{\vartheta} \\ \hat{M} \\ \hat{Q} \\ 1 \end{bmatrix}_{n+1}
=
\begin{bmatrix} N_y \\ \dfrac{2h_k}{l}\overline{N}_{xy} \\ \dfrac{4h_k{}^2}{l^2}\overline{N}_x \\ \dfrac{192}{5}\dfrac{E d_k h_k{}^3}{l^4}\bar{v} \\ \dfrac{192}{5}\dfrac{E d_k h_k{}^3}{l^4}w \\ \dfrac{192}{5}\dfrac{E d_k h_k{}^4}{l^4}\vartheta \\ \dfrac{1}{h_k}M \\ Q \\ 1 \end{bmatrix}
=
$$

$$
\begin{bmatrix}
1 & \dfrac{h}{h_k} & -\left(\dfrac{h}{h_k}\right)^2 & -\dfrac{1}{3}\left(\dfrac{h}{h_k}\right)^3\dfrac{d}{d_k} & 0 & 0 & 0 & 0 & 0 \\[2mm]
0 & 1 & -2\left(\dfrac{h}{h_k}\right) & -\left(\dfrac{h}{h_k}\right)^2\dfrac{d}{d_k} & 0 & 0 & 0 & 0 & 0 \\[2mm]
0 & 0 & 1 & \dfrac{h}{h_k}\dfrac{d}{d_k} & 0 & 0 & 0 & 0 & 0 \\[2mm]
0 & 0 & 0 & 1 & 0 & 0 & 0 & 0 & 0 \\[2mm]
0 & 0 & 0 & 0 & 1 & \dfrac{h}{h_k} & -\dfrac{\alpha}{2}\left(\dfrac{h}{h_k}\right)^2\left(\dfrac{d_k}{d}\right)^3 & -\dfrac{\alpha}{6}\left(\dfrac{h}{h_k}\right)^3\left(\dfrac{d_k}{d}\right)^3 & 0 \\[2mm]
0 & 0 & 0 & 0 & 0 & 1 & -\alpha\left(\dfrac{h}{h_k}\right)\left(\dfrac{d_k}{d}\right)^3 & -\dfrac{\alpha}{2}\left(\dfrac{h}{h_k}\right)^2\left(\dfrac{d_k}{d}\right)^3 & 0 \\[2mm]
0 & 0 & 0 & 0 & 0 & 0 & 1 & \dfrac{h}{h_k} & 0 \\[2mm]
0 & 0 & 0 & 0 & 0 & 0 & 0 & 1 & 0 \\[2mm]
0 & 0 & 0 & 0 & 0 & 0 & 0 & 0 & 1
\end{bmatrix}
\begin{bmatrix} \hat{N}_y \\ \hat{N}_{xy} \\ \hat{N}_x \\ \hat{v} \\ \hat{w} \\ \hat{\vartheta} \\ \hat{M} \\ \hat{Q} \\ 1 \end{bmatrix}_{n}
\qquad (22\text{-}100)^*
$$

---

* From "Analysis of Folded Plate Structures by Transfer Matrices," by Ivar Holand, *Proceedings of the International Colloquium on Simplified Calculation Methods*, Brussels, 1961. Courtesy of North-Holland Publishing Co., Amsterdam.

In abbreviated form,

$$\{Z\}_{n+1} = [T]\{Z\}_n \tag{22-101}$$

where $[T]$ is the transfer matrix. The last row of the matrix is meant to take care of loads.

To transform the actions of plate $n - (n + 1)$ at its end$(n + 1)$ to the actions of the plate $(n + 2) - (n + 3)$ at its end $(n + 2)$, we need a *point matrix*. This matrix has also to take care of the changes in the actions brought about by line load and moment applied at the fold. The pairs of quantities $\hat{v}$ and $\hat{w}$, $\hat{N}_y$ and $\hat{Q}$ transform according to the rule developed in (22-75). The line load P applied at the junction $n + 1$, $n + 2$ (Fig. 22-12) contributes a component of $P \cos \theta$ in the direction of $N_{y,n+2}$ and $-P \sin \theta$ in the direction of Q_{n+2}. The

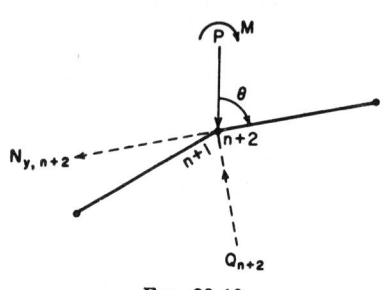

FIG. 22-12

moment M applied at the fold is added on as M/h_k to M_{n+1} to give M_{n+2}. The shear force N_{xy} has to be the same for both the plates at their junction, and the longitudinal stress

$$\frac{N_{x,n+1}}{d_{n+1}} = \frac{N_{x,n+2}}{d_{n+2}}$$

With these explanations, it is now possible to write down the relation between $\{Z\}_{n+1}$ and $\{Z\}_{n+2}$ by means of a *point matrix*.

$$
\begin{bmatrix} \hat{N}_y \\ \hat{N}_{xy} \\ \hat{N}_x \\ \hat{v} \\ \hat{w} \\ \vartheta \\ \hat{M} \\ Q \\ 1 \end{bmatrix}_{n+2}
=
\begin{bmatrix}
\cos\gamma & 0 & 0 & 0 & 0 & 0 & 0 & \sin\gamma & P\cos\theta \\
0 & 1 & 0 & 0 & 0 & 0 & 0 & 0 & 0 \\
0 & 0 & \dfrac{d_{n+2}}{d_{n+1}} & 0 & 0 & 0 & 0 & 0 & 0 \\
0 & 0 & 0 & \cos\gamma & \sin\gamma & 0 & 0 & 0 & 0 \\
0 & 0 & 0 & -\sin\gamma & \cos\gamma & 0 & 0 & 0 & 0 \\
0 & 0 & 0 & 0 & 0 & 1 & 0 & 0 & 0 \\
0 & 0 & 0 & 0 & 0 & 0 & 1 & 0 & \dfrac{M}{h_k} \\
-\sin\gamma & 0 & 0 & 0 & 0 & 0 & 0 & \cos\gamma & -P\sin\theta \\
0 & 0 & 0 & 0 & 0 & 0 & 0 & 0 & 1
\end{bmatrix}
\begin{bmatrix} \hat{N}_y \\ \hat{N}_{xy} \\ \hat{N}_x \\ \hat{v} \\ \hat{w} \\ \vartheta \\ \hat{M} \\ Q \\ 1 \end{bmatrix}_{n+1}
$$

$$\tag{22-102}*$$

* From "Analysis of Folded Plate Structures by Transfer Matrices," by Ivar Holand, *Proceedings of the International Colloquium on Simplified Calculation Methods*, Brussels, 1961. Courtesy of North-Holland Publishing Co., Amsterdam.

In abbreviated form

$$\{Z\}_{n+2} = [F]\{Z\}_{n+1} \tag{22-103}$$

22-13 The Procedure in Outline

We start at edge 1 (Fig. 22-13). If it is a free edge, four out of the eight actions, namely, Q, N_y, N_{xy}, and M, are zero. Hence we

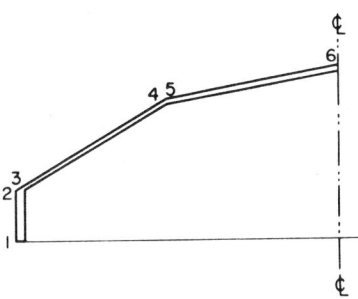

Fig. 22-13

have only four unknown actions. The eight actions at the other end 2 of the first plate are now expressed in terms of these four unknowns by making use of the transfer matrix. Thus

$$\{Z\}_2 = [T]\{Z\}_1 \tag{22-104}$$

We now encounter the first fold. The actions of the next plate at this junction are now expressed in terms of $\{Z\}_2$ by premultiplying it by the point matrix $[F]$. Hence

$$\{Z_3\} = [F]\{Z_2\} \tag{22-105}$$

$$= [F][T]\{Z_1\} \tag{22-106}$$

Thus we have succeeded in writing the action matrix $\{Z_3\}$ in terms of the original four unknowns. This process is continued until we reach the line of symmetry or the other end of the structure where four conditions will be available for evaluating the four unknowns involved. When once these unknowns are found, the actions at any point may be easily determined. To start with, all plates are considered to be rigidly clamped. The resulting unbalanced moments at the fold are applied as external moments M. The line loads P transferred to a fold are easily computed as in normal folded-plate analysis. In Design Example 22-3, the procedure is explained in detail.

Design Example 22-3

ANALYSIS OF A FOLDED PLATE BY THE METHOD OF TRANSFER MATRICES

1. DATA

Geometry

Span = 70 ft

Plate No.	Width h, ft	Horizontal projection, ft	Thickness d, ft	Inclination θ to the vertical
1-2	3	0	0.5000	0°
3-4	10	8.660	0.3333	60°
5-6	10	9.848	0.3333	80°

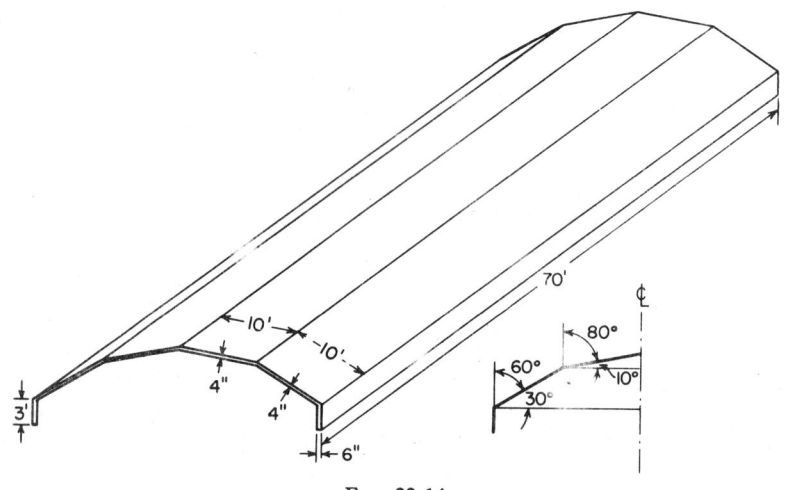

FIG. 22-14

Loading

Dead weight = 12.5 psf
Live load   = 15.0 psf of surface area on inclined plates

The total load per foot length of each plate is calculated as shown below.

Plate No.	Dead weight, kip	Live load, kip	Total load per ft length, kip
1-2	0.225	0	0.225
3-4	0.500	0.150	0.650
5-6	0.500	0.150	0.650

2. TRIGONOMETRIC RATIOS AND OTHER QUANTITIES

Joint or fold	Angle between adjacent plates, γ	$\cos \gamma$	$\sin \gamma$
2-3	60°	0.50000	0.86603
4-5	20°	0.93969	0.34202
6-6′	...	...	...

The width and thickness of plate 3-4 are chosen as the arbitrary dimensions h_k and d_k, respectively. Therefore, $h_k = 10$ ft and $d_k = 0.3333$ ft.

Table of Coefficients

Quantity	Coefficient	
N_y	1.0	1.0
N_{xy}	$\dfrac{2h_k}{l}$	0.28571
N_x	$\dfrac{4h_k^2}{l^2}$	0.08163
v	$\dfrac{192}{5} \dfrac{Ed_k h_k^3}{l^4}$	0.00053306E
w	$\dfrac{192}{5} \dfrac{Ed_k h_k^3}{l^4}$	0.00053306E
ϑ	$\dfrac{192}{5} \dfrac{Ed_k h_k^4}{l^4}$	0.0053306E
M	$\dfrac{1}{h_k}$	0.1
Q	1.0	1.0
1	1.0	1.0

$$\alpha = \frac{7.2(2h_k)^6}{d_k^2 l^4} = 172.76258$$

3. JOINT LOADS AND MOMENTS

The plates are first assumed to be fixed at the joints 2-3, 4-5, 6-6′, 5′-4′, and 3′-2′. The joint loads and the fixed-end moments are calculated on this basis. The total joint loads and the unbalanced moments at the joints are given below.

Joint or fold	Load P, kip	Unbalanced moment M, kip ft
2-3	$\left(0.225 + \dfrac{0.650}{2}\right) = 0.550$	$-0 + \dfrac{(0.650 \times 8.660)}{12} = +0.46908$
4-5	$\left(\dfrac{0.650}{2} + \dfrac{0.650}{2}\right) = 0.650$	$-\dfrac{0.650 \times 8.660}{12} + \dfrac{0.650 \times 9.848}{12} = +0.06435$
6	$\dfrac{0.650}{2} = 0.325$	$-\dfrac{0.650 \times 9.848}{12} = -0.53343$

It may be noted that at fold 6, which is on the line of symmetry, the load and the moment corresponding only to plate 5-6 are written down. This is because the boundary conditions are applied at the line of symmetry.

4. FORMULATION OF MATRICES

Matrix at Edge 1

This being a free edge, four of the stress resultants, namely, N_y, N_{xy}, M, and Q, are known to be zero. The other four stress resultants N_x, v, w, and ϑ are the unknowns. Hence the state of stress at this edge is conveniently expressed in terms of the unknowns in matrix form as given below.

$$
\begin{bmatrix} \hat{N}_y \\ \hat{N}_{xy} \\ \hat{N}_x \\ \hat{v} \\ \hat{w} \\ \hat{\vartheta} \\ \hat{M} \\ \hat{Q} \\ 1 \end{bmatrix}
=
\begin{bmatrix}
0 & 0 & 0 & 0 & 0 \\
0 & 0 & 0 & 0 & 0 \\
1 & 0 & 0 & 0 & 0 \\
0 & 1 & 0 & 0 & 0 \\
0 & 0 & 1 & 0 & 0 \\
0 & 0 & 0 & 1 & 0 \\
0 & 0 & 0 & 0 & 0 \\
0 & 0 & 0 & 0 & 0 \\
0 & 0 & 0 & 0 & 1
\end{bmatrix}
\begin{bmatrix} \hat{N}_x \\ \hat{v} \\ \hat{w} \\ \hat{\vartheta} \\ 1 \end{bmatrix}
$$

In this treatment, we shall be dealing with the reduced quantities $\hat{N}_y$, $\hat{N}_{xy}$, $\hat{N}_x$, etc., instead of the actual quantities N_y, N_{xy}, N_x, etc. The reduced and the actual quantities are related to each other by the respective coefficients tabulated in step 2.

The stress resultants at edge 2 are obtained from those at edge 1 by multiplying the above matrix by transfer matrix 1-2. The stress resultants at edge 3 are obtained by multiplying the matrix of stress resultants at edge 2 by the point matrix 2-3. This process of multiplying alternatively by transfer and point matrices is followed till the edge 6, which is on the line of symmetry, is reached. The entire procedure is arranged in tabular form as shown in Table 22-1.

5. BOUNDARY CONDITIONS

The following boundary conditions are applied at edge 6:

(i) Shear is zero, i.e.,

$$\hat{N}_{xy} = 0$$
$$-2.499\,\hat{N}_x - 2.894\,\hat{v} - 3.583\,\hat{w} - 1.417\,\hat{\vartheta} - 3.305 = 0 \tag{1}$$

(ii) Horizontal displacement is zero, i.e.,

$$v \sin \theta_{(\text{plate 5-6})} + w \cos \theta_{(\text{plate 5-6})} = 0$$

$$(-0.768 \, \hat{N}_x + 0.058 \, \hat{v} + 0.985 \, \hat{w} + 0.637 \, \hat{\vartheta} + 3.305) \sin 80°$$

$$+ (-30.60 \, \hat{N}_x - 9.629 \, \hat{v} - 2.669 \, \hat{w} + 1.139 \, \hat{\vartheta} + 112.685) \cos 80° = 0$$

or

$$-6.070 \, \hat{N}_x - 1.615 \, \hat{v} + 0.506 \, \hat{w} + 0.826 \, \hat{\vartheta} + 22.815 = 0 \qquad (2)$$

(iii) Rotation is zero, i.e.,

$$\hat{\vartheta} = 0$$

$$-65.275 \, \hat{N}_x - 21.955 \, \hat{v} - 8.529 \, \hat{w} - 1.559 \, \hat{\vartheta} + 208.197 = 0 \qquad (3)$$

(iv) The resultant vertical force is zero. i.e.,

$$\hat{Q}_{(\text{edge 6})} \sin \theta - \hat{N}_{v_{(\text{edge 6})}} \cos \theta - \text{applied load}_{(\text{edge 6})} = 0$$

$$(0.522\hat{N}_x + 0.219\hat{v} + 0.099\hat{w} + 0.030\hat{\vartheta} - 1.182) \sin 80°$$

$$-(-3.550\hat{N}_x - 2.622\hat{v} - 2.332\hat{w} - 0.813\hat{\vartheta} - 0.893) \cos 80° - 0.325 = 0$$

$$+1.131\hat{N}_x + 0.671\hat{v} + 0.503\hat{w} + 0.171\hat{\vartheta} - 1.334 = 0 \qquad (4)$$

The four unknown quantities $\hat{N}_x$, $\hat{v}$, $\hat{w}$, and $\hat{\vartheta}$ are found out by solving the above four equations. Their values are as given below.

$$\hat{N}_x = +7.3918$$

$$\hat{v} = -16.2671$$

$$\hat{w} = +11.9880$$

$$\hat{\vartheta} = -12.4632$$

6. CALCULATION OF STRESS RESULTANTS

The unknowns evaluated above are the reduced values of N_x, v, w, and ϑ. Dividing them by the corresponding coefficients tabulated in step 2, we get their actual values at the free edge. The values are

$$N_x = +90.553 \text{ kip/ft}$$

$$v = \frac{-30,516.49}{E} \text{ ft}$$

$$w = \frac{+22,488.99}{E} \text{ ft}$$

$$\vartheta = \frac{-2,338.04}{E}$$

The reduced values of stresses at any other edge are calculated by postmultiplying the matrix of reduced stress resultants (9×5 matrix) corresponding to that edge (given in the previous table) by a column matrix containing the values of the four unknowns

Table 22-1

Edge 1 — Initial boundary matrix

	$\hat{N}_z$	ϑ	$\hat{w}$	ϑ	s	1
$\hat{N}_v$	0	0	0	0	0	0
$\hat{N}_{zv}$	0	0	0	0	0	0
$\hat{N}_z$	1	0	0	0	0	0
ϑ	0	1	0	0	0	0
$\hat{w}$	0	0	1	0	0	0
s	0	0	0	0	1	0
$\hat{M}$	0	0	0	1	0	0
$\hat{Q}$	0	0	0	0	0	0
1	0	0	0	0	0	1

Transfer matrix 1‑2

	$\hat{N}_v$	$\hat{N}_{zv}$	$\hat{N}_z$	ϑ	$\hat{w}$	s	$\hat{M}$	$\hat{Q}$	1
$\hat{N}_v$	1	0.300	−0.090	−0.014	0	0	0	0	0
$\hat{N}_{zv}$	0	1	−0.600	−0.135	0	0	0	0	0
$\hat{N}_z$	0	0	1	0.450	0	0	0	0	0
ϑ	0	0	0	1	0	0.300	0	0	0
$\hat{w}$	0	0	0	0	1	0	−2.303	−0.230	0
s	0	0	0	0	0	1	−15.352	−2.303	0
$\hat{M}$	0	0	0	0	0	0	1	0.300	0
$\hat{Q}$	0	0	0	0	0	0	0	1	0
1	0	0	0	0	0	0	0	0	1

Edge 2 — Matrix of reduced stress resultants at edge 2

	$\hat{N}_z$	ϑ	$\hat{w}$	ϑ	s	1
$\hat{N}_v$	−0.090	−0.014	0	0	0	0
$\hat{N}_{zv}$	−0.600	−0.135	0	0	0	0
$\hat{N}_z$	1	0.450	0	0	0	0
ϑ	0	1	0	0	0.300	0
$\hat{w}$	0	0	1	0	0	0
s	0	0	0	0	1	0
$\hat{M}$	0	0	0	1	0	0
$\hat{Q}$	0	0	0	0	0	0
1	0	0	0	0	0	1

Point matrix 2‑3

	$\hat{N}_v$	$\hat{N}_{zv}$	$\hat{N}_z$	ϑ	$\hat{w}$	s	$\hat{M}$	$\hat{Q}$	1
$\hat{N}_v$	0.500	0	0	0	0	0	0	0.866	0.275
$\hat{N}_{zv}$	0	1	0	0	0	0	0	0	0
$\hat{N}_z$	0	0	0.667	0	0	0	0	0	0
ϑ	0	0	0	0.500	0.866	0	0	0	0
$\hat{w}$	0	0	0	−0.866	0.500	0	0	0	0
s	0	0	0	0	0	1	0	0	0
$\hat{M}$	0	0	0	0	0	0	1	0	0.047
$\hat{Q}$	−0.866	0	0	0	0	0	0	0.500	−0.476
1	0	0	0	0	0	0	0	0	1

Edge 3 — Matrix of reduced stress resultants at edge 3

	$\hat{N}_z$	ϑ	$\hat{w}$	ϑ	s	1
$\hat{N}_v$	−0.045	−0.007	0	0	0	0.275
$\hat{N}_{zv}$	−0.600	−0.135	0	0	0	0
$\hat{N}_z$	0.667	0.300	0	0	0	0
ϑ	0	0.500	0.866	0	0.260	0
$\hat{w}$	0	−0.866	0.500	0	0.150	0
s	0	0	0	0	1	0
$\hat{M}$	0	0	0	1	0	0.047
$\hat{Q}$	0.078	0.012	0	0	0	−0.476
1	0	0	0	0	0	1

Transfer matrix 3-4

Edge 4	$\hat{N}_v$	$\hat{N}_{zv}$	$\hat{N}_z$	$\hat{v}$	$\hat{w}$	ϑ	$\hat{M}$	$\hat{Q}$	1
$\hat{N}_v$	1	1	-1	1	-0.333	0	0	0	0
$\hat{N}_{zv}$	0	1	-2	1	-1	0	0	0	0
$\hat{N}_z$	0	0	1	1	-1	0	0	0	0
$\hat{v}$	0	0	0	1	0	0	0	0	0
$\hat{w}$	0	0	0	0	1	1	-86.381	-28.794	0
ϑ	0	0	0	0	0	1	-172.763	-86.381	0
$\hat{M}$	0	0	0	0	0	0	1	1	0
$\hat{Q}$	0	0	0	0	0	0	0	1	0
1	0	0	0	0	0	0	0	0	1

Matrix of reduced stress resultants at edge 4

-1.312	-0.608	-0.289	-0.087	0.275
-1.933	-1.235	-0.866	-0.260	0
0.667	0.800	0.866	0.260	0
0	0.500	0.866	0.260	0
-2.244	-1.203	0.500	1.150	9.663
-6.733	-1.010	0	1	33.040
0.078	0.012	0	0	-0.429
0.078	0.012	0	0	-0.476
0	0	0	0	1

Point matrix 4-5

Edge 5	$\hat{N}_v$	$\hat{N}_{zv}$	$\hat{N}_z$	$\hat{v}$	$\hat{w}$	ϑ	$\hat{M}$	$\hat{Q}$	1
$\hat{N}_v$	0.940	0	0	0	0	0	0	0.342	0
$\hat{N}_{zv}$	0	1	0	0	0	0	0	0	0
$\hat{N}_z$	0	0	1	0	0	0	0	0	0
$\hat{v}$	0	0	0	0.940	0.342	0	0	0	0
$\hat{w}$	0	0	0	-0.342	0.940	0	0	0	0
ϑ	0	0	0	0	0	1	0	0	0
$\hat{M}$	0	0	0	0	0	0	1	0	0
$\hat{Q}$	-0.342	0	0	0	0	0	0	0.940	0
1	0	0	0	0	0	0	0	0	1

Matrix of reduced stress resultants at edge 5

-1.206	-0.568	-0.271	-0.081	0.208
-1.933	-1.235	-0.866	-0.260	0
0.667	0.800	0.866	0.260	0
-0.768	0.058	0.985	0.637	3.305
-2.109	-1.301	0.174	0.992	9.080
-6.733	-1.010	0	1	33.040
0.078	0.012	0	0	-0.423
0.522	0.219	0.099	0.030	-1.182
0	0	0	0	1

Transfer matrix 5-6

Edge 6	$\hat{N}_v$	$\hat{N}_{zv}$	$\hat{N}_z$	$\hat{v}$	$\hat{w}$	ϑ	$\hat{M}$	$\hat{Q}$	1
$\hat{N}_v$	1	1	-1	1	-0.333	0	0	0	0
$\hat{N}_{zv}$	0	1	-2	1	-1	0	0	0	0
$\hat{N}_z$	0	0	1	1	-1	0	0	0	0
$\hat{v}$	0	0	0	1	0	0	0	0	0
$\hat{w}$	0	0	0	0	1	1	-86.381	-28.794	0
ϑ	0	0	0	0	0	1	-172.763	-86.381	0
$\hat{M}$	0	0	0	0	0	0	1	1	0
$\hat{Q}$	0	0	0	0	0	0	0	1	0
1	0	0	0	0	0	0	0	0	1

Matrix of reduced stress resultants at edge 6

-3.550	-2.622	-2.332	-0.813	-0.893
-2.499	-2.894	-3.583	-1.417	-3.305
-0.101	0.858	1.851	0.897	3.305
-0.768	0.058	0.985	0.637	3.305
-30.600	-9.629	-2.669	1.139	112.685
-65.275	-21.955	-8.529	-1.559	208.197
0.600	0.231	0.099	0.030	-1.605
0.522	0.219	0.099	0.030	-1.182
0	0	0	0	1

and 1. For example, the reduced values of the stress resultants at edge 4 are found as shown below.

$$
\begin{bmatrix} \hat{N}_v \\ \hat{N}_{zv} \\ \hat{N}_z \\ \hat{\vartheta} \\ \hat{w} \\ \vartheta \\ \hat{M} \\ \hat{Q} \\ 1 \end{bmatrix}_{\text{at edge 4}} = \begin{bmatrix} -1.312 & -0.608 & -0.289 & -0.087 & 0.275 \\ -1.933 & -1.235 & -0.866 & -0.260 & 0 \\ 0.667 & 0.800 & 0.866 & 0.260 & 0 \\ 0 & 0.500 & 0.866 & 0.260 & 0 \\ -2.244 & -1.203 & 0.500 & 1.150 & 9.663 \\ -6.733 & -1.010 & 0 & 1 & 33.040 \\ 0.078 & 0.012 & 0 & 0 & -0.429 \\ 0.078 & 0.012 & 0 & 0 & -0.476 \\ 0 & 0 & 0 & 0 & 1 \end{bmatrix} \begin{bmatrix} +7.3918 \\ -16.2671 \\ +11.9880 \\ -12.4632 \\ +1.0 \end{bmatrix}
$$

The actual values of the stress resultants are found out by dividing the reduced values by the corresponding coefficients tabulated in step 2. It is to be noted here that the value of the moment M obtained from this analysis is not its final value but corresponds to the unbalanced moment. The final value of M is obtained by adding to it the initial fixed-end moment. The details of calculation for N_v, N_{zv}, N_z, and M at edge 4 are given below for illustration.

$$
N_v = [+7.3918(-1.312) - 16.2671(-0.608) + 11.9880(-0.289)
$$
$$
- 12.4632(-0.087) + 1(+0.275)] = -1.904 \text{ kip/ft}
$$

$$
N_{zv} = \frac{1}{0.28571} [+7.3918(-1.933) - 16.2671(-1.235) + 11.9880(-0.866)
$$
$$
- 12.4632(-0.260) + 1(0)] = -4.705 \text{ kip/ft}
$$

$$
N_z = \frac{1}{0.08163} [+7.3918(+0.667) - 16.2671(+0.800) + 11.9880(+0.866)
$$
$$
- 12.4632(+0.260) + 1(0)] = -11.542 \text{ kip/ft}
$$

$$
M = \frac{1}{0.1} [+7.3918(+0.078) - 16.2671(+0.012) + 11.9880(0)
$$
$$
- 12.4632(0) + 1(-0.429)] - \frac{(-650 \times 8.660)}{12} = -0.904 \text{ kip ft/ft}
$$

The stress resultants at all edges are calculated as explained above and the values tabulated. The values of v, w, ϑ, and Q are not given as these quantities are not usually required for the design of reinforcement.

	Units	Edge 2	Edge 3	Edge 4	Edge 5	Edge 6
N_v	kip/ft	−0.446	+0.052	−1.904	−1.708	−2.290
N_{zv}	kip/ft	−7.836	−7.836	−4.705	−4.705	0
N_z	kip/ft	+0.872	+0.581	−11.542	−11.542	−4.924
M	kip ft/ft	0	0	−0.904	−0.904	−1.641

EXPERIMENTAL DESIGN

22-14 Mathematical versus Physical Models

Our discussion so far has been limited to analytical methods of approach to shell problems. The trend today is toward more complex shapes and boundary conditions which are not easily amenable to treatment by analytical means. The need to find an adequate method of analysis to deal with such complex structures has led to a great deal of research effort aimed at perfecting experimental techniques involving the use of scale models. Broadly speaking, there are two approaches available for studying the behavior of a structure—mathematical analysis or the use of models. In the mathematical approach, we idealize and simplify the physical problem, making a number of assumptions, and construct what may be termed a *mathematical model*. Alternatively, we may construct a *physical model* of the structure to scale and study its behavior. Each approach has its merits and limitations. When applied to shell problems, the latter approach has the advantage that complex shapes and boundary conditions which are not easily expressed analytically may be dealt with without undue difficulty.

22-15 Historical Notes

Although the extensive use of models as aids to structural design is a recent development, master builders down the centuries are known to have employed models of some kind or other to test their intuitive ideas. Thus Michelangelo is reported to have made use of a wooden model of St. Peter's Basilica to visualize the forces developed in it. The statement "A builder's most blessed expense is the money spent on models is attributed to him. The trite saying that "a test is worth a thousand opinions" also underlines the great value of experiments as aids to judgment.

A model analysis may be undertaken to supplement or confirm the results of a mathematical analysis or as a substitute for it. In some European countries, notably Italy and Spain, the results of a model analysis can be offered in lieu of a theoretical calculation. This practice is also sanctioned by the ACI Committee 334 on Concrete Shell Structures. Nervi and Torroja frequently resorted to this alternative in conceiving and designing some of their well-known structures. The Istituto Sperimentale Modelli E Strutture (ISMES) at Bergamo, Italy, the Laboratorio Nacional de Engenharia Civil at Lisbon, the Laboratorio Central de Ensayo de Materiales de Construccion at Madrid, and the Cement and Concrete Association Laboratory in Slough, England, are some of the European laboratories which are particularly well equipped for work on structural models. Similar facilities are now becoming available at M.I.T. and the PCA Research and Development Laboratories in the United States.

22-16   Direct and Indirect Models

Models are of two kinds—direct and indirect.   On a direct model, loads are applied and strains and displacements are observed.   In the indirect technique, a known displacement is given to the structure at a point and the resulting displacements at other points are observed. The object is to construct influence lines or influence surfaces.   An example of the indirect procedure is Begg's method for arriving at influence lines for statically indeterminate structures.   Direct models are more suitable for investigations on shell structures.

22-17   Types of Models

We may sometimes use models to simulate prototype behavior only within the elastic range.   Such models, known as elastic models, function very much in the same way as analog computers.   But in certain other instances, we may be interested in simulating structural behavior right up to and including failure.   This means that the model has to reproduce the prototype behavior faithfully even in the inelastic range.

22-18   Dimensional Analysis and the Theory of Models

If the model is to simulate the prototype, certain nondimensional numbers, known as π terms, have to be the same for the model as well as the prototype.   The π terms governing a physical phenomenon may be found by using the well-known π theorem first enunciated by Buckingham in 1914.[1]   The theorem states that *if a physical phenomenon is governed by m variables which may be described by n fundamental dimensions, the number of π terms is $(m - n)$*.   Buckingham stated his theorem without proof. In fact, it was no more than a rule of thumb.   Later writers such as Bridgman [159] and Van Driest [160] have amended and restated the theorem thus: *The number of π terms or dimensionless products is equal to the number of variables minus the rank of their dimensional matrix*.   In problems of mechanics, we shall be concerned with only three fundamental dimensions—force, length, and time, or mass, length, and time. Time will not figure in problems of statics; hence we need deal with only two fundamental dimensions.   Limitations of space forbid a detailed discussion of dimensional analysis.   Readers may refer to the book by Murphy [161] for a more extended presentation.   We shall now proceed

---

[1] Buckingham, E., "On Physically Similar Problems: Illustrations of the Use of Dimensional Equations," *Physical Review*, vol. IV, no. 4, p. 345, 1914.

to apply dimensional theory to the stress analysis of shells. In a model experiment on a shell, we are ultimately interested in arriving at the stresses and displacements in the prototype based on the observed results on the model. This means that we need relations connecting corresponding quantities—known as homologous quantities in dimensional theory—for the model and the prototype. The first step in the dimensional analysis of a physical problem is the careful identification and listing of the variables involved. Dimensional analysis cannot correct any omission on our part to include *all* the relevant variables. In our problem, the variables are σ, a, b, d, E, ν, y, P, and p, where

$$\sigma = \text{stress in the shell}$$

$$a, b = \text{linear dimensions of the shell}$$

$$d = \text{thickness of the shell}$$

$$E = \text{Young's modulus}$$

$$\nu = \text{Poisson's ratio}$$

$$y = \text{any displacement}$$

$$P = \text{a concentrated load}$$

$$p = \text{a uniformly distributed load}$$

The next step is to compile a dimensional matrix of the variables as shown below.

Dimension \ Variable	a	b	d	P	p	E	$-\nu$	σ	y
F	0	0	0	1	1	1	0	1	0
L	1	1	1	0	-2	-2	0	-2	1

The number of variables is 9 and the rank of the dimensional matrix is 2. Hence there are seven π terms. These may be written by inspection. Because Poisson's ratio is dimensionless, it must appear as one of the π terms. We may thus write $\pi_1 = \nu$, $\pi_2 = a/b$, $\pi_3 = a/d$, $\pi_4 = P/Ed^2$, $\pi_5 = p/E$, $\pi_6 = \sigma/E$, and $\pi_7 = y/a$. The two characteristics of π terms are that they are dimensionless and independent. It may be easily verified that any other nondimensional number that may be formed out of the variables can be derived from these seven π terms. For instance, $\sigma/p = \pi_6/\pi_5$, $P/\sigma d^2 = \pi_4/\pi_6$, and so on. The π terms, π_1

to π_7, have to be the same for the model and the prototype. This means that

$$\nu_m = \nu_p \tag{22-107}$$

$$\frac{a_m}{b_m} = \frac{a_p}{b_p} \tag{22-108}$$

$$\frac{a_m}{d_m} = \frac{a_p}{d_p} \tag{22-109}$$

$$\frac{P_m}{E_m d_m{}^2} = \frac{P_p}{E_p d_p{}^2} \tag{22-110}$$

$$\frac{p_m}{E_m} = \frac{p_p}{E_p} \tag{22-111}$$

$$\frac{\sigma_m}{E_m} = \frac{\sigma_p}{E_p} \tag{22-112}$$

and

$$\frac{y_m}{a_m} = \frac{y_p}{a_p} \tag{22-113}$$

Relation (22-107) implies that Poisson's ratio has to be the same for both the model and prototype. This condition cannot be met in practice as Poisson's ratio for concrete ranges between 0 and 0.16 whereas its value for Perspex or plastics such as Plexiglas lies between 0.35 and 0.38. Cowan [162] estimates that the error introduced as a result is negligible. Even in an extreme instance where $\nu = 0.45$, the error is only 26 % for that part of the deformation caused by shear, which is usually only a negligible part of the total deformation. Relations (22-108) and (22-109) are just a statement of *geometrical similarity*, which means that all the geometrical dimensions of the prototype are scaled down in the same proportion in building the model. Let this scale be λ so that $a_p = \lambda a_m$, $b_p = \lambda b_m$, and $d_p = \lambda d_m$. Next, let us consider the relation

$$\frac{\sigma_m}{E_m} = \frac{\sigma_p}{E_p}$$

or

$$\sigma_p = \sigma_m \frac{E_p}{E_m} \tag{22-114}$$

This equation enables us to relate the stress in the model to that in the prototype. The relation $y_m/a_m = y_p/a_p$ may be recast as

$$y_p = y_m \frac{a_p}{a_m} = \lambda y_m \tag{22-115}$$

This relation is used to arrive at any deflection of the prototype from the corresponding deflection of the model. Relations (22-110) and (22-111) show how loading is to be simulated on the model. Thus, from Equation (22-110) we may write

$$P_m = P_p \frac{E_m}{E_p} \left(\frac{d_m}{d_p}\right)^2 \tag{22-116}$$

Similarly, the relationship between prototype and model loadings when the load is uniformly distributed may be arrived at from the equation

$$p_m = p_p \frac{E_m}{E_p} \tag{22-117}$$

Any two of the three scales, $a_p/a_m = \lambda$, E_p/E_m, or P_p/P_m, may be chosen independently. Thus, if the material of the model and its linear scale are specified, λ and E_p/E_m become fixed and the loading scale alone remains to be worked out.

22-19 Fabrication of Elastic Models

Elastic models are usually made out of plastics which are easily machined, sawed, and drilled, have a low modulus of elasticity, and are relatively inexpensive. Creep and sensitivity to temperature are their chief disadvantages as model materials. The most widely used material is Perspex or Plexiglas. Celluloid polyvinyl chloride, plaster of paris, or mixtures of plaster of paris and diatomite have also been occasionally employed. The properties of Perspex and Plexiglas are given below.

Modulus of elasticity	420–460 $\times$ 10^3 psi
Poisson's ratio	0.35–0.38
Tensile strength	8,000–10,000 psi
Specific gravity	1.19
Softening point	115–150°C
Coefficient of linear expansion	9 $\times$ 10^{-5} per °C

These materials are acrylic resins belonging to the *thermoplastic* group. Plastics of this group undergo no permanent change on heating. On being heated to temperatures in the range of 150 to 300°C, they become rubbery in consistency and may be formed into any desired shape by pressing against a mold. Plastics of the *thermosetting* group do not possess this property. They maintain their rigid shape after the thermoset has taken place. The *thermoplastic* resins are more suitable for model work. For information on other plastics, the reader may consult the book by Preece and Davies [163]. Properties of Perspex are described in detail in a series of handbooks issued by I.C.I. Ltd., Plastics Division, Welwyn Garden City, Herts, England. Similar data on

Plexiglas are given in a manual published by Rohm and Haas Co. [164] in the United States.

Singly curved or developable shapes are easily formed. The mold is inserted in the oven with the rigid sheet placed over it. On being heated, the sheet takes the desired shape without any application of pressure.

For forming surfaces of double curvature, the method of vacuum forming developed at M.I.T. and described by Litle and Hansen [165] appears to be the most suitable. The technique consists in heating the plastic and forcing it against a mold by vacuum pressure, when it is still in the rubbery state. Some thickness variation is inevitable when doubly curved shapes are formed by this process. The effect of this variation on the structural behavior of the model has not yet been adequately assessed.

Machine-shop facilities required for working with plastics must include at least a large engine lathe and a milling machine. Machining and working with plastics calls for special care. The precautions to be taken have been discussed in *Bulletin* D76 issued by PCA [166].

Two kinds of cements are available for assembly by cementing of Plexiglas models. One dissolves the Plexiglas in the neighborhood of the joint to develop bond; the action of the other, known as a polymerizable cement, depends on an internal chemical reaction. Acetic acid and chloroform may be used as solvents. Polymerizable cements include PS-18 cement and its variations. Tensol cement No. 7 is suitable for use with Perspex. Preece and Davies [163] give more details of this cement and the manner in which it is to be used. When solvent cements are used, there is a local reduction in the modulus of elasticity. Hence precautions need to be taken to ensure that the solvent cements do not come in contact with parts of the model which are not being bonded.

It is essential that plastic models are tested under controlled conditions of temperature and humidity.

22-20 Loading of Plastic Models

Plastic models are usually loaded by hanging weights or by means of lever systems. The weight of the model is not adequate to simulate the action of dead weight on the prototype. This difficulty is usually got over by hanging weights at a number of discrete points on the model.

22-21 Planning of a Model Study within the Elastic Range

The various steps involved in planning a model study within the elastic range on a shell structure are best explained by means of a concrete example. Suppose we are interested in the stresses and deflections in a short cylindrical shell of the following dimensions under the action of its dead weight (Fig. 22-15).

Data

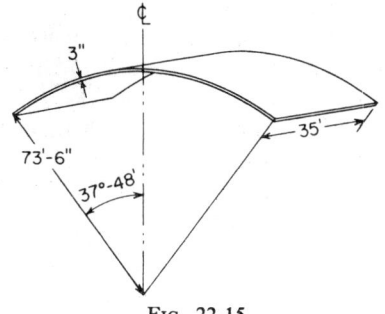

Length *l*	= 35 ft
Radius *a*	= 73.5 ft
Thickness *d*	= 3 in.
Semicentral angle ϕ_c	= 37°48′

Fig. 22-15

Let us select Perspex sheet 1/16 in. thick as the model material. Its modulus of elasticity is 420×10^3 psi and its Poisson's ratio is 0.38. As the model is to be geometrically similar and the sheet thickness of 1/16 in. is to represent the shell thickness of 3 in., the linear scale works out to 48. The prototype shell being 3 in. thick, its dead weight per square foot works out to 37.5 psf. From relation (22-117), the model

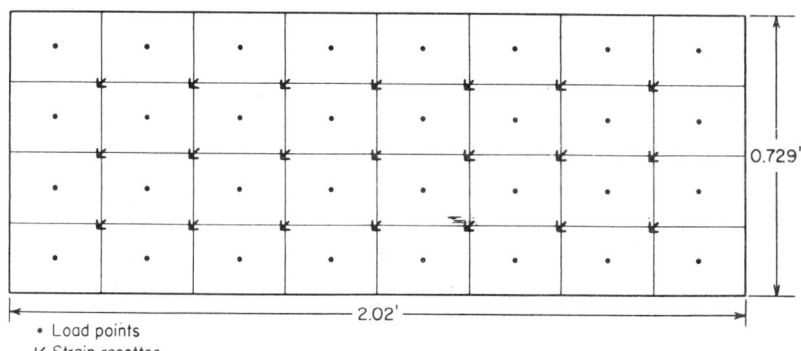

• Load points
ㅂ Strain rosettes

Developed plan of the model showing loading points
and positions of strain gauges

Fig. 22-16

load intensity $= 37.5 \times (420 \times 10^3/3 \times 10^6) = 5.25$ psf. If we decide to load the model by hanging weights at 32 points, the load at each point works out to 0.242 lb. The loading points and the electrical resistance strain-gage rosette locations are indicated in the developed plan of the shell (Fig. 22-16). The manner in which this load is applied is schematically shown in Fig. 22-17. The maximum load to which the model could be loaded was estimated at 42 lb. An approximate stress analysis indicated that at this load, the stress in the model would reach 6,000 psi. This is the limiting stress up to which the model may be expected to behave elastically.

Foil gages may be fixed by means of Tensol 7 or other cement at the locations indicated on the plan. Johnson and Homewood [167] report that Duco cement is satisfactory for installing SR-4 gages on Plexiglas. Foil gages are found to be more suitable for model analysis because of their superior heat-dissipation characteristics and lower transverse sensitivity.

The traverse shown in Fig. 22-18 may be fixed to the shell by using Tensol 7 or other cement.

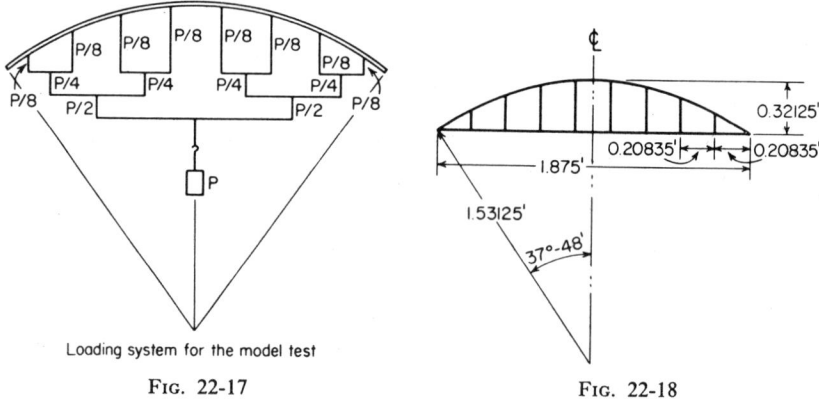

Loading system for the model test

Fig. 22-17   Fig. 22-18

A Savage and Parsons 50-channel null-balancing-type bridge is suitable for strain measurement using an accumulator as a steady source of power supply. For zones of low strain it is advisable to use a Sanborn recorder.

22-22 Model Simulation in the Inelastic Range

In certain studies we may wish the model to simulate prototype behavior right up to failure. If this is the case, the material of the model must have a stress-strain curve which is similar to that of the prototype material. This means that the stress-strain curve of the material can be converted into that of the prototype by a simple change of scale along the stress and strain axes (Fig. 22-19). Moreover, Poisson's ratio for the two materials must be the same. It is evident that

Stress in prototype $= \alpha$ times the stress in model

Strain in prototype $= \beta$ times the strain in model

For models of this type, we may proceed to find the scales involved as follows: Let us consider the π term $P/\sigma d^2$ which has to have the same value for the model as well as the prototype. It is clear that

$$\frac{P_m}{\sigma_m d_m{}^2} = \frac{P_p}{\sigma_p d_p{}^2}$$

Or, the load scale for concentrated loads

$$P_p = P_m \left(\frac{\sigma_p}{\sigma_m}\right) \left(\frac{d_p}{d_m}\right)^2$$

$$= P_m \, \alpha\lambda^2 \qquad\qquad (22\text{-}118)$$

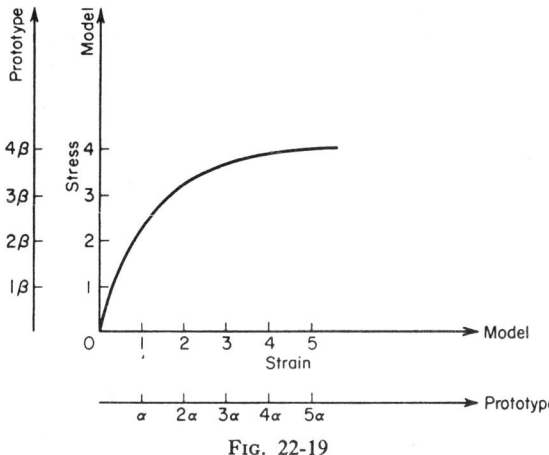

FIG. 22-19

The scale ratio for uniformly distributed loads will be the same as the stress scale, i.e.,

$$p_p = p_m\alpha \qquad\qquad (22\text{-}119)$$

Since any displacement may be thought of as strain times a length,

$$y_p = \beta\lambda y_m \qquad\qquad (22\text{-}120)$$

The scale for stress resultants, such as N_x (force per unit length), is easily verified to be

$$(N_x)_p = (N_x)_m\alpha\lambda \qquad\qquad (22\text{-}121)$$

and for moment stress resultants, such as M_x (force), it is given by

$$(M_x)_p = (M_x)_m\alpha\lambda^2 \qquad\qquad (22\text{-}122)$$

Mixtures of plaster of paris and diatomite may be employed to simulate concrete. The α of this material lies in the range of 5 to 10 and its β value is around 0.40. It is essential to have $\beta = 1$ in modeling structures whose deformations are large enough to influence their structural behavior. Suitable modeling materials for such applications are mortars and concretes made out of pumice aggregate. The ISMES at Bergamo has perfected this technique. The aggregate employed is a pumice stone from the island of Lipari [168]. The value of α may be adjusted to be between 2 and 12. The value of β, of course, is unity. Another model material which has sometimes been employed is micro-concrete with both α and β equal to 1.

If the prototype material is reinforced concrete, we have to find two model materials so that the α and β values of concrete and the model material chosen to simulate it are the same as those of steel and the model material selected to simulate it. It may often be difficult to find such a pair. One would then be obliged to use steel with the same mechanical properties in the model as in the prototype. But the area of model reinforcement will no longer be $1/\lambda^2$ of the prototype reinforcement. We proceed to find the scale ratio for the area of reinforcement as follows: Because strain in the prototype has to be β times the strain in the model

$$\frac{P_p}{A_p E_p} = \beta \frac{P_m}{A_m E_m}$$

But, since we are using steel to model steel, $E_p = E_m$. Hence the desired scale ratio

$$\frac{A_p}{A_m} = \frac{1}{\beta} \frac{P_p}{P_m} = \frac{\alpha \lambda^2}{\beta} \tag{22-123}$$

Steel threaded rod annealed for 1 hr at 600°C used in conjunction with wet gypsum plaster is reported [169] to be a satisfactory means of simulating reinforced concrete. The powder recommended for use is of the autoclave or alpha hemihydrate variety. The powder is to be added to the water contained in a rubber bowl. After the paste is allowed to stand for a minute, it is mixed with a flexible steel spatula and poured into a mold. A powder which sets 10 min after it is added to water is reported to be suitable.

Reinforced concrete still remains the best means of simulating reinforced concrete. But this, however, means the scale of the model has to be reasonably large so that the scaled thickness is not too small. Photo 22-1 shows a half-full-scale model of a prototype conoid shell of 50 ft span and 40 ft chord width under test at the Structural Engineering Research Laboratory at Roorkee, India. Most civil engineering structures, unlike aircraft structures, tend to be one of a kind. Hence testing a large-scale model of this kind cannot be economically justified unless the project involves the building of a large number of identical units of a particular type of shell roof. In this instance, since 100 units of identical size were to be built, the cost of the testing program amounted to only a negligible fraction of the total cost of the structures planned.

Photo 22-1. A test to destruction in progress on a half-scale model of the conoid shell roofs shown in Photos 17-8 and 17-9.

22-23 Photoelasticity Applied to Shells

The principal load acting on a shell is its dead weight. If the model shell is made of Araldite, its own weight is insufficient to produce fringes at room temperature. Hence photoelasticity techniques can be applied to shell structures only if we take recourse to centrifuging and stress freezing. The application of these techniques to shell structures has been investigated by Sharma [170].

22-24 Other Materials and Techniques

The *Proceedings of the Symposium on Shell Research* [171] held at Delft in 1961 is a storehouse of information on various model techniques developed for the stress analysis of shells of various kinds. The reader seeking more information on this subject is well advised to consult this publication.

STABILITY PROBLEMS

22-25 Need for Research

We know very little at present about the stability problems relating to reinforced-concrete shells. Experienced designers are aware [172], however, that safety against buckling assumes great importance and largely influences the selection of thickness and curvatures when the size of the shell is large. Most of the information that we have at present is derived from investigations carried out on metal shells. Even here, there are wide divergences between the results of analytical and experimental investigations. There is no other area of shell design where research is more urgently needed.

22-26 Stability of Cylindrical Shells

Causes of Instability

Instability in a cylindrical shell may be caused by

(*a*) Local buckling in zones submitted to compressive stress.
(*b*) Flattening of shells, known as the Brazier effect, which occurs particularly in long shells without edge beams.
(*c*) Instability caused by the combined effect of bending and torsion in the shell as a whole. This can occur in asymmetric shells.

The Brazier effect has been discussed by Lundgren [2]. He suggests that in flat shells where the length is large compared with the rise, the edges of the shell should at least be partially supported either horizontally by adjacent shells or vertically by the provision of edge beams.

We have already seen in connection with our discussion on northlight shells that the torsion on the cross section may be eliminated if the center of gravity of the cross section is made to coincide with the shear center.

In what follows, we shall be primarily concerned with buckling caused by compressive stresses.

Long Shells ($\rho < 7$; $\kappa < 0.12$)

In long shells, the tendency to buckle is confined to a zone near the crown where the longitudinal compressive stresses are high. Theoretical studies based on the theory of small deflections indicate [173] that the critical stress for an axially compressed cylinder is

$$\sigma_{cr} = \frac{1}{\sqrt{3(1 - \nu^2)}} \frac{Ed}{a} \tag{22-124}$$

For a steel cylinder,

$$\sigma_{cr} \cong 0.605 \frac{Ed}{a} \tag{22-125}$$

where σ_{cr} = elastic critical stress
$\quad\quad d$ = thickness of the shell
$\quad\quad a$ = radius of the shell

Experiments on metal shells, however, show that shells fail at a load ranging from 10 to 60 % of the classical critical load predicted by the small-deflection theory. Barring a few experimental results reported by Lundgren [2], there is hardly any information available on the behavior of axially compressed reinforced-concrete shells. Until more extensive and reliable tests are carried out, we may, following Lundgren, estimate the critical stress in a long reinforced-concrete shell subjected to axial compression as

$$\sigma_{cr} = 0.20 \frac{Ed}{a} \tag{22-126}$$

We may visualize two extreme cases—a thick shell wherein failure by crushing of the concrete occurs before the elastic critical stress is reached and a thin shell wherein the elastic critical stress is reached before the concrete is crushed. Shells which are neither too thick nor too thin lie between these two limits. Their carrying capacity is less than both the theoretical buckling load as well as the load corresponding to the crushing strength of the concrete. To cover all the requirements of the three cases enumerated above, we may write

$$\frac{1}{\sigma_{pr}} = \frac{1}{\sigma_{cr}} + \frac{1}{f_c'} \tag{22-127}$$

where σ_{pr} = practical ultimate stress
$\quad\quad f_c'$ = cylinder strength of the concrete at 28 days

The above relation may be recast in the form

$$\sigma_{pr} = \frac{f_c'}{1 + \dfrac{f_c'}{\sigma_{cr}}} \tag{22-128}$$

We may now substitute for σ_{cr} from (22-126) and introduce a factor of safety of 4 to get

$$\sigma_{per} = \frac{0.25 f_c'}{1 + \dfrac{5a}{Ed} f_c'} \tag{22-129}$$

where σ_{per} = permissible stress in concrete. This formula may be employed to check the safety of long shells against buckling. This approach to buckling problems has also been adopted by Paduart [174]. We may also use charts such as those given by Bradshaw [175].

Short Shells $(\rho > 10; \kappa > 0.15)$

In short shells, transverse stresses tend to be critical. The critical stress may be estimated on the basis of the available theoretical results for a cylinder subjected to external pressure.   These results are given below without proof.   The reader interested in the derivation may consult the book by Gerard [176].

$$\sigma_{cr} = \frac{\pi^2 kE}{12(1 - \nu^2)} \left(\frac{d}{l}\right)^2 \tag{22-130}$$

where

$$k = \frac{(1 + \beta^2)^2}{\beta^2} + \frac{12Z^2}{\pi^4 \beta^2 (1 + \beta^2)^2}$$

$$\beta = \frac{nl}{\pi a}$$

$$Z = \frac{L^2}{at} (1 - \nu^2)$$

$n =$ number of half wavelengths around the circumference

The buckling coefficient k is shown plotted against the parameter Z in Fig. 22-20.   A few important particular cases that arise may be noted.

Case 1   *Very Short Shells for which* $Z \to 0$

In this case $k = 4$.   Substituting this value in (22-130) and assuming $\nu = 0.16$,

$$\sigma_{cr} = 3.3763 \, E \left(\frac{d}{l}\right)^2 \tag{22-131}$$

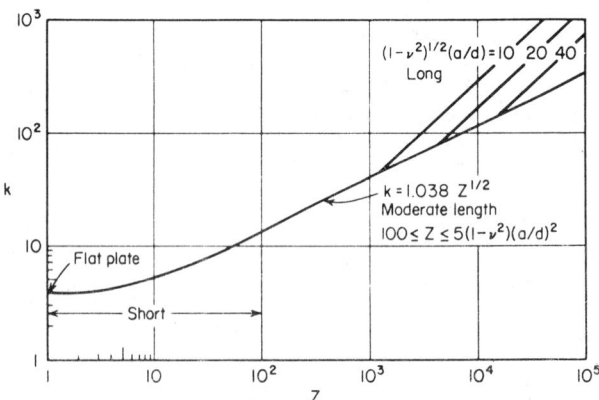

Buckling coefficients for simply supported circular
cylinders under external pressure

Fɪɢ. 22-20.   *(From "Introduction to Structural Stability Theory" by George Gerard, copyright © 1962, McGraw-Hill, Inc.   Used by permission of McGraw-Hill Book Company.)*

For shells lying in about the same range, Lundgren [2] gives the formula

$$\sigma_{cr} = E\left[3.4\left(\frac{d}{l}\right)^2 + 0.025\left(\frac{l}{a}\right)^2\right] \tag{22-132}$$

This formula holds good for $l < 2.3\sqrt{da}$.

Case 2 Shells of Moderate Length

From Fig. 22-20, it is seen that $k = 1.038\,Z^{1/2}$ when $Z > 100$. In this region, the graph is practically a straight line. Substituting this value of Z in (22-130),

$$\sigma_{cr} = \frac{0.855}{(1 - \nu^2)^{3/4}}\left(\frac{d}{a}\right)^{3/2}\frac{a}{l}$$

Observing that $\nu = 0.16$ for concrete,

$$\sigma_{cr} = 0.8718\left(\frac{d}{a}\right)^{3/2}\frac{a}{l}$$

Lundgren [2] gives the following formulas applicable to shells with $l > 2.3\sqrt{da}$:

$$\sigma_{cr} = E\frac{0.89\dfrac{d}{l}\sqrt{\dfrac{d}{a}}}{1 - 1.18\dfrac{\sqrt{da}}{l}} \tag{22-133}$$

for $2.3\sqrt{da} < l < 4.0\sqrt{da}$

and

$$\sigma_{cr} = 1.1\,E\frac{d}{l}\sqrt{\frac{d}{a}} \tag{22-134}$$

for $l > 4.0\sqrt{da}$.

Formula (22-134) would be found applicable in most cases. As before, the critical stress σ_{cr} is to be substituted in (22-128) and the resulting σ_{pr} reduced by a factor of 4 to find σ_{per}.

Intermediate Shells

Shells with ρ values between 7 and 10 are of relatively infrequent occurrence. For such shells, σ_{per} may be calculated on the basis of both axial and circumferential compression and the lower of the two values adopted.

Procedure Recommended by ACI Committee 334

The procedures outlined in the previous paragraphs for arriving at σ_{per} in long, short, and intermediate shells are based on the recommendations of Lundgren [2], Paduart [174], and the Indian Standard Criteria [30]. The procedure recommended by the ACI Committee 334 is somewhat different. Where the influences of creep, deformations, and decentering conditions are not expected to be abnormal, the Committee recommends that the critical buckling stress be computed in the two directions—axial and circumferential—with the aid of graphs available for that purpose using the unreduced value of the modulus of elasticity at the time of decentering. The factors of safety F_1 and F_2 in the two directions are found by dividing the corresponding critical stress by the working stress. The overall factor of safety F is then related to the factors of safety F_1 and F_2 by the equation

$$\frac{1}{F} = \frac{1}{F_1} + \frac{1}{F_2} \tag{22-135}$$

Hence

$$F = \frac{F_1 F_2}{F_1 + F_2} \tag{22-136}$$

22-27 Discrepancies between Theory and Experiment

The discrepancy that exists between theoretical and experimental results relating to axially compressed cylinders has already been noted.

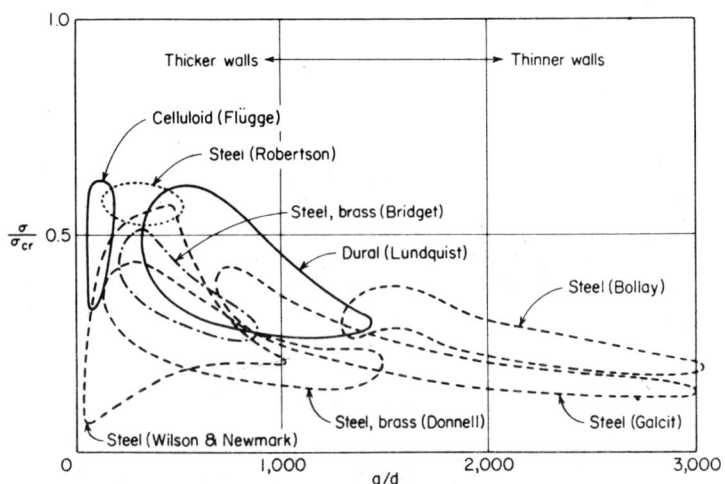

Comparison of experimental buckling stress for a cylinder under axial
compression with that given by the classical linear theory

FIG. 22-21. *(From "Effect of Imperfections on Buckling of Thin Cylinders and Columns under Axial Compression" by Donnell and Wan, Journal of Applied Mechanics, vol. 17, no. 1, 1950. Used by permission of American Society of Mechanical Engineers.)*

In Fig. 22-21 the areas in which the experimental points obtained by various investigators lie are indicated. The experimental results not only show wide scatter but are substantially lower than the theoretical values.

Various large-deflection theories have been advanced in recent years in an effort to resolve the observed divergence between theory and experiment. Two competing theories hold the field at present. The first is the initial-imperfection theory of Donnell and Wan [177]. This theory attributes lower experimental failure loads to initial geometrical imperfections. Material imperfections are not considered. The second explanation, again based on large deflections, is that offered by von Kármán and Tsien [18]. They were the first to establish the existence of finite equilibrium configurations at loads considerably lower than the classical critical load. It was suggested that the small amount of energy required to push the shell "over the hump" could come from the small vibrations that occur in the testing machine. According to this theory, random impulses present during testing trigger the jump into finite-deflection equilibrium configurations, before the classical critical load predicted by the small-deflection eigenvalue theory is reached. This phenomenon is known as "snap through" or "oil canning." The post-buckling behavior of an axially compressed perfect cylinder, according to the "jump theory," and cylinders with various degrees of initial imperfections, according to the Donnell-Wan theory, is shown in Fig. 22-22.

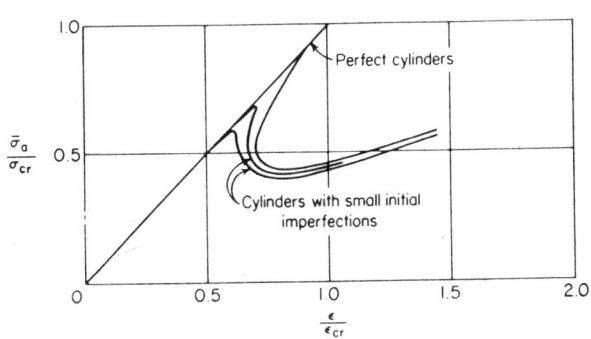

$\bar{\sigma}_0$ = average axial stress

ϵ = axial strain

ϵ_{cr} = strain corresponding to the critical stress

Postbuckling behavior of moderate-length
cylinders in axial compression

FIG. 22-22. *(From "Introduction to Structural Stability Theory" by George Gerard, copyright © 1962, McGraw-Hill, Inc. Used by permission of McGraw-Hill Book Company.)*

22-28 Stability of Shells of Double Curvature

Barring the spherical cap which has been the subject of several theoretical and experimental investigations, there is hardly any literature on the buckling of shells of double curvature. It is generally known that shells of double curvature, being nondevelopable surfaces, are more resistant to buckling than cylindrical shells. Again, anticlastic surfaces are generally more stable, because one of the membrane stresses that develop in them is tensile.

According to Wansleben's theory[1] the intensity of load p_k at which a doubly curved shell of constant curvature would buckle is given by

$$p_k = \frac{2Ed^2}{\sqrt{3(1 - v^2)}} k_1 k_2 \qquad (22\text{-}137)$$

where k_1 and k_2 are the two principal curvatures. If $k_1 = k_2 = 1/R$, we arrive at the well-known Zoelly's formula for a spherical shell, which may be written as

$$p_k = \frac{2E}{\sqrt{3(1 - v^2)}} \left(\frac{d}{R}\right)^2$$

Experimental values of the buckling load are found to be only a small fraction of this theoretical result based on the small-deflection theory. Tsien [178], using large-deflection theory, arrived at the formula

$$p_k = 0.312 \frac{Ed^2}{R^2} \qquad (22\text{-}138)$$

for a spherical shell. This is really the value corresponding to "snap through" or "oil canning." Based on tests on eight aluminum doubly curved translational shells, Schmidt [179] arrived at the formula

$$p_k = 0.32E \frac{d^2}{R_1 R_2} \qquad (22\text{-}139)$$

where R_1 and R_2 are the principal radii of curvature. Thus Tsien's "snap through" value is in close agreement with experimental results.

Because all the available experimental results relate to metal shells, they are not directly applicable to reinforced-concrete shells. Schmidt recommends a coefficient of 0.15 in his formula for checking the stability of reinforced-concrete shells. The Indian Standard Criteria [30] recommends a coefficient of 0.10. The only report available on the collapse

---

[1] Girkmann, K., "Flächentragwerke," 4th ed., pp. 516–529, Springer-Verlag OHG, Vienna, 1956.

of a shell of double curvature due to buckling is that due to Csonka [180]. Basing his calculations on this shell, he recommends the formula

$$p_k = 0.06E \frac{d^2}{R_1 R_2}$$

It is not yet known if the classical "snap through" and empirical formulas discussed in the preceding paragraph would apply to anti-clastic shells such as the hyperbolic paraboloid. Ralston [181] and Dayaratnam and Gerstle [182] have developed methods for determining the buckling loads applicable to hyperbolic paraboloids based on the small-deflection theory. A large-deflection analysis and experimental results are not yet available for such shells.

Tests on reinforced-concrete shells of various geometries need to be taken up by laboratories all over the world if the present gap in our knowledge is to be filled.

OTHER UNSOLVED PROBLEMS

In this brief account, it is not possible to do justice to all the problems that still await solution. Some of the areas where further studies may be fruitful are identified and listed below.

(i) Structural dynamics of shells—their response to vibration, shocks, blasts, and earthquakes

(ii) Analysis and design of shells for thermal stresses

(iii) Analysis of shell structures to carry heavy live loads

(iv) Study of shrinkage and creep effects

(v) Analysis of shells of double curvature with realistic boundary conditions

(vi) Experiments on reinforced-concrete shells to arrive at criteria for stability

(vii) Work on longitudinally continuous cylindrical shells

(viii) Establishment of rational criteria for the design of reinforcement

(ix) Tests on reinforced-concrete shells to establish buckling loads and criteria for design

(x) Limit analysis of shells

This list is merely suggestive and is by no means exhaustive.

APPENDIX

Membrane Forces in Semielliptic Shells* (Refer to Art. 6-10)

	θ	t_2	s	t_1	t_2'	s'	t_1'
$\dfrac{b}{a} = 0.1$	0°	−10.0000	0	−29.9000	−10.0000	0	−30.000
	7°30′	−2.2514	−14.4440	−12.8970	−2.2321	−14.4500	+13.191
	15°	−0.4582	−9.9152	−61.1100	−0.4425	−9.8274	+62.612
	30°	−0.0663	−5.3263	−132.9100	−0.0574	−5.0448	+143.365
	45°	−0.0197	−3.4935	−170.1500	−0.0140	−2.9703	+208.968
	60°	−0.0077	−2.5866	−59.3000	−0.0038	−1.7263	+258.511
	75°	−0.0029	−2.1377	−96.7700	−0.0007	−0.8033	+289.465
	90°	0	−2.0000	0	0	0	+300.000
$\dfrac{b}{a} = 0.2$	0°	−5.0000	0	−14.8000	−5.0000	0	−15.000
	7°30′	−2.9643	−6.8178	−7.0740	−2.9389	−6.8889	−7.039
	15°	−1.1469	−7.1851	+6.2660	−1.1078	−7.1903	+6.889
	30°	−0.2338	−4.8571	+27.7570	−0.2025	−4.6394	+31.182
	45°	−0.0754	−3.3724	+38.9880	−0.0533	−2.8846	+49.923
	60°	−0.0302	−2.5325	+37.5330	−0.0151	−1.7093	+63.662
	75°	−0.0114	−2.1310	+23.0280	−0.0030	−0.8015	+72.132
	90°	0	−2.0000	0	0	0	+75.000
$\dfrac{b}{a} = 0.3$	0°	−3.3333	0	−9.7000	−3.3333	0	−10.000
	7°30′	−2.6038	−3.5810	−7.2021	−2.5815	−3.6798	−7.330
	15°	−1.4822	−4.8847	−1.7772	−1.4317	−4.9683	−1.457
	30°	−0.4357	−4.2244	+8.7658	−0.3773	−4.0915	+10.796
	45°	−0.1582	−3.1852	+14.8150	−0.1118	−2.7523	+20.544
	60°	−0.0663	−2.4972	+15.0070	−0.0331	−1.6816	+27.590
	75°	−0.0256	−2.1200	+9.3745	−0.0066	−0.7987	+31.886
	90°	0	−2.0000	0	0	0	+33.333
$\dfrac{b}{a} = 0.4$	0°	−2.5000	0	−7.1000	−2.5000	0	−7.500
	7°30′	−2.1707	−2.1159	−5.0359	−2.16!1	−2.2272	−6.297
	15°	−1.5366	−3.3314	−3.4326	−1.4843	−3.4679	−3.318
	30°	−0.6157	−3.5541	+2.5529	−0.5332	−3.5109	−4.007
	45°	−0.2561	−2.9503	+6.4784	−0.1811	−2.5862	+10.340
	60°	−0.1129	−2.4227	+7.1517	−0.0570	−1.6444	+14.977
	75°	−0.0452	−2.1046	+4.5988	−0.0117	−0.7947	+17.801
	90°	0	−2.0000	0	0	0	−18.750

* From "Scienza delle Costruzioni," vol. III, pp. 484–485, 1958, by Belluzzi, O. Used ll permission of Nicola Zanichelli, Editore, Bologna, Italy.

PART V

NEW DEVELOPMENTS

CHAPTER 23

FINITE ELEMENT ANALYSIS OF THIN SHELLS

23-1 Introduction

A famous mathematician once observed that the Laws of Nature are written in the language of mathematics. These often take the form of ordinary or partial differential equations. The Electronic Digital Computer is an amazingly fast calculating tool; but, it can handle only arithmetic. The differential equations governing physical phenomena, such as the behaviour of beams, plates and thin shells have therefore to be reduced to a system of linear simultaneous equations, before the computer can solve them by a series of arithmetical operations. Finite Element Techniques do precisely this. Applied to thin shells, they can handle complex shapes, difficult boundary conditions and arbitrary loading. Variations in thickness, cutouts and non-linearity stemming from geometry as well as the constitutive laws of materials and anisotropy are easily accommodated. Cracks can be modelled and the analysis may be continued right up to collapse. It is this versatility that gives this technique an edge over older methods such as Finite Differences.

23-2 Finite Element Discretization

Finite element analysis involves the replacement of the continuum by discrete elements. Thus, the loaded plate in Fig. 23-1, representing a problem in plane-stress is replaced by an assemblage of rectangular elements with nodes at their four corners. Sometimes, additional nodes are also located at points on the sides of the element as indicated in Fig. 23-2.

Instead of rectangular elements, one could have used triangular or quadrilateral elements to model the loaded plate. The shape of elements chosen is often dictated by the geometric configuration of the structure under study. Provided certain conditions are satisfied, the stresses and displacements resulting from a finite element analysis may be expected to converge towards their exact values as we progress from a coarser to a finer sub-division of the structure. The conditions will be spelt out later.

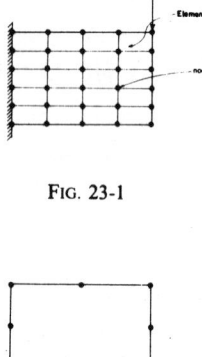

FIG. 23-1

FIG. 23-2

Line elements are suitable for beam problems. Problems of plane stress and flat plate bending are handled by discretization involving triangular, rectangular or quadrilateral elements. Thin shells may be idealized by an assemblage of flat faceted, curved or solid elements in the form of 'bricks' or tetrahedrons.

23-3   Historical Notes

Two landmark papers to which the origin of the Finite Element Method is generally traced are those due to Turner, Clough, Martin and Topp [183] and Argyris and Kelsey [184]. A brief historical account of the developments leading to the Finite Element Technique is given by Gallagher [185]. The limited purpose of this chapter is to introduce the reader to this powerful and versatile technique and summarize the state of the art in applying it to the analysis of thin shells. The reader interested in a more detailed exposition of the method is referred to the following excellent texts on the subject [186] [187] [188] [189] [190] and [185].

23-4   Application to Beam Bending

The problem of beam bending has been chosen for detailed exposition, because the simplicity of the problem permits the sequences involved to be clearly seen. The beam may be replaced by line elements of the type shown in Fig. 23-3.

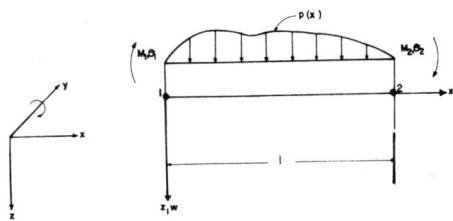

FIG. 23-3

The nodes are numbered 1 and 2. Let the deflexions at 1 and 2 (positive downward) be w_1 and w_2 and let the slopes of the neutral axis be θ_1 and θ_2 at 1 and 2 respectively. The right hand rule is followed for defining positive moments and rotations at the nodes. The element has four degrees of freedom consisting of two deflexions and two rotations. The nodal displacements may be collected together and written as:

$$[\Delta]^T = [w_1 \quad \theta_1 \quad w_2 \quad \theta_2] \qquad (23\text{-}1)$$

The following step by step procedure will be employed to set up the Element Stiffness Matrix.

Step 1

The first step is to select a polynomial displacement function to represent the displacement w at a distance x from the origin. Drawing on our knowledge of elementary strength of materials, a cubic is considered suitable and we may write:

$$w = a_1 + a_2 x \pm a_3 x^2 + a_4 x^3 \qquad (23\text{-}2)$$

Or, in matrix form:

$$[w] = [1 \quad x \quad x^2 \quad x^3] \begin{Bmatrix} a_1 \\ a_2 \\ a_3 \\ a_4 \end{Bmatrix} = [P]\{A\} \qquad (23\text{-}3)$$

where,

$$[P] = [1 \quad x \quad x^2 \quad x^3]$$

and,

$$\{A\} = \begin{Bmatrix} a_1 \\ a_2 \\ a_3 \\ a_4 \end{Bmatrix}$$

Because positive rotations result in positive w,

$$\theta = \frac{dw}{dx} = a_2 + 2a_3x + 3a_4x^2 \tag{23-4}$$

We may apply (23-3) and (23-4) to the ends of the element to get:

$$\begin{Bmatrix} w_1 \\ \theta_1 \\ w_2 \\ \theta_2 \end{Bmatrix} = \begin{bmatrix} 1 & 0 & 0 & 0 \\ 0 & 1 & 0 & 0 \\ 1 & \ell & \ell^2 & \ell^2 \\ 0 & 1 & 2\ell & 3\ell^2 \end{bmatrix} \begin{Bmatrix} a_1 \\ a_2 \\ a_3 \\ a_4 \end{Bmatrix}$$

$$\begin{array}{ccc} \{\Delta\} & C & \{A\} \\ 4 \times 1 & 4 \times 4 & 4 \times 1 \end{array}$$

Or,

$$\{\Delta\} = [C]\{A\} \tag{23-5}$$

where $\{\Delta\}$ are the *nodal displacements* of the element. The matrix $[C]$ may be inverted by solving the four simultaneous equations represented by (23-5) to determine a_1, a_2, a_3 and a_4. C^{-1} so found is given below:

$$C^{-1} = \begin{bmatrix} \ell^3 & 0 & 0 & 0 \\ 0 & \ell^3 & 0 & 0 \\ -3\ell & -2\ell^2 & 3\ell & \ell^2 \\ 2 & \ell & -2 & \ell \end{bmatrix} \tag{23-6}$$

From 23-5,

$$\{A\} = [C]^{-1}\{\Delta\} \tag{23-7}$$

The substitution of (23-7) in (23-3) leads to:

$$\underset{1 \times 1}{w} = \underset{1 \times 4}{[P]} \underset{4 \times 4}{[C]^{-1}} \underset{4 \times 1}{\{\Delta\}} = \underset{1 \times 1}{[N]\{\Delta\}} \tag{23-8}$$

In what follows, $[P][C]^{-1}$ will be denoted by $[N]$.

What we have achieved in Step 1 is to find an expression for the displacements $\{w\}$ at any point in the beam element in terms of the nodal displacement $\{\Delta\}$.

Step 2

We next set up an expression for the strain ϵ. For a beam, the strain is the curvature. i.e.

$$\epsilon = -\frac{d^2 w}{dx^2}$$

Or,

$$\epsilon = -\frac{d^2}{dx^2}[P][C]^{-1}\{\Delta\} \tag{23-9}$$

Note that $[C]^{-1}$ and $\{\Delta\}$ involve only constants and only the matrix $[P]$ need to be differentiated. Denoting $-\frac{d^2}{dx^2}[P]$ by $[M]$, we have:

$$\epsilon = [M][C]^{-1}\{\Delta\}$$
$$= [0 \quad 0 \quad -2 \quad -6x][C]^{-1}\{\Delta\} \tag{23-10}$$

Step 3

An expression is now set up for stress σ. For a beam 'the stress' corresponding to curvature is an internal moment $= [D]\epsilon$ where $D = EI$.

$$Moment = [D][M][C]^{-1}\{\Delta\} \tag{23-11}$$

Step 4

An expression for the strain energy may now be written as:

$$U = \tfrac{1}{2} \int \{\Delta\}^T [C^{-1}]^T [M]^T [D][M][C]^{-1}\{\Delta\} \, dv \qquad (23\text{-}12)$$

The integration is to be carried over the volume of the element.

Step 5

Let the element be acted upon by a distributed load $p(x)$. The expression for the work V done by external forces takes the form:

$$V = - \int \Delta \cdot p(x) \, dx$$

Or,

$$V = - \int_0^l \{\Delta\}^T [N]^T p(x) \, dx \qquad (23\text{-}13)$$

Step 6

We may form the expression for potential energy as:

$$\Pi = U + V$$

The theorem of potential energy states that of all displacement functions satisfying boundary conditions, those that satisfy the equations of equilibrium must lead to a stationary value of the potential energy. Hence,

$$\frac{\partial \Pi}{\partial \{\Delta\}} = \int_v [C^{-1}]^T [M]^T [D][M][C]^{-1} \, dv \{\Delta\}$$

$$- \int_0^l [N]^T p(x) \, dx = 0 \qquad (23\text{-}14)$$

Let us introduce the following notation:

$$[k] = \int_v [C^{-1}]^T [M]^T [D][M][C]^{-1} \, dv \qquad (23\text{-}15)$$

and

$$\{Q\} = \int_0^l [N]^T p(x) \, dx \qquad (23\text{-}16)$$

$[k]$ and $[Q]$ are respectively known as the *Element Stiffness Matrix* and *the consistent load vector for the element*. The consistent load vector allocates the share of the external load acting on the element to nodes 1 and 2. When all the stiffness matrices of elements composing the structure are properly assembled, we arrive at the Global Stiffness Matrix of the structure denoted by $[K]$. Similarly the assembly of the consistent load vectors will result in the load vector for the structure designated by $[F]$.

Step 7

We now arrive at the following relationship for the element

$$\{Q\} = [k]\{\Delta\} \qquad (23\text{-}17)$$

Step 8

The element stiffness matrix is computed by evaluating integral (23-15). Evaluation of the volume integral in this instance merely involves integrating over the length of the element. Thus,

$$[k] = \int_0^l [C^{-1}]^T [M]^T [D][M][C]^{-1} \, dx$$

Noting that $D = EI$,

$$[k] = EI \int_0^l [C^{-1}]^T [M]^T [M][C]^{-1} \, dx$$

The steps involved are as follows:

First form $[M]^T[M]$ to get:

$$[M]^T[M] = \begin{bmatrix} 0 & 0 & 0 & 0 \\ 0 & 0 & 0 & 0 \\ 0 & 0 & 4 & 12x \\ 0 & 0 & 12x & 36x^2 \end{bmatrix}$$

Next premultiply by $[C]^{-1}$ to get:

$$\frac{1}{l^3}\begin{bmatrix} 0 & 0 & 0 & 0 \\ 0 & 0 & 0 & 0 \\ 0 & 0 & 4 & 12x \\ 0 & 0 & 12x & 36x^2 \end{bmatrix}\begin{bmatrix} l^3 & 0 & 0 & 0 \\ 0 & l^3 & 0 & 0 \\ -3l & -2l^2 & -3l & -l^2 \\ 2 & l & -2 & l \end{bmatrix}$$

$$= \frac{1}{l^3}\begin{bmatrix} 0 & 0 & 0 & 0 \\ 0 & 0 & 0 & 0 \\ -12l+24x & -8l^2+12lx & \begin{matrix}-12l\\-24x\end{matrix} & -4l^2+12lx \\ \begin{matrix}-36lx\\+72x^2\end{matrix} & \begin{matrix}-24l^2x\\+36lx^2\end{matrix} & \begin{matrix}-36lx\\-72x^2\end{matrix} & \begin{matrix}-12l^2x\\+36lx^2\end{matrix} \end{bmatrix}$$

Next premultiplication by $[C^{-1}]^T$ and EI are carried out. The result is integrated from 0 to l to give

$$[k] = \frac{EI}{l^3}\begin{bmatrix} 12 & 6l & -12 & 6l \\ & 4l^2 & -6l & 2l^2 \\ \text{symmetric} & & 12 & -6l \\ & & & 4l^2 \end{bmatrix} \qquad (23\text{-}18)$$

23-5 Shape Functions

The Stiffness Matrix for the beam element can be set up in a more elegant manner by utilizing *shape functions*. There are two overriding advantages to be gained by following this alternative procedure. The inversion of $[C]$ which can be quite lengthy and tedious for plate bending and thin shell problems is circumvented. The integrals involved in setting up the element stiffness matrix and the consistent load vector can be evaluated by simple numerical procedures such as Gauss Quadrature. For these reasons, the shape function procedure will be followed in the rest of this chapter.

As already outlined, the basic problem in the finite element procedure is to express the displacements at any point on the element in terms of the nodal displacements and to arrive at the relationship given in (23-8). The expression for w for a beam may be built up by superimposing four shape functions each multiplied by the corresponding nodal displacement. We have to find four functions collectively denoted by $[N]$ so that

$$[N] = [N_1 \quad N_2 \quad N_3 \quad N_4].$$

The shape function N_1 is such that it makes $w_1 = 1$ and $w_2 = \theta_2 = \theta_1 = 0$. Similar remarks apply to N_2, N_3 and N_4. The four shape functions are shown sketched in Fig. 23-4.

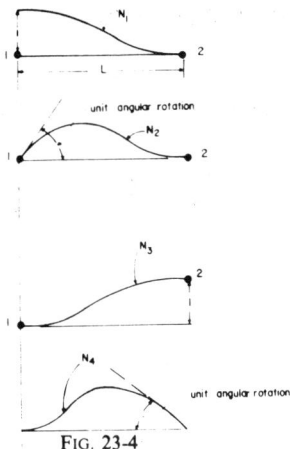

FIG. 23-4

Referring to the beam element in Fig. 23-3 and introducing the non-dimensional co-ordinate $s = \dfrac{x}{l}$, the shape functions may be found by using the following procedure.

Let us suppose that the shape function corresponding to w_1 is to be found. Let $N_1 = As^3 + Bs^2 + Cs + d$, where the undetermined coefficients A, B, C and D are found from the four following conditions:

$$N_1 = 1 \text{ at } s = 0; \qquad N_1 = 0 \text{ at } s = 1$$

Also, $\dfrac{dN_1}{ds} = 0$ at $s = 0$ and at $s = 1$.

It is easily verified that:

$$A = 2; B = -3 \quad C = 0 \text{ and } D = 1.$$

Consequently,

$$N_1 = 1 - 3s^2 + 2s^3$$

Similarly,

$$N_2 = ls(1 - 2s + s^2)$$
$$N_3 = s^2(3 - 2s) \quad \text{and}$$
$$N_4 = ls^2(s - 1)$$

Hence,

$$w(x) = N_1 w_1 + N_2 \theta_1 + N_3 w_2 + N_4 \theta_2 \qquad (23\text{-}19)$$

Or $w = [N]\{w\}$

For finding slopes and curvatures, we need derivatives which are:

$$\frac{dw}{dx} = \theta = \frac{dw}{ds} \frac{ds}{dx}$$

Noting that $s = \dfrac{x}{l}$, $\dfrac{ds}{dx} = \dfrac{1}{l}$. Or

$$\frac{dw}{dx} = \frac{1}{l} \frac{dw}{ds} \qquad (23\text{-}20)$$

$$\frac{d^2 w}{dx^2} = \frac{1}{l^2} \frac{d^2 w}{ds^2}$$

$$= \frac{1}{l^2} \frac{d^2}{ds^2} [N]\{\Delta\} = [B]\{\Delta\} \qquad (23\text{-}21)$$

We may proceed as indicated in Steps 2 to 7 of Art 23-4 to arrive at the element stiffness matrix as:

$$[k] = \int_v [B]^T [D][B] \, dv$$

which for the beam problem takes the form:

$$[k] = EIl \int_0^1 [B]^T [B] \, ds, \text{ noting that } [D] = EI \text{ and } dx = l \, ds.$$

Similarly,

$$\{Q\} = l \int_0^1 [N]^T p \, ds.$$

The element stiffness matrix $[k]$ may now be evaluated as follows:

$$[B] = \frac{1}{l^2} \frac{d^2}{ds^2}[N]$$

$$= \frac{1}{l^2} [6(2s - 1) \quad 2l(3s - 2) \quad -6(2s - 1) \quad 2l(3s - 1)]$$

$$[k]_{(4 \times 4)} = EIl \int_0^1 \left\{ \begin{array}{c} \dfrac{6(2s - 1)}{l^2} \\[2mm] \dfrac{2(3s - 2)}{l} \\[2mm] \dfrac{-6(2s - 1)}{l^2} \\[2mm] \dfrac{2(3s - 1)}{l} \\[2mm] {\scriptstyle (4 \times 1)} \end{array} \right\} \underbrace{\left[\dfrac{6(2s - 1)}{l^2} \quad \dfrac{2(3s - 2)}{l} \quad \dfrac{-6(2s - 1)}{l^2} \quad \dfrac{2(3s - 1)}{l} \right]}_{(1 \times 4)} ds.$$

$$\tag{23-22}$$

The matrix $[k]$ will be (4×4). By way of illustration, let us compute the element k_{11}.

$$k_{11} = \frac{EI}{l} \int_0^1 \frac{36}{l^4} (2s - 1)^2 \, ds.$$

The integrations involved are more easily performed by Gauss-Quadrature if the limits are from -1 to $+1$. Such a change can be effected by the transformation:

$$s = \frac{1}{2}(1 + k) \qquad \text{and} \qquad ds = \frac{1}{2} \, dk.$$

Hence,

$$k_{11} = \frac{18EI}{l^3} \int_{-1}^{+1} k^2 \tag{23-23}$$

23-6   Gauss-Quadrature

Integrals such as (23-23), often encountered in Finite Element Analysis, are most easily evaluated by a numerical procedure known as Gauss-Quadrature.

This procedure is useful in finding a definite integral $\int_{-1}^{+1} F(x)\ dx$. According to this rule,

$$\int_{-1}^{+1} F(x)\ dx = \sum_{i=1}^{n} C_i F(x_i).$$

The abscissa points at which evaluation of the function is to be carried out are designated by x_i and the C_i are weighting coefficients. The points x_i are sometimes referred to as *Gauss points*. The number of sampling points x_i to be used will depend on the function involved and the accuracy desired. In Table 23-1, the values of x_i and C_i are given.

Table 23-1 Gauss Quadrature

No. of Points n	Coefficients C_i	Abiscissas x_i
2	$C_1 = C_2 = 1.0$	$-x_1 = x_2 = 0.5773502962$
3	$C_1 = C_3 = 0.55555$ $C_2 = 0.888888$	$-x_1 = x_3 = 0.7745966692$ $x_2 = 0.0$
4	$C_1 = C_4 = 0.3478548451$ $C_2 = C_3 = 0.6521451549$	$x_1 = x_4 = 0.86113\ 63116$ $-x_2 = x_3 = 0.33998\ 10436$
5	$C_1 = C_5 = 0.2369268851$ $C_2 = C_4 = 0.4786286705$ $C_3 = 0.568888$	$-x_1 = x_5 = 0.9061798459$ $-x_2 = x_4 = 0.5384693101$ $x_3 = 0.0$
6	$C_1 = C_6 = 0.1713244924$ $C_2 = C_5 = 0.3607615730$ $C_3 = C_4 = 0.4679139346$	$-x_1 = x_6 = 0.93246\ 95142$ $-x_2 = x_5 = 0.66120\ 93865$ $-x_3 = x_4 = 0.2386191861$

For evaluating a low order integral involving polynomials such k^2 in (23-24), the use of two Gauss points is adequate. From Table 23-1, the appropriate x_i values are $\pm \dfrac{1}{\sqrt{3}}$ and $C_i = 1.0$. Hence,

$$k_{11} = \frac{18EI}{l^3} \int_{-1}^{+1} k^2 = \frac{18EI}{l^3} \left[\frac{1}{3} + \frac{1}{3} \right] = \frac{12EI}{l^3}$$

Alternately, the numerical integrations involved can be carried out on a programmable calculator such as HP-34C. Using either of these methods, the element stiffness matrix [k] is evaluated as:

$$[k] = \begin{bmatrix} 12 & 6l & -12 & 6l \\ & 4l^2 & -6l & 2l^2 \\ & & 12 & -6l \\ \text{symmetric} & & & 4l^2 \end{bmatrix} \qquad (23\text{-}24)$$

Comparison with equation (23-18) will show that the value of $[k]$ is identical with what was arrived at previously.

23-7 Illustrative Problem

The application of the Finite Element method to beams is best explained by means of a fully worked example.

Example 23-1

A cantilever beam of span L (Fig. 23-5) carries a uniformly varying load whose intensity varies from p at the fixed end to zero at the free end.

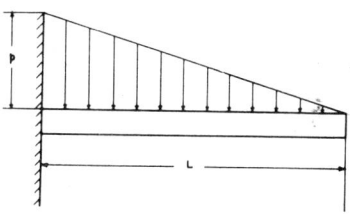

FIG. 23-5

Arrive at its deflexions and slopes at the free end and $\dfrac{L}{2}$ from the fixed end.

The beam has a uniform flexural rigidity of EI.

SOLUTION

STEP 1

The beam is conveniently divided into two elements each of which is of length l so that $L = 2l$. The element numbers are enclosed in squares and the node numbers are circled (Fig. 23-6).

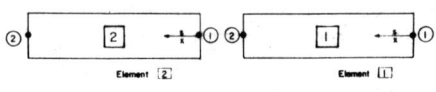

FIG. 23-6

STEP 2

Co-ordinate x and non-dimensional co-ordinate $s = \dfrac{x}{l}$ are chosen as indicated. $l \, ds = dx$.

$[\Delta]^T = [w_1 \quad \theta_1 \quad w_2 \quad \theta_2]$ for each element.

$[N] = [(1 + 2s^3 - 3s^2), \quad l(s - 2s^2 + s^3), \quad (3s^2 - 2s^3), \quad l(s^3 - s^2)]$

The consistent load vector has to be set up for element ①. To this end, the load intensity at a non-dimensional distance s needs to be found as $p(s) = \dfrac{p}{2}s$ from Fig. 23.7.

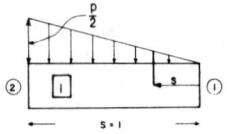

FIG. 23-7

The consistent load vector

$$\{Q\} = l \int_0^1 \frac{p}{2}s \, ds \, [N]^T$$

$$= \frac{lp}{2} \int_0^1 \begin{Bmatrix} s(1 - 3s^2 + 2s^3) \, ds \\ ls^2(1 - 2s + s^2) \, ds \\ s^3(3 - 2s) \, ds \\ ls^3(s - 1) \, ds \end{Bmatrix}$$

Simplifying,

$$\{Q\} = \begin{Bmatrix} \dfrac{3}{40} pl \\[2mm] \dfrac{1}{60} pl^2 \\[2mm] \dfrac{7}{40} pl \\[2mm] -\dfrac{1}{40} pl^2 \end{Bmatrix} \tag{23-25}$$

In a similar manner, we proceed to set up the consistent load vector for element ② as follows. At a distance s from node (1) of element ②, the load intensity (Fig. 23.8) is found as:

$$\frac{p}{2} + \frac{p}{2}s = \frac{p}{2}(1 + s).$$

Consistent load vector for element ② therefore takes the form:

$$\{Q\} = \frac{lp}{2} \int_0^1 \begin{Bmatrix} (1 + s)(1 - 3s^2 + 2s^3) \, ds \\ l(1 + s)s(1 - 2s + s^2) \, ds \\ s^2(1 + s)(3 - 2s) \, ds \\ ls^2(1 + s)(s - 1) \, ds \end{Bmatrix}$$

Simplifying,

$$\{Q\} = \left\{ \begin{array}{c} \dfrac{13}{40}\,pl \\[2ex] \dfrac{7}{120}pl^2 \\[2ex] \dfrac{17}{40}\,pl \\[2ex] -\dfrac{pl^2}{15} \end{array} \right\} \tag{23-26}$$

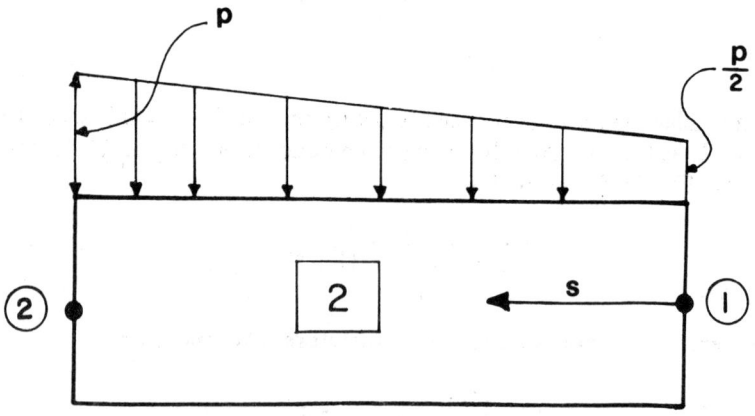

Fig. 23-8

We are now in a position to formulate the following relationships for elements ⑴ and ⑵.

For element $\boxed{1}$,

$$
\frac{EI}{l^3}
\begin{bmatrix}
12 & 6l & -12 & 6l \\
6l & 4l^2 & -6l & 2l^2 \\
-12 & -6l & 12 & -6l \\
6l & 2l^2 & -6l & 4l^2
\end{bmatrix}
\begin{Bmatrix}
w_1 \\ \theta_1 \\ w_2 \\ \theta_2
\end{Bmatrix}
=
\begin{Bmatrix}
\dfrac{3}{40}pl \\[2mm]
\dfrac{1}{60}pl^2 \\[2mm]
\dfrac{7}{40}pl \\[2mm]
-\dfrac{1}{40}pl^2
\end{Bmatrix}
\tag{23-27}
$$

For element $\boxed{2}$,

$$
\frac{EI}{l^3}
\begin{bmatrix}
12 & 6l & -12 & 6l \\
6l & 4l^2 & -6l & 2l^2 \\
-12 & -6l & 12 & -6l \\
6l & 2l^2 & -6l & 4l^2
\end{bmatrix}
\begin{Bmatrix}
w_1 \\ \theta_1 \\ w_2 \\ \theta_2
\end{Bmatrix}
=
\begin{Bmatrix}
\dfrac{13}{40}pl \\[2mm]
\dfrac{7}{120}pl^2 \\[2mm]
\dfrac{17}{40}pl \\[2mm]
-\dfrac{pl^2}{15}
\end{Bmatrix}
\tag{23-28}
$$

Transition from local to Global Co-ordinates and assembly of Global Stiffness Matrix and Load Vector

The global numbering of the nodes is indicated in Fig. 23-9. It is clear that in so far as element $\boxed{1}$ is concerned the local and global numbers attached to

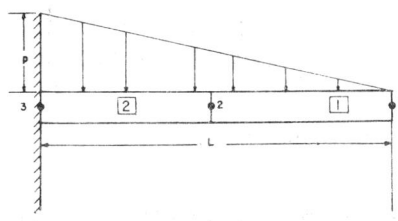

Fig. 23-9

its nodes happen to be the same. For element ② its local node 1 becomes global node 2 and its local node 2 is now labelled global node 3. To arrive at the Global Stiffness Matrix and the global load vector, element stiffness matrices and consistent load vectors have to be superimposed as schematically shown in (23-29).

$$
\begin{bmatrix}
12 & 6l & -12 & 6l & - & \\
6l & 4l^2 & -6l & 2l^2 & - & \\
-12 & -6l & \begin{matrix}12 \\ +12\end{matrix} & \begin{matrix}-6l \\ +6l\end{matrix} & -12 & 6l \\
6l & 2l^2 & \begin{matrix}-6l \\ +6l\end{matrix} & \begin{matrix}4l^2 \\ +4l^2\end{matrix} & -6l & 2l^2 \\
 & & -12 & -6l & 12 & -6l \\
 & & 6 & 2l^2 & -6l & 4l^2
\end{bmatrix}
\begin{Bmatrix}
w_1 \\ \theta_1 \\ w_2 \\ \theta_2 \\ w_3 \\ \theta_3
\end{Bmatrix}
=
\begin{Bmatrix}
\dfrac{3}{40}pl \\[2mm]
\dfrac{1}{60}pl^2 \\[2mm]
\dfrac{7}{40}pl + \dfrac{13}{40}pl \\[2mm]
-\dfrac{1}{40}pl^2 + \dfrac{7}{120}pl^2 \\[2mm]
\dfrac{17}{40}pl \\[2mm]
-\dfrac{pl^2}{15}
\end{Bmatrix}
$$

Node 2 being common to elements ① and ②, the contributions of the two are to be added as indicated.

Simplifying and inserting boundary conditions $w_3 = \theta_3 = 0$, we get:

$$
\begin{bmatrix}
12 & 6l & -12 & 6l \\
6l & 4l^2 & -6l & 2l^2 \\
-12 & 6l & 24 & 0 \\
6l & 2l^2 & 0 & 8l^2
\end{bmatrix}
\begin{Bmatrix}
w_1 \\ \theta_1 \\ w_2 \\ \theta_2
\end{Bmatrix}
=
\begin{Bmatrix}
\dfrac{3}{40}pl \\[2mm]
\dfrac{1}{60}pl^2 \\[2mm]
\dfrac{1}{2}pl \\[2mm]
\dfrac{pl^2}{30}
\end{Bmatrix}
\dfrac{l^3}{EI}
\tag{23-29}
$$

The four resulting linear simultaneous equation may be written as:

$$
12w_1 + 6l\theta_1 - 12w_2 + 6l\theta_2 = \frac{3}{40}pl
$$

$$
6lw_1 + 4l^2\theta_1 - 6lw_2 + 2l^2\theta_2 = \frac{1}{60}pl^2
$$

$$-12w_1 + 6l\theta_1 + 24w_2 \quad = \frac{pl}{2}$$

and

$$6lw_1 + 2l^2\theta_1 + 8l^2\theta_2 \quad = \frac{pl^2}{30} \qquad (23\text{-}30)$$

Solution of the equations will give w_1, θ_1, w_2 and θ_2. The results obtained by Finite Element Analysis (F.E.M.) are shown compared with those arrived at the Elementary Strength of Materials (S.M.). The are identical.
Note that $L = 2l$

F.E.M. Solutions   *S.M. Solutions*

$$w_1 = 0.0334\frac{pL^4}{EI} \qquad\qquad w_1 = 0.0334\frac{pL^4}{EI}$$

$$w_2 = 0.0128\frac{pL^4}{EI} \qquad\qquad w_2 = 0.0128\frac{pL^4}{EI}$$

$$\theta_1 = -0.0417\frac{pL^3}{EI} \qquad\qquad \theta_1 = -0.0417\frac{pL^3}{EI}$$

$$\theta_2 = -0.0391\frac{pL^3}{EI} \qquad\qquad \theta_2 = -0.0391\frac{pL^3}{EI} \qquad (23\text{-}31)$$

23-8 Requirements for Convergence

Certain conditions have to be satisfied by chosen displacement functions, if the results are to converge monotonically towards correct answers, as we progress from a coarser to finer mesh of finite elements. These requirements may be stated as follows:

(i) Interelement compatability requirements have to be met. This condition is satisfied if interelement compatability is ensured to the order of $n - 1$, where n is the order of highest derivative occurring in the potential energy function. For a beam, the highest derivative found in the potential energy function is $\frac{d^2w}{dx^2}$. Hence, interelement compatability with respect to both w and $\frac{dw}{dx}$ has to be achieved. The same requirements apply to plate bending problems. In problems of plane stress, the highest derivative present in the potential energy function is of the first order. It will therefore be sufficient if compatability is maintained up to a zero order of the derivatives. In other words, it is enough if the displacements u and v in the x and y directions respectively are inter-element compatible. Displacement function satisfying the foregoing compatability conditions are described as *conformable*. Depending upon the order to derivations up to which interelement compatability is achieved, displacement functions are characterized as C^0, C^1, etc., the superscript denoting the order.

(ii) The selected displacement function must be capable of representing all possible rigid body motions properly.

(iii) The displacement function must include terms capable of describing all relevant constant states of strain to which the element will be subjected when eventually shrunk to infinitesimal size as a result of successive mesh refinement. It has been demonstrated that the absence of such terms may cause the results to converge to a wrong answer in plate-bending problems [191].

(v) If the selected displacement function is a polynomial, it has to be complete. For instance, if it is a cubic, it must include all the cubic, quadratic, linear and constant terms appearing in the Pascal Triangle up to the third degree (Fig. 23-10). This, strictly speaking, is not a requirement.

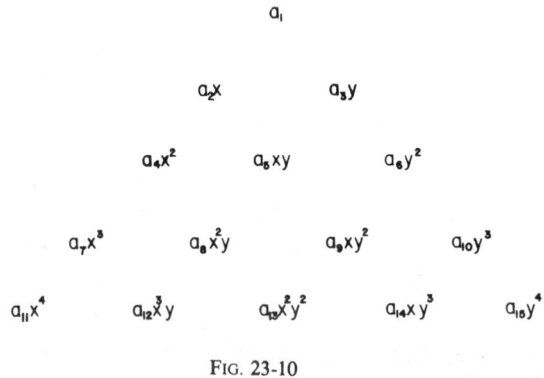

FIG. 23-10

Let us, for example, consider the choice of displacement functions u and v for a plane stress problem. If rectangular elements are used, there are altogether eight degrees of freedom, two at each of the four nodes. Thus the displacement function for u and v must have four terms. The first three are easily chosen as a_1, a_2x and a_3y. The fourth term has to be chosen from among the three quadratic terms of the Pascal Triangle. We will be violating the condition of completeness by selecting only one of second degree terms. The natural choice is a_5xy; it will at least preserve geometrical isotropy because the inclusion of this term will not imply a bias towards x or y.

Although the polynomial displacement functions for u and v are thus incomplete, they have been found to give satisfactory results. It has been pragmatically demonstrated that acceptable accuracy and convergence can sometimes by achieved even if one or more of the foregoing conditions are violated. Thus, the 12 term polynomial displacement function presented in the next article for plate bending problems yields satisfactory results, although it is incomplete and continuous only up to C^0 against a rigorous C^1 requirement. The inclusion of the required rigid body terms is not always possible, unless extremely complex displacement functions are selected. Haisler and Stricklin have shown [192] that the omission of rigid body terms for shells of revolution does not always affect the accuracy of the results. Sometimes, the rigorous inclusion of all rigid body terms may not turn out to be an unmixed blessing. Pecknold and Schnobrich [193] have shown that the inclusion of all the five rigid body terms in the displacement functions for shallow shells leads to some inplane interelement discontinuities. Cantin [194] shows how solutions deficient in rigid body terms can be corrected *a posteriori* by the addition of extra terms.

The subject of convergence requirements is bedevilled by numerous complexities. The reader interested in a more detailed study of the subject is directed to texts on the Finite Element Method [185] [195] [190].

23-9 Application to Plate Bending Problems

Analogy with beams would suggest that the so-called 'crossed-beam' displacement function

$$w = (a_1 + a_2x + a_3x^2 + a_4x^3)(b_1 + b_2x + b_3x^2 + b_4x^3)$$

might prove suitable for plate-bending problems utilizing rectangular elements. However, it will be seen on recasting the expression into shape function form that it lacks the simple twist term necessary to represent constant shearing strain. [185] The rectangular element has 12 apparent degrees of freedom (d.o.f.) in the form of a displacement w and two angular displacements θ_x and θ_y at each of its 4 nodes. It is therefore logical to look for a 12 term polynomial. Looking at the Pascal Triangle, the terms up to the third degree will provide ten terms. The terms x^3y and xy^3 provide the most logical choice because the term x^2y^2 cannot be paired and the inclusion of x^4 and y^4 will lead to quartic variation of w along the edges leading to less manageable discontinuities of displacements along interelement boundaries. The chosen displacement function therefore takes the form:

$$w = a_1 + a_2x + a_3y + a_4x^2 + a_5xy + a_6y^2 + a_7x^3$$
$$+ a_8x^2y + a_9xy^3 + a_{10}y^3 + a_{12}x^3y + a_{14}xy^3$$

(23-32)

It can be readily recast into shape function form as follows by choosing the origin of co-ordinates at the centre of the rectangular element and introducing non-dimensional co-ordinates s and t (Fig. 23-11),

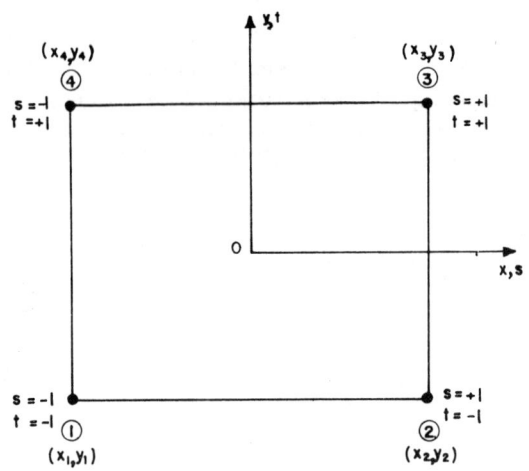

Fig. 23-11

where,

$$s = \frac{x}{x_2} ; \qquad t = \frac{y}{y_3}$$

$dx = x_2\ ds$ and $dy = y_3\ dt$.

Before we proceed to rewrite w in the shape function form, it is relevant to note some of its characteristics. It satisfies the governing differential equation for equilibrium. The necessary rigid body and constant strain terms are present. It is an incomplete quartic polynomial providing continuity of the order of C^0. Consequently, the angular displacements θ_x and θ_y are not interelement compatible. Positive senses and directions of concentrated moments forces slopes and displacements are as indicated in Fig. 23-12 for the rectangular element sketched in Fig. 23-11. It is to be noted that $\theta_x = \frac{\partial w}{\partial y}$ and $\theta_y = -\frac{\partial w}{\partial x}$. The minus sign indicates that a positive rotation about the y axis causes a negative w displacement. With these preliminaries, we now proceed to set up the displacement w in shape function form written as:

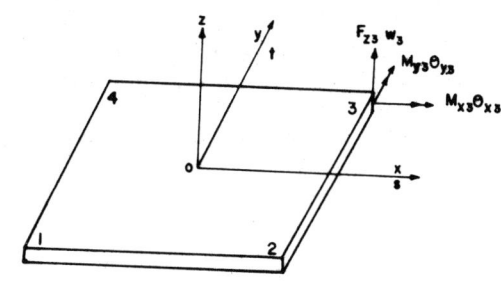

FIG. 23-12

$$w = [[Nw] \quad [N\theta x] \quad [N\theta y]] \begin{Bmatrix} \{w\} \\ \{\theta_x\} \\ \{\theta_y\} \end{Bmatrix} = [N]\{\Delta\} \qquad (23\text{-}33)$$

$$\underset{1 \times 1}{} \qquad \underset{1 \times 12}{} \qquad \underset{12 \times 1}{}$$

where,

$$[Nw] = \{[N_1(x)N_1(y)][N_2(x)N_1(y)][N_2(x)N_2(y)][N_1(x)N_2(y)]\}$$

$$[N\theta x] = \left\{ \left[\frac{1}{2}(1-s)N_3(y)\right]\left[\frac{(1+s)}{2}N_3(y)\right]\left[-\frac{(1+s)}{2}N_4(y)\right] \right.$$

$$\left. \left[-\frac{(1-s)}{2}N_4(y)\right] \right\}$$

$$[N\theta y] = \left\{ \left[-\frac{1}{2}(1-t)N_3(x)\right]\left[\frac{(1-t)}{2}N_4(x)\right]\left[\frac{(1+t)}{2}N_4(x)\right] \right.$$

$$\left. \left[-\frac{(1+t)}{2}N_3(x)\right] \right\} \qquad (23\text{-}34)$$

The functions involved are:

$$N_1(x) = \frac{1}{4}(s^3 - 3s + 2) \qquad \text{This function satisfies the conditions:}$$

$$N_1(x) = 1 \text{ at } s = -1 N_1(x) = 0 \text{ at } s = +1$$

and

$$\frac{dN_1(x)}{dx} = 0 \text{ at } s = \pm 1$$

$$N_2(x) = \frac{1}{4}(3s - s^3 + 2) \text{ satisfying conditions:}$$

$$N_2(x) = 1 \text{ at } s = +1 \text{ and } N_2(x) = 0 \text{ at } s = -1.$$

Also,

$$\frac{dN_2(x)}{dx} = 0 \text{ at } s = \pm 1$$

$$N_1(y) = \frac{1}{4}(t^3 - 3t + 2) \text{ satisfying conditions:}$$

$$N_1(y) = 1 \text{ at } t = -1 \text{ and } N_1(y) = 0 \text{ at } t = +1$$

Also,

$$\frac{dN_1(y)}{dy} = 0 \text{ at } t = \pm 1$$

$$N_2(y) = \frac{1}{4}(3t - t^3 + 2) \text{ satisfying conditions:}$$

$$N_2(y) = 1 \text{ at } t = +1 \text{ and } N_2(y) = 0 \text{ at } t = -1$$

Also,

$$\frac{dN_2(y)}{dy} = 0 \text{ at } t = \pm 1$$

$$N_3(x) = \frac{x}{4}\left[(s + 1 - s^2) - \frac{x_2}{x}\right] \text{ satisfying conditions:}$$

$$N_3(x) = 0 \text{ at } s = \pm 1$$

Also,

$$\frac{-dN_3(x)}{dx} = 0 \text{ at } s = +1 \text{ and } \frac{-dN_3(x)}{dx} = 1 \text{ at } s = -1$$

$$N_4(x) = \frac{x}{4}\left[(1 - s - s^2) + \frac{x_2}{x}\right] \text{ satisfying conditions:}$$

$$N_4(x) = 0 \text{ at } s = \pm 1$$

Also,

$$\frac{-dN_4(x)}{dx} = 0 \text{ at } s = -1 \text{ and } \frac{-dN_4(x)}{dx} = 1 \text{ at } s = +1$$

$$N_3(y) = \frac{y}{4}\left[(t + 1 - t^2) - \frac{y_3}{y}\right] \text{ satisfying conditions:}$$

$$N_3(y) = 0 \text{ at } t = \pm 1$$

Also,

$$\frac{\partial N_3(y)}{\partial y} = 1 \text{ at } t = -1 \text{ and } \frac{\partial N_3(y)}{\partial y} = 0 \text{ at } t = +1$$

$$N_4(y) = -\frac{y}{4}\left[(1 - t - t^2) + \frac{y_3}{y}\right] \text{ satisfying conditions:}$$

$$N_4(y) = 0 \text{ at } t = \pm 1$$

Also,

$$\frac{\partial N_4(y)}{\partial y} = 1 \text{ at } t = +1 \text{ and } \frac{\partial N_4(y)}{\partial y} = 0 \text{ at } t = -1. \quad (23\text{-}35)$$

It will be verified that the functions chosen satisfy the definition of interpolation functions.

F.E.M.: APPLICATIONS TO THIN SHELLS

23-10 Historical Notes

Thin shells may be modelled by flat-faceted, curved thin shell or solid finite elements. Two comprehensive surveys, tracing the development of these classes of elements, have been presented by Gallagher in 1969 [191] and 1974 [194].

In the following sections, a brief account is given of their respective merits and shortcomings.

23-11   Flat-Faceted Elements

As soon as sophisticated elements become available for representing plane stress and plate bending behaviour, efforts began to be made to account for membrane and bending actions in thin shells by combining them and carrying out a co-ordinate transformation to arrive at the Stiffness Matrix of the complete shell in global co-ordinates [191]. Chu and Schnobrich [197] have successfully used a flat-faceted triangular element with 27 degrees of freedom to model translational shells. Acceptable accuracy is reported. The loss of accuracy resulting from the absence of curvature in the model is sought to be offset by using an element with a large number of degrees of freedom. The shortcomings of representing the smooth continuous surface of a thin shell by flat elements are rather obvious. These are:

(i) There is no coupling between membrane and bending behaviour within the flat elements, although this is present in the shell being modelled on account of its curvature.

(ii) It may happen that along the edge of an element, the lateral displacement function chosen gives a cubical variation while the in-plane displacements normal to the edge are linear. These will match those of the adjoining element only if the two elements are coplanar.

However, errors on this account will tend towards zero as the elements become more nearly coplanar with progressive mesh refinement. [190]

(iii) Another shortcoming is the appearance of 'discontinuity' moments at the junctions of elements, when none are present at those points in the shell being modelled. Gallagher [196] illustrates this situation by referring to the problem of a pressurized hemisphere modelled by truncated cone elements. 'Discontinuity' bending moments show up at the junctions of the elements, although it is known that the structure is subjected only to hoop-tension and is free from bending moments. Gallagher [196] suggests that this problem may be overcome by substituting the computed displacements in the stress displacement relations of the curved shell. Also, numerical experimentation discloses that this effect is less severe for general shells than axisymmetric shells.

The shortcomings of approximating thin shells by flat elements are remediable by the use of either refined elements or a large number of simpler elements. The approach has the merit of extreme simplicity. Moreover, flat elements have the capability of representing rigid body motions without strain [190]. For these reasons, they still find favour with many design offices.

23-12 Curved Thin Shell Elements

It is logical to think in terms of modelling thin shells by means of curved thin shell finite elements. But this seemingly attractive proposition is not without its own problems. Gallagher [196] mentions four difficulties. These may be summarised as:

(i) Difficulties associated with the choice of an appropriate shell theory. Theories available range in sophistication from the simpler shallow shell theories of Marguerre-Vlasov and Novozhilav at one end of the spectrum to the non-shallow theories of Love and Novozhilov. We have also to decide whether element stiffnesses are to be formulated in local or global co-ordinates.

(ii) Difficulties relating to the description of geometry. Enough attention is not always paid in prescribing the geometry, especially the curvature characteristics of the surface, with adequate precision. The resulting geometrical approximations can sometimes more than offset the advantages to be gained by using refined displacement functions.

(iii) Problems involved in incorporating terms in the chosen displacement functions to represent rigid body motions. In curved shell elements, such rigid body motions can be incorporated only by a coupling of the individual displacement terms. Moreover, displacement functions so modified to incorporate rigid body terms may guarantee continuity only at the nodes; but this does not always mean that the performance of the element does not improve [198] [199]. Cantin has pointed out that constructing displacement functions for curved shell elements to meet the requirements of Article 23-8 *apriori* will prove to be a formidable task. Instead, he suggests the addition of a correction *a posteriori* [194].

(iv) Difficulties arise in ensuring C^1 continuity of interelement displacements.

23-13 Solid Elements

Before we proceed to discuss solid elements, two new concepts have to be introduced. These are 'natural' co-ordinate systems and isoparametric elements. Elements in which the same interpolation functions are used to define the shape of the element as well as the displacement functions are described as isoparametric. For example, if isoparametric co-ordinates are employed, the x and y co-ordinates at any point in the quadrilateral shown in (Fig. 23-13a) can be expressed in terms of its nodal co-ordinates $x_1\, y_1, x_2$

y_2, x_3 y_3 and x_4 y_4 and interpolation functions N_i ($i = 1, 2 \ldots 4$) and natural co-ordinates ξ and ν so that:

$$x = \sum_{i=1}^{4} N_i x_i \quad \text{and} \quad y = \sum_{i=1}^{4} N_i y_i \qquad (23\text{-}36)$$

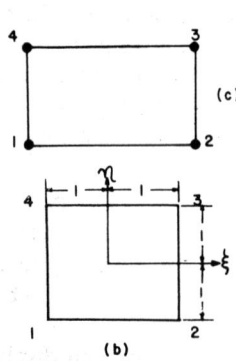

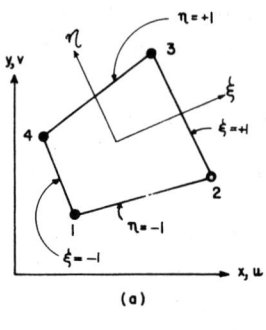

Fig. 23-13

In effect, a unit square in natural co-ordinates (Fig. 23-13b) ξ and ν gets mapped as a quadrilateral in the xy system (Fig. 23-13a). The 'parent element' is a rectangle (Fig. 23-13c). If the quadrilateral is in a state of plane-stress with two degrees of freedom (u_i, v_i) at each of its four nodes, the u and v displacements at any point in the quadrilateral may be written as:

$$u = \sum_{i=1}^{4} N_i u_i \qquad v = \sum_{i=1}^{4} N_i v_i \qquad (23\text{-}37)$$

where,

u_i and v_i are nodal co-ordinates.

On comparing (23-36) and (23-37) it is seen that the same interpolation functions have been employed to describe the shape of the element as well as its displacements. The element just now described is known as a *linear iso-parametric element* because the sides of the element are straight and so is the distribution of displacements.

It may easily be verified that the appropriate interpolation functions are:

$$N_1 = \frac{(1 - \xi)(1 - \eta)}{4} \quad ; \quad N_2 = \frac{(1 + \xi)(1 - \eta)}{4}$$

$$N_3 = \frac{(1 + \xi)(1 + \eta)}{4} \quad ; \quad N_4 = \frac{(1 - \xi)(1 + \eta)}{4} \quad \text{(23-38)}$$

Also,

$$\begin{Bmatrix} x \\ y \end{Bmatrix} = \begin{bmatrix} N_1 & 0 & N_2 & 0 & N_3 & 0 & N_4 & 0 \\ 0 & N_1 & 0 & N_2 & 0 & N_3 & 0 & N_4 \end{bmatrix} \begin{Bmatrix} x_1 \\ y_1 \\ x_2 \\ y_2 \\ x_3 \\ y_3 \\ x_4 \\ y_4 \end{Bmatrix} \quad \text{(23-39)}$$

$$\begin{Bmatrix} u \\ v \end{Bmatrix} = \begin{bmatrix} N_1 & 0 & N_2 & 0 & N_3 & 0 & N_4 & 0 \\ 0 & N_1 & 0 & N_2 & 0 & N_3 & 0 & N_4 \end{bmatrix} \begin{Bmatrix} u_1 \\ v_1 \\ u_2 \\ v_2 \\ u_3 \\ v_3 \\ u_4 \\ v_4 \end{Bmatrix} \quad \text{(23-40)}$$

For the quadrilateral element with eight nodes (Fig. 23-14) the interpolation functions are (190):

$$N_i = \frac{1}{4}(1 + \xi\xi_i)(1 + \eta\eta_i)(\xi\xi_i + \eta\eta_i - 1) \quad \text{for } i = 1, 3, 5, 7$$

$$N_i = \frac{1}{2}(1 - \xi^2)(1 + \eta\eta_i) \quad \text{for } i = 2, 6$$

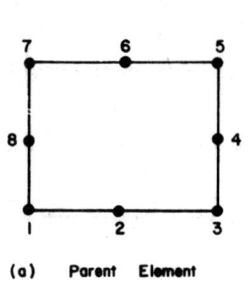

(a)   Parent Element

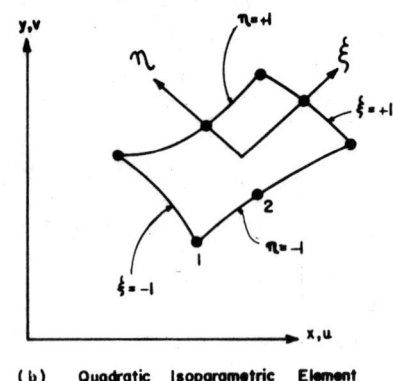

(b)   Quadratic Isoparametric Element

FIG. 23-14

$$N_i = \frac{1}{2}(1 + \xi\xi_i)(1 - \eta^2) \qquad \text{for } i = 4 \text{ and } 8$$

with i taking on values 1 to 8. The consecutive values of ξ_i and η_i are

$$\xi_i = -1, 0, 1, 1, 1, 0, -1, -1 \text{ and}$$
$$\eta_i = -1, -1, -1, 0, 1, 1, 1, 0 \qquad (23\text{-}41)$$

Isoparametric elements satisfy constant strain and rigid body requirements but ensure only C^0 continuity.

A 'degenerate' solid element of the shape sketched in (Fig. 23-15a) is useful in modelling thin shell structures. There are 20 nodes and the eight lines connecting the top and bottom surfaces are straight but not normal. The global co-ordinates x, y and z vary linearly with ζ and so do the direct strains in the ξ and η directions. It is assumed that no strain occurs across the element thickness.

The shape of the element may be defined as:

$$\begin{Bmatrix} x \\ y \\ z \end{Bmatrix} = \sum_{i=1}^{8} N_i \frac{1+\zeta}{2} \begin{Bmatrix} x_{it} \\ y_{it} \\ z_{it} \end{Bmatrix} + \sum_{i=1}^{8} N_i \frac{1-\zeta}{2} \begin{Bmatrix} x_{ib} \\ y_{ib} \\ z_{ib} \end{Bmatrix} \qquad (23\text{-}42)$$

where x_{it}, y_{it}, z_{it} and x_{ib}, y_{ib} and z_{ib} are respectively the cartesian global co-ordinates of the eight top and eight bottom nodes and x, y, z are the

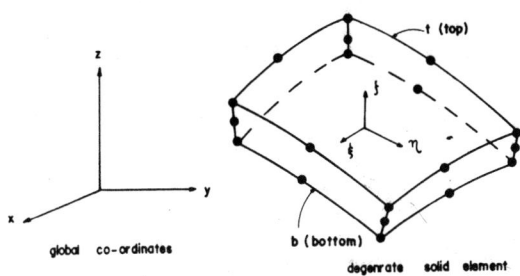

FIG. 23-15*a*

co-ordinates of any point on the element. The top and bottom surfaces are defined by $\zeta = \pm 1$ respectively; clearly the mid-surface is represented by $\zeta = 0$. The following artifice will permit x, y and z to be written in terms of the mid-surface normal co-ordinates. It is seen from (23-42) that co-ordinates x, y, z on the mid-surface with $\zeta = 0$ are the average of the co-ordinates at the top and bottom surfaces for which $\zeta = \pm 1$ respectively. Let us define:

$$2\{x_i \quad y_i \quad z_i\} = \{x_{it} \quad y_{it} \quad z_{it}\} + \{x_{ib} \quad y_{ib} \quad z_{ib}\}$$

$$\bar{V}_{3i} = \{x_{it} \quad y_{it} \quad z_{it}\} - \{x_{ib} \quad y_{ib} \quad z_{ib}\} \qquad (23\text{-}43)$$

Inserting (23-43) in (23-42),

$$\begin{Bmatrix} x \\ y \\ z \end{Bmatrix} = \sum_{i=1}^{i=8} N_i \begin{Bmatrix} x_i \\ y_i \\ z_i \end{Bmatrix} + \sum_{i=1}^{8} N_i \frac{\zeta}{2} \bar{V}_{3i} \qquad (23\text{-}44)$$

We have thus expressed the cartesian global co-ordinates x, y, z at any point on the element in terms of the mid-surface nodal co-ordinates x_i, y_i and z_i.

Element displacements in x, y, z global co-ordinates will now be expressed in terms of the mid-surface displacements u_i, v_i, w_i and by two rotations α_i and β_i of a line initially normal to the mid-surface about two axes parallel to the mid-surface. Referring to Fig. 23-15(b), the vectors $\bar{V}_{1i}$, $\bar{V}_{2i}$ and $\bar{V}_{3i}$ are mutually perpendicular and the rotations α_i and β_i are collinear with $\bar{V}_{2i}$ and V_{1i} respectively. The rotations α_i and β_i cause inplane displacements of a point on the vector $\bar{V}_{3i}$ of magnitudes

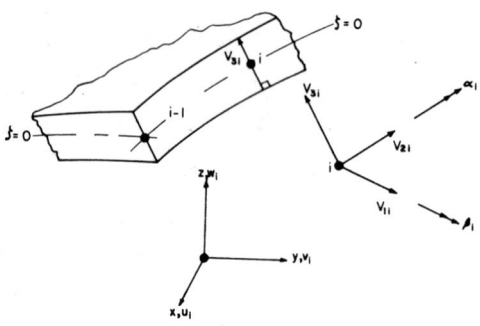

Fig. 23-15b

$$u^1{}_i = \frac{\zeta d_i \alpha_i}{2} \quad \text{and} \quad v^1{}_i = - \frac{\zeta d_i \beta_i}{2}$$

in the direction of $\bar{V}_{1i}$ and $\bar{V}_{2i}$ respectively. d_i is the thickness of the shell at point i. The xyz components of these displacements may be found by making use of the direction cosines of $\bar{V}_{1i}$ and $\bar{V}_{2i}$. If these direction cosines are respectively $(l_{1i} \quad m_{1i} \quad n_{1i})$ and $(l_{2i} \quad m_{2i} \quad n_{2i})$, the displacements at any point of the element in xyz system may be written as:

$$\begin{Bmatrix} u \\ v \\ w \end{Bmatrix} = \sum_{i=1}^{8} N_i \begin{Bmatrix} u_i \\ v_i \\ w_i \end{Bmatrix} + \sum_{i=1}^{i=8} N_i \zeta [\mu_i] \frac{d_i}{2} \begin{Bmatrix} \alpha_i \\ \beta_i \end{Bmatrix} \tag{23-45}$$

The above expression may be seen to consist of two parts. The first part is contributed by the mid-surface nodal displacements u_i, v_i w_i and the second part by the rotations α_i and β_i. A comparison of (23-44) and (23-45) will reveal that the shape as well as the displacements of the solid isoparametric element are described by the same shape functions N_i.

It has been found that the isoparametric solid element described in the foregoing paragraph is too stiff in flexure. When subjected to flexure, it deforms in the shear mode (Fig. 23-16). One of two artifices may be employed to remedy the situation. The first involves the use of a reduced order of numerical integration in arriving at the Stiffness Matrix. Apart from improving the performance of the element, the reduction of the order of integration from $3 \times 3 \times 3$ to $2 \times 2 \times 2$ also leads to considerable reduction in computational labour [200]. Alternatively, nonconforming displacement modes can be added to eliminate transverse displacement as suggested by Wilson et al [201]. Choi and Schnobrich [200] have demonstrated the improvement that results by means of examples.

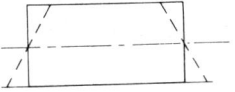

Fig. 23-16

One of the principal advantages inherent in the use of solid elements is that it circumvents the need to select a shell theory.

General purpose computer programs incorporating isoparametric solid elements are available from the ASAS (Atkins Stress Analysis System) Library. These are GCS6, an aribtrarily curved thin shell triangular element with varying thickness (parabolic variation) and GCS8 an arbitrarily curved thin shell quadrilateral element with varying thickness (parabolic variation). Details of these elements may be found in reference [202]. The practising engineer may use these 'black box' programs fairly easily without going into their theory. For example, the element GCS8 has eight mid-surface nodes (Fig. 23-17). It has 32 degrees of freedom. Twelve of these are at the corner nodes i.e. u, v, w at each of the four nodes and twenty at the mid-side nodes consisting of u, v, w at nodes 2, 4, 6 and 8 and rotations θ_x and θ_y at the two Gauss-points situated at distances of $\dfrac{L}{2\sqrt{3}}$ on either side of mid node on each side. A curved beam element GCB3 is also available from ASAS Library to be used in conjunction with elements of the GCS family for handling shells with edge beams.

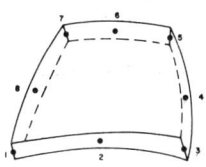

Fig. 23-17

23-12   A Simple Shallow Curved Shell Element

In selecting an element for thin shell analysis, one has to work out the trade-off between accuracy on the one hand and the complexity and cost of computation on the other. A relatively simple shallow curved shell element which gives results of acceptable accuracy is that due to Connor and Brebbia [203]. The stress-strain and moment-curvature relations used in the development of the element are those underlying the shallow shell theory of Vlasov developed in Chapter 19. If a global co-ordinate system is chosen as in Fig. 23-18(a), the stress-resultants acting on a element of the shell will be as shown in Fig. 23-18(b) and (c). In the selected co-ordinate system, the strains take the following form:

$$\epsilon_x = \frac{\partial u}{\partial x} - rw = \frac{\partial u}{\partial x} - \frac{\partial^2 z}{\partial x^2} w$$

$$\epsilon_y = \frac{\partial v}{\partial y} - tw \qquad = \frac{\partial v}{\partial y} - \frac{\partial^2 z}{\partial y^2} w$$

$$\gamma_{xy} = \frac{\partial u}{\partial y} + \frac{\partial v}{\partial x} - 2sw = \frac{\partial u}{\partial y} + \frac{\partial v}{\partial x} - 2 \frac{\partial^2 z}{\partial x \partial y} w \qquad (23\text{-}46)$$

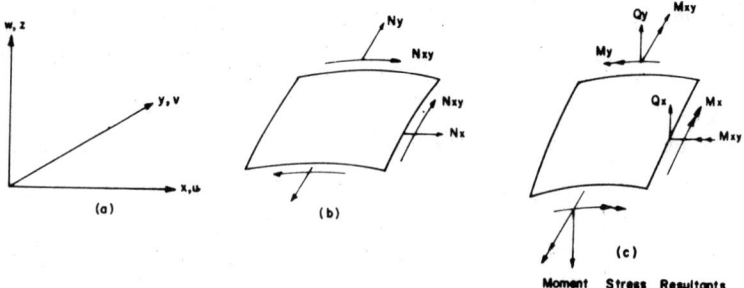

(a)   (b)

(c)

Moment Stress Resultants

FIG. 23-18

The expressions have also been written in their expanded form avoiding the use of the symbols r, s, t because we propose to use s and t to denote non-dimensional local co-ordinates in the subsequent development. Curvatures may be written as:

$$X_x = -\frac{\partial^2 w}{\partial x^2}$$

$$X_y = -\frac{\partial w}{\partial y^2}$$

$$\tau = -\frac{\partial^2 w}{\partial x \partial y} \qquad (23\text{-}47)$$

The stress-strain and moment-curvature relations take the form:

$$N_x = \frac{E_d}{1 - v^2}(\epsilon_x + v\epsilon_y)$$

$$N_y = \frac{E_d}{1 - v^2}(\epsilon_y + v\epsilon_x)$$

$$N_{xy} = N_{yx} = \frac{E_d}{2(1 + v)}\gamma_{xy} = \frac{E_d}{1 - v^2}\frac{(1 - v)}{2}\gamma_{xy}$$

$$M_x = D(\chi_x + v\chi_y)$$

$$M_y = D(\chi_y + v\chi_x)$$

$$M_{xy} = D\frac{(1 - v)}{2}2\tau \qquad (23\text{-}48)$$

Connor and Brebbia introduced the following displacement functions for the element which is assumed to have five degrees of freedom at each of its four nodes:

$$u = a_1 + a_2\bar{x} + a_3\bar{y} + a_4\bar{x}\bar{y}$$

$$v = a_5 + a_6\bar{x} + a_7\bar{y} + a_8\bar{x}\bar{y}$$

$$w = a_9 + a_{10}\bar{x} + a_{11}\bar{y} + a_{12}\bar{x}^2 + a_{13}\bar{x}\bar{y}$$

$$\qquad + a_{14}\bar{y}^2 + a_{15}\bar{x}^3 + a_{16}\bar{x}^2\bar{y} + a_{17}\bar{x}\bar{y}^2$$

$$\qquad + a_{18}\bar{y}^3 + a_{19}\bar{x}^3\bar{y} + a_{20}\bar{x}\bar{y}^3 \qquad (23\text{-}49)$$

It will be seen that the expressions assumed for u and v are those of the plane stress-element and the expression for w is that of the plate-bending element developed in Article 23-11. Unlike in the flat-faceted element, coupling between strain caused by the in-plane displacements (u, v) and curvature caused by w exists because of the curvature of the shell appearing in the expressions for ϵ_x, ϵ_y and γ_{xy}. The five degrees of freedom at each node are represented by u, v, w, θ_x and θ_y. The manner in which the expression for w can be recast into shape function form has already been explained in detail under Article 23-11. Referring to the non-dimensional co-ordinates s and t of (Fig. 23-11), the expressions for u and v can be transformed into shape function form and expressed as:

$$u = [N_1 \quad N_2 \quad N_3 \quad N_4] \begin{Bmatrix} u_1 \\ u_2 \\ u_3 \\ u_4 \end{Bmatrix}$$

$$v = [N_1 \quad N_2 \quad N_3 \quad N_4] \begin{Bmatrix} v_1 \\ v_2 \\ v_3 \\ v_4 \end{Bmatrix}$$

where,

$$N_1 = \frac{1}{4}(1-s)(1-t)$$

$$N_2 = \frac{1}{4}(1+s)(1-t)$$

$$N_3 = \frac{1}{4}(1+s)(1+t)$$

$$N_4 = \frac{1}{4}(1-s)(1+t)$$

and u_1, u_2, u_3, u_4 and v_1, v_2, v_3, v_4 are nodal displacements.   (23-50)

The expressions (23-46) for the strains involve $\dfrac{\partial u}{\partial x}$, $\dfrac{\partial v}{\partial y}$ etc. These may be found as follows:

$$\frac{\partial u}{\partial x} = \begin{bmatrix} \dfrac{\partial N_1}{\partial x} & \dfrac{\partial N_2}{\partial x} & \dfrac{\partial N_3}{\partial x} & \dfrac{\partial N_4}{\partial x} \end{bmatrix} \begin{Bmatrix} u_1 \\ u_2 \\ u_3 \\ u_4 \end{Bmatrix}$$

(23-51)

$$\frac{\partial N_1}{\partial x} = \frac{\partial N_1}{\partial s} \frac{\partial s}{\partial x}. \text{ But } \frac{\partial s}{\partial x} = \frac{1}{x_2}.$$

Hence $\dfrac{\partial N_1}{\partial x} = -\dfrac{1}{4}(1 - t)\dfrac{1}{x_2} = \dfrac{1}{4x_2}(t - 1).$

(Note that s is a function of x only).
Similarly,

$$\frac{\partial N_2}{\partial x} = \frac{1}{4x_2}(1 - t)$$

$$\frac{\partial N_3}{\partial x} = \frac{1}{4x_2}(1 + t)$$

$$\frac{\partial N_4}{\partial x} = -\frac{1}{4x_2}(1 + t)$$

Inserting these into (23-51), we may write:

$$\frac{\partial u}{\partial x} = \left[\frac{1}{4x_2}(t - 1) \quad \frac{1}{4x_2}(1 - t) \quad \frac{1}{4x_2}(1 + t) \quad -\frac{1}{4x_2}(1 + t)\right]\begin{Bmatrix} u_1 \\ u_2 \\ u_3 \\ u_4 \end{Bmatrix}$$

$$(23\text{-}52)$$

Similar expressions may be set up for $\dfrac{\partial u}{\partial y}$, $\dfrac{\partial v}{\partial y}$ and $\dfrac{\partial v}{\partial x}$.

By way of illustration, the Connor and Brebbia element will be now used to analyze a paraboloidal shallow shell on a square base with clamped boundaries under the action of a central concentrated load. The weight of the shell itself will be ignored. In the next section, the illustrative example is worked out by hand computation to provide an insight into the sequential procedure involved.

Example 23-2

PROBLEM

Find the central deflexion of a clamped paraboloid of revolution with a square ground plan of 40' × 40' and central rise of $f = 4'$ when submitted to the action of a central concentrated load of 1 lb. The thickness of the shell $d = 3''$ (Fig. 23-19(a)). The Modulus of Elasticity of concrete $E = 2 \times 10^6$ p.s.i. and Poisson's Ratio $\nu = 0.15$.

STEP 1—GEOMETRY

Referring to Fig. 23-19(a), the equation of the paraboloid may be set up as:

$$z = \frac{2f}{a^2}(x^2 + y^2) \qquad (23\text{-}53)$$

$$\frac{\partial^2 z}{\partial x^2} = \frac{\partial^2 z}{\partial y^2} = \frac{4f}{x^2} = \frac{1}{R} \text{ where } R \text{ is the radius of curvature.}$$

$$R = \frac{a^2}{4f} = \frac{40 \times 40}{4 \times 4} = 100' \qquad \frac{\partial^2 z}{\partial x \partial y} = 0.$$

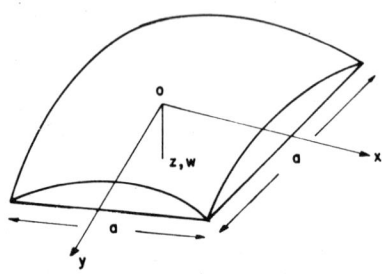

Fig. 23-19*a*

STEP 2—EXPRESSION FOR *w*

Taking advantage of symmetry, it will be enough to consider one quadrant of the shell. We will use a single element ① to represent the quadrant. The shell being clamped, displacements and rotations at nodes 2, 3 and 4 are zero. Because of symmetry, the displacements u, v and the rotations are zero at the centre. We are thus left with one unknown i.e. w_1 the displacement w_1 at the centre. Hence, the expressions for w at any point on the element will involve only the nodal displacement w_1.

Thus, from (23-34),

$$w = \frac{1}{16}(s^3 - 3s + 2)(t^3 - 3t + 2)w_1$$

where,

$$s = \frac{4x}{a} \quad ; \quad t = \frac{4y}{a}$$

and

$$ds = \frac{4}{a} dx \qquad dt = \frac{4}{a} dy$$

$$u = v = 0 \tag{23-53}$$

STEP 3—EXPRESSIONS FOR STRAINS AND CURVATURES

$\epsilon_x = \dfrac{\partial u}{\partial x} - \dfrac{w}{R}$. Noting that $u = 0$ at all the nodes.

$$\epsilon_x = -\frac{1}{16R} (s^3 - 3s + 2)(t^3 - 3t + 2)$$

Similarly,

$$\epsilon_y = -\frac{1}{16R} (s^3 - 3s + 2)(t^3 - 3t + 2)$$

$\gamma_{xy} = 0$ because u and v displacements are zero at all the nodes.

$$\chi_x = -\frac{\partial w^2}{\partial x^2}$$

$$-\frac{\partial w}{\partial x} = \frac{1}{16} (3s^2 - 3)(t^3 - 3t + 2) \frac{\partial s}{\partial x} w_1$$

$$-\frac{\partial^2 w}{\partial x^2} = -\frac{6s}{16} (t^3 - 3t + 2)\left(\frac{\partial s}{\partial x}\right)^2 w_1$$

$$= -\frac{6s}{16} \times \frac{16}{a^2} (t^3 - 3t + 2)w_1$$

Similarly,

$$\chi_y = -\frac{\partial^2 w}{\partial y^2} = -\frac{6s}{a^2} (s^3 - 3s + 2)w_1$$

$$2\tau = -2\frac{\partial^2 w}{\partial x \partial y} = -\frac{2}{16} \times \frac{16}{a^2} \times 9(s^2 - 1)(t^2 - 1)$$

These may be assembled together as Matrix [B].

$$
\begin{Bmatrix} \epsilon_x \\ \epsilon_y \\ \gamma_{xy} \\ \chi_x \\ \chi_y \\ 2\tau \end{Bmatrix} = \begin{bmatrix} -\dfrac{1}{16R}(s^3 - 3s + 2)(t^3 - 3t + 2) \\[2mm] -\dfrac{1}{16R}(s^3 - 3s + 2)(t^3 - 3t + 2) \\[2mm] 0 \\[2mm] -\dfrac{6s}{a^2}(t^3 - 3t + 2) \\[2mm] -\dfrac{6s}{a^2}(s^3 - 3s + 2) \\[2mm] -\dfrac{18}{a^2}(s^2 - 1)(t^2 - 1) \end{bmatrix} [w_1]
$$

$$(6 \times 1) \hspace{5cm} (23\text{-}54)$$

STEP 4—ELASTICITY MATRIX

The Elasticity Matrix [D] for converting strains and curvatures into stress and moment resultants may be written as:

$$
[D] = \frac{E_d}{1 - \nu^2} \begin{bmatrix} 1 & \nu & 0 & 0 & 0 & 0 \\[2mm] \nu & 1 & 0 & 0 & 0 & 0 \\[2mm] 0 & 0 & \dfrac{1-\nu}{2} & 0 & 0 & 0 \\[2mm] 0 & 0 & 0 & \dfrac{d^2}{12} & \dfrac{\nu d^2}{12} & 0 \\[2mm] 0 & 0 & 0 & \dfrac{\nu d^2}{12} & \dfrac{d^2}{12} & 0 \\[2mm] 0 & 0 & 0 & 0 & 0 & \dfrac{1-\nu}{2}\,\dfrac{d^2}{12} \end{bmatrix}
$$

$$(6 \times 6) \hspace{5cm} (23\text{-}55)$$

STEP 5

Form matrix $[D][B]$

$$[D][B] = \frac{E_d}{1 - \nu^2} \begin{bmatrix} -\dfrac{(1+\nu)}{16}(s^3 - 3s + 2)(t^3 - 3t + 2) & [w_1] \\[3mm] -\dfrac{(1+\nu)}{16}(s^3 - 3s + 2)(t^3 - 3t + 2) \\[3mm] 0 \\[3mm] -\dfrac{6\lambda}{a^2}\{s(t^3 - 3t + 2) + \nu t(s^3 - 3s + 2)\} \\[3mm] -\dfrac{6\lambda}{a^2}\{t(s^3 - 3s + 2) + \nu s(t^3 - 3t + 2)\} \\[3mm] -\dfrac{18\lambda}{a^2}\dfrac{(1-\nu)}{2}(s^2 - 1)(t^2 - 1) \end{bmatrix} \qquad (23\text{-}56)$$

where, $\lambda = \dfrac{d^2}{12}$

STEP 6—ELEMENT STIFFNESS MATRIX

We are now ready to form the element stiffness matrix by forming $[B]^T[D][B]$ which will turn out to be a 1×1 matrix. This matrix will have to be integrated over the volume of the element—in this instance the area of the element—to arrive at the element stiffness matrix $[k]$.

$$[k] = \int\int [B]^T[D][B] \, dx \, dy$$

$$= \int_{-1}^{+1}\int_{-1}^{+1} [B]^T[D][B] \, ds \, dt. \times \frac{a^2}{16}$$

The term $\dfrac{a^2}{16}$ arises from the replacement of $dx\,dy$ by $ds\,dt$.

$$\int\int [B]^T[D][B] \, ds \, dt \times \frac{a^2}{16}$$

$$= \frac{2(1 + \nu)}{256R^2} \times \frac{Ed}{(1 - \nu^2)} \times \frac{a^2}{16} \int_{-1}^{+1} (s^3 - 3s + 2)^2 \, ds \int_{-1}^{+1} (t^3 - 3t + 2)^2 \, dt$$

$$+ \frac{36\lambda}{a^4} \cdot \frac{Ed}{(1 - \nu^2)} \times \frac{a^2}{16} \left[\int_{-1}^{+1}\int_{-1}^{+1} \{s^2(t^3 - 3t + 2)^2 \right.$$

$$+ 2\nu st(s^3 - 3s + 2)(t^3 - 3t + 2) + t^2(s^3 - 3s + 2)^2$$

$$\left. + \frac{9(1 - \nu)}{2}(s^2 - 1)^2(t^2 - 1)^2\} \, ds \, dt \right]$$

The first term represents the contribution of membrane action and the second that of flexure. The first term may be recast and simplified as:

$$\frac{2 \times 1.15}{256} \times \frac{12}{16} \times \left(\frac{a}{d}\right)^2 \times \left(\frac{a}{R}\right)^2 \times D \times \frac{1}{a^2} \times \left(\frac{2}{7} + 6 + 8 - \frac{12}{5}\right)^2$$

$$= 3899.0576 \frac{D}{a^2} \quad \text{where } D = \frac{Ed^3}{12(1 - \nu^2)}$$

Noting that $\dfrac{\lambda Ed}{(1 - \nu^2)} = D$, the second term representing the contribution of flexure may be simplified as $47.1771 \dfrac{D}{a^2}$. Hence the stiffness matrix $[k]$ of the element becomes

$$(3899.0576 + 47.1771) \frac{D}{a^2} \,.$$

The load at node 1 of the element is $\dfrac{1}{4}$. The nodal deflexion is w_1. Hence,

$$\frac{1}{4} = 3946.2347 \frac{D}{a^2} w_1$$

$$= 3946.2347 \times \frac{2 \times 10^6 \times (3.00)^3}{12(1 - 0.15^2)} \times \frac{w_1}{480 \times 480}$$

Simplifying,

$$w_1 = \frac{1}{4} \times \frac{480 \times 480 \times 12(1 - 0.15^2)}{2 \times 10^6 \times (3)^3 \times 3946.2347}$$

$$= 3.1706 \times 10^{-6} \text{ inches}$$

To check the results, the problem was run using the solid GCS8 element of the ASAS library with 25 elements covering a quadrant of the shell. The central deflexion was found to be 2.524×10^{-6} inches. Computer outputs of the contours of displacement z, and stresses σ_{xx}, σ_{yy} and σ_{xy} (corresponding to stress-resultants N_x, N_y, and N_{xy} respectively) are shown in Figs. 23-20(a), (b), (c) and (d).

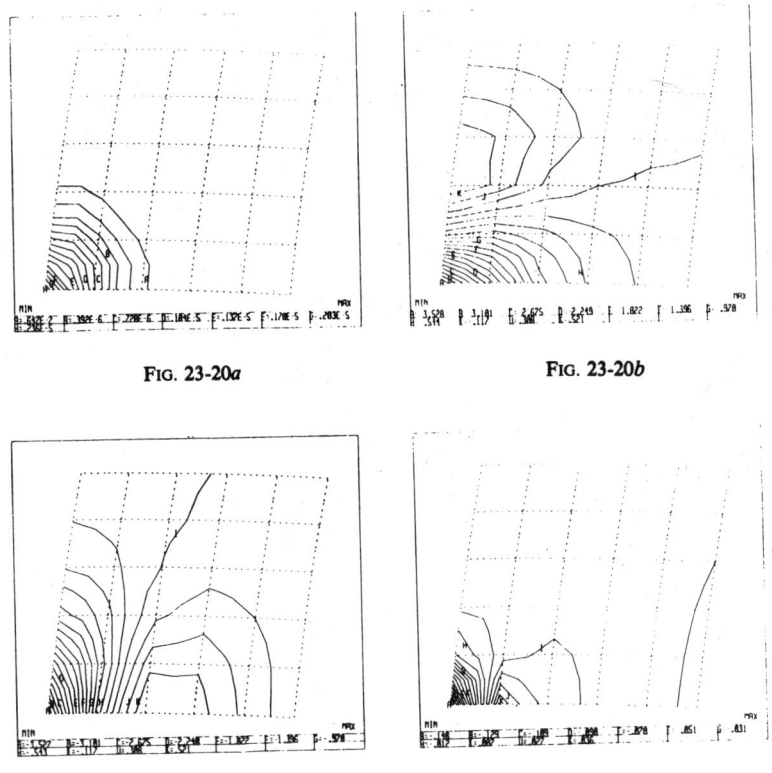

Fig. 23-20a

Fig. 23-20b

Fig. 23-20c

Fig. 23-20d

23-15 Extension to Shells of Variable Curvature

Bhattacharya and Ramaswamy [204] have modified the constant curvature formulation of Connor and Brebbia to make it applicable to shells of variable curvature. As a test example, they compared the FEM answers with the results reported by Ramaswamy and Jain [205] for a clamped funicular shell of rectangular ground plan submitted to the action of uniform vertical loading solved by using the Galerkin method. The results obtained by the two methods, presented in Fig. 23-21 (a) to (d), show very close agreement.

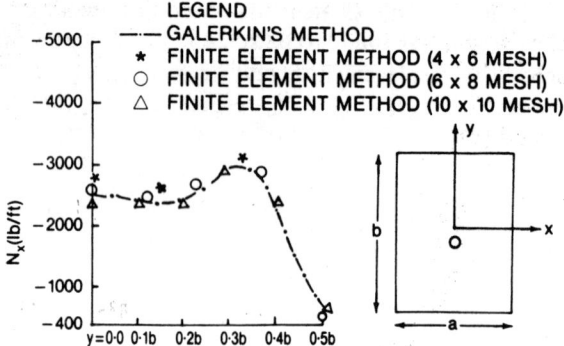

Fig. 23-21*a*

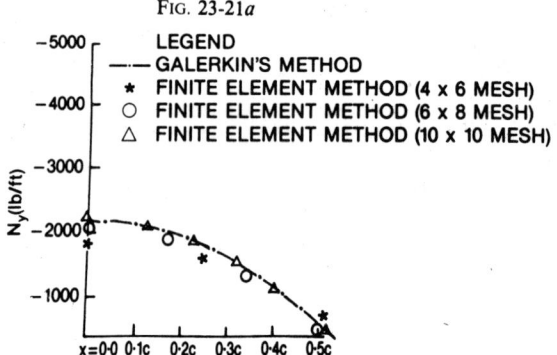

Fig. 23-21*b*

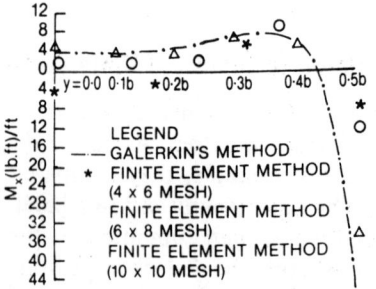

Fig. 23-21*c*

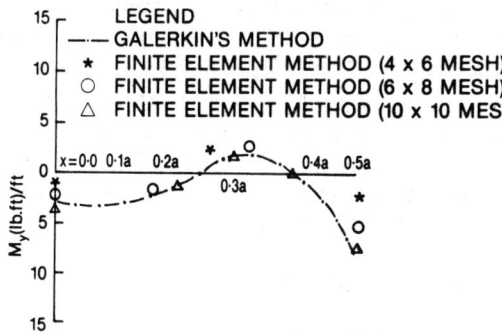

Fɪɢ. 23-21*d*

23-16 Analysis of Funicular Shells by Solid Elements

A clamped funicular shell, square in ground plan, was analyzed by using the quadrilateral solid element GCS8. The displacement pattern and the variation of the moment and stress-resultants are shown in Fig. 23-22 (a),

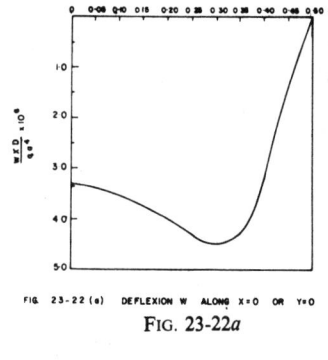

FIG. 23-22 (a)   DEFLEXION W ALONG X=0   OR   Y=0

Fɪɢ. 23-22*a*

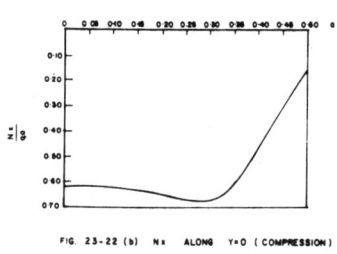

FIG. 23-22 (b)   Nx   ALONG   Y=0 (COMPRESSION)

Fɪɢ. 23-22*b*

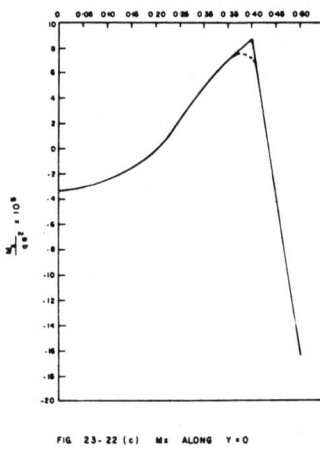

FIG 23-22 (c)   Mx   ALONG   Y=0

Fɪɢ. 23-22*c*

(b), (c). The ordinates z of the shell for use in the program were computed by the approximate formula (18-4) of Chapter 18, Article 18-6. This approximation is not, in fact, necessary. The ordinates z can be computed to any desired degree of accuracy by using the series solution given in Table 18-3. A program for computing five terms of the series for a funicular shell of rectangular ground plan by using H.P. 34-C is given in Appendix II. The values of z so computed can be used in the GCS8 program. This would mean that within the element itself, the edges will be approximated by second degree curves.

23-17   Shells with Edge Beams

Most of the published papers on the analysis of thin shells by the Finite Element Method relate to shells with idealized boundary conditions. Authors, for the most part, demonstrate the merits of the approaches they advocate by applying them to carefully selected test examples. On the other hand, shells that the practising engineer is likely to encounter most frequently in practice are those with edge beams which in turn are supported on columns. Bhattacharya and Ramaswamy [204] have shown how such problems can be formulated and solved by using F.E.M. techniques. They make use of the modified Connor and Brebbia element with five degrees of freedom at each node. The beam element has, however, six degrees of freedom. At the junction, one of the degrees of freedom of the beam element had to be suppressed. Shells with edge beams can also be solved by using the GCS8 solid element for idealizing the shell and the GCB3 element from the sams ASAS library for modelling the edge member.

Schnobrich [206] has shown how hipped roof hyperbolic paraboloid structures with edge members can be analyzed.

23-18   Convergence of Results

A question that often arises is whether a particular idealization by finite elements will lead to results that will converge towards a shallow or deep shell solution. Jones [207] discusses this problem in some depth by referring to case studies. The conclusions reached may be summarized as follows. The flat-faceted idealization leads to results which converge towards the deep shell solution. The isoparametric parabolic element [208] with 2×2 Gauss point integration gives results that tend to converge to the shallow shell solution. On the other hand, the cubic element with modified transverse shears [208] led to answers which converge towards the deep shell solution. The element of Cowper, Lindberg and Olson (1970) is essentially a shallow shell formulation in global co-ordinates. As is only to be expected, the results, in this instance, converged towards the shallow shell solution. Where convergence towards the deep shell solution is desired, use may be made of the method developed by Olson [209] wherein he formulates a shallow shell triangular conforming element in local cartesian co-ordinates and effects a transformation to deep shell global cartesian co-ordinates. Continuity of normal displacement and its first two derivatives and the tangential displacements u and v and their first derivatives are preserved *at the nodes* while assembling the elements. Displacement continuity between nodes is lost. Acceptable accuracy is however achieved with a relatively small number of elements as demonstrated in the test cases reported by the author.

23-19   Computer Software for Shell Analysis

The analysis of thin shells by the Finite Element Method is now within the reach of all design offices because of the availability of software for this purpose from a number of organizations. The more important of these and the agencies concerned are listed in Table 23-2.

Table 23-2

No.	NAME OF THE PROGRAM	AGENCY
1	ASKA (Automatic System for Kinematic Analysis)	J.H. Argyris & H. Parish Institut für Statik und Dynamik der Luft-und Raumfahrtkonstruktionen Universität Stuttgart 7 Stuttgart 80 Paffenwalding 27, West Germany.
2	SAP IV and NONSAP	E.L. Wilson & K. Wong NISEE 720 Davis Hall University of California Berkeley, CA 94720, USA
3	STRUDL II	ICES User's Group Inc. P.O. Box 8243 Cranston RI 02920, USA
4	NASTRAN	Cosmic Information Services 112 Barrow's Hall The University of Georgia Athens, Georgia GA 30602, USA
5	ASAS (Atkins Stress Analysis System)	Atkins R & D Woodcote Grove Ashley Road, Epsom, UK
6	MARC	P. Marcal and H.D. Hibbitt Marc Analysis Research Corporation
7	EASE 2	Engineering Analysis Corporation 1611 South Pacific West Highway Redondo Beach CA 90277, USA
8	ANSYS (Analysis System)	John S. Swanson Swanson Analysis Systems Inc.

CHAPTER 24

FUNICULAR SHELL FLOORS

24-1 Concrete Shells for Floors

Concrete shells are seldom considered for intermediate floors, their use being largely restricted to roofs. Apart from the description of such floors designed and built in India under the supervision of the author and his colleagues [210] [211], there is hardly any literature on such applications in the English language. The potential of concrete shells for use as intermediate floors of offices, apartments, warehouses and multi-storey parking garages is still to be fully exploited. The general reluctance to use concrete thin shells for intermediate floors is understandable because the curvature they present at the top and boundaries can lead to awkward problems. A structural form which particularly lends itself to such applications is the shallow precast funicular shell already described in Chapter 18. In the course of a recent literature survey to locate any available information on the use of shells as floors in the past, the author came across an interesting account given by Torroja [212] of solid tile groined vaults that the traditional craftsmen of Estremadura in Spain built with the help of simple tools (Fig. 24-1). The manner in which funicular shells may be employed for intermediate floors is illustrated in Fig. 24-2. The haunch filling on the top of the shell surface need not be of structural grade concrete. Earth or lightweight concrete can be substituted if so desired. If structural concrete is used for this purpose, it can be assumed to function compositely with the rib. The undersurface of such floors present a pleasing waffle pattern (Fig. 24-18).

24-2 Waffle-Slab versus Waffle-Shells

Waffle-slab intermediate floors are often favoured for office buildings, apartments, and multi-storey parking garages, mainly because they eliminate the need for complex and expensive formwork. Waffle shells offer all the advantages inherent in waffle-slab construction. In addition, a waffle-shell panel tends to be much stiffer than a waffle-slab of the same dimensions for

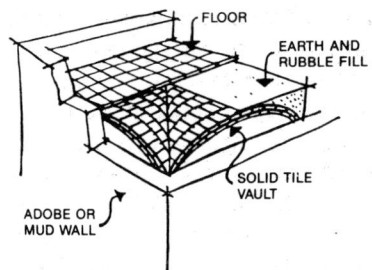

FIG. 24-1

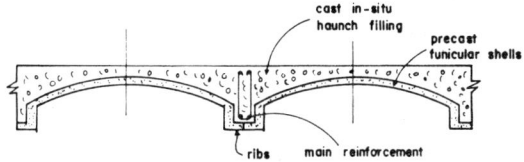

FIG. 24-2

FIG. 24-18   Waffle funicular shell floor (By courtesy of the Structural Engineering Research Centre, Madras, India)

the same equivalent thickness. But the most overriding advantage that this form of construction offers is that there is no need to remove the domes after the floor is formed as in the waffle-slab technique. The precast funicular shells serve as 'lost-forms' which are incorporated in the structure and participate in the load carrying function. In the absence of a Code governing this form of construction, the superior stiffness characteristics of a waffle shell panel cannot be fully exploited in design at present. Accepted procedures relating to waffle-slabs described in ACI 318-77 [213] may be used to arrive at conservative designs of such structures until more refined methods are developed on the basis of extensive analytical and experimental studies similar to those carried out on flat plates [214] [215]. In the next section, a worked example illustrates the design procedure relating to a typical waffle-shell floor. The precast funicular shell forms are temporarily propped when in-situ concrete is poured.

Design Example 24.1

PROBLEM

Design an intermediate waffle-shell floor of a building of three storeys with storey heights of 12′ to carry a Live Load of 50 p.s.f. Floor finish may be estimated at 10 p.s.f Columns are spaced 32′ apart, centre to centre, in both directions. Solid heads are provided over columns.

$$f'_c = 5000 \text{ psi} \qquad f_y = 60,000 \text{ psi}$$

SOLUTION

The layout is shown in Figure 24-3. We will adopt 24″ × 24″ square interior columns. The edge column shall be 30″ × 30″. The precast funicular shells will be 1″ thick and 72″ × 72″ in plan. The details of the shell module may be seen in Figure 24-4. They are placed 72″ centre to centre.

PROPERTIES OF SHELL MODULE

(1) *Load per square foot*

The load per square foot of the floor may be computed as follows (Fig. 24-4):

Volume of solid 72″ × 72″ × 15″ = 6 × 6 × 1.25 = 45 cft.

Deduct Voids:

Volume of prism $\qquad = \dfrac{62}{12} \times \dfrac{62}{12} \times \dfrac{8}{12} \qquad = 17.80$

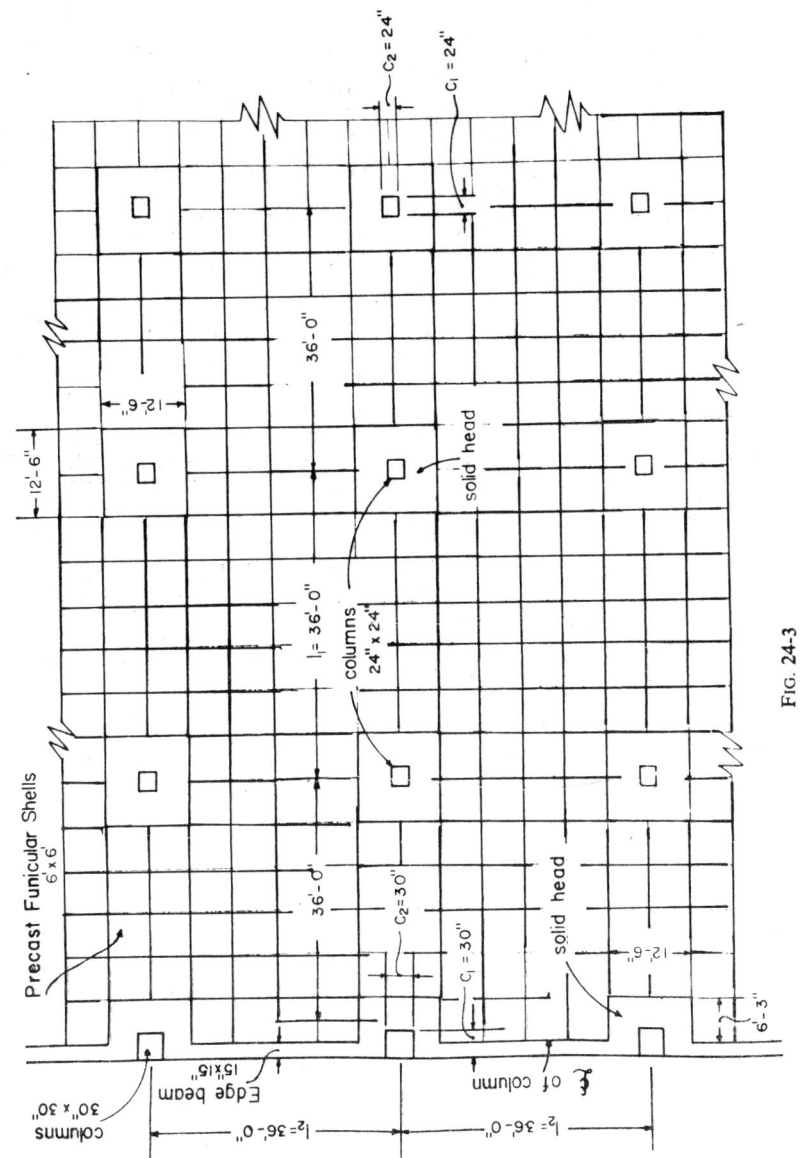

$C_2 = 24''$

$C_1 = 24''$

$36'-0''$

$12'-6''$

$12'-6''$

$l_1 = 36'-0''$

solid head

columns
$24'' \times 24''$

Precast Funicular Shells
$6' \times 6'$

$36'-0''$

$C_2 = 30''$

$C_1 = 30''$

solid head

$12'-6''$

$6'-3''$

Edge beam
$15'' \times 15''$

℄ of column

columns
$30'' \times 30''$

$l_2 = 36'-0''$

$l_2 = 36'-0''$

Fig. 24-3

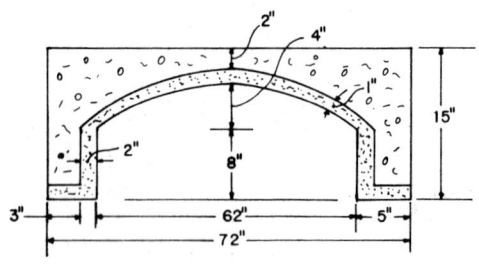

FIG. 24-4

$$\text{Volume of shell cavity} = \frac{4}{9} \times \frac{62}{12} \times \frac{62}{12} \times \frac{4}{12} = 3.95$$

Total deductions	$\underline{21.75}$ cft
Net volume = (45 − 21.75)	= 23.25 cft
Weight per square foot = 23.25 × 150	= 96.9 lbs.

PROPERTIES OF COMPOSITE RIB (Figure 24-5)

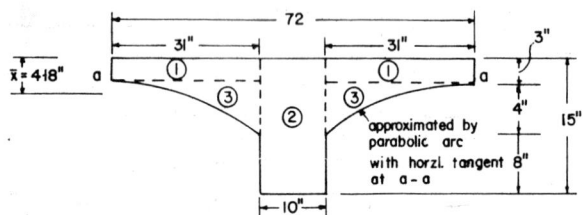

FIG. 24-5

Area

Areas of (1) = 2 × 31 × 3		= 186 in²
Area of (2) = 15 × 10		= 150 in²
Areas of (3) = $2 \times \dfrac{1}{3} \times 31 \times 4$ =		82.67 in²
Total area		$\underline{418.67}$ in²

Position of Centroid

The distance of the centroid from the top is computed as:

$$\bar{x} = \frac{(186 \times 1.5) + (150 \times 7.5) + (82.67 \times 4.2)}{418.67} = 4.18''$$

Note that the C.G. of the parabolic parts labelled 3 lie at a distance of $\left(3 + \dfrac{3}{10} \times 4\right) = 4.2$ in from the top.

MOMENT OF INERTIA

$\left.\begin{array}{l}\text{Moment of inertia of areas (1) about}\\ \text{about own centroidal axis}\end{array}\right\} = \dfrac{1}{12} \times 62 \times 27 = 139.5 \text{ in}^4$

Add Ad^2 for (1) $= (62 \times 3)(4.18 - 1.5)^2$

$= 1335.92 \text{ in}^4$

$\left.\begin{array}{l}\text{Moment of inertia of piece (2)}\\ \text{about own centroidal axis}\end{array}\right\} = \dfrac{1}{12} \times 10 \times 15^3 = 2812.50 \text{ in}^4$

Add Ad^2 for (2) $= (15 \times 10)(7.5 - 4.18)^2 = 1653.36 \text{ in}^4$

$\left.\begin{array}{l}\text{Moment of inertia of pieces (3)}\\ \text{about axes } a\text{-}a\end{array}\right\} = 94.48 \text{ in}^4$

$\left.\begin{array}{l}\text{Moment of inertia of piece (3)}\\ \text{about own centroidal axes}\end{array}\right\} = 94.48 - \dfrac{1}{3}\left(\dfrac{31}{4}\right)(1.2)^2 = 34.96 \text{ in}^4$

$\left.\begin{array}{l}\text{Hence moment of inertia of 2}\\ \text{pieces (3) about own centroidal axis}\end{array}\right\} = 69.92 \text{ in}^4$

Ad^2 for pieces (3) $= 82.67 \times (4.2 - 4.1)^2 \simeq 0$

Total I $= 139.50 + 1335.92 + 2812.50 + 69.92$

$= 6011.2 \text{ in}^4$

Hence, a section of the floor 36' wide, taken between solid heads, comprising six ribs will have a moment of inertia of $6 \times 6011.2 = 36067 \text{ in}^4$

MOMENT OF INERTIA OF SECTION TAKEN AT SOLID HEAD

A section of the floor 36' wide taken at the solid head will comprise a solid head 12'-6" wide and 15" deep with 1½ ribs on either side making up a total of three ribs and a solid head (Figure 24-6).

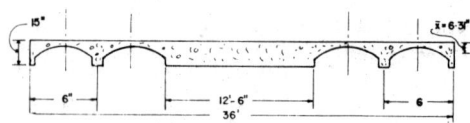

Fig. 24-6

POSITION OF C.G. OF SECTION

The distance $\bar{x}$ of the C.G. from the top is computed as

$$\bar{x} = \frac{(150 \times 15 \times 7.5) + (3 \times 418.67 \times 4.18)}{(2250 + 3 \times 418.67)} = 6.31''$$

MOMENT OF INERTIA

$$\left.\begin{array}{l}\text{Moment of inertia of solid head}\\ \text{about own centroidal axis}\end{array}\right\} = \frac{1}{12} \times 150 \times 15^3 = 42187.5 \text{ in}^4$$

Add $Ad^2 = (150 \times 15)(7.5 - 6.31)$ $\qquad = 3186.23 \text{ in}^4$

Moment of inertia of 3 composite
ribs about own centroidal axis $= 3 \times 6011.2 \qquad = 18033.6 \text{ in}^4$

Add $Ad^2 = 3 \times 418.67(6.31 - 4.18)^2 \qquad\qquad = \underline{4775.97 \text{ in}^4}$

Total I $\qquad\qquad\qquad\qquad\qquad\qquad\qquad = \underline{\underline{68183 \text{ in}^4}}$

We will design a typical intermedicate panel in the 1_1 direction comprising an edge panel and a first interior panel. In the analysis and design of the panels, the procedure given in the CRSI Handbook [216] will be closely followed. The notations used are those given in ACI 318-77. The Direct Design Method described in the Code is followed.

LOADS

(i) *Where no solid head is provided*

Load per square foot of waffle floor	=	96.9 lb
Floor finish	=	10.0 lb
Total dead load	=	106.9 lb

Design dead load $= 1.4 \times$ dead load	$= 149.66$ lbs
Live load	$= 50$ p.s.f.
Design live load $= 1.7 \times$ live load	$= 85$ p.s.f.
Hence total ultimate design load $= 149.66 + 85$	$= 234.66$ lb/sq. ft.

(ii) *Load where solid heads are provided*

Weight of solid head $\times 1.4 = 1.4 \times 1.25 \times 150 = 262.50$ p.s.f.	
Finishes $\times 1.4$	$= 14$ p.s.f.
$1.7 \times$ live load $= 1.7 \times 50$	$= 85.00$ p.s.f.
Total ultimate design load	$= 361.50$ p.s.f.

STIFFNESS PROPERTIES OF COLUMNS AND SLABS

To compute moments by the Direct Design Method, it is necessary to determine the flexural stiffness K_c and K_s of the columns and slabs, the torsional stiffness K_t, combined stiffness K_{ec} and stiffness ratios α_{min}, α_c and α_{ec}.

α_{min}

$(L) =$ Live load $= 50$ p.s.f.

$(D) =$ Dead load $= 106.9$ p.s.f.

$$\frac{D}{L} = 2.12 > 2.0.$$

Hence according to the Code, there is no α_{min} requirement.

EXTERIOR COLUMN

The Analogous column corresponding to the exterior column is shown in Figure 24-7. The moment of inertia of the column is regarded as infinite within the slab. It can be shown by using the Column Analogy that $\dfrac{K_c}{E} = k_c \cdot \dfrac{I_g}{l_c - h}$, where I_g is the gross moment of inertia of the column, l_c is the storey height and h is the plate thickness.

$$k_c = 1 + 3\left(\frac{l_c}{l_c - h}\right)^2$$

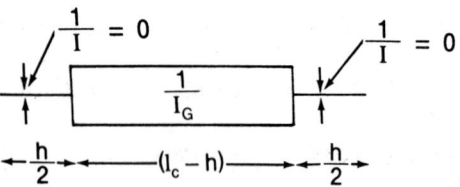

FIG. 24-7

In this instance, $l_c = 144$; $h = 15$; $k_c = 4.75$.

Hence $\dfrac{K_c}{E_c} = \dfrac{4.75(67.500)}{129} = 2485.$

K_s for Exterior Panel Slab

K_s for the exterior panel slab will be computed by Column Analogy. Within the columns (i.e. from the centre line to the face of a column, $\dfrac{l}{I}$ is to be multiplied by a factor $= \left(1 - \dfrac{c_2}{l_2}\right)^2$.

This factor will be computed for the edge and interior columns as follows.

For edge column,

$$\left(1 - \frac{c_2}{l_2}\right)^2 = \left(1 - \frac{30}{432}\right)^2 = 0.8659$$

For the interior column

$$\left(1 - \frac{c_2}{l_2}\right)^2 = \left(1 - \frac{24}{432}\right)^2 = 0.8920$$

The Analogous Column is shown sketched in Figure 24-8. The $\dfrac{l}{I}$ values relevant to the five segments into which the Analogous Column is divided are computed as follows:

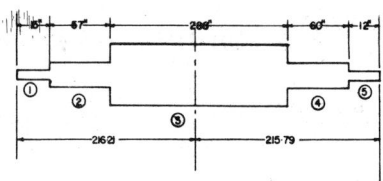

Fig. 24-8

Segment (1) $\dfrac{1}{I} = \dfrac{0.8659}{68183} = 0.000012700$

Segment (2) $\dfrac{1}{I} = \dfrac{1}{68183} = 0.000014666$

Segment (3) $\dfrac{1}{I} = \dfrac{1}{36067} = 0.000027726$

Segment (4) $\dfrac{1}{I} = \dfrac{1}{68183} = 0.000014666$

Segment (5) $\dfrac{1}{I} = \dfrac{0.8920}{68183} = 0.000013082$

Area and Centroidal distance $\bar{x}$ from A-A

No.	Area	Moment about A-A	
(1)	0.0000127×15	$= 0.00019050 \times 7.5$	$= 0.00142875$
(2)	0.000014666×57	$= 0.00083596 \times 43.5$	$= 0.0363435$
(3)	0.000027726×288	$= 0.007985088 \times 216$	$= 1.724779$
(4)	0.000014666×60	$= 0.00087996 \times 390$	$= 0.34318440$
(5)	0.000013082×12	$= 0.00015698 \times 426$	$= 0.06687518$
	Area $= \boxed{0.01004894}$		$\Sigma \boxed{2.172631689}$

$$\bar{x} = \dfrac{2.172631689}{0.0010048494} = 216.21'' \text{ from AA}$$

Moment of Inertia of Analogous Column

$$\frac{0.00001270}{3}(216.21)^3 = 42.78679897$$

$$\frac{0.00000196}{3}(201.21)^3 = 5.32210575$$

$$\frac{0.00001306}{3}(144.21)^3 = 13.05593721$$

$$\frac{0.000013082}{3}(215.79)^3 = 43.81742330$$

$$\frac{0.00000158}{3}(203.79)^3 = 4.45742906$$

$$\frac{0.00001306}{3}(143.79)^3 = 12.942196$$

$$I = \Sigma = \overline{122.38189033}$$

$$\frac{K_s}{E_c} = \frac{1}{0.010048494} + \frac{216^2}{122.38189033} = \underline{480.75}\ \text{in}^3$$

Computation of Torsional Stiffness for Exterior Column

At solid heads, the attached torsional member has a section as shown in Figure 24-9.

$$C_d = \left[1 - 0.63\frac{x}{y}\right]\frac{x^3 y}{3} = \left[1 - \frac{0.63 \times 15}{30}\right]\frac{15^3 \times 30}{3} = 23118.75$$

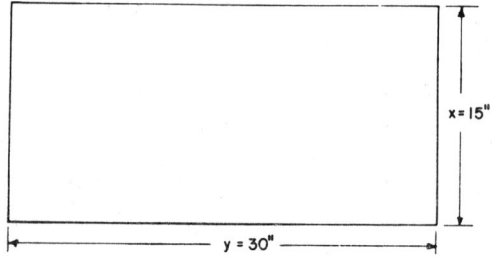

$$x = 15''$$

$$y = 30''$$

FIG. 24-9

Between solid heads, the attached torsional member has a section as shown in Figure 24-10.

$$C_s = \left[1 - \frac{0.63 \times 15}{18}\right] \times \frac{15^3 \times 18}{3} + \left[1 - \frac{0.63 \times 3}{12}\right] \times \frac{3^3 \times 12}{3}$$

$$= 9709.74$$

$$\frac{1}{K_t} = \frac{l_2\left(1 - \frac{C_2}{l_2}\right)^5}{20E_cC_d} + \frac{l_2\left(1 - \frac{2x_2}{l_2}\right)^5}{20E_c}\left(\frac{1}{C_s} - \frac{1}{C_d}\right)$$

where, x_2 = distance from column centre to the edge of the solid head.

$$\frac{1}{K_t} = \frac{432\left(1 - \frac{30}{432}\right)^5}{20 \times 23119E_c} + \frac{432}{20E_c}\left(1 - \frac{2 \times 72}{432}\right)\left(\frac{1}{9710} - \frac{1}{23119}\right)$$

$$= \frac{0.000082183}{E_c}$$

Hence, $\dfrac{K_t}{E_c} = 1217 \text{ in}^3$

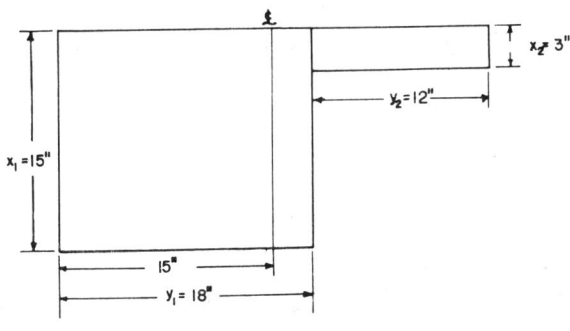

FIG. 24-10

K_{ec} for the exterior column is worked out from the formula:

$$\frac{1}{K_{ec}} = \frac{1}{\Sigma K_c} + \frac{1}{K_t}$$

$$K_{ec} = \left(\frac{2 \times 2485.46}{1 + \dfrac{2 \times 2485.46}{1217}} \right) = 977.65 \text{ in}^3$$

$$\alpha_{ec} = \frac{978}{481} = 2.033$$

TOTAL FACTORED STATIC MOMENTS

Exterior Panel

The loads acting on the exterior span, the reactions, moments and shears as a simple beam are shown in Figure 24-11. The total factored static moment is 1239 kip. ft.

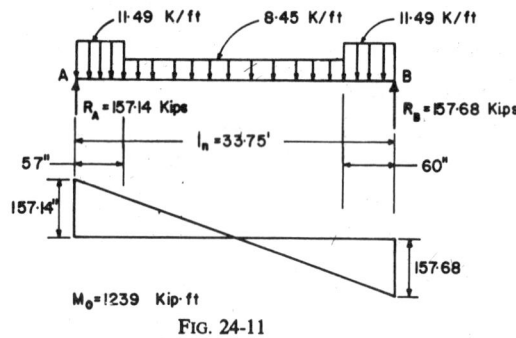

FIG. 24-11

First Interior Span

The loads, reactions, shears and moments acting on the first interior span, computed as a simply supported beam are shown in Figure 24-12. The total factored static moment is 1259 kip. ft.

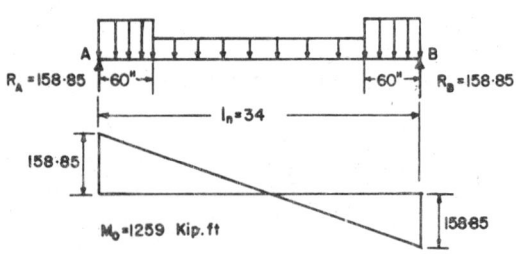

FIG. 24-12

Negative and Positive Factored Moments

(i) *Exterior Span*

Negative Moment Exterior $= -\left(\dfrac{0.65 \times 1239}{1 + \dfrac{1}{2.033}}\right) = 540$ kip. ft.

Positive Moment $= \left(0.63 - \dfrac{0.28}{1 + \dfrac{1}{2.033}}\right) = +548$ kip. ft.

Negative Moment Interior $= \left(0.75 - \dfrac{0.10}{1 + \dfrac{1}{2.033}}\right) = -846$ kip. ft.

(ii) *First Interior Span*

Negative Factored Moment $= 0.65 \times 1259 = -818$ kip. ft.

Positive Factored Moment $= 0.35 \times 1259 = +441$ kip. ft.

Reactions (Figure 24-13)

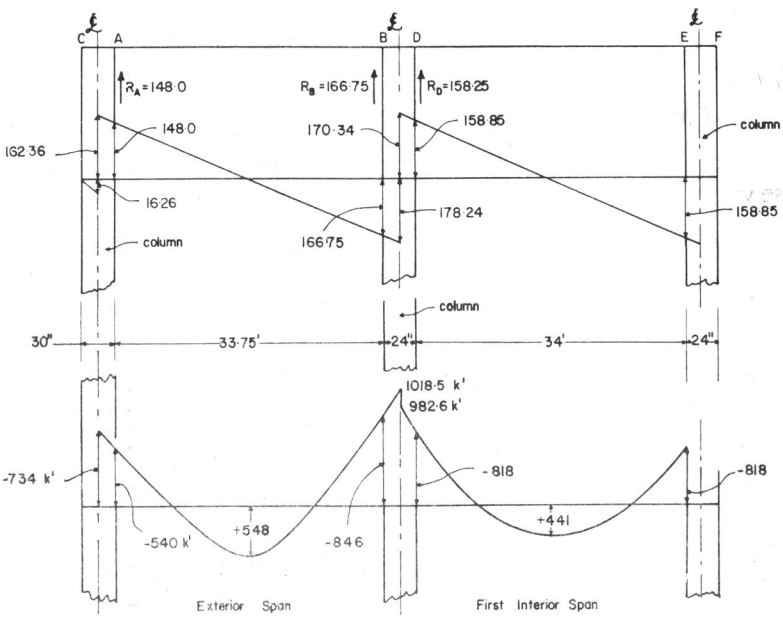

Fɪɢ. 24-13

$$R_A = 157.14 - \frac{(846 - 540)}{33.75} = 148 \text{ kips.}$$

$$R_B = 157.68 + \frac{(846 - 540)}{33.75} = 166.75 \text{ kips.}$$

$$\left.\begin{array}{l}\text{Shear at centre line} \\ \text{exterior column}\end{array}\right\} = 166.75 + \frac{11.49 \times 15}{12} = 162.36 \text{ kips}$$

$$\left.\begin{array}{l}\text{Shear left of center} \\ \text{line}\end{array}\right\} = \frac{15}{12} \times 1.25 \times 0.3615 \times 36 = 16.26 \text{ kips}$$

$$\text{Reaction} \qquad = 162.36 + 16.26 = \underline{178.62 \text{ kips}}$$

$$\left.\begin{array}{l}\text{Shear to the left of} \\ \text{the centre line of} \\ \text{first interior column}\end{array}\right\} = 166.75 + 11.49 = 178.24 \text{ kips}$$

$$\left.\begin{array}{l}\text{Shear to the right of} \\ \text{the centre line of} \\ \text{the first interior} \\ \text{column}\end{array}\right\} = 158.85 + 11.49 = 170.34 \text{ kips}$$

$$\left.\begin{array}{l}\text{Reaction at first} \\ \text{interior column}\end{array}\right\} = (178.24 + 170.34) = \underline{348.58 \text{ kips}}$$

BENDING MOMENTS

The factored moments previously found are at the face of the columns. These values have to be extrapolated to the centre of the columns as follows:

EXTERIOR COLUMN

$$\text{Moment at } A \qquad = -540 \ k' \text{ feet}$$

$$\left.\begin{array}{l}\text{Moment at centre} \\ \text{line of exterior} \\ \text{column}\end{array}\right\} = -540 - 11.94 \times \frac{15}{12} \times \frac{15}{2 \times 12}$$

$$= -734 \ k'$$

FIRST INTERIOR COLUMN

Proceeding from the face B, the bending moment at the centre line is found as:

$$-846 - (166.75 \times l) - (11.49 \times l \times 0.5) = -1018.5 \ k'$$

Proceeding from the face D, we arrive at a value of the bending moment at the same central line as:

$$-818 - (158.85 \times l) - (11.49 \times l \times 0.5) = -982.6 \ k'$$

Obviously, at this central line, there is an unbalanced moment of $1018.5 - 982.6 = 35.9 \ k'$ which will be later distributed to the members meeting at this point in the ratio if their stiffnesses.

$\dfrac{K_s}{E_c}$ for Interior Slab

The $\dfrac{K_s}{E_c}$ for the interior slab can be found by Column-Analogy by following the procedure already explained in detail for the exterior panel slab. The calculations are self-explanatory.

The Analogous Column of the first interior panel slab is shown in Figure 24-14. Factor by which $\dfrac{l}{I}$ is to be multiplied in segment (1) is $\left(1 - \dfrac{24}{432}\right)^2 =$ 0.89. The I at a section through the solid head is 68183 in^4; a section between solid heads will have $I = 36067$ in^4.

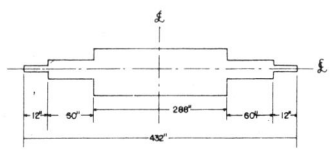

FIG. 24-14

Segment	$\dfrac{l}{I}$	Area	
Segment (1)	0.00001305	0.00001305×12	$= 0.00015660$
Segment (2)	0.00001467	0.00001467×12	$= 0.00088020$
Segment (3)	0.00002773	0.00002773×288	$= 0.00798624$
Segment (4)	0.00001467	0.00001467×60	$= 0.0008802$
Segment (5)	0.00001305	0.00001305×12	$= 0.00015660$
	Area of Analogous Column		$= \underline{0.01005984}$

Moment of Inertia of Analogous Column

$$\frac{1}{12} \times 0.00001305(432)^3 \qquad\qquad = \quad 87.6760$$

$$\frac{1}{12} \times (0.00001467 - 0.00001305) \times (408)^3 = \quad 9.1688$$

$$\frac{1}{12} \times (0.00002773 - 0.00001467) \times (288)^3 = \quad 25.9980$$

I of Analogous Column $\qquad\qquad\qquad = \overline{122.8427}$

$$\frac{K_s}{E_c} = \frac{1}{0.01005984} + \frac{216^2}{(122.8427)} = \underline{479.20}\ \text{in}^3$$

TORSIONAL CONSTANT FOR INTERIOR COMLUMN

The section of the attached torsional member between solid heads is shown in Figure 24-15.

$x_1 = 3 \qquad y_1 = 7$ (two areas)

$x_2 = 10 \qquad y_2 = 15$ (one area)

$$C_s = \left[1 - \frac{0.63 \times 3}{7}\right] \frac{3^3 \times 7}{3} \times 2 + \left[1 - \frac{0.63 \times 10}{15}\right] \times \frac{10^3 \times 15}{3}$$

$$= 2192$$

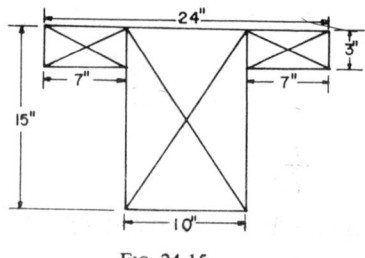

FIG. 24-15

The section of the attached torsional member at the solid section is shown in Figure 24-16.

$$C_d = \left[1 - \frac{0.63 \times 15}{24}\right] \times \frac{15^3 \times 24}{3} = 16,369$$

$$\frac{1}{K_t} = \frac{432\left(1 - \frac{24}{432}\right)^5}{20 \times E_c \times 16369} + \frac{432\left(1 - \frac{2 \times 72}{432}\right)^5}{20E_c}\left(\frac{1}{2192} - \frac{1}{16369}\right)$$

Solving, $\dfrac{K_t}{E_c} = 473$ in^3

Hence, $K_{ec} = \dfrac{2 \times 1018}{1 + \dfrac{2 \times 1018}{473}} = 384$ in^3

$\dfrac{K_s}{E_c}$ for the interior slab has already been found to be 479; the $\dfrac{K_s}{E_c}$ of the exterior slab is 480.

Hence, $\alpha_{ec} = \dfrac{384}{479 + 480} = 0.40$ at the first interior support.

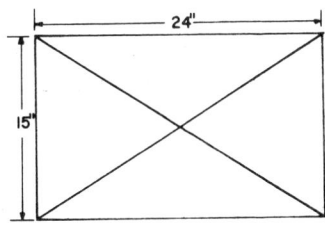

Fig. 24-16

CHECKING EXTERIOR EDGE COLUMN FOR SHEAR

The critical section to be examined for shear is shown in Figure 24-17. Effective depth $= 15 - 1.5 = 13.5$ in. Area A_{sh} resisting shear $= (43.5 + 2 \times 36.75) \times 13.5 = 1579.5$ in^2 say 1580 in^2. The distance of the centroid of this area from the outside edge of the column is found as:

$$\bar{x} = \frac{(43.5) \times (36.75) + (2 \times 36.75 \times 18.38)}{(2 \times 36.75 + 4.35)} = 24.20 \text{ in.}$$

Hence, the distance of the centroid from the centre line of the column $= 25.20 - 15 = 10.20$ in. The distance $C_{AB} = (36.75 - 25.20) = 11.55$ in. $C_{CD} = 25.20$.

The polar moment of inertia J of the shear-resisting section is found as:

$$J = \left(c_1 + \frac{d}{2}\right)\frac{d^3}{6} + \frac{2}{3}d(C_{AB}^3 + C_{CD}^3) + (c_2 + d)d(C_{AB})^2$$

$$c_1 = c_2 = 30 \text{ in}$$

$$J = \frac{36.75 \times 13.5^3}{6} + \frac{2}{3} \times 13.5(11.55^3 + 25.2^3) + 43.5 \times 13.5(11.55)^2$$

$$= 251,305 \text{ in}^4$$

Reaction at centre line of the column = 178.62 kips.

To arrive at V_u, we have to deduct from this the load acting on the area (43.5″ × 36.75″) defining the critical section. Hence, V_u =
$$178.62 - \frac{(36.75 \times 43.5)}{144} \times 0.3615 = 174.61 \text{ kips.}$$

We must next arrive at the unbalanced moment at the centroid of the shear-resisting section.

M_{un} = Bending moment at the centre line of the column −
$\qquad$ (178.62 × 10.2) − (moment of forces on the left of the centre line)

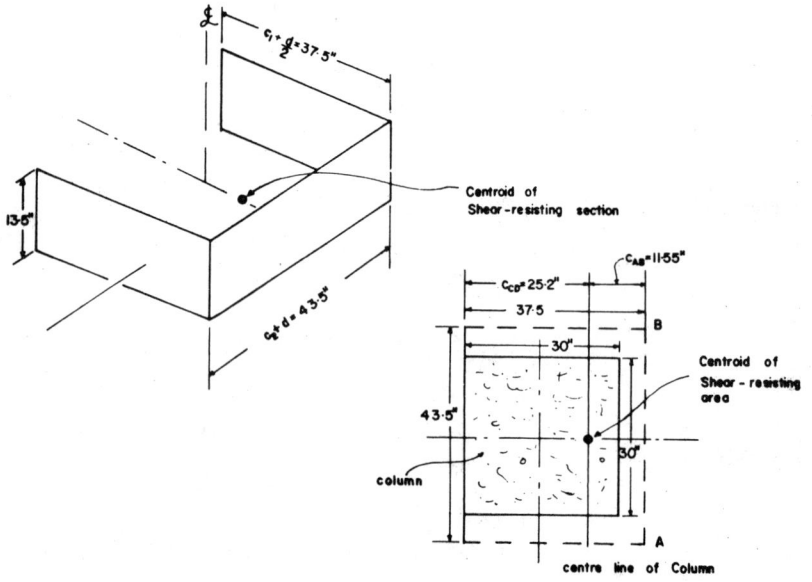

Fig. 24-17

$$= 734 - \frac{178.62 \times 10.2}{12} - 16.26 \times \frac{15}{2} \times \frac{1}{12}$$

$$= 572 \ k'$$

The ACI Code specifies that a fraction of this unbalanced moment will be transferred in shear. This fraction for the exterior column becomes

$$F = \left(1 - \frac{1}{1 + \frac{2}{3}\sqrt{\dfrac{c_1 + \dfrac{d}{2}}{c_2 + d}}}\right)$$

substituting $c_1 = c_2 = 30$ and $d = 13.5$, the factor becomes 0.38.

$$\text{Shear stress} = \frac{V_u}{\phi A_{sh}} + \frac{F \times M_{un} \times C_{AB}}{\phi J}$$

$$= \left(\frac{174610}{0.85 \times 1580}\right) + \frac{0.38 \times 572000 \times 11.55}{0.85 \times 251{,}305}$$

$$= 271 \text{ psi} < 283 \text{ psi}$$

Allowable shear stress $= 4\sqrt{f'c} = 4\sqrt{5000} = 283$ psi. Hence, the column size os o.k.

Checking First Interior Column for Shear

The area A_{sh} resisting shear is found as:

$$A_{sh} = 2 \times (c_1 + c_2 + 2d)d = (24 + 24 + 27) \times 2 \times 13.5 = \underline{2025} \text{ in}^2$$

$$J = \frac{d(c_1 + d)^3}{6} + \frac{d^3}{6}(c + d) + \frac{d(c_2 + d)(c_1 + d)^2}{2}$$

$$c_1 = c_2 = 24 \qquad d = 13.5 \qquad C_{AB} = 18.75''$$

$$J = \underline{489.987} \text{ in}^4$$

$$\text{Fraction } F = \left(1 - \frac{1}{1 + \dfrac{2}{3}}\right) = 0.40$$

Reaction at the first interior support $= 348.58$ kips.

To find V_u, we have to deduct the load on the critical area from the reaction.

$$V_u = 348.58 - \left(\frac{37.5 \times 37.5}{144}\right) \times 0.3615 = 345 \text{ kips}$$

We had previously noted that there is an unbalanced moment of 3.59 kips over the first interior support. The part of the unbalanced moment that is transferred to the columns

$$= \left[\frac{1}{1 + \dfrac{1}{\alpha_{ec}}}\right] \times 3.59 = \left[\frac{1}{1 + \dfrac{1}{0.40}}\right] \times 3.59 = 0.29 \times 3.59$$

$$= 10.26 \text{ kip ft.}$$

Shear stress $v_u = \dfrac{345000}{2025 \times 0.85} + \dfrac{0.40 \times 10.26 \times 1000 \times 18.75}{0.85 \times 489,987}$

$$= 200.6 \text{ psi} < 283 \text{ psi O.K.}$$

Design of Ribs

Before we proceed to the design of the ribs, the bending moments have to be apportioned to the column strips and middle strips in accordance with ACI 318-77. Thus, at the face of the exterior column, the entire bending moment of −540 kip feet will be assigned to the column-strip. The positive moment of 548 kip feet in the exterior span will be shared by the column and middle strips in the ratio of 60 to 40. Thus, the shares of the column and middle strips works out to 328.8 kip feet and 219.2 kip feet respectively. The negative bending moment of −846 kip feet at the first interior support will be apportioned to the column and middle strips in the ratio of 75:25; hence, the column-strip receives a bending moment of −643.5 kip feet and the middle strip −211.5 kip feet. Similar sharing of bending moments between column and middle strips will also take place in the first interior span.

The column-strip consists of the solid heads, wherever these are provided. Elsewhere, three ribs with their flanges form the column-strip. The middle-strips consist of three ribs and the appropriate width of flange. T beam action can be assumed only where the bending moments are positive causing compression in the flanges. Because of the large flange widths available, the neutral axis is likely to fall within the flange. In such cases, the ribs can be analysed by using principles applicable to the ultimate strength design of rectangular sections, the effective flange width being regarded as the breadth of the beam. In Table 24-1, the areas of steel, number of bars and other relevant details relating to various critical sections are tabulated.

Table 24-1

Section	M_u	d	b	j_d	A_s	Number of bars and Size
Face of Ext. Column Column Strip Middle Strip	$-540\ k'$ 0	$13.5''$	$150''$	$13.10''$	$9.16\ \text{in}^2$	21 #6 bars
Positive Moment *Ext. Span* Column Strip Middle Strip	$-328.8\ k'$ $+219.2\ k'$	$13.5''$ $13.5''$	$216''$ $216''$	$13.33''$ $13.33''$	$5.5\ \text{in}^2$ $3.65\ \text{in}^2$	15 #6 bars (5 in each rib) 9 #6 bars (3 in each rib)
First Int. Support Column Strip Middle Strip	$-643.5\ k'$ $-211.5\ k'$	$13.5''$ $13.5''$	$150''$ $216''$	$13.0''$ $13.33''$	$11\ \text{in}^2$ $3.5\ \text{in}^2$	25 #6 bars 9 #6 bars (3 in each rib)
First Int. Span *Positive Moment* Column Strip Middle Strip	$+264.6\ k'$ $+176.4\ k'$	$13.5''$ $13.5''$	$216''$ $216''$	$13.20''$ $13.33''$	$4.45\ \text{in}^2$ $2.94\ \text{in}^2$	9 #7 bars (3 in each rib) 9 # bars (3 in each rib)
First Int. Span *& Int. Support* Column Strip Middle Strip	$-613.5\ k'$ $-204.5\ k'$	$13.5''$ $13.5''$	$150''$ $216''$	$13.0''$ $13.33''$	$10.48\ \text{in}^2$ $3.4\ \text{in}^2$	25 #6 bars 9 #6 bars (3 in each rib)

An example will illustrate the design procedure. Let us suppose that the reinforcement to be provided in the column strip to carry its share of the positive moment in the exterior span (amounting to 328.8 k') is to be found. We proceed as follows:

$$\text{Bending moment} = +328,800\ \text{lb ft}$$

Being a positive moment, the flange will be effective in compression. The share of the moment that each rib needs to carry

$$= \frac{328{,}800 \times 12}{3} = 1{,}315{,}200 \text{ in lb.}$$

The effective flange width appropriate in this case is the centre to centre spacing of the ribs $= 72''$ Average thickness of flange $= \frac{7+3}{2} = 5''$.

Overhang permitted $= 5 \times 8 = 40'' < 31''$ (actual)

Effective depth after allowing for adequate cover $= 13.5''$

Hence, $b = 72''$.

Let us assume a trial value of $J_d = 0.95$

$$A_s = \frac{1315{,}200}{0.90 \times 60{,}000 \times 0.95 \times 13.5} = 1.9 \text{ in}^2$$

Let us find 'a', the depth of the equivalent compression stress block as:

$$a = \frac{1.9 \times 60{,}000 \times 0.90}{0.85 \times 5000 \times 72} = 0.34''$$

$$J_d = 13.5 - \frac{a}{2} = 13.5 - 0.17 = 13.33''$$

Using this refined value of J_d, the area of steel is recomputed as:

$$A_s = \frac{1315{,}200}{0.90 \times 60{,}000 \times 13.33} = 1.83 \text{ in}^2$$

5 #6 bars in each rib will give 2.20 in^2 > 1.83 in^2.

The embeddment lengths and other details must conform to Code requirements.

Design of Ribs for Shear

Let us for example examine the ribs for shear at the first interior support. The critical section for shear is situated at a distance of $d = 13.5''$ from the edge of the solid head. The ultimate shear at the critical section is:

$$V_u = 0.235 \times 0.50 \times (36^2 - 14.75^2) + \left(\frac{1018.5 - 734}{36} \right)$$

$$= 134.62 \text{ kips.}$$

At this section, 6 ribs are available bd for each $= 10 \times 13.5 = 135 \text{ in}^2$.

$$v_u = \frac{134,620}{6 \times 135} = 166.20 \text{ psi.}$$

$$v_c = 2.2\sqrt{f'}_c = 155.56 \text{ psi } v_u > v_c$$

$$v_u - v_c = 166 - 156 = 10 \text{ psi} < 4\sqrt{f'}_c$$

Hence stirrups are required.

Let stirrups be #4 single vertical bars apaced 6" apart.

$$\text{Capacity of stirrup} = \frac{0.20(60,000)}{6 \times 10}$$

$$= 200 \text{ psi} > 10 \text{ psi}$$

24-3 Notes on Construction

Funicular shells required for intermediate floors are best cast on accurately profiled moulds. The ordinates of the surface may be computed by using the program given in Appendix II. The shells themselves need no reinforcement except a #4 bar placed in the edge beams. It is however good practice to place mesh reinforcement in the 2" screed on the top of the shell to avoid cracks that may be formed during transport and erection. Such reinforcement, if provided, will also arrest hair cracks, caused by temperature and shrinkage.

Handling hooks need to be provided in the edge beams. The funicular shells are required to be adequately propped until the composite floor consisting of the precast funicular shell and the cast in place concrete gain their desired strength.

CHAPTER 25

SHELL FOUNDATIONS

25-1 Introductory Remarks

Spread footings for columns, transmitting heavy loads to weak soils, tend to be massive. If rafts are provided, they need to be excessively thick to be rigid enough to control settlements within tolerable limits. A shell foundation, only a few inches thick, can provide the same rigidity as a much thicker raft. The substitution of shell foundations for spread footings and rafts can therefore lead to considerable savings in concrete and reinforcing steel. The consequent cost reduction can amount to as much as 50% in countries where construction materials are expensive and labour is relatively cheap. It is not therefore surprising that much of the pioneering research, development and practical applications relating to shell foundations has been carried out in developing countries, notably Mexico and India [217] [218] [219]. Studies have demonstrated that the cost savings increase with increasing column loads and decreasing allowable soil pressures, with greater sensitivity to the latter [219]. Felix Candela [217] is generally accorded the credit for pioneering the use of hyperbolic paraboloid for foundation footings. It will be shown later that the hyperbolic paraboloid with straight edges which he introduced for such applications is not necessarily the best structural form for foundation footings. It may be mentioned in passing that the use of inverted masonry barrel shells for foundations predates the advent of reinforced concrete as a construction material.

25-2 Choice of Shell Forms

Many forms of shells and folded plates are available for use as foundations. Inverted barrel shells and folded plates are suitable for use as continuous footings. The cone and hyperbolic paraboloid footing (consisting of four quadrants joined together) can act as foundations for a single column. The cone can be used in two different ways—with either its cavity facing up and vertex down [220] or its cavity in contact with the soil (Figure 25-1a, b, c). An inverted spherical dome (Figure 25-1d) can provide a foundation for a group of columns located on the circumference of a ring beam [221]. The funicular shell may be used either with its convexity facing up ⌢ or with its cavity facing up ∪ as a foundation. It is particularly suitable for use in the form of a multiple shell raft (Figure 25-1e). Hypar shells are also suitable for combined footings (Figure 25-1f).

25-3 Characteristics of Foundation Shells

Although foundation shells may be visualized as inverted roofs with soil pressures providing the loads, it must be appreciated that the following characteristics set them apart from thin shells used for roofing:

i) The loads to be carried being much heavier, the shells are relatively thicker and possess more flexural rigidity.

ii) These shells are generally smaller in size.

iii) If f is the rise and d the thickness, their $\left(\dfrac{f}{d}\right)$ ratios are much smaller than those of roof shells. Hence, they deflect more.

These characteristics have to be borne in mind in designing shell foundations.

25-4 Selection of Shell Dimensions

Apart from structural considerations, soil parameters have also to be considered when arriving at the demensions of shell foundations. From a structural point of view, a deep shell is to be preferred, because it is stiffer and develops smaller stresses. But it will increase the surface area and cost of forming. Moreover, tests carried out on funicular shells resting on sand with their cavities facing up have shown that the settlement increases as the depth of the shell at its centre increases. A trade-off between structural and settlement considerations has to be worked out to arrive at a suitable central depth. Nicholls and Izadi [222] have arrived at optimum semi-vertex angles for conical shells to minimize cost, assuming a range of values for the cost ratio of reinforcing steel to concrete. The dimensions of edge-members and the amount of reinforcement in them have also to be selected judiciously because it often happens that the premature cracking or failure of the edge-members does not permit the strength of the shell to be fully exploited. Tests on funicular shells have shown that the provision of deeper edge beams not only raises the ultimate load but also the load at which the first crack is formed. Moreover, as the depth of the edge beam is increased, the location of the first crack tends to move towards the mid-point of the beam [223].

25-5 Hypar Shell Footings: State of the Art

Practice is far ahead of theory. Conservative designs of hypar footings based on simple design procedures have been successfully used in practice. Their performance has been proven by full-scale load tests [224] carried out at Madras, India in connection with their proposed use for a four-storeyed

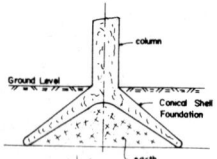

FIG. 25-1*a*

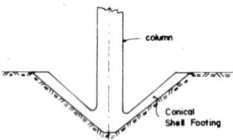

FIG. 25-1*b*

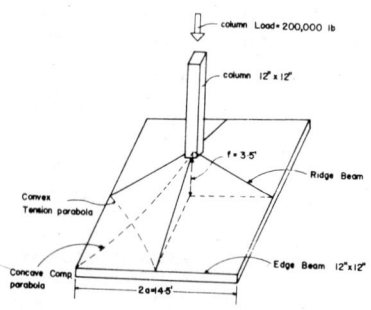

FIG. 25-1*c*

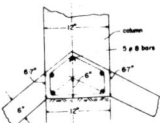

FIG. 25-1c

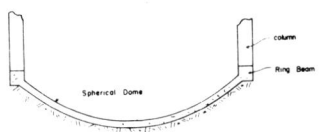

FIG. 25-1d

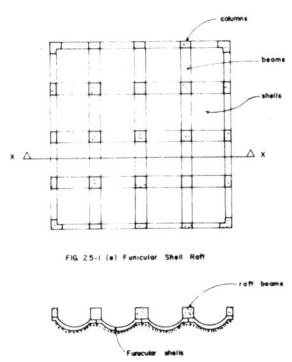

FIG. 25-1 (e) Funicular Shell Raft

FIG. 25-1e

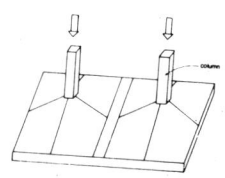

FIG. 25-1f

housing project. The combined footing tested was 18′3″ × 9′6″ in plan and consisted of two hypar footings each designed to carry a column load of 65 tons. The rise of the hypars was 2′6″. Each hypar footing in fact consisted of four hyperbolic paraboloids.

The soil at the site was dense sand with a safe bearing capacity of 0.75 ton per square foot. The footings were designed on the basis of the membrane theory, assuming a uniformly distributed soil reaction. The design was carried out by the working stress method allowing 600 psi in direct compression in the concrete and 16,000 psi in tension in the reinforcing steel. The shell 6″ thick was reinforced with #4 bars at 4½″ centres, in two directions parallel and perpendicular to the edges. The combined footing was loaded to 1.5 times the design load by means of two loading jacks positioned centrally over the columns and reacting against two loading beams carrying sand bags heaped on a platform. The measured stresses in the concrete and steel were far below the levels assumed in the design indicating ample margins of safety. The following conclusions may drawn from the reported test results:

i) The soil reaction under hypar footings may be assumed to be uniform despite the arching observed in experiments [222], indicating higher pressures at the edges than at the centre.

ii) The membrane theory may be used to simplify the design procedure, although it is well-known that in hyperbolic paraboloids with boundaries consisting of straight line generators, a certain amount of flexure is unavoidable. The favourable performance in load tests of hypar footings designed on the basis of the membrane theory is at least partially attributable to the conservative design practice of ignoring the tensile strength of concrete in the shell and designing the reinforcement to take all of the tension.

iii) The working stress method is to be preferred, because the object is to ensure that neither the shell nor edge and ridge beams crack under service loads, permitting the reinforcement to rust in contact with moist earth.

A simple design office procedure based on the foregoing assumptions for arriving at the dimensions of hypar footings and the reinforcement to be provided in the shell and the edge and ridge members will now be outlined with the aid of an example.

Design Example 25-1

PROBLEM

Design a Reinforced Concrete hypar footing for a 12″ × 12″ square column carrying a service load of 200 kips, assuming that the safe soil bearing pressure is 1,500 psf at 3′6″ below the ground level. Adopt the Working Stress Method.

$$f'_c = 5,000 \text{ psi} \qquad f_y = 50 \text{ ksi}$$

SOLUTION

The earth-fill over the footing and the weight of the filling transmitted to the soil will be estimated at an average density of 125 pounds per cubic foot.

Pressure on soil $= 3.5 \times 125 = 438$ psf

$$\left.\begin{array}{l}\text{Allowable soil bearing}\\ \text{pressure available for}\\ \text{service loads}\end{array}\right\} = 1{,}500 - 438 = 1{,}050 \text{ psf}$$
$$\text{say } 1{,}000 \text{ psf}$$

Plan area of footing $= \dfrac{200{,}000}{1{,}000} = 200 \text{ ft}^2$.

A footing $14'6'' \times 14'6''$ will suffice.
Hence $2a = 14.5$ and $a = 7.25$.
Assume a rise $f = 3.5'$ for the footing at the centre.

Design of the Shell

Referring to Chapter 17, and assuming the shell to be shallow, membrane shear $N_{xy} = -\dfrac{1{,}000 \times (7.25)^2}{2 \times 3.5} = -7{,}509$ lb/ft.

The membrane shear will cause tension in the directions of the convex parabola and compression in the directions of the concave parabola of a magnitude equal to the membrane shear. Instead of placing the reinforcing steel along the direction of the convex parabola, it is convenient, from a practical point of view, to arrange the steel in the form of an orthogonal grid with the reinforcement running parallel and perpendicular to the edges of the hypars. Steel is assumed to take all of the tension. The contribution of the concrete is ignored.

$$A_s = \frac{7{,}509}{20{,}000} = 0.375 \text{ in}^2/\text{foot}.$$

Using #4 bars, spacing works out to:

$$\frac{0.196 \times 12}{0.375} = 6.28''$$

Provide #4 bars at 6" centres (horizontal distance) is both the directions.

The tensile stress in the shell is next checked by dividing the tensile force per foot by the transformed area per foot to give:

$$\text{tensile stress} = \frac{7,509}{(12 \times 6 \times 9 \times 0.392)} = 99 \text{ psi}$$

This is very much less than the ultimate tensile stress of concrete in tension which may be computed as $6\sqrt{f'_c} = 6\sqrt{5,000} = 420$ psi. Hence the concrete in the shell will not crack under service loads.

Design of Edge Beam

$$\text{Tensile force in edge beam} = a \times N_{xy} = 7.25 \times 7,509$$
$$= 51,173 \text{ lb}$$

$$A_s = \frac{7,509 \times 7.25}{20,000} = 2.72 \text{ in}^2$$

Use 4 #8 bars to give 3.14 in² > 2.72 in²

Try an edge beam $12'' \times 12''$.
Transformed area $= (144 + 9 \times 4 \times 0.7854)$
$$= 172 \text{ in}^2$$

$$\text{Tensile stress in concrete} = \frac{7,509 \times 7.25}{172} = 316 < 420 \text{ psi. OK}$$

Design of the Ridge Beam

Length of ridge beam $= \sqrt{a^2 + f^2}$
$$= \sqrt{3.5^2 + 7.25^2} = 8.05'$$

Compressive force in ridge member $= 8.05 \times 7,509 = 120,894$ lbs.

At the apex of the shell a triangular rib 6" deep and as wide as the column will be formed. The ridge member will therefore have a section at this point as indicated in Figure 25-1c.

$$\text{Area of rib} = \left(\frac{6.7 + 12.7}{2}\right) \times 2 \times 6 = 116.4 = A_g.$$

$$A_{st} = 5 \times 0.7854 = 3.93 \text{ in}^2$$

$$(A_g - A_{st}) = (116.4 - 3.93) = (112.47 \text{ in}^2).$$

Design ultimate axial load $= 0.80 \times 0.70$

$$[0.85 \times 5,000 \times 112.47 + 50,000 \times$$

$$3.93]$$

$$= 377,718$$

$\left.\begin{array}{l}\text{Permissible working load} \\ \text{under axial compression}\end{array}\right\} = 377,718 \times 0.4$

$$= 151,087 > 120,894 \text{ lb.}$$

Hence provide a ridge beam as indicated in Figure 25-1c with five #8 bars as reinforcement.

Detailing of Reinforcement

The reinforcement from the ridge beams need to be properly anchored into the edge beams. For more information on detailing, the reader is referred to reference [225]. Wherever practicable, the edge and ridge beams can be avoided and the reinforcement can be accommodated in the shell itself. This practice, followed by Candela, will facilitate the casting of the shell on profiled mounds without interference from projecting beams. It is good practice to cover the profiled earth mound with a 3″ layer of lean concrete to obtain an even surface against which the concrete can be cast.

In countries where labour-intensive technologies are inappropriate because of high wage structures, it will be advantageous to precast hypar footings and install them in precut trenches. Preforming of profiled earth mounds to match the undersurface of the hyperbolic paraboloid may not ensure uniform contact. It is reported that better results may be achieved by filling the cavity of the hyperbolic paraboloid after installation by pouring sand into it through a hole provided in the column. The sand so poured is thoroughly compacted by centrifugal blasting. A simple centrifugal vane rotor that can be operated by a 2 H.P. motor has been developed for this purpose by Kurian and Shah [226].

25-6 Funicular Shell Footings and Rafts

Because of its remarkable stiffness, the funicular shell is particularly suitable for use in foundations, especially in the form of a raft. Bhattacharya and Ramaswamy [227] have studied the behaviour of such shells resting on elastic foundations. They discarded the Winkler and Elastic half space foundation models as unsuitable, because the former is too crude and the latter is inappropriate, unless the shell is resting on rock. Instead, they used a single layer two parameter model proposed by Vlasov and Leontev [228]. They compared the performance of a funicular shell with its convex surface pointing upwards with a hyperbolic paraboloid and a plate resting on the same foundation when acted upon by a point load at the crown. The superior characteristics of the funicular shell as a foundation will be readily apparent from Tables 25-1 and 25-2.

Table 25-1

DEFLEXION AND STRESS RESULTANTS	SHELL RISE = 3"		SHELL RISE = 6"		SHELL RISE = 9"	
	FUNICULAR SHELL	HYPAR	FUNICULAR SHELL	HYPAR	FUNICULAR SHELL	HYPAR
Deflexion—ft	0.000012	0.000026	0.000006	0.000016	0.000004	0.000011
Moment—lbft/ft	0.205	0.289	0.167	0.266	0.143	0.259
Shear lb/ft	0.00	−0.006	0.00	−0.003	0.00	−0.002

The shells were 8′ × 8′ in plan; the Soil Modulus was assumed as 600 lbs/ft^2. The value γ, the soil parameter governing the distribution of the displacement with the depth of the soil is taken as 2.0.

Table 25-2

DESCRIPTION	MAXIMUM DEFLEXION IN FT.	MAXIMUM MOMENT IN FT LB PER FOOT	MAXIMUM THRUST IN LBS/FOOT
1. Plate without foundation	0.167×10^{-3}	0.3312	—
2. Plate with foundation	0.101×10^{-3}	0.2930	—
3. Shell without foundation	0.62×10^{-5}	0.1800	−5.20
4. Shell with foundation	0.61×10^{-5}	0.1670	−5.15

It is evident from Table 25-2 that the foundation itself plays only an insignificant role because of the high stiffness of the shell. This would point to the important conclusion that the funicular shells will perform satisfactorily as a foundation even on every weak soils because its structural behaviour is only marginally influenced by the support that it receives from the soil on which it rests.

Two interesting field investigations reported by Ayyar et al [229] [223] seem to provide experimental confirmation of the bahaviour predicted by theory. In the first of these studies [229] single square shells of size 110 cm with a central rise of 7.35 cm were tested by applying loads in three modes:

i) as a uniformly distributed load along the edge beams

ii) as a concentrated load at the mid-points of the edge beams

iii) as a concentrated load at the corners of the shell.

The site chosen at Monkompu in Kerala State, India where a soft clayey deposit is underlain by a stiff clay 30 to 35 metres thick. The test results indicated that a funicular shell loaded as a raft functions as a rigid foundation for nearly half its ultimate load. The pressure distribution on the clayey soil tended to be slightly convex in the initial stages with higher soil pressures at the centre than at the edges, possibly because the shape of the shell increases the resistance of the soil at the centre. With increasing loads, a near uniform pressure distribution was reached, indicating that the shell behaved as a rigid foundation. The shell did not appear sensitive to the manner in which it was loaded.

A more comprehensive field test was carried out on a multiple shell raft [223] at Kuttonad, Kerala State, India. The site, notorious for its foundation problems, is overlain by a marine clay. Even single storeyed buildings resting on strip foundations have shown signs of distress. The details of the raft and loading arrangements used may be seen in Figure 25-2. The structure was hydraulically loaded by filling the tank with water. Measurements of settlements and soil pressures were recorded during the construction and as the loading proceeded over a period of time. Settlements were observed

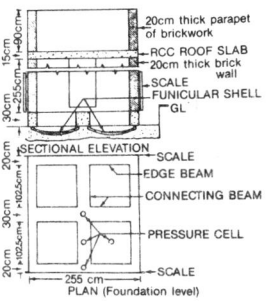

FIG. 25-2

by reading scales fixed to the walls using a precision levelling instrument. The variation of the settlement with time and the distribution of pressures on the soil are shown in Figures 25-3 and 25-4. The water loading of the structure began on August 15, 1978. By the end of March 1979, the total settlement observed was only 7.5 mm. The differential settlement was less than 1 mm. An observation of considerable interest is that the confinement of the soil in the interstices between the shells tended to increase the bearing capacity, besides reducing the settlement. The provision of a grid of beams preempted premature failure of the shell on account of the failure of the edge beams. The test results point to promising possibilities that multiple funicular shell rafts offer in transmitting heavy loads to weak soils.

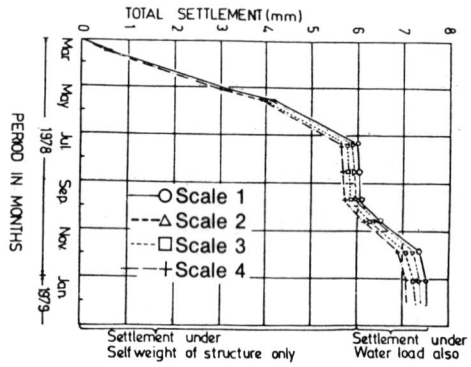

FIG. 25-3

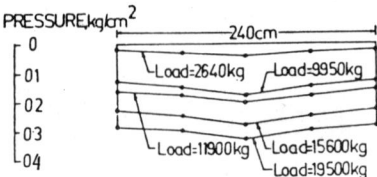

FIG. 25-4

CHAPTER 26

SHELLS OF FERROCEMENT

26-1 Historical Notes

When Jean Louis Lambot built a boat of plastered mesh for use in his estate at Miraval in 1843, he was not fully aware of the magnitude of the contribution that he was making to concrete technology. He had, in fact, discovered, without fully realizing it, that the secret of improving the extensibility and ductility of concrete (and hence its tensile strength) lies in embedding several layers of small diameter steel meshes in rich cement mortar. This accidental discovery anticipated to a remarkable extent the concepts of micro-reinforcement and crack arrest. Lambot filed French and Belgian patents for the new material 'Ferciement'—the forerunner of modern ferrocement. This early contribution was all but forgotten until Nervi resurrected the technique under the name of ferrocemento in the early forties [230]. He used the new material, in the form of corrugated precast units, to dramatic advantage in roofing long span structures, the most notable of which is the 300 feet span Turin Exhibition Hall. Realizing that its light-weight, impermeability and resistance to impact and marine-borers made it an ideal material for building boats, the firm of Nervi and Bartoli pioneered building a motor sailor *Irene* with a hull of ferrocement, just 1.37″ thick. This development triggered interest in several countries such as New Zealand, Canada, Thailand and India in the use of ferrocement for building fishing trawlers and sailing boats. As a result, ferrocement is today an accepted material for the fabrication of marine craft. The use of ferrocement in building construction has not, however, progressed at the same rapid pace, except perhaps in the Soviet Union and Eastern Europe.

GOSSTROY, the standardization organization in the U.S.S.R. has a Code of Practice for ferrocement. It has been claimed that approximately three million square feet of roofing has been built in the Soviet Union utilizing armocement (ferrocement) components for a variety of structures such as metro underground stations, farm buildings and industrial sheds [231]. A very large proportion of these roofs are thin shells. Developing countries such as China, India and Thaliand use labour-intensive technologies to produce ferro-cement building components, water-tanks, grain storage bins and gas-holders for biogas plants. On the other hand, in the U.S.S.R. and Eastern Europe, the production of ferro-cement components has been mechanized to a large extent. Techniques such as vibro-stamping, vibro-pressing and vibro-moulding are employed [232]. Ferrocement is thus a unique material which at once lends itself to both highly industrialized and labour-intensive technologies. It therefore finds acceptance in the developing and the developed parts of the world. Successful practice all over the world has leap-frogged ahead of theory which is limping far behind.

In the United States, interest in ferrocement appears to have been sparked by a report published in 1973 by the National Academy of Sciences [233]. But this study focussed on the possible applications of ferrocement in the developing countries of the world. As a sequel to the report, the International Ferrocement Information Centre came into being at the Asian Institute of Technology, Bangkok in 1976. Among developments in North America mention needs to be made of the setting up of ACI Committee 549 in 1974 "to study and report on the engineering properties, construction practices and practical applications of ferrocement and similar materials and to develop standards and safeguards for ferrocement construction."

26-2 Composition

Ferrocement consists of several layers of woven wire or welded wired galvanized mesh or expanded metal covered by rich cement mortar with a cement to sand ratio by weight ranging from 1 to 2.5. The mesh wires are of 0.5 to 1.5 mm diameter (0.020 to 0.62") and are spaced 6 to 25 mm apart (¼ to 1") in both the directions to form a grid. In addition, there may be a skeletal layer of regular reinforcing bars of small diameter (ranging from #1 to #3 bars). An important characteristic of ferrocement is the specific surface of the wires used as micro-reinforcement. The specific surface is defined as the ratio of bond surface to the volume of the reinforcement. A value of the specific surface of up to 10 in^2/in^3 is acceptable. Some of the properties of ferrocement and specifications for its constituent materials have been presented by Naaman [234]. The thickness of ferrocement members normally ranges between ¼" to 2". Steel contents of up to 40 lbs/ft^3 are generally prescribed. It has been observed that the extensibility of concrete increases five fold when the steel content is in the range of 25 to 32 lbs/ft^3 (400 to 500 kg/m^3) [230]. The cross-section of a typical ferrocement member may be seen in Figure 26-1.

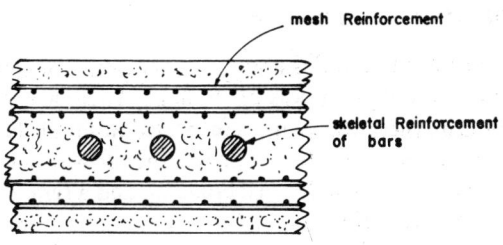

FIG. 26-1

26-3 Properties of Ferrocement

The attributes of ferrocement which are particularly favourable for its use in thin shell construction may be summarised as follows:

(i) It is very light compared to reinforced concrete because of the very thin sections that are practicable.

(ii) It is moisture-proof [235] and needs no additional protection in the form of water-proofing felt.

(iii) It has the mouldability of concrete; but unlike concrete, it has considerable tensile strength and ductility. While reinforced concrete thin shells have to be designed as membranes carrying loads primarily in compression, with some secondary flexure, it is practicable to design suspended shells of ferrocement which are in tension. Buckling inherent in compression forms which limit their spans is absent in tension solutions. Hence, much longer spans are conceptually possible.

(iv) The need for expensive formwork can be obviated altogether or minimized because the meshes supported on skeletal reinforcing bars can be plastered in-situ or shotcreted [236].

(v) The material lends itself to mass production of components if mechanized techniques such as vibro-stamping, vibro-moulding or vibro-pressing are employed. The lightweight components pose no serious problems of transportation, handling and erection.

(vi) A reinforced concrete thin shell is normally analyzed on the basis of elastic behaviour. But when we proceed to design reinforcement, the concrete is assumed to be cracked and all of the tension is assigned to the steel. The design of a ferrocement thin shell structure does not involve this dichotomy, because the composite material has an ultimate tensile strength of about 5,000 p.s.i. and a Modulus of Rupture of 8,000 p.s.i. The permissible tensile stresses are of the order of 1,500 p.s.i. Hence the structure can be designed and analyzed on the same basic assumptions.

We will deal with each of these aspects in greater detail in later sections.

26-4 Definition of Ferrocement

Experts are not unanimous about the definition of ferrocement. Broadly speaking, there are two schools of thought. There are those who hold that ferrocement is a synergistic composite. Stated in simple language, this merely means that the presence of steel improves the properties of concrete—mainly its extensibility reflected in delayed cracking and a remarkable increase in its Modulus of Rupture. This school favours the definition of ferrocement as a thin highly reinforced body of rich mortar in which steel reinforcement is distributed throughout the section so that when stressed it behaves more or less as a *homogenous material*. This definition seems to be closer to the intuitive ideas of Nervi and his uncanny deductions about the role of closely spaced small diameter reinforcement in improving the extensibility of concrete. The other school tends to regard ferrocement as improved reinforced concrete.

26-5 Gaps between Theory and Practice

Reference has already been made to the gap that separates theory from practice. However, of late, there have been a spurt of published papers which promise to close some of the gaps in our knowledge relating to the behaviour of ferrocement in tension, flexure and tension combined with flexure. Study of behaviour of ferrocement in tension is important, because the thin shells of water tanks are in hoop tension. Behaviour in flexure is of paramount interest in so far as roofing elements are concerned. In hybrid structures, such as hanging ribbons of ferrocement, behaviour under combined flexure and tension needs to be understood. Some of the recent published papers in these areas of interest need special mention. Balaguru, Naaman and Shah have investigated the behaviour of ferrocement in flexure [237]. The accent in this study is on deflexion and cracking. Johnston et al and Naaman have studied behaviour in tension as well as flexure [238] [234]. Both Balaguru et al and Johnston et al have presented models of ferrocement by refining available models for reinforced concrete. An extensive series of tests on flexure were also undertaken at the Structural Engineering Research Centre (SERC), Madras, India involving twenty specimens with varying steel contents and using three different types of meshes (hexagonal, woven square mesh and expanded metal). [239].

The author would propose the following hypothesis to understand the behaviour in flexure of ferrocement, regarding it as a homogenous material. From intuitive considerations, this theory may be stated thus:

(i) When a beam specimen has a certain volume content of steel (stated in cubic feet/cubic feet), it will behave as a homogenous material.

(ii) The presence of uniformly distributed micro-reinforcement considerably enhances the modulus of rupture of the material.

(iii) The beam specimen fails when this enhanced Modulus of Rupture is reached.

(iv) To take into account the contributions of steel mesh and mortar, one may use the non-dimensional index $\dfrac{\lambda f_u}{f'_m}$ where,

λ = the volume content of steel expressed in cubic feet/cubic feet.

f_u = ultimate tensile strength of the material of the mesh

f'_m = ultimate compressive strength of mortar.

The foregoing considerations may be incorporated in the following expression for the ultimate flexural moment M_u in the form:

$$M_u = k_1 f'_m \left(1 + k_2 \frac{\lambda f_u}{f'_m}\right) \frac{bD^2}{6}$$

where,

$k_1 f'_m$ = a measure of the ultimate tensile strength in flexure of the unreinforced mortar

k_2 = a coefficient to be determined

b = breadth of the member

D = total depth of the member

The author carried out a linear regression analysis of the SERC and Johnston test results to evaluate k_1 and k_2. This led to the following relationship:

$$\frac{M_u}{bD^2} = 265.83 + 1090.73 \frac{\lambda f_u}{f'_m} \tag{26-1}$$

If this equation of physically interpreted, it is seen that, in the limit, when no reinforcement is present i.e. $\lambda = 0$, the modulus of rupture of unreinforced mortar given by $\dfrac{6M_u}{bD^2}$ will be as high as $6 \times 265.83 = 1,595$ p.s.i. Obviously, the extrapolation of the best fit straight line right up to $\lambda = 0$ is not valid.

Intuitively, one would expect a minimum steel content defining the dividing line between reinforced concrete and ferrocement behaviour. To find this limit, a regression analysis of the test results relating to low steel contents was next carried out. This led to the relationship:

$$\frac{M_u}{bD^2} = 74 + 1{,}888 \, \frac{\lambda f_u}{f'_m} \qquad (26\text{-}2)$$

The intersection of these two lines defines a value of $\lambda = 0.24$ corresponding to a steel content of 12.5 lbs/cubic foot. Above this steel content, homogenous ferrocement behvaiour may be expected and equation (26-1) may be used to predict ultimate bending moments that a given section can carry or alternatively to arrive at the steel required to resist a given ultimate moment.

It is desirable to design thin shells by the working stress method. The ultimate flexural capacity of the section may be reduced by a suitabie factor to arrive at the safe bending moment that a section can carry under working loads.

In so far as sections in axial tension are concerned studies have shown that the ultimate strength of a member is the ultimate strength of the mesh reinforcement [234]. The ultimate strength may be reduced by a suitable factor to arrive at the safe working load that a ferrocement member can carry.

In addition, a tension member needs to be checked to ensure that the tensile stress in the composite section does not exceed 1000 p.s.i. The allowable compressive stress in ferrocement members may be taken as $0.45 f'_c$ where f'_c is the compressive strength of the mortar which is generally of the order of 5000 p.s.i.

Thin shells are subjected to either axial tension combined with flexure or axial compression accompanied by flexure. The principles outlined above provide a suitable basis for designing such members until more refined procedures become available. For purposes of design Equation (26-1) may be more meaningful recast as follows to permit the Modulus of Rupture corresponding to various values of $\dfrac{\lambda f_u}{f'_m}$ to be readily calculated:

$$\frac{6M_u}{bD^2} = 1595 + 6544 \, \frac{\lambda f_u}{f'_m} \qquad (26\text{-}3)$$

$$\lambda > 0.24$$

26-6   Design of Ferrocement Thin Shells

Design Example 26-1

PROBLEM

At a point in a cylindrical shell $M\phi$ = 150 ft lb/foot

$$N\phi = 4000 \text{ lb/foot (tension)}$$

$$Nx = 4000 \text{ lb/foot (tension)}$$

The shell is of ferrocement and is 1″ thick

$$f_u = 65{,}000 \text{ p.s.i.} \qquad f'_m = 5{,}000 \text{ p.s.i.}$$

Welded wire mesh available is 0.042″ diameter with a mesh spacing of ½″ in both directions.

SOLUTION

Let the skeletal reinforcement in the longitudinal direction (x direction) consist of #3 bars at 6″ centres. Let there be four layers of mesh reinforcement (two at top and two at the bottom of the skeletal #3 reinforcing bars) draped to suit the radius of curvature of the shell in the ϕ direction.

Design in the circumferential (ϕ) direction

Consider 1 foot length of the shell.

$$\text{Area of concrete} = 12 \times 1 = 12 \text{ in}^2$$

There will be 24 bars available in each layer.

$$\left. \begin{array}{l} \text{Area of steel furnished by four} \\ \text{layers in both the directions} \end{array} \right\} = 2 \times 4 \times \frac{\pi}{4} \times (0.042)^2 \times 24$$

$$= 0.27 \text{ in}^2$$

$$\lambda = \frac{0.27}{12}; \qquad \frac{\lambda f_u}{f'_m} = \frac{0.27}{12} \times \frac{65{,}000}{5{,}000} = 0.29$$

The skeletal #3 bars in the x direction will be ignored for design in the ϕ direction.

The modulus of rupture may now be found (26-3) as

$$(1595 + 0.29 \times 6544) = 3493 \text{ p.s.i.}$$

Allowable tensile stress in flexure may be conservatively estimated as ½ × 3493 = 1746.5 p.s.i.

Superimposing the effects of $M\phi$ and $N\phi$, the maximum tensile stess is

$$= \frac{6M\phi}{bD^2} + \frac{N\phi}{bD} + \frac{6 \times 150 \times 12}{12 \times 1^2} + \frac{4000}{12 \times 1} = 900 + 333 = 1233 \text{ p.s.i.}$$

$$< 1746.5$$

Hence the four meshes provided are adequate.

Design in the x direction

The reinforcement available in the x direction consists of the four layers of mesh (with 24 bars in each available in a one foot length) and the skeletal #3 bars at 6″ as providing an area of 2×0.11 per foot length of the shell measured in the circumferential direction.

$$\left.\begin{array}{l}\text{Area of steel available in the}\\ x \text{ direction/foot of shell}\end{array}\right\} = 0.22 + 4 \times 24(.042)^2 \times \frac{\pi}{4}$$

$$= 0.22 + 0.13 = 0.35 \text{ in}^2$$

At an allowable stress of 20,000 p.s.i., safe axial tension $= 0.35 \times 20,000$

$$= 7,000 \text{ lbs} > N_x$$

Hence the reinforcement provided is adequate.

$$\text{Tension in the composite} = \frac{4,000}{12 \times 1} = 333 \text{ p.s.i.} < 1,000 \text{ p.s.i.}$$

$$\text{OK}$$

The reinforcement scheme is shown in Figure 26-2.

26-7 Notes on Construction

As already mentioned, ferrocement thin shells for roofs can be built without the use of forms. The skeletal steel is first laid in position, propping it if necessary. Next the mesh reinforcement is tied to it. The mason can now deposit the mortar in position and finish the surfaces with a trowel.

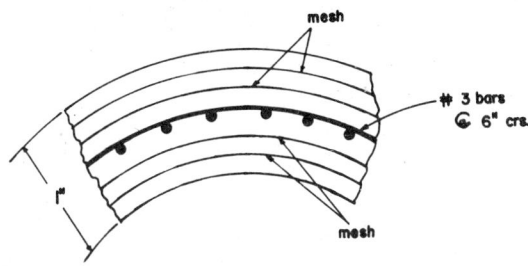

Fig. 26-2

(Photographs 26-1a, b, c.) Alternatively, the mortar can be deposited by shotcreting.

A funicular ferrocement shell of an unusual shape recently built in India may be seen in Photo-26.2a. Its plan resembles the English letter W. It was built to provide a covered work space of 8,000 square feet for mentally handicapped trainees attached to the Occupational Therapy Department of the B.M. Institute, Ahmedabad. The shape of the roof was generated by a hanging chain net model—a technique pioneered by Frei Otto for tension structures—and inverting the surface so found to arrive at the shape of funicular shell. The skeletal funicular frame was formed out of 190 mm (7.48″) diameter steel tubes. The ferrocement skin, 1″ thick, was designed to act integrally with the lattice frame. The sequences involved may be understood by looking at Photos 26-2a, b, c, d. A paper on this project is expected to be published in the near future.

26-8 Hanging Ribbons of Ferrocement

Shell roofs—with the exception of shells of revolution—are economical only up to a span of about 100 feet. Being primarily in compression, shells in excess of 100′ span pose problems of elastic stability. For larger spans, it is therefore logical to look for tension shapes in the form of suspended shells.

If suspended shells are too light, they give rise to problems of flutter. It is also axiomatic that large span structures must be as light as possible; other-

PHOTO 26-1*a*   Ferrocement shell under construction.

PHOTO 26-1*b*   The application of mortar on the skeleton.

PHOTO 26-1c   A view of completed ferrocement shell.

wise, one will be obliged to design the roof primarily to carry itself. These considerations would point to a tension structural form which can be built with a meterial which is neither too light nor too heavy and has adequate tensile strength. Moreover, it will be an advantage if the material is waterproof by itself. Ferrocement satisfies all these requirements. Hanging ribbons of ferrocement just about 1″ thick have been shown to be practicable for a span of up to 60 m. The author had the opportunity to inspect such a roof in Czechoslovakia. Such roofs can be designed by regarding them as beam cable structures [240] [241]. The concept is shown schematically in Figure 26-3. The roof is given a gentle curvature in the longitudinal direction to facilitate drainage. The double curvature also helps in stabilizing the roof. The ribbon is cast on the ground at the site of construction in pieces equal to the developed span of the roof and in widths of about three feet. It is reinforced with several layers of mesh and a central layer of regular small diameter bars. The ribbon is temporarily hung from the bottom chord of a truss profiled to the shape of the hanging roof and hoisted into position using two cranes. The regular reinforcing bars projecting from the ribbon are welded to dowels projecting from the edge beams. The truss is then relieved and the joints between the ribbon and the edge beam are concreted. Where a high degree of thermal comfort is desired, sandwich ribbons consisting of two ferrocement skins and a core of styrofoam can be designed.

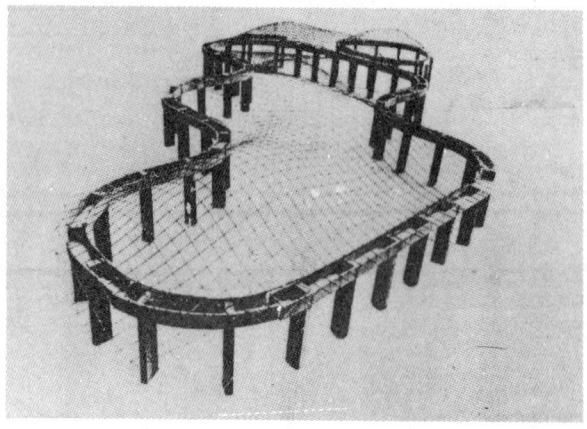

PHOTO 26-2*a*   Model of the ferrocement shell for the B.M. Institute, Ahmedabad, India. *(By courtesy of Mr. Gautam and Miss Gira Sarabhai)*

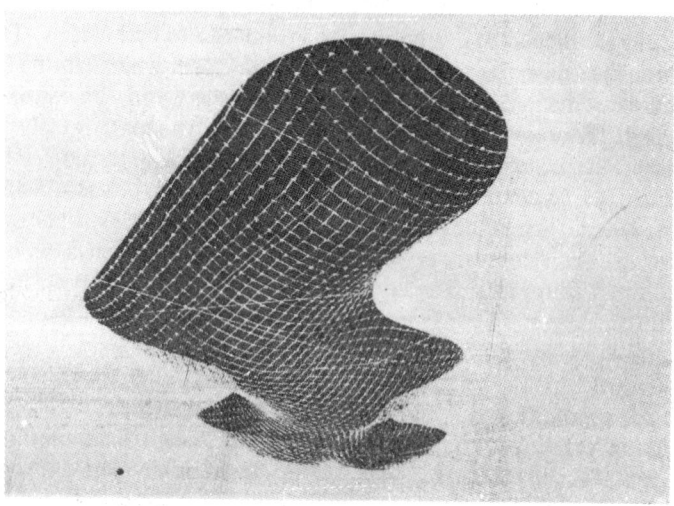

PHOTO 26-2*b*   Chain model of ferrocement shell for the B.M. Institute, Ahmedabad, India. *(By courtesy of Mr. Gautam and Miss Gira Sarabhai)*

PHOTO 26-2c Ferrocement shell for B.M. Institute, Ahmedabad, India *(By courtesy of Mr. Gautam and Miss Gira Sarabhai)*

PHOTO 26-2d A view of the completed ferrocement shell roof for B.M. Institute, Ahmedabad, India *(By courtesy of Mr. Gautam and Miss Gira Sarabhai)*
Architectural & Construction details—Gautam Sarabhai
Structural Design—Structural Engineering Research Centre, India.
Construction—Hiralal P. Mistry
Contractors—H.C. Shah B.H. Shah
Ganon Dunkerly & Co. Ltd.
Ahmedabad, India.

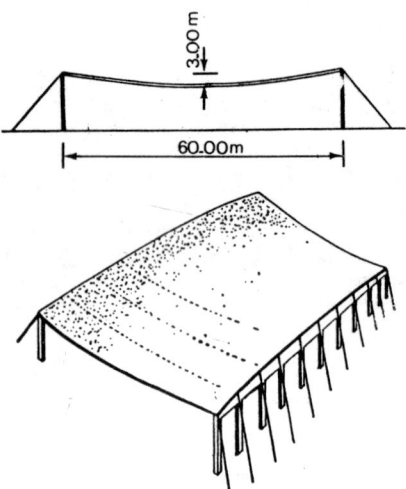

FIG. 26-3

Sealing strips are inserted between adjacent ribbons. Photo 26-3 shows an experimental ferrocement ribbon of 25 m span ready for testing at the Structural Engineering Research Centre at Madras. The possibilities of using ferrocement imaginatively in tension structures opens up exciting new frontiers which await to be fully explored.

PHOTO 26-3   Ferrocement Ribbon ready for testing *(By courtesy of Structural Engineering Research Centre, Madras, India)*

APPENDIX

Membrane Forces in Semielliptic Shells* (Refer to Art. 6-10)

	θ	t_2	s	t_1	t_2'	s'	t_1'
$\dfrac{b}{a} = 0.1$	0°	−10.0000	0	−29.9000	−10.0000	0	−30.0000
	7°30′	−2.2514	−14.4440	−12.8970	−2.2321	−14.4500	+13.1915
	15°	−0.4582	−9.9152	−61.1100	−0.4425	−9.8274	+62.6127
	30°	−0.0663	−5.3263	−132.9100	−0.0574	−5.0448	+143.3653
	45°	−0.0197	−3.4935	−170.1500	−0.0140	−2.9703	+208.9684
	60°	−0.0077	−2.5866	−59.3000	−0.0038	−1.7263	+258.5110
	75°	−0.0029	−2.1377	−96.7700	−0.0007	−0.8033	+289.4656
	90°	0	−2.0000	0	0	0	+300.0000
$\dfrac{b}{a} = 0.2$	0°	−5.0000	0	−14.8000	−5.0000	0	−15.0000
	7°30′	−2.9643	−6.8178	−7.0740	−2.9389	−6.8889	−7.0394
	15°	−1.1469	−7.1851	+6.2660	−1.1078	−7.1903	+6.8892
	30°	−0.2338	−4.8571	+27.7570	−0.2025	−4.6394	+31.1820
	45°	−0.0754	−3.3724	+38.9880	−0.0533	−2.8846	+49.9230
	60°	−0.0302	−2.5325	+37.5330	−0.0151	−1.7093	+63.6628
	75°	−0.0114	−2.1310	+23.0280	−0.0030	−0.8015	+72.1328
	90°	0	−2.0000	0	0	0	+75.0000
$\dfrac{b}{a} = 0.3$	0°	−3.3333	0	−9.7000	−3.3333	0	−10.0000
	7°30′	−2.6038	−3.5810	−7.2021	−2.5815	−3.6798	−7.3303
	15°	−1.4822	−4.8847	−1.7772	−1.4317	−4.9683	−1.4570
	30°	−0.4357	−4.2244	+8.7658	−0.3773	−4.0915	+10.7961
	45°	−0.1582	−3.1852	+14.8150	−0.1118	−2.7523	+20.5443
	60°	−0.0663	−2.4972	+15.0070	−0.0331	−1.6816	+27.5906
	75°	−0.0256	−2.1200	+9.3745	−0.0066	−0.7987	+31.8866
	90°	0	−2.0000	0	0	0	+33.3333
$\dfrac{b}{a} = 0.4$	0°	−2.5000	0	−7.1000	−2.5000	0	−7.5000
	7°30′	−2.1707	−2.1159	−5.0359	−2.1611	−2.2272	−6.2979
	15°	−1.5366	−3.3314	−3.4326	−1.4843	−3.4679	−3.3180
	30°	−0.6157	−3.5541	+2.5529	−0.5332	−3.5109	−4.0072
	45°	−0.2561	−2.9503	+6.4784	−0.1811	−2.5862	+10.3403
	60°	−0.1129	−2.4227	+7.1517	−0.0570	−1.6444	+14.9777
	75°	−0.0452	−2.1046	+4.5988	−0.0117	−0.7947	+17.8011
	90°	0	−2.0000	0	0	0	−18.7500

* From "Scienza delle Costruzioni," vol. III, pp. 484–485, 1958, by Belluzzi, O. Used by permission of Nicola Zanichelli, Editore, Bologna, Italy.

	θ	t_2	s	t_1	t_2'	s'	t_1'
$\dfrac{b}{a} = 0.5$	0°	−2.0000	0	−5.5000	−2.0000	0	.−6.0000
	7°30′	−1.8400	−1.3596	−4.9715	−1.8243	−1.4774	−5.3537
	15°	−1.4678	−2.3273	−3.5901	−1.4178	−2.4980	−3.6412
	30°	−0.7482	−2.9286	+0.0205	−0.6479	−2.9692	+1.1338
	45°	−0.3578	−2.6870	+2.7392	−0.2530	−2.4000	+5.6921
	60°	−0.1707	−2.3316	+3.5449	−0.0853	−1.5988	+9.1525
	75°	−0.0699	−2.0851	+2.3916	−0.0181	−0.7897	+11.2822
	90°	0	−2.0000	0	0	0	+12.0000
$\dfrac{b}{a} = 0.6$	0°	−1.6667	0	−4.4000	−1.6667	0	−5.0000
	7°30′	−1.5801	−0.9252	−4.1138	−1.5666	−1.0467	−4.6089
	15°	−1.3599	−2.6685	−3.3356	−1.3135	−1.8616	−3.5303
	30°	−0.8314	−2.3846	−1.0995	−0.7200	−2.4982	−0.2311
	45°	−0.4540	−2.4125	+0.8156	−0.3210	−2.2059	+3.2338
	60°	−0.2338	−2.2269	+1.6148	−0.1169	−1.5465	+6.0009
	75°	−0.0995	−2.0616	+1.9600	−0.0258	−0.7836	+7.7419
	90°	0	−2.0000	0	0	0	+8.3333
$\dfrac{b}{a} = 0.7$	0°	−1.4286	0	−3.5857	−1.4286	0	−4.2857
	7°30′	−1.3795	−0.6547	−3.4246	−1.3677	−0.7785	−4.0281
	15°	−1.2472	−1.2225	−2.9754	−1.2047	−1.4309	−3.2996
	30°	−0.8745	−1.9291	−1.5883	−0.7574	−2.1037	−0.9154
	45°	−0.5388	−2.1403	−0.2507	−0.3810	−2.0134	+1.8087
	60°	−0.3006	−2.1117	+0.4793	−0.1503	−1.4889	+4.1129
	75°	−0.1336	−2.0344	+0.4785	−0.0346	−0.7765	+5.6079
	90°	0	−2.0000	0	0	0	+6.1224
$\dfrac{b}{a} = 0.8$	0°	−1.2500	0	−2.9500	−1.2500	0	−3.7500
	7°30′	−1.2217	−0.4755	−2.8602	−1.2113	−0.6008	−3.5692
	15°	−1.1422	−0.9103	−2.6050	−1.1033	−1.1293	−3.0493
	30°	−0.8886	−1.5548	−1.7692	−0.7696	−1.6609	−1.2618
	45°	−0.6095	−1.8799	−0.8633	−0.4310	−1.8293	+0.9317
	60°	−0.3686	−1.9890	−0.2309	−0.1843	−1.4275	+2.8991
	75°	−0.1718	−2.0035	−0.0164	−0.0445	−0.7685	+4.2237
	90°	0	−2.0000	0	0	0	+4.6875
$\dfrac{b}{a} = 0.9$	0°	−1.1111	0	−2.4333	−1.1111	0	−3.3333
	7°30′	−1.0950	−0.3510	−2.3874	−1.0857	−0.4774	−3.2000
	15°	−1.0484	−0.6849	−2.2546	−1.0127	−0.9116	−2.8123
	30°	−0.8834	−1.2493	−1.7922	−0.7651	−1.5149	−1.4298
	45°	−0.6653	−1.6369	−1.2168	−0.4704	−1.6575	·+0.3698
	60°	−0.4357	−1.8616	−0.6927	−0.2178	−1.3638	+2.0777
	75°	−0.2137	−1.9692	−0.2970	−0.0553	−0.7597	+3.2755
	90°	0	−2.0000	0	0	0	+3.7037
$\dfrac{b}{a} = 1.0$	0°	−1.0000	0	−2.0000	−1.0000	0	−3.0000
	7°30′	−0.9914	−0.2611	−1.9829	−0.9830	−0.3862	−2.8977
	15°	−0.9659	−0.5176	−1.9319	−0.9330	−0.7500	−2.5980
	30°	−0.8660	−1.0000	−1.7321	−0.7500	−1.2990	−1.5000
	45°	−0.7071	−1.4142	−1.4142	−0.5000	−1.5000	0
	60°	−0.5000	−1.7321	−1.0000	−0.2500	−1.2990	+1.5000
	75°	−0.2588	−1.9319	−0.5176	−0.0670	−0.7500	+2.5980
	90°	0	−2.0000	0	0	0	+3.0000

APPENDIX II

PROGRAM FOR COMPUTING ORDINATES OF A FUNICULAR SHELL OF RECTANGULAR GROUND PLAN

[f] CLEAR [PRGM]

h LBL A	STO 7	STO.1	k
STO 1	RCL 5	RCL 3	×
$x \rightleftharpoons y$	ge^x	h$\frac{1}{x}$	h π
STO 2	h LST x	3	3
1	CHS	hy^x	hy^x
STO 3	ge^x	STO.2	÷
h LBL B	+	RCL 3	STO 9
RCL 3	0.5	1	RCL 9
h π	×	−	h PAUSE
×	RCL 7	0.5	RCL 3
2a	ge^x	×	ENTER
÷	h LST x	STO.3	9
STO 4	CHS	1	$f\,x=y$
RCL 4	ge^x	CHS	hRTN
RCL 1	+	RCL.3	2
×	0.5	hy^x	STO + 3
STO 5	×	RCL.2	GTOB
RCL 4	÷	×	Switch to run
RCL 2	CHS	RCL.1	x
×	1	×	ENTER
STO 6	+	STO + 8	y
RCL 4	RCL 6	RCL 8	Press button A
b	f cos	16 a^2	
×	×	×	

Notes: In run mode, registers have to be cleared and calculator set in radian mode.

$$k = \frac{g}{N}$$

Refer to Chapter 18 for notation used

BIBLIOGRAPHY

1. Ramaswamy, G. S., and M. Ramaiah: "Characteristic Equation of Cylindrical Shells—A Simplified Method of Solution," *Journal of the American Concrete Institute*, vol. 58, no. 4, p. 471, October, 1961.
2. Lundgren, H.: "Cylindrical Shells," vol. 1, Cylindrical Roofs, The Danish Technical Press, The Institution of Danish Civil Engineers, Copenhagen, 1949 (third printing, January, 1960).
3. "Design of Cylindrical Concrete Shell Roofs," ASCE Manual of Engineering Practice, no. 31, The American Society of Civil Engineers, New York, 1952.
4. Ramaswamy, G. S., and A. Carbone: "Design and Construction of Cylindrical Shells," National Building Organisation, Government of India, New Delhi, December, 1960.
5. Ramaswamy, G. S., M. Ramaiah, and B. Y. Ballal: "Matrix Methods in Structural Analysis—2 and 3," *Indian Concrete Journal*, December, 1963, and February, 1964.
6. Morice, P. B.: "Linear Structural Analysis," The Ronald Press Company, New York, 1959.
7. McMinn, S. J.: "Matrices for Structural Analysis," John Wiley & Sons, Inc., New York, 1962.
8. Argyris, J. H., and S. Kelsey: "Energy Theorems and Structural Analysis," Butterworth & Co. (Publishers), Ltd., London, 1960.
9. Sokolnikoff, Ivan S., and E. S. Sokolnikoff: "Higher Mathematics for Engineers and Physicists," 2d ed., McGraw-Hill Book Company, New York, 1941.
10. Weatherburn, C. E.: "Differential Geometry of Three Dimensions," vol. 1, Scientific Book Co., Patna, India, 1955.
11. Struik, Dirk J.: "Lectures on Classical Differential Geometry," 2d ed., Addison-Wesley Publishing Company, Inc., Reading, Mass., 1961.
12. Novozhilov, V. V.: "The Theory of Thin Shells" (translated from Russian by P. G. Lowe), Erven P. Noordhoff, NV, Groningen, Netherlands, 1964.
13. Belluzzi, O.: "Scienza delle costruzioni," vol. III, pp. 484–485, 1958; edited by Nicola Zanichelli, Editore, Bologna, Italy.
14. Finsterwalder, U.: "Die Theorie der Kreiszylindrischen Schalengewölbe System Zeiss-Dywidag und ihre Anwendung auf die Grossmarkthalle in Budapest," vol. 1, *Internationale Vereinigung für Brückenbau und Hochbau (IABSE)*, Abh., pp. 127–152, Zürich, 1932.
15. Finsterwalder, U.: "Die Querversteiften Zylindrischen Schalengewölbe mit Kreissegmentförmigem Querschnitt," vol. 4, *Ing.-Arch.*, 1933.
16. Donnell, L. H.: "The Stability of Thin Walled Tubes under Torsion," *National Advisory Committee for Aeronautics Report* 479, Washington, D.C., 1933.
17. Donnell, L. H.: "A New Theory for the Buckling of Thin Cylinders under Axial Compression and Bending," *Transactions of the American Society of Mechanical Engineers*, vol. 56, p. 108, 1934.

18. Kármán, Th. Von, and H. S. Tsien: "The Buckling of Thin Cylindrical Shells under Axial Compression," *Journal of Aeronautical Science*, vol. 8, no. 8, pp. 303–312, June, 1941.

19. Jenkins, R. S.: "Theory and Design of Cylindrical Shell Structures," *The O. N. Arup Group of Consulting Engineers, Bulletin* 1, London, 1947.

20. Flügge, W.: "Stresses in Shells," Springer-Verlag OHG, Berlin, 1960.

21. Schorer, H.: "Line Load Action on Thin Cylindrical Shells," *Proceedings of the American Society of Civil Engineers*, vol. 61, p. 281, 1935.

22. Rabich, R.: "Die Statik der Schalenträger. Die Berechnung der Randstörung am Randträger," *Bauplanung—Bautechnik*, H. 1, 10 Jg, January, 1956.

23. Aas-Jakobsen, A.: "Zylinderschalen mit veränderlichem Krümmungshalbmesser und veränderlicher Schalenstärke," *Bauingenieur*, vol. 18, Berlin, 1937.

24. Arya, A. S.: "Elastic Theories of Cylindrical Shells," vol. 5, no. 3, *India in Industries, Proceedings of the Seminar on Shell Structures* held at the Bengal Engineering College, Shibpur, January, 1962, Oxford Book and Stationery Co., Calcutta 16.

25. Dischinger, F.: "Die Strenge Theorie der Kreiszylinderseehale in ihrer Anwandung auf die Zeiss-Dywidag-Schalen," vol. 34, *Beton und Eisen*, 1935.

26. Holand, I.: "Design of Circular Cylindrical Shells," Oslo University Press, Oslo, 1957.

27. Moe, J.: "On the Theory of Cylindrical Shells," *Publication of the International Association for Bridge and Structural Engineering*, vol. 13, pp. 283–296, Zürich, 1953.

28. McNamee, J. J.: "Existing Methods for the Analysis of Concrete Shell Roofs," *Proceedings of the Symposium on Concrete Shell Roof Construction*, July, 1952, Cement and Concrete Association, London, 1954.

29. Fischer, L.: "Design of Cylindrical Shells with Edge Beams," *Journal of the American Concrete Institute*, vol. 52, p. 481, December, 1955.

30. "Criteria for the Design of Reinforced Concrete Shell Structures and Folded Plates," Indian Standard 2210–1962, Indian Standard Institution, New Delhi, 1963.

31. "Directives for the Design, Calculation and Execution of Cylindrical Shells in Reinforced Concrete," *Rapport* 12, *Report of the Commission Schaaldaken*, Commission for Conducting Research (C.U.R.), Gravenhage, Holland, December, 1956.

32. Hotzler, Herbert: "Experience with the Typification of Shell Roof Constructions in the German Democratic Republic," *The International Association for Shell Structures, Bulletin* 9, Madrid 7, Spain.

33. Zalewski, W., and J. Krzeminski: "Shell and Spacial Structure Shapes Applied in Poland," a-11, *International Colloquium on Construction Processes of Shell Structures*, Madrid, 1959.

34. Same as reference 14, p. 128.

35. Aas-Jakobsen, A.: "Einzellasten auf Kreiszylinderschalen," *Bauingenieur*, vol. 22, p. 343, 1941.

36. Mast, Paul E.: "Design and Construction of Northlight Barrel Shells," *Journal of the American Concrete Institute*, vol. 59, no. 4, April, 1962.

37. Haas, A. M.: "Research on a Prestressed Concrete Northlight Shell Structure," p. 659, *F.I.P. Second Congress*, Amsterdam, 1955, Session IIIb, Paper 1.

38. Ramaswamy, G. S., and M. Ramaiah: "Analysis of Northlight Shells," Paper A 5, Central Building Research Institute Symposium on Shell Structures, *Indian Concrete Journal*, December, 1959.

39. Rüdiger, D., and J. Urban: "Circular Cylindrical Shells," B. G. Teubner Verlagsgesellschaft mbH, Leipzig, 1955.

40. Wilby, C. B.: "A Method of Designing Northlight Shell Roofs," *Indian Concrete Journal*, vol. 35, no. 1, pp. 6–10, January, 1961.

41. Mihailescu, M., and I. Ungureanu: "A New Shell Form for Prestressed Sheds," Paper a-13, *International Colloquium on Construction Processes of Shell Structures,* Madrid, 1959.

42. Chronowicz, A.: "Approximate Analysis of the Northlight Shell," *Civil Engineering and Public Works Review,* vol. 52, no. 614, p. 880, August, 1957.

43. Kirkland, C. W., and A. Goldstein: "The Design and Construction of a Large Span Prestressed Concrete Shell Roof," *The Structural Engineer,* vol. 29, pp. 107–127, April, 1951.

44. Marshall, W. T.: "The Elimination of Moments in Shell Roofs by Prestressing," *Proceedings of the Institution of Civil Engineers,* vol. 3, Part III, pp. 276–282, April, 1954.

45. Haas, A. M.: "Prestressed Shells and Precast Units," A19, *Proceedings of the World Conference on Prestressed Concrete,* San Francisco, July, 1957.

46. Same as reference 37, p. 664.

47. "Important Shell Structures in India," Central Building Research Institute Symposium on Shell Structures, *Indian Concrete Journal,* vol. 33, p. 469, December, 1959.

48. Sexton, C. G.: "Prestressed Reinforced Concrete Hangar at the Civil Airport of Karachi," *Journal of the Institution of Civil Engineers,* vol. 29, pp. 109–130, December, 1947.

49. Dabrowski, Ryszard: "Analysis of Prestressed Cylindrical Shell Roofs," *Journal of the Structural Division, Proceedings of the American Society of Civil Engineers,* October, 1963, ST-5.

50. Hajnal-Kónyi, K., et al.: Discussion on Paper, "Design and Construction of a Large Span Prestressed Concrete Shell Roof," by Kirkland, G. W., and A. Goldstein, *The Structural Engineer,* vol. 29, pp. 306–311, November, 1951.

51. Cretu, Mircea: "Prestressed Concrete Shell Roofs," Central Building Research Institute Symposium on Shell Structures, December, 1959, *Indian Concrete Journal,* October, 1960.

52. "Tentative Recommendations for Prestressed Concrete," reported by ACI-ASCE Joint Committee 323, *Journal of the American Concrete Institute,* vol. 54, no. 7, pp. 545–578, January, 1958.

53. Ehlers, G.: "Ein Neues Konstruktions Prinzip," *Bauingenieur,* vol. 11, no. 8, p. 125, Berlin, 1930.

54. Gruber, E.: "Berechnung Prismatischer Scheibenwerke," vol. 1, *Publication of the International Association for Bridge and Structural Engineering,* Zürich, 1932.

55. Winter, G., and M. Pei: "Hipped Plate Construction," *Journal of the American Concrete Institute,* vol. 43, p. 505, January, 1947.

56. Gaafar, I.: "Hipped Plate Analysis, Considering Joint Displacements," Paper 2696, *Transactions of the American Society of Civil Engineers,* vol. 119, pp. 743–784, 1954.

57. Simpson, H.: "Design of Folded Plate Roofs," *Journal of the Structural Division, Proceedings of the American Society of Civil Engineers,* vol. 84, January, 1958.

58. Whitney, C. S., B. G. Anderson, and H. Birnbaum: "Reinforced Concrete Folded Plates Construction," *Journal of the Structural Division, Proceedings of the American Society of Civil Engineers,* vol. 85, ST8, October, 1959.

59. Girkmann, K.: "Flächentragwerke," Springer-Verlag OHG, Vienna, 1948.

60. Yitzhaki, D.: "The Design of Prismatic and Cylindrical Shell Roofs," North-Holland Publishing Company, Amsterdam, 1959.

61. Vlasov, V. Z.: "Handbook für Platten und Schalen," Moskau, 1939.

62. Phase I Report on Folded Plate Construction, Report of the Task Committee on Folded Plate Construction, *Journal of the Structural Division, Proceedings of the American Society of Civil Engineers,* vol. 60, p. 365, ST6, December, 1963.

63. Brielmaier, A. A.: "Prismatic Folded Plates," *Journal of the American Concrete Institute*, vol. 59, p. 407, March, 1962.

64. Rao, G. S.: "Analysis of Folded Plate Roofs by Iteration," *Indian Concrete Journal*, vol. 36, no. 10, p. 365, October, 1962.

65. Morice, P. B.: "An Approximate Solution to the Problem of Longitudinally Continuous Shells," *Magazine of Concrete Research*, vol. 9, no. 26, pp. 95–104, August, 1957.

66. Jenkins, R. S.: "A Variational Method for Design of Cylindrical Shells," *Proceedings of the Second Symposium on Concrete Shell Roof Construction*, Oslo, July, 1957.

67. Olsen, O.: "Continuous Shells," *Proceedings of the Second Symposium on Concrete Shell Roof Construction*, Oslo, July, 1957.

68. Inglis, C. E.: "The Determination of Critical Speeds, Natural Frequencies and Modes of Vibration by Means of Basic Functions," *Transactions of the North East Coast Institution of Engineers & Ship Builders*, vol. 61, pp. 111–136, 1944

69. Geckeler, J.: "Über die Festigkeit Achsensymmetrischer Schalen" (Strength of Axially Symmetric Shells), *Forschungsarbeiten auf dem Gebiet des Ingenieurwesens*, no. 276, 1926.

70. Kolykunov, E. V.: "Fundamentals of Calculation of Elastic Shells" (in Russian), pp. 187–194.

71. Fornerod, M. F.: "Prestressed Concrete Shell Roof Construction," *Publication of International Association for Bridge and Structural Engineering*, vol. 8, pp. 91–103, Zürich, 1947.

72. Soare, Mircea: "Application des équations aux différences finies au calcul des coques," Éditions de l'académie de la République Populaire Roumaine, Bucarest, Éditions Eyrolles, Paris, 1962.

73. Singhal, A. C., A. Villaveces, and S. Utku: "A Computer Analysis of Thin Shells of Revolution," *Proceedings of the World Conference on Shell Structures*, Oct. 1–4, 1962, San Francisco, Calif., 1962, National Academy of Sciences-National Research Council, Publication 1187, p. 189, Washington D.C.

74. Pucher, A.: "Ueber den Spannungszustand in Doppelt Gekrümmten Flächen," vol. 33, p. 298, *Beton und Eisen*, 1934.

75. Arup, O. N., and R. S. Jenkins: "The Design of a Reinforced Concrete Factory at Brynmawr, South Wales," *Proceedings of the Institution of Civil Engineers*, vol. 2, Part III, p. 345, December, 1953.

76. Doganoff, I.: "Betrachtungen über die Berechnungsverfahren Rechteckiger Kugelcalotten," *International Colloquium on Simplified Calculation Methods*, International Association for Shell Structures, Brussels, Sept. 4–6, 1961.

77. Parme, A. L.: "Hyperbolic Paraboloids and Other Shells of Double Curvature," *Journal of the Structural Division, Proceedings of the American Society of Civil Engineers*, vol. 82, ST5, pp. 1057–1 to 32, September, 1956.

78. Fischer, L.: "Determination of Membrane Stresses in Elliptic Paraboloids Using Polynomials," *Journal of the American Concrete Institute*, vol. 32, no. 4, pp. 433–441, October, 1960.

79. Csonka, P.: "Calotte Shell over Rectangular Base," *Acta Technica*, Academiae Scientiarum Hungaricae, Tomus XIII, Fasciculi 1-2, Budapest, 1955.

80. Csonka, P.: "Calculation of Calotte Shells over Rectangular Bases," *Acta Technica*, Academiae Scientiarum Hungaricae, Tomus XI, Fasciculi 3-4, Budapest, 1955.

81. Beles, Aurel A., and Mircea Soare: "Paraboloidul Eliptic si Hiperbolic in Constructii" (in Rumanian), Editura Academiei Republicii Populare Romine, 1964.

82. McKelvey, K. K., and Maxey Brooke, "The Industrial Cooling Tower with Special

Reference to the Design, Construction, Operation and Maintenance of Water Cooling Towers," Elsevier Publishing Company, Amsterdam, 1959.

83. Hajnal-Kónyi, K.: "Recent Developments in Shell Concrete Construction," vol. 9, pp. 194–223, Architects' Year Book.

84. Kulasinghe, A. N. S.: "Some Precast Shells in Ceylon," *Proceedings of the World Conference on Shell Structures*, Oct. 1–4, 1962, San Francisco, National Academy of Sciences-National Research Council, Publication 1187, p. 177, Washington D.C.

85. Candela, F.: "Structural Applications of Hyperbolic Paraboloidal Shells," *Journal of the American Concrete Institute*, vol. 26, Title 51-20, p. 397, December, 1954.

86. Vreedenburgh, C. G. J.: "The Hyperbolic Paraboloidal Shell and Its Mechanical Properties," The Philips Pavilion at the 1958 Brussels World Fair, *Philips Technical Review* (Eindhoven, Netherlands), vol. 20, no. 1, 1958–1959.

87. Bouma, A. L., and F. K. Ligtenberg: "Model Tests for Proving the Construction of the Pavilion," The Philips Pavilion at the 1958 Brussels World Fair, *Philips Technical Review* (Eindhoven, Netherlands), vol. 20, no. 1, 1958–1959.

88. Bouma, A. L.: "Some Applications of the Bending Theory Regarding Doubly Curved Shells," *Proceedings of the Symposium on the Theory of Thin Elastic Shells* (Delft, Aug. 24–28, 1959), North-Holland Publishing Company, Amsterdam, 1960.

89. Aimond, F.: "Étude statique des voiles minces en paraboloide hyperbolique travaillant sans flexion," 4-e vol., pp. 1–112, *Association Internationale des Ponts et Charpentes, Mémoires*, Zürich, 1936.

90. Booth, L. G., and J. E. Gallagher: "Wind Tunnel Tests on Four Linked Timber Hyperbolic Paraboloids," *Proceedings of the World Conference on Shell Structures*, Oct. 1–4, 1962, San Francisco, Calif., 1962, National Academy of Sciences-National Research Council, Publication 1187, p. 279, Washington, D.C.

91. Ramaswamy, G. S., and M. N. Keshava Rao: "The Membrane Theory Applied to Hyperbolic Paraboloid Shells," *Indian Concrete Journal*, May, 1961.

92. Meng, Ching-Hung, and Louis M. Laushey: "Stresses in Hyperbolic Paraboloid Shells Loaded Laterally," *Bulletin of the International Association for Shell Structures*, no. 15, Madrid, Spain, 1960.

93. Yu, C. W., and Ladislav B. Kriz: "Tests of a Hyperbolic Paraboloid Reinforced Concrete Shell," *Proceedings of the World Conference on Shell Structures*, Oct. 1–4, 1962, San Francisco, Calif., 1962, National Academy of Sciences-National Research Council, Publication 1187, p. 261, Washington, D.C.

94. Rosati, L.: "Contributo allo Studio Statico dello Volte Sottili Sghembe Senza Rigidezza a Flessione," *Atti E Rassegna Technica*, Torrino, pp. 157–164, May, 1951.

95. Parme, A. L.: "Elementary Analysis of Hyperbolic Paraboloid Shells," *The International Association for Shell Structures, Bulletin* 4, Madrid 7, Spain, 1960.

96. Hajnal-Kónyi, K.: "Construction of a Hyperbolic Paraboloid Shell Roof over the Garage of the Lincolnshire Motor Company Ltd., Lincoln," a-8, *International Colloquium on Construction Processes of Shell Structures*, Madrid, 1959.

97. Jones, L. L.: "Tests on a One-tenth Scale Model of a Hyperbolic Paraboloid Shell Roof," *Cement and Concrete Association Technical Report* TRA/334, 52 Grosvenor Gardens, London, August, 1960.

98. Tedesko, Anton: "Shell at Denver—Hyperbolic Paraboloidal Structure of Wide Span," *Journal of the American Concrete Institute*, vol. 32, no. 4, pp. 403–412, October, 1960.

99. De Cossio, R. D.: "Discussion of the Paper, Hyperbolic Paraboloidal Umbrella Shells under Vertical Loads, by Harrenstien, H.P.," *Journal of the American Concrete Institute*, vol. 32, no. 12, p. 1603, June, 1961.

100. Soare, M.: "Membrane Theory of Conoidal Shells" (translation—no. 82—of the

article in German that appeared in *Der Bauingenieur*, vol. 33, no. 7, pp. 256–265, July, 1958), Cement and Concrete Association, 52 Grosvenor Gardens, London, 1959.

101. Rao, G. S.: "Membrane Analysis of a Conoidal Shell with a Parabolic Directrix," *Indian Concrete Journal*, p. 325, September, 1961.

102. Bhise, V. M., V. P. Apte, and B. S. Phadke: "Membrane Analysis of Catenary and Circular Conoidal Shells," *Indian Concrete Journal*, p. 179, May, 1962.

103. Ramaswamy, G. S.: "Polynomial Stress Function for Parabolic Conoids," *Indian Concrete Journal*, vol. 35, no. 8, August, 1961.

104. Gol'denveizer, A. L.: "Theory of Elastic Thin Shells," pp. 474–483, Pergamon Press, New York, 1961.

105. Billig, Kurt: "Corrugated Concrete Shell Roofs," *Bulletin of the Central Building Research Institute*, vol. 1, no. 3, Roorkee, India, November, 1953, reprinted by the Concrete Association of India, Bombay, 1955.

106. Waller, J. H. de W., and A. C. Aston: "Corrugated Concrete Shell Roofs," *Proceedings of the Institution of Civil Engineers*, vol. 2, Part III, pp. 153–182, August, 1953.

107. Narayana, S. K.: "Construction of a Twin Ctesiphon Shell at the Central Building Research Institute, *Indian Concrete Journal*, vol. 33, no. 12, p. 429, December, 1959.

108. Ramaswamy, G. S., and S. M. K. Chetty: "A New Form of Doubly Curved Shell for Roofs and Floors," a-4, International Colloquium on Construction Processes of Shell Structures, Madrid, *The International Association for Shell Structures, Bulletin* 1.

109. Ramaswamy, G. S.: "The Theory of a New Funicular Shallow Shell of Double Curvature," *Indian Concrete Journal*, p. 336, September, 1960.

110. Harrenstien, H. P.: "Configuration of Shell Structures for Optimum Stresses," *Proceedings of the Symposium on Shell Research, Delft*, Aug. 30–Sept. 2, 1961, p. 232, North-Holland Publishing Company, Amsterdam, 1961.

111. Isler, Heinz: "Experimental Shell Design," *Proceedings of the Symposium on Shell Research, Delft*, Aug. 30–Sept. 2, 1961, p. 356, North-Holland Publishing Company, Amsterdam, 1961.

112. Caminos, Horacio: "Studies on Models of a Type of Membranal Structure," no. 10, *International Colloquium on Construction Processes of Shell Structures*, Madrid, 1959.

113. Saether, K.: "The Structural Membrane," *Journal of the American Concrete Institute*, vol. 32, no. 7, p. 827, January, 1961.

114. Vlasov, V. Z.: "Allgemeine Schalentheorie und Ihre Anwendung in der Technik," Akademie-Verlag GmbH, Berlin, 1958.

115. Marguerre, K.: "Zur Theorie der Gekrümmten Plate Grosser Formänderung," *Proceedings of the Fifth International Congress of Applied Mechanics, Cambridge, Mass.*, pp. 93–101, 1938.

116. Vlasov, V. Z.: "The Basic Differential Equations in the General Theory of Elastic Shells," *Prikladnaya Matematika i Mekhanika*, vol. 8, 1944 (translated from Russian into English by TNACA, Technical Memorandum 1241, February, 1951).

117. Reissner, E.: "On Some Aspects of the Theory of Thin Elastic Shells," *Journal of Boston Society of Civil Engineers*, vol. 42, no. 2, 1955.

118. Keshava Rao, M. N., and S. P. Sharma: "Application of the Methods of Finite Differences for the Analysis of a Conoid Shell Using the Bending Theory," *Indian Concrete Journal*, vol. 36, no. 3, p. 84, March, 1962.

119. Hadid, H. A.: "An Analytical and Experimental Investigation into the Bending Theory of Elastic Conoidal Shells," Thesis, University of Southampton, March, 1964.

120. Apeland, K., and E. P. Popov: "Analysis of Bending Stresses in Translational Shells," *Proceedings of the Colloquium on Simplified Calculation Methods*, p. 9, Brussels, September, 1961, North-Holland Publishing Company, Amsterdam, 1962.

726   Bibliography

121. Timoshenko, S., and S. Woinowsky-Krieger: "Theory of Plates and Shells," 2d ed., p. 109, McGraw-Hill Book Company, New York, 1959.

122. Flügge, W., and D. A. Conrad: "A Note on the Calculation of Shallow Shells," *Journal of Applied Mechanics*, p. 683, December, 1959.

123. Noor, A. K., and A. S. Veletsos: "A Study of Doubly Curved Shallow Shells," Thesis, Civil Engineering Studies, Structural Research Series 274, University of Illinois, November, 1963.

124. Loof, W. H.: "Discussion on the paper, Analysis of Bending Stresses in Translational Shells, by Apeland, K., and Popov, E. P.," *Proceedings of the Colloquium on Simplified Calculation Methods*, Brussels, September, 1961, p. 34, North-Holland Publishing Company, Amsterdam, 1962.

125. Zalewski, W.: "A Simplified Manner of Calculating Shell Diaphragms Traverses," *Proceedings of the Colloquium on Simplified Calculation Methods*, Brussels, International Association for Shell Structures, Sept. 4–6, 1961, North-Holland Publishing Company, Amsterdam, 1962.

126. Pauw, Adrian, and Daniel P. Jenny: "Lightweight Aggregate for Thin Concrete Shells," *Proceedings of the World Conference on Shell Structures*, Oct. 1–4, 1962, San Francisco, National Academy of Sciences-National Research Council, Publication 1187, p. 159, Washington, D.C.

127. Esquillan, N.: "General Report," Colloquium on Non-traditional Construction Processes of Shell Structures, *The International Association for Shell Structures, Bulletin 1*, Madrid.

128. Coppel, Th., and J.-J. Coulon: "Échafaudages Tubulaires," pp. 154–155 (French), Prix du Livre Technique du Batiment 1961, Dunod, Paris, 1963.

129. Hruban, I., and K. Hruban: "Northlight Shell Roofs—New Constructional Methods," *Indian Concrete Journal*, December, 1959, pp. 423–426.

130. Hahn, M.: "Usine Dunlop à Amiens Sheds en Béton Armé," Paper a-1, Colloquium on Non-traditional Construction Processes of Shell Structures, *The International Association for Shell Structures, Bulletin 1*, Madrid.

131. Gvozdev, A.: "Construction Récentes de Coques Minces en Béton Armé et en Béton Précontraint en Union Soviétique" (French), *Proceedings of the Second Symposium on Concrete Shell Roof Construction*, July, 1957, Teknisk Ukeblad, Oslo.

132. Haas, A. M.: "Reinforced or Prestressed Concrete Béton Armé et Précontraint," Colloquium on Non-traditional Construction Processes of Shell Structures, *The International Association for Shell Structures, Bulletin 1*, Madrid.

133. Shastri, N. V.: "Northlight Shell Roofs—Practical Aspects of Construction," Central Building Research Institute Symposium on Shell Structures, *Indian Concrete Journal*, vol. 33, no. 12, December, 1959.

134. Chatterjee, B. K.: "Economics of Shell Roof Construction in India," Central Building Research Institute Symposium on Shell Structures, *Indian Concrete Journal*, vol. 33, no. 12, p. 456, December, 1959.

135. Same as reference 128, p. 265.

136. Ramaswamy, G. S., Z. George, and B. V. Srinivasa Rao: "Movable Shuttering for Folded Plate Roof of Museum Building at C. B. R. I., Roorkee," *Indian Concrete Journal*, August, 1961.

137. Hurd, M. K.: "Formwork for Concrete," *American Concrete Institute Special Publication 4*, p. 289, Detroit, 1963.

138. Same as reference 128, p. 263.

139. Ramaswamy, G. S.: "Economical Roofs for Industrial and Storage Structures," *Cement and Concrete*, vol. 3, no. 3, p. 51, New Delhi, October–December, 1962.

140. Same as reference 137, p. 291.

141. Rosenlund, Jack E.: "Construction of the Dallas Memorial Auditorium," *Journal of the American Concrete Institute*, Title 54–17, p. 329, October, 1957.
142. Mokk, L.: "Prefabricated Concrete for Industrial and Public Structures," p. 398, Akadémiai Kiado, Publishing House of the Hungarian Academy of Sciences, Budapest, 1964.
143. Same as reference 142, p. 427.
144. Harry, C. Walter: "Precast Folded Plates Become Standard Products," *Journal of the American Concrete Institute*, vol. 60, no. 10, p. 1375, October, 1963.
145. "Couverture en coques précontraintes préfabriquées," *Batir* 128, p. 32, March, 1964.
146. Same as reference 142, p. 448.
147. Same as reference 142, p. 415.
148. Billington, David P.: "An American Engineer Views Precast and Prestressed Concrete in the Soviet Union," *Civil Engineering (ASCE)*, vol. 28, no. 10, pp. 53–57, October, 1958.
149. Haas, A. M.: "Design of Thin Concrete Shells," vol. 1, p. 108, John Wiley & Sons, Inc., New York, 1962.
150. Same as reference 137, p. 296.
151. Flint, A. R., and A. E. Low: "The Construction of Hyperbolic Paraboloid Type Shells without Temporary Formwork," Colloquium on Non-traditional Construction Processes of Shell Structures, *The International Association for Shell Structures, Bulletin*, Madrid, 1959.
152. "Concrete Shell Structures, Practice and Commentary," Report of ACI Committee 334, *Journal of the American Concrete Institute*, vol. 61, no. 9, p. 1104, September, 1964.
153. Waling, J. L., Earl E. Ziegler, and Harry G. Kemmer: "Hy-par Shell Construction by Offset Wire Methods," *Proceedings of the World Conference on Shell Structures*, Oct. 1–4, 1962, San Francisco, National Academy of Sciences-National Research Council Publication 1187, p. 453, Washington, D.C.
154. Same as reference 137, p. 44.
155. Tottenham, H.: "The Matrix Progression Method in Structural Analysis," chap. 7, Introduction to Structural Problems in Nuclear Reactor Engineering, International Series of Monographs on Nuclear Energy, Pergamon Press, New York, 1962.
156. Desai, J. R., and H. Tottenham: "Approximate Solutions in the Shell Theory of Arch Dams," International Symposium on the Theory of Arch Dams, Southampton University, Pergamon Press, 1964.
157. Pestel, E., and F. Leckie: "Matrix Methods in Elastomechanics," McGraw-Hill Book Company, New York, 1963.
158. Holand, I.: "Analysis of Folded Plate Structures by Transfer Matrices," *Proceedings of the Colloquium on Simplified Calculation Methods*, Brussels, Sept. 4–6, 1961, North-Holland Publishing Company, Amsterdam, 1962.
159. Bridgman, P. W.: "Dimensional Analysis," Yale University Press, New Haven, 1922.
160. Van Driest, E. R.: "On Dimensional Analysis and the Presentation of Data in Fluid Flow Problems," *Journal of Applied Mechanics*, vol. 13, no. 1, p. A-34, March, 1946.
161. Murphy, G.: "Similitude in Engineering," The Ronald Press Company, New York, 1950.
162. Cowan, H. J.: "Some Applications of the Use of Direct Model Analysis in the Design of Architectural Structures," *Journal of the Institution of Engineers*, Australia, July–August, 1961.
163. Preece, B. W., and J. D. Davies: "Models for Structural Concrete," C. R. Books Ltd., London, 1964.

164. Plexiglas Sheet-cementing, *Plexiglas Design, Fabrication and Molding Data Bulletin* 7d, Rohm and Haas Company, Philadelphia, 1959.

165. Litle, William A., and Robert J. Hansen: "The Use of Models in Structural Design," *Journal of the Boston Society of Civil Engineers*, vol. 50, no. 2, p. 59, April, 1963.

166. Carpenter, J. E., D. D. Magura, and N. W. Hanson: "Structural Model Testing— Techniques for Models of Plastic," *Portland Cement Association, Bulletin* D76, vol. 6, no. 2, May, 1964.

167. Johnson, A. E., and R. H. Homewood: "Stress and Deformation Analysis from Reduced-scale Plastic-model Testing," *Proceedings of the Society for Experimental Stress Analysis*, vol. 18, no. 2, p. 81, September, 1961.

168. Oberti, G.: "Arch Dams: Development of Model Research in Italy," Istituto Sperimentale Modelli E Strutture (ISMES), Bergamo, Italy, December, 1957.

169. Brock, G.: "Direct Models as an Aid to Reinforced Concrete Design," *Engineering*, p. 468, April, 1959.

170. Sharma, S. P.: "Application of Photoelasticity to Shells of Revolution," Thesis, University of London, July, 1959.

171. Haas, A. M., and A. L. Bouma: "Shell Research," *Proceedings of the Symposium on Shell Research*, Delft, August–September, 1961, North-Holland Publishing Company, Amsterdam, 1961.

172. Esquillan, N.: "The Design and Construction of the Shell Roof of the Exhibition Palace of the National Centre of Industries and Technology, Paris," Cement and Concrete Association, December, 1958.

173. Fung, Y. C., and E. E. Sechler: "Instability of Thin Elastic Shells," Structural Mechanics, *Proceedings of the First Symposium on Naval Structural Mechanics*, p. 115, Pergamon Press, New York, 1960.

174. Paduart, A.: "Introduction au casul et à l'execution des voiles minces en Béton Armé" (in French), Centre D'Information de L'Industrie Cimentière Belge, Eyrolles, Paris, 1961.

175. Bradshaw, Richard R.: "Some Aspects of Concrete Shell Buckling," *Journal of the American Concrete Institute*, vol. 60, no. 3, p. 313, March, 1963.

176. Gerard, George: "Introduction to Structural Stability Theory," McGraw-Hill Book Company, New York, 1962.

177. Donnell, L. H., and C. C. Wan: "Effect of Imperfections on Buckling of Thin Cylinders and Columns under Axial Compression," *Journal of Applied Mechanics*, vol. 17, no. 1, March, 1950.

178. Tsien, H. S.: "A Theory for the Buckling of Thin Shells, *Journal of the Aeronautical Sciences*, vol. 9, no. 10, pp. 373–384, August, 1942.

179. Schmidt, H.: "Ergebnisse von Beulversuchen mit Doppelt Gekrümmten Schalen- modellen aus Aluminium," *Proceedings of the Symposium on Shell Research*, Delft, August–September, 1961, pp. 159–181, North-Holland Publishing Company, Amsterdam, 1961.

180. Csonka, P.: "The Buckling of a Spheroidal Shell Curved in Two Directions," *Acta Technica*, vol. 14, no. 3–4, pp. 425–437, Academiae Scientiarum Hungaricae, Budapest, 1956.

181. Ralston, A.: "On the Problem of Buckling of Hyperbolic Paraboloidal Shell Loaded by Its Own Weight," *Journal of Mathematical Physics*, vol. 35, no. 1, 1956.

182. Dayaratnam, P., and Kurt H. Gerstle: "Buckling of Hyperbolic Paraboloids," *Proceedings of the World Conference on Shell Structures*, Oct. 1–4, 1962, San Francisco, National Academy of Sciences-National Research Council, Publication 1187, p. 289, Washington, D.C.

183. Turner, M.R., Clough R., Martin H., and Topp L.: "Stiffness and Deflexion Analysis of Complex Structures", *J. Aero. Sci.*, 23. No. 9, Sep. 1956, PP. 805-823.

184. Argyris, J. and Kelsey S.: "Energy Theorems and Structural Analysis", Butterworth Scientific Publications, London, 1960.

185. Gallagher, Richard H.: "Finite Element Analysis Fundamentals", Prentice Hall Inc., Englewood Cliffs, New Jersey, USA, 1975. Chapter 1.

186. Desai, C. S. and Abel J. K.: "Introduction to the Finite Element Method" Van Nostrand Reinhold, New York, 1972.

187. Zienkiewicz, O. C.: "The Finite Element Method in Engineering Science", McGraw-Hill, London, 1971. 2nd Ed.

188. Zienkiewicz, O. C. and Cheung, T. K.: "The Finite Element Method in Structural and Continuum Mechanics", McGraw-Hill, London, 1967.

189. Segerlind, Larry J.,: "Applied Finite Element Analysis", John Wiley and Sons, Inc. New York/London/Sydney/Toronto, 1976.

190. Cook, Robert D.,: "Concepts and Applications for Finite Element Analysis", John Wiley and Sons Inc. 1974.

191. Proceedings of Symposium on "Applications of Finite Element Methods in Civil Engineering", School of Engineering, Vanderbilt Univ. and American Society of Civil Engineers, Edited by William H. Rowan Jr. & Robert M. Hackett, 1969. Keynote address by R. H. Gallagher on "Analysis of Plate and Shell Structures" pp. 168-170.

192. Haisler, W. and Stricklin, J.: "Rigid Body Displacement of Curved Elements in the Analysis of Shells by the Matrix Displacement Method", AIAA Journal Vol. 5, No. 8, Aug. 1967, pp 1525-27.

193. Pecknold, D. A. W. and Schnobrich, W. C.: "Finite Element Analysis of Skewed Shallow Shells", Civil Engineering Studies Structural Research Series, No. 332, University of Illinois, Urbana Illinois, January 1968.

194. "Finite Elements for Thin Shells and Curved Members" Edited by Ashwell, D. G. and Gallagher, R. H. John Wiley and Sons, 1976, Chapter 4, "Rigid-Body Motions and Equilibrium in Finite Elements" by Gilles Cantin.

195. Desai, C. S.,: "Elementary Finite Element Method", Printice-Hall Inc. Englewood Cliffs, New Jersey 07623, 1979.

196. "Finite Elements for Thin Shells and Curved Members" Edited by D. G. Ashwell and R. H. Gallagher John Wiley and Sons 1976. Chapter 1, "Problems and Progress in Thin Shell Finite Element Analysis".

197. Chu, T. C. and Schnobrich, W. C.: "Finite Element Analysis of Translational Shells", Computers & Structures, Vol. 2, pp. 197-222, Pergamon Press, 1972.

198. Cantin, G. and Clough, R. W.,: "A Curved Cylindrical Shell Finite Element", AIAA J., 6, 1057 (1968).

199. Fonder, G. A., Clough, R. W.: "Explicit Addition of Rigid Body Motions in Curved Finite Elements", AIAA J., 11, 305 (1972).

200. Choi, Chang Koon and Schnobrich, W. C.,: "Non-conforming Finite Element Analysis of Shells", *Journal of the Engineering Mechanics Division*, ASCE, Aug. 1975, pp 447-464.

201. Wilson, E. L. et al: "Incompatible Displacement Models", International Symposium on Numerical Computer Methods in Structural Mechanics, University of Illinois, Urbana, Illinois, Sept. 1971.

202. "Experience of Finite Element Analysis of Shell Structures", by N. C. Knowles, A. Razzaque and J. B. Spooner in "Finite Elements for Thin Shells and Curved

Members", Chapter 13. Edited by D. G. Ashwell and R. H. Gallagher, John Wiley and Sons, 1976.

203. Connor Jr., Jerome J. and Brebbia, Carlos,: "Stiffness Matrix for Shallow Rectangular Shell Element", *Journal of the Engineering Mechanics Division*, ASCE, Oct. 1967, pp. 43-65.

204. Bhattacharya, B. and Ramaswamy, G. S.,: "Analysis of Funicular Shells by the Finite Element Method" *Journal of Structural Eng.*, Vol. 6, No. 3, 1978 pp 158-164. Published by the Structural Engineering Research Centre, Madras-600020, India.

205. Ramaswamy, G. S. and Jain, R. D.: "Analysis of Funicular Shells with Clamped Boundaries using Bending Theory", Bulletin of the International Association for Shell Structures, No. 32, Madrid, Spain.

206. Schnobrich, William C.: "Analysis of Hipped Roof Hyperbolic Paraboloid Structures", *Journal of the Structural Eng. Div.*, Proceedings of the ASCE, July 1972 pp 1576-83.

207. Jones, Rombert F. Jr.: "Shell and Plate Analysis by Finite Elements", *Journal of the Structural Division* ASCE, May 1973, pp. 889-901.

208. Zienkiewicz, O. C., Taylor, R. L. and Too, J. M.: "Reduced Integration Technique in General Analysis of Plates and Shells" *International Journal of Numerical Methods in Engineering*, Vol. 3, 1971. pp. 275-290.

209. Olson, Mervyn D.,: "Analysis of Arbitrary Shells", Thin Shell Structures, Theory Experiment and Design. Edited by Y. C. Fung and E. E. Sechler. Prentice Hall Inc. Inglewood, Calif. 1974 pp 403-433.

210. Zacharia, George and Deshmukh, R. S.: "Precast Concrete Doubly-Curved Funicular Shells Form Grid Floors for Concrete Testing Laboratory", *Indian Concrete Journal*, June 1973, pp 1-7.

211. Ramaswamy, G. S., Chetty, S. M. K., Bhargava, R. N. and Zacharia, George,: "Funicular Shell for Transit Shed Floor for Heavy Industrial Loading" Bulletin of the International Association for Shell Structures, No. 30.

212. Torroja, Eduardo,: "Philosophy of Structures" English version, University of California Press, Berkley, and Los Angeles, 1958.

213. ACI Standard Building Code Requirements for Reinforced Concrete (ACI 318-77). Detroit, Michigan, U.S.A.

214. Corley, W. G., Sozen, M. A., Seiss, C. P.,: "The Equivalent Frame Analysis for Concrete Slabs", Univ. of Illinois, Civil Engineering Studies, Structural Research Series No. 218, June 1961.

215. Sozen, Mete A. and Seiss, C. P.: "Investigations on Multipanel Reinforced Concrete Slabs", Design methods—their evolution and comparison, Journal ACI 60, No. 8, Aug. 1963.

216. CRSI Handbook, Concrete Reinforcing Steel Institute, 180 North Lasalle Street, Chicago, Illinois.

217. Faber, Colin,: "Candela/The Shell Builder", Reinhold Publishing Corporation, New York, 1963, 240 pages.

218. Kaimal, S. S.: "Hypar Footings for a Housing Project in India", Bulletin No. 32 of the International Association for Shell and Spatial Structures, Dec. 1967.

219. Kurien, Nainan, P.: "Economy of Hyperbolic Paraboloidal Shell Footings", Technical Note, Geotechnical Engineering, Vol. 8, 1977, pp. 53-39.

220. Sharma, A. K. and Mawal, M. B.: "Conical Shells as Foundations" Concrete Construction and Architecture, April 1972, Vol. III No. 10, pp 14-17

221. Sharma, A. K. and Mawal, M. B.: "A New Foundation for Tower-Shaped Structures", *Indian Geotechnical Journal*, Vol. 1, No. 2, April 1971, pp. 172-184.

222. Nicholls, R. L., Izadi, M. V.: "Design and Testing of Cone and Hypar Footings", Proceedings of ASCE, *Journal of the Soil Mechanics and Foundation Division*, Vol. 94 No. S. M. 1 pp. 47-72.

223. Ramanatha Ayyar, T. S., Ramachandra Rao, V. and Gopalakrishnan Nair, T. K.: "Funicular Shell Foundations on Soft Soils", Proceedings of the International Association for Housing Science Conference, Miami, 1979. pp 705-715.

224. Varghese, P. G. and Kaimal, S. S.: "Field Test on a Combined Hyperbolic Paraboloid Shell Foundation", Bulletin of the International Association of Shell Structures, No. 44, pp. 43-55.

225. Kurian, Nainan P.: "Ultimate Strength Design of Hyperbolic Paraboloid Shell Footings", Bulletin of the International Association for Shell and Spatial Structures, Bulletin No. 51, pp. 71-77.

226. Kurian, Nainan P., Syed, Shah H.: "A Novel Technique for the Installation of Precast Hypar Footings", Bulletin of the International Association for Shell and Spatial Structures, Bulletin No. 56, pp. 6.

227. Bhattacharya, B. and Ramaswamy, G. S.,: "A Finite Element Analysis of Funicular Shells on a Two Parameter Foundation Model" Bulletin No. 65, International Association for Shell and Spatial Structures, pp. 45-54.

228. Vlasov, V. Z.,: "Beams, Plates and Shells on Elastic Foundations", Israle Program for Scientific Transalations, Jerusalem, 1966.

229. Ramanatha, Ayyar. T. S., Ramachandra Rao, N., Gopalakrishnan, Nair, T. K.: "Stress Distribution Below Funicular Shell Foundations", Bulletin No. 69 of the International Association for Shell and Spatial Structures. pp. 35-38.

230. Nervi, P. L.,: "Structures", Translation by Giusappina and Salvadori, F. W. Dodge Corporation, New York, 1956.

231. Khaidukov, G. K.,: "Development of Armocement Structures", Bulletin of International Association of Structures, No. 36, pp. 85-97.

232. Walkus, B. R., Maria Kamariska, and Edward Szejkowski,: "Mechanicazed Method of Manufacture of Prefabricated Shells".

233. "Ferrocement Applications in Developing Countries", National Academy of Sciences, Washington D.C., 1973.

234. Naaman, Antoine E.,: "Performance Criteria for Ferrocement", *Journal of Ferrocement*, Vol. 9, No. 2, 1979.

235. Lachance, L., and Fugeure, P.,: "Ferro Shot-crete for Thin Shell Structures", Bulletin of IASS No. 44, Dec. 1970, pp. 13-18.

236. Lachance, L. and Picard, A.,: "Construction of Ferrocement Shells", Bulletin of IASS, No. 70, pp. 69-76.

237. Balaguru, P. N., Naaman, Antoine E. and Shah, S. P.: "Analysis and Behaviour of Ferrocement in Flexure", *Journal of the Structural Division*, ASCE Oct. 1977, pp. 1938-1951.

238. Johnston, Coiln D., Mowat, Dallas N.,: "Ferrocement—Material Behaviour in Flexure", *Journal of the Structural Division*, ASCE, October 1974, pp. 2053-69.

239. Abdul Karim, E. and Paul Joseph G.,: "Investigation on Flexural Behaviour of Ferrocement and its Application to Long Span Roofs", *Journal of Ferrocement*, Vol. 8, No. 1, Jan 1978, pp. 1-21.

240. Ramaswamy, G. S., and Al Adeeb, A. M.: "Ferrocement: A New Material for Long Spans", Research Paper CP5/72, Scientific Research Foundation, Building Research Centre, Baghdad, 1974.

241. Subrahmanyam, B. V., Karim, Abdul E., Ganesh Babu, K., and Ramaiah, M.,: "Ferrocement Ribbon Roofs for Long. Spans", Structural Engineering Research Centre, Madras-600020, India (paper under publication) RILEM International Symposium on Ferrocement, July 1981.

NAME INDEX

SUBJECT INDEX